Dictionary of
PARASITOLOGY

DICTIONARY OF PARASITOLOGY

Peter J. Gosling

CRC Press is an imprint of the
Taylor & Francis Group, an **informa** business
A TAYLOR & FRANCIS BOOK

Cover images are from DPDx: Laboratory Identification of Parasites of Public Health Concern, http://www.dpd.cdc.gov/dpdx/default.htm

CRC Press
Taylor & Francis Group
6000 Broken Sound Parkway NW, Suite 300
Boca Raton, FL 33487-2742

First issued in paperback 2019

ISBN-13: 978-0-415-30855-7 (hbk)
ISBN-13: 978-0-367-39254-3 (pbk)

Library of Congress Cataloging-in-Publication Data

Catalog record is available from the Library of Congress

Visit the Taylor & Francis Web site at
http://www.taylorandfrancis.com

and the CRC Press Web site at
http://www.crcpress.com

About the Author

Dr. Peter J. Gosling is an internationally renown scientist and science writer. He is a former Senior Government Scientist with the Department of Health in the United Kingdom, where he held responsibility for leading policy in zoonoses and in the infection risks of xenotransplantation, providing additional scientific input to policy development in the microbiology of water and the environment business areas. He has worked for many years in public health in the United Kingdom and is considered an expert on the genus *Aeromonas*. He has also worked internationally having helped to develop a diagnostic medical microbiology service in the University Hospital in Jeddah, Saudi Arabia, during the 1980s.

During his career he has been an active member of many scientific committees and has served as part of the secretariat for the U.K. Advisory Committee on Dangerous Pathogens and the Group of Experts on Cryptosporidium in water supplies. He has written successful copy for many agencies including the British Dental Association (BDA), Department of Health (DoH), Drinking Water Inspectorate (DWI), Health and Safety Executive (HSE), Organisation for Economic Co-operation and Development (OECD) and the World Health Organisation (WHO). He has also provided expert guidance and written copy for a variety of U.K. Government expert advisory committees including the Advisory Committee on Dangerous Pathogens (ACDP), the United Kingdom Xenotransplantation Interim Regulatory Authority (UKXIRA) and the Group of Experts on Cryptosporidium in Water Supplies.

Dr. Gosling is the author/editor of over 50 publications, including the popular science book *Pasteur — A Beginner's Guide* (Hodder & Stoughton); the textbook *The Genus Aeromonas* (John Wiley & Sons); a reference book covering aspects of visually communicating scientific information, the *Scientist's Guide to Poster Presentations* (Kluwer Academic/ Plenum Publishers); and a previous dictionary, *The Dictionary of Biomedical Sciences* (Taylor & Francis). His works have been published in many countries including the United Kingdom, the United States of America, Japan, Saudi Arabia and in Spain where he now resides.

Preface

Parasitic diseases are responsible for considerable morbidity and mortality in humans and animals throughout the world, and often present with nonspecific signs and symptoms. They are also of great economical importance in agriculture and horticulture. However, although many books have been published that cover various aspects of human, animal and plant parasitology, and public health problems associated with parasites that contaminate food and water supplies, none to date has provided a comprehensive guide for the beginner who is baffled by parasitology jargon. One of the main aims of the proposed *Dictionary of Parasitology* is to clarify this confusion.

This dictionary contains over 11,500 entries that define all the basic principles of parasitology, together with a wealth of other information. The dictionary reflects current practice in all aspects of parasitology and includes spellings, punctuation, abbreviations, acronyms, symbols, nomenclature, prefixes, and suffixes. It covers the field of parasitology in a concise, clear, authoritative, up-to-date manner, addressing the need for understanding and accurate use of terms, particularly when communicating scientific information.

The dictionary covers terms associated with human, veterinary, plant, insect, and fish parasitology, and entries are broadly assigned to one of these divisions (although, as is the nature of parasitology, there are many occasions when a variety of labels could be applied to any one entry). Entries cover control measures, immunology, physiology, pharmacology, etc., and are additionally demarcated.

It was my intention that the *Dictionary of Parasitology* should provide the depth and breadth of knowledge to make it both an informative and useful volume for students and experts alike. To provide these, together with science writers and editors of scientific texts, with an extensive guide to the terms used in the field of parasitology. I hope I have fulfilled my aims and that this dictionary will have international appeal and become a frequently consulted, informative, practical guide to the subject.

Peter J. Gosling

Dedication

To my wife, Deanna, without whose constant encouragement, care and support this work would not have been accomplished.

Notes on use

The order of headings is based on the alphabetical sequence of letters in the term and no headings are inverted, *e.g.* there is an entry under 'multilocular hydatid', but not under 'hydatid, multilocular'.

Headings defined appear in bold type and the term 'abbreviation' includes contracted and shortened forms of words and phrases, and acronyms and initials. I have followed the common practice of omitting the full point after each initial in abbreviations but have used them when a word has been truncated. (Thus, 'AGD' has no full points but 'path.' does).

Throughout the dictionary I have supplied American spellings to words where they differ from those in current use in the United Kingdom.

A

a- (terminology) A word element. [Latin] Prefix denoting *without, not.*

Aa (chemistry) An abbreviation of 'A'bsolute 'a'lcohol.

AAAS (society) An abbreviation of 'A'merican 'A'ssociation for the 'A'dvancement of 'S'cience.

AAB (society) An abbreviation of 'A'ssociation of 'A'pplied 'B'iologists.

AAFS (education) An abbreviation of 'A'merican 'A'academy of 'F'orensic 'S'ciences.

AAM (society) An abbreviation of 'A'merican 'A'ssociation for 'M'icrobiology.

AAPS (society) An abbreviation of 'A'merican 'A'ssociation for the 'P'romotion of 'S'cience.

a/b (epidemiology) An abbreviation of 'a'ir'b'orne. Also termed abn.

ab- (terminology) A word element. [Latin] Prefix denoting *from, off, away from.*

abattoir fever (human parasitology) *See* Q fever.

Abbreviata (veterinary parasitology) A genus of nematodes in the subfamily Physalopterinae, parasites of reptiles (especially saurians) or more rarely in amphibians and primates.

abdom. (anatomy) An abbreviation of 'abdom'en or 'abdom'inal.

abdomen (abdom.) (anatomy) The part of the trunk between the thorax and the pelvis. It is separated above from the thorax by a muscular partition called the diaphragm, and below by the pelvic floor. The structures which make up the abdominal wall are skin, fat of varying thickness, muscle, another layer of fat, and the thin, slippery membrane called the peritoneum which lines the abdominal cavity. The peritoneum covers all the internal organs in the abdominal cavity, *i.e.* the stomach, and most of the organs of the digestive system including the small and large intestine with the appendix and caecum (*USA*: cecum), and the liver and pancreas. Other abdominal organs include the kidneys, the adrenal glands, the spleen, and some large and important blood vessels including the aorta and the inferior vena cava. The prostate gland in the male and the uterus, ovaries and Fallopian tubes in the female are in the pelvis below the abdominal cavity.

abdomin(o)- (terminology) A word element. Prefix denoting *abdomen.*

abdominal infusion (parasitology) Infusion of saline into the peritoneal cavity, usually through a catheter inserted through the abdominal wall, for diagnostic purposes. The fluid returned may be examined for parasites, red blood cells, bacteria, enzymes, etc. Also termed abdominal lavage; peritoneal lavage.

abdominal lavage (parasitology) *See* abdominal infusion.

abdominal palpation (human/veterinary parasitology) Palpation of the contents of the abdomen either through the abdominal wall or per rectum.

abdominal paracentesis (methods) Insertion of a trocar through a small incision or a needle into the abdominal cavity, to remove ascitic fluids, inject a therapeutic agent, or to collect a sample for cytological and chemical examination. It is an important technique in the investigation of causes of ascites in all species. Also termed abdominocentesis.

abdominal respiration (human/veterinary parasitology) Inspiration and expiration accomplished mainly by the abdominal muscles and diaphragm. May occur in tick paralysis due to paralysis of the intercostal muscles.

See also tick paralysis.

abdominal section (methods) Laparotomy; incision of the abdominal wall.

abdominal threadworm (parasitology) *See Setaria equina.*

aberrant migration (parasitology) The migration of larvae of a parasite into sites not typically found in the life cycle. This frequently occurs in accidental hosts.

aberration (1 physics; 2 genetics) 1 The production of an image with coloured (*USA*: colored) fringes that occurs because the focal length of a lens is different for different colours of light (chromatic aberration), so that the lens focuses different colours in different planes; or the production of an image that is not sharp when a beam of light falls on the edges as well as on the centre (*USA*: center) of a lens (sperical aberration). 2 Change in the number or the structure of chromosomes in a cell..

Seeo also mutation.

abiotic (biology) Nonliving; used primarily for the nonliving parts of ecosystems, or of the environment in general.

ablastin (parasitological immunology) Serum antibodies that inhibit protozoan replication.

ablation (biology) The removal or destruction of any part of an organism.

abn (epidemiology) An abbreviation of 'a'ir'b'or'n'e. Also termed a/b.

abnormal (general terminology) Departing from normality. The term is usually used for departures from normality in an undesirable direction rather than in a desirable one.

abomasum (biology) The glandular compartment of a ruminant stomach.

aboral (terminology) Away from or opposite the mouth in those groups of animals that have no clear-cut dorsal or ventral surfaces.

abrasion (medical) The area from which the surface layer of the skin has been rubbed.

abs. (general terminology) An abbreviation of 'abs'olute/ 'abs'orbent/ 'abs'tract.

abscess (parasitology) A localized collection of pus in a cavity formed by the disintegration of tissue. Most abscesses are formed by invasion of tissues by bacteria, but some are caused by protozoa or even helminths. For specific abscesses see under anatomical sites, *e.g.* brain abscess.

abscissa (mathematics) The horizontal co-ordinate of a graph, conventionally called the *x-axis*. *See* Cartesian co-ordinates.

absgiute value (mathematics) A number or value expressed regardless of its sign.

absol. (general terminology) An abbreviation of 'absol'ute.

absolute alcohol (chemistry) Ethanol (ethyl alcohol) that contains no more than 1 per cent water.

absolute scale (measurement) A temperature scale with zero at the absolute zero of temperature.

absolute temperature (measurement) Temperature reckoned from absolute zero (-459.67°F or -273.15°C).

absolute value (measurement) The size of an observation or measurement regardless of its sign.

absolute zero (measurement) The lowest possible temperature, designated 0 on the kelvin or Rankine scale, the equivalent of -273.15°C or -459.67°F.

absorbance (chemistry) A measure of the amount of light absorbed by a solution. The difference between the amount of incident light and transmitted light expressed mathematically is referred to as the absorbance of the solution.

absorbed dose (pharmacology) The amount of a chemical that enters the body of an exposed organism.

absorbent gauze (parasitology) White cotton cloth of various thread counts and weights, supplied in various lengths and widths. Often used as a rough filter in preparing faecal (*USA*: fecal) specimens for parasitological investigation.

absorption factor (biochemistry/pharmacology) That fraction of a chemical coming into contact with an organism that is absorbed via the skin, respiratory system or gastrointestinal tract and enters the body of the organism.

abstr. (general terminology) An abbreviation of 'abstr'act.

abt (general terminology) An abbreviation of 'ab'ou't'.

Abyssinian tick typhus (human parasitology) *See* boutonneuse fever.

Ac (chemistry) Chemical symbol for actinium.

ac. (general terminology) An abbreviation of 'ac'tivity.

Acad. sci. (education) An abbreviation of 'Acad'emy of 'sci'ence.

academic education (education) Education in the theoretical principles of a subject or course.

Acanthamoeba (parasitology) A genus of small amoebae (*USA*: amebae), found in soil and water; they have been found in sporadic cases of pneumonia, general systemic infection and have produced meningoencephalitis after experimental administration. Possibly associated with granulomatous encephalitis in greyhounds. Species includes *Acanthamoeba castellani*, and *Acanthamoeba culbertsoni*. (*USA*: acanthameba).

acanthamoebiasis (parasitology) An infection by amoebae (*USA*: amebae) of the genus *Acanthamoeba*, such that has been observed rarely in dogs. (*USA*: acanthamebiasis) *See Acanthamoeba*.

acanthella (parasitological physiology) The developmental stage of an acanthocephalan parasite in which the larva develops definitive organ systems, that occurs between the acanthor and cystacanth stages.

acanth(o)- (terminology) A word element. [German] Prefix denoting *sharp spine*, *thorn*.

Acanthobdellida (fish parasitology) A primitive order of leech in the class Hiudinea of the phylum Annelida that is parasitic on salmon.

Acanthocephala (parasitology) A phylum of elongate, mostly cylindrical organisms with an anterior proboscis covered with many hooks, parasitic in the intestines of all classes of vertebrates. Acanthocephalans are not encountered as commonly as parasitic flatworms (trematodes and tapeworms) or nematodes, but are found in many species of fishes, amphibians, birds, and mammals. Several morphological characteristics serve to separate acanthocephalans from other parasitic worms, but probably the most notable is the presence of an anterior, protrusible proboscis that is usually covered with hooks. It is this characteristic that gives the acanthocephalans their common name, the thorny-headed worms. Complete life cycles are known for only approximately 25 species of acanthocephalans, but all species appear to follow the same basic pattern. The adult acanthocephalans occur in the intestine of the definitive host. The sexes are separate *i.e.* they are dioecious, and the females produce eggs that are passed in the host's faeces (*USA*: feces). The eggs are ingested by an intermediate host, an arthropod, and in the intermediate host the parasite goes through several developmental or juvenile stages. The definitive host is infected with it when it eats an intermediate host containing the infective juvenile stage called a cystacanth. Also termed: the thorny-headed worms.

acanthocephalans (parasitology) Members of the phylum Acanthocephala.

acanthocephaliasis (parasitology) Infestation with worms of the phylum Acanthocephala.

acanthocephalid (parasitology) *See acanthocephalans.*

Acanthocephalus (fish parasitology) A genus of thorny-headed worms of the family Echinorhynchinae. Species includes *Acanthocephalus jacksoni*, a parasite of trout.

Acanthocheilonema (parasitology) *See Dipetalonema.*

***Acanthophorides* species** (insect parasitology) A species of the insect order Diptera in the group Phoridae, a parasite of the ants *Eciton* and *Labidus*. *Acanthophorides* species attacks the adult stage of the ant.

acanthor (parasitology) The stage of acanthocephalan parasite which hatches from the egg.

Acarapis (insect parasitology) A species of mite in the phylum Prostigmata that are parasites to the bees, Apinae.

Acari (human/plant/veterinary parasitology) Members of the Arthropod class Chelicerata, the subclass Acari is divided into two superorders containing a total of seven orders between them. Superorder Actinotrichida contains the orders: Prostigmata; Astigmata; Oribatida; and the Superorder Anactinotrichida contains the Orders: Notostigmata; Holothyrida; Ixodida; Mesostigmata. Acari comprise the mites; the largest group within the arthropod class Arachnida, with over 48,000 described species. This number is misleading since it is estimated that only between five to ten per cent of all mite species have been formally described. Mites are distinctive in both their small size (adult body length ranging from 0.1–30 millimeters), and ecological diversity. Some mites are predators like almost all other arachnids, but mites may also feed on plants, fungi, microorganisms, or as parasites on or in the bodies of other animals. Many species are serious pests of agricultural crops, either through direct damage or indirectly as vectors of plant pathogens. Other species are parasitic on domestic animals and cause losses in meat, egg, and fiber production. Others, such as the human scabies mite, are direct agents of human disease, or, as in the case of chiggers and ticks, vectors of pathogens. Other mites may affect humans by infesting stored food products. Many species of Astigmata are known as 'stored product mites' because they have moved from their ancestral rodent nest habitats into human food stores. Such mites may also cause damage in animal feed by causing allergic reactions in livestock, and are also known to cause skin irritation in humans handling infested material. A related group of astigmatid mites, also ancestrally nest inhabiting, is the family Pyroglyphidae. These mites have colonized human habitations from bird nests and are the primary source for allergens in house dust. Commonly known as 'house dust mites', such species, particularly those in the genus *Dermatophagoides*, produce many proteins that induce allergic responses in sensitive individuals. House dust allergy may take the form of respiratory distress or skin irritation. Mites typically inhabit beds, chairs and carpets in houses, and their shed skins and faeces (*USA*: feces) provide the bulk of the allergens in house dust extracts. Some mites are beneficial to humans in their role as biological control agents against agricultural pests.

acariasis (parasitology) Infestation with arthropod parasites of the order Acarina including the ticks and mites. Also termed acarinosis.

acaricide (control measures) Pesticide that destroys ticks and mites. Common examples includes the organophosphorus compounds, the synthetic pyrethroids, and the carbamates.

acarid (parasitology) A tick or a mite of the order Acarina.

Acarina (parasites) An order of arthropods (class Arachnida), including mites and ticks.

acarine (parasitology) Pertaining to or of the nature of members of the order Acarina including the ticks and mites.

acarinosis (parasitology) Any disease caused by mites. Also termed acariasis.

acarodermatitis (parasitology) Skin inflammation due to bites of parasitic mites (acarids).

Acaroidea (parasitology) A superfamily of the order Acarina.

Acarus (parasitololgy) A genus of mites which often live as parasites on the external surface of larger animals. They cause various skin diseases such as itch and mange. They are also parasites to the wasps, *Dolichovespula*, where they feed on larval haemolymph (*USA*: hemolymph). Species include *Acarus fairnae*, *Acarus longion*, *Acarus siro*, *Acarus tawinae*. *See Tyroglyphus.*

acceptable daily intake (ADI) (pharmacology) The estimated amount of a substance that can be ingested daily during life without causing appreciable adverse effects. It is expressed in mg/kg body-weight/day.

acceptable risk (epidemiology) Risk for which the benefits rank larger than the potential hazards.

access (1 computing; 2 medical) 1 Computer term for getting access to a file or other collection of data.

2 Surgical term for the ease of reaching the target organ or site in an operation.

access time (computing) Period of time required for reading out of, or writing into, a computer's memory.

accessibility (general terminology) Ease of gaining access.

accessory (general terminology) A term referring to a supplementary item or one affording aid to another similar and generally more important thing.

accidental host (parasitology) An animal host to a parasite which is not the usual host species for that parasite.

accidental parasite (parasitology) A parasite that parasitizes an organism other than the usual host.

acclimatization (physiology) Way in which an organism adapts to a different or changing environment.

accolé (parasitological physiology) Early ring form of *Plasmodium falciparum* found at margin of red cells.

accreditation (quality standards) Certification of the expertise of individuals and establishments working in biomedical sciences including parasitology. Accreditation is performed by several organisations who establish standards and accredit on the basis of education, experience, and accomplishments.

accuracy (quality standards) *See* quality control.

ACDP (advisory committee) An abbreviation of 'A'dvisory 'C'ommittee on 'D'angerous 'P'athogens.

acellular (biology) Having a body which is not composed of cells. Acellular organisms may have a complex structure differentiated into specialized areas and organelles. Such organisms are also described as unicellular.

acentric chromosome (genetics) A chromosome with no centromere. This may be merely a large part of a chromosome that has broken off from another one. Acentrics tend to get lost because they cannot be manoeuvred (*USA*: maneuvered) during the normal cell-division process.

acet. (chemistry) An abbreviation of 'acet'one.

acetabulum (parasitological physiology) (plural: acetabula) 1 The posterior or ventral attachment organ of digenetic trematodes. 2 One of the four suction cups surrounding the head of cyclophyllidean cestodes which are used to maintain position in the gastrointestinal tract.

acetaldehyde (chemistry) CH_3CHO Colourless (*USA*: colorless) liquid organic compound with a pungent odour (*USA*: odor); a simple aldehyde made by the oxidation of ethanol (ethyl alcohol). Also termed: acetic aldehyde; ethanal.

acetamide (chemistry) CH_3CONH_2 Colourless (*USA*: colorless) deliquescent crystalline organic compound, with a 'mousy' odour (*USA*: odor). Also termed: ethanamide.

acetarsol (pharmacology) An organic arsenical used as an antiprotozoal agent, especially in turkeys and geese. Also termed: acetarsone.

acetarsone (pharmacology) *See* acetarsol.

acetate (chemistry) A salt of acetic acid.

acetate base (methods) Cellulose acetate sheet used as a support or base for X-ray film.

acetate tape slide (methods) A method of collecting ectoparasites such as mites, lice or fleas, and their eggs, for diagnostic purposes by pressing the sticky side of the tape against the skin and haircoat and applying the tape to a glass slide which is then examined microscopically.

acetic acid (chemistry) CH_3COOH Colourless (*USA*: colorless) liquid carboxylic acid, with a pungent odour (*USA*: odor) and acidic properties. It is the acid in vinegar. Also termed: ethanoic acid.

***Acetodextra* species** (fish parasitology) Trematodal fish parasites which may be found in the swim bladder of channel catfish (*Ictalurus punctatus*), and in adult form in the mesentaries, liver, and spleen. Species *Acetodextra ameiuri* may also be found in the ovaries of parasitized fish.

acetone (chemistry) CH_3COCH_3 Colourless (*USA*: colorless), volatile liquid organic compound, formed during the degradation of fats as well as a widely manufactured solvent. It is a simple ketone. Also termed: 2-propanone.

acetonitrile (chemistry) CH_3CN Colourless liquid organic compound, with a pleasant odour (*USA*: odor). Also termed: methyl cyanide.

acetonuria (biochemistry) The presence of acetone in the urine, which gives it a sweet characteristic smell.

acetophenone (chemistry) $C_6H_5COCH_3$ Colourless liquid organic compound, with a sweet pungent odour (*USA*: odor). Also termed: phenyl methyl ketone.

acetyl chloride (chemistry) CH_3COCl Colourless (*USA*: colorless), highly refractive liquid organic compound, with a strong odour (*USA*: odor). Also termed: ethanoyl chloride.

acetyl Co-A (physiology) An abbreviation of acetyl 'Co'enzyme A.

acetyl coenzyme A (acetyl Co-A) (physiology/biochemistry) A derivative of pantothenic acid, acetyl coenzyme A is an intermediate in many key metabolic processes, particularly in the transfer of the products of glycolysis to the Kreb's cycle and in fatty acid metabolism.

acetylethylenimine (control measures) One of a group of related alkylating agents used in the preparation of inactivated vaccines.

ACFAS (education) An abbreviation of '*A'ssociation 'C'anadienne 'F'rannçaise pour l'A'vancement des 'S'ciences*. [French-Canadian, meaning 'Association for the Advancement of Science'].

Acholeplasma (parasitology) A genus of the class Mollicutes and very closely related to the genus *Mycoplasma*.

Acholeplasma laidlawii (veterinary parasitology) A species of *Acholeplasma* which is commomly found in the lungs of calves with enzootic pneumonia but of doubtful significance.

Acholeplasma oculi (veterinary parasitology) A species of *Acholeplasma* which is commomly found in the conjunctivae of sheep with contagious ophthalmia but which is of unproven pathogenicity in this situation.

Achroia grisella (insect parasitology) Arthropods in the Lepidoptera that, similarly to *Ephestia cautella*, attacks the brood comb of the bees *Apis mellifera* and possibly *Bombus*.

achromatic (1, 2 microscopy) 1 Having no colour (*USA*: color). 2 Not easily coloured (*USA*: colored) by staining agents.

achromatic lens (microscopy) Combination of two or more lenses that has a focal length that is the same for two or more different wavelengths of light. This arrangement largely overcomes chromatic aberration.

Achtheres (fish parasitology) A genus of the class Crustacea which parasitize freshwater fish.

acid (chemistry) Member of a class of chemical compounds whose aqueous solutions contain hydrogen ions. Solutions of acids have a pH of less than 7. Strong acids dissociate completely (into ions) in solution; weak acids only partly dissociate. An acid neutralizes a base to form a salt, and reacts with most metals to liberate hydrogen gas.

acid anhydride (chemistry) Member of a group of organic compounds, general formula RCOOCOR# (where R and R# are alkyl radicals), that can be regarded as a carboxylic acid from which a molecule of water has been removed.

acid chloride (chemistry) Member of a group of volatile pungent-smelling organic compounds, general formula RCOCl (where R is an alkyl radical), much used in organic synthesis.

acid dyes (microscopy) Those dyes that contain an acidic organic component which stains materials such as cytoplasm and collagen when combined with a metal.

acid fuchsin (microscopy) A mixture of sulphonated (*USA*: sulfonated) fuchsins; used in various complex stains.

acid methyl green stain (microscopy) Stains protozoal nuclei a bright green and is recommended for the detection of *Balantidium coli* in faecal (*USA*: fecal) smears.

acid phosphatase (biochemistry) A lysosomal enzyme that hydrolyzes phosphate esters liberating phosphate, showing optimal activity at a pH between 3 and 6; found in eryhrocytes, prostatic tissue, spleen, kidney and other tissues.

acid picro mallory stain (microscopy) A staining technique that may be used to make connective tissue visible under the microscope.

acid salt (chemistry) Salt formed when not all the replaceable hydrogen atoms of an acid are substituted by a metal or its equivalent; *e.g.* sodium hydrogencarbonate (bicarbonate), $NaHCO_3$, diammonium hydrogenphosphate, (NH_4)-$2HPO_4$.

acid stain (microscopy) A stain in which the colouring (*USA*: coloring) agent is in the acid radical.

acid-base balance (biology) The proportion of acid and base required to keep the blood and body fluids neutral.

acid-fast (microscopy) Characteristic of certain microorganisms, which involves resistance to decolourisation (*USA*: decolorization) by acids when stained by an aniline dye, such as carbol fuchsin.

acidophil granules (microscopy) Granules staining with acid dyes.

Acinogaster (insect parasitology) Mites in the Prostigmata that, similarly to *Acinogaster marianae*, are phoretic on the ants *Eciton*, *Labidus*, *Neivamyrmex* and *Nomamyrmex*.

Acinogaster marianae (insect parasitology) *See Acinogaster*.

ack. (literature terminology) An abbreviation of 'ack'nowledge. Also termed ackt.

ackt (literature terminology) *See* ack.

ACLAM (society) An abbreviation of 'A'merican 'C'ollege of 'L'aboratory 'A'nimal 'M'edicine.

ACME (computing) An abbreviation of 'A'dvanced 'C'omputer for 'ME'dical research.

ACMT (education) An abbreviation of 'A'merican 'C'ollege of 'M'edical 'T'echnologists.

acoelomate (anatomy) Lacking a coelom (*USA*: celom), the true body cavity, which occurs in some invertebrate groups such as nematodes. (*USA*: acelomate).

Acomatacarus (human/veterinary parasitology) A trombidiform mite of the family Trombiculidae. The larvae are parasitic. Species includes *Acomatacarus australiensis* which may be found in humans and dogs, and *Acomatacarus galli* which may be found in chickens, mice, rats, and rabbits. Also termed: chiggers; scrub itch-mite.

aconitase (biochemistry) The enzyme that reversibly interconverts citric acid, *cis*-aconitic acid and isocitric acid in the citric acid (Krebs) cycle by dehydration and hydration reactions.

Acontylus (plant parasitology) A parasitic nematode of plants, in the family Rotylenchulinae. Species include *Acontylus vipriensis*.

ACOP (quality standards) An abbreviation of 'A'pproved 'C'ode 'O'f 'P'ractice.

ACOS (advisory committee) An abbreviation of 'A'dvisory 'C'ommittee 'O'n 'S'afety.

Acotyledon (insect parasitology) *See Ascarus*.

ACP (society) An abbreviation of 'A'ssociation of 'C'linical 'P'athologists.

ACPV (society) An abbreviation of 'A'merican 'C'ollege of 'P'oultry 'V'eterinarians.

acquired (epidemiology) Incurred as a result of factors acting from or originating outside the organism; not inherited.

acquired anaphylaxis (immunology) That in which sensitization is known to have been produced by administration of a foreign antigen.

acquired character (biology) A noninheritable modification produced in an animal as a result of its own activities or of environmental influences.

Acquired Immune Deficiency Syndrome (AIDS) (medical) Virus disease more commonly known by its abbreviation AIDS. A disease caused by the HIV (human immunodeficiency) virus, transmitted through the exchange of body fluids such as blood or semen. AIDS predisposes individuals to infection by organisms that include parasites, *e.g.* approximately 75 per cent of AIDS patients suffer from a type of pneumonia due to the protozoan parasite *Pneumocystis carinii*.

acquired immunity (immunology) Antigen specific immunity attributable to the production of antibody and of specific immune T lymphocytes (responsible for cell-mediated immunity), following exposure to an antigen, or passive transfer of antibody or immune lymphoid cells (adoptive immunity).

acridine (pharmacology) A mutagenic compound occurring in coal tar that has been used in the manufacture of dyes and intermediates. Derivatives are used as antiseptics, *e.g.* acriflavin, and antimalarial drugs. Acridine is a strong irritant to mucous membranes and skin, and it causes sneezing on inhalation.

acridine orange (microscopy) An orange stain that binds nonspecifically to nucleic acids, proteins, polysaccharides and glycosaminoglycans. Together with fluorescent microscopy, it is reportedly more sensitive than conventional staining methods for demonstrating *Haemobartonella felis* in blood smears. It fluoresces at 530 nm when intercalated into double-stranded DNA, or at 640 nm when ionically bound to single-stranded DNA. It produces mutations, some involving reading frame shifts, although its carcinogenicity is uncertain.

acroanaesthesia (medical) Loss of sensation in one or more of the bodily extremities. (*USA*: acroanesthesia).

acrocentric chromosome (genetics) A chromosome whose centromere is towards one end.

acromio- (terminology) A word element. [German] Prefix denoting *acromion*.

Acrostichus (insect parasitology) A genus of nematode in the Diplogasteridae that are parasites to the bees *Halictus farinosus*, and that may be found in the reproductive tract, possibly being a sexually transmitted disease.

acrylamide (chemistry/methods) A substance used in the preparation of clear gel for use in acrylamide gel electrophoresis. Humans exposed to acrylamide monomer, but not its polymers, are vulnerable to neurotoxic injury. Although acute high doses can result in an encephalopathy that is apparently reversible, repeated smaller doses are cumulative and result in a distal sensorimotor axonopathy.

acrylamide gel electrophoresis (methods) A technique using an acrylamide gel for the electrophorectic separation of proteins and RNA according to their molecular weight. The monomer can be cast in the form of sheets or cylinders by polymerisation *in situ* to give a clear gel.

acrylic resin (chemistry/optics) Transparent thermoplastic formed by the polymerization of ester or amide derivatives of acrylic acid. The resins are used to make artificial fibres (*USA*: fibers) and for optical purposes, such as making lenses (*e.g.* Acrilan, Perspex).

acrylonitrile (chemistry) $CH_2=CHCN$ Colourless (*USA*: colorless) liquid nitrile used in manufacture of acrylic resins. Also termed: vinyl cyanide, propenonitrile.

ACTH (biochemistry) An abbreviation of 'A'dreno-'C'ortico'T'rophic 'H'ormone. (*USA*: Adrenocorticotropic hormone).

actinide (chemistry) Member of a series of elements in Group IIIB of the periodic table, of atomic numbers 90 to 103 (actinium, at. no. 89, is sometimes also included). All actinides are radioactive. Also termed: actinoid.

actinium (Ac) (chemistry) Radioactive element in Group IIIB of the periodic table (usually regarded as one of the actinides); it has several isotopes, with half-lives of up to 21.7 years. It results from the decay of uranium-235. At. no. 89; r.a.m. 227 (most stable isotope).

actino- (terminology) A word element. [German] Prefix denoting *ray, radiation*.

Actinocleidus (fish parasitology) A genus of flukes that are parasitic in fish.

activated charcoal (chemistry) Charcoal that, after pyrolysis during manufacture, has been subjected to steam or air at high temperature, which makes it an effective absorber of substances.

activated lymphocyte (immunology) A lymphocyte that has reacted on exposure to antigen or to a mitogen.

active anaphylaxis (immunology) *See* acquired anaphylaxis.

active immunity (immunology) An immunity to a disease that has been acquired through previous exposure to it. *Contrast* passive immunity.

See also immune system.

active immunization (immunology) Stimulation with a specific antigen to promote an immune response. Any of a vast number of foreign substances may induce an active immune response. Since active immunization induces the body to produce its own antibodies and specifically reactive cells and to go on producing them, protection against disease will last several years, in some cases for life.

active mass (chemistry) Concentration of a substance that is involved in a chemical reaction.

active principle (pharmacology) Any constituent of a drug that helps to confer upon it a medicinal property.

active sensitization (immunology) The sensitization that results from the injection, ingestion or inhalation, of antigen into an animal.

See also active immunity, active immunization.

active surveillance (epidemiology) Sampling, including necropsy examination, of clinically normal samples of the population; important in the surveillance of diseases in which subclinical cases and carriers predominate.

active transport (biochemistry/physiology) The various energy-requiring processes that permit the movement of chemicals/ions across biological membranes. All active transport systems require metabolic energy; can be inhibited by chemicals that affect energy metabolism; are selective in terms of the molecules transported; are saturable; and can transport chemicals against a concentration gradient. *Contrast* diffusion.

actuarial methods (statistics) Statistical techniques relating to preparation of mortality and other analytical tables.

Acuaria (parasites) A genus of nematode parasites. Species includes: *Acuaria hamulosa* (*See Cheilospirura hamulosa*), *Acuaria spiralis* (*Synhimantus spiralis*; *See Synhimantus*), *Acuaria uncinata* (*Echinuria uncinata*; *See Echinuria*).

Aculops fuchsiae (plant parasitology) The fuchsia gall mite. *See* gall.

acute (1 medical; 2 mathematics) 1 Any process which has a sudden onset and has a relatively brief duration. *Contrast* chronic. 2 Describing an angle of less than 90 degrees.

acute abdomen (medical) The condition in which the patient is suffering from acute severe pain in the abdomen.

acute infection (medical) An infection of short duration, of the order of several days.

acute inflammation (medical) Inflammation, usually of sudden onset, marked by the classic signs of heat, redness, swelling, pain and loss of function, and in which vascular and exudative processes predominate.

acute moist dermatitis (parasitology) A superficial bacterial infection of the skin, usually caused by self-trauma, *e.g.*, scratching, rubbing, biting, which in some animals such as dogs, ectoparasites are common pre-cipitating causes. Affected skin is moist, weeping, and has a covering of matted haircoat and dried exudate. Staphylococcus spp. are usually present. Also termed pyotraumatic dermatitis, 'hot spots'.

acute tolerance (pharmacology) Development of tolerance to a drug after only one or a few doses. *Contrast* chronic tolerance.

acyl group (chemistry) Part of an organic compound that has the formula RCO; where R is a hydrocarbon group. (*e.g.* acetyl group CH_3CO-).

ad hoc (general terminology) Unprincipled, particularly of explanations constructed *post hoc* to explain data.

adaptation (1, 2 biology; 3 physiology) 1 Characteristic of an organism which improves that organism's chance of survival in its environment. 2 Change in the behaviour (*USA*: behavior) of an organism in response to environmental conditions. 3 Change in the sensitivity of a sensory mechanism after it has been exposed to a particular continuous stimulus. This allows the mechanism to adjust the sensitivity scale according to the level of the stimulus.

adaptive value (genetics) The amount of increase or decrease in reproductive fitness that a given genotype or mutation confers, relative to other genotypes and in a particular environment.

adder (computing) Part of a computer that adds digital signals (addend, augend and a carry digit) to produce the sum and a carry digit.

additive effect (biochemistry) The combined effect of two or more chemicals, this being equal to the sum of the individual effects.

additive variance (genetics) The component of genetic variance that is due to the additive effect of two or more genes on the trait in question, *i.e.* excluding any dominance or interaction.

address (1, 2 computing) 1 Identity of a location's position in a memory or store. 2 Specification of an operand's location.

adenine (molecular biology) One of the four nitrogenous bases which form the core of both DNA and RNA. It pairs with thymine (T) in DNA, and uracil (U) in RNA. Also termed: 6-aminopurine.

See also replication.

adeno- (terminology) A word element. [German] Prefix denoting *gland*.

adenomatous (biology) A term describing a growth which has glandular structure.

adenomatous nodules (biology) A lump, bump or growth which resembles glandular material.

Adenophorea (plant parasitology) A class of parasites of plants that includes the nematodes *Longidorus africanus*, the needle nematode, *Paratrichodorus minor*, the stubby root nematode and *Xiphinema index*, the dagger nematode.

adenosine (biochemistry) Compound consisting of the base adenine linked to the sugar ribose. Its phosphates are important energy carriers in biochemical processes.

adherent cell (methods) A cell that adheres to the glass or plastic container in cell cultures, to form the monolayer.

See also tissue culture.

adhesive tape (methods) A strip of fabric or other material evenly coated on one side with a pressure-sensitive adhesive material which may be used as a method of collecting ectoparasites. *See* acetate tape slide.

ADI (pharmacology) An abbreviation of 'A'cceptable 'D'aily 'I'ntake.

adip(o)- (terminology) A word element. [Latin] Prefix denoting *fat*.

adolescaria (parasitological physiology) A developmental stage in the life cycle of trematodes.

adoptive immunity (immunology) Passive immunity of the cell-mediated type conferred by the administration of sensitized lymphocytes from an immune donor to a naive recipient.

adren(o)- (terminology) A word element. [Latin] Prefix denoting *adrenal glands*.

adsorbent (methods) Substance on which adsorption takes place.

adsorption chromatography (methods) Chromatography technique in which the stationary phase is an adsorbent.

adult (parasitological physiology) The stage in which a parasite, or other organism, has fully developed and grown to full size and strength within their life cycle.

adulticide (pharmacology/control measures) An agent which kills the adult stage of a parasite, *e.g.* a drug used to kill adult heartworms (*Dirofilaria immitis*) as distinct from that used to kill the microfilaria. Some agents are capable of both effects.

adverse (general terminology) Pertaining to going against something, or being harmful.

Advisory Committee on Dangerous Pathogens (ACDP) (advisory committee) A UK Government committee which advises the Health and Safety Commission, the Health and Safety Executive and Health and Agriculture Ministers, as required, on all aspects of hazards and risks to workers and others from exposure to pathogens including parasites. The ACDP was set up in 1981 following a second outbreak of laboratory-acquired smallpox and was the successor to the Dangerous Pathogens Advisory Group.

Aedes (parasitology) A genus of the family Culicidae, the mosquitoes. *Aedes* mosquitoes occur around the world and there are over 950 species. They can cause a serious biting nuisance to people and animals, both in the tropics and in cooler climates. In tropical countries *Aedes aegypti* is an important vector of dengue, dengue haemorrhagic fever, yellow fever and other viral diseases. A closely related species, *Aedes albopictus*, can also transmit dengue. They are known to transmit the *Plasmodium* spp. of bird malaria, and the arboviruses of Rift Valley Fever and Japanese B encephalitis. Some species are vectors of equine encephalomyelitis virus. In some areas *Aedes* species transmit filariasis. Heartworm (*Dirofilaria immitis*) uses *Aedes* spp. as intermediate hosts for the development of microfilariae. *Aedes* also cause insect worry in animals and some species, *e.g. Aedes vigilex*, and may cause fatalities amongst puppies and piglets.

Aedes aegypti (parasitology) *See Aedes*.

Aedes albopictus (parasitology) *See Aedes*.

Aegyptianella (parasitology) A genus of the family of rickettsia-like organisms called Anaplasmataceae.

Aegyptianella pullorum (avian parasitology) The species of rickettsia-like organisms that cause aegyptianellosis in birds.

aegyptianellosis (avian parasitology) The disease of birds caused by *Aegyptianella pullorum*, and possibly other organisms of the same genus, that is characterized by anorexia, diarrhoea (*USA*: diarrhea), fever and paralysis.

Aelurostrongylus (veterinary parasitology) A genus of lungworms of the family Angiostrongylidae which cause chronic cough and weight loss. Species includes *Aelurostrongylus abstrusus* found in cats, *Aelurostrongylus falciformis* found in badgers, and *Aelurostrongylus pridhami* found in wild mink.

-aemia (terminology) A word element. [German] Suffix pertaining to *the blood*. (*USA*: -emia).

Aenigmatis dorni (insect parasitology) A species of the insect order Diptera in the group Phoridae, a parasite of the ants *Formica* species, *Lasius* and *Myrmica*. Like

Aenigmatis franzi and *Aenigmatis lubbocki*, *Aenigmatis dorni* attacks the pupa stage of the ant.

Aenigmatis franzi (insect parasitology) A species of the insect order Diptera in the group Phoridae, a parasite of the ants *Formica* species, *Lasius* and *Myrmica.* Like *Aenigmatis dorni and Aenigmatis lubbocki*, *Aenigmatis franzi* attacks the pupa stage of the ant.

Aenigmatis lubbocki (insect parasitology) A species of the insect order Diptera in the group Phoridae, a parasite of the ants *Formica* species, *Lasius* and *Myrmica.* Like *Aenigmatis franzi and Aenigmatis dorni*, *Aenigmatis lubbocki* attacks the pupa stage.

aer(o)- (terminology) A word element. [German] Prefix denoting *air*, *gas*.

aerobic (physiology) Pertaining to an organism that requires free oxygen, or to the uptake of oxygen by an organism. Aerobic cells or organisms require oxygen and depend upon the pathways of aerobic respiration *i.e.* glycolysis to pyruvate, tricarboxylic acid cycle and electron transport system, for their energy generation. *See* aerobic respiration.

aerobic respiration (physiology) Process by which cells obtain energy from the oxidation of fuel molecules by molecular oxygen with the formation of carbon dioxide and water. This process yields more energy than anaerobic respiration.

See also glycolysis; Krebs cycle.

aerogenic (physiology) Gas producing. *Contrast* anaerogenic.

Aeroglyphus (insect parasitology) *See Ascarus.*

aerosol transmission (epidemiology) Person-to-person or animal-to-animal transmission of disease agents via aerosolised droplets. Respiratory diseases in particular may be transmitted in this way.

aetiological (epidemiology) Pertaining to aetiology (*USA*: etiology).

aetiological factors (epidemiology) Risk factors contributing to the cause of a disease. (*USA*: etiological factors).

aetiology (1, 2 epidemiology) 1 The cause or causes of a disease. 2 Study of the causes of a disease. (*USA*: etiology).

See also epidemiology.

AFAS (society) An abbreviation of '*A'ssociation 'F'rançaise pour l''A'vancement des 'S'ciences.* [French, meaning 'Association for the Advancement of the Sciences'.]

AFD (methods) An abbreviation of 'A'ccelerated 'F'reeze 'D'ried.

Afenestrata (plant parasitology) A genus of nematodal plant parasites in the family Heteroderinae. Species include *Afenestrata koreana* and hosts include *Phyllostachys* species (bamboo).

afferent (anatomy) Leading towards (*e.g.* towards an organ of the body); the term is particularly used of various nerves and blood vessels.

See also efferent.

affinity (1 chemistry; 2 biology) 1 The tendency of some substances to combine chemically with others (*e.g.* the likelihood of a substrate binding to an enzyme, a ligand binding to a receptor, or an antibody combining site with a single antigenic determinant). 2 A similarity of structure or form between different species of plants or animals.

affinity chromatography (methods) A method of chromatography that utilizes the biologically important binding interactions that occur on protein surfaces. *E.g.* an enzyme substrate is covalently coupled to an inert matrix such as a polysaccharide bead. The enzyme can be bound to the bead and thereby separated when present in very low concentration in a very complex mixture of other macromolecules.

affinity labelling (methods) A procedure for identifying or quantifying receptors by covalently binding a high affinity ligand to the receptor. Binding can be achieved by either using a ligand possessing a chemical grouping capable of covalently binding to the receptor directly, or by allowing a cross-linking reagent to covalently bind to both the ligand and the receptor.

AFNOR (quality standards) An abbreviation of '*A'ssociation 'F'rançaisede 'Nor'malisation,* French Standards Association.

African sleeping sickness (human/veterinary parasitology) A fatal infection of the nervous and lymphatic systems that is endemic in certain parts of Africa and is caused by a flagellate protozoan called *Trypanosoma*, particularly *Trypanosoma brucei gambiense* in West Africa and *Trypanosoma brucei rhodesiense* in East Africa. The vector of the flagellate is the tsetse fly *Glossina*, which also feeds on cattle, the latter acting as a reservoir for the parasite. African sleeping sickness is not to be confused with encephalitis which is caused by a virus. Also termed: African trypanosomiasis.

African swine fever (veterinary parasitology) A peracute, highly contagious, highly fatal disease of pigs caused by African swine fever virus, previously a member of the family *Iridoviridae*, now an unclassified virus. The virus is carried in wart hogs in which it produces no disease and is transmitted to European pigs via the tick *Ornithodoros moubata porcinus.* The disease was originally confined to southern Africa, but has spread to North Africa, to Europe, including Spain, Portugal and Belgium, and also to Cuba and the Dominican Republic.

African trypanosomiasis (human/veterinary parasitology) *See* African sleeping sickness.

Afrina wevelli (plant parasitology) *Afrina wevelli* was described from South Africa in 1985, recovered from seed galls on *Eragrostis curvula*, weeping love grass. The diagnostic morphological features of this species include: many longitudinal lines in the lateral field, the procorpus narrowing more appreciably before the median bulb, excretory pore situated more posterior, absence of convoluted isthmus, and the genital primordium of the juvenile situated far forward. Also termed: seed gall nematode.

Afrocypholaelaps (insect parasitology) A species of mites in the Mesostigmata, that are phoretic and cleptoparasitic on the bees *Apinae* and Meliponinae.

AFTM (society) An abbreviation of 'A'merican 'F'oundation of 'T'ropical 'M'edicine.

Ag (immunology) An abbreviation of 'A'nti'g'en.

Agamomermis (insect parasitology) A genus of nematodes in the Mermithidae that are parasites to the bees *Apis mellifera*. They may be found in workers, queens, and drones. It is possible that workers take up eggs with water from foliage.

Agamomermis pachysoma (insect parasitology) A species of *Mermithidae* in the phylum Nematoda and genus *Agamomermis* that are parasites to the wasps, *Vespula* species.

agar (chemistry) A complex polysaccharide obtained from seaweed, especially that of the genus *Gelidium*. It resists the digestive action of bacteria, and is therefore used by bacteriologists to make a medium on which cultures can be grown. It is commonly combined with broth or blood for this purpose. Also termed: agar-agar.

AGD (fish parasitology) An abbreviation of 'A'moebic (*USA*: amebic) 'G'ill 'D'isease.

ageing (physiology) The phenomenon of progressive failure of self-repair of an organism.

agent (epidemiology) Any factor whose excessive presence or relative absence is essential for the occurrence of disease.

agglutination (immunology) Aggregation of separate particles into clumps or masses; especially the clumping of organisms or cells by antibody specific to, or directed against, surface antigenic determinants. Agglutination is commonly used as an end point in immunological tests.

See also agglutinin.

agglutination titre (immunology) The highest dilution of a serum that causes clumping of organisms or other particulate antigens. (*USA*: agglutination titer).

agglutinin (immunology) Antibody that causes agglutination.

aggr. (general terminology) An abbreviation of 'aggr'egate.

aggregate selection criteria (statistics) The sum of all of the criteria to be used in a selection programme (*USA*: program) with each criterion multiplied by its relative importance to the other criteria.

agit. (general terminology) Abbreviation of '*agit'atum*. [Latin, meaning 'shaken'].

Aglenchus (parasitology) A genus of the family Atylenchinae of nematode worms which includes *Aglenchus agricola* which is parasitic to grasses and clover. The body of *Aglenchus* species is 0.4–0.7mm long, and is straight to slightly curved. *Aglenchus agricola* is common and widespread in Europe. Other species include *Aglenchus dakotensis* found in USA, *Aglenchus mardanensis* found in Pakistan and *Aglenchus muktii* found in India.

aglm (general terminology) An abbreviation of 'ag'g'l'o'm'erate.

-agogue (terminology) A word element. [German] Suffix denoting *something that leads* or *induces*.

-agra (terminology) A word element. [German] Suffix denoting attack, seizure.

agreement method (epidemiology) When a particular factor is common to all occurrences of the disease, it may be that the common factor is the cause of the disease.

agricultural chemical (control measures) Chemical used in agriculture. These include pesticides, anthelmintics, and the like.

Agriolimax meticulatus (parasitology) The common garden slug which is intermediate host of the sheep lungworm, *Cystocaulus ocreatus*.

Agriostomum (parasitology) A genus of the subfamily Chabertinae of nematode worms which includes *Agriostomum vryburgi* parasitic to *Bos indicus* cattle.

agw (measurement) An abbreviation of 'a'ctual 'g'ross 'w'eight.

AHF (chemistry) An abbreviation of 'A'nhydrous 'H'ydro'F'luoric acid.

ahr (epidemiology) An abbreviation of 'a'cceptable 'h'azard 'r'ate.

AIBS (society) An abbreviation of 'A'merican 'I'nstitute of 'B'iological 'S'ciences.

AIDS (medical) An abbreviation of 'A'cquired 'I'mmune 'D'eficiency 'S'yndrome.

AIDS-related complex (ARC) (medical) The medical condition of a debilitated HIV-positive person not diagnosed with AIDS.

Aino virus (parasitology) A bunyavirus transmitted by insects.

air (chemistry) Mixture of gases that forms the Earth's atmosphere. Its composition varies slightly from place to place, particularly with regard to the amounts of carbon dioxide and water vapour it contains, but the average composition of dry air is (percentages by volume): nitrogen 78.1 per cent; oxygen 20.9 per cent; agon 0.9 per cent; other gases 0.1 per cent.

air sac disease (veterinary parasitology) *See* airsacculitis.

air sacs (anatomy) Sacs that communicate with the respiratory, air-filled membranous system in birds and primates.

air sacs inflammation (veterinary parasitology) *See* airsacculitis.

air sacs mite (veterinary parasitology) *See Cytodites*.

air temperature (epidemiology) The temperature of the surrounding air as measured by a dry bulb thermometer.

airborne transmission (epidemiology) Spread of infection by droplet nuclei or dust through the air. Without the intervention of winds or drafts the distance over which airborne infection takes place is short, of the order of 10 to 20 feet.

airsacculitis (veterinary parasitology) Inflammation of the air sacs, which may be a result of air sac mite, *Cytodites nudus*, infestation.

AIS (education) An abbreviation of 'A'ssociate of the 'I'nstitute of 'S'tatisticians.

akabane virus (parasitology) A bunyavirus transmitted by insects, including *Culicoides brevitarsis*, and the cause of arthrogryposis and hydranencephaly, recognized in newborn calves and lambs following in utero infection in the early months of gestation.

Al (chemistry) The chemical symbol for element 13, aluminum.

-al (terminology) Suffix usually denoting *that an organic compound is an aldehyde*.

al. (chemistry) An abbreviation of 'al'cohol.

Ala (chemistry) An abbreviation of 'Ala'nine.

ala (parasitological anatomy) Plural: alae. [Latin] A winglike process, *e.g.* the cervical and caudal alae of nematodes; the wing-like structures at the oesophageal (*USA*: esophageal) and tail regions of the worm.

Alaeuris brachylophi (veterinary parasitology) Oxyurid worms which may be found in the intestines of lizards and turtles.

alanine (Ala) (biochemistry) $CH_3C(NH_2)$- COOH Amino acid commonly found in proteins. Also termed: 2-aminopropanoic acid.

Alaria (veterinary parasitology) A genus of the family Diplostomatidae of digenetic trematodes. Species include *Alaria alata* that may be found in intestines of wild and domestic carnivores but which is generally nonpathogenic.

alariasis (veterinary parasitology) Infestation with the intestinal fluke *Alaria* spp. Many affected animals show no symptoms but very heavy infestations may cause an increase in faecal (*USA*: fecal) mucus and occasionally haemorrhagic (*USA*: hemorrhagic) enteritis.

albendazole (pharmacology/control measures) A benzimidazole anthelminic with high efficiency in sheep and cattle against all intestinal nematodes, except *Trichuris* spp., and intestinal tapeworms and lungworms and, with increased dose rates, will kill adult liver fluke.

albumen (1 parasitological physiology; 2 biochemistry.) 1 Substance produced by vitelline glands of flukes; also forms outer coating of ascarid eggs. 2 White of an egg.

albumin (biochemistry) A group of proteins characterised by their heat coagulation characteristics and solubility in dilute salt solutions. Although found in most tissues and species, they are best known from mammalian blood. Serum albumins together form the most abundant class of blood proteins, and are mainly responsible for keeping the osmotic pressure constant in the circulation, and facilitates the transport of fats, hormones, haptens and drugs. Albumins vary in their composition, but all contain carbon, hydrogen, nitrogen, oxygen and sulphur (*USA*: sulfur). Albumins are colloidal and do not pass through parchment membranes or the membranes of normal living cells; they are coagulated by heat, following which they become insoluble in water until treated with caustic alkalis (*USA*: caustic alkalies) or mineral acids; they are precipitated by various chemicals such as alcohol, tannin, nitric acid and mercury perchloride.

Alcataenia cerorhincae (avian parasitology) A cestode found in the bird, the rhinoceros anklet (*Cerorhinca monocerata*).

Alcataenia fraterculae (avian parasitology) A cestode found in horned puffin (*Fratercula coniculata*).

alcian blue stain (methods) Staining technique for acid mucins.

alcian blue/PAS stain (methods) Staining technique for acid and neutral mucins.

alcian blue/safranine stain (methods) Staining technique for different types of mast cell.

alcohol (1, 2 chemistry) 1 Member of a large class of organic compounds that contain one or more hydroxyl (-OH) groups. A primary alcohol has the general formula R - CH_2OH, where R is an alkyl group (or H in the case of methanol). A secondary alcohol has the formula RR'CHOH, and a tertiary alcohol has the formula RR'R" OH. Alcohols with two -OH groups are diols. Alcohols react with acids to form esters. 2 The term alcohol is often used to refer specifically to ethanol.

aldehyde (chemistry) Member of a large class of organic compounds that have the general formula RCHO, where R is an alkyl group or aryl group. Aldehydes may be made by the controlled oxidation of alcohols. Their systematic names end with the suffix *-al* (*e.g.* the systematic name of acetaldehyde, CH_3CHO, is ethanal). All are reducing agents, and undergo addition, condensation and polymerisation reactions. When oxidised, they form carboxylic acids.

aldehyde fuchsin stain (microscopy) Staining technique for pancreatic islet cells.

aldicarb (control measures) A carbamate pesticide.

aldimine (biochemistry) *See* Schiff's base.

aldol (chemistry) $CH_3CHOHCH_2CHO$ Viscous liquid organic compound. It is an aldehyde, a condensation product of acetaldehyde (ethanal). Also termed: acetaldol; beta-hydroxy-butyraldehyde.

aldol reaction (chemistry) Condensation reaction between two aldehyde or two ketone molecules that produces a molecule containing an aldehyde (-CHO) group and an alcohol (-OH) group, hence *ald-ol* reaction.

aldose (chemistry) Type of sugar whose molecules contain an aldehyde (-CHO) group and one or more alcohol (-OH) groups.

aldoxycarb (control measures) A carbamate pesticide.

aldrin (insect parasitology/control measures) An insecticide. See chlorinated hydrocarbons.

Aleppo boil (human/veterinary parasitology) A common name for cutaneous leishmaniasis. Also termed Aleppo button. *See* leishmaniasis.

Aleppo button (human/veterinary parasitology) Also termed: Aleppo boil. *See* leishmaniasis.

Aleppo gall (plant parasitology) A gall resembling a nut referred to as a gallnut or nutgall found on oaks in the Middle East. The Aleppo gall is a source of material from which a high-quality ink is manfactured. *See* gall.

alfalfa root nematode (plant parasitology) *See Heterodera.*

-algia (terminology) A word element. [German] Suffix denoting *pain.*

algo- (terminology) A word element. [German] Prefix denoting *pain, cold.*

ALGOL (computing) Acronym for the early high-level computer programming language 'Algo'rithmic 'L'anguage, used for manipulating mathematical and scientific data.

algorithm (computing) Operation or set of operations that are required to effect a particular calculation or to manipulate data in a certain way, usually to solve a specific problem. The term is commonly used in the context of computer programming.

alicyclic compound (chemistry) Member of a class of organic chemicals that possess properties of both aliphatic compounds and cyclic compounds.

alimentary canal (anatomy) Tube in the body of animals along which food passes, and in which the food is subjected to physical and chemical digestion and is absorbed. Also termed: alimentary tract; digestive tract; enteric canal; gut.

aliphatic (1, 2 chemistry) 1 Fatty or oily. 2 Pertaining to a hydrocarbon that does not contain an aromatic ring.

aliphatic compound (chemistry) An organic compound (hydrocarbon) whose molecules contain a main chain of carbon atoms, as opposed to ring structures. The chain of carbon atoms may be straight or branched, and the componds saturated or unsaturated. The class includes alkanes, alkenes and alkynes and their derivatives.

alizarin red stain (microscopy) A staining technique for calcium deposits.

alk. (chemistry) An abbreviation of 'alk'aline.

alkali (chemistry) A chemical compound, usually oxides, hydroxides, or carbonates and bicarbonates of metals. Alkalis (*USA*: alkalies) that combine with acid in water to form salts, hence neutralising the acid. [pl. alkalis (*USA*: alkalies)] *See* pH.

alkali metal (chemistry) One of the elements in Group IA of the periodic table. They are lithium, sodium, potassium, rubidium, caesium (*USA*: cesium) and francium, which are all soft silvery metals that react vigorously with water and occur naturally only as their compounds.

alkaline (chemistry) Having a pH between 7 and 14.

alkaline congo red stain (microscopy) A tissue staining technique for amyloid.

alkaline earth (chemistry) One of the metallic elements in Group IIA of the periodic table. They are beryllium, magnesium, calcium, strontium, barium and radium.

alkyl group (chemistry) Group formed by removing a hydrogen atom from an aliphatic compound. *e.g.* methane (CH_4) and ethane (C_2H_6) form methyl (CH_3 -) and ethyl (C_2H_5 -) groups, respectively. It is a hydrocarbon radical.

all(o)- (terminology) A word element. [German] Prefix denoting *other, deviating from normal.*

allele (genetics) One of a pair of genes situated at the same location on homologous chromosomes that controls a specific inherited trait. If the two alleles are identical, the individual is a homozygote, in respect of that locus, and if they are different, the individual is a heterozygote. Each gene may have a very large number of alternative alleles, since an allele need only differ from the original form by one base-pair out of thousands. Also termed 'allelomorph'. *See* mutation.

allergen (immunology) Any substance that is capable of eliciting an allergic response. Allergens are usually proteins and may be derived from insects, worms etc.

See also allergy, allergic response.

allergen extract (immunology) An extract usually containing protein of any substance including that derived from insects, worms etc. to which an animal may be allergic.

allergenicity (immunology) The potential of a compound to provoke an allergic response.

allergic (immunology) Pertaining to or caused by allergy.

allergic response (immunology) A physiological response occurring when an antigen makes contact with mast cells of the immune system. This contact triggers the release of histamine, which induces local inflammation and produces symptoms of asthma, hay fever or the swelling of insect bites. *See* allergy.

allergic rhinitis (medical) Any allergic reaction of the nasal mucosa, occurring perennially (nonseasonal allergic rhinitis) or seasonally.

allergic shock (immunology) *See* anaphylaxis.

allergy (immunology) Abnormal sensitivity of the body to a substance known as an allergen. Contact with the allergen causes symptoms such as skin rashes, watery eyes, rhinitis, etc.

allethrin (control measures) A first-generation pyrethroid used to control ectoparasites on dogs, and may be used in insecticide vapourizers (*USA*: vaporizers) to help protect humans against mosquitoes and biting flies.

alloantibody (immunology) An antibody produced by one individual that reacts with alloantigens of another individual of the same species.

alloantigen (immunology) An antigen existing in alternative (allelic) forms in a species, thus inducing an immune response when one form is transferred to members of the species who lack it.

Allodermanyssus (veterinary parasitology) A genus of mites, members of the family Dermanyssidae. Species includes *Allodermanyssus sanguineus*, which is a blood-sucker in rodents.

allogenic (biology) Affected by outside factors.

allogenic antigen (immunology) An antigen occurring in some but not all individuals of the same species, formerly called isoantigen.

Allomermis lasiusi (insect parasitology) A species of *Mermithidae* in the phylum Nematoda that are parasites to the ants, *Lasius* species, which like *Allomermis myrmecophila* is associated with changes in morphology.

Allomermis lasiusi (insect parasitology) A species of nematode in the Mermithidae that, similarly to *Allomermis myrmecophila*, are parasites to the ants *Lasius* spp. The infection is associated with changes in morphology.

Allomermis myrmecophila (insect parasitology) A species of *Mermithidae* in the phylum Nematoda that are parasites to the ants, *Lasius* species, which like *Allomermis lasiusi* is associated with changes in morphology.

Allomermis myrmecophila (insect parasitology) *See Allomermis lasiusi.*

allotypic determinant (immunology) An antigenic determinant which varies among members of the same species.

almond cyst nematode (plant parasitology) *See Heterodera.*

Alouattamyia (veterinary parasitology) A parasitic fly of which the larval grub causes cutaneous cysts in *Alouatta* species, the howling monkeys of New World primates.

ALP (chemistry) An abbreviation of 'AL'kaline 'P'hosphatase.

alphacypermethrin (control measures) A pyrethroid insecticide commonly employed in the control of cockroaches.

alphanumeric (computing) Describing characters or their codes that represent letters of the alphabet or numerals, particularly in computer applications.

Alphitobius diaperinus (parasitology) A beetle common to the litter in most poultry houses, and suspected of being involved in transmission of Marek's disease. Larval stages are termed the lesser mealworm. *See* beetles.

alternate host (parasitology) Intermediate host.

alternating current (physics) A current that flows in opposite directions sinusoidally.

Alu family (genetics) A set of highly repetitive DNA elements dispersed throughout the human genome; each member is about 300 base-pairs in length, and is repeated 300,000 times, the copies being scattered around the chromosomes. The Alu sequence copies make up some 3 per cent of the total human DNA. The copies exist in slightly varying forms, and it is the combination of these variants that gives the individuality analysed in genetic fingerprinting. The Alu sequence got its name from that of the restriction endonuclease, *Alu I*, by which it was isolated.

Alucha floridensis (insect parasitology) A species of the insect order Hymenoptera in the group *Eulophidae*, a parasite of the ants *Camponotus* and *Crematogaster*, that attacks the larvae and pupa stages.

alum. (chemistry) An abbreviation of 'alum'inium.

aluminium (Al) (chemistry) Silvery-white metallic element in Group IIIA of the periodic table. It occurs (as aluminosilicates) in many rocks and clays and in bauxite, its principal ore, from which it is extracted by electrolysis. It is a light metal, protected against corrosion by a surface film of oxide. At. no. 13; r.a.m. 26.9815. (*USA*: aluminum).

aluminum Al (chemistry) *See* aluminium.

alveolar clearance (physiology) Particles that reach the alveoli are cleared by two principal routes: 1 by phagocytosis and removal either via the mucociliary process or via the lymphatic system; 2 by dissolution of the particles, with the dissolved material passing either to the blood stream or the lymphatic system.

alveolar cyst (veterinary parasitology) Hydatid cysts, metacestode stage of taeniid tapeworms, which contains multiple vesicles, daughter cysts, found in microtine rodents.

alveolar hydatid disease (medical) *See* hydatid disease.

alveolus (anatomy) 1 Minute sac in the vertebrate lung. There are vast numbers of alveoli, and most of the exchange of gases between air and blood takes place within them. *See* respiration. 2 In zoology the term is also used for any small indent or sac in the surface of an organ.

Am (1 chemistry; 2 laboratory equipment) 1 Chemical symbol for americium. 2 An abbreviation of 'am'meter.

AMA (1 general terminology; 2, 3 governing body) 1 An abbreviation of 'A'gainst 'M'edical 'A'dvice. 2 An abbreviation of 'A'ustralian 'M'edical 'A'ssociation. 3 An abbreviation of 'A'merican 'M'edical 'A'ssociation.

amastigote (parasitological physiology) A term used in relation to flagellates. A stage in the life cycle that is round or oval in shape without any free flagellum. Also termed: 'leishmanial'stage.

amber codon (genetics) One of the three nonsense codons in DNA that code for 'stop' and therefore signal the termination of the protein being synthesised (*USA*: synthesized). Amber is UAG. *See* protein synthesis.

ambient temperature (parasitology) Temperature of the immediate environment.

Ambiphyra (fish parasites) A genus of the phylum Ciliophora. A ciliate protozoan which may cause heavy mortality in young fish. They are sessile organisms with a cylindrical to conical body with oral cilia and a permanent motionless equatorial ciliary fringe. They range in size from approximately 60–100 mm and adhere to the epithelium of the skin and/or gills. Disease and death of fish have been associated with chronic infections of the gills due to mechanical blockage of respiratory epithelium. Diagnosis of this parasite is dependent upon identification of this organism within the skin or gill scrapings or histopathology.

Amblyomma (parasitology) A genus of ticks of the family Ixodidae that includes *Amblyomma americanum*, the lone star tick.

Amblyomma americanum (human/veterinary parasitology) A three-host tick which may cause painful bites and tick paralysis and transmits Q fever, tularemia and Rocky Mountain spotted fever of humans. Also termed the lone star tick.

Amblyomma cajennense (human parasitology) A three-host tick that may transmit spotted fever of humans and leptospirosis due to *Leptospira pomona*. Also termed cayenne tick.

Amblyomma hebraeum (human/veterinary parasitology) A three-host tick that may cause severe bite wounds and transmit *Rickettsia ruminantium*, the cause of heartwater. Also termed bont tick.

Amblyomma maculatum (human/veterinary parasitology) A three-host tick which causes paralysis but does not transmit disease. Also termed Gulf coast tick.

Amblyomma pomposum (human/veterinary parasitology) A three-host tick, a vector for *Rickettsia ruminantium*, the cause of heartwater.

Amblyomma variegatum (human/veterinary parasitology) A three-host tick which transmits *Rickettsia ruminantium*, the cause of heartwater, *Coxiella burnetii* the cause of Q fever and Nairobi sheep disease. Also termed variegated or tropical bont tick.

American Academy of Forensic Sciences (AAFS) (society) A society to encourage the study of all aspects of forensic science including parasitology.

American Association for the Advancement of Science (society) A USA professional society whose members are from all branches of science, medicine and engineering.

American dagger nematode (plant parasitology) *See Xiphinema.*

American dog tick (veterinary parasitology) *See Dermacentor variabilis.*

American screwworm fly (veterinary parasitology) *See Cochliomyia.*

American Standard Code for Information Interchange (ASCII) (computing) A standard format for representing characters in differing computer programmes (*USA*: programs).

American trypanosomiasis (human/veterinary parasitology) A disease of humans caused by *Trypanosoma cruzi* in which many animal species act as carriers. The disease in dogs includes anaemia (*USA*: anemia), debility and splenomegaly; in cats there are posterior paralysis and convulsions. The disease is transmitted by reduviid bugs. Also termed Chagas' disease.

americium (Am) (chemistry) Radioactive element in Group IIIB of the periodic table (one of the actinides). It has several isotopes, with half-lives of up to 7,650 years. It is used as a source of alpha-particles. At. no. 95; r.a.m. 243 (most stable isotope).

amicarbalide (pharmacology) An aromatic diamidine used as an antiprotozoal in the treatment of babesiosis, particularly in cattle, and also may have been used in ehrlichia infections.

amide (chemistry) Member of a group of organic chemical compounds in which one or more of the hydrogen atoms of ammonia (NH_3) have been replaced by an acyl group (-RCO). One hydrogen atom in primary amides, two hydrogen atoms in secondary amides and three hydrogen atoms in tertiary amides have been so replaced. Also termed: alkanamides.

amido (chemistry) The monovalent radical $-NH_2$ united with an acid radical.

amidostomatosis (avian parasitology) Parasitic infestation with *Amidostomum* species which may occur under the cuticular lining of the gizzard of ducks, geese and other aquatic birds.

Amidostomum (parasitology) A genus of the helminth family of Amidostomatidae.

Amidostomum anseris (avian parasitology) A parasite of ducks and geese where it may be found in the gizzard. Resultant blood-sucking and mucosal damage may be fatal to young birds.

aminase (biochemistry) One of a group of enzymes that can catalyse the hydrolysis of amines.

amination (chemistry) Transfer of an amino group ($-NH_2$) to a compound.

amine (chemistry) Member of a group of organic chemical substances in which one or more of the hydrogen atoms of ammonia (NH_3) have been replaced by a hydrocarbon group. One hydrogen atom in primary amines, two hydrogen atoms in secondary amines and three hydrogen atoms in tertiary amines have been so replaced.

amino (chemistry) The monovalent radical $-NH_2$, when not united with an acid radical.

amino acid (biochemistry) The building blocks of proteins, amino acids are organic compounds that contain an acidic carboxyl group (-COOH) and a basic amino group ($-NH_2$). Twenty amino acids are commonly found in proteins in animals, and about 100 more that are rare and found only in plants. Those that can be synthesized by a particular organism are known as 'non-essential'; 'essential' amino acids must be obtained from the environment, usually from food. The essential amino acids for humans are: isoleucine, leucine, lysine, me-thionme, phenylalanine, threonine, tryptophan and valine.

amino acid sequencer (methods) An automatic machine for determining the amino acid sequence of a protein.

amino acid transamidation (chemistry) *See* transamination.

amino group (chemistry) Chemical group with the general formula -NRR', where R and R' may be hydrogen atoms or organic radicals; the commonest form is $-NH_2$. Compounds containing amino groups include amines and amino acids.

aminoacetic acid (chemistry) Glycine.

aminocarb (control measures) A carbamate pesticide.

aminoisovaleric acid (chemistry) An alternative term for valine.

aminotoluene (chemistry) An alternative term for toluidine.

amitraz (pharmacology/control measures) A formamidine used as a topical acaricide on cattle, sheep, pigs and fruit crops and miticide in the treatment of generalized demodectic mange in dogs. In horses it may cause fatal impaction of the intestine.

See also colon impaction colic.

ammeter (laboratory equipment) An instrument for measuring electric current, usually calibrated in amperes. The common moving-coil ammeter is a type of galvanometer.

ammonia (chemistry) (NH_3) A colourless (*USA*: colorless) gas with a pungent odour (*USA*: odor), detectable by humans down to concentrations of 53 ppm. Inhalation of the vapour (*USA*: vapor) arising from a solution of ammonia, while uncomfortable, is not dangerous, but inhalation of the gas can cause oedema (*USA*: edema) of the lungs and bronchitis but this is rarely serious. It is formed naturally by the bacterial decomposition of proteins, purines and urea; made in the laboratory by the action of alkalis (*USA*: alkalies) on ammonium salts; or synthesized commercially by fixation of nitrogen. Liquid ammonia is used as a refrigerant. The gas is the starting material for making nitric acid and nitrates.

ammonia solution (chemistry) An alternative term for ammonium hydroxide.

ammonium (chemistry) A hypothetical radical, NH_4, forming salts analogous to those of the alkaline metals.

See also ammonia.

ammonium hydroxide (chemistry) NH_4OH Alkali made by dissolving ammonia in water, giving a solution that probably contains hydrates of ammonia. It is used for making soaps and fertilizers. 880 ammonia is a saturated aqueous solution of ammonia (density 0.88 g cm^{-3}). Also termed: ammonia solution.

ammonium ion (chemistry) The ion NH_4^+, which behaves like a metal ion.

Amnicola limosa porosa (veterinary parasitology) A snail, the intermediate host of *Metorchis conjunctus*, a liver fluke of cats and dogs.

amnion (anatomy) Innermost membrane that envelops an embryo or foetus (*USA*: fetus) (in mammals, reptiles and birds) and encloses the fluid-filled amniotic cavity.

amniotic fluid (anatomy) Liquid that occurs in the amnion surrounding a foetus (*USA*: fetus).

Amoeba (parasitology) A genus of the subphylum Sarcodina. It is a single-celled mass of protoplasm which changes shape by extending cytoplasmic processes called pseudopodia by which it moves and absorbs nourishment. The majority of amoebae (*USA*: amebas) are free-living in soil and water. (*USA*: amebic).

See also amoebic.

amoebiasis (human/veterinary parasitology) An infection caused by the genus *Amoeba* or related genera, especially in the intestine with *Entamoeba histolytica*. Saprophytic amoebae (*USA*: amebas) also occasionally cause disease.

See also amoebic dysentery; amoebic meningoencephalitis; acanthamoebiasis.

amoebic (parasitology) Pertaining to, caused by, or of the nature of, amoebae (*USA*: amebas). (*USA*: amebic).

amoebic dysentery (human/veterinary parasitology) An acute amoebic (*USA*: amebic) dysentery of humans and nonhuman primates. A mild to severe necroulcerative enterocolitis caused by *Entamoeba histolytica* which rarely occurs spontaneously in dogs and cats. A similar disease of reptiles and amphibians is caused by *Entamoeba invadens* and *Entamoeba ranarum*. (*USA*: amebic dysentery).

amoebic gill disease (fish parasitology) A disease caused by *Paramoeba* spp. It is an important disease in sea-caged salmonids; manifested by lethargy, flared opercula, and rapid death that is encouraged by high temperatures. (*USA*: amebic gill disease).

amoebic meningoencephalitis (human parasitology) A meningoencephalitis caused by *Naegleria fowleri* that is restricted in occurrence to humans. (*USA*: amebic meningoencephalitis).

amoebicide (control measures) A substance that is destructive to amoebas (*USA*: amebas). (*USA*: amebicide).

amoebocyte (cytology) Cell that demonstrates amoeboid movement.

See also amoeba.

amoeboid (parasitology) Resembling an amoeba (*USA*: ameba). (*USA*: ameboid).

amoeboid movement (parasitology) Movement like that of an amoeba (*USA*: ameba), by means of pseudopodia, a pseudopod being a projection of the surface of a cell caused by alternate formaion of sol and gel protoplasm in the region of the temporarily formed 'false foot'. (*USA*: ameboid movement).

Amoebotaenia (parasitology) A genus of tapeworms of the family Dilepididae.

Amoebotaenia cuneata (avian parasitology) A species of tapeworm in the genus *Amoebotaenia*. A parasite of poultry, heavy infestations may cause a haemorrhagic (*USA*: hemorrhagic) enteritis, and chronic infestions may cause wasting and reduced growth rate.

Amoebotaenia sphenoides (avian parasitology) A species of tapeworm in the genus *Amoebotaenia*. Like *Amoebotaenia cuneata*, *Amoebotaenia sphenoides* is a parasite of poultry and heavy infestations may cause a haemorrhagic (*USA*: hemorrhagic) enteritis, and chronic infestions may cause wasting and reduced growth rate.

amorphia (biology) The state of being amorphous.

amorphous (biology) Without clear shape or structure.

amoyphism (biology) The state of being amorphous.

AMP (biochemistry) An abbreviation of 'A'denosine 'M'ono'P'hosphate.

amp. (1 measurement; 2 equipment) 1 An abbreviation of 'amp'ere. 2 An abbreviation of 'amp'oule.

amperage (measurement) Strength of an electric current in amperes or milliamperes.

ampere (A) (measurement) SI unit of electric current strength, the current yielded by one volt of electromotive force against one ohm of resistance. It was named after the French physicist André Marie Ampère (1775–1836).

amphi- (terminology) A word element. [German] Prefix denoting *both*, *on both sides*.

Amphibia (biology) A class of animals containing the amphibians.

amphibians (biology) Members of the animal class Amphibia which includes frogs, newts, salamanders and toads all capable of living on land or in water.

amphid (parasitological anatomy) One of a pair of sense organisms, minute papillae on the anterior end, present on parasitic nematodes.

Amphilinidea (fish parasitology) An order of flatworms in subclass Cestodaria, class Cestoidea and the phylum Platyhelminthes. Amphilinideans are secondary monozoic cestodes found in the body cavity of teleost and chondrostean fishes and also in turtles. They have a status independent of gyrocotylideans and caryphyllideans. *Gyrometra* is an example.

Amphimerus (avian parasitology) Avian digenetic trematodes of the family Opisthorchiidae.

Amphimerus elongatus (avian parasitology) Avian digenetic trematode which blocks biliary and pancreatic ducts in poultry, ducks and pigeons. *See Amphimerus*.

Amphimerus pseudofelineus (avian parasitology) Avian digenetic trematode that is a liver fluke in cats and coyotes. *See Amphimerus*.

amphinomids (parasitology) Members of the family Amphinomidae, a group of browsing marine worms with bristles that cause skin irritation.

amphipathic (chemistry) Molecules containing both polar and non-polar regions in their structure.

Amphistomata (parasitology) A suborder of Trematodes that includes the genus *Gastrodiscoider* and *Watsonius*.

amphistomes (parasitology) *See* paramphistomes.

amphistomiasis (parasitology) Also termed stomach fluke infestation. *See* paramphistomiasis.

amphistomous (anatomy) Having a sucker at each end of the body, as in leeches.

amphitrichous (anatomy) Pertaining to having flagella at each end.

amphixozes (parasitology) Diseases that affect humans and other animals with equal facility.

amphophilic (microscopy) Pertaining to staining with either acid or basic dyes.

amphoteric (chemistry) Capable of acting as both an acid and a base and able to neutralize either bases or acids.

amplifier (epidemiology) A host that epidemiologically serves the function of increasing the size of the pathogen population.

ampoule (equipment) One of a range of small tubular glass vessels, closed at one or both ends, used for storage of, *e.g.* a freeze-dried cell suspenion. Also termed vial.

amprolium (control measures) An antiprotozoal parasiticide.

amputation (medical) The removal of part of the body, usually a limb.

amu-darya nematode (plant parasitology) *See Heterodera.*

amyl(o)- (terminology) A word element. [German] Prefix denoting *starch.*

amylase (chemistry) An enzyme that catalyzes the hydrolyis of starch into simpler compounds. The alpha-amylases occur in animals and include pancreatic and salivan amylase; the beta-amylases occur in higher plants.

Amyloodinium (fish parasitology) A protozoan, a member of the Sarcomastigophora, the cause of velvet disease of aquarium fish. The gills are the primary site of inflammation, but the lesions may spread to the skin.

See also velvet disease; *Oodinium*.

amylose (chemistry) Polysaccharide, a polymer of glucose, that occurs in starch.

amylum (chemistry) An alternative term for starch.

Amyrsidea dedonsai (avian parasitology) A louse of guinea fowl.

An (chemistry) Chemical symbol for actinon.

ana- (terminology) A word element. [German] Prefix denoting *upward, again, backward, excessively.*

***Anacamptomyia* species** (insect parasitology) A species of the insect order Diptera in the group Tachinidae, a parasite of the wasps *Belonogaster, Icaria, Polistes and Ropalidida* that attacks the pupa stage.

anadromous fish (biology) Fish which live most of their lives in the sea, but return to fresh water to spawn.

anaerobic culture (parasitology) A culture carried out in the absence of air.

anaesthetic (pharmacology) Drug that induces overall insensibility (general anaesthetic (*USA*: general anesthetic) or loss of sensitivity in one area (local anaesthetic [*USA*: local anesthetic]). (*USA*: anesthetic).

anal (biology) Pertaining to the anus.

anal groove (parasitological anatomy) Microscopic anterior or posterior groove to anus of ticks, which may be used to aid identification.

Analges (avian parasitology) A genus of the Analgesidae family of mites. Species includes *Analges passerinus* found in the plumage of passerine birds.

analgesic (pharmacology) Drug that relieves pain without causing loss of consciousness.

analog computer (computing) Computer that represents numerical values by continuously variable physical quantities (*e.g.* voltage, current). *Contrast* digital computer.

analogidigital converter (computing) A device that converts the output of an analog computer into digital signals for a digital computer.

analysis (methods) The separation into component parts. *See* chromatography; qualitative analysis; quantitative analysis; volumetric analysis.

analysis of variance (ANOVA) (statistics) A statistical procedure for testing hypotheses about population means in which the overall variability in samples of data is split into additive components associated with the independent variables.

analyte (methods) A substance or material determined by a chemical analysis.

analytic (methods) Pertaining to or emanating from analyasis.

analytic reagent (AR) (chemistry) Grade of chemical.

analytical (methods) Pertaining to or emanating from analysis.

analytical balance (laboratory equipment) A laboratory balance sensitive to very small variations of the order of 0.001 mg.

analytical epidemiology (epidemiology) Statistical analysis of epidemiological data in an attempt to establish relationships between causative factors and incidence of disease.

analytical methods (statistics/epidemiology) Techniques used to draw statistical inferences including multiple regression, path analysis, discriminate analysis and logistic analysis.

analytical study (epidemiology) A method for testing a hypothesis as part of an investigation of the association between a disease and possible causes of the disease.

anaphase (cytology/genetics) Stage in mitosis and moiosis (cell division) in which chromosomes migrate to opposite poles of the cell by means of the spindle.

anaphylactic (immunology) Pertaining to anaphylaxis.

anaphylactic antibody (immunology) Antibody, usually IgE, formed after the first exposure to certain allergens and responsible for the signs of anaphylaxis following subsequent exposures to the same allergen.

anaphylactic reaction (immunology) *See* anaphylaxis.

anaphylactic rhinitis (medical) *See* allergic rhinitis.

anaphylaxis (immunology) An unusual or exaggerated allergic reaction of an animal to foreign protein or other substances. Anaphylaxis is an immediate or antibody-mediated hypersensitivity reaction (type I) produced by the release of vasoactive agents such as histamine and serotonin. Release is a consequence of the binding of IgE antibodies to Fc receptors on the surface of particularly mast cells and basophils. Antigen binding to two adjacent IgE molecules causes perturbation of the cell membrane leading to the release of vasoactive substances. Anaphylaxis may be localized, usually cutaneous, or generalized. Also termed: anaphylactic shock.

See also hypersensitivity.

Anaplasma (veterinary parasitology) A genus of organisms in the family Anaplasmataceae. They are gram-negative rod-shaped, spherical, coccoid or ring-shaped bodies that multiply only in the erythrocytes of ruminants. They are transmitted by arthropods and are susceptible to tetracyclines.

Anaplasma centrale (veterinary parasitology) A species of *Anaplasma* that causes a mild form of anaplasmosis in cattle and that has been used as a vaccine against *Anaplasma marginale*.

Anaplasma marginale (veterinary parasitology) A species of *Anaplasma* that causes anaplasmosis in ruminants. The infection is transmitted mechanically by many biting insects, *Boophilus* spp. and *Dermacentor* spp. *e.g.* being biological vectors.

Anaplasma ovis (veterinary parasitology) A species of *Anaplasma* that causes anaplasmosis in sheep.

Anaplasmataceae (parasitology) The family of organisms containing the genera *Anaplasma*, *Paranaplasma*, *Aegyptianella*, *Haemobartonella*, and *Eperythozoon*.

anaplasmosis (veterinary parasitology) A disease caused by *Anaplasma* species. In cattle it is a chronic, often remitting disease characterized by fever, jaundice, emaciation and anaemia (*USA*: anemia) but never haemoglobinuria (*USA*: hemoglobinuria). It is transmitted by a number of insect vectors, especially ticks. In sheep and goats the disease is subclinical.

See also babesiosis.

anat. (anatomy) An abbreviation of 'anat'omy/ 'anat'omical.

Anaticola (avian parasitology) A genus of bird lice in the superfamily Ischnocera. Species include the duck lice *Anaticola anseris* and *Anaticola crassicornis*.

Anatidae (biology) Family of aquatic birds, including the ducks, geese and swans.

Anatoecus (parasitology) A genus of lice in the suborder Mallophaga. Species includes the duck lice *Anatoecus dentatus* and *Anatoecus icterodes*.

anatomic (biology) Pertaining to anatomy, or to the structure of the body.

anatomical (biology) Pertaining to anatomy, or to the structure of the body.

anatomist (biology) One skilled in anatomy.

anatomy (biology) The science dealing with the form and structure of living organisms.

Anatrichosoma (parasitology) A genus of the nematode subfamily Capillariinae.

Anatrichosoma cutaneum (veterinary parasitology) A species in the genus *Anatrichosoma* that parasitizes most Old World nonhuman primates.

Anatrichosoma cuteum (veterinary parasitology) A species in the genus *Anatrichosoma* that parasitizes most Old World nonhuman primates.

anatriptic (pharmacology) A medicine applied by rubbing.

anchor (parasitological anatomy) A 'hook' in the haptor of a monogenetic trematode which aids attachment.

anchor worm (fish parasitology) *See Lernaea*.

Ancryocephalus (fish parasitology) A fluke genus of monogenetic nematodes in the family Dactylogyridae that is a parasite of fish.

ancyl(o)- (terminology) A word element. [German] Prefix denoting *bent*, *crooked*, *in the form of a loop*, *adhesion*, *fixed*.

Ancylostoma (parasitology) A genus of nematode parasites (hookworms) belonging to the family Ancylostomatidae.

Ancylostoma brasiliense (human/veterinary parasitology) A species of the genus *Ancylostoma* parasitic in dogs and cats in tropical and subtropical regions, whose larvae may cause creeping eruption in humans.

Ancylostoma caninum (veterinary parasitology) A species of the genus *Ancylostoma*; the common hookworm of dogs and cats.

Ancylostoma ceylanicum (veterinary parasitology) A species of the genus *Ancylostoma*; a hookworm of dogs and cats in Asia that resembles *Ancylostoma brasiliense*.

Ancylostoma duodenale (human parasitology) A species of the genus *Ancylostoma*; the Old World hookworm of man, a species found chiefly in temperate areas.

Ancylostoma tubaeforme (veterinary parasitology) A species of the genus *Ancylostoma*; a hookworm of cats.

Ancylostomatidae (parasitology) A family of nematode parasites having teeth or two ventrolateral cutting plates at the entrance to a large buccal capsule, and small teeth at its base; the hookworms.

ancylostomiasis (parasitology) Infection by worms of the genus *Ancylostoma* or by other hookworms such as *Necator Americanus*. The symptoms caused by hookworm infestation includes melaena (*USA*: melena), pallor of mucosae, hypernoea (*USA*: hyperpnea), poor exercise tolerance, oedema (*USA*: edema), and anaemia (*USA*: anemia). Also termed hookworm disease.

See also hookworm.

Ancystropus zeleborii (veterinary parasitology) A parasitic mite which may be found under the eyelids and in plugged meibomian glands of Chiroptera bats.

And gate (computing) Computer logic element that combines two binary input signals to produce one output signal according to particular rules. Also termed: AND element.

Andersonstrongylus milksi (veterinary parasitology) A metastrongyloid nematode that may be found in the respiratory tract of dogs.

andr(o)- (terminology) A word element. [German] Prefix denoting *male*, *masculine*.

-ane (chemistry/terminology) Suffix usually denoting that an organic compound is an alkane or cycloalkane; *e.g.* methane, cyclohexane.

anergy (immunology) Absence of reaction to antigens or allergens.

anes. (pharmacology) An abbreviation of 'anes'thetic (*UK*: anaesthetic).

aneurin (biochemistry) An alternative term for thiamine (vitamin B_1).

angiogram (methods) X-ray picture produced following the injection of an X-ray opaque dye into an artery.

angiostrongyliasis (parasitology) Infection by nematodes of the genus *Angiostrongylus*.

Angiostrongylidae (parasitology) A family of nematodes that includes parasites in the genus *Angiostrongylus*.

angiostrongylosis (parasitology) *See* angiostrongyliasis.

Angiostrongylus (parasitology) A genus of worms of the family Angiostrongylidae.

Angiostrongylus cantonensis (human/veterinary parasitology) A species of worms in the genus *Angiostrongylus*. The rat lungworm which may cause eosinophilic meningitis in humans and other species including dogs.

Angiostrongylus costaricensis (human/veterinary parasitology) A species of worms in the genus *Angiostrongylus* that parasitizes the blood vessels of the alimentary canal of wild rodents and may infect humans causing eosinophilic granulomas in the intestine.

Angiostrongylus mackerrasae (human/veterinary parasitology) A species of worms in the genus *Angiostrongylus*. A rat lungworm which may also cause eosinophilic meningitis in humans.

Angiostrongylus vasorum (veterinary parasitology) A species of worms in the genus *Angiostrongylus* that may occur in the pulmonary artery and right ventricle of dogs and foxes, resulting in pulmunary emphysema and fibrosis which may be accompanied by congestive heart failure.

angstrom (Å) (measurement) Unit of wavelength for electromagnetic radiation, including visible light and X-rays, equal to 10^{-10} m, or 0.1 nm. It was named after the Swedish physicist Anders Ångstrom (1814–74).

Anguillicola (fish parasitology) Flukes of the swim bladder of eels which cause devastating losses in eels in Europe. Species include *Anguillicola australiensis*, *Anguillicola crassus*, and *Anguillicola novaezeelandiae*.

anguillicolosis (fish parasitology) Parasitic infestation by members of the *Anguillicola* genus of flukes in the swim badder of eels.

Anguina (plant parasitology) A genus of nematodes in the order Tylenchida and family Anguinidae. It contains the first three recorded plant parasites: *Anguina tritici* in 1743 by Needham, *Anguina agrostis* from seed galls of bentgrass by Steinbuch in 1799, and *Anguina graminis* from galls on grass leaves by Hardy in 1850. They are obese, 1.5 to 5 mm long, and adults are found only in plant stem or inflorescence galls. *See Anguina tritici.*

Anguina agrostis (plant parasitology) A grass nematode that parasitizes grass seedheads and converts the seed into a gall. A toxin is produced when *Clavibacter toxicus* (*Corynebacterium raythi*) is present and if ingested by animals the clinical effects are principally nervous ones especially incoordination, convulsions and death. The grasses susceptible to the worm are annual or Wimmera rye (*Lolium rigidum*) and Chewing's fescue (*Festuca rubra commutata*). Also termed *Anguina lolii*; *Anguina funesta*; bentgrass nematode; seed-gall nematode. *See* gall.

Anguina graminis (plant parasitology) Also termed: fescue leaf gall nematode.

Anguina tritici (plant parasitology) A species of nematodes in the genus *Anguina* whose hosts are wheat and rye. Oats and barley have also been recorded as hosts but little or no reproduction occurs on them although oats may be attacked and severely deformed at seedling stage. They are sedentary endoparasites, diploid and amphimictic. During their life cycle each female lays up to 2,000 eggs within galls. The life cycle follows the standard pattern with four moults (*USA*: molts), the final moult (*USA*: molt) in both sexes occurring only after galls have formed. The adults die soon after oviposition; the eggs hatch quickly and first stage juveniles moult (*USA*: molt) to second stage. By harvest-time galls contain only second stage juveniles which are very resistant to dessication and have been revived after 28 years storage. If galls are sown with seed, second stage juveniles emerge in moist soil, invade host seedlings and feed ectoparasitically on the tissues of young leaves near the growing point. The galls are shed from the ears more readily than the grains so many fall out during harvest and infest the soil. Each gall contains at first up to 40 or more adults of each sex which produce up to 30,000 or more eggs and/or juveniles. Infested seedlings are more or less severely stunted and show characteristic rolling, twisting and crinkling of the leaves with consequent growth distortion. Infested ears are generally undersized, shorter and thicker than healthy ones. Some or all of the grains are replaced by galls. The most effective control is by modern mechanical seed cleaning which eliminates the galls. Also termed: seed and leaf gall nematode.

anhyd. (chemistry) An abbreviation of 'anhyd'rous. Also termed anhydr.

anhydride (chemistry) Chemical compound formed by removing water from another compound (usually an acid); *e.g.* an acid anhydride or acidic oxide.

anhydrite (chemistry) Naturally occurring calcium sulphate (*USA*: calcium sulfate), used to make fertilizers.

See also gypsum.

anhydrous (chemistry) Describing a substance that is devoid of moisture, or lacking water of crystallization.

aniline (chemistry) $C_6H_5NH_2$ Colourless (*USA*: colorless) oily liquid organic compound, one of the basic chemicals (feedstock) used in the manufacture of dyes, pharmaceuticals and plastics. Also termed: aminobenzene; phenylamine.

animal (biology) Member of a large kingdom of organisms that feed heterotrophically on other organisms or organic matter. Animals are usually capable of movement and react quickly to stimuli (because they have sense organs and a nervous system). Animal cells have limited growth, no chlorophyll, and are surrounded by a cell membrane. In some classifications, certain unicellular organisms such as protozoans are also included. The clasification of animals into major groups is now generally agreed upon by most biologists, unlike the situation in the plant kingdom where there are several alternative systems of classification.

animal bite (parasitology) Trauma caused by teeth and usually heavily contaminated with microorganisms.

animal model (methods) Any condition in an animal that has enough similarities to a condition in humans that studies of the animal disease will assist in understanding the human disorder.

animal starch (chemistry) An alternative term for glycogen.

anion (physics/methods) A negatively charged particle, which in an electric field will migrate toward the positive pole. *See* anode; electrolysis.

anionic detergent (chemistry) A substance that, when dissolved, contributes a hydrophobic ion which carries a negative charge to the solution.

anis(o)- (terminology) A word element. [German] Prefix denoting *unequal*.

anisakiasis (human/avian parasitology) The disease in humans caused by infestation with worms of the genus *Ainsakis*. The symptoms often resemble that of peptic ulcer in cases of gastric involvement, and diagnosis may be made from a biopsy specimen of stomach or intestine and gastroscopy may also be of value in some cases. The larva is characterised by the prominent lateral chords which split at their anterior end. The parasite may be found in salt-water fish but only rarely in fresh-water fish and survives smoking but is killed by heat and freezing. Also termed: herring worm disease.

See also Phocanema, *Terranova* and *Contracaecum*.

Anisakidae (human/avian/fish/veterinary parasitology) A family of nematode parasites of marine mammals, fish, birds, reptiles and, by accidental infestation, humans.

See also Phocanema, *Terranova* and *Contracaecum*.

Anisakis (human/fish/veterinary parasitology) A genus of nematodes that parasitize the stomachs of fish-eating marine mammals such as dolphins, porpoises and whales. The eggs passed in the faeces (*USA*: feces) of these animals produce second stage larvae which develop in crustaceans into third stage larvae. Further development occurs in fish, which act as a source of the parasite to humans causing - anisakiasis, and marine mammals. *See* anisakiasis.

anisogamy (genetics) The production of two unequal-sized types of gametes in a species. It is the system used by all higher eukaryotes, including humans, in which the ovum is much larger than the sperm. *Contrast* isogamy.

ankyroid (terminology) Hooklike; like the fluke of an anchor.

annealing (molecular biology) The reassociation of complementary strands of DNA after they have been denatured, usually by melting at a high temperature. The annealed strands can either be original complementary strands or they can be any two strands that have enough base-pairs matching for reassociation to occur. This hybridization of DNA is one technique used in genetic engineering. Annealing can also occur between a strand of DNA and a strand of RNA. *See* denature.

annelid (parasitology) A member of the phylum Annelida. A group containing the segmented worms and leeches. Annelids are characterized by the presence of metameric segmentation, a coelom (*USA*: celom), well-developed blood and nervous systems, and nephridia.

Annelida (parasitology) A phylum of metazoan invertebrates, the segmented worms, including leeches.

annual (1, 2 terminology) 1 Pertaining to being conducted once each year. 2 Any plant that germinates from seed, grows to maturity and produces new seed all within one year or growing season.

***Anocentor* species** (parasitology) *See Dermacentor*.

anode (physics/methods) An electrode that is positively charged relative to another with a current flowing between the two; it attracts anions. *Contrast* cathode.

See also electrolysis.

Anoetus (insect parasitology) A species of mite in the phylum Asigmata that are parasites to the bees, Halictinae and *Apis*.

Anoetus myrmicarum (insect parasitology) A species of mite in the phylum Asigmata that are parasites to the termites, *Reticulitermes lucifugus*, and that has caused high mortality in laboratory colonies.

Anomala orientalis (plant parasitology) The oriental beetle. *See* beetles.

Anomotaenia borealis (insect parasitology) *See Chaenotania craterformis*.

Anomotaenia brevis (insect/avian parasitology) A species of helminth in the phylum *Cestoda* that are parasites to the ant, *Leptothorax* species, that are infected as larvae, cause aberrant colours (*USA*: colors) and whose final hosts are woodpeckers.

anon. (literary terminology) An abbreviation of 'anon'ymous, used with capital as attribution of unknown authorship.

Anopheles (human/veterinary parasitology) A genus of the Culicidae family of mosquitoes. About 380 species of *Anopheles* occur around the world, some 60 species being sufficiently attracted to humans to act as vectors of malaria. A number of *Anopheles* species are also vectors of filariasis and viral diseases. Many *Anopheles* species feed on both humans and animals. They differ, however, in the degree to which they prefer one over the other. Some species feed mostly on animals while others feed almost entirely on humans, the latter species being the more dangerous as vectors of malaria. Species includes: *Anopheles balabacensis*, *Anopheles barbirostris*, *Anopheles culicifacies*, *Anopheles darlingi*, *Anopheles maculatus*, *Anopheles minimus* and *Anopheles stephensi*.

anopheline (parasitology) Pertaining to the *Anopheles* genus of mosquitoes.

Anophryocephalus ochotensis (veterinary parasitology) A tapeworm of the Californian sea lion.

Anopleura (parasitology) A suborder of sucking lice that includes *Haematopinus*, *Linognathus*, and *Solenoptes* spp. (*USA*: Anoplura).

Anoplocephala (veterinary parasitology) A genus of the tapeworm family of Anoplocephalidae. Species includes *Anoplocephala magna*, a parasite of horses and donkeys, and *Anoplocephala perfoliata* a parasite of horses.

Anoplocephaloides (veterinary parasitology) A genus of the tapeworm family Anoplocephalidae. Species includes *Anoplocephaloides mammillana* found in the small intestine of the horse where heavy infestations may cause unthriftiness.

anoplocephalosis (parasitology) An infestation caused by tapeworms of the family Anoplocephalidae.

Anoplophora chinensis (plant parasitology) The citrus longhorn beetle. *See* beetles.

Anoplophora malasiaca (plant parasitology) The white-spotted longicorn beetle. *See* beetles.

Anoplotaenia dasyuri (veterinary parasitology) A metacestode that may be found in heart, lungs and skeletal muscles of the Australian marsupials, wallabies and pademelons.

anoscope (methods) A speculum or endoscope used in direct visual examination of the anal canal.

anoscopy (methods) The examination of the anal canal with an anoscope.

anosigmoidoscopy (methods) The endoscopic examination of the anus and distal colon.

ANOVA (statistics) An abbreviation of 'AN'alysis 'O'f 'VA'riance.

ANP (physiology/biochemistry) An abbreviation of 'A'trial 'N'atriuretic 'P'eptide.

ANS (anatomy) An abbreviation for 'A'utonomic 'N'ervous 'S'ystem.

ANSL (quality standards) An abbreviation of 'A'ustralian 'N'ational 'S'tandards 'L'aboratory.

ant (insect parasitology) A ubiquitous insect, which may act as intermediate hosts for the trematode *Dicrocoelium dendriticum* and cestodes *Raillietina* spp.

ant bite conjunctivitis (veterinary parasitology) A form of conjunctivitis recorded in calves and recumbent animals attacked by fire ants (*Solenopsis invicta*).

antacid (pharmacology) Substance used medicinally to combat excess stomach acid (*e.g.* various compounds of magnesium).

antagonism (biology) The inhibiting or nullifying action of one substance or organism on another, *e.g.* the exhaustion of a food supply by one organism at the expense of another.

Antarctophthirus (veterinary parasitology) A genus of the insect family Echinophthiriidae, lice which infest pinnipeds such as sea lions. Species includes *Antarctophthirus microchir*.

antemortem (medical) Before death.

antenna (parasitological anatomy) (plural: antennae) Usually one of a pair of many-jointed, whip-like structures present on the head of many arthropods, particularly insects, on the first appendage on the head, and crustaceans, on the second appendage. Antennae have a sensory function, though in some crustaceans they are used for attachment or swimming.

Antennophours (insect parasitology) A species of mites in the Mesostigmata that, similarly to *Bisternalis*, are cleptoparasitic on the bees Meliponinae.

anterior (biology) Near the head of an animal.

See also posterior.

anterior station trypanosomes (parasitology) A section of the genus *Trypanosoma* in which the infectious stages accumulate in the mouthparts and salivary glands of the intermediate host so that the parasite is transmitted when the insect vector takes a blood meal. Also termed *Salivaria*.

antero- (terminology) A word element. [Latin] Prefix denoting *anterior, in front of*.

anthelminthic (pharmacology/control measures) Anthelmintic.

anthelmintic (1, 2 pharmacology/control measures) 1 Destructive to worms. 2 An agent destructive to worms. They are classified as antinematicidal, antitrematicidal, and anticesticidal. Also termed: dewormer.

anthelmintic poisoning (pharmacology/control measures) *See* under individual anthelmintics.

anthelmintic resistance (pharmacology/control measures) Frequent dosing of animals, especially ruminants running at pasture, selectively retains worms with innate resistance to a particular anthelmintic, and a population of resistant worms may result. Side resistance to other compounds in the same chemical group may occur as is observed in the benzimidazole group of compounds and in the levamisole-morantel group.

Anthonomus bisignifer (plant parasitology) The strawberry weevil.

Anthonomus grandis (plant parasitology) The cotton boll weevil.

Anthonomus signatus (plant parasitology) The strawberry blossom weevil.

anthropo- (terminology) A word element. [German] Prefix denoting *human being*.

anthropophilic (parasitology) Preferring human beings to animals, an occurrence that may be said of certain mosquitoes.

anthropozoonosis (parasitology) A disease of either animals or humans that may be transmitted from one species to the other.

anthropozoophilic (parasitology) Attracted to both human beings and animals. May be said of certain mosquitoes.

anti- (terminology) A word element. [German] A prefix denoting *effective against, counteracting*.

antiamoebic (1, 2 pharmacology/control measures) 1 Destroying or suppressing the growth of amoebae (*USA*: amebae). 2 A substance that destroys or suppresses the growth of amoebae (*USA*: amebas). (*USA*: antiamebic).

antibabesials (pharmacology/control measures) Drugs that are effective against red blood cell parasites of the genus *Babesia*. Older drugs are trypan blue and quinuronium. Modern treatments are aromatic diamidines and carbanilides such as imidocarb, amicarbalide, Berenil (Ganaseg) and phenamidine.

antibiosis (biology) An association between two populations of organisms that is detrimental to one of them, or between one organism and an antibiotic produced by another.

antibody (immunology) Highly specific molecule produced by the immune system in response to the presence of an antigen, which it neutralizes.

See also acquired immunity.

antibody classes (immunology) *See* immunoglobulin.

antibody-antigen reaction (immunology) The specific combination of antigen with homologous antibody resulting in the reversible formation of antibody-antigen complexes that differ in composition according to the antibody-antigen ratio.

See also antigen.

antibody-mediated hypersensitivity (immunology) Types I, II and III hypersensitivity reactions. Also termed immediate hypersensitivity.

antibody-mediated immunity (immunology) humoral immunity.

anticestodal (1, 2 pharmacology/control measures) 1 Destructive to cestodes. 2 A substance destructive to tapeworms.

anticestodal drugs (pharmacology/control measures) Drugs that are destructive to cestodes and tapeworms, which include the natural organic compound arecoline and the synthetic compounds bunamidine, niclosamide, dichlorophen, praziquantel, uredofos and resorantel.

anticoagulant (pharmacology/control measures) An agent that prevents blood from clotting. Anticoagulants are often used in the collection of blood for testing.

anticoagulant rodenticide (control measures) Anticoagulant substances effective in the control of rodents that includes warfarin, pindone, diphacinone, phentolacin, and Valone.

anticoccidials (pharmacology/control measures) Chemicals effective in the treatment and control of coccidiosis in birds and animals. Examples include clopidol, quinolones, monensin, lasalocid, robenidine, amprolium, dinitolmide, nicarbazin and sulphonamide (USA: sulfonamide). Also termed coccidiostatic drugs.

anticodon (genetics) The three adjacent nucleotides in a transfer-RNA molecule that are complementary to, and base pair with, the three complementary nucleotides of a codon in a messenger-RNA molecule. *See* protein synthesis; genetic code.

antidiarrhoea (1, 2 pharmacology) 1 Counteracting diarrhoea (*USA*: diarrhea). 2 A substance that counteracts diarrhoea (*USA*: diarrhea). (*USA*: antidiarrheal).

antidysenteric (pharmacology) Counteracting dysentery.

antiemetics (pharmacology) Drugs that suppress vomiting. The chief antiemetics are the phenothiazines, such as chlorpromazine, perphenazine, prochlorperazine or promethazine.

antigen (immunology) A substance that elicits an immune response. Usually a protein molecule that the body's immune system recognizes as 'foreign', but they may be other macromolecules, *e.g.* bacteria.

See also antibody.

antigenemia (immunology) The presence of antigen in the blood.

antigenic (immunology) Having the properties of an antigen.

antigenic determinant (immunology) A structural component of an antigen against which immune responses are made and to which antibody or T cell receptor binds; an antigen such as a protein has many antigenic determinants. Also termed epitope.

antigenic mimicry (immunology) Similarities between sequences found in non-host proteins and host proteins which may result in cross-reacting immune responses and autoimmune disease.

antigenicity (immunology) The capacity to stimulate the production of antibodies or cell-mediated immune responses.

antihelminthic (pharmacology) Anthelmintic.

antimalarial (pharmacology) A drug used to treat or prevent malaria. The class of drug most commonly used to treat or prevent infection are the quinidines, of which chloroquine is the standard. However, in some parts of the world certain forms of the protozoan that causes malaria are resistant to chloroquine. In such cases, quinine, the traditional remedy for malaria, may be used.

antimony potassium tartrate (pharmacology/control measures) A substance which may be used as an antiparasitic agent in schistosomiasis, trypanosomiasis and leishmaniasis. Also termed tarter emetic.

antimony (Sb) (chemistry) Blue-white semimetallic element in Group VA of the periodic table. At. no. 51; r.a.m. 121.75. Trivalent and pentavalent antimony compounds are used in medicine as anti-infective agents in the treatment of tropical diseases, especially those of protozoan origin. All antimony compounds are potentially poisonous and must be used with caution.

antinematodal (pharmacology/control measures) *See* nematocide.

antinematodal drugs (pharmacology/control measures) Drugs active against nematodes that includes piperazine, imidazothiazoles and tetrahydropyrimidines (*e.g.* tetramizole, levamisole, morantel, pyrantel), benzimidazole and pro-benzimidazoles (*e.g.* thiabendazole, mebendazole, parbendazole, fenbendazole, oxfendazole, cambendazole, flubendazole, febantel, thiophanate, netobimin), macrocyclic lactones (*e.g.* ivermectin, moxidectin, doramectin), organophosphorus compounds (*e.g.* dichlorvos, haloxon, trichlorfon), salicylanilides (*e.g.* closantel, nitroscanate).

antiparallel (genetics) The orientation of DNA strands in a duplex. One strand runs 5' to 3' in one direction whereas the complementary strand runs 5' to 3' in the opposite direction. By virtue of the polarity of DNA, the two strands are therefore antiparallel.

antiparasitic (1, 2 pharmacology/control measures) 1 Destroying parasites. 2 A substance that destroys parasites.

antiparasitic agent (pharmacology/control measures) Agents active against parasites that include insecticides, acaricides and anthelmintics. Their suitability depends on their efficiency in reducing parasite loads, especially of the immature forms; the breadth of their therapeutic spectrum; safety; and their ease of administration; cost and freedom from tissue residue problems and the development of resistance.

antipediculotic (1,2 pharmacology/control measures) 1. Effective against lice and in treatment of pediculosis. 2 A substance that is effective against lice.

Antipolistes anthracella (insect parasitology) A species of the order Arthropods in the group Lepidoptera, a parasite of the wasps *Polistes* species.

antiprotozoal (1, 2 pharmacology/control measures) 1 Destroying protozoa, or checking their growth or reproduction. 2 A substance with that effect.

antiprotozoal agent (pharmacology/control measures) *See* anticoccidials; trypanocidal.

antisense RNA (genetics) An RNA transcript having the polarity complementary to that of normally transcribed RNA.

antiseptic (pharmacology) Substance that prevents sepsis by killing bacteria or preventing their growth. Also termed: germicide.

antiserum (immunology) Blood serum containing antibodies, used in vaccines to treat or prevent a disease or to combat animal venom (*e.g.* snakebite).

antispasmodic drugs (pharmacology) Drugs used to relieve spasm in smooth muscle such as those in the respiratory tract or intestinal wall.

Antispila **species** (insect parasitology) A species of the order Arthropods in the group Lepidoptera, a parasite of the wasps *Polistes* species.

antitick serum (veterinary parasitology/control measures) Hyperimmune serum produced in dogs by exposing naturally resistant animals to a continuous and heavy infestation with the tick *Ixodes holocyclus*. Used in treatment and temporary protection of dogs against tick paralysis.

antitoxin (immunology) Type of antibody against a toxoid produced in the body by a disease organism or by vaccination.

antitrematodal (pharmacology/control measures) Drugs having efficiency in the treatment of trematodes. Because of the lower susceptibility to injury of immature flukes the measure of efficiency of antitrematodal drugs is measured by their efficiency against immature flukes. Antitrematodal drugs includes albendazole, bithionol suloxide, bromsalans, carbon tetrachloride, clioxanide, diamfenetide, tetrachlorodifluorethane, hexachloroethane, hexachloroparaxylene, hexachlorophene, niclofolan, nitroxynil, oxyclozanine, rafoxanide, and triclabendazole.

antrum (1, 2 anatomy) 1 Cavity in a bone (*e.g.* a sinus). 2 Part of the stomach next to the pyloris.

anus (anatomy) Posterior opening of the alimentary canal, through which the undigested residue of digestion is passed.

anx (literary term) An abbreviation of 'an'ne'x'.

ANZAAS (education) An abbreviation of 'A'ustralian and 'N'ew 'Z'ealand 'A'ssociation for the 'A'dvancement of 'S'cience.

Aonidiella citrina (plant parasitology) Yellow scale.

aorta (anatomy) Principal artery that takes oxygenated blood from the heart to all parts of the body other than the lungs.

Aotiella aotophilus (veterinary parasitology) Biting lice found on New World primates.

aperture (microscopy) Effective diameter of a lens or lens system. Its reciprocal is the f-number.

APHA (society) An abbreviation of 'A'merican 'P'ublic 'H'ealth 'A'ssociation.

Aphaniptera (parasitology) The insect order containing the fleas, whose members are laterally compressed, blood-sucking forms that lack wings but have well developed hind legs modified for jumping. Also termed: Siphonaptera.

Aphanius (control measures) A larvivorous fish belonging to the tooth carp family Cyprinodontidae originally distributed in tropical and subtropical Africa, that feeds on mosquito larvae that has been used around the world in attempts to control malaria, other mosquito-borne diseases and mosquito nuisance.

Aphelenchina (plant parasitology) A suborder of plant parasites in the order Tylenchida that includes *Aphelenchoides fragariae*, the foliar nematode.

Aphelenchoides (plant parasitology) A genus of nematodes in the order Tylenchida and family Aphelenchoididae. Females are slender, usually under 1 mm long and when killed by heat the male tail curls ventrally into a 'walking stick' shape. *See Aphelenchoides fragariae.*

aphelenchoides besseyi (plant parasitology) Also termed: rice white-tip nematode; spring dwarf nematode; strawberry bud nematode; white-tip nematode.

aphelenchoides composticola (plant parasitology) Also termed: mushroom nematode.

Aphelenchoides fragariae (plant parasitology) A species of nematodes in the genus *Aphelenchoides* whose hosts cover a wide range of over 250 hosts in 47 families, including fern, lily, begonia, African violet, strawberry, and many aquatic plants in temperate and tropical regions. They are a migratory endoparasite in leaves, but also feed ectoparasitically on leaf and flower buds in strawberry, for example. The nematode enters leaves through stomata or directly, and are sexually reproducing, thus requiring males. Their life cycle is completed in 10–13 days preferring moist, cool situations, and the female produces approximately 30 eggs. Nematode feeding causes blotches and necrotic lesions between veins on leaves which start as water-soaked spots and then turn brown. In strawberries, the nematode causes malformed leaves with crinkled edges. Control measures include foliar or soil treatments with systemic chemicals. Also termed: foliar nematode.

aphelenchoides olesistus (plant parasitology) Also termed: fern nematode.

aphelenchoides oryzae (plant parasitology) Also termed: rice nematode.

aphelenchoides ribes (plant parasitology) Also termed: currant nematode.

aphelenchoides ritzemabosi (plant parasitology) Also termed: black currant nematode; chrysanthemum foliar nematode; chrysanthemum leaf nematode; chrysanthemum nematode.

Aphelenchoidinae (plant parasitology) A subfamily in the suborder Aphelenchina that includes *Aphelenchoides fragariae*, the foliar nematode.

aphid (plant parasitology) An insect of the order Hemiptera that parasitize many plants during the warm seasons. Aphids feed on plant juices by using piercing and sucking mouthparts and are of considerable economic importance because they can act as vectors of plant viruses. Of interest in animal health are the black aphids, *Aphis craccivora*, which can exist in very large numbers on burr trefoil, *Medicago polymorpha*, and may stimulate the production of phytoallexins in the plant, which then photosensitize animals grazing the infested pasture. Also termed: green fly; plant louse.

apholate (control measures) A chemical sterilant for insects which may cause congenital ocular defects in sheep.

Aphomia sociella (insect parasitology) A species of the order Arthropods in the group Lepidoptera, a parasite of the wasps *Dolichovespula* species. They are also a parasite of the bees *Bombus* species that usually destroys entire colonies, attacking the brood and locating the host nest by smell.

Aphyosemion (control measures) A larvivorous fish belonging to the tooth carp family Cyprinodontidae originally distributed in tropical and subtropical Africa, that feeds on mosquito larvae that has been used around the world in attempts to control malaria, other mosquito-borne diseases and mosquito nuisance.

Apicomplexa (parasitology) A phylum of protozoa, including the coccidia.

Apicystis (Mattesia) bombi (insect parasitology) A protozoal species of gregarinida that are parasites to the bees *Bombus* spp., and *Psithyrus*, including the queens.

Aplocheilus panchax (control measures) A larvivorous fish, belonging to the tooth carp family Cyprinodontidae originally distributed in India and south-east Asia, that feeds on mosquito larvae that has been used around the world in attempts to control malaria, other mosquito-borne diseases and mosquito nuisance.

apneustic (parasitological anatomy) Pertaining to the aquatic larvae of certain insects where spiracles are closed or absent.

apo- (terminology) A word element. [German] Prefix denoting *away from*, *separated*.

Apocephalus (insect parasitology) A genus of the insect order Diptera in the group Phoridae, a parasite to adults of the ants *Acromyrmex*, *Atta*, *Azteca*, *Camponotus*, *Eciton*, *Formica*, *Iridomyrmex*, *Labidus*, *Nomamyrmex*, *Paraponera*, *Pheidole* and *Solenopsis*, that attacks the adult stage. Typically the female attacks the host ant from the air outside the nest and oviposits on the head and thorax or inside the gaster.

Apocephalus borealis (insect parasitology) A species of the insect order Diptera in the group Phoridae, a parasite of the bees *Apis cerana* and *Bombus* species, that attacks the adult stage and feeds on thorax muscles. They are also a parasite of the wasps *Vespula* species, and *Vespa*, that like *Gymnoptera vitripennis* attacks the adult stage.

apochromatic lens (microscopy) A lens corrected for both chromatic and spherical aberration.

Apodicrania termilophila (insect parasitology) A species of the insect order Diptera in the group Phoridae, a parasite to larvae of the ants *Solenopsis species*, the parasite being tended by the ants.

apodous (biology) Without legs.

apoenzyme (biochemistry) Protein part of an enzyme that consists of a protein and a non-protein portion.

Aponomma (veterinary parasitology) A genus of ticks occurring almost entirely on reptiles. Resembles *Amblyomma* spp. except that they have no eyes. Species includes *Aponomma aruginans* parasites of wombats, and *Aponomma concolor* parasites of echidnas.

Apophallus (veterinary parasitology) A genus of intestinal flukes (digenetic trematodes) of the family Heterophyidae. Species includes *Apophallus muhlingi* and *Apophallus donicum* parasites of dogs and cats.

apoplast (biology) 1 A plastid which lacks chromatophores. The adjective form – apoplastic is applied to individual protozoans that lack colour (*USA*: color) in a group which is generally coloured (*USA*: colored). 2 Those areas of the plant that are outside the symplast, comprising the parts outside the plasmalemma, such as cell walls, intercellular spaces and the dead tissues of xylem.

apoptosis (physiology) The process of programmed cell death.

apostome (parasitology) A member of the order Apostomatida of ciliated protozoa.

apothecaries' weights and measures (measurements) A generally superseded system used for measuring and weighing drugs and solutions. It is gradually being replaced by the metric system.

apothecary (pharmacology) A pharmacist; a person who compounds and dispenses drugs.

apothecary ounce (measurements) Unit of weight, 1/12 of a pound (apothecary). 1 oz ap. = 1.0971 oz (avoirdupois) = 31.103481 grams. Also termed: ounce, troy.

app. (1, 2, 3, 4, general terminology) 1 An abbreviation of 'app'aratus (also termed appar.; apparat.). 2 An abbreviation of 'app'arent. 3 'app'lied (also termed appl.). 4 'app'roximate (also termed approx.).

apparat. (general terminology) *See* app.

apparatus (1 equipment; 2 biology) 1 An arrangement of a number of parts acting together to perform a special function. 2 Certain organ systems such as respiratory or digestive apparatus.

appd. (general terminology) An abbreviation of 'app'rove'd'.

append (computing) Adding a character to the end of an existing string of characters or writing data after the end of existing data in a file.

appendage (1, 2 biology) 1 A protuberant outgrowth, such as a tail, a limb or limblike structure. 2 A thing or part appended.

appendix (anatomy) Vestigial outgrowth of the caecum (*USA*: cecum) in some mammals. In human beings its full name is vermiform appendix.

appl. (1, 2 general terminology) 1 An abbreviation of 'appl'icable. 2 An abbreviation of 'appl'ied.

apple cyst nematode (plant parasitology) The nematode *Globodera mali.*

apple root-knot nematode (plant parasitology) The nematode *Meloidogyne mali. See Meloidogyne.*

applications programme (computing) Computer programme (*USA*: program) written by the user for a specific purpose, *e.g.* record-keeping or stock control. (*USA*: applications program).

applicator (pharmacology) A device, *e.g.* a flat wooden stick, used to apply medicaments to a small area of exposed tissue.

approx. (general terminology) *See* app.

apps (literary terminology) An abbreviation of 'app'endice's'.

appx (literary terminology) An abbreviation of 'app'endi'x'.

Aproctella stoddardi (avian parasitology) (synonym: Microfilaria fallisi) A tissue nematode in turkeys, doves, quail, and grouse. Necropsy lesions include granulomatous pericarditis.

APST (society) An abbreviation of 'A'ssociation of 'P'rofessional 'S'cientists and 'T'echnologists.

Apterophora attophila (insect parasitology) A species of the insect order Diptera in the group Phoridae, a parasite of the ants *Atta, Camponotus, Eciton* species, *Iridomyrmex, Pachycondyla,* and *Solenopsis. Apterophora attophila,* like *Apterophora caliginosa, Auxanommatidia myrmecophila, Borgmeierphora kempfi, Borgmeierphora multisetosa* and *Cataclinusa* species attacks the adult stage of the ant. Also a parasite of the termites *Nasutitermes* species, that, like *Apterophtora caliginosa,* attacks the adult stage of the termite.

Apterophora caliginosa (insect parasitology) A species of the insect order Diptera in the group Phoridae, a parasite of the ants *Atta, Camponotus, Eciton* species, *Iridomyrmex, Pachycondyla,* and *Solenopsis. Apterophora caliginosa,* like *Apterophora attophila, Auxanommatidia myrmecophila, Borgmeierphora kempfi, Borgmeierphora multisetosa* and *Cataclinusa* species, attacks the adult stage of the ant. Also a parasite of the termites *Nasutitermes* species, that, like *Apterophtora attophila,* attacks the adult stage of the termite.

apyogenous (medical) Not caused by pus.

apyretic (medical) Without fever.

apyrexia (medical) Absence of fever.

apyrogenic (medical) Not producing fever.

aq. (chemistry/terminology) An abbreviation of 'aq'ueous/ '*aq'ua.* [Latin, meaning 'water'].

AQL (quality standards) An abbreviation of 'A'cceptable 'Q'uality 'L'evel.

aquatic (biology) Living or growing in, happening in, or connected with water.

aquatic respiration (physiology) A process in which freshwater and marine organisms carry out gas exchange with the water that surrounds them. Although fully saturated water contains only about 0.02 percent as much oxygen per unit volume as air, as the tissues are in direct contact with the gas-carrying medium flowing past, the system is very efficient and often a relatively small respiratory surface is required. Many aquatic organisms such as protozoa and nematodes, have no special respiratory organs, relying on a large surface area for gas exchange.

aqueous (chemistry) Dissolved in water, or chiefly consisting of water.

aqueous habitat (biology) Any habitat in which water is the medium in which the organisms live.

aqueous humour (anatomy) The liquid between the lens and cornea of the eye.

aqueous solution (chemistry) Solution in which the solvent is water.

Ar (chemistry) The chemical symbol for Argon.

arabinose (chemistry) $C_5H_{10}O_5$ Crystalline pentose sugar derived from plant polysaccharides (such as gums).

arachnid (parasitology) A member of the class Arachnida in the phylum Arthropoda. The class contains, amongst others, ticks and mites.

Arachnida (parasitology) A class of organism in the phylum Arthropoda that contains, amongst others, the ticks and mites.

arachnidism (parasitology) Poisoning from a spider bite.

aragonite (chemistry) Fairly unstable mineral form of calcium carbonate ($CaCO_3$).

Aranaeomorpha (parasitology) A suborder of spiders with horizontal fangs; includes the red-backed spider.

Araneida (parasitology) An order of the class Arachnida which contains the spiders.

arch(i)- (terminology) A word element. [German] Prefix denoting *ancient, beginning, first, original.*

Archaeopsylla erinacei (veterinary parasitology) A species of flea whose principal host is hedgehogs.

Archiacanthocephala (parasitology) A class in the phylum Acanthocephala. Most archiacanthocephalans are parasites of predaceous birds and mammals. *See* Acanthocephala.

archiving (computing) Removal of infrequently needed data from a computer to a less accessible storage.

Arcyophora (veterinary parasitology) A genus of moths in the order Lepidoptera. Species includes *Arcyophora longivalvis* and *Arcyophora patricula,* which are nocturnal moths that feed on the secretions of cows' eyes and physically transmit infections of the conjunctival sac between cattle.

area-incidence ratio (epidemiology) The number of new cases of a specific disease in a population during a specified time period, divided by the geographic area size in which the observations are made, multiplied by the time elapsed, *e.g.* cases per acre per month.

arecoline (pharmacology) An alkaloid obtained from the nut of the tree *Areca catechu* that is previously a preferred treatment for cestodes in dogs. The acetarsol, hydrobromide and carboxyphenylstilbonate salts have been used for this purpose. Oral administration causes paralysis of the worms and catharsis, so the worms are expelled alive and intact.

Arg. (chemistry) An abbreviation of 'Arg'inine.

Argas (avian/human parasitology) A genus of ticks parasitic on poultry and other birds and occassionally humans.

Argas miniatus (parasitology) A species in the genus *Argas* that is similar to *Argas persicus* and originally classified with that species.

Argas persicus (avian/human parasitology) A species in the genus *Argas* that causes tick worry of birds and paralysis of poultry and which may transmit *Aegyptianella pullorum* and *Borrelia anserina.*

Argas radiatus (parasitology) A species in the genus *Argas* that is similar to *Argas persicus* and originally classified with that species.

Argas reflexus (avian/human parasitology) A species in the genus *Argas* that infests pigeons and may transmit *Borrelia anserina* to poultry.

Argas robertsi (parasitology) A species in the genus *Argas* that is similar to *Argas persicus* and originally classified with that species.

Argas sanchezi (parasitology) A species in the genus *Argas* that is similar to *Argas persicus* and originally classified with that species.

argasid tick (parasitology) A tick belonging to the family Argasidae.

Argasidae (parasitology) A family of arthropods made up of the softbodied ticks.

Arge pullata (veterinary parasitology) Scandinavian birch sawfly whose larvae, which contains lophyrotomin, causes severe liver necrosis in sheep that eat them.

Argentine (terminology) Having some relationship with the country Argentina.

Argentine tick (parasitology) *See Margaropus winthemi.*

argentum (chemistry) [Latin] Silver (symbol Ag).

arginine (chemistry) $C_6H_{14}N_4O_2$ Colourless (*USA*: colorless) crystalline essential amino acid of the alpha-ketoglutaric acid family.

argon (Ar) (chemistry) An inert gas element in Group 0 of the periodic table (the rare gases). It makes up 0.93 per cent of air (by volume), from which it is extracted. It is used to provide an inert atmosphere in electric lamps and discharge tubes, and for welding reactive metals (such as aluminium). At. no. 18; r.a.m. 39.948.

Argulus (fish parasitology) A genus of freshwater fish lice in the class Crustacea that cause cutaneous ulceration leading to secondary infection. They may also transmit spring viremia of carp, and heavy infestations cause poor growth and erratic swimming. Species include *Argulus brachiuran.*

arista (1, 2 anatomy) 1 A bristle at the base of the antenna in insects. 2 A bristle-like structure in the flowering glumes of grasses.

arithmetic mean (statistics) A measure of central tendency, obtained by dividing the sum of a set of numbers by the number in the set. *Contrast* geometric mean; median; mode.

arithmetic mean average (statistics) of a collection of numbers obtained by dividing the sum of the numbers by the quantity of numbers.

arithmetic unit (computing) Part of a computer's central processor that performs arithmetical operations (addition, subtraction, multiplication, division).

armamentarium (parasitology) The entire equipment of a practitioner, such as medicines, instruments, books etc.

armed scolex (parasitological anatomy) A descriptive term for the rostral end of a cestode which contains hooks.

armed tapeworm (parasitology) *See Taenia solium.*

Armigeres (parasitology) Mosquito species reputed to be capable of carrying Japanese encephalitis virus.

Armillifer (parasitology) A genus of the class Pentastomida consisting of organisms of uncertain taxonomy. Usually considered to be aberrant crustaceans.

armour (biology) (*USA*: armor). *See* exoskeleton.

armyworm (plant parasitology) The larvae of the owlet moth (*Laphygma exempta*) that migrates in large groups ravaging crops.

aromatic compound (chemistry) Member of a large class of organic chemicals that exhibit aromaticity, the simplest of which is benzene.

array (computing) An ordered pattern of symbol strings having one or more dimensions.

arrested larval development (parasitology) *See* hypobiosis.

arrheno- (terminology) A word element. [German] Prefix denoting *male, masculine.*

Arrhenodes minutus (plant parasitology) The oak timberworm beetle. *See* beetles.

arsenate (chemistry) Salt or ester of arsenic acid (H_3AsO_4). Also termed: arsenates.

arsenic acid (chemistry) H_3AsO_4 Tribasic acid from which arsenates are derived. An aqueous solution of arsenic(V) oxide, As_2O_5. Also termed: orthoarsenic acid; arsenic(V) acid.

arsenic (As) (chemistry) Silver-grey semimetallic element in Group VA of the periodic table which exists as several allotropes. It is extracted from sulphide (*USA*: sulfide) ores (*e.g.* arsenopyrite) and is used in insecticides and drugs, alloys, and as a donor impurity in semiconductors. Some of the arsenicals are used for infectious diseases, especially those caused by protozoa. Arsenic compounds are very poisonous. The substance known as white arsenic is arsenic(III) oxide (arsenious oxide), As_4O_6. At. no. 33; r.a.m. 74.9216.

arsenical (1, 2 chemistry) 1 Pertaining to arsenic. 2 A compound containing arsenic.

arsenites (control measures) Soluble arsenic compounds, *e.g.* sodium arsenite. The most toxic of forms of arsenic which may be used as insecticides.

artefact (microscopy) Something that appears during preparation or examination of material which is not present in the natural state. (*USA*: artifact).

arterio- (terminology) A word element [Latin/German] Prefix denoting *artery.*

arteriol(o)- (terminology) A word element. [Latin] Prefix denoting *arteriole.*

arteriole (anatomy) Small artery.

artery (anatomy) Blood vessel that carries oxygenated blood from the heart to other tissues. An exception is the pulmonary artery, which carries deoxygenated blood to the lungs.

arthr(o)- (terminology) A word element. [German] Prefix denoting *joint*, articulation.

arthropod (parasitology) An individual of the phylum Arthropoda, containing joint-limbed organisms that possess a hard exoskeleton.

arthropod transmission Transmission by insect, either mechanically via a contaminated proboscis or feet, or biologically when there is growth or replication of the organism in the arthropod.

Arthropoda (parasitology) A phylum of the annual kingdom including bilaterally symmetrical animals with hard, segmented bodies bearing jointed appendages; embracing the largest numher of known animals, with at least 740,000 species, divided into 12 classes. It includes the arachnids, crustaceans and insects.

arthropodal (parasitology) Pertaining to or emanating from arthropods.

arthropodal allergy (parasitological immunology) *See* allergy.

arthropodal ectoparasites (parasitology) Insects that parasitize the skin of animals.

arthropod-borne encephalomyelitis (parasitology) Arthropods are the transmitting agents for a number of viral agents responsible for causing encephalomyelitis.

arthropodic disease (parasitology) Disease caused by infestation with arthropod parasites, *e.g.* tick worry, tick paralysis, lousy.

Arthula flavofasciata (insect parasitology) A species of the insect order Hymenoptera in the group Ichneumonidae, a parasite of the wasps *Polistes* species and *Ropalidia* species. *Arthula flavofasciata* like *Arthula formosana* attacks the larvae stage of the wasp.

Arthula formosana (insect parasitology) A species of the insect order Hymenoptera in the group Ichneumonidae, a parasite of the wasps *Polistes* species and *Ropalidia* species. *Arthula formosana* like *Arthula flavofasciata* attacks the larvae stage of the wasp.

artificial selection (statistics) A selection based on human decisions.

aryl group (chemistry) Radical that is derived from a compound exhibiting aromaticity by the removal of one hydrogen atom. *e.g.* phenyl, C_6H_5 derived from benzene.

arylation (chemistry) The chemical addition of an aryl group to another molecule.

As (chemistry) The chemical symbol for Element 33, arsenic.

asap (general terminology) An abbreviation of 'a's 's'oon 'a's 'p'ossible.

asb (chemistry) An abbreviation of 'asb'estos.

asbestos (chemistry) Fibrous variety of a number of rock-forming silicate minerals that are heat-resistant and chemically inert.

ASc. (education) An abbreviation of 'A'ssociate in 'Sc'ience.

ASc.W (society) An abbreviation of 'A'ssociation of 'Sc'ientific 'W'orkers.

Ascarapis (insect parasitology) A species of mites in the Prostigmata, that are parasitic to the bees *Apinae*. A pest to honeybees.

ascaricide (parasitology/control measures) A substance destructive to ascarids.

ascarid (parasitology) Any of the phasmid nematodes of the Ascaridoidea, which includes the genera *Ascaris*, *Parascaris*, *Toxocara* and *Toxascaris*.

ascaridata (parasitology) Ascarids.

Ascaridia (parasitology) A genus of nematode worms of the family Heterakidae which includes *Ascaridia columbae*, *Ascaridia dissimilis*, *Ascaridia compar*, *Ascaridia numidae* and *Ascaridia razia*.

Ascaridia galli (avian parasitology) A parasite of domesticated and wild birds which causes enteritis with diarrhoea (*USA*: diarrhea), and unthriftiness especially in birds up to 3 months of age.

Ascaridida (fish parasitology) An order of roundworms in subclass Secernentea, and the phylum Nematoda. There are two superfamilies of this order, the Ascaridoidea and Seuratoidea, each with many genera of fish nematodes. *Goezia*, *Cucullanus* and *Pseudanisakis* are representatives of these two superfamilies.

ascaridiosis (parasitology) Parasitic infestation with any of the phasmid nematodes of the Ascaridoidea, which includes the genera *Ascaris*, *Parascaris*, *Toxocara* and *Toxascaris*.

***Ascaris* species** (human/veterinary parasitology) Nematode parasites found in the intestine of many animal species. The organisms are members of the family Ascarididae. Species includes: *Ascaris columnaris*, a parasite of wild animals; *Ascaris equorum*, a parasite of horses; *Ascaris lumbricoides*, a parasite of humans; *Ascaris schroederi* and *Ascaris suum*, parasites of pigs; and *Ascaris vitulorum*, a parasite of cattle, buffalo, sheep and goats. *Ascaris lumbricoides* can reach a length of 30 cm and in large numbers can block the gut and kill the host.

Ascarops (veterinary parasitology) A genus of nematode worms of the family Spirocercidae. Species include *Ascarops strongylina* and *Ascarops dentata*, which may be found in pigs.

Ascarus (insect parasitology) A species of mites in the Astigmata, which similar to *Acotyledon* and *Aeroglyphus*, are parasites to the bees *Apis*, *Apis mellifera*.

ascending mesocolon (anatomy) *See* mesocolon.

Aschistonyx eppoi (plant parasitology) The Japanese gall midge. *See* gall.

ASCII (computing) An abbreviation of 'A'merican 'S'tandard 'C'ode for 'I'nformation 'I'nterchange, a standard format for representing characters on a computer.

ascites (medical) Abnormal accumulation of fluid from the blood in the peritoneal cavity, occurring in heart, liver and kidney failure.

ascitic fluid (medical) Serous fluid in peritoneal cavity.

aseptic (parasitology) Referring to being free from pathogenic microorganisms.

aseptic fever (parasitology) Fever associated with aseptic wounds, presumably due to the disintegration of leukocytes or to the absorption of avascular or traumatized tissue.

asexual reproduction (biology) Reproduction that does not involve gametes or fertilization. There is a single parent, and all offspring are genetically identical. This commonly occurs in prokaryotes, and also fungi, Protozoa and some plants.

See also vegetative propagation.

ASF (parasitology) An abbreviation of 'A'frican 'S'wine 'F'ever.

ASGBI (society) An abbreviation of 'A'natomical 'S'ociety of 'G'reat 'B'ritain and 'I'reland.

Ashmeadopria (insect parasitology) A species of the insect order Hymenoptera in the group Diapriidae, a parasite of the ants *Ecitonini*, *Plagiolepis*, *Solenopsis*, *and Tetramorium*, that attacks the larvae and pupa stages.

Asn (biochemistry) An abbreviation of 'As'paragi'n'e.

ASN (statistics) An abbreviation of 'A'verage 'S'ample 'N'umber.

ASP (parasitology) An abbreviation of 'A'ncylostoma-'S'ecreted 'P'roteins.

Asp. (biochemistry) An abbreviation of 'Asp'artic acid.

Aspiculuris (veterinary parasitology) A nematode genus of the oxyurid family Heteroxynematidae. Species includes *Aspiculuris tetraptera*, which is found in rodents.

aspidium (pharmacology/control measures) The dried products of a genus of plants known as male fern (*Dryopteris filixmas*) that contains an oleoresin capable of causing liver damage and in the past was used as an anthelmintic.

Aspidogastrea (fish parasitology) A subclass of flatworms in the class Trematoda and phylum Platyhelminthes. Two orders are accepted in the subclass Aspidogastrea, the Aspidogastrida with one family, Aspidogastridae, and the Stichocotylida with three, Stichocotylidae, Multicalycidae and Rugogastridae. Members of the three sub-families of the Aspidogastridae, which occur in molluscs, teleost fish and turtles, have as a holdfast a ventral disc divided into a marginal ring of alveoli surrounding medial alveoli. In the Stichocotylida, one family (Stichocotylidae) has a holdfast composed of isolated suckers; in another (Multicalycidae) it is composed of fused suckers; and in the third (Rugogastridae), it is a series of raised transverse septa (ruga).

aspirate (1, 2 methods) 1 To withdraw fluid by negative pressure, or suction. 2 The fluid obtained by aspiration.

aspirating needle (equipment) A long, hollow needle for removing fluid from a cavity.

aspiration (methods) The removal of fluids by suction.

aspiration biopsy (parasitology) Biopsy in which tissue is obtained by application of suction through a needle attached to a syringe.

aspiration biopsy needle (equipment) A needle to which suction can be applied in order to withdraw a core of tissue from a solid organ.

aspirator (equipment) Apparatus that produces suction in order to draw a gas or liquid from a vessel or cavity.

aspirin (pharmacology) $CH_3COO.C_6H_4$-COOH Drug that is commonly used as an analgesic, antipyretic (to reduce fever) and anti-inflammatory. Also termed: acetylsalicyllic acid.

assassin bug (parasitology) *See* reduviid bug.

assay (methods) Determination of the purity of a substance or the amount or activity of any particular constituent of a mixture.

assessment (medical) The critical analysis and evaluation or judgment of the status or quality of a particular condition, situation, or other subject of appraisal.

Association of the British Pharmaceutical Industry (ABPI) (association) A UK trade association that aims to ensure that medicinal and related products are of the highest quality and are readily available for the treatment of human and animal disease.

astatine (At) (chemistry) Radioactive element in Group VIIA of the periodic table (the halogens). It has several isotopes, of half-lives of 2×10^{-6} sec to 8 hr. Because of their short half-lives, they are available in only minute quantities. At. no. 85; r.a.m. 210 (most stable isotope).

aster phase (genetics) An alternative term for metaphase.

asthen(o)- (terminology) A word element. [German] Prefix denoting *weak*, *weakness*.

astigmatism (microscopy) Failure of an optical system, such as a lens or a mirror, to focus the image of a point as a single point.

asymptomatic (medical) Not showing any signs or symptoms of disease whether disease is present or not.

at. no. (physics) An abbreviation of 'at'omic number.

at. wt. (physics) An abbreviation of 'at'omic 'w'eigh't'.

atel(o)(io)- (terminology) A word element. [German] Prefix denoting *incomplete*, *imperfectly developed*.

Athesmia (veterinary parasitology) A genus of trematode parasites of the family Dicrocoeliidae. Species include *Athesmia foxi* which may be found in South American monkeys.

atom (chemistry) Fundamental particle that is the basic unit of matter. An atom consists of a positively charged nucleus surrounded by negatively charged electrons restricted to orbitals of a given energy level. Most of the mass of an atom is in the nucleus, which is composed principally of protons (positively charged) and neutrons (electrically neutral); hydrogen is exceptional in having merely one proton in its nucleus. The number of electrons is equal to the number of protons, and this is the atomic number. The chemical behaviour (*USA*: behavior) of an atom is determined by how many electrons it has, and how they are transferred to, or shared with, other atoms to form chemical bonds.

See also relative atomic mass; isotope; molecule; subatomic particle; valence.

atomic mass (chemistry) An alternative term for atomic weight.

See also relative atomic mass.

atomic mass unit (AMU) (chemistry) Arbitrary unit that is used to express the mass of individual atoms. The standard is a mass equal to 1/12 of the mass of a carbon atom (the carbon-12 isotope). A mass expressed on this standard is called a relative atomic mass (r.a.m.), symbol A_r. Atomic mass unit is also termed dalton.

atomic number (at. no.) (chemistry) Number equal to the number of protons in the nucleus of an atom of a particular element, symbol Z. Also termed: proton number.

atomic orbital (chemistry) Wave function that characterizes the behaviour (*USA*: behavior) of an electron when orbiting the nucleus of an atom. It can be visualized as the region in space occupied by the electron.

atomic weight (at. wt.) (chemistry) Relative mass of an atom given in terms of atomic mass units. *See* relative atomic mass.

atomicity (chemistry) Number of atoms in one molecule of an element; *e.g.* argon has an atomicity of 1, nitrogen 2, ozone 3.

atopic reagin (immunology) The antibody responsible for hypersensitivity reactions to specific substances with signs of atopy resulting.

ATP (biochemistry) An abbreviation of 'A'denosine 'T'ri'P'hosphate.

atrichia (parasitology) An absence of hair or of flagella or cilia.

Atriotaenia (veterinary parasitology) A genus of tapeworms of the family Linstowiidae. Species include *Atriotaenia megastoma* and *Atriotaenia procyonis*, which may be found in mustelids and procyonids, *e.g.* in skunk, badger, marten, ermine and mink.

atrium (anatomy) One of the thin-walled upper chambers of the heart. Also termed: auricle.

atrophy (parasitology) A wasting away, or decrease in the size and function of a cell, tissue, or organ.

attack (parasitology) An episode or onset of illness.

attack rate (epidemiology) The proportion of a population affected by the disease during a prescribed, usually short, period of time.

attenuated (parasitology) Weakened, reduced in virulence.

attenuated vaccine (control measures) A vaccine prepared from live microorganisms that have lost their virulence but retained their ability to induce protective immunity.

attribute (1 statistics; 2 computing) 1 A qualitative variable which cannot be expressed in numerical terms, *e.g.* Insectivorous as a behavioural (*USA*: behaviorial) attribute. 2 An element which determines the appearance of a character, *e.g.* italics, bold.

attribute-specific rate (parasitology) The rate of occurrence of a specific attribute.

Au (chemistry) The symbol for the element gold. [From the Latin, 'aurum'].

AUC (statistics) An abbreviation of 'A'rea 'U'nder 'C'urve.

Auchmeromyia (parasitology) A genus of flies of the family Calliphoridae.

Auchmeromyia luteola (human/veterinary parasitology) An African fly, the larva of which is a bloodsucker that parasitizes pigs and people. Also termed Congo floor maggot.

audi(o)- (terminology) A word element. [Latin] Prefix denoting *hearing*.

auditory (medical) Relating to the ear or hearing.

auditory canal (anatomy) Tube that leads from the outer ear to the ear drum (tympanum).

Aulonocephalus lindquisti (avian parasitology) A caecal (*USA*: cecal) and colonic nematode of uncertain pathogenicity in quail.

aurantiactinomyxon (fish parasitology) A myxozoan parasite that is considered to be the causative agent of 'proliferative gill disease' of catfish. This parasite causes a rapid, severe, proliferation of the gill epithelium which results in impairment of respiration and osmoregulatory function. The intermediate host is thought to be a microscopic aquatic oligochaete worm, *Dero digitata*, which is found in the mud and sediment in the ponds of affected catfish. Diagnosis is dependent upon the presence of the parasite within swollen, clubbed proliferative gill lamellae, many of which are fractured due to the associated chondrodysplasia.

auric (chemistry) Trivalent gold. Also termed: gold(III); gold(3+).

auricular (medical) Pertaining to or emanating from the ear.

auricular mange (veterinary parasitology) Infestation with ear mites. *See* otodectic mange.

aurous (chemistry) Monovalent gold. Also termed: gold(I); gold(1+).

auscultation (medical) Listening to sounds produced in the body, usually used to hear sounds produced by heart and lungs.

Austeucharis (insect parasitology) A species of the insect order Hymenoptera in the group of parasitic wasps the Eucharitidae. A parasite of the ants *Myrmecia* species, that attacks the larvae and pupa stages.

Australorbis (parasitology) An important genus of snail that are hosts of schistosomes, *Schistosoma mansoni*.

Austrobilharzia (parasitology) A genus of digenetic trematodes of the family Schistosomatidae.

Austrobilharzia variglandis (veterinary parasitology) A species of digenetic trematodes in the genus *Austrobilharzia* that are principally a parasite of waterfowl but that can also cause dermatitis in mammals that frequent shallow aquatic environments.

Austrosimulium (parasitology) *See* black fly.

aut(o)- (terminology) A word element. [German] Prefix denoting *self*.

autoantigen (immunology) A tissue constituent that stimulates production of autoantibodies or self-reactive T lymphocytes in the animal in which it occurs.

autocatalysis (chemistry) Catalytic reaction that is started by the products of a reaction that was itself catalytic. *See* catalyst.

autoclave (laboratory equipment) Airtight container that heats and sometimes agitates its contents under high-pressure steam. Autoclaves are usually used for sterilization or industrial processing.

autoclave tape (laboratory equipment) Special masking tape used to close packages of surgical materials to be autoclaved. It includes a heat sensitive dye in diagonal stripes. The appearance of the dye can be misunderstood; it does not indicate that the package has been sterilized only that it has been exposed to some heat.

autoecious (parasitology) Completing the entire life cycle on a single species of host.

autogamy (1 parasitology physiology; 2 plant physiology) 1 The process by which the two parts of a divided cell nucleus reunite, as in some protozoans. 2 Self-fertilization in plants.

autogenous vaccine (control measures) A vaccine prepared from cultures of material derived from a lesion of the animal to be vaccinated.

autoinfection (parasitology) Spread of infection from one part of the body to another, in parasitology used to describe a host which is both intermediate and definitive host without parasite transmission from other animals, which may be particularly damaging due to the large number of offspring many parasites produce.

autolysis (biochemistry) Breakdown of the contents of an animal or plant cell by the action of enzymes produced within that cell.

autonomously replicating sequence (molecular biology) Usually plasmids that replicate independently of chromosomal DNA.

autopsy (medical) The examination of a body after death, so that physicians can arrive at a true diagnosis of the cause of death.

autosome (genetics) Any chromosome other than one of the sex chromosomes.

autotroph (biology) Organism that lives using autotrophism.

autotrophism (biology) In bacteria and green plants, the ability to build up food materials from simple substances, *e.g.* by photosynthesis.

See also heterotrophism.

autumn fly (parasitology) *See Musca autumnalis*. Also termed face fly.

Auxanommatidia myrmecophila (insect parasitology) A species of the insect order Diptera in the group Phoridae, a parasite of the ants *Atta*, *Camponotus*, *Eciton* species, *Iridomyrmex*, *Pachycondyla*, and *Solenopsis*. Like *Apterophora attophila*, *Apterophora caliginosa*, *Borgmeierphora kempfi*, *Borgmeierphora multisetosa* and *Cataclinusa* species, *Auxanommatidia myrmecophila* attacks the adult stage of the ant.

Auxopaedeutes lyriformis (insect parasitology) A species of the insect order Hymenoptera in the group Diapriidae, a parasite of the ants *Formica*, *Paratrechina* and *Solenopsis*, that attacks the larvae and pupa stages.

Auxopaedeutes sodalis (insect parasitology) A species of the insect order Hymenoptera in the group Diapriidae, a parasite of the ants *Formica*, *Paratrechina*, and *Solenopsis* that like *Auxopaedeutes lyriformis*, attacks the larvae and pupa stages.

AV (chemistry) An abbreviation of 'A'cid 'V'alue.

AV valves (physiology) An abbreviation of 'A'trioV'entricular 'valves' of heart.

AVA (society) An abbreviation of 'A'ustralia 'V'eterinary 'A'ssociation.

average (statistics) Number that is representative of a collection of numbers; *e.g.* an arithmetic mean, geometric mean, mode or median.

avermectins (pharmacology/control measures) A group of chemically related anthelmintics belonging to the macracytic lactones, and produced by fermenting *Streptomyces avermitilis*. A number of compounds with anthelmintic activity are produced by this process. The combination of two dehydrogenated avermectins is called ivermectin.

Aves (biology) A class comprising all of the birds. Any living organism that has feathers is classed as a bird. Besides the diseases that afflict them as birds, they are also of importance as vectors of disease for other species.

Avg (statistics) An abbreviation of 'av'era'g'e.

avian (biology) Pertaining to or emanating from members of the class Aves. *See* Aves.

avian canker (avian parasitology) Disease of birds caused by *Trichomonas gallinae* and characterized hy accumulations of caseous material in the throat.

avian diseases (biology) Diseases affecting birds.

avian malaria (avian parasitology) A disease transmitted by mosquitoes affecting most species of birds and caused by *Plasmodium* spp., *e.g. Plasmodium* *gallinaceum* in fowl, *Plasmodium juxanucleare* in fowl and turkeys, *Plasmodium durae* and *Plasmodium griffithsi* in turkeys. The disease is characterized by anaemia (*USA*: anemia) which may be fatal.

avian trichomoniasis (avian parasitology) A disease of the upper digestive tract of young birds, especially pigeon squabs, caused by *Trichomonas gallinae*. Necrotic lesions are present in the mouth, pharynx, oesophagus (*USA*: esophagus), crop and sometimes proventriculus and conjunctival sac. Also termed canker, frounce, roup.

avicide (control methods) Agents, such as avitrol (4-aminopyridine), used to poison bird pests.

avidity (immunology) An imprecise measure of the strength of antibody-antigen binding based on the rate at which the complex is formed.

Avioserpens (parasitology) A genus of nematodes of the family Dracunculidae. Species includes *Avioserpens taiwana* and *Avioserpens mosgovoyi*, which may be found in ducks and other waterfowl.

Avitellina (parasitology) A nonpathogenic tapeworm genus of the family Thysanosomatidae. Species includes *Avitellina centripunctata* which may be found in sheep and goats.

AVMA (society) An abbreviation of 'A'merican 'V'eterinary 'M'edical 'A'ssociation.

avoir. (measurements) An abbreviation of 'Avoir'dupois.

avoirdupois (measurements) System of weights based on a pound (symbol lb), equivalent to 2.205kg, and subdivided into 16 ounces, or 7,000 grains. In science and medicine it has been almost entirely replaced by SI units. *See* avoirdupois ounce.

avoirdupois ounce (measurement) Unit of weight, 1/16 of a pound. 1 oz = 28.349527 grams. *See* avoirdupois.

A/W (measurement) An abbreviation of 'A'ctual 'W'eight.

aw (chemistry) An abbreviation of 'a'tomic 'w'eight.

awl nematode (plant parasitology) *See Dolichodorus heterocephalus*.

axiom (philosophy) Principle that is taken as true, without needing proof.

See also hypothesis; law; theorem.

axis (1 mathematics; 2 biology; 3 anatomy) 1 A line of significant reference for a graph or figure; *e.g.* *x*- and *y*-axes in Cartesian co-ordinates. 2 Central line of symmetry of an organism. 3 The second (cervical) vertebra, which articulates with the atlas and allows the head to turn from side to side.

axoneme (1 parasitological anatomy; 2 genetics) 1 The central core of a cilium or flagellum, consisting of two central microtubules surrounded by nine peripheral microtubule pairs. 2 The gene-string or filament forming the basis of a chromosome.

axostyle (parasitological anatomy) A stiff rod of protoplasm acting as an internal skeletal support in certain protozoa such as *Trichomonas*, which extends from the kinetosome to the posterior end of the organism.

Azadirachta indica (insect parasitology/control measures) Toxic Indian plant, known for its insecticidal properties, in the family Meliaceae. Also termed neem.

azamethiphos (insect parasitology/control measures) An organophosphorus compound that may be used as an insecticide in toxic baits for fly control.

azeotrope (chemistry) Mixture of two liquids that boil at the same temperature.

Azgiidae (parasitology) A family of digenetic trematodes in the phylum plathyhelminthes that includes the *Proterometra* species.

azide (chemistry) One of the acyl group of compounds, or salts, derived from hydrazoic acid (N_3H). Most azides are unstable, and heavy metal azides are explosive.

azinphos-methyl (control measures) An organophosphorus, nonsystemic insecticide and acaricide. Poisoning with the compound causes typical signs for organophosphates.

azo compound (chemistry) Organic compound of general formula R-N=N-R', in which R and R' are usually aromatic groups.

azo dye (chemistry) Member of a class of dyes that are derived from compounds containing an amino group and have an -N=N- linkage in their molecules. They are intensely coloured (*USA*: colored) (usually red, brown or yellow) and account for a large proportion of the synthetic dyes produced. Azo dyes can be made as acid, basic, direct or mordant dyes.

Azolla (control measures) A free-floating fern that may in certain areas be used to completely cover water surfaces and prevent breeding by mosquitoes.

azomethine (biochemistry) *See* Schiff's base.

azure (microscopy) One of three metachromatic basic dyes (azures A, B and C).

azurophil (microscopy) A tissue constituent staining with azure or a similar metachromatic thiazine dye.

B

B (1 chemistry; 2 haematology) 1 The chemical symbol for element 5, boron. 2 A blood type.

B App. Sc. (education) An abbreviation of 'B'achelor of 'App'lied 'Sc'ience.

B cell (immunology/haematology) An abbreviation of 'b'eta cell.

B Hy. (education) An abbreviation of 'B'achelor of 'Hy'giene/Public Health.

B Med Biol. (education) An abbreviation of 'B'achelor of 'Med'ical 'Biol'ogy.

B Med Sc. (education) An abbreviation of 'B'achelor of 'Med'ical 'Sc'ience.

B Sc. App. (education) An abbreviation of 'B'achelor of 'App'lied 'Sc'ience.

b. pt (physics) An abbreviation of 'b'oiling 'p'oin't'.

B. Sc. (education) Bachelor of Science; - (LA) Bachelor of Science (Laboratory Assistant); - (Med.) Bachelor of Science in Medicine; - (Med. Lab. Tech.) Bachelor of Science in Medical Laboratory Technology; - (Med. Sci.) Bachelor of Science in Medical Sciences; - (M.L.S.) Bachelor of Science in Medical Laboratory Science; - (Nutr.) Bachelor of Science in Nutrition; - (Pharm.) Bachelor of Science in Pharmacy; - (R.T.) Bachelor of Science in Radiologic Technology; - (Vet.) Bachelor of Science in Veterinary Science, - (Vet.Sc. & A.1A.) Bachelor of Science in Veterinary Science and Animal Husbandry.

Ba (chemistry) The chemical symbol for element 56, barium.

BA Sc. (education) An abbreviation of 'B'achelor of 'A'pplied 'Sc'ience.

BAAS (education) An abbreviation of 'B'ritish 'A'ssociation for the 'A'dvancement of 'S'cience.

Babesia (parasitology) A genus of large, round to pyriform protozoa, of the family Babesiidae which includes piroplasms. These protozoa pass part of their life cycle in erythrocytes and transmission between animals is by ticks.

Babesia bigemina (veterinary parasitology) A species of protozoa in the genus *Babesia* that may cause babesiosis of cattle and some wild ruminants.

Babesia bovis (veterinary parasitology) A species of protozoa in the genus *Babesia* that may cause babesiosis of cattle and some wild ruminants.

Babesia caballi (veterinary parasitology) A species of protozoa in the genus *Babesia* that may cause a mild form of babesiosis in horses.

Babesia canis (veterinary parasitology) A species of protozoa in the genus *Babesia* that may cause babesiosis in dogs.

Babesia cati (veterinary parasitology) A species of protozoa in the genus *Babesia* that may be found in cats.

Babesia divergens (veterinary parasitology) A species of protozoa in the genus *Babesia* that may cause a mild form of babesiosis of cattle and some wild ruminants.

Babesia equi (veterinary parasitology) A species of protozoa in the genus *Babesia* that may cause babesiosis in horses.

Babesia felis (veterinary parasitology) A species of protozoa in the genus *Babesia* that may cause babesiosis of cats.

Babesia gibsoni (veterinary parasitology) A species of protozoa in the genus *Babesia* that may cause babesiosis in dogs.

Babesia herpaiuri (veterinary parasitology) A species of protozoa in the genus *Babesia* that may be found in cats.

Babesia hylomysci (veterinary parasitology) A species of protozoa in the genus *Babesia* that may be found in red deer.

Babesia major (veterinary parasitology) A species of protozoa in the genus *Babesia* that may cause a mild form of babesiosis of cattle.

Babesia motasi (veterinary parasitology) A species of protozoa in the genus *Babesia* that may cause acute babesiosis in sheep and goats.

Babesia ovis (veterinary parasitology) A species of protozoa in the genus *Babesia* that may cause a mild form of babesiosis in sheep and goats.

Babesia pantherae (veterinary parasitology) A species of protozoa in the genus *Babesia* that may be found in cats, including leopards.

Babesia rodhaini (veterinary parasitology) A species of protozoa in the genus *Babesia* that may be found in red deer.

Babesia vogeli (veterinary parasitology) A species of protozoa in the genus *Babesia* that may be found in dogs.

babesiasis (medical/veterinary parasitology) *See* babesiosis.

babesicide (pharmacology/control measures) Destructive to *Babesia* spp.

babesiosis (medical/veterinary parasitology) A group of diseases caused by the protozoan *Babesia* spp. and transmitted by blood-sucking ticks. Clinically they are all characterized by fever and intravascular haemolysis (*USA*: hemolysis) manifested by a syndrome of anaemia (*USA*: anemia), haemoglobinuria (*USA*: hemoglobinuria) and jaundice. Also termed tick fever, Texas fever, redwater fever.

bacillary band (parasitological anatomy) A term used in relation to nematodes. A row of longitudinal cells formed by the hypodermis. Seen in Trichinelloidea.

back emf (physics) Electromotive force produced in a circuit that opposes the main flow of current; *e.g.* in an electrolytic cell because of polarization or in an electric motor because of electromagnetic induction.

backbone (anatomy) An alternative term for vertebral column.

background radiation (radiology) Radiation from natural sources, including outer space (cosmic radiation) and radioactive substances on Earth (*e.g.* in igneous rocks such as granite).

background rate (epidemiology) The rate, often low, at which some agent or event occurs, at a particular time or in a particular place, in the absence of a specific hazard.

backing store (computing) Computer store that is larger than the main (immediate access) memory, but with a longer access time.

backup (computing) The act of saving computer files to ensure that a copy survives a major problem in the computer. Achieved by periodic copying of data in one computer to another unit of the same medium or to a different medium so that the data may be recovered in the event of its loss.

bact. (bacteriology) An abbreviation of 'bact'eria/ 'bact'erial/ 'bact'eriology.

bacteraemia (bacteriology) The condition of having bacteria in the bloodstream. (*USA*: bacteremia).

bacteria (bacteriology) A large group of single-celled microscopic organisms. They can occur either free-living in air, water, soil, etc; or they may be parasitic upon a host plant or animal. Only a minority cause any disease, and more are actually useful to their hosts, *e.g.* in breaking down food as part of the digestive process. The three basic shapes are *bacillus* (rod), *spirillum* (spiral) and *coccus* (spherical), but some bacteria may link up to form chains or clumps. Multiplication is fast, usually by fission. Between them, the various species can utilize almost any type of organic molecule as food and inhabit almost any environment. Some can even use inorganic elements such as sulphur (*USA*: sulfur).The activities of some bacteria are of great significance to man, *e.g.* fixation of nitrogen in certain plants, as agents of decay, and their medical importance as disease-causing agents (pathogens). Much genetical research is done on bacteria because, being prokaryotes, they have simple genetic machinery. Sing.: 'bacterium'.

bacterial (bacteriology) Refering to bacteria or caused by bacteria.

bacterial canker of stone fruits (plant parasitology) *See* *Criconemella xenoplax*.

bactericidal (bacteriology) The action of a substance which destroys bacteria.

bactericide (pharmacology) A substance which kills bacteria.

See also bacteriostatic.

bacteriocins (bacteriology) Protein antibiotic-like substances produced by bacteria, which are lethal to other bacteria. These are generally strains closely related to the producer organism.

bacteriological (bacteriology) Pertaining to bacteriology.

bacteriologist (bacteriology) A scientist who specialises (*USA*: specializes) in the study of bacteria.

bacteriology (microbiology) The scientific study of bacteria, their effects on organisms and their uses in agriculture and industry (*e.g.* in biotechnology).

bacteriolysin (bacteriology) A protein, usually an immunoglobulin, which destroys bacterial cells.

bacteriolysis (bacteriology) The destruction of bacterial cells.

bacteriolytic (bacteriology) A substance which can destroy bacteria.

bacteriophage (bacteriology) Virus that infects bacteria. When inside a cell it replicates using its host's enzymes; the release of new viruses may disintegrate the cell. Bacteriophages have been used extensively in research on genes. Also termed: phage.

bacteriostatic (bacteriology) Substance that inhibits the growth of bacteria without killing them.

See also bactericide.

bacterium (bacteriology) The singular form of the term 'bacteria'. *See* bacteria.

Baermann technique (parasitology methods) A laboratory method for separating parasite larvae from faeces (*USA*: feces), soil or herbage for counting or identification.

Baghdad boil (medical/veterinary parasitology) *See* leishmaniasis.

Bakerdania (insect parasitology) Mites in the Prostigmata that are parasitic on a variety of ants and that may be phoretic on *Reticulitermes* and other various termites.

baking soda (chemistry) An alternative term for sodium hydrogencarbonate (sodium bicarbonate).

balance (1 physics; 2 physiology) 1 Apparatus for weighing things accurately; types include a beam balance, spring balance, torsion balance and substitution balance. 2 Sense supplied by organs within the semicircular canals of the inner ear.

balantidiasis (medical/veterinary parasitology) Infection by protozoa of the genus *Balantidium*.

Balantidium (human/veterinary parasitology) A genus of ciliated protozoa, including many species found in the intestine in vertebrates and invertebrates, predominantly in primates, pigs and humans. Species include *Balantidium coli*.

Balmer series (physics) Visible atomic spectrum of hydrogen, consisting of a unique series of energy emission levels, which appears as lines of red, blue and blue-violet light. It is the key to the discrete energy levels of electrons.

bals. (microscopy) An abbreviation of 'bals'am.

balsam (microscopy) Vegetable resin combined with oil, *e.g.* Canada balsam, used in preparing slides for microscopy, which comes from the balsam fir of North America.

bandicoot (parasitology) Small nocturnal burrowing marsupial that is insectivorous and a host for *Ixodes holocyclus*. Also termed *Perameles* spp.

bar (measurement) Unit of pressure defined as 10^5 newtons per square metre; equal to approximately one atmosphere.

bar chart (statistics) Graph that has vertical or horizontal bars whose lengths are proportional to the quantities they represent.

See also pie chart.

bar code (general terminology) Information about a product coded as a series of thick and thin parallel lines printed on it (or on a label attached to it), which can be scanned and 'read' by a machine.

barber's pole worm (parasitology) *See Haemonchus*.

barium (Ba) (chemistry) Silver-white metallic element in Group IIA of the periodic table (the alkaline earths), obtained mainly from the mineral barytes (barium sulphate [*USA*: barium sulfate]). Its soluble compounds are poisonous, and used in fireworks; its insoluble compounds are used in pigments and medicine. At. no. 56; r.a.m. 137.34.

barium sulphate (chemistry) $BaSO_4$ White crystalline insoluble powder, used as a pigment and as the basis of 'barium meal' to show up structures in X-ray diagnosis (because it is opaque to X-rays). It occurs as the mineral barytes (also called barite). (*USA*: barium sulfate).

barograph (physics) Type of barometer that produces a chart recording changes in atmospheric pressure over a period of time.

barometer (physics) Instrument for measuring atmospheric pressure, much used in meteorology.

Barr body (genetics) Condensed X-chromosome, seen in female cells of mammals, due to one or the other of the two X-chromosomes in each cell being inactivated.

BAS (education) An abbreviation of 'B'achelor of 'A'pplied 'S'cience.

basal ganglion (anatomy) Region of grey matter within the white matter that forms the inner part of the cerebral hemispheres of the brain.

basal metabolic rate (BMR) (physiology) Minimum amount of energy on which the body can survive, measured by oxygen consumption and expressed in kilojoules per unit body surface.

basal nucleus (anatomy) Also known as basal ganglia, a body of grey matter (nerve tissue) located deep within the cerebral hemispheres of the brain, and concerned with movement.

base (1 molecular biology; 2,3,4 mathematics; 5 chemistry) 1 One of the nucleotides of DNA or RNA (*i.e.* adenine, cytosine, guanine, thymine or uracil). 2 The horizontal line upon which a geometric figure stands. 3 The starting number for a numerical or logarithmic system; *e.g.* binary numerals have the base 2, common logarithms are to the base 10. 4 Number on which an exponent operates; *e.g.* in 5^2, 5 is the base (and 2 the exponent). 5 A member of a class of chemical compounds whose aqueous solutions contain OH^- ions. A base neutralizes an acid to form a salt.

See also alkali.

base metals (chemistry) Metals that corrode, oxidize or tarnish on exposure to air, moisture or heat; *e.g.* copper, iron, lead.

base pair (Bp) (molecular biology) Used as a unit of length to describe a sequence of DNA.

base pairing (molecular biology) Specific pairing between complementary nucleotides in double-stranded DNA or RNA by hydrogen bonding; *e.g.* in DNA guanine pairs with cytosine, and adenine pairs with thymine.

See also purine; pyrimidine.

basic (1 computing; 2 chemistry) 1 An abbreviation of 'b'eginners 'a'll-purpose 's'ymbolic 'i'nstruction 'c'ode. A computer language. 2 Having a tendency to release hydroxide (OH^-) ions. *See* base.

basic input output system (BIOS) (computing) That part of the computer operating system which communicates between the keyboard, the screen, the printer and other peripheral devices.

basic oxide (chemistry) Metallic oxide that reacts with an acid to form a salt and water.

basic salt (chemistry) Salt that contains hydroxide (OH^-) ions; *e.g.* basic lead carbonate, $2PbCO_3.Pb(OH)_2$.

bass tapeworm (fish parasitology) *See Proteocephalus ambloplitis.*

batch file (computing) A file which contains a number of commands which are carried out sequentially when the file is executed. Also termed script file.

batch processing (computing) A method of processing in which all of the input transactions of a given type that are to be processed on a computer are collected together and processed at one time.

Bathmostomum (veterinary parasitology) A genus of the Ancylostomatidae family of blood-sucking nematodes. Species include *Bathmostomum sangeri* which may be found in the caecum (*USA*: cecum) of the Indian elephant.

battery (physics) Device for producing electricity (direct current) by chemical action. Also termed: cell. *See* Daniell cell; dry cell; primary cell; secondary cell.

See also solar cell.

Baylisascaris (veterinary parasitology) A genus in the Ascarididae family of nematodes which cause cerebrospinal nematodiasis. Species include *Baylisascaris columnaris* found in dogs, *Baylisascaris transfuga* found in captive and zoo bears and *Baylisascaris procyonis* found in rodents.

BBC (chemistry) An abbreviation of 'B'romo-'B'enzyl 'C'yanide.

BBT (physiology) An abbreviation of 'B'asal 'B'ody 'T'emperature.

BDH (chemistry) An abbreviation of 'B'ritish 'D'rug 'H'ouses.

Be (chemistry) Chemical symbol for beryllium.

beak overgrowth (avian parasitology) A condition found in birds resulting most commonly from infestation by the mite *Cnemidocoptes pilae*.

beaker (equipment) A round laboratory vessel of various materials, usually with parallel sides and often with a pouring spout.

Beckmann thermometer (equipment) Mercury thermometer used for accurately measuring very small changes or differences in temperature. The scale usually covers only 6 or 7 degrees. It was named after the German chemist Ernst Beckmann (1853–1923).

becquerel (Bq) (measurement) SI unit of radioactivity, equal to the number of atoms of a radioactive substance that disintegrate in one second. It was named after the French physicist Antoine Henri Becquerel (1852–1908). 1 Bq = 2.7×10^{-11} curies (the former unit of radioactivity).

bed bugs (parasitology) *See Cimex lectularius.*

beef tapeworm (parasitology) *Taenia saginata.*

Beer's law (physics) Concerned with the absorption of light by substances, it states that the fraction of incident light absorbed by a solution at a given wavelength is related to the thickness of the absorbing layer and the concentration of the absorbing substance. Also termed: Beer Lambert law.

beet sugar (chemistry) An alternative term for sucrose.

beetles (insect parasitology) Members of the insect order Coleoptera. They are common intermediate hosts for tapeworms.

beggar tick (parasitology) *Bidens frondosa.*

behaviour (medical) Way of acting. (*USA*: behavior).

behavioural scientist (medical) A person who specialises (*USA*: specializes) in the study of behaviour. (*USA*: behavioral scientist).

bel (B) (measurement) Unit representing the ratio of two amounts of power, *e.g.* of sound or an electronic signal, equal to 10 decibels. It was named after the American inventor Alexander Graham Bell (1847–1922).

Belonolaimus (plant parasitology) A genus of nematodes in the order Tylenchida and family Belonolaimidae. They are large nematodes that are usually 2 to 3 mm in length. *See Belonolaimus longicaudatus.*

Belonolaimus longicaudatus (plant parasitology) A species of nematodes in the genus *Belonolaimus* whose hosts cover a wide range, including turf grasses, cereals, vegetables, forage crops, fruits, trees, ornamentals, weeds, peanut, corn, cotton, tomato, squash, soybeans, grasses, and pine roots. They have not been reported outside of the USA. They are migratory ectoparasites, feeding at root tip and along the sides of roots and prefer light, sandy soils. Reproduction is indeed inhibited in soils with large amounts of organic matter. Soil moisture is also important; optimum reproduction occurs at seven per cent soil moisture; some reproduction occurs at three per cent soil moisture; reproduction and development are inhibited at 30 per cent moisture unless aeration is supplied. Optimum temperatures for reproduction seem to be about 30°C. Males and females are present, but life history details are not well understood. In general, nematodes are confined to the upper 30 cm of soil, but within this zone there is vertical and seasonal fluctuation depending on temperature. High air temperatures force these nematodes deeper into the soil, while lower surface temperatures result in a higher concentration of nematodes in the surface layers, with little or no increase below. Infected plants show an increased tendency to wilt in dry conditions, severe stunting, leaf chlorosis and plant death may occur. Infested areas vary in size and shape, but the boundary between diseased and healthy plants is usually well-defined. They render *Fusarium* wilt-resistant cotton susceptible. Feeding causes devitalized and stubby roots, while necrosis and discolouration (*USA*: discoloration) occur less frequently. Their long stylet which is 100–160 µm in length, penetrates to inner cortex and endodermis, causing root tip damage, resulting in reduced root system with short, stubby branches. Gall-like swellings may occur at root tips in corn due to repeated production of new branches, and feeding causes lesions on cotton roots, followed by root girdling. Control measures include the use of nematicides. Also termed: sting nematode.

benchmark programme (computing) A computer programme (*USA*: program) designed to assess efficiency of hardware or software.

Benedenia (fish parasites) A genus of the class of monogenetic Trematoda in the order Capsalidae that is an important oral and cutaneous fluke parasite of aquarium, cultured and marine fish.

Benedict's test (biochemistry) Test used to detect the presence of a reducing sugar by the addition of a solution containing sodium carbonate, sodium citrate, potassium thiocyanate, copper sulphate (*USA*: copper sulfate) and potassium ferrocyanide. A change in colour (*USA*: color) from blue to red or yellow on boiling indicates a positive result. It was named after the American chemist S. Benedict (1884–1936).

See also Fehling's test.

benign theileriosis (veterinary parasitology) A condition caused by infestation with the protozoan parasites of the species *Theileria orientalis*.

benzaldehyde (chemistry) C_6H_5CHO Colourless (*USA*: colorless) aromatic aldehyde, with a smell of almonds, used as a flavouring (*USA*: flavoring), in perfumes and as an intermediate in making dyes.

benzene (chemistry) C_6H_6 Colourless (*USA*: colorless) inflammable liquid hydrocarbon, simplest of the aromatic compounds. It is used as a solvent and in the manufacture of plastics.

benzene ring (chemistry) Cyclic (closed-chain) arrangement of six carbon atoms, as in a molecule of benzene. Molecules containing one or more benzene rings display aromatic character.

benzene-1,3-diol (chemistry) An alternative term for resorcinol.

benzocaine (pharmacology) A local anaesthetic (*USA*: anesthetic) drug administered by topical application for the relief of pain in the skin surface or mucous membranes.

benzoic acid (chemistry) C_6H_5COOH White crystalline organic compound, used as a food preservative because it inhibits the growth of yeasts and moulds (*USA*: molds).

benzole (chemistry) An alternative term for benzene.

benzpyrene (chemistry) Cyclic organic compound, found in coal-tar and tobacco smoke, which has strong carcinogenic properties.

benzpyrrole (chemistry) An alternative term for indole.

BER (physiology) An abbreviation of 'B'asic 'E'lectric 'R'hythm.

berkelium (Bk) (chemistry) Radioactive element in Group IIIB of the periodic table (one of the actinides). It has several isotopes, with half-lives of up to 1,400 years, made by alpha-particle bombardment of americium-241. At. no. 97; r.a.m. 247 (most stable isotope).

Bernoulli's theorem (physics) At any point in a tube through which a liquid is flowing the sum of the potential, kinetic and pressure energies is constant. Also termed: Bernoulli's principle. It was named after the Swiss mathematician and physicist Daniel Bernoulli (1700–82).

Bertiella (veterinary parasitology) A genus of nonpathogenic tapeworm of the family Anoplocephalidae. Species include *Bertiella mucronata* and *Bertiella studeri* found in primates, and *Bertiella obesa* found in koalas.

bertielliasis (human/veterinary parasitology) Infestation with *Bertiella* spp. of tapeworms. Occurs in primates, possums, koalas and occasionally in humans and dogs.

beryllium (Be) (chemistry) Silver-grey metallic element in Group IIA of the periodic table (the alkaline earths). It is used for windows in X-ray tubes and as a moderator in nuclear reactors. At. no. 4; r.a.m. 9.0122.

Besnoitia (veterinary parasitology) A genus of sporozoan parasites in the family Sarcocystidae. There are a number of species that are found only in wild animals. Horses and cattle are affected by disease in their role as intermediate hosts. In many of the species the definitive host is the cat but the definitive host in the others remains unidentified.

Besnoitia bennetti (veterinary parasitology) A species of tapeworm in the genus *Besnoitia* that causes besnoitiosis of horses and donkeys.

Besnoitia besnoiti (veterinary parasitology) A species of tapeworm in the genus *Besnoitia* that causes besnoitiosis of cattle.

Besnoitia darlingi (veterinary parasitology) A species of tapeworm in the genus *Besnoitia* that may be found in opossums, and possibly lizards.

Besnoitia wallacei (veterinary parasitology) A species of tapeworm in the genus *Besnoitia* that may be found in cats.

besnoitiosis (veterinary parasitology) The cutaneous form of besnoitiosis in horses and burros, caused by *Besnoitia bennetti* and characterized by a widespread, serious dermatitis. The disease in cattle, caused by *Besnoitia besnoiti*, is a systemic one manifested by swelling of the lymph nodes, subcutaneous swellings, diarrhoea (*USA*: diarrhea) and abortion.

beta decay (radiology) Disintegration of an unstable radioactive nucleus that involves the emission of a beta particle. It occurs when a neutron emits an electron and is itself converted to a proton, resulting in an increase of one proton in the nucleus concerned and a corresponding decrease of one neutron. This leads to the formation of a different element (*e.g.* beta decay of carbon-14 produces nitrogen).

beta particle (radiology) High-velocity electron emitted by a radioactive nucleus undergoing beta decay.

beta radiation (radiology) Radiation, consisting of beta particles (electrons), due to beta decay.

beta ray (radiology) Stream of beta particles.

bethanechol chloride (pharmacology) A parasympath-omimetic drug administered orally to stimulate motility in the intestines and to treat urinary retention.

betz cells (cytology) The large pyramidal cells of the motor cortex of the brain. Named after the Ukrainian anatomist Vladimir Betz (1834–94).

bezafibrate (pharmacology) A lipid-lowering drug administered orally to reduce levels, or change the proportions of various lipids in the bloodstream of patients with hyperlipidaemia (*USA*: hyperlipidemia).

BFO (physics) An abbreviation of 'B'eat-'F'requency 'O'scillator.

BGP (physiology/biochemistry) An abbreviation of 'B'one 'G'la 'P'rotein.

bibl. (literary terminology) An abbreviation of 'bibl'iographer/bibl'iographical/bibl'iography. Also termed bibliog.

bibliographical (literary terminology) Pertaining to the literature of a subject.

bicarb. (chemistry) An abbreviation of 'bicarb'onate of soda.

bicarbonate (chemistry) An alternative term for hydrogen carbonate.

Bicaulus (veterinary parasitology) A genus of nematode worms in the family Protostrongylidae. Now included in the genus *Varestrongylus*. Species include *Bicaulus sagitattus* and *Bicaulus schulzi* which may be found in the lungs of deer, goat, and sheep.

biceps (anatomy) Two-headed muscle in the arm or thigh.

bichromate (chemistry) An alternative term for dichromate.

biconcave (optics) Describing a lens that is concave on both surfaces.

biconvex (optics) Describing a lens that is convex on both surfaces.

bilateral symmetry (physiology) Type of symmetry in which a shape is symmetrical about a single axis or plane (each half being a mirror image of the other). *E.g.* most vertebrates are bilaterally symmetrical.

See also radial symmetry.

bile (biochemistry) Alkaline mixture of substances produced by the liver and stored in the gall bladder, which passes it to the duodenum, where it emulsifies fats (preparing them for digestion) and neutralizes acid. Its (yellowish) colour (*USA*: color) is due to bile pigments (*e.g.* bilirubin).

bile duct (anatomy) Tube that carries bile from the gall bladder to the duodenum.

bilharzia (human/veterinary parasitology) Disease that affects human beings and domestic animals in some subtropical regions, which results from an infestation by one of the parasitic blood flukes belonging to the genus *Schistosoma*. The larvae of the flukes develop inside freshwater snails and become free-swimming organisms which can attach themselves to a wading mammal, penetrating the skin and entering the bloodstream. Also termed: schistosomiasis.

bilharziasis (medical/veterinary parasitology) *See* schistosomiasis.

Bilharziella (parasitology) A genus of the family Schistosomatidae.

Bilharziella polonica (avian parasitology) A trematode parasite found in the abdominal blood vessels of ducks.

bilharziosis (medical/veterinary parasitology) *See* schistosomiasis.

biliary (anatomy) To do with bile or the gall bladder.

bilirubin (biochemistry) Major pigment in bile, formed in the liver by the breakdown of haemoglobin (*USA*: hemoglobin). Its accumulation causes the symptom jaundice.

billion (measurement) Number now generally accepted as being equivalent to 1,000 million (10^9). Formerly in Britain a billion was regarded as a million million (10^{12}).

binary code (mathematics) An alternative term for binary notation.

binary compound (chemistry) Chemical that consists of two elements.

binary fission (physiology) Method of reproduction employed by many single-celled organisms such as protozoa, in which the so-called mother cell divides in half (by mitosis), forming two identical, but independent, daughter cells. It is a type of asexual reproduction.

binary notation (mathematics) Number system to the base 2, involving only two digits, 0 and 1. Instead of the units tens, hundreds, etc. of the decimal system, units twos, fours, eights, etc. are used. Thus, *e.g.* 8 is given as 1000 and 9 as 1001. Binary notation is important in electronics and computers because the 0 and 1 can be represented by a circuit being 'on' or 'off'. Also termed: binary code; binary system.

binary system (mathematics) An alternative term for binary notation.

binding energy (physics) Energy required to cause a nucleus to decompose into its constituent neutrons and protons.

binocular (optics) 1 Pertaining to both eyes. 2 Having two eyepieces, as in a microscope.

binocular microscope (equipment) A microscope with two eyepieces, permitting use of both eyes simultaneously.

binomial (mathematics) polynomial that has two variables *e.g.* $(x + 2y)^2$.

binomial nomenclature (biology) System by which organisms are identified by two Latin names. The first is the name of the genus (generic name), the second is the name of the species (specific name), *e.g. Homo sapiens.* Also termed: Linnaean system (after the Swedish biologist Carolus Linnaeus, 1707–78).

See also classification.

binomial theorem (mathematics) Formula for calculating the power of a binomial without complicated multiplication.

bioassay (biochemistry) Method for quantitatively determining the concentration of a substance by its effect on living organisms, *e.g.* its effect on the growth of bacteria.

biochemistry (biochemistry) Study of the chemistry of living organisms.

biocide (pharmacology/control measures) Destructive to organisms including amoebae (*USA*: amebae) (amebicide).

biodegradation (biology) Breakdown or decay of substances by the action of living organisms, especially saprophytic bacteria and fungi. Through biodegradation, organic matter is recycled.

bioenergetics (physiology) Study of the transfer and utilization of energy in living systems.

See also ATP.

bioengineering (biology) Application of engineering science and technology to living systems.

biofeedback (physiology) Method of controlling a bodily process (*e.g.* heartbeat) that is not normally subject to voluntary control by making the person concerned aware of measurements from instruments monitoring that process.

biogenesis (biology) Theory that living organisms may originate only from other living organisms, as opposed to the theory of spontaneous generation.

biogeography (epidemiology) Scientific study of the geographic distribution of living organisms.

biological clock (biology) Hypothetical mechanism in plants and animals that controls periodic changes in internal functions and behaviour (*USA*: behavior) independently of environment, *e.g.* diurnal rhythms and hibernation patterns.

biological vector (parasitology) An arthropod vector in whose body the infecting organism develops or multiplies before becoming infective to the recipient individual.

biologist (biology) A specialist in biology.

biology (biology) Scientific study of living organisms.

bioluminescence (physiology) Emission of visible light by living organisms, *e.g.* certain bacteria, fungi, fish and insects. The light is produced by enzyme reactions in which chemical energy is converted to light.

biomass (biology) Total mass of living matter in a given environment or food chain level.

biome (epidemiology) A large, distinct, easily differentiated community of organisms in a major ecological region.

biomechanics (biology) The application of mechanical laws of living structures.

biomedicine (medical/veterinary parasitology) Clinical medicine based on the principles of the natural sciences, such as biology and biochemistry.

biomembrane (anatomy) Any membrane, *e.g.* the cell membrane, of an organism.

biometry (biology) Application of mathematical and statistical methods to the study of living organisms.

bionic (bioengineering) Describing an artificial device or system that has the properties of a living one.

biophysics (biology) Use of ideas and methods of physics in the study of living organisms and processes.

biopsy (histology) Small sample of cells or tissue removed from a living subject for laboratory examination (usually as an aid to diagnosis).

bioptome (equipment) A cutting instrument for taking biopsy specimens.

BIOS (computing) An abbreviation of 'B'asic 'I'nput 'O'utput 'S'ystem.

biosafety (methods) The safe handling of biological materials, particularly infectious agents which are classified on the basis of degree of risk to humans working with them and includes definition of biosafety levels for handling such agents.

bioscience (biology) The study of biology wherein all the applicable sciences such as physics, chemistry, etc. are applied.

biosecurity (parasitology) Security from transmission of infectious diseases, parasites and pests.

biostatics (biology) The science of the structure of organisms in relation to their function.

biostatistics (statistics) Vital statistics.

biosynthesis (biology) Formation of the major molecular components of cells, *e.g.* proteins, from simple components.

biota (biology/epidemiology) All the living organisms of a particular area; the combined flora and fauna of a region.

biotechnology (biotechnology) Utilization of living organisms for the production of useful substances or processes, *e.g.* the use of recombinant DNA technology and genetic engineering to manufacture a wide variety of biologically useful substances such as vaccines and hormones by expression of cloned genes in various host cell systems including bacteria, yeast and insect cells.

biotic (biology) 1 Pertaining to life or living organisms. 2 Pertaining to the biota.

biotic community (biology/epidemiology) The assemblage of living things, including animals, plants and bacteria, which inhabit a specific biotope.

biotin (biochemistry) Coenzyme that is involved in the transfer of carbonyl groups in biochemical reactions, such as the metabolism of fats; one of the B vitamins.

biotope (biology/epidemiology) An area of land surface that provides uniform conditions over its entire surface for animal and plant life.

biotoxicology (biology/parasitology) Scientific study of poisons produced by living organisms, their cause, detection and effects, and treatment of conditions produced by them.

biotoxin (biology/parasitology) A poisonous substance produced by a living organism.

biotype (biology) A group of individuals having the same genotype.

See also biovar.

biovar (parasitology) A group of strains of a species of microorganisms having differentiable biochemical or physiological characteristics.

biparental (biology) Derived from two parents, male and female.

bipolar (biology) 1 Having two poles. 2 Pertaining to both poles.

bipotentiality (biology) Ability to develop or act in either of two different ways.

birch sawfly (parasitology) *See Arge pullata.*

bird bug (avian parasitology) A number of bugs in the family Cimicidae (order Hemiptera) which infest birds. *See Haematosiphon*; *Oeciacus vicarius*; *Ornithodorus.*

bird flea (avian parasitology) *See Ceratophyllus.*

bird lice (avian parasitology) Members of the order Mallophaga and includes *Amyrsidea*, *Anaticola*, *Anatoecus*, *Bonomiella*, *Campanulotes*, *Chelopistes*, *Ciconiphilus*, *Clayia*, *Colocerus*, *Colpocephalum*, *Columbicola*, *Cuclogaster*, *Goniocles*, *Goniodes*, *Hohorstiella*, *Holomenopon*, *Lagopoecus*, *Liperus*, *Menacanthus*, *Menopon*, *Numidicola*, *Ornithobius*, *Oxylipeuris*, *Physconelloides*, *Somaphantus* and *Trinoton.*

bird malaria (avian parasitology) *See* avian malaria.

bird tick (avian parasitology) *See Haemaphysalis chordeilis*; *Argas.*

bisacodyl (pharmacology) A stimulant laxative administered orally or topically as suppositories to promote defecation and relieve constipation. It can be used to evacuate the colon prior to rectal examination or surgery.

bismuth (Bi) (chemistry) Silvery-white metallic element in Group VA of the periodic table. It is used as a liquid metal coolant in nuclear reactors and as a component of low-melting point lead alloys. Its soluble compounds are poisonous; its insoluble ones are used in medicine. At. no. 83. r.a.m. 208.9806.

Bisternalis (insect parasitology) *See Antennophours.*

bisulphate (chemistry) An alternative term for hydrogensulphate (*USA*: hydrogensulfate). (*USA*: bisulfate).

bisulphite (chemistry) An alternative term for hydrogensulphite (*USA*: hydrogensulfite). (*USA*: bisulfite).

bit (computing) An abbreviation of 'bi'nary digi't'; either 1 or 0, the only two digits in binary notation. Amount of information that is required to express choice between two possibilities. The term is commonly applied to a single digit of binary notation in a computer.

See also byte.

bite biopsy (methods) Instrumental removal of a fragment of tissue.

bithionol (pharmacology) A bacteriostatic agent especially effective against gram-positive cocci that also has anthelmintic properties.

bithionol sulphoxide (pharmacology) An effective cestocide, which is also used as a fasciolicide, usually in combination with other compounds because of its poor efficiency against immature flukes. It has been largely superseded as an anthelmintic. (*USA*: bithionol sulfoxide).

Bithynia (parasitology) A genus of snails that act as intermediate hosts for the miracidia of the *Opisthorchis* spp. of bile duct flukes.

biting (parasitology) Pertaining to the characteristic behaviour (*USA*: behavior) of performing a bite.

biting louse (parasitology) *See* species of the insect suborder Mallophaga.

biting midge (parasitology) Insects of the family Ceratopogonidae. Also termed punkies, no-see-ums, sandflies.

biting pattern (parasitology) The pattern of distribution of bites, or of diseases transmitted by insect bites, which may suggest the identity of the biter.

biuret test (biochemistry) Test used to detect peptides and proteins in solution by treatment of biuret (NH_2 - $CONHCONH_2$) with copper sulphate (*USA*: copper sulfate) and alkali to give a purple colour (*USA*: color).

bivalent (chemistry) Having a valence of two. Also termed: divalent.

Bivitellobilharzia (veterinary parasitology) A genus of the family Schistosomatidae of blood flukes. Species include *Bivitellobilharzia loxodontae* and *Bivitellobilharzia nairi* which may be found in the portal vein of elephants.

bkgd (general terminology) An abbreviation of 'b'ac'kg'roun'd'.

black blowfly (parasitology) *See Phormia.*

black fly (parasitology) A term which covers a number of genera and species that play a part in the transmission of *Onchocerca* spp. Also termed: buffalo gnat; sandfly; *Cnephia*; *Austrosimulium pestilens*; *Austrosimulium bancrofti.*

See also Simulium.

black gill disease (fish parasitology) Nonspecific term for dark gills in shrimp, that may be caused by amongst others, ciliate (apostome) infection.

black grub (fish parasitology) Metacercariae of digenetic flukes in the skin and/or musculature of finfish.

black scour worm (parasitology) *See Trichostrongylus.*

black soil itch (veterinary parasitology) *See Eutrombicula sarcina.*

blackhead (avian parasitology) A disease of birds, especially turkeys, caused by the protozoan parasite *Histomonas meleagridis* and characterized by necrotic lesions in the liver and inflammation and distension of the caecum (*USA*: cecum). Earth worms are important vectors in the life cycle of the parasite.

blackhead disease of bananas (plant parasitology) *See Radopholus similis.*

blackleaf 40 (control measures) A commercially available concentrate of nicotine sulphate (*USA*: nicotine sulfate) used as a parasiticide in horticulture and containing 40 percent of alkaloidal nicotine.

black-legged tick (parasitology) *See Ixodes.*

blacktail condition (fish parasitology) A disease of fish caused by a protozoan parasite, *Myxosoma* spp., which destroys cartilage in the vertebral column, causing abnormal activity of caudal pigment cells and black discolouration (*USA*: discoloration) of the tail.

blackwater fever (medical/veterinary parasitology) *See* babesiosis.

bladder (anatomy) A membranous sac containing gas or fluid, *e.g.* the gall bladder and urinary bladder.

See also vesicle.

bladder worm (parasitology) *See Cysticercus tenuicollis.*

blade (equipment) Scalpel.

blanket therapy (pharmacology/control measures) Treatment of all animals in the group used usually as a protective measure against infestation or infection, or because a large proportion are suspected to be infested or infected and it is more cost-effective to treat all of them than to test and treat selectively.

blast(o)- (terminology) A word element. [German] Prefix denoting a bud, budding.

blastocyst (biology) A stage in the embryogenesis of mammals, prior to implantation. It is the first stage at which there is any differentiation among the ball of cells. The blastocyst consists of an outer layer of protective cells (the trophectoderm) surrounding a fluid-filled cavity containing the inner cell mass that goes on to develop as the embryo. The human blastocyst is about 0.15 mm in diameter.

blastula (histology) Hollow sphere composed of a single layer of cells produced by cleavage of a fertilized ovum in animals.

bleach (chemistry) Substance used for removing colour (*USA*: color) from, *e.g.* cloth, paper and straw. A common bleach is a solution of sodium chlorate(I) (sodium hypochlorite), NACIO, although hydrogen peroxide, sulphur dioxide (*USA*: sulfur dioxide), chlorine, oxygen and even sunlight are also used as bleaches. All are oxidizing agents.

bleaching powder (chemistry) White powder containing calcium chlorate(I) (calcium hypochlorite), $Ca(OCl)_2$, made by the action of chlorine on calcium hydroxide (slaked lime). When treated with dilute acid it generates chlorine, which acts as a bleach.

blind spot (biology) Area on the retina of the vertebrate eye which, because it is at the point of entry of the optic nerve, is without light sensitive cells and is thus blind.

blister beetle (parasitology) *See Epicauta vittata.*

blister fly (parasitology) *Cantharis vesicatoria.*

block (computing) To highlight a block of text which will then be modified by the next command.

blood (haematology) Fluid in the bodies of animals that circulates and transports oxygen and nutrients to cells, and carries waste products from them to the organs of excretion. It also transports hormones and the products of digestion. Essential for maintaining uniform temperature in warmblooded animals, blood is made up mainly of erythrocytes, leucocytes, platelets, water and proteins.

See also haemocyanin; haemoglobin.

blood cells (haematology) *See* erythrocytes; leucocytes.

blood clotting (pharmacology) *See* haemostasis.

blood count (haematology) A count of the number of red and white blood cells and platelets.

blood fluke (parasitology) *See Schistosoma.*

blood parasites (parasitology) A general term used in referring to those parasites which pass most of their lives in the vascular system, *e.g. Babesia, Theileria, Trypanosoma* and *Dirofilaria.*

blood plasma (haematology) Fluid part of blood in which blood cells are suspended. It contains plasma proteins, urea, sugars and salts.

blood poisoning (medical) An alternative term for septicaemia (*USA*: septicemia).

blood pressure (physiology) Pressure of blood flowing in the arteries. It varies between the higher value of the systolic pressure (when the heart's ventricles are contracting and forcing blood out of the heart) and the lower value of the diastolic pressure (when the heart is filling with blood). It is affected by exercise, emotion and various drugs.

blood sampling (parasitology) Laboratory examination of samples of blood. These are usually collected from an appropriate vein, but arterial samples are required for some special techniques. Collection is by a closed method, either a hypodermic syringe or a vacuumized container. The site of collection varies with the species and the circumstances. The sample may be clotted, for serum estimations, or collected with an anticoagulant for examinations on plasma or whole blood.

blood serum (haematology) Fluid part of blood from which all blood cells and fibrin have been removed. It may contain antibodies, and such serums are used as vaccines.

blood sugar (biochemistry) The energy-generating sugar glucose, whose level in the blood is controlled by the hormone insulin.

blood vascular system (anatomy) System consisting of the heart, arteries, veins and capillaries. The heart acts as a central muscular pump which propels oxygenated blood from the lungs along arteries to the tissues; deoxygenated blood is carried along veins back to the heart and lungs.

blood vessel (anatomy) An artery, vein, or capillary. Many blood vessels have muscular walls, whose contraction and relaxation aids blood flow.

blood worms (parasitology) The large strongyles of horses. *See Strongylus vulgaris*; *Strongylus edentatus*; *Strongylus equinus.*

bloodworm (fish parasite) A parasitic worm of cyprinid fish. Also termed: *Philometra abdominalis.*

blowfly (parasitology) A member of the family Calliphoridae of insects.

blowfly myiasis (parasitology) *See* cutaneous myiasis.

blowfly strike (parasitology) Invasion of skin or exposed mucosae by blowfly larvae.

See also cutaneous myiasis.

blowing (parasitology) Infestation of dead or living material with blowfly maggots.

blue louse (parasitology) *Linognathus ovillus.*

blue tick (parasitology) *See Boophilus decoloratus.*

blue vitriol (chemistry) An alternative term for the hydrated form of copper(II) sulphate (*USA*: copper (II) sulfate).

BLWA (society) An abbreviation of 'B'ritish 'L'aboratory 'W'are 'A'ssociation.

BM (medical) An abbreviation of 'b'owel 'm'ovement.

BMA (governing body) An abbreviation of 'B'ritish 'M'edical 'A'ssociation.

BMJ (literary terminology) An abbreviation of 'B'ritish 'M'edical 'J'ournal.

BMR (physiology) An abbreviation of 'B'asal 'M'etabolic 'R'ate.

BMS (society) An abbreviation of 'B'ritish 'M'ycological 'S'ociety.

BNP (physiology/biochemistry) An abbreviation of 'B'rain 'N'atriuretic 'P'eptide.

body louse (parasitology) *See Menacanthus.*

body mange (parasitology) *See* psoroptic mange.

boiling point (b.p.) (physics) Temperature at which a liquid boils, freely turning into a vapour (*USA*: vapor) its vapour pressure then equals the external pressure.

bomb fly (parasitology) *See Hypoderma.*

bonamiasis (fish parasitology) An important protozoan disease affecting haemocytes (*USA*: hemocytes) of flat oysters worldwide.

bond (chemistry) Link between two atoms in a molecule. *See* co-ordinate bond; covalent bond; ionic bond.

bond energy (chemistry) Energy involved in bond formation.

bond length (chemistry) Distance between the nuclei of two atoms that are chemically bonded.

bone (anatomy) Skeletal substance of vertebrates. It consists of cells (osteocytes) distributed in a matrix of collagen fibres (*USA*: collagen fibers) impregnated with a complex salt (bone salt), mainly calcium phosphate, for hardness. Cells are connected by fine channels that permeate the matrix. Larger channels contain blood vessels and nerves. Some bones are hollow and contain bone marrow. There are about 206 bones in an adult human skeleton.

bone marrow (haematology) The soft haematopoietic (*USA*: hematopoietic) tissue that has stem cells capable of differentiating into erythrocytes, leucocytes (*USA*: leukocytes) or platelets (thrombocytes).

Bonomiella columbae (veterinary parasitology) A species of lice in the genus *Bonomiella* that may be found on pigeons.

bont tick (parasitology) *See Amblyomma hebraeum.*

book louse (parasitology) The intermediate hosts of *Avitellina* spp. tapeworms.

Boophilus (parasitology) A genus of ticks in the family Ixodidae.

Boophilus annulatus (parasitology) A one-host tick which may transmit *Babesia bigemina.* Also termed North American tick.

Boophilus calcaratus (parasitology) A one-host tick which may transmit *Anaplasma marginale, Babesia bigemina* and *Babesia berbera.*

Boophilus decoloratus (parasitology) A one-host tick which may transmit *Anaplasma marginale, Babesia bigemina, Babesia ovis, Babesia trautmanni* and *Babesia theileri.*

Boophilus microplus (parasitology) A one-host tick which may transmit *Anaplasma marginale, Babesia bigemina* and *Babesia bovis* and *Coxiella burnettii.*

Booponus (parasitology) A genus of flies of the family Calliphoridae.

Booponus intonsus (veterinary parasitology) A species of fly whose maggots may cause myiasis and skin damage on the lower parts of the legs of ruminants. Also termed foot maggot.

boot (computing) To start up a computer and bring it to a state of readiness to receive commands. Also termed: initializing.

boracic acid (chemistry) An alternative term for boric acid.

borax (chemistry) $Na_2B_4O_7.\ 10H_2O$ White amorphous compound, soluble in water, which occurs naturally as tincal. It is used in the manufacture of enamels and heat-resistant glass. Also termed: disodium tetraborate; sodium borate.

boric acid (control measures/chemistry) H_3BO_3 White crystalline compound, soluble in water, which occurs naturally in volcanic regions of Italy. It has antiseptic properties and may be used as an insecticide. Also termed: boracic acid.

boron (B) (chemistry) Amorphous, non-metallic element in Group IIIA of the periodic table. Because of its high neutron absorption, it is used for control rods in nuclear reactors. Important compounds include borax and boric acid. At. no. 5; r.a.m. 10.81.

borosilicate glass (chemistry) Heat-resistant glass of low thermal expansion, made by adding boron oxide (B_2O_3) to glass during manufacture.

Bos (parasitology) A genus of cattle of the family Bovidae which includes buffalo, bison and many other wild ruminants.

Bos indicus (parasitology) The zebu species, much prized for their resistance to tick infestation and because of their hardiness in hot climates. Also termed Brahman (USA), Afrikaner (Africa).

bot fly (parasitology) The flies that produce the maggots known as bots and the diseases referred to as gasterophilosis and nasal bot fly infestation.

bot fly infestation (parasitology) *See* gasterophilosis.

Bothriocephalus (fish parasites) A genus of tapeworms in the order Pseudophyllidae that occur, but appear to have little pathogenicity, in many wild fish. Species include *Bothriocephalus acheilognathi* found in carp and goldfish.

bothrium (parasitological anatomy) A term used in relation to cestodes. A longitudinal groove in the scolex of pseudophyllideans, *e.g. Diphyllobothrium* spp.

Bourgelatia (veterinary parasitology) A genus of the family Chabertiidae of nematodes. Species include *Bourgelatia diducta*, which may be found in pig intestines.

boutonneuse fever (human/veterinary parasitology) A tick-borne rickettsial disease of humans, endemic in the Mediterranean area caused by *Rickettsia conorii*. Dogs are sometimes infected and may be a reservoir for the disease.

Bovicola (veterinary parasitology) A genus of lice of the superfamily Ischnocera. Species includes *Bovicola painei* that may be found on goats. Also termed: *Damalinia*.

bovine protozoal abortion (veterinary parasitology) *See* neosporosis.

bovine trichomoniasis (veterinary parasitology) A contagious venereal disease of cattle caused by *Trichomonas foetus* and characterized by infertility, abortion and pyometra.

bowel (anatomy) An alternative term for intestine.

Bowman's capsule (anatomy) Dense ball of capillary blood vessels that cover the closed end of every nephron in the kidney. From it leads the uriniferous tubule. It was named after the British physician William Bowman (1816–92).

Boyle's law (physics) At constant temperature the volume of a gas V is inversely proportional to its pressure p; *i.e.* $p\,V$ = a constant. It was named after the Irish chemist Robert Boyle (1627–91).

bp (genetics) An abbreviation of 'b'ase-'p'air.

Brachylaemidae (avian/veterinary parasitology) A family of trematodes that infest birds and mammals.

Brachylecithum mosquensis (insect parasitology) A species of helminth, Trematodes, that are parasites to the ants *Camponotus* spp. Typically one or two metacercaria are found in the host's brain, resulting in sluggish behaviour (*USA*: behavior), the ants preferring open spaces. They have two other hosts in land snails and a bird.

Bradybaena (parasitology) Terrestrial snail, the intermediate host for the pancreatic flukes *Eurytrema* spp.

bradyzoite (parasitology) Small comma-shaped form of *Toxoplasma gondii* found in tissues enclosed in a pseudocyst, and assumed to be slow-growing form of the parasite.

brain (anatomy) Principal collection of nerve cells that form the anterior part of the central nervous system, consisting (in mammals) of a moist pinkish-grey mass protected by the bones of the cranium (skull). It receives, mostly via spinal nerves from the spinal cord but also via cranial nerves from organs in the head, sensory information through afferents carrying the output of sense organs and it sends out instructions along efferents (*e.g.* motor nerves) to the effector organs (*e.g.* muscles). It is also the centre (*USA*: center) of intellect and memory (so that behaviour [*USA*: behavior] can be based on past experience), and is responsible for the coordination of the whole body.

See also cerebellum; cerebrum; cortex.

brain cestodal cyst (parasitology) *See* coenurosis.

brain stem (anatomy) Part of the brain at the end of the spinal cord, consisting of the midbrain, medulla oblongata and pons.

branched chain (chemistry) Side group(s) attached to the main chain in the molecule of an organic compound.

breastbone (anatomy) An alternative term for sternum.

breathing (physiology) An alternative term for respiration, but often restricted to the physical actions of inhalation and exhalation.

breeze fly (parasitology) *See Tabanus*.

BRI (education) An abbreviation of 'B'iological/'B'rain 'R'esearch 'I'nstitute.

bright-field microscope (microscopy) The standard bench microscope.

brine (chemistry) Concentrated solution of common salt (sodium chloride).

british dog tick (veterinary parasitology) *See Ixodes canisuga*.

British thermal unit (BTU) (measurement) Amount of heat required to raise the temperature of 1 pound of water through 1°F. 1 BTU = 1,055 joules.

broad fish tapeworm (fish parasites) *See Diphyllobothrium latum*.

broad tapeworm (parasitology) *See Diphyllobothrium latum*.

Broca's area (anatomy) Speech centre (*USA*: center) of the brain. It was named after the French surgeon Paul Broca (1824–80).

brom. (chemistry) An abbreviation of 'brom'ide.

bromide (chemistry) Binary compound containing bromine; a salt of hydrobromic acid.

bromide paper (photography) Photographic (light-sensitive) paper coated with an emulsion containing silver bromide, used for making black-and-white prints and enlargements.

bromine (Br) (chemistry) Dark red liquid non-metallic element in Group VIIA of the periodic table (the halogens), extracted from sea-water. It has a pungent smell. Its compounds are used in photography and as anti-knock additives to petrol. At. no. 35; r.a.m. 79.904.

bromocyclen (pharmacology) A halogenated hydrocarbon used as an acaricide. Also termed: bromociclen.

bromophenophos (veterinary parasitology) An organophosphorus compound used to treat *Fasciola hepatica* infections in cattle. Also termed: bromofenofos.

bromophos-ethyl (pharmacology) An organophosphate insecticide used principally as an acaricide.

bromsalans (pharmacology) Biphenolic compounds used as fasciolicides; includes dibromsalan and tribromsalan. They are very effective against juvenile flukes.

bronchiole (anatomy) Terminal air-conducting tube (1mm in diameter) of mammalian lungs, arising from secondary subdivision of a bronchus and terminating in alveoli.

bronchus (anatomy) One of the two air-carrying divisions of the trachea (windpipe) into the lungs. The bronchi become inflamed in the disorder bronchitis.

brood capsule (parasitological anatomy) A term used in relation to cestodes. A small cyst attached to the germinal layer of the hydatid, containing many protoscoleces.

Brooklynella (fish parasites) An ectoparasitic ciliated protozoon that causes severe lesions on the gills of marine fish kept in aquaria.

brooklynellosis (fish parasites) Disease of marine fish caused by infestation with *Brooklynella* spp., a unicellular ciliate. Characterized by hyperplasia of the gill epithelium and clinically by respiratory impairment.

brown chicken louse (avian parasitology) *See Goniodes dissimilis.*

brown dog tick (veterinary parasitology) *See Rhipicephalus sanguineus.*

brown ear tick (parasitology) *See Rhipicephalus appendiculatus.*

brown ring test (chemistry) Laboratory test for nitrates in solution. An acidic solution of iron(II) sulphate (*USA*: iron(II) sulfate) (ferrous sulphate [*USA*: ferrous sulfate]) is added to the suspected nitrate solution in a test-tube, and concentrated sulphuric acid (*USA*: sulfuric acid) carefully poured down the inside of the tube. A brown ring at the junction of the liquids indicates the presence of a nitrate.

brown stomach worm (parasitology) *See Ostertagia ostertagi; Teladorsagia.*

brown winter tick (parasitology) *See Dermacentor nigrolineatus.*

Brownian movement (physics) Random motion of particles of a solid suspended in a liquid or gas, caused by collisions with molecules of the suspending medium. It was named after the British botanist Robert Brown (1773–1858). Also termed: Brownian motion.

browse (computing) Term for scanning through information in a computer programme (*USA*: program).

Brugia (human/veterinary parasitology) A genus of the family Onchocercidae of worms. Species includes *Brugia celonensis*, *Brugia malayi*, *Brugia patei*, *Brugia pahangi*, *Brugia timori*. In tropical countries *Brugia* spp. are parasitic in the lymphatics of primates, carnivores and insectivores, and parasitized animals may act as reservoirs for human infection. Transmission is via mosquitoes.

Brumptia bicanda (veterinary parasitology) Intestinal trematode in the rhinoceros.

brush biopsy (parasitology) Removal of cells and tissue fragments using a brush with stiff bristles (introduced through an endoscope), which is effective in obtaining tissue samples from inaccessible places.

BS (1 education; 2 society; 3 quality standards) 1 An abbreviation of 'B'achelor of 'S'cience/'S'urgery. 2 An abbreviation of 'B'iochemical/'B'iometric/'B'iological 'S'ociety. 3 An abbreviation of 'B'ritish 'S'tandard.

BSAVA (society) An abbreviation of 'B'ritish 'S'mall 'A'nimals 'V'eterinary 'A'ssociation.

BSc (education) An abbreviation of 'B'achelor of 'Sc'ience.

bsc (general terminology) An abbreviation of 'b'a's'i'c'.

BSCC (society) An abbreviation of 'B'ritish 'S'ociety for 'C'linical 'C'ytology.

BSCP (quality standards) An abbreviation of 'B'ritish 'S'tandard 'C'ode of 'P'ractice.

BSI (quality standards) An abbreviation of 'B'ritish 'S'tandards 'I'nstitution.

BSIRA (society) An abbreviation of 'B'ritish 'S'cientific 'I'nstrument 'R'esearch 'A'ssociation.

BSP (society) An abbreviation of 'B'achelor of 'S'cience in 'P'harmacy.

BTPS (physics) An abbreviation of 'B'ody 'T'emperature and 'P'ressure, 'S'aturated with water vapour (*USA*: vapor).

BTX (chemistry) An abbreviation of 'b'enzene, 't'oluene and 'x'ylene.

bty (physics) An abbreviation of 'b'at't'er'y'.

buccal (anatomy) Pertaining to the cheek and mouth.

buccal capsule (parasitological anatomy) A term used in relation to nematodes. The mouth cavity of the nematode.

buccal cavity (anatomy) Mouth cavity in front of the opening of the pharynx, where mammals food is usually chewed and subjected to the initial stages of digestion.

Bucephalus (fish parasitology) Trematode parasite causing parasitic castration of clams and scallops.

budding (biology) Method of asexual reproduction employed by yeasts and simple animal organisms.

budesonide (pharmacology) A corticosteroid drug with anti-inflammatory, anti-allergic and anti-asthmatic properties. It may be administered by inhalation to prevent attacks of asthma and rhinitis and topically to treat severe inflammatory skin disorders such as psoriasis and eczema.

buffalo fly (veterinary parasitology) *See Haematobia exigua.*

buffalo gnat (veterinary parasitology) *See* black fly.

buffalo louse (veterinary parasitology) *See Haematopinus tuberculatus.*

buffer (1 chemistry; 2 computing) 1 A solution that resists changes in pH on dilution or on the addition of acid or alkali. 2 A temporary store for data in transit between the central processor and an output device.

buffy coat (haematology) Layer of white blood cells and platelets above red blood cell mass when blood is sedimented.

bug (1 parasitology; 2 computing) 1 A member of the family Cimicidae in the order Hemiptera and includes the blood-sucking bugs. *See Haematosiphon*; *Oeciacus vicarius*. 2 An error in computer hardware or software.

bulb (parasitological anatomy) A term used in relation to nematodes. Posterior portion of the oesophagus.

Bulinus (parasitology) Water snail, intermediate host for paramphistomes and other trematodes.

bumetanide (pharmacology) A powerful diuretic drug which can be administered orally or by injection and is used to treat oedema (*USA*: edema), particularly pulmonary oedema. It is one of the class of loop diuretics.

BUN (biochemistry/haematology) An abbreviation of 'B'lood 'U'rea 'N'itrogen.

bunamidine (veterinary parasitology) A cestocide used extensively for the treatment of tapeworms in dogs and cats. Used usually as the hydrochloride in dogs and cats and as the hydroxynaphthoate in sheep and goats.

bunostomiasis (veterinary parasitology) Infestation with the hookworms *Bunostomum phlebotomum* in cattle and *Bunostomum trigonocephalum* in sheep causing anaemia (*USA*: anemia) and anasarca due to blood loss, together with poor growth. Also termed: hookworm disease.

Bunostomum (veterinary parasitology) A genus of hookworms of the family of Ancylostomatidae. Species includes *Bunostomum phlebotomum* that may be found in *Bos indicus* and *Bos taurus* cattle, and *Bunostomum trigonocephalum* that may be found in sheep, goat, and deer.

Bunsen burner (laboratory equipment) Gas burner that efficiently mixes air with the fuel gas, commonly used for heating in laboratories. It was named after the German chemist Robert Bunsen (1811–99).

bupivacaine hydrochloride (pharmacology) A local anaesthetic drug administered by injection that has a long duration of action. It is an amide and chemically related to lignocaine hydrochloride.

buquinolate (pharmacology) A quinolone cocciostat that may be used in poultry.

Burenella dimorpha (insect parasitology) A protozoal species of microsporidia that are parasites to the ants *Solenopsis geminata*, *Solenopsis invicta*, *Solenopsis quinquecuspis*, *Solenopsis richteri*, and *Solenopsis saevissima*-complex.

burette (laboratory equipment) Long vertical graduated glass tube with a tap, used for the addition of controlled and measurable volumes of liquids (*e.g.* in making titrations in volumetric analysis).

burrowing nematode (plant parasitology) *See Radopholus similis.*

bursa (anatomy) A closed sac lined with synovial membrane containing fluid placed where there may be friction between structures such as tendons and bones so that they can move easily over each other. Sometimes they become inflamed, as the result of either injury or infection, a condition called bursitis.

bursa of fabricius (immunology) Lymphoid organ which processes B-lymphocytes.

Bursaphelenchus (plant parasitology) A genus of nematodes in the order Tylenchida and family Aphelenchoididae. The females are similar to Aphelenchoides but some species have a cuticular flap extending posteriorly over the vulva. There are some 49 described species of *Bursaphelenchus*, most of which have a phoretic relationship with insects, especially bark beetles and wood borers. All species feed on fungi. *See Bursaphelenchus xylophilus.*

Bursaphelenchus xylophilus (plant parasitology) A species of nematodes in the genus *Bursaphelenchus* whose host is the pine trees and which may be found in USA and Japan. Species of pine vary in their resistance. Sexually reproducing, the nematodes are vectored by cerambycid longhorn beetles, also known as sawyers, of the genus *Monochamus*. The beetles are wood borers in the larval stage and *Monochamus alternatus* is the primary vector of *Bursaphelenchus xylophilus* in Japan. In a damaged pine forest more than 75 percent of pine sawyer adults may have dauer larvae, averaging 15,000/insect. The insect lays eggs in the bark of weakened tree, and larvae hatch after 1 week burrowing into wood and moult (*USA*: molt), creating a U-shaped tunnel back toward the surface, ending in a pupal chamber. Nematode development requires 5 days to reach the adult form, followed by 28 days of oviposition of 80 eggs per day. With unlimited resources a single female could give rise to 260,000 offspring in 15 days. The nematode has a resting stage in its life cycle. It appears in wood after the nematode population reaches its highest level and is regarded as the third stage, but different in its morphological and biological features from the usual third-stage juvenile. It is adapted to surviving unfavourable (*USA*: unfavorable) conditions, such as dry conditions, low temperatures, and lack of food. They are called the dispersal third-stage juvenile to differentiate them from the ususal third-stage juveniles. They have densely packed materials in their bodies, which are lipid droplets in the intestine. They also have a thicker cuticle than other stages. They moult (*USA*: molt) to dauer larvae in wood. This is then the fourth stage juvenile or dauer larvae which are adapted to being carried by the insect vector to a new habitat. The dauer larvae begin to moult (*USA*: molt) immediately after crawling off the insect's body onto pine twigs, which are the feeding site of the vector. Morphologically dauer larvae can be differentiated by a dome-shaped head, lack of stylet, degenerate oesophagus (*USA*: esophagus) and oesophageal (*USA*: esophageal) glands, and subcylindrical tail with a digitate terminus. Their bodies are covered with a protective sticky substance that seems to play a role in attaching these nematodes to the insect's body and to help them move on the insect's body, and to leave the insect. Transmitted dauer larvae invade trees through wounds on twigs created by feeding of *Monochamus alternatus*. They moult (*USA*: molt) to the adult stage, begin to feed on epithelial cells, and reproduce in the resin canals. Death of the parenchyma cells is the first visible pathological response. This is followed by cessation of oleoresin exudation. This occurs 6 to 9 days after infection and before nematode populations have increased or spread from the original infection site. After cessation of oleoresin exudation, the nematode population increases rapidly. Dead or dying trees contain tens of millions of nematodes in wood throughout the tree. From winter to spring, dispersal third stage nematodes accumulate around pupal chambers of *Monochamus alternatus*. In late spring, the dispersal third-stage juveniles moult (*USA*: molt) to become dauer larvae coinciding with the time of pupation of *Monochamus alternatus*. Dauer larvae appear on the wall surface of the pupal chamber, and climb up the long perithecial neck of blue stain fungi, *Ceratocystis* spp. At the tip of the neck they attach to spore masses of blue stain fungi with sticky substances. Contamination with dauer larvae occurs only on adults just after emerging from pupae. Dry conditions greatly reduce the number of dauer larvae which attach to beetles. Adults become infested with nematodes as they emerge from pupa and burrow to the surface. Nematodes usually remain on the surface and in the tracheae of beetles. The insect then flies to a healthy tree and feeds on young tissues. Trees already infected decline, thus attracting more insects to weakened trees for oviposition. A new generation of insects then becomes infected in the diseased trees. Drought and other stress may hasten the decline of trees. Control measures include treating insect vectors with insecticides and burning infected trees. Trunk injection of nematicides has also been used, but must be performed before symptoms occur. Also termed: pine wood nematode.

bus (computing) A pathway connecting the components of a computer, usually shared by several components.

bush fly (parasitology) *See Musca.*

bushel (measurement) Unit for measuring dry goods by volume, equal to 4 pecks and equivalent to 8 gallons or 36.369 litres (in Britain) or 35.238 litres (United States).

butamisole (pharmacology) An anthelmintic used as the hydrochloride. which may be administered by injection in dogs for the treatment of *Trichuris vulpis* and *Ancylostoma caninum*.

butane (chemistry) C_4H_{10} Gaseous alkane, used as a portable supply of fuel.

butanedioic acid (chemistry) An alternative term for succinic acid.

butanol (chemistry) C_4H_9OH Colourless (*USA*: colorless) liquid alcohol that exists in four isomeric forms. Also termed: butyl alcohol.

butopyronoxyl (control measures) An insect repellent effective against ticks.

butyl (chemistry) A hydrocarbon radical, C_4H_9.

butyl alcohol (chemistry) An alternative term for butanol.

butyl chloride (pharmacology) A substance that has been used as an anthelmintic but that may cause poisoning similar to that caused by carbon tetrachloride.

BV (haematolgy) An abbreviation of 'B'lood 'V'olume.

BVA (society) An abbreviation of 'B'ritish 'V'eterinary 'A'ssociation.

BW (biology) An abbreviation for 'B'ody 'W'eight.

bwd (general terminology) An abbreviation of 'b'ack'w'ar'd'.

bypass (medical) An auxiliary flow; a shunt; a surgically created pathway circumventing the normal anatomical pathway, as an intestinal bypass.

by-product (biotechnology) Incidental or secondary product of manufacture.

byte (computing) A number of binary digits that function as a unit, *e.g.* in the store of a computer. The term is sometimes limited to a unit consisting of eight binary digits.

C

C and G (education) An abbreviation of 'C'ity and 'G'uilds.

C Biol. (education) An abbreviation of 'C'hartered 'Biol'ogist.

c cm (measurement) An abbreviation of 'c'ubic 'c'enti'm'etre (*USA*: cubic centimeter). Also termed cc or millilitre (*USA*: milliliter).

c ft (measurement) An abbreviation of 'c'ubic 'f'ee't'/ 'f'oo't'.

C peptide (physiology/biochemistry) An abbreviation of 'C'onnecting 'peptide'.

C terminal (physiology/biochemistry) A term signifying an end to peptide or protein having a free -COOH group.

C (1 chemistry; 2 measurement) 1 chemical symbol for carbon. 2 An abbreviation of 'C'elsius.

c.cm. (measurement) An abbreviation of 'c'ubic 'c'enti 'm'etre (*USA*: cubic centimeter).

c.p.m. (measurement) An abbreviation of 'c'ounts 'p'er 'm'inute.

c.p.s. (measurement) An abbreviation of 'c'ycles 'p'er 's'econd.

C_{19} steroids (physiology/biochemistry) A symbol for steroids containing 19 carbon atoms.

C_{21} steroids (physiology/biochemistry) A symbol for steroids containing 21 carbon atoms.

CA (1 medical; 2, 3 biochemistry) An abbreviation for 1 'C'hronological 'A'ge. 2 An abbreviation for 'ca'techolamine. 3 An abbreviation of 'C'ellulose 'A'cetate.

Ca (chemistry) The chemical symbol for calcium.

ca. (1 physics; 2 Literature terminology) 1 An abbreviation of 'ca'thode. 2 An abbreviation of 'c'irc'a'. [Latin, meaning about].

Ca^{2+} (chemistry) Calcium ions.

Caballonema (parasitology) A genus of the subfamily Cyathostominae, small strongyles, cyathostomes of equids that has a distribution limited to the USSR and China.

cac(o)- (terminology) A word element. [German] Prefix denoting bad, ill.

cache (computing) A section of random access memory in which recently accessed information can be stored in anticipation of early future use.

cad (medical) An abbreviation of 'cad'aver. Also termed cadav.

CADE (computing) An abbreviation of 'C'omputer 'A'ssisted 'D'ata 'E'valuation.

cadmium (Cd) (chemistry) Silvery-white metallic element in Group IIB of the periodic table. It is used, amongst other things, as corrosion-resistant electroplating on steel articles. Its compounds are used as yellow or red pigments. At. no. 48; r.a.m. 112.40.

cadmium oxide (pharmacology) A toxic compound used at one time as an anthelmintic for pigs.

cADPR (physiology/biochemistry) An abbreviation of 'c'yclic 'A'denosine 'D'is'P'hosphate 'R'ibose.

caecal blackhead (parasitology) *See* *Histomonas meleagridis*. (*USA*: cecal blackhead).

caecal coccidiosis (parasitology) *See* coccidiosis. (*USA*: cecal coccidiosis).

caecum (anatomy) Pouch or pocket, such as one at the junction of the small and large intestines from which hangs the appendix.

Caenocholax fenyesi (insect parasitology) Arthropods in the Strepsiptera that are endoparasitic in the larvae of the ants *Solenopsis invicta*.

Caenorhabditis dolichura (insect parasitology) A species of nematode in the Rhabditidae that are parasites to the ants *Acanthomyops*, *Camponotus*, *Formica* spp., *Lasius* spp., and *Tetramorium*. They may be found in the pharyngeal glands of the head capsule.

caesium clock (physics) Atomic clock used in the SI unit definition of the second. (*USA*: cesium clock).

caesium (Cs) (chemistry) Soft reactive metallic element in Group IA of the periodic table (the alkali metals), a major fission product of uranium. It is used in photoelectric cells. The isotope caesium-137 is used in radiotherapy. At. no. 55; r.a.m. 132.9055. (*USA*: cesium)

CAF (medical) An abbreviation of 'C'ardiac 'A'ssessment 'F'actor.

cage-side tests (parasitology) Various clinicopathologic testing procedures which may be carried out at the location of an animal's cage; the equivalent of bed-side tests for human patients.

cal (measurement) An abbreviation of 'cal'orie (small).

Cal (measurement) An abbreviation of 'Cal'orie.

Calaenostanus (insect parasitology) A species of mites in the Mesostigmata, that similarly to *Edbarellus* and

Eumellitiphis are phoretic and cleptoparasitic on the bees *Apinae* and Meliponinae.

calamine (chemistry) Zinc ore whose main constituent is zinc oxide.

calciferol (biochemistry) Fat-soluble vitamin formed in the skin by the action of sunlight, which controls levels of calcium in the blood. Also termed: vitamin D.

calcitonin (biochemistry) Hormone secreted in vertebrates that controls the release of calcium from bone. In mammals it is secreted by the thyroid gland. Also termed: thyrocalcitonin.

calcium (Ca) (chemistry) Silver-white metallic element in Group IIA of the periodic table (the alkaline earth). The fifth most abundant element on Earth, it occurs mainly in calcium carbonate minerals; it also occurs in bones and teeth. At. no. 20; r.a.m. 40.08.

calcium carbonate (chemistry) $CaCO_3$ White powder or colourless (*USA*: colorless) crystals, the main constituent of chalk, limestone and marble.

calcium chloride (chemistry) $CaCl_2$ White crystalline compound, which forms several hydrates, used to control dust, as a de-icing agent and as a refrigerant. The anhydrous salt is deliquescent and is employed as a desiccant.

calcium hydrogencarbonate (chemistry) $CaHCO_3$ White crystalline compound, stable only in solution and the cause of temporary hardness of water. Also termed: calcium bicarbonate.

calcium hydroxide (chemistry) $Ca(OH)_2$ White crystalline powder which gives an alkaline aqueous solution known as limewater, used as a test for carbon dioxide (which turns it cloudy). Also termed: calcium hydrate; hydrated lime; caustic lime; slaked lime.

calcium oxide (chemistry) CaO White crystalline powder made commercially by roasting limestone (calcium carbonate). It is used to make calcium hydroxide. Also termed: lime, quicklime.

calcium phosphate (chemistry) $Ca_3(PO_4)_2$ White crystalline solid which makes up the mineral component of bones and teeth, and occurs as the mineral apatite. It is produced commercially as bone ash and basic slag. Treated with sulphuric acid (*USA*: sulfuric acid) it forms the fertilizer known as superphosphate.

calcium sulphate (chemistry) $CaSO_4$ White crystalline compound which occurs as the minerals anhydrite and (as the dihydrate) gypsum, used to make plaster of paris. It is the cause of permanent hardness of water. It is used in making ceramics, paint, paper and sulphuric acid (*USA*: sulfuric acid), and is the substance in blackboard 'chalk'. (*USA*: calcium sulfate).

caleancus (anatomy) Heel-bone; the major bone of the foot. Also termed: calcaneum.

caliber (measurement) *See* calibre.

calibration (1, 2 measurements) 1 Measuring scale on a scientific instrument or apparatus. 2 Determination of the accuracy of an instrument, usually by measurement of its variation from a standard, to ascertain necessary correction factors.

calibrator (measurement) An instrument for dilating a tubular structure or for determining the calibre (*USA*:caliber) of such a structure.

calibre (measurement) The diameter of the lumen of a canal or tube. (*USA*: caliber).

Calicophoron (veterinary parasitology) Stomach flukes of ruminants. Species include *Calicophoron calicophorum*, *Calicophoron cauliorchis*, *Calicophoron ijimai* and *Calicophoron raja*.

See also Paramphistomosis.

California black-legged tick (parasitology) *See Ixodes pacificus.*

California eyeworm (parasitology) *See Thelazia californiensis.*

californium (Cf) (chemistry) Radioactive element in Group IIIB of the periodic table (one of the actinides), produced by alpha-particle bombardment of curium-242; it has several isotopes, with half-lives of up to 800 years. At. no. 98; r.a.m. 251 (most stable isotope).

callipers (equipment) Instrument with two bent or curved legs used to measure thickness or diameter of a solid, or the internal dimensions of a hollow object. (*USA*: calipers).

Calliphora (veterinary parasitology) A genus of flies which includes *Calliphora augur*, *Calliphora australis*, *Calliphora erythrocephala*, *Calliphora fallax*, *Calliphora hilli*, *Calliphora novica*, *Calliphora stygia* and *Calliphora vomitoria*, that may initiate blowfly strike in sheep but which mainly assume importance in sheep that are already infested.

calliphorid (parasitology) Pertaining to blowflies.

calliphorid flies (parasitology) *See* blowfly.

Calliphoridae (parasitology) The family containing most of the important blowflies, includiug *Calliphora*, *Chrysomyia*, *Lucilia*, *Callitroga* and *Phormia* spp.

Calliphorinae (parasitology) The blowfly subfamily.

calliphorine myiasis (parasitology) *See* cutaneous myiasis.

Callitroga (parasitology) A genus of screw-worms which includes *Callitroga americana*, *Callitroga hominivorax* and *Callitroga macellaria*. (Also termed *Cochliomyia hominivorax* and *Cochliomyia macellaria*).

See also cutaneous myiasis.

Caloglyphus (insect parasitology) A species of mites in the Astigmata, that similarly to *Carpoglyphus* are cleptoparasitic to the bees *Apinae*.

calomel (chemistry) An alternative term for mercury(I) chloride.

calomel half-cell (physics) Electrode consisting of mercury, mercury(I) chloride and potassium chloride solution, employed as a reference electrode because it has a known constant potential. Also termed: calomel electrode; calomel reference electrode.

calorie (measurement) Amount of heat required to raise the temperature of 1 g of water by 1°C at one atmosphere pressure; equal to 4.184 joules.

See also Calorie.

Calorie (measurement) Amount of heat required to raise the temperature of 1 kg of water by 1°C at one atmosphere pressure; equal to 4.184 kilojoules. It is used as an unit of energy of food (when it is sometimes spelled with a small *c*). Also termed: kilocalorie; large calorie.

See also calorie.

calorific (measurement) Generating heat measurable in calories.

calorific value (physics) Quantity of heat liberated on the complete combustion of a unit weight or unit volume of fuel.

calorimeter (physics/chemistry) Apparatus for measuring heat quantities generated in or evolved by materials in processes such as chemical reactions, changes of state and salvation. The technique is known as calorimetry.

calorimetry (measurement) Measurement of the heat eliminated or stored in any system.

CAM (physiology/biochemistry) An abbreviation of 'C'ell 'A'dhesion 'M'olecule.

Camallanus (fish parasites) A nematode genus which infests freshwater turtles and aquarium fish.

cambendazole (pharmacology) An efficient broad-spectrum anthelmintic.

See also albendazole.

Camelostrongylus (veterinary parasitology) A genus of the family Trichostrongylidae of intestinal helminths. Species include *Camelostrongylus* mentulatus which may be found in sheep, camel, and wild ruminants.

cAMP (biochemistry) An abbreviation of 'c'yclic 'a'denosine 'm'ono'p'hosphate.

Campanulotes (avian parasitology) A genus of bird lice. Species include *Campanulotes bidentatus compar,* also termed small pigeon louse.

camphor (chemistry) Naturally-occurring organic compound with a penetrating aromatic odour (*USA*: odor). Also termed: gum camphor; 2-camphanone.

Camponotus (parasitology) An ant genus which is second intermediate host to the flukes *Dicrocoelium* spp.

candela (cd) (measurement) SI unit of luminous intensity.

candle power (cp) (measurement) Luminous intensity expressed in candelas (formerly in international candles).

cane-sugar (chemistry) An alternative term for sucrose.

canine babesiosis (veterinary parasitology) A haemolytic (*USA*: hemolytic) disease of dogs caused by *Babesia canis* or *Babesia gibsoni*, transmitted by a tick, and characterized by anaemia (*USA*: anemia) and haemoglobinuria (*USA*: hemoglobinuria). Also termed tick fever, malignant jaundice.

canine nasal mites (veterinary parasitology) *Pneumonyssus caninum.*

canker (veterinary parasitology) Ulceration, especially (1) of the lip or oral mucosa; (2) in horses of the horn of the sole of the foot; (3) often used erroneously to describe otitis externa.

Cannizzaro reaction (chemistry) The formation of an alcohol and an acid salt by the reaction between certain aldehydes and strong alkalis (*USA*: alkalies). It was named after the Italian chemist Stanislao Cannizzaro (1826–1910).

canonical form (chemistry) *See* mesomerism.

cap. (general terminology) An abbreviation of 'cap'acity.

capacitance (physics) Ratio of the charge on one of the conductors of a capacitor (there being an equal and an opposite charge on the other conductor) to the potential difference between them. The SI unit of capacitance is the farad.

capacitor (physics) Device that can store charge and introduces capacitance into an electrical circuit. Also termed: condenser; electrical condenser.

capacity (terminology) The power to hold, retain, or contain, or the ability to absorb; usually expressed numerically, as the measure of such ability.

Capillaria (fish/avian/veterinary parasitology) A genus of parasitic nematodes of the subfamily Capillariinae and most commonly parasitic in birds. They cause capillariasis. Species include *Capillaria anatis*, *Capillaria annulata* (*Capillaria contorta*), *Capillaria caudinflata*, *Capillaria obsignata*, which may be found in birds; *Capillaria aerophila*, (*Capillaria bilobata*, *Capillaria bovis*, *Capillaria brevipes*, *Capillaria didelphis*, *Capillaria entomelas*, *Capillaria erinacea*, *Capillaria feliscati*, *Capillaria hepatica*, *Capillaria megrelica*, *Capillaria mucronata*, *Capillaria philippinensis*, *Capillaria plica*, *Capillaria putorii*, which may be found in mammals. There are others which occur in small rodents and in fish.

capillariasis (avian/veterinary parasitology) Infection with nematodes of the genus capillaria. In birds the disease is manifested by chronic gastroenteritis and the affected birds are emaciated. The disease in mammals may be enteritis with diarrhoea (*USA*: diarrhea) (*Capillaria bovis*, *Capillaria entomelas*), cystitis (*Capillaria feliscati*, *Capillaria plica*), hepatitis (*Capillaria hepatica*) or bronchopneumonia (*Capillaria aerophila*, *Capillaria didelphis*).

capillarid (parasitology) A member of the family Capillariidae.

capillariomotor (physiology) Pertaining to the functional activity of the capillaries.

capillaritis (medical) Inflammation of the capillaries.

capillarity (physics) A phenomenon resulting from surface tension that causes low-density liquids to flow along narrow (capillary) tubes or soak into porous materials. Also termed: capillary action.

capillary (1 physics; 2 anatomy) 1 Any narrow tube, in which capillarity can occur. 2 Finest vessel of the blood vascular system in vertebrates. Large numbers of capillaries are present in tissues. Their walls are composed of a single layer of cells through which exchange of substances, *e.g.* oxygen, occurs between the tissues and blood.

capitulum (parasitological anatomy) A term used in relation to arthropods. The 'false-head', found in Arachnida. The basal portion of this structure is 'basis capituli'.

caprylic acid (chemistry) An alternative term for octanoic acid.

capsalid fluke (parasitology) *See Neobenedinia.*

capsule (1 bacteriology; 2 anatomy) 1 In certain bacteria, gelatinous extracellular envelope, in some cases conferring protective properties on the cell. 2 Sheath of membrane that surrounds an organ or area of tissue; *e.g.* the synovial capsule that surrounds the moving parts of joints.

carbamide (chemistry) An alternative term for urea.

carbanion (chemistry) Transient negatively charged organic ion that has one more electron than the corresponding free radical.

carbene (chemistry) Organic radical that contains divalent carbon.

carbocyclic (chemistry) Describing a cyclic organic compound consisting of aromatic rings containing only carbon and hydrogen atoms. Also termed: homocyclic.

carbofuran (pharmacology) A carbamate acaricide and nematocide.

carbohydrase (biochemistry) Any of a group of enzymes that catalyze the hydrolysis of higher carbohydrates to lower forms.

carbohydrate (chemistry) Compound of carbon; hydrogen and oxygen that contains a saccharose group or its first reaction product, and in which the ratio of hydrogen to oxygen is 2:1 (the same as in water). Cellulose, starch and all sugars are common carbohydrates. Digestible carbohydrates in the diet are a good source of energy.

carbolic acid (chemistry) An alternative term for phenol.

carbon (C) (chemistry) Non-metallic element in Group IVA of the periodic table which exists as several allotropes (including diamond and graphite). It occurs in all living things and its compounds are the basis of organic chemistry. It is the principal element in coal and petroleum. Its nonorganic compounds include the oxides carbon monoxide (CO) and carbon dioxide (CO_2), carbides and carbonates. Carbon is used for making electrodes, brushes for electric motors, carbon fibres (*USA*: carbon fibers) and steel. Diamonds are used as gemstones and industrially as abrasives. Its isotope carbon-14 is the basis of radiocarbon dating. At. no. 6; r.a.m. 12.001.

carbon black (chemistry) Finely divided form of carbon obtained by the incomplete combustion or thermal decomposition of natural gas or petroleum oil.

carbon dioxide CO_2 (chemistry) Colourless (*USA*: colorless) gas formed by the combustion of carbon and its organic compounds, by the action of acids on carbonates, and as a product of fermentation and respiration. It is a raw material of photosynthesis. It is used in fire extinguishers, in fizzy (carbonated) drinks, as a coolant in nuclear reactors and, as solid carbon dioxide (dry ice), as a refrigerant. The accumulation of carbon dioxide in the atmosphere creates the greenhouse effect.

carbon disulphide (chemistry) CS_2 Liquid chemical, used as a solvent for oils, fats and rubber and in paint-removers. (*USA*: carbon disulfide).

carbon monoxide (chemistry) CO Colourless (*USA*: colorless) odourless (*USA*: odorless) poisonous gas produced by the incomplete combustion of carbon or its compounds. It is used in the chemical industry as a reducing agent.

carbon tetrachloride (chemistry) An alternative term for tetrachloromethane.

carbonate (chemistry) Salt of carbonic acid (H_2CO_3), containing the ion CO_3^{2-}. Carbonates commonly occur as minerals (*e.g.* calcium carbonate) and are readily decomposed by acids to produce carbon dioxide.

See also hydrogencarbonate.

carbonation (chemistry) Addition of carbon dioxide under pressure to a liquid.

carbonic acid (chemistry) H_2CO_3 acid formed by the combination of carbon dioxide and water. Its salts are carbonates.

carbonic acid anhydrase (biochemistry) *See* carbonic anhydrase.

carbonic anhydrase (biochemistry) An enzyme which catalyzes the reversible conversion of carbon dioxide to bicarbonate ions and thus facilitates the transport and elimination of carbon dioxide from tissues.

carbonium ion (chemistry) Positively charged fragment that arises from the heterolytic fission of a covalent bond involving carbon.

carbonyl chloride (chemistry) An alternative term for phosgene.

carbonyl compound (chemistry) Chemical containing the radical CO (carbonyl group), formed when carbon monoxide combines with a metal (*e.g.* nickel carbonyl, $Ni(CO)_4$).

carbonyl group (chemistry) The group = CO. *See* carbonyl compound.

carboxyl group (chemistry) The organic group COOH, characteristic of carboxylic acids. Also termed: carboxy group, oxatyl group.

carboxylic acid (chemistry) Organic acid that contains the carboxyl group, COOH; *e.g.* (ethanoic) acetic acid, CH_3COOH.

carbylamine (chemistry) An alternative term for isocyanide.

carcinogen (toxicology) Substance capable of inducing a cancer. Most carcinogens are also mutagens, and this is thought to be their principal mode of action.

carcinogenic (toxicology) Causing cancer or capable of causing cancer.

card punch (computing) Machine for punching coded sets of holes in punched cards, to be fed through a card reader for inputting data into a computer.

card reader (computing) Computer input device that reads data off punched cards.

cardi(o)- (terminology) A word element. [German] Prefix denoting *heart.*

cardiac (anatomy) Relating to the heart.

See also cardiac muscle.

cardiac muscle (anatomy) Specialized striated muscle, found only in the heart, that continually contracts rhythmically and automatically (*i.e.* it is myogenic).

cardinal number (mathematics) Number of elements in a set (*e.g.* all sets with 5 elements have the cardinal number 5); in everyday terms, an ordinary counting number.

See also ordinal number.

cardiogram (physiology) An alternative term for an electrocardiogram.

cardiovascular system (anatomy) Heart and the network of blood vessels that circulate blood around the body.

Carmyerius (veterinary parasitology) A genus of rumen (digenetic trematode) flukes which are members of the family Paramphistomatidae that are found in Asia. Species includes *Carmyerius gregarius*, *Carmyerius spatiosus* (small rumen flukes of Asia).

Caro's acid (chemistry) An alternative term for peroxomonosulphuric(VI) acid (*USA*: peroxomonosulfuric acid), H_2SO_5 (persulphuric acid [*USA*: persulfuric acid]).

carotene (biochemistry) Pigment that belongs to the carotenoids, a precursor of retinol (vitamin A).

carotenoid (biology/chemistry) Member of a group of pigments found in fruits and vegetables (*e.g.* carrots), which can absorb light during photosynthesis. They also aid in the protection of prokaryotes from damage by light.

See also carotene.

carotid artery (anatomy) Either of two main arteries that carry blood from the heart to the head.

carotid body (physiology) Receptor located between the internal and external carotid arteries. It is sensitive to changes in the carbon dioxide and oxygen content of the blood, and sends impulses to the respiratory and blood-vascular centres (*USA*: centers) in the brain to adjust these levels, if necessary.

carpal (anatomy) Bone in the foot and wrist of tetrapods. There are 10–12 carpals in most animals; in human beings there are eight.

Carpoglyphus (parasitology) A genus of the insect family of Acaridae. Species includes *Carpoglyphus lactis*, the cause of dried fruit mite dermatitis.

carpus (anatomy) The wrist, consisting of eight bones in human beings.

carrier (1 epidemiology; 2 medical) 1 Somebody who carries pathogens (*e.g.* bacteria, viruses, parasites) and can pass them on to others, although not necessarily themselves having symptoms of the disease. 2 Vector (in medicine).

See also coenzyme.

Cartesian co-ordinates (mathematics) System for locating a point, P, by specifying its distance from axes at right-angles, which intersect at a point, O, called the origin. For a point on a plane, the distance from the horizontal or *x-axis* is called the ordinate of P; the distance from the *y-axis* is called the abscissa. The point's Cartesian co-ordinates are (x, y). They were named after the French mathematician Rene Descartes (1596–1650).

cartilage (anatomy) In animals that have a bony skeleton, pre-bone tissue occurring in young animals that becomes hard through calcification. It also occurs in some parts of the skeleton of adults, where it is structural (*e.g.* forming ear flaps and the larynx) or provides a cushioning or lubricating effect during movement (*e.g.* intervertebral discs, cartilage at the ends of bones in joints). In some animals, *e.g.* sharks, rays and related fish (Chondrichthyes), the whole skeleton is composed of cartilage even in the adult. Also termed: gristle.

Caryophyllaeus (fish parasitology) A genus of the family of cestodes Caryophyllaeidae. Species includes *Caryophyllaeus fimbriceps*, and *Caryophyllaeus laticeps* found in carp.

Caryophyllidea (fish parasitology) An order of flatworms in subclass Eucestoda, class Cestoidea and the phylum Platyhelminthes. Caryophyllideans are a large group of cestodes of fresh water siluriform and cypriniform fish. The monozoic body, one set of reproductive organs, and many features of their biology, set them apart from all other members of the subclass Eucestoda.

Caryospora (veterinary parasitology) Apicomplexan protozoan parasites affecting mostly reptiles and raptors.

Caryospora bigenetica (veterinary parasitology) A species of ampicomplexan protozoan parasite in the genus *Caryospora* that is an occasional cause of pyogranulomatous dermatitis in young dogs.

caryosporiosis (parasitology) Infection by a species of coccidia in the genus *Caryospora.*

case (epidemiology) An animal which has the specified disease or condition which is under investigation.

casein (biochemistry) Protein that occurs in milk which serves to store amino acids as nutrients for the young of mammals.

castor bean tick (parasitology) *See Ixodes ricinus.*

CAT (radiology) An abbreviation of 'C'omputerised 'A'xial 'T'omography.

cat flea (veterinary parasitology) *See Ctenocephalides felis.*

cat fur mite (veterinary parasitology) *Lynxacarus radovsky. See Lynxacarus.*

catabolism (biochemistry) Part of metabolism concerned with the breakdown of complex organic compounds into simple molecules for the release of energy in living organisms.

catalysis (chemistry) Action of a catalyst.

catalyst (chemistry) Substance that increases the rate of a chemical reaction without itself undergoing any permanent chemical change. Some reactions take place so slowly without a catalyst that they are rendered virtually impossible.

See also enzyme.

Catatropis (avian/veterinary parasitology) A genus of intestinal flukes (digenetic trematodes) of the family Notocotylidae. Species includes *Catatropis* verrucosa found in avian caeca (*USA*: ceca).

catecholamine (biochemistry) Member of a group of amines that include neurotransmitters (*e.g.* dopamine) and hormones (*e.g.* adrenaline [*USA*: epinephrine]).

caterpillar (insect parasitology) The larval stage of insects of the Lepidoptera family.

Cathaemasiidae (avian parasitology) A family of alimentary canal flukes (digenetic trematodes) of birds.

cathode (physics) Negatively-charged electrode in an electrolytic cell or battery.

cathode rays (physics) Stream of electrons produced by the cathode (negative electrode) of an evacuated discharge tube (such as a cathode-ray tube).

cathode-ray oscilloscope (CRO) (physics) Apparatus based on a cathode-ray tube which provides a visible image of one or more rapidly varying electrical signals.

cathode-ray tube (physics) Vacuum tube that allows the direct observation of the behaviour (*USA*: behavior) of cathode rays. It is used as the picture tube in television receivers, radar displays, visual display units (VDUs) of computers and in cathode-ray oscilloscopes.

cation (chemistry) Positively charged ion, which travels towards the cathode during electrolysis.

Catostomus commersoni (fish/veterinary parasitology) The common sucker fish, a secondary host for *Metorchis conjunctus*, a fluke of cats and dogs.

cattle grub (veterinary parasitology) *See Hypoderma.* Also termed warble fly.

cattle louse (veterinary parasitology) *See Haematopinus*; *Linognathus.*

cattle maggot (veterinary parasitology) *See Hypoderma.*

cattle tick (veterinary parasitology) Any one of a large variety of tick species, the title being used locally to designate the preponderant species.

cattle tick fever (veterinary parasitology) *See* babesiosis. Also termed Texas fever; redwater fever.

cauda equina (anatomy) Sheaf of nerve roots that arise from the lower end of the spinal cord and serve the lower parts of the body.

caudal vertebra (anatomy) One of the set of vertebrae nearest to the base of the spinal column.

caustic (chemistry) Describing alkaline substances that are corrosive towards organic matter.

caustic lime (chemistry) An alternative term for calcium hydroxide.

caustic potash (chemistry) An alternative term for potassium hydroxide.

caustic soda (chemistry) An alternative term for sodium hydroxide.

cavography (methods) Radiography of the vena cava.

cavum Plural: Cava [Latin] cavity.

cavus (terminology) [Latin] hollow.

cayenne tick (parasitology) *See Amblyomma cajennense.*

CBC (haematology) An abbreviation of 'C'omplete 'B'lood and differential 'C'ount.

CBF (physiology) An abbreviation of 'C'erebral 'B'lood 'F'low.

CBG (physiology/biochemistry) An abbreviation of 'C'orticosteroid-'B'inding 'G'lobulin.

CBR (statistics) An abbreviation of 'C'rude 'B'irth 'R'ate.

CC (medical) An abbreviation of 'C'hief 'C'omplaint.

cc (measurement) An abbreviation of 'c'ubic 'c'entimetre (*USA*: cubic centimeter).

cc hr (measurement) An abbreviation of 'c'ubic 'c'entimetres per 'h'ou'r' (*USA*: cubic centimeters per hour).

cc sec (measurement) An abbreviation of 'c'ubic 'c'entimetres per 'sec'ond (*USA*: cubic centimeters per second).

CCD (physics) An abbreviation of 'c'harge-'c'oupled 'd'evice.

CCE (chemistry) An abbreviation of 'C'arbon-'C'hloroform 'E'xtract.

cckw (general terminology) An abbreviation of 'c'ounter'c'loc'kw'ise. Also termed ccw.

CCU (medical) An abbreviation of 'C'ritical 'C'are 'U'nit.

CD (1 haematology; 2 medical) 1 An abbreviation of 'C'luster 'D'ifferentiating type antigens. 2 An abbreviation of 'C'aesarean 'D'elivery.

Cd (chemistry) Chemical symbol for cadmium.

cd (measurement) An abbreviation of 'c'an'd'ela.

CD antigen (immunology) A group of cell surface molecules which act as markers on T lymphocytes.

CD4 (immunology) A cell surface protein, usually on helper T-lymphocytes, that recognises foreign antigens on antigen-presenting cells.

CDC (governing body) An abbreviation of 'C'enters for 'D'isease 'C'ontrol and Prevention.

CDNA (genetics) An abbreviation of 'C'omplementary 'D'eoxyribo'N'ucleic 'A'cid.

CDR (statistics) An abbreviation of 'C'rude 'D'eath 'R'ate.

CD-ROM (computing) An abbreviation of 'C'ompact 'D'isk-'R'ead 'O'nly 'M'emory.

CE (1 chemistry; 2 statistics) 1 An abbreviation of 'C'arbon 'E'quivalent. 2 An abbreviation of 'C'onstant 'E'rror.

Ce (chemistry) Chemical symbol for cerium.

Cebalges (veterinary parasitology) Members of Psoroptidae family of mange mites which includes *Cebalges gaudi* a parasite of primates.

-cele (terminology) A word element. [German] Suffix denoting *tumour* (*USA*: *tumor*), *hernia*; in American English, may also mean *cavity*.

cell (biology) Fundamental unit of living organisms. It consists of a membrane bound compartment, often microscopic in size, containing protoplasm. Although a cell is not independent of its environment, all biochemical reactions of metabolism necessary for life (within a favourable ecological niche) are located either within the protoplasm, within or associated with the cell membrane, or within the local environment surrounding the cell (which may be modified by secretory substances). In all cells, the nucleus contains nucleic acids (*e.g.* DNA) essential for the synthesis of new proteins. Cells may be highly specialized for their particular function and grouped into tissues, *e.g.* muscle cells.

cell body (cytology) Part of a nerve cell that contains the cell nucleus and other cell components from which an axon extends.

cell differentiation (cytology) Way in which previously undifferentiated cells change structurally and take on specialized roles during growth and development (*e.g.* becoming liver cells or bone cells).

cell division (cytology) Splitting of cells in two by division of the nucleus (mitosis) and division of the cytoplasm after duplication of the cell contents.

See also meiosis.

cell membrane (cytology) *See* plasma membrane.

cellulose (biology) Major polysaccharide of plants found in cell walls and in some algae and fungi. It is composed of glucose units aligned in long parallel chains, and gives cell walls their strength and rigidity.

Celsius scale (measurement) Temperature scale on which the freezing point of water is 0°C and the boiling point is 100°C. It is the same as the formerly used centigrade scale, and a degree Celsius is equal to a unit on the kelvin scale. To convert a Celsius temperature to kelvin., add 273.15 (and omit the degree sign). To convert a Celsius temperature to a Fahrenheit one, multiply by 9, divide by 5 and add 32. It was named after the Swedish astronomer Anders Celsius (1701–44).

cent. (measurement) An abbreviation of 'cent'igrade.

centigrade scale (measurement) Former name for the Celsius scale.

central nervous system (CNS) (anatomy) Concentration of nervous tissue responsible for co-ordination of the body. In vertebrates it is highly developed to form the brain and spinal cord. The CNS processes information from the sense organs and effects a response, *e.g.* muscle movement.

central processor (computing) Heart of a digital computer which controls and coordinates all the other activities of the machine, and performs logical processes on data loaded into it according to program instructions it holds.

centre of curvature (optics) Geometric centre of a spherical mirror. (*USA*: center of curvature).

centre of mass (physics) Point at which the whole mass of an object may be considered to be concentrated. Also termed: centre of gravity (*USA*: center of gravity). (*USA*: center of mass)

centrifugal force (physics) Outward force on any object moving in a circular path.

See also centripetal force.

centrifuge (laboratory equipment) Instrument used for the separation of substances by sedimentation through rotation at high speeds, *e.g.* the separation of components of cells. Sedimentation varies according to the size of the component.

centriole (cytology) Cylindrical body present in the microtubule organizing centre (*USA*: center) of most animal cells. During mitosis it forms the poles of the spindle.

centripetal force (physics) Force, directed towards the centre (*USA*: center), that causes a body to move in a circular path. For an object of mass m moving with a speed v in a curve of radius of curvature r, it is equal to mv^2/r.

centromere (genetics) Region on a chromosome that attaches to the spindle during cell division.

Centrorhynchus (veterinary parasitology) Thorny-headed worms of reptiles.

See also Macracanthorhynchus.

cephal(o)- (terminology) A word element. [German] Prefix denoting *head.*

Cephalopsis (parasitology) A member of the genus of flies in the family Oestridae.

Cephalopsis titillator (veterinary parasitology) A species of flies in the genus *Cephalopsis.* Nasal bot fly of camels; the larvae inhabit the nasal sinuses.

Cephenemyia (veterinary parasitology) A genus of bot flies in the family Oestridae. Species includes *Cephenemyia apicata, Cephenemyia auribarbis, Cephenemyia jellisoni, Cephenemyia phobifer, Cephenemyia pratti, Cephenemyia stimulator, Cephenemyia trompe,* and *Cephenemyia ulrichi,* which may be found in the nasal cavities of wild mammals.

Ceratomyxa (parasitology) A genus in the class Myxosporea, which may or may not be true protozoa.

Ceratomyxa shasta (fish parasites) A species of the genus *Ceratomyxa* that are an important parasite of young salmonids limited in occurrence to the Columbia river basin. Signs include swelling at vent, distended abdomen and subcutaneous boils and they may cause severe losses in young fish in culture ponds.

Ceratophyllus (avian/veterinary parasitology) A genus of fleas in the order Siphonaptera. Species includes *Ceratophyllus columbae, Ceratophyllus (Nosopsyllus) fasciatus, Ceratophyllus gallinae, Ceratophyllus garei,* and *Ceratophyllus* niger (Western chicken flea), that may be found in rodents and wild birds.

Ceratophyllus columbae (avian parasitology) A species of flea whose principal hosts are pigeons.

Ceratophyllus fasciatus (veterinary parasitology) A species of flea whose principal hosts are house mouse and rat. Also termed: *Nosopsyllus fasciatus.*

Ceratophyllus gallinae (avian parasitology) A species of flea whose principal hosts are chickens (European chicken flea).

Ceratophyllus garei (avian parasitology) A species of flea whose principal hosts are water fowl.

Ceratophyllus niger (avian parasitology) A species of flea whose principal hosts are chickens. The Western chicken flea.

Ceratopogonidae (parasitology) A family of biting midges; the most important genus is *Culicoides.*

cercaria (parasitology) Plural: cercariae [German] The final, free-swimming larval stage of a trematode parasite. Young flukes which develop from germ cells in sporocyst and redia.

cercarial (parasitology) Pertaining to or emanating from cercariae.

cercarial dermatitis (parasitology) *See Trichobilharzia.*

cerci (parasitological anatomy) A term used in relation to arthropods. Two slender appendages at the terminal abdominal segment.

cereal mite (parasitology) *See Tyroglyphus.*

cerebellum (anatomy) Front part of hindbrain. It is important in balance and muscular co-ordination.

cerebral cortex (anatomy) Outer region of the cerebral hemispheres of the brain that contains densely packed nerve cells which are interconnected in a complex manner.

cerebral hemisphere (anatomy) Paired expansions of the anterior end of the forebrain. The forebrain is enormously enlarged by these expansions.

cerebral nematodiasis (veterinary parasitology) Sporadic occurrence of chance invasion of brain by migrating nematode larvae, *e.g.* strongyles in horses. Also termed: kumri.

See also cerebrospinal nematodiasis.

cerebrospinal fluid (CSF) (anatomy) Liquid that fills the cavities of the brain, which are continuous with each other and the central canal of the spinal cord. The fluid nourishes and removes secretions from the tissues of the central nervous system (CNS).

cerebrospinal nematodiasis (veterinary parasitology) Invasion of the central nervous system by the microfilaria of *Setaria labiatopapillosa* (*Setaria digitata*) in most species causes an acute focal encephalomyelomalacia. The clinical picture is one of incoordination, then paralysis of the limbs. *Setaria equina* may cause endophthalmitis in horses by similar invasion.

cerebrum (anatomy) Part of the forebrain that expands to form the cerebral hemispheres.

Cerenkov counter (radiology) Method of measuring radioactivity by utilizing the effect of Cerenkov radiation.

Cerenkov radiation (radiology) Light emitted when charged particles pass through a transparent medium at a velocity that is greater than that of light in the medium. It was named after the Soviet physicist Pavel Cerenkov (1904–).

cerium (Ce) (chemistry) Steel-grey metallic element in Group IIIB of the periodic table (one of the lanthanides). It is used in lighter flints, tracer bullets, catalytic converters for car exhausts and gas mantles. At. no. 58; r.a.m.140.12.

cerumen (biochemistry) Wax that forms in the ear.

cervical (anatomy) Relating to the neck or to the cervix of the womb.

cervical vertebra (anatomy) Vertebra in the neck region of the spinal column, concerned with movements of the head.

cervix (anatomy) Neck, usually referring to the neck of the uterus (womb) at the inner end of the vagina.

cesarean (medical) Caesarean.

cesium (chemistry) *See* caesium.

cestocidal (pharmacology/control measures) Destructive to cestodes.

cestocide (pharmacology/control measures) A substance that destroys cestodes.

cestodal cysts (parasitological physiology) The larval (metacestode) stage of cestodes in mammal hosts, *e.g. Echinococcus* spp. cysts in humans, *Cyticercus tenuicollis* cysts in sheep.

Cestodaria (fish parasitology) A subclass of flatworms in the class Cestoidea and phylum Platyhelminthes. The subclass Cestodaria has two orders, the Gyrocotylidae and Amphilinidea, which are found in fish.

cestode (parasitology) 1 Any individual of the class Eucestoda. 2 Cestoid.

cestodiasis (parasitology) Infestation with tapeworms.

cestodology (parasitology) The scientific study of cestodes.

cestoid (parasitology) Resembling a tapeworm.

Cestoidea (fish parasitology) A class of flatworms in the phylum Platyhelminthes. The class Cestoidea has two subclasses, namely the Cestodaria and Eucestoda. *See* Cestodaria; Eucestoda.

Ceylonocotyle (parasitology) A genus of flukes (digenetic trematodes) in the family Paramphistomatidae. Species includes *Ceylonocotyle scoliocoelium*, and *Ceylonocotyle streptocoelium* that may be found in the rumen and reticulum. *See* paramphistomosis.

Cf (measurement) An abbreviation of 'c'ubic 'f'eet.

CFBS (society) An abbreviation of 'C'anadian 'F'ederation of 'B'iological 'S'ciences.

CFC (chemistry) An abbreviation of 'c'hloro'f'luoro'c'arbon.

CFF (physiology) An abbreviation of 'C'ritical 'F'usion 'F'requency.

CFU (biology) An abbreviation of 'C'olony 'F'orming 'U'nit.

cg (measurement) An abbreviation of 'c'enti'g'ram. Also termed cgm.

CGI (education) An abbreviation of 'C'ity and 'G'uilds 'I'nstitute.

cgm (measurement) *See* cg.

cGMP (physiology/biochemistry) An abbreviation of 'c'yclic 3',5'-'G'uanosine 'M'ono'P'hosphate.

CGP (physiology/biochemistry) An abbreviation of 'C'horionic 'G'rowth hormone-'P'rolactin.

CGRP (physiology/biochemistry) An abbreviation of 'C'alcitonin 'G'ene-'R'elated 'P'eptide.

cgs (measurement) An abbreviation of 'c'entimetre 'g'ram 's'econd. (*USA*: centimeter gram second).

C_{H2O} (physiology) A symbol for free water 'C'learance.

Chabertia (veterinary parasitology) A genus of nematodes in the superfamily Strongyloidea. Species includes *Chabertia ovina*, which may be found in the colon of ruminants. *See* chabertiasis.

chabertiasis (veterinary parasitology) Infestation with *Chabertia ovina*. Characterized by weight loss and the passage of soft faeces (*USA*: feces) containing much mucus. Also termed chabertiosis.

chabertiosis (veterinary parasitology) *See* chabertiasis.

Chaenotania craterformis (insect parasitology) A species of helminth in the Cestoda that similarly to *Chaenotania musculosa*, *Chaenotania unicornata* and possibly *Anomotaenia borealis*, are parasites to the ants *Crematogaster*, *Harpagoxenus* and *Leptothorax* spp., that cause aberrant colours (*USA*: colors) and sluggish behaviour (*USA*: behavior). The final hosts are woodpeckers.

Chaenotania musculosa (insect parasitology) *See Chaenotania craterformis.*

Chaenotania unicornata (insect parasitology) *See Chaenotania craterformis.*

Chagas' disease (medical/parasitology) South American trypanosomiasis, a disease spread by the cone-nosed or assassin reduviid bug. The organism is *Trypanosma cruzi* and the disease is a major cause of heart damage and heart failure in endemic areas. Named after the Brazilian physician Carlos Chagas (1879–1934).

chagoma (parasitology) A skin tumour (*USA*: tumor) in trypanosomiasis due to *Trypanosoma cruzi.*

chain of infection (epidemiology) A series of infections that are directly or immediately connected to a particular source.

chain reaction (physics/chemistry) Nuclear or chemical reaction in which the products ensure that the reaction continues (*e.g.* combustion).

Chalcoela iphitalis (insect parasitology) Arthropods in the Lepidoptera that are a parasitic to the wasps *Mischocyttarus* spp. and *Polistes* spp., attacking the larvae, and which are a common cause of colony mortality.

***Chalcura* sp.** (insect parasitology) Hymenoptera in the Eucharitidae that may be found in small pupa of *Formica and Odontomaches*.

chalk (chemistry) Mineral form of calcium carbonate ($CaCO_3$), a soft white limestone derived from the skeletons of microscopic marine animals.

chancre (medical) The 2 to 4 inch, hard, hot, painful lesion which develops at the site of tsetse-fly bites when the flea is a transmitter of trypanosomiasis.

Chandlurella quiscali (veterinary parasitology) A nematode parasite that may be found in the emu.

character (genetics) Variation caused by a gene; an inherited trait.

characteristic (mathematics) Integer part of a logarithm; *e.g.* the logarithm (to the base 10) of 200 is 2.3010, in which 2 is the characteristic.

Charchesium polysinum (veterinary parasitology) A protozoa that parasitizes the skin of tadpoles. Lesions may cover the gills and cause asphyxia.

charcoal (chemistry) Form of carbon made from incomplete burning of animal or vegetable matter.

charge (physics) *See* electric charge; electron charge.

charge/mass ratio (physics) Fundamental physical constant, the ratio of an electron's charge to its mass e/m_e, equal to 1.758796×10^{11} C kg^{-1} (coulombs per kilogram).

charge-coupled device (CCD) (physics) Semiconductor device used in television cameras, video cameras and astronomical telescopes to produce an electronic signal from an optical image.

Charles's law (physics) At a given pressure, the volume of an ideal gas is directly proportional to its absolute temperature. It was named after the French physicist Jacques Charles (1746–1823).

See also kinetic theory.

cheese fly (biology) *Piophila casei.* A fly that infests cheese but that is of aesthetic (*USA*: esthetic) importance only. Also termed cheese skipper.

cheese mite (parasitology) *See Tyroglyphus siro.*

cheil(o)- (terminology) A word element. [German] Prefix denoting *lip*.

Cheilospirura (parasitology) A genus of nematodes of the family Acuariidae.

Cheilospirura hamulosa (avian parasitology) A species of nematodes of the genus *Cheilospirura* that may be found in the gizzard of fowls and turkeys. Heavy infestations cause emaciation, weakness and anaemia (*USA*: anemia). Also termed: *Acuaria.*

chelate (chemistry) Chemical compound in which a central metal ion forms part of one or more organic rings of atoms. The formation of these compounds is useful in many contexts, *e.g.* medicine, in which chelating agents are administered to counteract poisoning by certain heavy metals. They can also act to buffer the concentration of metal ions (*e.g.* iron and calcium) in natural biological systems.

chelicerae (parasitology anatomy) Pair of moveable oral appendages adapted for cutting, carried by acarids, including ticks.

Chelopistes (avian parasitology) A genus of the superfamily Ischnocera of lice. Species within the genus includes *Chelopistes* meleagridis, which may be found in turkeys.

chem(o)- (terminology) A word element. [German] Prefix denoting *chemical, chemistry.*

chemical bond (chemistry) Linkage between atoms or ions within a molecule. A chemical reaction and the input or output of energy are involved in the formation or destruction of a chemical bond.

See also covalent bond; ionic bond.

chemical combination (chemistry) Union of two or more chemical substances to form a different substance or substances.

chemical equation (chemistry) Way of expressing a chemical reaction by placing the formulas of the reactants to the left and those of the products to the right, with an equality sign or directional arrow in between. The number of atoms of any particular element are the same on each side of the equation (when the reaction is in equilibrium). *E.g.* the chemical equation for the combination of hydrogen and oxygen to form water is written as: $2H_2 + O_2 \equiv 2H_2O$

chemical equilibrium (chemistry) Balanced state of a chemical reaction, when the concentration of reactants and products remain constant.

See also equilibrium constant.

chemical formula (chemistry) Method of representing the chemical composition of a single substance that uses the chemical symbols of the atoms, and indicates the numbers of those atoms, in a molecule of the substance *e.g.* the chemical formulae of water, calcium carbonate and ethane are H_2O, $CaCO_3$ and C_2H_6.

chemical hazard (toxicology) A chemical capable of causing poisoning is on the premises and represents a potential threat.

chemical kinetics (chemistry) The scientific study of the rates and mechanisms of chemical reactions.

chemical potential (chemistry) Measure of the tendency of a chemical reaction to take place.

chemical reaction (chemistry) Process in which one or more substances react to form a different substance or substances.

chemical symbol (chemistry) Letter or pair of letters that stand for an element in chemical formulae and equations; *e.g.* the symbols of carbon and chlorine are C and Cl.

chemiluminescence (chemistry/biochemistry) Luminescence that results from a chemical reaction (*e.g.* the oxidation of phosphorus in air). In living organisms it is termed bioluminescence.

chemistry (chemistry) The study of elements and their compounds, particularly how they behave in chemical reactions.

chemo. (medical) An abbreviation of 'chemo'therapy.

chemoreceptor (physiology) Cell that fires a nerve impulse in response to stimulation by a specific type of chemical substance; *e.g.* the taste buds on the tongue and olfactory bulbs in the nose contain chemoreceptors that provide the senses of taste and smell.

chemotaxis (biology) Response of organisms to chemical stimuli; *e.g.* the movement of bacteria towards nutrients.

chemotherapy (pharmacology) Treatment of a disorder with drugs that are designed to destroy pathogens or cancerous tissue.

Chenopodium ambrosioides (pharmacology) A plant of the *Chenopodium* genus in the family Chenopodiaceae, which contains wormseed oil which is used as an anthelmintic. Also termed *Chenopodium anthemminticum* var *ambrosioides*; wormseed.

Cheyletiella (human/veterinary parasitology) A genus of mites in the family Cheyletidae which causes a mild, scaling dermatitis in dogs and cats, and a more severe, pruritic dermatitis, mainly on the back of rabbits, and intensely pruritic vesicles in humans. Host specificity is not certain but generally *Cheyletiella blakei* infests cats, *Cheyletiella parasitovorax* infests rabbits and hares, and *Cheyletiella yasguri* infests dogs.

Cheyletiella dermatitis (parasitology) *See Cheyletiella.*

cheyletiellosis (parasitology) Infestation by *Cheyletiella* spp. Also termed: *Cheyletiella* dermatitis; 'walking' dandruff.

chiasma (1 genetics; 2 anatomy) 1 Point along the chromatid of a homologous chromosome at which connections occur during crossing-over or exchange of genetic material in meiosis. 2 Crossing-over point of the optic nerves in the brain.

chicken body louse (avian parasitology) *See Menacanthus stramineus.*

chicken caecal worms (avian parasitology) *See Heterakis*; *Subulura* spp.; *Strongyloides avium*; *Trichostrongylus tenuis*; *Aulonocephalus lindquisti*. (*USA*: chicken cecal worms).

chicken fluff louse (avian parasitology) *See Goniocotes gallinae.*

chicken head louse (avian parasitology) *See Cyclotogaster heterographa.*

chicken louse (avian parasitology) *See Menopon pallidum*; *Menacanthus stramineus.*

chicken mite (avian parasitology) *See Dermanyssus gallinae.*

chiclero ulcer (medical) Leishmaniasis ulcer, caused particularly by *Leishmania braziliensis.*

chigger (human parasitology) The six-legged red larva of mites of the family Trombiculidae, some species of which are vectors of the rickettsiae of scrub typhus of humans. Also termed: chigger mite; harvest mite; red bug. *See Trombicula*; trombiculosis.

chigger mite (parasitology) *See* chigger.

chigoe (parasitology) The sand flea, *Tunga penetrans*, of tropical and subtropical America and Africa, which should not to be confused with chigger mites. Also termed: chigger; chigoe flea.

See also chigger.

Chile saltpetre (chemistry) Old name for impure sodium nitrate.

Chilodenella (fish parasites) A genus of ciliates in the phylum Ciliophora which occur in fish and amphipods.

Chilodenella cyprini (fish parasites) A species of ciliates in the genus *Chilodenella* which may be found in fish and that causes a serious loss of skin at the gills.

Chilodenella hexasticha (fish parasites) A species of ciliates in the genus *Chilodenella* which may be found in fish.

chilodenellosis (fish parasitology) Disease of freshwater fish caused by the unicellular ciliate *Chilodenella* spp. Characterized by hyperplasia of the epithelium of the gill and difficult respiration, and infection of the skin.

Chilomastix (parasitology) A genus of parasitic protozoa in the order Trichomonadida that may be found in the intestines of vertebrates.

Chilomastix bettencourti (veterinary parasitology) A species of nonpathogenic protozoa in the genus *Chilomastix* that may be found in the caeca (*USA*: ceca) of rodents.

Chilomastix caprae (veterinary parasitology) A species of nonpathogenic protozoa in the genus *Chilomastix* that may be found in the caeca (*USA*: ceca) of goats.

Chilomastix cuniculi (veterinary parasitology) A species of nonpathogenic protozoa in the genus *Chilomastix* that may be found in the caeca (*USA*: ceca) of rabbits.

Chilomastix equi (veterinary parasitology) A species of nonpathogenic protozoa in the genus *Chilomastix* that may be found in the caeca (*USA*: ceca) of horses.

Chilomastix gallinarum (avian parasitology) A species of nonpathogenic protozoa in the genus *Chilomastix* that may be found in the caeca (*USA*: ceca) of chickens.

Chilomastix intestinalis (avian parasitology) A species of nonpathogenic protozoa in the genus *Chilomastix* that may be found in the caeca (*USA*: ceca) of turkeys.

Chilomastix wenrichi (veterinary parasitology) A species of nonpathogenic protozoa in the genus *Chilomastix* that may be found in the caeca (*USA*: ceca) of guinea pigs.

Chilomitus (parasitology) A genus of nonpathogenic protozoa of the order Trichomonadida.

Chilomitus caviae (veterinary parasitology) A species of nonpathogenic protozoa in the genus *Chilomitus* that similarly to *Chilomitus connexus* may be found in the caecum (*USA*: cecum) of the guinea pig.

Chilomitus connexus (veterinary parasitology) A species of nonpathogenic protozoa in the genus *Chilomitus* that, similarly to *Chilomitus caviae*, may be found in the caecum (*USA*: cecum) of guinea pigs.

chimaera (genetics) Genetic mosaic or organism composed of genetically different tissues arising from mutation or mixing of cell types of different organisms. It can be achieved by incorporating donor cells at an early stage of embryonic development of the recipient.

chinese blister fly (parasitology) *See Mylabris phalerata.*

chinese blistering beetle (parasitology) *See Mylabris phalerata.*

chip (computing) Complicated electronic circuits etched on silicon chips; the building blocks of modern electronics.

See also computer.

chirality (chemistry) Property of a molecule that has a carbon atom attached to four different atoms or groups, and which can therefore exist as a pair of optically active stereoisomers (whose molecules are mirror images of each other). Commonly called 'handedness', chirality is significant in the biological activity of molecules, in some of which the right-handed version is active and the left-handed version is not, or vice versa (*e.g.* many pheromones).

Chirodiscoides (veterinary parasitology) A genus of cutaneous mite of the suborder Sarcoptiformes. Species includes *Chirodiscoides caviae* which may be found in guinea pigs.

chitin (biochemistry) Polysaccharide found in the exoskeleton of arthropods, giving it hard, waxy properties.

Chlamydonema (parasitology) *See Physaloptera praeputialis.*

chloral (chemistry) An alternative term for trichloroethanal (trichloroacetaldehyde).

chloral hydrate (chemistry) An alternative term for trichloroethanediol.

chlorate (chemistry) Salt of chloric acid ($HClO_3$).

chlorfenethol (control measures) An acaricide used on agricultural crops and trees. It has a low toxicity but can cause depression, diarrhoea (*USA*: diarrhea), dyspnea, salivation and lacrimation.

chlorfenvinphos (control measures) An organophosphorus insecticide used in the control of ectoparasites in large and small animals.

chloride (chemistry) Binary compound containing chlorine; a salt of hydrochloric acid (HCl).

chlorinated hydrocarbons (control measures) Insecticidal substances which are no longer recommended for use on food animals because of their persistence in animal tissues and entry into the human food chain. Examples include aldrin, benzene hexachloride, chlordane, DDD, DDT, heptachlor, isodrin, lindane, and methoxychlor.

chlorination (1 chemistry; 2 control measures) 1 Reaction between chlorine and an organic compound to form the corresponding chlorinated compound; *e.g.* the chlorination of benzene produces chlorobenzene, C_6H_5Cl. 2 Treatment of a substance with chlorine; *e.g.* to disinfect it.

chlorine (Cl) (chemistry) Gaseous non-metallic element in Group VIIA of the periodic table (the halogens), obtained by the electrolysis of sodium chloride (common salt). It is a green-yellow poisonous gas with an irritating smell, used as a disinfectant and bleach and to make chlorine-containing organic chemicals. At. no. 17; r.a.m. 35.453.

Chloriopsoroptes (veterinary parasitology) A genus of mange mites in the family Psoroptidae. Species includes *Chloriopsoroptes kenyensis* which may be found in the African buffalo.

chlorofluorocarbon (chemistry) Fluorocarbon that has chlorine atoms in place of some of the fluorine atoms. Chlorofluorocarbons and fluorocarbons have similar properties, and are used as aerosol propellants and refrigerants.

chloroform (chemistry) An alternative term for trichloromethane.

chloroquine (pharmacology) An antimalarial/ antiprotozoal drug which can be administered orally, by injection or infusion to prevent and treat contraction of malaria in humans, and in the treatment of avian malaria, anaplasmosis and theileriosis in cattle and amoebiasis (*USA*: amebiasis) in non-human primates.

Choanotaenia (avian parasitology) A genus of non-pathogenic tapeworms in the family Dilepididae. Species includes *Choanotaenia infundibulum* which may be found in fowl and turkey intestines.

chol(o)- (terminology) Word element. [German] Prefix denoing *bile*.

cholangiohepatitis (veterinary parasitology) Inflammation of the biliary system and, by extension, of the periportal hepatic parenchyma, which in large animals, is nearly always the result of parasitic infestation.

chole- (terminology) A word element. [German] Prefix denoting *bile*.

choledocho- (terminology) A word element. [German] Prefix denoting *common bile duct*.

cholesterol (biochemistry) Sterol found in animal tissues, a lipid-like substance that occurs in blood plasma, cell membranes and nerves, and may form gallstones. High levels of cholesterol in the blood are connected to the onset of atheroselerosis, in which fatty materials are deposited in patches on artery walls and can restrict blood flow. Many steroids are derived from cholesterol.

choline (biochemistry) $HOC_2H_4N(CH_3)_3OH$ Organic compound that is a constituent of the neurotransmitter acetylcholine and some fats. It is one of the B vitamins.

chondr(o)- (terminology) A word element. [German] Prefix denoting *cartilage*.

chondrio- (terminology) A word element. [German] Prefix denoting *cartilage*, *granule*.

chondriosome (cytology) An alternative term for mitochondrion.

Choniangium (veterinary parasitology) A genus of roundworms of the family Strongylidae. Species includes *Choniangium epistomum* and *Choniangium magnostomum* that may be found in the caecum (*USA*: cecum) of Indian elephants.

chorion (biology) Membrane that surrounds an implanted blastocyst and the embryo and foetus (*USA*: fetus) that develop from it.

chorionic gonadotrophin (biochemistry) Hormone produced by the placenta during pregnancy. Its presence in the urine is the basis of many kinds of pregnancy testing.

chorionic villus sampling (biochemistry) Testing during early pregnancy of small samples of tissue taken from the chorion for the presence of foetal (*USA*: fetal) abnormalities.

Chorioptes (parasitology) A genus of mange mites of the family Psoroptidae.

Chorioptes bovis (veterinary parasitology) A species of mange mites in the genus *Chorioptes* which may be found on the pasterns of horses and cattle, on the scrotum of sheep and on the perineal region of cattle. They cause chorioptic mange.

Chorioptes texanus (veterinary parasitology) A species of mange mites in the genus *Chorioptes* which may be found on goats and reindeer.

chorioptic mange (veterinary parasitology) The common form of mange in cattle and horses caused by *Chorioptes bovis*. Cattle show small scabs on the perineum, tailhead and back of the udder without irritation. Horses show severe dermatitis behind the pastern with severe itching at first and then soreness. Sheep show a scaly dermatitis on the legs and on the scrotum of the ram, and may suffer from infertility. Also termed: leg mange; tail mange; symbiotic mange.

choroid (histology) Layer of pigmented cells, rich in blood vessels, between the retina and sclerotic of the eye.

chrom(o)- (terminology) A word element. [German] Prefix denoting colour (*USA*: color).

chromat(o)- (terminology) Word element. [German] Prefix denoting *colour* (*USA*: color), *chromatin*.

chromatic aberration (physics/genetics) *See* aberration.

chromatid (genetics) One of two thread-like parts of a chromosome, visible during prophase of meiosis or mitosis, when the chromosome has duplicated. Chromatids are separated during anaphase.

chromatin (biochemistry) Basic protein that is associated with eukaryotic chromosomes, visible during certain stages of cell duplication.

chromatography (chemistry) Method of separating a mixture by carrying it in solution or in a gas stream through an absorbent material. The separated substances may be extracted by elution.

chromium (Cr) (chemistry) Silver-grey metallic element in Group VIB of the periodic table (a transition element), obtained mainly from its ore chromite. It is electroplated onto other metals (particularly steel) to provide a corrosion-resistant decorative finish, and alloyed with nickel and iron to make stainless steels; its compounds are used in pigments and dyes. At. no. 24; r.a.m. 51.996.

chromophore (chemistry) Chemical grouping that causes compounds to have colour (*USA*: color) (*e.g.* the - N = N - group in an azo dye). Also termed: colour (*USA*: color) radical.

chromosome (genetics) Structure within the nucleus of a cell that contains protein and the genetic DNA. Chromosomes occur in pairs in diploid cells (ordinary body cells); haploid cells (gametes or sex cells) have only one chromosome in their nuclei. The number of chromosomes varies in different species. During cell division, each chromosome doubles and the two duplicates separate into the two new daughter cells.

chromosome map (genetics) Diagram showing the positions of various genes on appropriate chromosomes.

chron(o)- (terminology) A word element. [German] Prefix denoting *time*.

chronic (medical) Describing a condition or disorder that is long-standing (and often difficult to treat).

See also acute.

chronic murine pneumonia (veterinary parasitology) *See* murine respiratory mycoplasmosis.

chronic murine respiratory disease (veterinary parasitology) *See* murine respiratory mycoplasmosis.

chrys(o)- (terminology) A word element. [German] Prefix denoting *gold*.

Chrysomya (parasitology) A genus of flies of the family Calliphoridae. Species includes *Chrysomya albiceps*, *Chrysomya bezziana*, *Chrysomya chloropyga*, *Chrysomya mallochi*, and *Chrysomya rufifacies* which cause cutaneous myiasis.

See also screw-worm; *Lucilia*; *Callitroga*; *Phormia*.

Chrysomyia (parasitology) *See Chrysomya*.

Chrysops (human/veterinary parasitology) A genus of blood-sucking tropical flies of the family Tabanidae. Species include *Chrysops discalis* (deer fly), a vector of tularemia in the western USA, and *Chrysops silacea*, an intermediate host of *Loa loa*, a filarial parasite. These flies cause painful bites and worry livestock when they are about. They also mechanically transmit anaplasmosis, amongst others, and the larvae of the filariid parasite *Loa loa*. A number of trypanosome species are also transmitted mechanically by this important means. Also termed chrysops flies.

chyle (biochemistry) Milky fluid resulting from the absorption of fats in the lacteals of the small intestine; it is removed by the lymphatic system.

chyme (physiology) Partly digested food that passes from the stomach into the duodenum and small intestine.

Ci (measurement) An abbreviation of 'C'ur'i'e.

Ciconiphilus pectiniventris (avian parasitology) A species of louse in the genus *Ciconiphilus* which may be found on duck and geese.

-cide (terminology) A word element. [Latin] Suffix denoting *destruction or killing*; *an agent that kills or destroys*.

cili(o)- (terminology) A word element [Latin] Prefix denoting *cilia, ciliary*.

ciliary body (histology) Ring of tissue that surrounds the lens of the eye. It generates the aqueous humour and contains ciliary muscles, which are used in accommodation.

ciliary muscle (histology) Muscle that controls the lens of the eye and thus achieves accommodation.

Ciliata (parasitology) A class of protozoa of the subphylum Ciliophora, whose members possess cilia during the life cycle and of which a few species are parasitic.

ciliate (parasitology) 1 Having cilia. 2 Any member of the class Ciliata.

ciliate diarrhoea (veterinary parasitology) Colitis caused by Troglodytella in primates. (*USA*: ciliate diarrhea).

ciliated (parasitological anatomy) Provided with cilia.

Ciliophora (parasitology) A subphylum of Protozoa, which includes two major groups, the ciliates and suctorians, and distinguished from the other subphyla by the presence of cilia at some stage during the organism's life cycle.

cilium (cytology) Small hair-like structure that moves rhythmically on the surface of a cell or the whole epithelium. Cilia usually cover the surface of a cell and cause movement in the fluid surrounding it. They are used for locomotion by some single-celled aquatic organisms.

See also flagellum.

cimetidine (pharmacology) A histamine$_2$-antagonist which can be administered orally, by injection or infusion as an ulcer-healing drug.

Cimex (insect parasitology) A genus of the insect family Cimicidae.

Cimex hemipterus (avian parasitology) A species of parasitic bugs in the genus *Cimex* which may be found on poultry. Also termed: *Cimex rotundatus*.

Cimex lectularius (human/avian parasitology) A species of blood-sucking insects in the genus *Cimex* which may be found parasitizing humans for the most part but that can also affect pigeons and poultry. Also termed: bed bugs.

cine- (terminology) A word element. [German] Prefix denoting *movement*.

Cionella lubrica (parasitology) A terrestrial snail that is the first intermediate host of *Dicrocoelium dendriticum*.

circ. (physics) An abbreviation of 'circ'uit.

circadian rhythm (physiology) Cyclical variation in the physiological, metabolic or behavioural (*USA*: behavioral) aspects of an organism over a period of about 24 hours; *e.g.* sleep patterns. It may arise from inside an organism or be a response to a regular cycle of some external variation in the environment. Also termed: diurnal rhythm.

See also biological clock.

circuit (physics) An electrical term for the path that an electric current flows along.

circuit breaker (physics) Safety device in an electric circuit that interrupts the current flow in the event of a fault.

circulatory system (physiology) In animals, transport system that maintains a constant flow of tissue fluid in sealed vessels to all parts of the body (*e.g.* in the blood vascular system, oxygen and food materials dissolved in blood diffuse into each cell; waste products, including carbon dioxide, diffuse out of the cells and into the blood).

See also lymphatic system.

circum- (terminology) A word element. [Latin] Prefix denoting *around*.

circum. (measurement) An abbreviation of 'circum'ference.

circumference (measurement) Boundary of a circle, equal in length to π times the diameter (or 2π times the radius).

cirrus (parasitological anatomy) A term used in relation to trematodes. The male copulatory organ.

cirrus pore (parasitological anatomy) A term used in relation to trematodes. The opening through which the cirrus is protruded.

cirrus pouch (parasitological anatomy) A term used in relation to trematodes. A hollow organ surrounding the introverted cirrus.

cisapride (pharmacology) A motility stimulant which can be administered orally to release the neurotransmitter acetylcholine from the nerves of the stomach and intestine to treat conditions such as oesophageal reflux.

cistron (molecular biology) Functional unit of a DNA chain that controls protein manufacture.

cit. (literary terminology) An abbreviation of 'cit'ed.

citric acid (chemistry) $C_6H_8O_7$ Hydroxy-tricarboxylic acid, present in the juices of fruits and made by fermenting residues from sugar refining. It is important in the Krebs cycle, and is much used as a flavouring and in medicines.

citric acid cycle (biochemistry) An alternative term for Krebs cycle.

citrus nematode (plant parasitology) *See Tylenchulus semipenetrans*.

Cittotaenia (parasitology) A tapeworm genus of the family Anoplocephalidae.

Cittotaenia denticulata (veterinary parasitology) A species of tapeworms in the genus *Cittotaenia*, heavy infestations of which may cause digestive disturbances, emaciation and some deaths in rabbits and hares.

CK (biochemistry) An abbreviation of 'C'reatine 'K'inase.

Cl (chemistry) Chemical symbol for chlorine.

Cladotaenia (avian/veterinary parasitology) A genus of cestodes which may be found in wild birds, and the larval stages in the livers of rodents.

Claisen condensation (chemistry) Chemical reaction in which two molecules combine to give a compound containing a ketone group and an ester group.

clam digger's itch (medical parasitology) Schistosome dermatitis.

class (biology) In biological classification, one of the groups into which a phylum is divided, and which is itself divided into orders; *e.g.* Mammalia (mammals), Aves (birds).

classical physics (physics) Physics prior to the introduction of quantum theory and a knowledge of relativity.

classification (biology) The placing of living organisms into a series of groups according to similarities in structure, physiology, biochemistry, and other characteristics. The smallest group is the species. Similar species are placed in a genus, similar genera are grouped into families, families into orders, orders into classes, classes into phyla (or divisions in plants), and phyla into kingdoms. Modern classification is usually intended to reflect degrees of evolutionary relationship, although not all experts agree on single classification schemes for animals or for plants.

See also binomial nomenclature.

clathrate (chemistry) Chemical structure in which one atom or molecule is 'encaged' by a structure of other molecules, and not held by chemical bonds.

See also chelate.

Clavibacter toxicus (parasitology) A bacterial species (formerly *Corynebacterium rathei*) that grows in galls caused by the grass nematodes *Anguina* spp. on the seedheads of some temperate zone pasture grasses, *e.g.* *Agrostis avenacea*, and produces a highly potent toxin.

clavicle (anatomy) In vertebrates, the anterior bone of the ventral side of the shoulder girdle. Also termed: collarbone.

claw (parasitological anatomy) A term used in relation to arthropods. A sharp curved appendage at the distal end of the leg.

Clayia theresae (avian parasitology) A species of lice in the genus *Clayia* which may be found on guinea fowl.

cleavage (1 biology; 2 chemistry; 3 biochemistry) 1 A series of mitotic divisions of a fertilized ovum. 2 The splitting of a crystal structure along a certain plane parallel to a potential crystal face. 3 The splitting of chemical bonds (*e.g.* protease enzymes cleave peptide bonds from proteins to release amino acid residues).

Cleiodiscus (fish parasites) A genus of the family Dactylogyridae of monogenetic flukes which infest the gills of fish.

clinical thermometer (medical) Type of (mercury) thermometer used for taking body temperature.

Clinostomum complanatum (fish/avian parasitology) A species of flukes in the Clinostomidae family that similarly to *Clinostomum marginatum* parasitize fauces of piscivorous birds. The first hosts are snails and the second are fish, which allows the infestation of birds when they are eaten.

Clinostomum marginatum (fish/avian parasitology) A species of flukes in the Clinostomidae family that, similarly to *Clinostomum complanatum*, parasitize fauces of piscivorous birds. The first hosts are snails and the second are fish, which allows the infestation of birds when they are eaten.

clioxanide (pharmacology) An anthelmintic used specifically as a flukicide that because of its large dose size and greater variability in efficacy, depending on whether the dose goes into the rumen or not, has been largely superseded.

CLIP (physiology/biochemistry) An abbreviation of 'C'orticotropin-'L'ike 'I'ntermediate-lobe 'P'olypeptide.

CLL (haematology) An abbreviation of 'C'hronic 'L'ymphocytic 'L'eukaemia. (*USA*: Chronic Lymphocytic Leukemia).

cloaca (parasitological anatomy) A common opening for the rectum and the genital tract. A term used for example in relation to nematodes.

cloacal prolapse (avian parasitology) A condition caused in companion birds by parasitic enteritis.

clone (genetics) One of many descendants produced by asexual reproduction. Members of a clone have identical genetic constitution.

cloning vector (molecular biology) A DNA molecule used to transfer an inserted DNA segment into a host cell. Includes other viruses, phages and bacterial plasmids. Also termed cloning vehicle.

Clonorchis (parasitology) A genus of liver flukes in the family Opisthorchiidae.

Clonorchis sinensis (human/veterinary parasitology) (synonym: *Opisthorchis sinensis*) A species of liver flukes in the genus *Clonorchis* which may be found in bile ducts, ocassionally pancreatic ducts and duodenum in dogs, cats, pigs, some small wild mammals and humans. This may cause diarrhoea (*USA*: diarrhea), abdominal pain, jaundice and ascites in humans. Also termed Oriental or Chinese liver fluke.

clopidol (pharmacology) A pyridinol coccidiostat that may be used in poultry.

clorsulon (pharmacology) A benzenesulfonamide anthelmintic and flukacide that may be used in cattle and sheep.

closantel (pharmacology) An anthelmintic effective against *Fasciola hepatica*, *Haemonchus contortus* and nasal bots.

close (computing) A command to exit from a dialog box, a window or an application.

clotting factor (haematology) Protein-like structure, *e.g.* thrombin and fibrinogen, that induces blood coagulation when a blood vessel is broken.

cluster (epidemiology) A naturally occurring group of similar units, *e.g.* a group of cases of a single disease in time or space.

cluster fly (parasitology) *Pollenia rudis*.

clustering (epidemiology) The gathering together of disease events. The clustering may be in space (geographical clustering) or in time (temporal clustering).

See also cluster.

Cm (chemistry) Chemical symbol for curium.

cm (measurement) An abbreviation of 'c'enti'm'etre (*USA*: centimeter).

CMA (1 education; 2 society) 1 An abbreviation of 'C'ertified 'M'edical 'A'ssistant. 2 An abbreviation of 'C'hemical 'M'anufacturers 'A'ssociation.

cmpd (chemistry) An abbreviation of 'c'o'mp'oun'd'.

$CMRO_2$ (physiology/biochemistry) An abbreviation of 'C'erebral 'M'etabolic 'R'ate for 'O'xygen.

Cnemidocoptes (parasitology) A genus of the family Sarcoptidae of mites. Also termed *Knemidocoptes*.

Cnemidocoptes jamaicensis (avian parasitology) A species of mites in the genus *Cnemidocoptes* which causes scaly leg in some Jamaican wild birds.

Cnemidocoptes laevis gallinae (avian parasitology) A species of mites in the genus *Cnemidocoptes* which causes depluming itch of fowl, pheasants and geese.

Cnemidocoptes laevis laevis (avian parasitology) A species of mites in the genus *Cnemidocoptes* which causes depluming scabies in pigeons.

Cnemidocoptes mutans (avian parasitology) A species of mites in the genus *Cnemidocoptes* which causes scaly leg in fowls and turkeys.

Cnemidocoptes pilae (avian parasitology) A species of mites in the genus *Cnemidocoptes* which causes scaly leg and a crumbly, honeycomb-like mass at the cere (the firm, fleshy bond lying across the base of the beak) of the budgerigar or parakeet, called scaly face.

cnemidocoptic mange (avian parasitology) A series of diseases of birds including scaly leg of poultry, caused by *Cnemidocoptes mutans*, depluming itch of poultry caused by *Cnemidocoptes laevis gallinae* and scaly face and tassel foot of cage birds caused by *Cnemidocoptes pilae*.

Cnephia (veterinary parasitology) Small black flies which cause intense livestock worry. Also termed black fly. Species includes *Cnephia pecuarum*.

CNS (medical) An abbreviation of 'C'entral 'N'ervous 'S'ystem.

CNTF (physiology) An abbreviation of 'C'iliary 'N'euro'T'rophic 'F'actor.

Co (chemistry) Chemical symbol for cobalt.

co. (anatomy) An abbreviation of 'co'lon.

CoA (physiology/biochemistry) An abbreviation of 'Co'enzyme 'A'.

coagulation (1 haematology; 2 biochemistry; 3 chemistry) 1 Process by which bleeding is arrested. Thrombin is produced in the absence of antithrombin from the combination of prothrombin and calcium ions. This interacts with soluble fibrinogen to precipitate it as the insoluble blood protein fibrin, which forms a mesh of fine threads over the wound. Blood cells become trapped in the mesh and form a clot. 2 Irreversible setting of protoplasm on exposure to heat or poison. 3 Precipitation of colloids, *e.g.* proteins, from solutions.

cobalt (Co) (chemistry) Silver-white magnetic metallic element in Group VIII of the periodic table, used in alloys to make cutting tools and magnets. The radioactive isotope Co-60 is used in radiotherapy. At. no. 27; r.a.m. 58.9332.

Cobboldia (veterinary parasitology) A genus of parasitic flies in the family Gasterophilidae whose maggots inhabit the alimentary canal or tissues of mammalian hosts. Species include *Cobboldia elephantis* which may be found in the Indian elephant, and *Cobboldia loxodontis* which may be found in the African elephant and the rhinoceros.

COBOL (computing) Acronym of 'CO'mmon 'B'usiness 'O'riented 'L'anguage, a computer programming language designed for commercial use.

Coccidia (parasitology) A group of sporozoa in the family Eimeriidae that are commonly parasitic in epithelial cells of the intestinal tract, but also found in the liver and other organs. It includes three genera, *Eimeria*, *Isospora* and *Cystoisospora*.

coccidia (parasitology) Plural of coccidium.

coccidial (parasitology) Of, or pertaining to, or caused by Coccidia.

coccidian (parasitology) 1 Pertaining to Coccidia. 2 Any member of the Coccidia.

coccidiosis (fish/avian/veterinary parasitology) Infection by Coccidia causes enteritis in all species, the clinical picture varying between species. In calves it is a serious diarrhoea (*USA*: diarrhea) and dysentery and death may occur because of the blood and protein loss and the dehydration. In sheep the effects are poor production and poor weight gain, although diarrhoea (*USA*: diarrhea) and dysentery may occur. The clinical disease is rare in pigs and horses, but outbreaks, similar clinically to those in cattle, may occur in young animals. In dogs and cats, infection is most common in young puppies and kittens where it can be the cause of severe diarrhoea (*USA*: diarrhea) and even death. Adults usually experience only mild and self-limiting infections. All poultry species suffer severe outbreaks of the disease characterized by diarrhoea (*USA*: diarrhea) and dysentery. Subclinical infections causing reduced productivity are a feature of the disease in birds. Affected fish are cachectic and trail long mucoid faecal (*USA*: fecal) casts. The disease in all species except fish is caused by *Eimeria*, *Isospora* or *Cystoisospora*. In fish the species involved is *Eimeria* (*Epieimeria*), *Goussia*, and *Cryptosporidium* spp.

coccidiostatic drugs (pharmacology) Drugs which control coccidiosis. The greatest importance of coccidiosis is in the chicken industry and many agents have been developed in an attempt to reduce losses. The important drugs or groups of drugs for this purpose include clopidol, quinolones, monensin, lasalocid, salinomycin, robenidine, amprolium, dinitolmide, nicarbazin, sulphonamides (*USA*: sulfonamides) and halofuginone.

coccidiostats (pharmacology) *See* coccidiostatic drugs.

Coccidium (parasitology) *See Eimeria*; *Isospora*; *Sarcocystis*; *Toxoplasma*; *Hammondia*; *Besnoitia*; *Cryptosporidium*.

coccyx (anatomy) Bony structure in primates and amphibians, formed by fusion of tail vertebrae; *e.g.* in human beings it consists of three to five vestigial vertebrae at the base of the spine.

cochlea (anatomy) Spirally coiled part of the inner ear in mammals. It translates sound-induced vibrations into nerve impulses that travel along the auditory nerve to the brain, where they are interpreted as sounds.

Cochliomyia (parasitology) A genus of the fly family of insects Calliphoridae. Species includes *Cochliomyia hominivorax* and *Cochliomyia macellaria. See Callitroga.*

Cochlosoma (parasitology) A protozoan parasite of the family Cochlosomatidae.

Cochlosoma anatis (avian parasitology) A species of protozoan parasite in the genus *Cochlosoma* which may be found in the large intestine of domestic and wild ducks and turkeys, that may cause catarrhal enteritis and diarrhoea (*USA*: diarrhea).

cockle (veterinary parasitology) A dermatitis of sheep consisting of inflammatory nodules, especially on the neck and shoulders, that is thought to be of parasitic cause.

Codiostomum (parasitology) A genus of the nematode family of Strongylidae.

Codiostomum struthionis (veterinary parasitology) A species of strongylid nematode in the genus *Codiostomum* that may be found in the large intestine of the ostrich.

coefficient (1 mathematics; 2 physics) 1 In an algebraic expression, the numerical factor by which the variable is multiplied; *e.g.* in $5xy$, the coefficient of xy is 5. 2 Number or parameter that measures some specified property of a given substance; *e.g.* coefficient of friction, coefficient of viscosity.

-coele (terminology) A word element [German] Suffix denoting *cavity, space.*

coelomyarian (parasitological cytology) A term used in relation to nematodes. Somatic muscle cells in which the muscle fibres extend along the sides of the cell in addition to lying perpendicular to the hypodermis.

coelozoic (parasitology) Inhabiting the intestinal canal of the body.

coenurosis (veterinary parasitology) Infection with the intermediate stage of *Taenia multiceps* which invades the brain and spinal cord of sheep and causes a variety of syndromes characteristic of slowly developing space-occupying lesions of the nervous system. Ataxia, head-pressing, somnolence and occasional convulsions are common signs of brain involvement. Paralysis and recumbency are the usual signs in spinal cord involvement. Also termed: gid; sturdy.

Coenurus (parasitology) A metacestode, a larval stage of a tapeworm belonging to the genus *Taenia.*

Coenurus cerebralis (veterinary parasitology) Metacestode of Taenia *multiceps* which may be found in the brain and spinal cord of sheep, but in other organs of goats.

See also coenurosis.

Coenurus serialis (veterinary parasitology) Metacestode of the tapeworm *Taenia serialis* of dogs and foxes found in the subcutaneous tissues and muscles of the intermediate host, a lagomorph.

coenzyme (biochemistry) Organic compound essential to catalytic activities of enzymes without being utilized in the reaction. Coenzymes usually act as carriers of intermediate products, *e.g.* ATP.

COHb (physiology/biochemistry) An abbreviation of 'C'arbonmon'O'xy'H'aemoglo'b'in (*USA*: carbon oxyhemoglobin).

coherent units (measurement) System of units in which the desired units are obtained by multiplying or dividing base units, with no numerical constant involved.

See also SI units.

cohesion (physics) Attraction between similar particles (atoms or molecules of the same substance); *e.g.* between water molecules to create surface tension.

cohort (epidemiology) A group of individuals who share a characteristic acquired at the same time.

Coleoptera (insect parasitology) An order in the class Insecta. The beetles.

collagen (biochemistry) Fibrous protein connective tissue that binds together bones, ligaments, cartilage, muscles and skin.

collarbone (anatomy) An alternative term for the clavicle.

colloid (physics) Form of matter that consists of small particles, about 10^{-4} to 10^{-6} mm across, dispersed in a medium such as air or water. Common colloids include aerosols (*e.g.* fog, mist) and gels (*e.g.* gelatin, rubber). A non-colloidal substance is termed a crystalloid.

Collyriclum (avian parasitology) A trematode genus of the family Troglotrematidae. Species includes *Collyriclum faba* which causes subcutaneous cysts in fowls, turkeys, and wild birds.

colo- (terminology) A word element. [German] Prefix denoting *colon.*

Coloceras (avian parasitology) A genus of feather-eating lice of the family Philopteridae which infest pigeons and doves. Species includes *Coloceras damicorne* which may be found in pigeons.

colon (anatomy) Large intestine, in which the main function is the absorption of water from faeces (*USA*: feces).

colony (bacteriology) Bacterial growth on a solid medium that forms a visible mass.

Colorado tick fever (human parasitology) A disease of humans caused by *Coltivirus*, transmitted from small mammals by the tick, *Dermacentor andersoni*.

colorimeter (physics/chemistry) Instrument for measuring the colour (*USA*: color) intensity of a medium such as a coloured solution (which can be related to concentration and therefore provide a method of quantitative analysis). The technique is termed colorimetry.

colostrum (biology) Yellowish milky fluid secreted from the mammalian breast immediately before and after childbirth. It contains more antibodies and leucocytes (and less fat and carbohydrates) than true milk, which follows within a few days.

colour (physics) Visual sensation or perception that results from the adsorption of light energy of a particular wavelength by the cones of the retina of the eye. There are two or more types of cone, each of which are sensitive to different wavelengths of light. The brain combines nerve impulses from these cones to produce the perception of colour. The colour of an object thus depends on the wavelength of light it reflects (other wavelengths being absorbed) or transmits. (*USA*: color)

colp(o)- (terminology) A word element. [German] Prefix denoting *vagina*.

Colpocephalum (parasitology) A genus of lice of the family Menoponidae.

Colpocephalum tausi (avian parasitology) A species of lice of the genus *Colpocephalum* that may be found in turkeys.

Colpocephalum turbinatum (avian parasitology) A species of lice of the genus *Colpocephalum* that may be found in pigeons.

Columbicola (avian parasitology) A genus of lice in the superfamily Ischnocera. Species includes *Columbicola columbae* that may be found on pigeons.

columbium (chemistry) Former name of niobium.

combination (mathematics) Any selection of a given number of objects from a set, irrespective of their order. The number of combinations of r objects that can be obtained from a set of n objects (usually written $_nC_r$) is $n!/[r!(n - r)!]$. (The symbol ! stands for factorial.)

combustion (chemistry) Burning; the rapid oxidation of a substance accompanied by heat, light and flame. Usually the oxidizing agent is oxygen (often from air).

command (computing) An instruction which starts or stops a computer programme (*USA*: program).

commensalism (biology) Close relationship between two organisms from which one benefits and the other neither benefits nor suffers.

See also, parasitism; symbiosis.

common denominator (mathematics) Same denominator assigned to fractions (formerly with different ones) so that they can be added or subtracted; *e.g.* to subtract 1/3 from 1/2 the fractions are first assigned the common denominator of 6, becoming 2/6 and 3/6, and subtracted to give 1/6; using the same method they can be added (to give 5/6).

common difference (mathematics) Difference between any two consecutive terms of an arithmetic progression.

common genital pore (parasitological anatomy) A term used in relation to trematodes. The common opening formed by the fusion of the male genital duct and the uterus. This is usually enclosed in a small chamber.

common liver fluke (parasitology) *See Fasciola hepatica.*

common logarithm (mathematics) Logarithm to the base 10.

common salt (chemistry) An alternative term for sodium chloride.

commutative (mathematics) Describing a mathematical operation of the type $a * b$ whose result does not depend on the order in which the operation is carried out. *E.g.* the addition 7 + 2 has the same result as 2 + 7, so addition is commutative. Subtraction is not; *e.g.* 7 – 2 is not the same as 2 – 7.

competition (biology) The struggle within a community between organisms of the same or different species for survival.

See also natural selection.

compiler (computing) Computer program that converts a source language into machine code (readable by the computer).

complementary deoxyribonucleic acid (CDNA) (genetics) Single-stranded DNA complementary to an RNA; it is synthesized *in vitro* by using the RNA as a template for the action of reverse transcriptase.

complete metamorphosis (parasitological physiology) A term used in relation to arthropods. Holometabolous. The cycle of development involves egg, larva, pupa and adult. Each stage is morphologically distinct from the other.

complex compound (chemistry) An alternative term for a co-ordination compound.

complex ion (chemistry) Cation bonded by means of a co-ordinate bond.

complex number (mathematics) Number written in the form $x + iy$, where x and y are real numbers and i is the square root of -1 (*i.e.* $i^2 = -1$).

component (1 mathematics; 2 chemistry) 1 The resolved part of a vector quantity in a particular direction, usually one of a pair at right-angles. 2 One of the substances in a system.

composite (physics) Material composed of two or more other materials that give a combination with better properties than any of the components individually; *e.g.*

glass-fibre (*USA*: glass-fiber) or carbon-fibre (*USA*: carbon-fiber) reinforced plastic resin.

compound (chemistry) Substance that consists of two or more elements chemically united in definite proportions by weight. Also termed: chemical compound.

compound eye (parasitological anatomy) A term used in relation to arthropods. A large number of visual elements grouped together on each side of the head.

compound microscope (microscopy) The standard laboratory microscope consisting of a two lens system whereby the image formed by the system near the object (objective) is magnified by the one nearer the eye (eyepiece).

Compton effect (physics) Reduction in the energy of a photon as a result of its interaction with a free electron. Some of the photon's energy is transferred to the electron. It was named after the American physicist Arthur Compton (1892–1962).

computer (computing) Electronic device that can accept data, apply a series of logical operations to it (obeying a program), and supply the results of these operations. *See* analog computer; digital computer.

computer output (computing) The response to a command such as print, list, display. Usually taken to mean printed material but does include other materials, *e.g.* magnetic tape recordings put out on other output media by various output devices.

computerized axial tomography (CAT) scan (radiology) A technique for producing X-ray pictures of cross-sectional 'slices' through the brain and other parts of the body.

COMT (physiology/biochemistry) An abbreviation of 'C'atechol-'O'-'M'ethyl'T'ransferase.

concave (optics) Describing a surface that curves inwards (as opposed to one that is convex).

concentration (chemistry) Strength of mixture or solution. Concentrations can be expressed in very many ways; *e.g.* parts per million (for traces of a substance), percentage (*i.e.* parts per hundred by weight or volume), gm or kg per litre of solvent or per litre of solution, moles per litre of solution (molarity), moles per kg of solvent (molality), or in terms of normality. *See* normal.

See also solubility.

Concinnum (parasitology) A genus of flukes of the family Dicrocoeliidae.

Concinnum brumpti (veterinary parasitology) A species of flukes in the genus *Concinnum* that may be found in the liver and pancreas of African apes.

Concinnum procyonis (veterinary parasitology) - (synonym: *Eurytrema proctonis*) A species of flukes in the genus *Concinnum* that may be found in the pancreatic and biliary ducts of cats, and foxes.

Concinnum ten (veterinary parasitology) A species of flukes in the genus *Concinnum* that may be found in wild carnivores.

condensation (1 physics; 2 chemistry) 1 Change of a gas or vapour into a liquid or solid by cooling. 2 Condensation reaction.

condensation pump (laboratory equipment) An alternative term for diffusion pump.

condensation reaction (chemistry) Chemical reaction in which two or more small molecules combine to form a larger one, often with the elimination of a simpler substance, usually water.

condenser (1 chemistry; 2 physics; 3 optics) 1 An apparatus for changing a vapour into a liquid (by cooling it and causing condensation). 2 An alternative term for a capacitor. 3 A lens or mirror that concentrates a light source.

conditioned reflex (physiology) Animal's response to a neutral stimulus that learning has associated with a particular effect; *e.g.* a rat may learn (be conditioned) to press a lever when hungry because it associates this action with receiving food.

conductance (physics) Ability to convey energy as heat or electricity. Electrical conductance, measured in siemens, is the reciprocal of resistance. Also termed: conductivity.

conduction band (physics) Energy range in a semiconductor within which electrons can be made to flow by an applied electrical field.

conductor (1, 2 physics) 1 Material that allows heat to flow through it by conduction. 2 Material that allows electricity to flow through it; a conductor has a low resistance.

cone (cytology) Light-sensitive nerve cell present in the retina of the eye of most vertebrates; it can detect colour (*USA*: color).

cone-nose bug (parasitology) *See* reduviid bug.

configuration (1 physics; 2 chemistry; 3 computing) 1 The arrangement of electrons about the nucleus of an atom. Also termed: electron configuration. 2 The arrangement in space of the atoms in a molecule. 3 The group of interconnected machines that make up a practical working unit. Comprises computer, disk drive(s), printer, visual display unit.

confocal scanning microscope (microscopy) A microscope that utilizes a focused laser beam to visualize fluorescent molecules in a single plane.

conformation (chemistry) Particular shape of a molecule that arises through the relative spacial arrangement of atoms.

congenital (medical) Dating from birth or before birth. Congenital conditions may be caused by environmental factors or be inherited.

Congo floor maggot (parasitology) *Auchmeromyia luteola.*

Congo virus disease (parasitology) A transient fever of cattle and goats caused by a not well characterized arbovirus transmitted by *Hyalomma* spp. ticks.

conical flukes (parasitology) *See Paramphistomes.*

conjugated (chemistry) Describing an organic compound that has alternate single and double or triple bonds, *e.g.* buta-1, 3-diene, $H_2C=CH-CH=CH_2$.

conjugation (biology) Form of reproduction that involves the permanent or temporary union of two isogametes, *e.g.* in certain green algae. In protozoa, two individuals partly fuse, exchanging nuclear materials. When separated, each cell divides further to give uninucleate spores which eventually develop into full adults.

conjunctiva (histology) Layer of protective mucus-secreting epidermis that covers the cornea on the eyeball and is continuous with the inner lining of the eyelids of vertebrates.

conjunctival habronemiasis (veterinary parasitology) Granulomatous lesions in horses caused by invasion by *Habronema* spp larvae occurring on the third eyelid, or the eyelid proper, or on the conjunctiva of the medial canthus. *See* habronemiasis.

connect time (computing) The time period for which a user is connected to a computer. Used as the basis for charging for computer use, particularly for networking and dial-up computer services.

connective tissue (histology) Strong tissue that binds organs and tissues together. It consists of a glycoprotein matrix containing collagen in which cells, fibres (*USA*: fibers) and vessels are embedded. The most widespread connective tissue is areolar tissue.

Conocephalus (parasitology) Grasshoppers, an intermediate host for the flukes in the genus *Eurytrema.*

conoid (parasitological anatomy) A term used in relation to *Toxoplasma* and *Sarcocystis.* The hollow cone-like structure at the anterior end of the cystozoites, endozoites and merozoites. It is covered by spirally arranged fibrillar structures. It is generally believed that this organ is used by the parasite to penetrate host cells.

conservation of energy (physics) Law that states that in all processes occurring in an isolated system the energy of the system remains constant.

See also thermodynamics.

conservation of mass (physics) Principle which states that the products of a purely chemical reaction have the same total mass as the reactants.

constant (measurement) Quantity that remains the same in all circumstances; *e.g.* in the expression $2y = 5x^2$, the numbers 2 and 5 are constants (and x and y are variables).

contagious agalactia (veterinary parasitology) Acute mastitis, arthritis and ophthalmitis with painful joint swelling in goats and sheep caused by *Mycoplasma agalactiae.* Abortion is common, the udder is permanently damaged and many animals die.

contagious caprine/ovine pleuropneumonia (veterinary parasitology) A highly infectious and fatal pleuropneumonia of goats and sheep generally considered to be caused by *Mycoplasma mycoides* var. *mycoides.*

continuous wave (physics) Electromagnetic wave of constant amplitude.

contra- (terminology) A word element. [Latin] Prefix denoting *against, opposed.*

Contracaecum (fish/avian veterinary parasitology) A genus of roundworms in the family Anisakidae. Species includes *Contracaecum microcephalum* and *Contracaecum spiculigerum* which may be found in wild birds, and *Contracaecum osculatum* which may be found in seals, other fish eaters.

contractile vacuole (biology) Membranous sac within a single-celled organism (*e.g.* amoeba (*USA*: ameba) and other protozoans) which fills with water and suddenly contracts, expelling its contents from the cell. It carries out osmoregulation and excretion.

contraindication (parasitology) Any condition that renders a particular line of treatment improper or undesirable.

control experiment (epidemiology) An experiment made under standard conditions, to test the correctness of other observations.

control matching (statistics) *See* matched groups.

convection (physics) Transport of heat by the movement of the heated substance (usually a fluid such as air or water).

converging lens (optics) Lens capable of bringing light to a focus; a convex lens.

convex (optics) Describing a surface that is rounded outwards (as opposed to one that is concave).

coolant (physics) Liquid or gas that removes heat by convection.

Cooperia (veterinary parasitology) A genus of intestinal worms in the family Trichostrongylidae. Principal infestations in sheep and goats, *Cooperia punctata, Cooperia oncophora, Cooperia curticei*; in cattle, *Cooperia oncohora, Cooperia pectinata, Cooperia punctata*; miscellaneous infections with other species in ruminants generally are *Cooperia bisonis, Cooperia spatulata, Cooperia surnabada* (synonym: *Cooperia mcmasteri*). The worms inhabit the small intestine.

co-ordinate bond (chemistry) Type of covalent bond that is formed by the donation of a lone pair of electrons from one atom to another. Also termed: dative bond.

co-ordination compound (chemistry) Chemical compound that has co-ordinate bonds. Also termed: complex; complex compound.

co-ordination number (chemistry) Number of nearest neighbours of an atom or an ion in a chemical compound.

COPD (medical) An abbreviation of 'C'hronic 'O'bstructive 'P'ulmonary 'D'isease.

copepod (parasitology) A member of the subclass Copepoda of marine invertebrate parasites. There are more than 4500 species which includes *Cyclops* spp. the intermediate host for *Spirometra erinacei* and *Diphyllobothrium latum*.

coplin jars (equipment) Wide-mouthed glass jars, usually with vertically grooved interior walls, used for the storage or staining of slides containing blood smears or tissue sections.

copper (Cu) (chemistry) Reddish metallic element in Group IB of the periodic table (a transition element) which occurs as the free metal (native) and in various ores, chief of which is chalcopyrite. The metal is a good conductor of electricity and is used for making wire, pipes and coins. Its chief alloys are brass and bronze. Its compounds are used as pesticides and pigments. Copper is an important trace element in many plants and animals. At. no. 29; r.a.m. 63.546.

copper sulphate (pharmacology/control measures) Used as a parasiticide in aquariums and in the treatment of foot rot in cattle. (*USA*: copper sulfate).

copper(I) (chemistry) An alternative term for cuprous.

copper(I) oxide (chemistry) Cu_2O Insoluble red powder, used in rectifiers and as a pigment. Also termed: cuprous oxide; copper oxide; cuprite; red copper oxide.

copper(II) (chemistry) An alternative term for cupric.

copper(II) carbonate (chemistry) $CuCO_3$ Green crystalline compound which occurs (as the basic salt) in the minerals azurite and malachite. It is also a component of verdigris, which forms on copper and its alloys are exposed to the atmosphere.

copper(II) chloride (chemistry) $CuCl_2$ Brown covalently bonded compound, which forms a green crystalline dihydrate. It is used in fireworks to give a green flame and to remove sulphur (*USA*: sulfur) in the refining of petroleum. Also termed: cupric chloride.

copper(II) oxide (chemistry) CuO Insoluble black solid, used as a pigment. Also termed: cupric oxide; copper oxide.

copper(II) sulphate (chemistry) $CuSO_4$ White hygroscopic compound, which forms a blue crystalline pentahydrate, $CuSO_4.5H_2O$, used as a wood preservative, fungicide, dyestuff and in electroplating. Also termed: cupric sulphate (*USA*: cupric sulfate); blue vitriol. (*USA*: copper(II) sulfate).

copra mite (parasitology) *See Tyrophagus longior.*

coproculture (methods) Culture of the faeces (*USA*: feces) for the purpose of hatching parasite eggs and obtaining larvae for morphological identification.

copulatory bursa (parasitological anatomy) The bursa that embraces the female nematode during copulation; the structure is useful for the identification of some species of nematodes.

coracidium (parasitological physiology) The first larval stage of a pseudophyllidean cestode. This motile stage consists of a ciliated embryophore containing the oncosphere.

Cordylobia (parasitology) A genus of blowflies of the family Calliphoridae.

Cordylobia anthropophaga (human/veterinary parasitology) A species of blowflies of the genus *Cordylobia*, the maggot of which parasitizes humans, rodents, monkeys and dogs causing cutaneous lumps. Also termed tumbu or skin-maggot fly.

Cordylobia rodhaini (human/veterinary parasitology) A species of blowflies of the genus *Cordylobia*, the maggot of which causes skin lesions in antelope, rodents and humans. Also termed Lund's fly.

core (1 physics; 2 computing) 1 Magnetic material at the centre (*USA*: center) of a solenoid or within the windings of a transformer. 2 Element in a computer memory consisting of a piece of magnetic material that can retain a permanent positive or negative electric charge until a current passes through it (when the charge changes polarity).

core(o)- (terminology) A word element. [German] Prefix denoting pupil of the eye.

corky ring spot of potatoes (plant parasitology) *See Paratrichodorus minor.*

cornea (histology) Transparent connective tissue at the front surface of the eye of vertebrates, overlying the iris. Togther with the lens, it focuses incoming light onto the retina.

cornification (histology) An alternative term for keratinization.

coronary vessels (anatomy) Arteries and veins that carry the blood supplying the heart muscle in vertebrates.

coronene (chemistry) Aromatic organic compound which consists of six (or seven) benzene rings fused to form a circle.

corpus (parasitological anatomy) A term used in relation to nematodes. Anterior portion of the oesophagus. In *Rhabditis* type it is divisible into procorpus and metacorpus.

corpus callosum (histology) Thick bundle of nerve fibres (*USA*: nerve fibers) in the middle of the brain that connects the two cerebral hemispheres.

corpuscle (biology) Any small mass or body.

corridor disease (veterinary parasitology) A disease of cattle caused by *Theileria lawrenci* that clinically resembles East Coast fever.

corrosive sublimate (chemistry) An alternative term for mercury(I1) chloride.

cortex (anatomy) Outer layer of a structure, *e.g.* the outer layer of cells of the adrenal gland or brain.

See also medulla.

corticosteroid (biochemistry) Hormone secreted by the adrenal cortex, which controls sodium and water metabolism as well as glycogen formation.

corticotrophin (biochemistry) An alternative term for adrenocorticotrophic hormone (ACTH).

cortisone (biochemistry) Hormone isolated from the adrenal cortex, used in the treatment of rheumatoid arthritis and other inflammatory conditions. Also termed: 17-hydroxy- 11-dehydrocorticosterone.

Corynetes caerulus (insect parasitology) Mites in the Orbibatidia (Cryptostigmata) that may be phoretic on the larvae of the wasps *Vespa crabro.*

Corynosoma (parasitology) A genus of acanthocephalan parasites of the family Polymorphidae. Also termed thornyheaded worm.

Corynosoma semerme (veterinary parasitology) A species of acanthocephalan parasites in the genus *Corynosoma* which may be found as nonpathogenic parasites in the intestines of dogs and foxes but that may cause enteritis in mink.

Corynosoma strumosum (veterinary parasitology) A species of acanthocephalan parasites in the genus *Corynosoma* which may be found as nonpathogenic parasites in the intestines of dogs and foxes but that may cause enteritis in mink.

Cosmocephalus obvelatus (parasites) A nematode found in the oesophagus (*USA*: esophagus) of newly captured rockhopper penguins.

cost(o)- (terminology) A word element. [Latin] Prefix denoting rib.

costa (parasitological anatomy) 1 A cytoplasmic thickening seen at the base of the undulating membrane in some flagellates. 2 Thickening of the anterior edge of the wing in arthropods.

costal (anatomy) Concerning the ribs.

Costia (fish parasitology) A genus of flagellated protozoa in the family Tetramitidae. Also termed *Ichthyoboda*. Species include *Costia necatrix* and *Costia pyriformis* which are parasites on the skin of freshwater fish.

Cotugnia (avian parasitology) A genus of tapeworms in the family Davaineidae. Species include *Cotugnia cuneata*, *Cotugnia digonopora*, and *Cotugnia fastigata* which may be found in the intestines of birds.

Cotugnia digonopora (insect parasitology) A species of helminth in the Cestoda that are parasites to the ants *Monomorium* spp.

Cotylurus (avian parasitology) A genus of digenetic trematodes in the family Strigeidae. Species include *Cotylurus cornutus*, *Cotylurus flabelliformis*, *Cotylurus platycephalus*, and *Cotylurus variegatus* which are parasitic in the intestines of wild birds.

coulomb (C) (measurement) SI unit of electric charge, defined as the quantity of electricity transported by a current of 1 ampere in 1 second. It was named after the French physicist Charles Coulomb (1736–1806).

coumaphos (pharmacology/control measures) An organophosphorus insecticide and anthelmintic, effective against *Haemonchus*, *Trichostrongylus*, *Osteragia*, *Cooperia* and *Tricuris* spp. in cattle and sheep.

coumatetralyl (control measures) A derivative of warfarin and used as a rodenticide.

count (statistics) A numerical computation or indication.

counter (1 physics; 2 computing) 1 An electronic apparatus for detecting and counting particles, usually by making them generate pulses of electric current; the actual counting circuit is a scaler. 2 Any device that accumulates totals (*e.g.* of repeated program loops or cards passing through a punched card reader).

counter immunoelectrophoresis (CIE) (methods) A laboratory technique in which an electric current is used to accelerate the migration of antibody and antigen through a buffered gel diffusion medium. Antigens in a gel medium in which the pH is controlled are strongly negatively charged and will migrate rapidly across the electric field toward the anode. The antibody in such a medium is less negatively charged and will migrate in an opposite or 'counter' direction toward the cathode. If the antigen and antibody are specific for each other, they combine and form a distinct precipitin line.

counter scintillation spectrometer (physics) Scintillation counter capable of measuring the energy and the intensity of gamma radiation emitted from a material.

counterstain (microscopy) A stain applied to tender the effects of another stain more discernible.

couple (physics) A pair of equal and parallel forces acting in opposite directions upon an object. This produces a turning effect (torque) equal to one of the forces times the distance between them.

coupling (physics) Connection between two oscillating systems.

covalent bond (chemistry) Chemical bond that results from the sharing of a pair of electrons between two atoms.

See also co-ordinate bond; ionic bond.

covalent compound (chemistry) Chemical compound in which the atoms are covalently bonded.

covalent radius (chemistry) Effective radius of an atom involved in a covalent bond.

coverglass (microscopy) A thin glass that covers a mounted microscopical object or a culture. Also termed coverslip.

coverslip (microscopy) *See* coverglass.

coxa (parasitological anatomy) A term used in relation to arthropods. The proximal segment of the leg.

CPA (medical) An abbreviation of 'C'ardio'P'ulmonary 'A'rrest.

CPBA (methods) An abbreviation of 'C'ompetitive 'P'rotein-'B'inding 'A'ssay.

CPD (1 education; 2 haematology) An abbreviation of 'C'ontinuing 'P'rofessional 'D'evelopment. 2 An abbreviation of 'C'itrate-'P'hosphate-'D'extrose anti-coagulant.

cpd (chemistry) An abbreviation of 'c'om'p'oun'd'.

CPE (biology) An abbreviation of 'C'yto'P'athic 'E'ffect.

CPI (computing) An abbreviation of 'C'haracters 'P'er 'I'nch.

CPK (biochemistry) An abbreviation of 'C'reatine 'P'hospho'K'inase.

CPS (computing) An abbreviation of 'C'haracters 'P'er 'S'econd, as a criterion of the speed of computer printers.

cps (measurement) An abbreviation of 'c'ycles 'p'er 's'econd.

CPT (control measures) An abbreviation of 3-'C'hloro-'P'-'T'oluidine, an avicide.

CPU (computing) An abbreviation of 'C'entral 'P'rocessing 'U'nit.

CR (physiology) An abbreviation of 'C'onditioned 'R'eflex.

Cr (chemistry) Chemical symbol, chromium.

Cr. (physiology/biochemistry) An abbreviation of 'Cr'eatinine.

crani(o)- (terminology) A word element. [Latin] Prefix denoting skull.

cranial (anatomy) Relating to the cranium and brain.

cranial nerve (histology) One of the 10 to 12 pairs of nerves connected directly with a vertebrate's brain and supplying the sense organs, muscles of the head and neck and abdominal organs. Together with the spinal nerves they make up the peripheral nervous system.

cranium (anatomy) Part of the skull that encloses and protects the brain, consisting of eight fused bones in human beings.

Craterostomum (veterinary parasitology) A genus of nematode worms of the family Strongylidae. Species include *Craterostomum acuticaudatum* and *Craterostomum tenuicauda* which may be found in the large intestine of equines.

cream of tartar (chemistry) An alternative term for potassium hydrogentartrate.

creatinase (biochemistry) An enzyme that catalyzes the decomposition of creatine into urea and ammonia.

creatine (biochemistry) $NH_2C(NH)N(CH_3)CH_2COOH$ White crystalline amino acid present in muscle, where it plays an important role in muscle contraction. It is broken down to creatinine.

creatinine (biochemistry) $C_4H_7N_3O$ Heterocyclic crystalline solid formed by the breakdown of creatine and excreted in urine.

creeping (medical) Gradual progression of a lesion or tissue growth.

creeping eruption (parasitology) *See* cutaneous larva migrans.

Crenosoma (veterinary parasitology) A genus of lungworms of the family Crenosomatidae. Species includes *Crenosoma mephiditis*, *Crenosoma petrowi*, *Crenosoma striatum* which may be found in wild animals, and *Crenosoma vulpis* which may be found in dogs and wild carnivores.

Crepidostomum (fish parasite) A genus of digenetic trematode of the family Allocreadiidae which are parasites of salmonid and other fish.

Criconemella (plant parasitology) A genus of nematodes in the order Tylenchida and family Criconematidae. There is strong sexual dimorphism: females are small, stout, nematodes under 1 mm long, while the degenerate males are much thinner than the females, and juveniles, although smaller, are somewhat similar to females but the cuticular ornamentation is usually more elaborate with longitudinal rows of scales or spines. *See Criconemella xenoplax.*

Criconemella xenoplax (plant parasitology) A species of nematodes in the genus *Criconemella* whose hosts cover a wide range, including generally woody plants, all *Prunus* species (including peach, almond, apricot, cherry, and plum, also lettuce, grape, carnation, and pine) and which may be found in North and South America, Europe, Africa, India, Australia, and Japan. They are migratory ectoparasites feeding on root tips or along more mature roots. During their life cycle females produce three to five eggs per day. Their first moult (*USA*: molt) occurs inside the egg, with the egg maturing in approximately 10–12 days at 20–22°C. All stages feed and the life cycle is complete in 24–30 days. Nematode feeding has a root-pruning effect on plants, resulting in reduction of feeder roots, reduction in ability to withstand stress and reduced nutrient uptake which may affect susceptibility to bacterial canker of stone fruits. Control measures includes the use of nematicides. Also termed: ring nematode.

crin- (terminology) Prefix meaning secrete.

Crithidia (insect parasitology) A genus, the trypanosomes, in the family Trypanosomatidae which may be found in arthropods and other invertebrates, some of which are parasites to the wasps, *Vespa squamosa*.

See also Crithidia bobi; Crithidia mellificae.

Crithidia bobi (insect parasitology) A protozoal species of flagellates that are highly infective parasites to the bees *Bombus* spp., and *Psithyrus vestalis*.

Crithidia mellificae (insect parasitology) A protozoal species of flagellates that are commonly found parasites to the bees *Apis mellifera* but with no apparent harmful effects.

critical (1 chemistry; 2 medical) 1 A point at which one property or state changes to another property or state. 2 Pertaining to a crisis in a disease.

critical angle (optics) The smallest angle of incidence at which total internal reflection occurs.

critical care patient (medical) Care pertaining to a crisis in a disease.

critical care medicine (medical) Emergency care for victims of trauma or disease.

critical pressure (physics) Pressure necessary to condense a substance at its critical temperature.

critical state (physics) Conditions of temperature and pressure at which the liquid and gas phases become one phase, *i.e.* they have the same density.

critical temperature (1, 2 physics) 1 Temperature above which a gas cannot be liquefied (no matter how high the pressure). 2 Temperature at which a magnetic material loses its magnetism. Also termed: Curie point.

critical volume (physics) Volume of one unit of mass of a substance at its critical pressure and critical temperature.

CRO (equipment) An abbreviation of 'c'athode-'r'ay 'o'scilloscope.

chromatin (parasitology) The substance in the cell nucleus which stains with basic dyes.

chromatoid body (parasitology) The RNA-protein complex which stains deeply with basic dyes but not with iodine. It is found in the genus *Entamoeba*. In the electron microscope it reveals a crystalline structure resembling virus particles.

cross linkage (chemistry) Cross-linking between chains within a polymer.

crossing over (genetics) Exchange of material between homologous chromosomes during meiosis (cell division). It is the mechanism that alters the pattern of genes in the chromosomes of offspring, giving the genetic variation associated with sexual reproduction.

crotamiton (pharmacology) An acaricide used in the topical treatment of scabies in dogs and enemidocoptic mange in budgerigars.

crotoxyphos (control measures) An insecticide used for fly control and externally on livestock; a cholinesterase inhibitor. *See* organophosphorus compound.

CrP (physiology/biochemistry) An abbreviation of 'Cr'eatinine 'P'hosphate.

CRT (physics) An abbreviation for 'C'athode 'R'ay 'T'ube.

crufomate (pharmacology) An organophosphorus compound used as an anthelmintic and in the control of warble fly in cattle. Also termed: Montrel; Ruelene.

cry(o)- (terminology) A word element. [German] Prefix denoting *cold*.

cryogenics (physics) Branch of physics that involves low temperatures and their effects.

crypt(o)- (terminology) A word element. [German] Prefix denoting *concealed, pertaining to a crypt*.

Crypto. (parasitology) An abbreviation of *'Cryto'sporidium.*

Cryptobia (fish parasitology) A genus of biflagellate protozoa that parasitize fish, amphibians and other aquatic creatures. Species include *Cryptobia borreli*, *Cryptobia brachialis* and *Cryptobia cyprini* that are parasitic haemoflagellates (*USA*: hemoflagellates) of fish that are transmitted by leeches.

Cryptocaryon (parasitology) A ciliate protozoan.

Cryptocaryon irritans (fish parasitology) A species of protozoa in the genus *Cryptocaryon* that causes 'white spot' skin and gill disease in saltwater fish.

Cryptocotyle (fish/avian/veterinary parasitology) A genus of flukes (digenetic trematodes) in the family Heterophyidae. Species include *Cryptocotyle concavium*, -*Cryptocotyle jejuna* and *Cryptocotyle lingua* which are intestinal flukes that can cause enteritis in fish-eating birds, dogs and other mammals, fish being the intermediate hosts.

cryptosporidiosis (avian/human/veterinary parasitology) Infection with *Cryptosporidium* spp. which in all species causes diarrhoea (*USA*: diarrhea) in the newborn. The protozoan may depend on the prior presence of another pathogen, *e.g.* rotavirus, to exert its pathogenicity. The critical lesion in the disease is villous atrophy causing a malabsorption defect. *Cryptosporidium* also infects trachea, cloaca, bursa of fabricius, and conjunctival sacs of birds, the stomach of mice and snakes and the bile duct of monkeys, and immunodeficient foals.

See also Cryptosporidium.

Cryptosporidium (parasitology) A coccidial protozoan parasite in most species. A member of the family Eimeriidae.

Cryptosporidium bayleyi (avian parasitology) A species of the coccidial protozoan parasite *Cryptosporidium* that may be found in birds.

Cryptosporidium crotalis (veterinary parasitology) A species of the coccidial protozoan parasite *Cryptosporidium* that may be found in reptiles.

Cryptosporidium meleagridis (avian parasitology) A species of the coccidial protozoan parasite *Cryptosporidium* that may be found in birds.

Cryptosporidium muris (veterinary parasitology) A species of the coccidial protozoan parasite *Cryptosporidium* that may be found in mammals.

Cryptosporidium nasorum (fish parasitology) A species of the coccidial protozoan parasite *Cryptosporidium* that may be found in fish.

Cryptosporidium parvum (human/veterinary parasitology) A species of the coccidial protozoan parasite *Cryptosporidium* that may be found in mammals.

Cryptosporidium serpentis (veterinary parasitology) A species of the coccidial protozoan parasite *Cryptosporidium* that may be found in reptiles.

crystal (chemistry) A naturally produced angular solid of definite form.

crystal violet (chemistry) A brilliant organic deep purple dye.

crystallog. (chemistry) An abbreviation of 'crystallog'raphy.

crystalloid (physics) Substance that is not a colloid and can therefore pass through a semipermeable membrane.

crystd (chemistry) An abbreviation of 'cryst'allise'd'.

crystn (chemistry) An abbreviation of 'cryst'allisatio'n'.

Cs (chemistry) Chemical symbol for caesium (*USA*: cesium).

CS (physiology/biochemistry) An abbreviation of 'C'onditioned 'S'timulus.

CSA (1 quality standards; 2 Immunology) 1 An abbreviation of 'C'anadian 'S'tandards 'A'ssociation. 2 An abbreviation of 'C'olony 'S'timulating 'A'ctivity.

CSF (1 anatomy; 2 physiology) 1 An abbreviation of 'C'erebro'S'pinal 'F'luid; 2 An abbreviation of 'C'olony-'S'timulating 'F'actor.

CSLT (society) An abbreviation of 'C'anadian 'S'ociety of 'L'aboratory 'T'echnologists.

CSM (medical) An abbreviation of 'C'erebro'S'pinal 'M'eningitis.

CSO (education) An abbreviation of 'C'hief 'S'cientific 'O'fficer.

CST (education) An abbreviation of 'C'ollege of 'S'cience and 'T'echnology.

CT (radiology) An abbreviation for 'C'omputerized 'T'omography.

ctenidium (parasitological anatomy) A spine which occurs in rows on the heads of fleas which is useful for morphological identification. Also termed combs.

Ctenocephalides (veterinary parasitology) A genus of fleas of the order Siphonaptera. Species include *Ctenocephalides canis* (dogs and related species), *Ctenocephalides felis*, *Ctenocephalides felis felis*, *Ctenocephalides felis strongylus*, *Ctenocephalides felis damarensis* and *Ctenocephalides felis orientalis*, which infest cats, dogs and related small animals.

Ctenocephalides canis (veterinary parasitology) A species of flea whose principal hosts are dog and fox.

Ctenocephalides felis (human/veterinary parasitology) A species of flea whose principal hosts are cat, dog, rarely humans, primates and rodents.

CTP (physiology/biochemistry) An abbreviation of 'Cytidine 'T'ri'P'hosphate.

Cu (chemistry) Chemical symbol for copper. (The symbol is derived from Latin, 'cuprum'.)

cu. (measurement) An abbreviation of 'cu'bic.

cu. ft (measurement) An abbreviation of 'cu'bit 'f'ee't'/'f'oo't'.

cu. in. (measurement) An abbreviation of 'cu'bic 'in'ch.

cubic (cu.) (measurement) Of three dimensions.

culdoscope (equipment) An endoscope used in culdoscopy.

Culex (veterinary parasitology) A genus of mosquitoes found throughout the world; cause insect worry and many species transmit various disease-producing agents, *e.g.* microfilariae and apicomplexan parasites.

Culex pipiens (parasitology) A species of mosquito in the genus *Culex* that transmits the virus fowlpox.

Culex pipiens quinquefasciatus (avian parasitology) A species of mosquito in the genus *Culex* that is a serious pest of poultry and carrier of a number of poultry diseases.

Culex tarsalis (veterinary parasitology) A species of mosquito in the genus *Culex* that transmits western equine encephalomyelitis.

Culex tritaeniohynchus (parasitology) A species of mosquito in the genus *Culex* that transmits Japanese encephalitis virus.

Culicidae (parasitology) *See Culex*; *Anopheles*.

culicide (control measures) A substance that destroys mosquitoes.

culicifuge (control measures) A substance that repels mosquitoes.

culicine (parasitology) 1 Any member of the genus *Culex* or related genera. 2 Pertaining to, involving, or affecting mosquitoes of the genus *Culex* or related species.

culicoid (parasitology) *See Culicoides*.

Culicoides (veterinary parasitology) A large genus of biting midges belonging to the family Ceratopogonidae. They act as vectors of bluetongue, ephemeral fever, Akabane virus of cattle, epizootic haemorrhagic (*USA*: hemorrhagic) disease of deer, and African horse sickness and many are intermediate hosts of filarioid nematodes. Many are causes of cutaneous hypersensitivity in horses (Queensland or sweet itch).

Culicoides brevitarsis (veterinary parasitology) A species of biting midges in the genus *Culicoides* that is one of the vectors of bluetongue virus and that causes hypersensitivity to its bite (Queensland itch). Also termed: *Culicoides robertsi*.

See also sweet itch.

Culicoides furens (parasitology) A species of biting midges in the genus *Culicoides* that is the intermediate host of *Mansonella ozzardi*.

Culicoides grahami (parasitology) A species of biting midges in the genus *Culicoides* that is the intermediate host of *Dipetalonema perstans* and *Dipetalonema streptocerca*.

culicoides hypersensitivity (parasitology) *See* sweet itch.

Culicoides nubeculosus (parasitology) A species of biting midges in the genus *Culicoides* that is the intermediate host of *Onchocerca cervicalis*.

Culicoides pungens (parasitology) A species of biting midges in the genus *Culicoides* that is the intermediate host of *Onchocerca gibsoni*.

Culiseta (avian/veterinary parasitology) A genus of mosquitoes which act as vectors for equine encephalomyelitis. Species include *Culiseta melanura* which transmits American encephalomyelitis virus, that does not feed on large mammals but does on water birds.

cuniculus (parasitology) Plural: cuniculi [Latin] A burrow in the skin made by the mange mite, *Sarcoptes scabiei*.

cupric (chemistry) Bivalent copper. Also termed: copper(II); copper(2+).

cuprite (chemistry) Mineral form of copper(I) oxide.

cuprous (chemistry) Monovalent copper. Also termed: copper(I); copper(1 +).

curie (Ci) (measurement) Measure of radioactivity. 1 Ci = 3.700×10^{10} disintegrations per second. It was named after the Polish-born French scientist Marie Curie (1867–1934). It has been replaced in si units by the becquerel.

curium (Cm) (chemistry) Radioactive element in Group IIIA of the periodic table (one of the actinides), made by the alpha-particle bombardment of plutonium-239. It has several isotopes, with half-lives up to 1.7×10^7 years. At. no. 96; r.a.m. 247 (most stable isotope).

cusp (1 mathematics; 2 anatomy) 1 In co-ordinate geometry, point on a curve where it crosses itself, the two branches being on opposite sides of a common tangent. 2 The pointed part of a tooth.

cutaneous habronemiasis (veterinary parasitology) A disease of horses that is manifested by granulomatous lesions caused by the invasion of skin wounds or excoriations by the larvae of *Habronema* spp. and *Draschia megastoma*. Also termed: summer sore, bursati, granular dermatitis and jack sores.

See also swamp cancer.

cutaneous larva migrans (human/veterinary parasitology) Creeping eruption; a convoluted, thread-like skin eruption in humans and other species which appears to migrate. It is caused by the burrowing beneath the skin of roundworm larvae, in particular those of *Ancylostoma*, *Strongyloides* and *Gnathostoma* spp. *Ancylostoma braziliense*, *Ancylostoma caninum*, *Bunostomum phlebotomum* can cause the disease.

cutaneous myiasis (veterinary parasitology) Infestation of devitalized skin, skin covered by hair or wool fouled by faeces (*USA*: feces) or urine, or skin wound, by maggots of *Lucila* spp., *Phormia* spp., or *Calliphora* spp. Sheep are especially susceptible and large areas of skin may be destroyed and the sheep die as a result. Also termed calliphorine myiasis, blowfly myiasis or strike and struck.

cutaneous stephanofilarosis (veterinary parasitology) Infestation with *Parafilaria multipapillosa* causes subcutaneous nodules in horses, *Parafilaria bovicola* causes similar lesions in cattle, *Suifilaria suis* does the same in pigs, *Stephanofilaria dedoesi* causes dermatitis in cattle (cascado). *Stephanofilaria kaeli* and *Stephanofilaria assamensis*, cause dermatitis in cattle (humpsore), *Stephanofilaria zaheeri* causes contagious otorrhea (earsore), a dermatitis around the ear in buffalo, *Stephanofilaria stilesi* and *Stephanofilaria okinawaensis* also cause dermatitis in cattle. The dermatitis is manifested by small papules that enlarge to form itchy, scabby lesions that suffer much rubbing. The lesions occur at various sites on the body depending on the species of worm. *See Parafilaria*, *Stephanofilaria*, *Elaeophora schneideri*.

Cuterebra (veterinary parasitology) Large flies whose larvae are parasitic, mostly on wild rodents but cats and dogs are occasionally infected. The larvae burrow under the skin and cause cyst-like cavities. Species includes *Cuterebra americana*, *Cuterebra buccata*, *Cuterebra emasculator*, *Cuterebra lepivora*. Also termed: cuterebra flies.

cuticle (parasitological anatomy) Cuticula. The external non-cellular hyaline layer covering the nematodes.

Cutifilaria (parasites) A genus of nematodes in the family Onchocercidae. Species includes *Cutifilaria wenki* which may be found in the skin of deer.

CV (1 measurement; 2 medical; 3 physiology) 1 An abbreviation of 'C'alorific 'V'alue. 2 An abbreviation of 'C'ardio'V'ascular. 3 An abbreviation of 'C'losing 'V'olume.

CVA (medical) An abbreviation of 'C'erebro-'V'ascular 'A'ccident.

CVR (physiology) An abbreviation of 'C'erebral 'V'ascular 'R'esistance.

cw (general terminology) An abbreviation of 'c'lock'w'ise.

CY (chemistry) An abbreviation of 'CY'anide.

cy (general terminology) An abbreviation of 'c'apacit'y'.

cyan(o)- (terminology) A word element. [German] Prefix denoting *blue*.

cyanacethydrazide (pharmacology) An anthelmintic used in the treatment of lungworms. Also termed cyacetazide.

cyanide (chemistry) Compound containing the cyanide ion (CN-); a salt of hydrocyanic acid (HCN). All cyanides are highly poisonous.

cyanoferrate (chemistry) *See* ferricyanide; ferrocyanide.

cyanogen (chemistry) NCCN Colourless (*USA*: colorless) highly poisonous gas made by the action of acids on cyanides.

Cyathocephalus (parasitology) A tapeworm genus of the family Cyathocephalidae.

Cyathocephalus truncatus (fish parasite) A species of tapeworm in the genus *Cyathocephalus* that may be found in fish and that can cause retardation of growth.

Cyathospirura (parasites) A spiruroid nematode that causes granulomatuus lesions in the stomachs of cats and foxes. Species include *Cyathospirura seurati*.

Cyathostoma (avian parasitology) A genus of the roundworm family Syngamidae. Species includes *Cyathostoma bronchialis*, *Cyathostoma brantae*, *Cyathostoma lari*, and *Cyathostoma variegation* which may be found in the respiratory tract of birds.

cyathostomiasis (parasitology) Massive infestations of small strongyles, including *Cyathostomum*, *Cylicocyclus*, *Cylicodontophorus*, *Cylindropharynx*, *Cylicostephanus*, *Poteriostomum*, *Gyalocephalus*, and *Caballonema*. They all cause profuse diarrhoea (*USA*: diarrhea), and in some cases death.

Cyathostominae (veterinary parasitology) A subfamily of the strongylid roundworms of horses which replaces the original genus of *Trichonema*. It includes the genera *Cyathostomum*, *Cylicocyclus*, *Cylicodontophorus*, *Cylicostephanus*, *Poteriostomum*, *Petrovinema*, *Gyalocephalus*, *Cylindropharynx* and *Caballonema*.

cyathostomosis (parasitology) The disease state caused by *Cyathostomum* spp. infection.

Cyathostomum (veterinary parasitology) A genus of strongylid worms of the family Strongylidae. Species includes *Cyathostomum catinatum*, *Cyathostomum coronatum*, *Cyathostomum labiatum*, *Cyathostomum labratum*, *Cyathostomum sagittatum*, *Cyathostomum tetracanthum* which may be found in the large intestine of horses and other equids. Heavy infestations can cause severe diarrhoea (*USA*: diarrhea).

cycl(o)- (terminology) A word element. [German] Prefix denoting *round, recurring, ciliary body of the eye*.

cycle (physics) One of a repeating series of similar changes, *e.g.* in a wave motion or vibration. One cycle is equal to the period of the motion; the number of cycles per unit time is its frequency. A frequency of 1 cycle per second = 1 hertz.

cyclic (chemistry) Describing the molecule of any (usually organic) compound whose atoms form a ring. Benzene and the cycloalkanes are simple cyclic compounds.

cyclicAMP (physiology/biochemistry) An abbreviation of 'cyclic''A'denosine3',5'-'M'onos'P'hosphate.

cycloalkane (chemistry) Saturated cyclic hydrocarbon whose molecule has a ring of carbon atoms, general formula C_nH_{2n}; *e.g.* cyclohexane, C_6H_{12}.

Cyclocoelidae (avian parasitology) A family of large flukes (digenetic trematodes) that are parasitic in body and nasal cavities, and air sacs of aquatic birds. Species include *Tracheophilus cymbius* (*Tracheophilus sisowi*) which may be found in ducks.

Cyclophyllidea (avian/veterinary parasitology) An important order of cestodes in the class Cestoda, that are parasitic in birds and mammals.

cyclopropagative transmission (parasitology) Transmission in which the agent undergoes both development and multiplication in the transmitting vehicle.

Cyclops (parasitology) A genus of minute crustaceans with terrestrial life cycles, some species of which act as hosts of *Diphyllobothrium* and *Dracunculus* spp. Also termed: water flea.

Cyclotogaster heterographa (avian parasitology) A species belonging to the order Ischocera (insects) that affects fowl and partridge. Also termed head louse.

cyclozoonosis (parasitology) A zoonotic disease that requires at least two species of vertebrates as definitive and intermediate hosts. Examples include hydatid disease (*Echinococcus granulosus*) and trichinosis (*Trichinella spiralis*).

cyfluthrin (control measures) A fluorinated pyrethroid, used in ear tags on cattle to control flies and ticks.

cyhalothrin (control measures) A synthetic pyrethroid used to control lice and ked on sheep.

cyhexatin (pharmacology) An acaricide used in horticulture and reputed to cause congenital defects in experimental animals.

Cylicocyclus (veterinary parasitology) One of the eight genera of the strongylid nematodes in the subfamily Cyathostominae. Species includes *Cylicocyclus elongatus*, *Cylicocyclus insignis*, *Cylicocyclus leptostomus*, *Cylicocyclus nassatus* and *Cylicocyclus ultrajectinus* which may be found in the large intestine of horses and other Equidae.

Cylicodontophorus (veterinary parasitology) One of the genera of nematodes within the strongylid subfamily Cyathostominae. Species includes *Cylicodontophorus bicoronatus*, *Cylicodontophorus euproctus* and *Cylicodontophorus mettami* which may be found in the large intestine of horses.

Cylicospirura felineus (veterinary parasitology) A spiruroid worm associated with the formation of nodules in the stomach of domestic and wild felids.

Cylicostephanus (veterinary parasitology) One of the genera of strongylid nematodes in the subfamily Cyathostominae, previously included in the genus *Trichonema*. Species includes *Cylicostephanus calicatus*, *Cylicostephanus goldi*, *Cylicostephanus hybridus*, *Cylicostephanus longibursatus* and *Cylicostephanus minutus* which may be found in the large intestine of horses and other Equidae.

cylicostomes (parasitology) *See* Cyathostominae.

Cylindropharynx (parasitology) A genus of strongylid, nematodes in the subfamily Cyathostominae.

cypermethrin (pharmacology/control measures) A synthetic pyrethroid used as an insecticide and acaricide.

Cyrnea (parasites) A genus of spiruroid worms in the family Habronematidae.

Cyrnea colini (avian parasitology) A species of spiruroid worms in the genus *Cyrnea* that similarly to *Cyrnea piliata* may be found in the proventriculus and gizzard of birds.

Cyrnea piliata (avian parasitology) A species of spiruroid worms in the genus *Cyrnea* that similarly to *Cyrnea colini* may be found in the proventriculus and gizzard of birds.

cyst (1 medical; 2 parasitological physiology) 1 A closed epithelium-lined sac or capsule containing a liquid or semi-solid substance. Most cysts are harmless but they occasionally may change into malignant growths, become infected, or obstruct a gland. There are four main types of cysts: retention cysts, exudation cysts, embryonic cysts and parasitic cysts. 2 A stage in the life cycle of certain parasites, during which they are enveloped in a protective wall. The cyst may be located in any organ of the body and is generally spherical in shape. The cyst wall is argyrophilic.

cyst mites (parasitology) *See Laminosioptes cysticola.*

cyst(o)- (terminology) A word element. [German] Prefix denoting cyst, bladder.

cystacanth (insect parasitology) An intermediate stage of *Macracanthorhyncus* spp. in arthropods.

cystectomy (medical) 1 Excision of a cyst. 2 Excision or resection of the urinary bladder.

cysteine (biochemistry) $HSCH_2CH(NH_2)COOH$ Amino acid found in proteins.

cystic (medical) 1 Pertaining to or containing cysts; *See* cyst. 2 Pertaining to the urinary bladder or to the gallbladder.

cysticercoid (parasitology) A stage in the life cycle of pseudophyllidean cestodes. One of the forms of metacestodes in which there is a single, retracted scolex withdrawn into a small vesicle which has a very small cavity, *e.g. Dipylidium caninum*, that may be found in invertebrates such as oribatid mites, fleas, etc.

cysticercosis (parasitology) Infection with cystercerci in the intermediate hosts of a number of species-specific cestodes. The disease is largely symptomless except for cysticercosis caused by *Cysticercus cellulosae*, but has some food hygiene importance. Also termed: sheep measles, beef measles. *See Cysticercus.*

cysticercus (parasitology) Plural: cysticerci [German] The larval form of certain tapeworms. The term is often used synonymously with bladderworm, but not all bladderworms are cysticerci.

See also cysticercosis.

Cysticercus bovis (parasitological physiology) The larval stage of *Taenia saginata*, a tapeworm of humans. The cysticerci are found in the muscles and other tissues of cattle and humans are infected by eating uncooked beef. Also termed: beef measles.

Cysticercus cellulosae (parasitological physiology) The larval stage of *Taenia solium*, a tapeworm of humans. The cysticerci are found in the skeletal and cardiac muscles of the pig and in the muscles and central nervous system of humans. Humans are infected by eating uncooked pork. Also termed: pork measles.

Cysticercus fasciolaris (parasitological physiology) The larval stage of *Taenia taeniaeformis*, a tapeworm of cats and wild Felidae and related species. The cysticerci may be found in the liver of rodents.

Cysticercus ovis (parasitological physiology) The larval stage of *Taenia ovis*, a tapeworm of dogs and wild carnivores. The cysticerci are found in the skeletal and cardiac muscles of sheep and goats. Dogs are infected by ingesting raw infected meat.

Cysticercus pisiformis (parasitological physiology) The larval stage of *Taenia pisiformis*, a tapeworm of dogs and wild carnivores. The cysticerci may be found in the peritoneal cavity of rabbits and hares.

Cysticercus tarandi (parasitological physiology) The larval stage of *Taenia krabbei*, a cestode parasite of dogs and wild carnivores. The cysticerci may be found in the muscles of wild ruminants.

Cysticercus tenuicollis (parasitological physiology) The larval stage of the tapeworm *Taenia hydatigena*, a tapeworm of dogs and wild carnivores. The cysticerci may be found in the liver and on the peritoneum in sheep but also in other ruminants, including wild ones and pigs. Infection in the dog occurs when infected offal is eaten raw. Also termed: long-necked bladderworm.

cystiform (biology) Resembling a cyst.

cystigerous (biology) Containing cysts.

cystine (biochemistry) $(SCH_2CH(NH_2)COOH)_2$ Dimeric form of the amino acid cysteine, found in keratin.

cystitis (medical) An inflammation of the urinary bladder.

cyt(o)- (terminology) A word element. [German] Prefix denoting *a cell*.

Cystocaulus (veterinary parasitology) A nematode genus of the family Protostrongylidae of worms that includes *Cystocaulus nigrescens* (*Cystocaulus ocreatus*) which may be found in the lungs of sheep and goats.

cystoid (parasitology) 1 Resembling a cyst. 2 A cystlike, circumscribed collection of softened material, having no enclosing capsule.

Cystoisospora (parasitology) A relatively new genus including many coccidia previously classified as *Isospora* spp. with indirect life cycles.

Cystoisospora burrowsi (veterinary parasitology) Coccidia of dogs which may cause enteritis. *See Cystoisospora.*

Cystoisospora canis (veterinary parasitology) Coccidia of dogs which may cause enteritis. *See Cystoisospora.*

Cystoisospora felis (veterinary parasitology) Coccidia of cats which may cause haemorrhagic enteritis (*USA*: hemorrhagic enteritis). *See Cystoisospora.*

Cystoisospora heydorni (veterinary parasitology) Coccidia of dogs which may cause enteritis. *See Cystoisospora.*

Cystoisospora ohioensis (veterinary parasitology) Coccidia of dogs which may cause enteritis. *See Cystoisospora.*

Cystoisospora rivolta (veterinary parasitology) Coccidia of cats which may cause haemorrhagic enteritis (*USA*: hemorrhagic enteritis). *See Cystoisospora.*

Cystoisospora wallacei (veterinary parasitology) Coccidia of dogs which may cause enteritis. *See Cystoisospora.*

cystozoites (parasitological physiology) A term used in relation to *Toxoplasma* and *Sarcocystis*. Stages found in the cysts. As these multiply slowly they are also known as bradyzoites. These are morphologically similar to endozoites but contain a large amount of polysaccharide material and, therefore, react strongly to staining by the periodic acid-Schiff (PAS) method.

Cytauxzoon (parasitology) A genus of protozoan parasites of the family Theileriidae.

Cytauxzoon felis (veterinary parasitology) A species of protozoan parasites in the genus *Cytauxzoon* that causes a fatal disease of cats.

Cytauxzoon strepsicerosi (veterinary parasitology) A species of protozoan parasites in the genus *Cytauxzoon* that, similarly to *Cytauxzoon sylvicaprae* and *Cytauxzoon taurotragi*, multiply by fission in erythrocytes and which have been observed in wild animals and in cats.

Cytauxzoon sylvicaprae (veterinary parasitology) A species of protozoan parasites in the genus *Cytauxzoon* that, similarly to *Cytauxzoon strepsicerosi* and *Cytauxzoon taurotragi*, multiply by fission in erythrocytes and which have been observed in wild animals and in cats.

Cytauxzoon taurotragi (veterinary parasitology) A species of protozoan parasites in the genus *Cytauxzoon* that, similarly to *Cytauxzoon sylvicaprae* and *Cytauxzoon strepsicerosi*, multiply by fission in erythrocytes and which have been observed in wild animals and in cats.

cytauxzoonosis (veterinary parasitology) A disease of cats caused by *Cytauxzoon* spp. that is transmitted probably by ixodid ticks and characterized by fever, anaemia (*USA*: anemia) and jaundice.

-cyte (terminology) A word element. [German] Suffix denoting *a cell*.

cythioate (veterinary parasitology) An organophosphorus anthelmintic administered orally to dogs and cats for control of fleas.

cytochrome (biochemistry) Respiratory pigment found in organisms that use aerobic respiration.

cytochrome b (biochemistry) A microsomal cytochrome involved in such metabolic activities as fatty acid desaturation.

cytochrome c (biochemistry) A respiratory chain enzyme having the same prosthetic group as haemoglobin (*USA*: hemoglobin).

cytochrome oxidase stain (histology) Enzyme histochemical staining technique.

Cytodites (avian parasitology) A genus of mites that are parasites in the air sacs of birds; the air sac mite. They do not usually have any apparent pathogenic effect. Species includes *Cytodites nudus*.

cytogenetics (genetics) The scientific study of the number, behaviour (*USA*: behavior) and morphology of chromosomes as a means of understanding the genetics of an organism.

cytokines (immunology) Substances, usually peptides or proteins, that act as growth factors, *i.e.* affect the growth, division or differentiation of cells. Examples include interferon and platelet derived growth factor.

cytologist (cytology) A specialist in cytology.

cytology (cytology) The study of living cells, their origin, structure, function and pathology.

cytoplasm (cytology) Protoplasm of a cell other than that of the nucleus.

cytosine (biochemistry) Colourless (*USA*: colorless) crystalline compound, derived from pyrimidine. It is a major constituent of DNA and RNA. Also termed: 4-amino-2(H)-pyrimidinone.

cytosol (histology/cytology) A watery material remaining when all the organelles and membranes have been removed from an homogenate of cells by centrifugation.

cytotoxic (1 toxicology; 2 pharmacology) 1 Poisonous to cells. 2 Describing a drug that destroys or prevents the replication of cells.

D

D & C (medical) An abbreviation of 'D'ilatation 'and' 'C'urettage.

D (1 chemistry; 2 biochemistry) 1 Chemical symbol for deuterium. 2 Symbol for aspartic acid.

d (1 measurement; 2 physics; 3 medical; 4 general terminology) 1 An abbreviation of prefix 'd'eci, 10^{-1}. 2 An abbreviation of 'd'euteron. 3 An abbreviation of 'd'ied. 4 An abbreviation of 'd'ay.

D- (physiology/biochemistry) A symbol for the geometric isomer of L- form of chemical compound.

Da (chemistry) Chemical symbol for davyum.

da- (measurement) Prefix meaning 10, used with units of measurement. Also termed deca-.

dacry(o)- (terminology) A word element. [German] Prefix denoting *tears* or the *lacrimal system*.

dactyl(o)- (terminology) A word element. [German] Prefix denoting *a digit*.

Dactylogyrida (fish parasitology) An order of flatworms in class Monogenea and the phylum Platyhelminthes. Six families are contained within the order Dactylogyrida: the Acanthocotylidae, which are normally parasites of the body surface of elasmobranchs and of which *Pseudacanthocotyla* is a member; the Pseudomurraytrematidae, found on the gills of freshwater catastomid fish and of which *Anonchohaptor* is a member; the Ancyrocephalidae, found typically but not exclusively on Perciformes and of which *Amphibdelloides*, a parasite of electric rays, is a member; the Capsalidae which are found on marine and bracish water teleosts and of which *Megalocotyle* and *Nitzchia* are members; the Tetraonchidae which are often found on gill of salmonids and esocids; and the Dactylogyridae, usually found on Cypriniformes and of which *Lamellodiscus* is a member.

Dactylogyrus (fish parasitology) A genus of monogenetic flukes of the family Dactylogyridae that infest fish.

Dactylogyrus extensus (fish parasitology) A species of monogenetic flukes of the genus *Dactylogyrus* that, similarly to *Dactylogyrus vastator*, cause an important parasitic disease of the gills of marine and freshwater fish.

Dactylogyrus vastator (fish parasitology) A species of monogenetic flukes of the genus *Dactylogyrus* that, similarly to *Dactylogyrus extensus*, cause an important parasitic disease of the gills of marine and freshwater fish.

DAG (biochemistry) An abbreviation of 'D'i'A'cyl'G'lycerol.

dag (measurement) An abbreviation of 'd'ek'ag'ram (10 grams). Also termed decagram.

dagger nematode (plant parasitology) *See Xiphinema index*.

daisy wheel (computing) The printing head of an early model of letter quality printer.

dalton (measurement) An alternative term for atomic mass unit.

Dalton's atomic theory (physics) Theory that states that matter consists of tiny particles (atoms), and that all the atoms of a particular element are exactly alike, but different from the atoms of other elements in behaviour (*USA*: behavior) and mass. The theory also states that chemical action takes place as a result of attraction between atoms, but it fails to account satisfactorily for the volume relationships that exist between combining gases. It was proposed by the British scientist John Dalton (1766–1844).

Dalton's law of partial pressures (physics) In a mixture of gases, the pressure exerted by one of the component gases is the same as if it alone occupied the total volume.

Damalinia (parasitology) A genus of mammal lice of the superfamily Ischnocera. Also termed *Bovicola*.

Damalinia bovis (veterinary parasitology) A species of lice of the genus *Damalinia* that may be found in cattle.

Damalinia caprae (veterinary parasitology) A species of lice of the genus *Damalinia* that, similarly to *Damalinia crassiceps*, may be found in goats.

Damalinia crassiceps (veterinary parasitology) A species of lice of the genus *Damalinia* that, similarly to *Damalinia caprae*, may be found in goats.

Damalinia equi (veterinary parasitology) (synonym: *Damalinia pilosus*) A species of lice of the genus *Damalinia* that may be found in horses.

Damalinia limbata (veterinary parasitology) A species of lice of the genus *Damalinia* that may be found in angora goats.

Damalinia ovis (veterinary parasitology) A species of lice of the genus *Damalinia* that may be found in sheep.

damping (physics) Reduction in the amplitude of a waveform or oscillation.

dander (immunology) Small scales of animal skin, hair or feathers. Dander commonly cause allergic effects, especially asthma.

Daniell cell (physics) Electrolytic cell that consists of a zinc half-cell and a copper half-cell, usually arranged as a zinc cathode and a copper anode dipping into an electrolyte of dilute sulphuric acid (*USA*: sulfuric acid).

danofloxacin (pharmacology) Antimicrobial agent of the fluoroquinolone group with a wide range of activity including against mycoplasmas.

DAP&E (education) An abbreviation of 'D'iploma in 'A'pplied 'P'arasitology and 'E'ntomology.

Daphnia magna (toxicology) A standard invertebrate used for aquatic toxicology tests. The common name is 'water flea'.

Daphnia pulex (avian parasitology) A water flea, one of the intermediate hosts of *Echinuria uncinata*, a roundworm of ducks.

DApp.Sc. (education) An abbreviation of 'D'octor of 'App'lied 'Sc'ience.

daraf (measurement) Unit of the elastance of an electrical component. it is the reciprocal of capacitance (the word is farad backwards).

darkfield microscope (microscopy) A microscope used for examining unstained, often living cells, in which light is only directed into the objective lens if it is deflected by an object in its path. The object is thus viewed as a white structure in an otherwise black (darkfield) background.

darkling beetle (avian parasitology) This and other mealworms are common inhabitants of poultry houses and are suspected of aiding in the transmission of Marek's disease and other virus diseases and of attacking the skin of all types of birds.

darwin (measurement/genetics) A unit of evolutionary change. A darwin is equal to a change, increase or decrease, in any one character by a factor of 2.7 per million years. As this is a very fast rate of change, most evolutionary change is expressed in millidarwins.

Darwinism (biology) Theory of evolution which states that living organisms arise in their different forms by gradual change over many generations, and that this process is governed by natural selection. It was proposed by the British naturalist Charles Darwin (1809–82).

Dasypsyllus gallinulae (avian parasitology) A species of flea in the order Siphonaptera that may be found on wild birds.

data (statistics) Collection of information, often referring to results of a statistical study or to information supplied to, processed by or provided by a computer. The plural of 'datum'.

data bank (computing) An alternative term for a data base.

data base (1, 2 computing) 1 Organized collection of data that is held on a computer, where it is regularly updated and can easily be accessed (often by many users). Also termed: data bank. 2 Applications program that controls and makes use of a data base.

data base management system (DBMS) (computing) Computer software designed to establish and maintain a data base.

data processing (DP) (computing) The process of inputting, storing, and manipulating large amounts of information, using a computer.

data transmission (computing) Transfer of data between outstations and a central computer or between different computer systems.

dative bond (chemistry) An alternative term for a co-ordinate bond.

dauer larva (parasitology) A developmentally arrested dispersal stage that may be formed under conditions of starvation or overcrowding. The word 'dauer' comes from the German: 'Dauer' (noun), meaning *endurance* or *duration*. Whereas adult worms have an average life span of two to three weeks, dauer larvae survive for months. If given food after such a period they develop into adults that have a normal life span. For this reason the dauer larva is considered non-aging. Dauers look different from the other larval stages in that they are thinner relative to their length and their bodies are darker due to nutrient storage granules in the intestinal and hypodermal cells. They also have relatively impermeable cuticles, and are non-feeding.

daughter cell (cytology) One of the two cells produced when a cell divides by, *e.g.* mitosis.

daughter cyst (parasitological physiology) A term used in relation to cestodes. A cyst formed by endogenous or exogenous budding from the germinal layer of a hydatid.

daughter nucleus (cytology) One of the two nuclei produced when the nucleus of a cell divides.

Davainea (parasitology) A genus of tapeworms of the family Davaineidae.

Davainea proglottina (avian parasitology) A species of tapeworm of the genus *Davainea* that causes severe enteritis in fowls and other gallinaceous birds.

dB (measurement) An abbreviation of 'd'eci'B'el.

db (measurement) An abbreviation of 'd'eci'b'el. Also termed dB.

dbl (general terminology) An abbreviation of 'd'ou'bl'e.

DBMS (computing) An abbreviation of 'D'ata 'B'ase 'M'anagement 'S'ystem.

DBP (physiology/biochemistry) An abbreviation of 'Vitamin 'D'-'B'inding 'P'rotein.

DC (physics) An abbreviation of 'D'irect 'C'urrent.

DCl. Sc. (education) An abbreviation of 'D'octor of 'Cl'inical 'Sc'ience.

dcm (measurement) An abbreviation of 'd'e'c'a'm'etre (*USA*: decameter).

DCP (education) An abbreviation of 'D'iploma of 'C'linical 'P'athology.

DCPath. (education) An abbreviation of 'D'iploma of the 'C'ollege of 'Path'ologists.

DCSO (governing body) An abbreviation of 'D'eputy 'C'hief 'S'cientific 'O'fficer.

DDAVP (biochemistry) An abbreviation of 1-'D'eamino-8-'D'-'A'rginine'V'aso'P'ression.

DDR (education) An abbreviation of 'D'iploma in 'D'iagnostic 'R'adiology.

DDT (toxicology) Abbreviation of 'D'ichloro-'D'iphenyl-'T'richloroethane.

de- (terminology) A word element. [Latin] Prefix denoting *down from*; sometimes negative or privative, and often intensive.

dead (medical) Deceased.

deaminase (biochemistry) Enzyme that catalyses the removal of an amino group (-NH_2) from an organic molecule.

deamination (biochemistry) Enzymatic removal of an amino group (-NH_2) from a compound. The process is important in the breakdown of amino acids in the liver and kidney. Ammonia formed by deamination is converted to urea and excreted.

debug (computing) To remove a fault in a computer program; hence, by analogy, to correct any form of fault.

Debye-Hückel theory (physics) Theory that explains variations from the ideal behaviour (*USA*: behavior) of electrolytes in terms of interionic attraction, and assumes that electrolytes in solution are completely dissociated into charged ions. It was named after the Dutch physicists Peter Debye (1884–1966) and Erich Hückel (1896–).

dec. (1 medical; 2 measurement) 1 An abbreviation of 'dec'eased. 2 An abbreviation of 'dec'imetre (*USA*: decimeter).

deca- (terminology) A word element. [German] Prefix meaning *10 times*, used with units of measurement. Also termed da-; deka-.

decalcification (1 biochemistry; 2 histology) 1 Loss of calcium and other mineral salts from the normally mineralized tissues, bone and teeth. 2 A histological laboratory technique.

decant (chemistry) Carefully pour away the liquid above a precipitate, once it has settled.

decay (1 biology; 2 radiology) 1 Natural breakdown of an organic substance; decomposition. 2 Breakdown, through radioactivity, of a radioactive substance. The rate of such decay is typically an exponential function.

See also half-life.

deceased (dec.) (medical) Pertaining to a person who has died.

deci- (terminology) A word element. [Latin] Prefix denoting *one-tenth*; used to indicate one-tenth (10^{-1}) of the unit designated by the root with which it is combined.

decibel (db or dB) (measurement) Unit used for comparing power levels (on a logarithmic scale); one-tenth of a bel. It is commonly used in comparisons of sound intensity.

See also bel.

decigram (dg) (measurement) One-tenth of a gram.

decim. (measurement) An abbreviation of 'decim'etre (*USA*: decimeter).

decimal system (mathematics) Number system that uses the base 10; *i.e.* it uses the digits 1 to 9 and 0.

See also binary notation.

decimetre (dm; decim.) (measurement) One-tenth of a metre (*USA*: meter). (*USA*: decimeter).

decomposer (biology) Saprophytic organism that breaks down organic materials into simple molecules.

decomposition (1 chemistry; 2 biology) 1 Breaking down of a chemical compound into its component parts (elements or simpler substances), often brought about by heat, light or electrolysis. *See also* double decomposition. 2 Rotting of a dead organism, often brought about by bacteria or fungi.

Decrusia (parasitology) A genus of strongylid worms from the family Strongylidae.

Decrusia additicta (veterinary parasitology) A species of strongylid worms of the genus *Decrusia* that may be found in the large intestine of Indian elephants.

deer fly (veterinary parasitology) *See Chrysops.*

defaecate (biology) The act of defaecation (*USA*: defecation). (*USA*: defecate).

defaecation (biology) Elimination of wastes and undigested food, as faeces (*USA*: feces), from the rectum. (*USA*: defecation).

definite integral (mathematics) Integral that has exact limits.

deg. (measurement) An abbreviation of 'deg'ree.

deglutition (physiology) Another term for swallowing.

degradation (1 chemistry; 2 physics) 1 The conversion of a molecule into simpler components. 2 The irreversible loss of energy available to do work.

degree (1, 2 measurement) 1 Unit of difference in temperature used in temperature scales. 2 Unit derived by dividing a circle into 360 segments, used to measure angles and describe direction. It is subdivided into minutes and seconds (of arc). Both types of degrees have the symbol °.

degree of freedom (physics) One of several variable factors, *e.g.* temperature, pressure and concentration, that must be made constant for the condition of a system at equilibrium to be defined.

dehalogenation (biochemistry) The removal of halogen substituents from organic chemicals.

dehydrating agent (chemistry) An alternative term for a desiccant.

dehydration (1 chemistry; 2 medical) 1 Elimination of water from a substance or organism. 2 Lack of water in the body.

dehydrogenase (biochemistry) An enzyme that activates oxidation-reduction reactions by the removal of a pair of hydrogen atoms from a molecule.

deionization (chemistry) Method of purifying or otherwise altering the composition of a solution using ion exchange.

deka- (measurement) Prefix meaning 10, used with units of measurement. Also termed da; deca; deka.

Delafondia (parasitology) *See Strongylus.*

Deletrocephalus (veterinary parasitology) A nematode genus of the superfamily Strongyloidea, in the family Deletrocephalidae, that may be found in the large intestine of the rhea.

Delhi boil (parasitology) *See leishmaniasis.*

deliquescence (chemistry) Gradual change undergone by certain substances that absorb water from the atmosphere to become first damp and then aqueous solutions.

deliquescent (chemistry) Describing a substance that exhibits deliquescence.

delocalization (chemistry) Phenomenon that occurs in certain molecules, *e.g.* benzene. Some of the electrons involved in bonding the atoms are not restricted to one particular bond, but are free to move between two or more bonds. The electrical conductivity of metals is due to the presence of delocalized electrons.

delta ray (physics) Electron that is ejected from an atom when it is struck by a high-energy particle.

demephion (pharmacology/control measures) An organophosphorus compound used as an insecticide and acaricide.

demeton (control measures) An organophosphorus insecticide used as a treatment for sucking lice. It consists of a mixture of two compounds and toxicity varies considerably depending on their relative concentrations.

demise (medical) Death.

demodectic mange (veterinary parasitology) Mange which in all species is caused by species-specific *Demodex* spp. Characterized by folliculitis with hair loss and often pustule formation anywhere on the body, although the head, face, neck and shoulders are must often affected. The disease is most common in dogs where it is associated with an abnormality of cell-mediated immunity. Most often, there are one or a few areas of localized hair loss in young dogs which will heal spontaneously, although treatment is often given. Occasionally the disease is generalized, severe and resistant to all treatment. Examination of a skin scraping confirms the diagnosis. Also termed follicular mange; red mange; demodicosis.

Demodex (parasitology) A genus of mites in the family Demodicidae that are parasitic within the hair follicles of the host.

Demodex aries (veterinary parasitology) A species of mite of the genus *Demodex* that may be found in sheep.

Demodex bovis (veterinary parasitology) A species of mite of the genus *Demodex* that, similarly to *Demodex ghanensis*, may infest cattle.

Demodex caballi (veterinary parasitology) A species of mite of the genus *Demodex* that, similarly to *Demodex equi*, may infest horses.

Demodex canis (veterinary parasitology) A species of mite of the genus *Demodex* that may infest dogs.

Demodex caprae (veterinary parasitology) (synonym: *Demodex capri*) A species of mite of the genus *Demodex* that may infest goats.

Demodex cati (veterinary parasitology) A species of mite of the genus *Demodex* that may infest cats.

Demodex criceti (veterinary parasitology) A species of mite of the genus *Demodex* that may infest hamsters.

Demodex equi (veterinary parasitology) A species of mite of the genus *Demodex* that, similarly to *Demodex caballi*, may infest horses.

Demodex folliculorum (human parasitology) A species of mite of the genus *Demodex* that may infest humans.

Demodex ghanensis (veterinary parasitology) A species of mite of the genus *Demodex* that, similarly to *Demodex bovis*, may infest cattle.

Demodex muscardini (veterinary parasitology) A species of mite of the genus *Demodex* that may infest dormice.

Demodex ovis (veterinary parasitology) A species of mite of the genus *Demodex* that may infest sheep.

Demodex phylloides (veterinary parasitology) A species of mite of the genus *Demodex* that may infest pigs.

demodicosis (parasitology) Demodectic mange.

demogram (epidemiology) A report of demographic results, usually in grid form.

demographics (epidemiology) The graphic representation of demographic results.

demography (eidemiology) The statistical science dealing with populations, including matters of health, disease, births and mortality. Strictly speaking, the word refers to human populations but common usage includes lower animal populations.

denature (chemistry) To unfold the structure of the polypeptide chain of a protein by exposing it to temperature or extremes of pH. This results in the loss of biological activity and a decrease in solubility.

dendr(o)- (terminology) A word element. [German] Prefix denoting *tree, treelike*.

dendrite (biology) An elongated process from a nerve cell that projects it towards another nerve cell. Also termed: dendron.

Dendritobilharzia (parasitology) A genus of digenetic trematodes in the family Schistosomatidae.

Dendritobilharzia pulverulenta (avian parasitology) A species of digenetic trematodes in the genus *Dendritobilharzia* that may be found in the dorsal aorta of swans.

dendron (biology) An alternative term for dendrite.

denitrification (biology) Process that occurs in organisms, *e.g.* bacteria in soil, which breaks down nitrates and nitrites, with the liberation of nitrogen.

denominator (mathematics) Lower number of a fraction (the upper number of the numerator); *e.g.* in the fractions 2/3, 7/12 and 83/100, the denominators are 3, 12 and 100.

See also common denominator.

dens. (physics) An abbreviation of 'dens'ity.

density (1 physics; 2 statistics) 1 Mass of a unit volume of a substance. For an object of mass m and volume V, the density d is m/V. It is commonly expressed in units such as g cm^{-3} (although the SI unit is kg m^{-3}). 2 Number of items in a defined surface area (*e.g.* population density, charge density).

dent(o)- (terminology) A word element. [Latin] Prefix denoting *tooth, toothlike*.

DEnt. (education) An abbreviation of 'D'octor of 'Ent'omology.

deoxyribonucleic acid (DNA) (molecular biology) Nucleic acid that is usually referred to by its abbreviation. The long thread-like molecule consists of a double helix of polynucleotides held together by hydrogen bonds. DNA is found chiefly in chromosomes, and is the material that carries the hereditary information of all living organisms (although most, but not all, viruses have only ribonucleic acid, RNA).

deoxyribonucleotides (molecular biology) Compounds that are the fundamental units of DNA. Each nucleotide contains a nitrogenous base, a pentose sugar and phosphoric acid. The four bases characteristic of deoxyribonucleotides are adenine, guanine, cytosine and thymine.

See also deoxyribonucleic acid.

deoxyribose (biochemistry) The sugar that forms part of the structure of DNA.

dependence (statistics) Changes in one variable occurring systematically with changes in another.

depluming (avian parasitology) Removing the feathers of birds.

depluming itch (avian parasitology) *Cnemidocoptes laevis gallinae*. Also termed depluming mite; depluming scabies.

depluming mite (avian parasitology) *See* depluming itch.

depluming scabies (avian parasitology) *See* depluming itch.

depolarization (physics/chemistry) Removal or prevention of electrical polarity or polarization.

depression of freezing point (physics) Reduction of the freezing point of a liquid when a solid is dissolved in it. At constant pressure and for dilute solutions of a non-volatile solvent, the depression of the freezing point is directly proportional to the concentration of the solutes.

derivative (1 chemistry; 2 mathematics) 1 A compound, usually organic, obtained from another compound. 2 A coefficient representing the rate of change of one quantity.

Dermacentor silvarum (parasitology) A species of tick in the genus *Dermacentor*, of little apparent importance to hosts.

Dermacentor (parasitology) A genus of ticks of the family Ixodidae that are parasitic on various animals, and vectors of disease-producing microorganisms.

Dermacentor albipictus (human/veterinary parasitology) A species of one-host tick in the genus *Dermacentor*, that parasitizes moose mostly but also other wild ruminants and pastured livestock and that may transmit anaplasmosis and possibly Rocky Mountain spotted fever. Also termed moose tick; winter tick.

Dermacentor andersoni (human/veterinary parasitology) (synonym: *Dermacentor venustus*) A species of tick in the genus *Dermacentor*, commonly found in the western USA, that is parasitic on numerous wild mammals, most domestic animals, and humans. It is a vector of Rocky Mountain spotted fever, tularemia, Colorado tick fever, and Q fever in the USA, and is also one of the causes of tick paralysis in the USA.

Dermacentor halli (parasitology) A species of tick in the genus *Dermacentor*, of little apparent importance to animals.

Dermacentor marginatus (parasitology) A species of tick in the genus *Dermacentor*, of little apparent importance to animals.

Dermacentor nigrolineatus (veterinary parasitology) A one-host tick in the genus *Dermacentor*, which may be found mostly on white-tailed deer, but also on pastured livestock. Also termed brown winter tick.

Dermacentor nitens (veterinary parasitology) A one-host tick in the genus *Dermacentor*, that parasitizes mostly horses and is the vector of equine piroplasmosis, and predisposes animals to screw-worm attack. Also termed tropical horse tick; *Anocentor nitens*.

Dermacentor nuttalli (parasitology) A species of tick in the genus *Dermacentor*, of little apparent importance to animals.

Dermacentor occidentalis (human/veterinary parasitology) A three-host tick in the genus *Dermacentor*, that may be found on many animals. Immature forms may be found on rodents. They transmit anaplasmosis, Colorado tick fever, Q fever, tularemia, and cause tick paralysis. Also termed Pacific Coast tick.

Dermacentor parumapterus (human/veterinary parasitology) A species of tick in the genus *Dermacentor*, that is a vector of Rocky Mountain spotted fever.

Dermacentor reticulatus (parasitology) A three-host tick in the genus *Dermacentor*, that transmits equine piroplasmosis.

Dermacentor variabilis (human/veterinary parasitology) A three-host tick in the genus *Dermacentor*, that transmits *Anaplasma marginale* in cattle, tularemia in humans, is the chief vector of Rocky Mountain spotted fever in the central and eastern USA and causes tick paralysis in the dog. The dog is the principal host of the adult forms, but they are also parasitic on cattle, horses, rabbits and humans. Also termed American dog tick.

Dermacentor venustus (parasitology) *See Dermacentor andersoni.*

dermal (medical) Pertaining to the skin.

Dermanyssidae (parasitology) A family of parasitic mites, including *Dermanyssus*, *Ornithonyssus* and *Pneumonyssus* spp.

Dermanyssus gallinae (human/avian/veterinary parasitology) A species of parasitic mites in the genus *Dermanyssus*, of the Dermanyssidae family of mites. A major parasite of domestic fowls but may also occur on other birds including aviary and wild colonies, and are a common chance parasite of humans from heavy infestations in urban bird colonies. They may cause death in birds due to anaemia (*USA*: anemia), and are also a vector for spirochetosis of fowls and possibly the equine encephalitides. Also termed red mite.

dermat(o)- (terminology) A word element. [German] Prefix denoting *skin*.

dermatitis (medical) Inflammation of the skin.

Dermatobia (veterinary parasitology) A genus of bot flies in the family Cuterebridae, that are a major parasite of cattle in South America.

Dermatobia hominis (human/veterinary/avian parasitology) A species of bot flies in the genus *Dermatobia*, the larvae of which are parasitic in the skin of humans, mammals and birds causing subcutaneous swellings with central holes.

Dermatophagoides (parasitology) A genus of mites of the family Epidermoptidae. Species include *Dermatophagoides farinae*, and *Dermatophagoides pteronyssinus* (housedust mites).

dermatorrhagiae parasitaire (parasitology) [French] *See* parafilariasis.

Dermatoxys (parasitology) A genus of nematodes in the family Oxyuridae.

Dermatoxys veligera (veterinary parasitology) A species of nematodes in the genus *Dermatoxys* that may be found in the large intestine of rabbits and hares.

dermatozoon (parasitology) Any animal parasite on the skin; an ectoparasite.

Dermestes (insect parasitology) A genus of beetles in the order Coleoptera.

Dermestes lardarius (avian parasitology) A species of beetle in the genus *Dermestes* that destroys stored grain, meat, hides and accumulated droppings of pigeons. The larvae may attack pigeon fledglings. Also termed larder beetle.

dermis (histology) Layer of skin that is the innermost of the two main layers. It is composed of connective tissue and contains blood, lymph vessels, sensory nerves, hair follicles, sweat glands and some muscle cells.

See also epidermis.

Dermocystidium (fish/veterinary parasitology) A protozoan parasite which causes cysts in the skin of fish and amphibians.

Dermocystidium ranae (veterinary parasitology) A protozoan parasite in the genus *Dermocystidium* that parasitizes the grass frog (*Rana temporaria*).

Dermoglyphus (avian parasitology) A genus of the Dermoglyphidae family of mites. Species includes *Dermoglyphus elongatus*, and *Dermoglyphus minor* that may be found in the feather shafts of birds but cause no apparent lesion.

DES (pharmacology) An abbreviation of 'D'i'E'thyl'S'tilbestrol.

descriptive statistics (statistics) The use of statistics to summarise a set of known data in a clear and concise manner. *Contrast* statistics.

desiccant (chemistry) Substance that absorbs water and can therefore be used as a drying agent or to prevent deliquescence (*e.g.* anhydrous calcium chloride, silica gel), often used in a desiccator. Also termed: dehydrating agent.

desiccator (laboratory equipment) Apparatus used for drying substances or preventing deliquescence, *e.g.* a closed glass vessel containing a desiccant.

desk top publishing (DTP) (computing) Technique that uses a microcomputer linked to a word processor (with access to various type founts and justification programs) and a laser printer to produce multiple copies of a document that rivals conventional printing in quality. The addition of a scanner allows the introduction of simple graphics (illustrations).

desorption (1, 2 chemistry) 1 Reverse process to adsorption. 2 Removal of an adsorbate from an adsorbent in chromatography.

detergent (chemistry) Surfactant that is used as a cleaning agent. Detergents are particularly useful for cleaning because they lower surface tension and emulsify fats and oils (allowing them to go into solution with water without forming a scum with any of the substances that cause hardness of water). Soaps act in a similar way, but form an insoluble scum in hard water.

determinant (1 immunology; 2 biology; 3 mathematics) 1 Region or regions of an antigen molecule required for its 'recognition' (binding) by a particle or antibody. The selective nature of this molecular interaction confers specificity on the immune reaction of the antibody-producer. 2 Factor that transmits inherited characteristics, *e.g.* a gene. 3 Quantity obtained by adding products of elements of a square matrix according to certain rules.

deuterated compound (chemistry) Substance in which ordinary hydrogen has been replaced by deuterium.

deuterium (chemistry) D_2 One of the three isotopes of hydrogen, having one neutron and one proton in its nucleus. R.a.m. 2.0141. Also termed: heavy hydrogen.

deuterium oxide (chemistry) D_2O Chemical name of heavy water.

deuteron (chemistry) Positively charged particle that is composed of one neutron and one proton; it is the nucleus of a deuterium atom.

deviation (mathematics) The amount by which one value in a set of values differs from the arithmetic mean.

See also standard deviation.

devitrification (optics) Loss of transparency, *e.g.* in glass, caused by crystallization.

deworming preparations (pharmacology) *See* anthelmintic.

dexamethasone (pharmacology) A corticosteroid and anti-inflammatory drug that, administered orally, topically or by injection, can be used for a variety of purposes including the suppression of allergic and inflammatory conditions.

dextr(o)- (terminology) A word element. [Latin] Prefix denoting *right*.

dextrin (chemistry) Polysaccharide of intermediate chain length produced from the action of amylases on starch.

dextronic acid (chemistry) An alternative term for gluconic acid.

dextrorotatory (chemistry) Describing an optically active compound that causes the plane of polarized light to rotate in a clockwise direction. It is indicated by the prefix (+)- or *d*-.

dextrose (chemistry) An alternative term for glucose.

df (statistics) An abbreviation for 'd'egrees of 'f'reedom.

DFP (biochemistry) An abbreviation of 'D'i-isopropyl 'F'luoro'P'hosphate.

dg (measurement) An abbreviation of 'd'eci'g'ram.

DHEW (governing body) An abbreviation of 'D'epartment of 'H'ealth, 'E'ducation and 'W'elfare.

DHHS (governing body) An abbreviation of 'D'epartment of 'H'ealth and 'H'uman 'S'ervices.

Dhobie itch (medical) Allergic contact dermatitis. *See* dermatitis.

DHP (education) An abbreviation of *'D'iplome en 'H'ygiene 'P'ublique.* [French, meaning 'Diploma in Public Health'.]

DHS (education) An abbreviation of 'D'octor of 'H'ealth 'S'ciences.

di- (terminology) A word element. [German] Prefix denoting *two*.

dia- (terminology) A word element. [German] Prefix denoting *through*, *between*, *apart*, *across*, *completely*.

diag. (1 medical; 2, 3 general terminology) 1 An abbreviation of 'diag'nose. 2 An abbreviation of 'diag'onal. 3 An abbreviation of 'diag'ram.

diakinesis (cytology) Phase of cell division that occurs at the final stage of prophase of the first division in meiosis. During this phase the chromosomes become short and thick, forming more chiasmata, the nucleoli and nuclear membrane disappear, and the spindle appears for the process of division.

dialysed iron (chemistry) Colloidal solution of iron(III) hydroxide, $Fe(OH)_3$. It is a red liquid, used in medicine.

dialysis (biochemistry) Separation of colloids from crystalloids using selective diffusion through a semipermeable membrane. It is the process by which globular proteins can be separated from low-molecular weight solutes, as in filtering ('purifying') blood in an artificial kidney machine: the membrane retains protein molecules and allows small solute molecules and water to pass through.

diam. (measurement) An abbreviation of 'diam'eter. Also termed dia.

diameter (dia. or diam.) (measurement) The length of a line segment intersecting the circle's centre (*USA*: center) between two points on the circle.

diamphenethide (pharmacology) A fasciolicide used in sheep, which is effective against immature flukes but with diminishing activity as the fluke ages. An effective compound for use in prophylactic programmes (*USA*: programs) against *Fasciola hepatica*. Also termed: diamfenetide.

diaph. (anatomy) An abbreviation of 'diaph'ragm.

diaphragm (anatomy) A sheet of muscle present in mammals, located below the lungs attached to the body wall at the sides and separating the thorax from the abdomen. During respiration it forms an important part of the mechanism for filling and expelling air from the lungs.

Diaptomus gracilis (parasitology) A diaptomid copepod, a first intermediate host of the cestode *Diphyllobothrium latum*.

diarrhoea (medical) The result of too rapid a transit of the bowel contents so that there is insufficient time for reabsorption of water from the faeces (*USA*: feces). Consequently, the stools are loose, liquid and are passed more frequently than normal. (*USA*: diarrhea).

diastase (biochemistry) An alternative term for amylase.

diastole (physiology) Phase of the heartbeat in which the heart undergoes relaxation and refills with blood from the veins. The term also applies to a contractile vacuole in a cell when it refills with fluid.

See also systole.

diatomic (chemistry) Describing a molecule that is composed of two identical atoms, *e.g.* O_2, H_2 and Cl_2.

See also atomicity.

diazinon (control measures) An organophosphorus insecticide, used in ear tags for cattle and in flea collars and rinses for dogs. Also termed: dimpylate.

See also organophosphorus compound.

diazo compound (chemistry) Organic compound that contains two adjacent nitrogen atoms, but only one attached to a carbon atom. Formed by diazotization, diazo compounds are very important in synthesis, being the starting point of various dyes and drugs.

diazonium compound (chemistry) Organic compound of the type $RN_2^+X^-$, where R is an aryl group. The compounds are colourless (*USA*: colorless) solids, extremely soluble in water, used for making azo dyes. Many of them (particularly the nitrates) are explosive in the solid state.

diazotization (chemistry) Formation of a diazo compound by the interaction of sodium nitrite, an inorganic acid, and a primary aromatic amine at low temperatures.

dibasic (chemistry) Describing an acid that contains two replaceable hydrogen atoms in its molecules, *e.g.* carbonic acid, H_2CO_3, sulphuric acid (*USA*: sulfuric acid), H_2SO_4. A dibasic acid can form two types of salts: a normal salt, in which both hydrogen atoms are replaced by a metal or its equivalent (*e.g.* sodium carbonate, Na_2CO_3), and an acidic salt, in which only one hydrogen atom is replaced, *e.g.* sodium hydrogensulphate (*USA*: sodium hydrogensulfate) (bisulphate [*USA*: bisulfate]), $NaHSO_4$.

Dibothriocephalus (parasitology) *See Diphyllobothrium.*

dibromsalan (pharmacology) One of two bromsalans, the other being tribromsalan, which, mixed in a variety of proportions, are used as treatments for liver flukes.

See also bromsalans.

dibutyltin dilaurate (pharmacology) A coccidiostat used in commercial poultry farming.

DIC (medical) An abbreviation of 'D'isseminated 'I'ntravascular 'C'oagulation.

dicarboxylic acid (chemistry) Organic acid that contains two carboxyl groups.

dichlorophen (pharmacology) A phenol derivative used as a teniacide which has been superseded, except in combination with piperazine and other compounds.

dichoptic (biology) A descriptive term said of eyes which are widely separated, *e.g.* in some insects.

dichroism (chemistry) Property of some substances that makes them transmit some colours (*USA*: colors) and reflect others, or which display certain colours when viewed from one angle and different colours when viewed from another.

dichromate (chemistry) Salt containing the dichromate(VI) ion ($Cr_2O_7^{\ 2-}$), an oxidizing agent; *e.g.* potassium dichromate, $K_2Cr_2O_7$. Also termed: bichromate.

dicrocoeliasis (parasitology) Hepatic fascioliasis due to infection with *Dicrocoelium dendriticum*.

Dicrocoelium (parasitology) A genus of flukes (digenetic trematodes) in the family Dicrocoeliidae.

Dicrocoelium dendriticum (human/insect/veterinary parasitology) (synonym: *Dicrocoelium lanceolatum*) A liver fluke in the genus *Dicrocoelium* that infects most domestic and many wild animals and has been reported in humans. Heavy infestation causes cirrhosis and biliary obstruction and a clinical syndrome of oedema (*USA*: edema), anaemia (*USA*: anemia) and emaciation. Similarly to *Dicrocoelium hospes* and *Dicrocoelium lanceolatum*, they are parasites to the ants *Cataglyphis*, *Camponotus*, *Formica* spp. and *Proformica*. Typically one or two metacercaria are found in the host's brain, resulting in sluggish behavioural (*USA*: behavioral) change; the infected ants climb and cling to grasses. The final host is an ungulate.

Dicrocoelium hospes (veterinary parasitology) A liver fluke in the genus *Dicrocoelium* that may be found in the bile ducts of cattle in Africa.

Dicrocoelium lanceolatum (insect parasitology) *See Dicrocoelium dendriticum*.

dictyate (parasitological physiology) A stage in the development of oocytes which are arrested at the same stage of meiotic prophase; the stage varies between species.

Dictyocaulus (parasitology) A genus of lungworms in the family Dictyocaulidae.

Dictyocaulus arnfieldi (veterinary parasitology) A species of lungworm in the genus *Dictyocaulus* that is primarily a lungworm of donkeys but which has little apparent clinical effect in them. The infestation when present in horses is also clinically innocuous and the worms do not usually mature in this species.

Dictyocaulus cameli (veterinary parasitology) A species of lungworm in the genus *Dictyocaulus* that may be found in camels.

Dictyocaulus eckerti (veterinary parasitology) A species of lungworm in the genus *Dictyocaulus* that may be found in deer.

Dictyocaulus filaria (veterinary parasitology) A species of lungworm in the genus *Dictyocaulus* that cause a chronic disease in small ruminants. Clinical signs are limited to persistent cough, moderate dyspnea and loss of condition. When occurring in young sheep and goats they may be severely affected, occasionally fatally.

Dictyocaulus viviparus (veterinary parasitology) A species of lungworm in the genus *Dictyocaulus*. The common lungworm of cattle, which may cause several serious diseases including verminous pneumonia, acute interstitial pneumonia and secondary bacterial pneumonia. The infestation is widespread and affects mainly young cattle. Massive larval intakes can cause a high mortality due to a peracute syndrome of interstitial pneumonia. More moderate infestations cause a syndrome of paroxysmal coughing, moderate dyspnea and loss of condition. On occassion, calves may die as a result of a secondary bacterial pneumonia. Warm moist autumnal conditions are most conducive to serious outbreaks.

Dicymolomia pegasalis (insect parasitology) Arthropods in the Lepidoptera that are a parasitic to the wasps *Polistes* spp., attacking the larvae, and suspected of having eliminated wasps from some islands.

1,1-diethoxyethane (chemistry) An alternative term for acetal.

dielectric (physics) Nonconductor of electricity in which an electric field persists in the presence of an inducing field (but opposes it). A dielectric is the insulating material in a capacitor.

dielectric constant (physics) An alternative term for relative permittivity.

dielectric strength (physics) Property of an insulator that enables it to withstand electric stress without breaking down.

dielectrophoresis (physics) Movement of electrically polarized particles in a variable electric field.

Dientamoeba (human/veterinary parasitology) A genus of amoebas (*USA*: amebas) that are commonly found in the colon and appendix of primates and of humans.

Dientamoeba fragilis (human/veterinary parasitology) A species of amoebas (*USA*: amebas) in the genus *Dientamoeba*, that occurs in the caecum (*USA*: cecum) of humans and monkeys. This species that has been associated with diarrhoea (*USA*: diarrhea) but its pathogenicity is unclear.

diethylcarbamazine (veterinary parasitology) An antifilarial agent used in dogs as the citrate salt for the prevention of heartworm. It is also used for the treatment of ascarids in dogs and immature lungworms in cattle and sheep.

didelphic (parasitological anatomy) A term used in relation to nematodes. Bicornate. Having a double set of reproductive system.

differential calculus (mathematics) Branch of mathematics that deals with continuously varying quantities. It uses differentiation for calculating rates of change, slopes of curves, maximum and minimum values, etc.

differentiation (mathematics) Method used in calculus to determine the derivative f' of a function f. If $f(x) = Ax^n$, $f'(x) = nAx^{n-1}$.

diffraction (physics) Bending of the path of a beam (*e.g.* of light or electrons) at the edge of an object.

diffraction grating (optics) Optical device that is used for producing spectra. It consists of a sheet of glass or plastic marked with closely spaced parallel lines (as many as 10,000 per centimetre). The spectra are produced by a combination of diffraction and interference.

diffusion of gases (physics) Phenomenon by which gases mix together, reducing any concentration gradient to zero, *e.g.* in gas exchange between plant leaves and air.

See also active transport.

diffusion of light (physics) Spreading or scattering of light.

diffusion of solutions (chemistry) Free movement of molecules or ions of a dissolved substance through a solvent, resulting in complete mixing.

See also osmosis.

diffusion pump (laboratory equipment) Apparatus used to produce a high vacuum. The pump employs mercury or oil at low vapour-pressure which carries along in its flow molecules of a gas from a low pressure established by a backing pump. Also termed: condensation pump; vacuum pump.

diflubenzuron (control measures) An ectoparasiticide similar to lufenuron.

di-form (chemistry) Term indicating that a mixture contains the dextrorotatory and the laevorotatory forms of an optically active compound in equal molecular proportions.

Digenea (fish parasitology) A subclass of flatworms in the class Trematoda and phylum Platyhelminthes. About seventy families of fish digeneans have been recognised and except for some species of Fellodistomidae, Azygiidae, Gorgoderidae, Sanguinicolidae, Syncoeliidae, Zoogonidae and Didymozoidae, and all species of Ptychogonimidae and Aphanhysteridae, they all occur exclusively in teleosts. Traditionally the Digenea may be divided into the Gasterostomata (Bucephalidae) and Prosostomata (other families), however the Bucephalidae are now associated with more conventional digenean groups and are now included amongst six other families as possible candidates for the most primative of extant digeneans.

digenean (parasitology) Pertaining to or of the nature of members of the subclass Digenea.

digenetic (1 parasitological physiology; 2 parasitology) 1 Having two stages of multiplication, one sexual in the mature forms, the other asexual in the larval stages. 2 Belonging to the subclass Digenea.

digenetic trematodes (parasitology) *See* Digenea.

digestion (physiology) Breakdown of complex substances in food by enzymes in the alimentary canal to produce simpler soluble compounds, which pass into the body by absorption and assimilation. Ultimately carbohydrates (*e.g.* starch, sugar) are broken down to glucose, proteins to amino acids, and fats to fatty acids and glycerol.

digit (1 mathematics/computing; 2 anatomy) 1 A single numeral; an integer under 10. 2 A finger or toe.

digital computer (computing) Computer that operates on data supplied and stored in digital or number form.

digital display (equipment) Display that shows readings of a measuring machine, clock, etc. by displaying numerals.

digital/analog converter (computing) Device that converts digital signals into continuously variable electrical signals for use by an analog computer.

dihydrate (chemistry) Chemical (a hydrate) whose molecules have two associated molecules of water of crystallization; *e.g.* sodium dichromate, $Na_2Cr_2O_7.2H_2O$.

2,3-dihydroxybutanedioic acid (chemistry) An alternative term for tartaric acid.

dihydroxypurine (chemistry) An alternative term for xanthine.

dihydroxysuccinic acid (chemistry) An alternative term for tartaric acid.

dilate (physiology) To widen; to produce dilation.

dilated (physiology) A state of dilation.

dilation (physiology) The widening or expansion of an organ, opening, passage or vessel. An alternative term: dilatation.

diluent (chemistry) Solvent used to reduce the strength of a solution.

dilute (1, 2 chemistry) 1 To reduce the strength of a solution by adding water or other solvent. 2 Describing a solution in which the amount of solute is small compared to that of the solvent.

dilution (chemistry) Process that involves the lowering of concentration.

dim. (general terminology) An abbreviation of 'dim'ension.

dimension (mathematics) Power to which a fundamental unit is raised in a derived unit; *e.g.* acceleration has the dimensions $[LT^{-2}]$, *i.e.* + 1 for length and - 2 for time, equivalent to length divided by the square of time.

dimensional analysis (mathematics) Prediction of the relationship of quantities. If an equation is correct the dimensions of the quantities on each side must be identical. It is an important way of checking the validity of an equation.

dimer (chemistry) Chemical formed from two similar monomer molecules.

dimethylbenzene (chemistry) An alternative term for xylene.

dimethyl sulphoxide (DMSO) (chemistry) An industrial solvent that has the ability to penetrate plant and animal tissues and to preserve living cells during freezing. (*USA*: dimethyl sulfoxide).

diminazene (pharmacology) A diamidine antiprotozoal agent used as the aceturate.

Dina parva (avian parasitology) A leech that may be found in the nasal cavity of aquatic birds.

dinitolmide (pharmacology) A nitrobenzamide anticoccidial agent used in poultry.

dinitrogen oxide (chemistry/medical) N_2O Colourless (*USA*: colorless) gas made by heating ammonium nitrate and used as an anaesthetic (*USA*: anesthetic). Also termed: nitrogen oxide; nitrous oxide; dental gas; laughing gas.

dinitro-o-toluamide (pharmacology) A toxic coccidiostat which may cause ataxia, torticollis and reduced growth.

Dinobdella (parasitology) A member of the genus *Limnatis* of leeches of the class Hirudinea.

Dinobdella ferox (human/veterinary parasitology) A species of leech that may be found in the pharynx of ruminants and the upper respiratory tract of dogs, monkeys and humans.

dinobdelliasis (parasitology) Infestation of all animal species with the nasal leeches *Dinobdella ferox*.

dinsed (pharmacology) An anticoccidial agent used in mixtures for the control of coccidiosis in chickens.

Dinychus (insect parasitology) Mites in the Mesostigmata that, similarly to *Ipiduropoda*, *Oodinychus*, *Phaulodinchyus*, *Prodynchius*, *Trematurella*, *Urobovella*, *Urodiscella*, and *Uroseius*, are commonly found in nests, some being parasitic and phoretic on emerging queens on various ants.

Dioctophyme (parasitology) A genus of nematode worms in the family Dioctophymidae.

Dioctophyme renale (veterinary parasitology) A species of very large nematode worms in the genus *Dioctophyme* which may be found in the kidneys and other organs of dogs and wild carnivores and occasionally ruminants. They may be severely destructive and cause fatal uremia.

dioctophymosis (parasitology) Infection with the kidney worm *Dioctophyme renale*.

diode (1, 2 physics) 1 Electron tube (valve) containing two electrodes, an anode and a cathode. 2 Rectifier made up of a semiconducting crystal with two terminals.

dioecious (parasitological anatomy) A term used in relation to trematodes. Condition where male and female gonads are found in different individuals.

diol (chemistry) Organic compound containing two hydroxyl groups and having the general formula $C_nH_{2n}(OH)_2$. Diols are thick liquids or crystalline solids, and some have a sweet taste. Ethane-1,2-diol (ethylene glycol) is the simplest diol, widely used as a solvent and as an antifreeze agent. Also termed: dihydric alcohol; glycol.

dioptre (measurement/microscopy) Unit that is used to express the power of a lens. It is the reciprocal of the focal length of the lens in metres. The power of a convergent lens with a focal length of one metre is said to be + 1 dioptre. The power of a divergent lens is given a negative value.

Diorchis nyrocae (avian parasitology) A common cestode of ducks in the family Hymenolepididae.

dioxan (chemistry) $(CH_2)_2O_2$ Colourless (*USA*: colorless) liquid cyclic ether. It is inert to many reagents and frequently used in mixtures with water to increase the solubility of organic compounds such as alkyl halides. Also termed: 1,4-dioxan.

dioxin (chemistry) $C_{12}H_4Cl_4O_2$ By-product of organic synthesis (*e.g.* of the disinfectant trichlorophenol) which can cause allergic skin reactions. Also termed: 2,3,7,8-tetrachlorodibenzo-*p*-dioxin.

dip. (education) An abbreviation of 'dip'loma.

Dip. Pharm. (education) An abbreviation of 'Dip'loma in 'Pharm'acy.

Dip. App. Sc. (education) An abbreviation of 'Dip'loma of 'App'lied 'Sc'ience.

Dip. Bac. (education) An abbreviation of 'Dip'loma in 'Bac'teriology.

Dip. BMS (education) An abbreviation of 'Dip'loma in 'B'asic 'M'edical 'S'ciences.

Dipetalonema (parasitology) A genus of filarioid parasites of the superfamily Filarioidea.

Dipetalonema dracunculoides (human/veterinary parasitology) A species of filarioid parasites in the genus *Dipetalonema* which may be found in peritoneal membranes of dogs and humans.

Dipetalonema evansi (veterinary parasitology) A species of filarioid parasites in the genus *Dipetalonema* which may be found in spermatic and pulmonary artery of camels.

Dipetalonema gracile (veterinary parasitology) A species of filarioid parasites in the genus *Dipetalonema* which similarly to *Dipetalonema marmosetae*, *Dipetalonema obtusa* and *Dipetalonema tamarinae* may be found in the peritoneal cavity of primates.

Dipetalonema grassii (veterinary parasitology) A species of filarioid parasites in the genus *Dipetalonema* which may be found in the subcutaneous tissue of dogs.

Dipetalonema loxodontis (veterinary parasitology) A species of filarioid parasites in the genus *Dipetalonema* which may be found in the African elephant. Also termed: *Loxodontofilaria loxodontis*.

Dipetalonema marmosetae (veterinary parasitology) A species of filarioid parasites in the genus *Dipetalonema* which similarly to *Dipetalonema gracile*, *Dipetalonema obtusa* and *Dipetalonema tamarinae* may be found in the peritoneal cavity of primates.

Dipetalonema obtusa (veterinary parasitology) A species of filarioid parasites in the genus *Dipetalonema* which similarly to *Dipetalonema marmosetae*, *Dipetalonema gracile* and *Dipetalonema tamarinae* may be found in the peritoneal cavity of primates.

Dipetalonema odendhali (veterinary parasitology) A species of filarioid parasites in the genus *Dipetalonema* which may be found in the subcutaneous and intermuscular tissues of California sea lion.

Dipetalonema perstans (human parasitology) A species of filarioid parasites in the genus *Dipetalonema* which similarly to *Dipetalonema streptocerca* are primarily parasitic in humans, with other primates serving as reservoir hosts.

Dipetalonema reconditum (veterinary parasitology) A species of filarioid parasites in the genus *Dipetalonema* which may be found in body cavities, connective tissues and kidneys of dogs.

Dipetalonema spirocauda (veterinary parasitology) A species of filarioid parasites in the genus *Dipetalonema* which may be found in the right heart and pulmonary artery of seals.

Dipetalonema streptocerca (human parasitology) A species of filarioid parasites in the genus *Dipetalonema* which similarly to *Dipetalonema perstans* are primarily parasitic in humans, with other primates serving as reservoir hosts.

Dipetalonema tamarinae (veterinary parasitology) A species of filarioid parasites in the genus *Dipetalonema* which similarly to *Dipetalonema marmosetae*, *Dipetalonema obtusa* and *Dipetalonema gracile* may be found in the peritoneal cavity of primates.

diphenhydramine hydrochloride (pharmacology) An antihistamine drug which can be administered orally for the symptomatic relief of allergic symptoms.

diphenoxylate hydrochloride (pharmacology) An opiod anti-diarrhoeal (*USA*: anti-diarrheal) drug used to treat chronic diarrhoea (*USA*: diarrhea).

Diphyllidea (fish parasitology) An order of flatworms in subclass Eucestoda, class Cestoidea and the phylum Platyhelminthes.

diphyllobothriasis (parasitology) Infection with *Diphyllobothrium* spp.

Diphyllobothrium (parasitology) A genus of long tapeworms in the family Diphyllobothriidae

Diphyllobothrium dalliae (human/fish/veterinary parasitology) A species of long tapeworms in the genus *Diphyllobothrium* which similarly to *Diphyllobothrium dendriticum*, *Diphyllobothrium pacificum*, *Diphyllobothrium strictum*, *Diphyllobothrium minus*, and *Diphyllobothrium ursi* may be found in fish-eating mammals including humans.

Diphyllobothrium dendriticum (human/fish/ veterinary parasitology) A species of long tapeworms in the genus *Diphyllobothrium* which similarly to *Diphyllobothrium dalliae*, *Diphyllobothrium pacificum*, *Diphyllobothrium strictum*, *Diphyllobothrium minus*, and *Diphyllobothrium ursi* may be found in fish-eating mammals including humans.

Diphyllobothrium erinacei (parasitology) *See Spirometra erinacei.*

Diphyllobothrium latum (human/veterinary parasitology) A species of long tapeworms in the genus *Diphyllobothrium*. The broad or fish tapeworm, this species may be found in the small intestines of humans, dogs, cats and other fish-eating mammals.

Diphyllobothrium minus (human/fish/veterinary parasitology) A species of long tapeworms in the genus *Diphyllobothrium* which similarly to *Diphyllobothrium dalliae*, *Diphyllobothrium dendriticum*, *Diphyllobothrium pacificum*, *Diphyllobothrium strictum* and *Diphyllobothrium ursi* may be found in fish-eating mammals including humans.

Diphyllobothrium pacificum (human/fish/ veterinary parasitology) A species of long tapeworms in the genus *Diphyllobothrium* which similarly to *Diphyllobothrium dalliae*, *Diphyllobothrium dendriticum*, *Diphyllobothrium minus*, *Diphyllobothrium strictum* and *Diphyllobothrium ursi* may be found in fish-eating mammals including humans.

Diphyllobothrium strictum (human/fish/ veterinary parasitology) A species of long tapeworms in the genus *Diphyllobothrium* which similarly to *Diphyllobothrium dalliae*, *Diphyllobothrium dendriticum*, *Diphyllobothrium minus*, *Diphyllobothrium pacificum* and *Diphyllobothrium ursi* may be found in fish-eating mammals including humans.

Diphyllobothrium ursi (human/fish/veterinary parasitology) A species of long tapeworms in the genus *Diphyllobothrium* which similarly to *Diphyllobothrium dalliae*, *Diphyllobothrium dendriticum*, *Diphyllobothrium minus*, *Diphyllobothrium pacificum*, *Diphyllobothrium strictum* and may be found in fish-eating mammals including humans.

Diplogaster aerivora (insect parasitology) A species of nematode in the Diplogasteridae that are parasites to the termites *Leucotermes lucifugus*, that may be found in the head around the mouth parts, the host becoming sluggish.

Diplogonoporus (parasitology) A genus of tapeworms in the family Diphyllobothriidae.

Diplogonoporus grandis (human/veterinary parasitology) A species of tapeworms in the genus *Diplogonoporus* which may be found in humans, probably as an final accidental host. The real host is probably a seal.

diploid (genetics) In a cell or organism, describing the existence of chromosomes in homologous pairs, *i.e.* twice the haploid number (2*n*). Apart from gametes, it is characteristic of all animal cells.

Diplopylidium (veterinary parasitology) A genus of tapeworms in the family Dipylidiidae which may be found in dogs and cats.

Diploscapter lycostoma *(Diploscapter coronata)* (insect parasitology) A species of nematode in the Rhabditidae that are parasites to the ants *Camponotus* spp., *Formica* spp., *Iridomyrmex* spp., *Lasius* spp., and *Myrmica*. They may be found in the pharyngeal glands.

Diplostomum (parasitology) A genus of digenetic trematodes of the family Diplostomatidae.

Diplostomum spathaceum (fish/avian parasitology) A species of digenetic trematodes in the genus *Diplostomum* which may be found in the intestines of gulls. They are also a cause of cataracts in fish and heavy infestations may cause severe mortalities.

diplotene (genetics) In meiosis, the stage in late prophase when the pairs of chromatids begin to separate from the tetrad formed by the association of homologous chromosomes. Chiasmata can be seen at this stage.

Diplozoon (parasitology) A genus of monogean trematode parasites in the family Diplozooidae.

Diplozoon barbi (fish parasitology) A species of monogean trematode parasites in the genus *Diplozoon* that, similarly to *Diplozoon paradoxus*, may be found in the gills of freshwater fish, where they cause damage and predispose to bacterial infection.

Diplozoon paradoxus (fish parasitology) A species of monogean trematode parasites in the genus *Diplozoon* that, similarly to *Diplozoon barbi*, may be found in the gills of freshwater fish, where they cause damage and predispose to bacterial infection.

dipole (physics) Pair of equal and opposite electric charges at a (short) distance from each other. Some asymmetric molecules act as dipoles.

dipole moment (physics) Product of one charge of a dipole and the distance between the charges.

Diptera (insect parasitology) An order of insects with two wings, including flies, gnats and mosquitoes.

dipterous (1 anatomy; 2 parasitology) 1 Having two wings. 2 Pertaining to insects of the order Diptera.

dipylidiasis (parasitology) Infection with *Dipylidium caninum*.

Dipylidium (parasitology) A genus of tapeworms of the family Dipylidiidae. Species includes *Dipylidium gracile*, *Dipylidium compactum*, *Dipylidium diffusum*, *Dipylidium buencaminoi* (synonym: *Dipylidium caninum*), and *Dipylidium sexcoronatum*.

Dipylidium caninum (human/veterinary parasitology) The dog tapeworm, parasitic in dogs and cats and occasionally found in humans. Aesthetically (*USA*: esthetically) unattractive but causes little apparent damage other than anal irritation.

dir. (governing body) An abbreviation of 'dir'ector.

direct (terminology) Without intervening steps.

direct calorimetry (measurement/physiology) Measurement of heat actually produced by an organism which is confined in a sealed chamber or calorimeter.

direct current (dc) (physics) Electric current that always flows in the same direction (as opposed to alternating current).

direct dye (chemistry) Dye that does not require a mordant.

director (dir.) (governing body) The head of a department or institution.

Dirofilaria (parasitology) A genus of nematode parasites of the superfamily Filarioidea.

Dirofilaria acutiuscula (veterinary parasitology) A species of nematode parasite in the genus *Dirofilaria* that causes swelling in the subcutaneous fascia of the dorsolumbar area of the peccary.

Dirofilaria conjunctivae (human/veterinary parasitology) A species of nematode parasite in the genus *Dirofilaria* that causes a zoonotic infection on the eyelids of humans, due to infestation of wildlife.

Dirofilaria corynodes (veterinary parasitology) A species of nematode parasite in the genus *Dirofilaria* that may be found in monkeys.

Dirofilaria immitis (human/veterinary parasitology) A species of nematode parasite in the genus *Dirofilaria* that may be found in dog, cat, fox and wolf and occassionally in humans and many other species. Transmitted by the intermediate hosts, *Culex*, *Aedes*, *Anopheles* and other mosquito genera. Found in the blood vessels, especially the heart and the pulmonary artery, they cause heartworm disease.

Dirofilaria reconditum (veterinary parasitology) A species of nematode parasite in the genus *Dirofilaria* that may be found in the subcutaneous tissues of dogs.

Dirofilaria repens (human/veterinary parasitology) A species of nematode parasite in the genus *Dirofilaria* that may be found in the subcutaneous tissues of the dog and cat, and occasionally humans.

Dirofilaria roemeri (parasitology) *See Pelecitus roemeri.*

Dirofilaria striata (veterinary parasitology) A species of nematode parasite in the genus *Dirofilaria* that may be found in the bobcat.

Dirofilaria tenuis (human/veterinary parasitology) A species of nematode parasite in the genus *Dirofilaria* that may be found in subcutaneous tissues of raccoons and humans.

Dirofilaria ursi (veterinary parasitology) A species of nematode parasite in the genus *Dirofilaria* that may be found in black bears.

dirofilariasis (parasitology) Infection with nematodes of the genus *Dirofilaria*, that includes subcutaneous swellings. *See* heartworm disease.

dis- (terminology) A word element. [Latin] Prefix denoting *reversal, separation.* [German] *duplication.*

dis. (medical) An abbreviation of 'dis'charge. Also termed disch.

disab. (medical) An abbreviation of 'disab'ility. Also termed disabl.

disaccharide (chemistry) One of the class of common sugars, including lactose and sucrose, that can be broken down by hydrolysis, under the action of enzymes, to yield two monosaccharides.

disch. (medical) *See* dis.

discharge (1, 2 physics) 1 High-voltage 'spark' (current flow) between points of large potential ditterence (*e.g.* lightning); 2 Removal of the charge between the plates of a capacitor by allowing current to flow out of it; 3 Removal of energy from an electrolytic cell (battery or accumulator) by allowing current to flow out of it. 4. In electrochemistry, the process by which ions are converted to neutral atoms at an electrode during electrolysis (by gain or loss of electrons).

Discocotyle (parasitology) A genus of monogenetic trematode parasites of the family Discocotylidae.

Discocotyle sagittata (fish parasite) A species of monogenetic trematode in the genus *Discocotyle* that is a significant parasite of fish causing serious mortalities due to damage to the gills.

discontinuous variable (statistics) *See* discrete variable.

discrete variable (statistics) One in which the possible values are not on a continuous scale.

discriminant (mathematics) Quantity $b^2 - 4ac$, derived from the coefficients of a quadratic equation of general formula $ax^2 + bx + c = 0$.

disease carrier (parasitology) *See* carrier; vector.

disintegration constant (radiology) Probability of a radioactive decay of an atomic nucleus per unit time. Also termed: decay constant; transformation constant.

disk (computing) Magnetic disc used to record data in computers.

See also floppy disk; hard disk.

diskette (computing) An alternative term for a floppy disk.

disodium oxide (chemistry) An alternative term for sodium monoxide.

disodium tetraborate (chemistry) An alternative term for borax.

disophenol (DNP) (pharmacology) An injectable anthelmintic for hookworms in cats and dogs and *Haemonchus contortus* in sheep.

See also nitrophenol.

disp. (pharmacology) An abbreviation of 'disp'ensary.

dispensary (pharmacology) A department of a hospital or clinic supplying drugs and other medical supplies on demand.

dispersion (physics) Splitting of an electromagnetic radiation (*e.g.* visible light) into its component wavelengths when it passes through a medium (because different wavelengths undergo different degrees of diffraction or refraction).

displacement (mathematics) Position of one point relative to another, including both the distance between the two points and the direction of the first point from the second point.

displacement reaction (chemistry) An alternative term for substitution reaction.

display (equipment) Short name for a liquid-crystal display (LCD) or a visual display unit (VDU).

dissociation (chemistry) Temporary reversible chemical decomposition of a substance into its component atoms or molecules, which often take the form of ions, *e.g.* it occurs when most ionic compounds dissolve in water.

dissociation constant (chemistry) Equilibrium constant of a dissociation reaction, and therefore a measure of the affinity of atoms or molecules in a compound.

distillate (chemistry) Condensed liquid obtained by distillation.

distillation (chemistry) Method for purification or separation of liquids by heating to the boiling point, condensing the vapour, and collecting the distillate. Formerly, the method was used to produce distilled water for chemical experiments and processes that required water to be much purer than that in the mains water supply. In this application distillation has been largely superseded by ion exchange.

distilled water (chemistry) Water that has been purified by distillation.

distomatosis (parasitology) *See* fascioliasis.

distomiasis (parasitology) Infection due to trematodes or flukes.

See also fascioliasis.

Distomum hepaticum (parasitology) *See Fasciola hepatica.*

distortion (physics) Change from the ideal shape of an object or image, or in the form of a wave pattern (*e.g.* an electrical signal).

disulphuric acid (chemistry) An alternative term for oleum. (*USA*: disulfuric acid).

DIT (biochemistry) An abbreviation of 'D'i'I'odo'T'yrosine.

dithiazanine (pharmacology) A broad-spectrum anthelmintic and microfilaricide used for the treatment of heartworm in dogs; also used for treatment of *Strongyloides* and *Spirocerca* spp. Also termed: dizan.

dithionate (chemistry) Salt derived from dithionic acid ($H_2S_2O_6$). Also termed: hyposulphate (*USA*: hyposulfate).

dithionic acid (chemistry) $H_2S_2O_6$ Strong acid that decomposes slowly in concentrated solutions and when heated. Also termed: hyposulphuric acid (*USA*: hyposulfuric acid).

dithionite (chemistry) Name that is given to any of the salts of dithionous acid, all of which are strong reducing agents.

dithionous acid (chemistry) $H_2S_2O_4$ Strong but unstable acid that is found only in solution.

Ditylenchus (plant parasitology) A genus of nematodes in the order Tylenchida and family Anguinidae. They are slender nematodes, straight or slightly curved when killed by heat, with a head skeleton and stylet similar to *Anguina*. *See Ditylenchus dipsaci.*

Ditylenchus dipsaci (plant parasitology) A species of nematodes in the genus *Ditylenchus* whose hosts cover a wide range with over 450 hosts complicated by there being eight to ten host races or biotypes, some with limited host range, and include oat race: polyphagous, most grains, rye, corn, and oats; alfalfa race: rather specific, but alfalfa, many weeds, clovers; bulb race: most bulbs, daffodil, narcissus, and tulip. Other hosts include onion, garlic, carrots, peas, potatoes, strawberry, sugarbeets; apples and peaches in nurseries; weeds. They are cosmopolitan, especially in temperate regions. These nematodes are migratory endoparasites. They are sexually reproducing, with a life cycle of 19–23 days at 15°C. Their first moult (*USA*: molt) occurs within the egg. Nematodes live 45–75 days when sexually mature, and females each produce 200–500 eggs. The fourth stage juvenile is a survival stage that can go into cryptobiosis on or below the surface of plant tissue. At the beginning of the crop season, fourth-stage juvenile enters young tissues, especially seedlings when below the soil surface. Feeding breaks down middle lamellae, the nematode probably secretes a pectinase enzyme, causing the plant parts to become 'crisp' and easily broken. Plants become distorted and stunted; infected tissues are spongy; damage can predispose plants to other problems. Migration on plant parts above ground requires free water, and may occur after rain or sprinkler irrigation in alfalfa. Nematode enters through stomata or by direct penetration and survives in soil without a host for as long as two years, probably feeding on fungi. Control measures includes the use of systemic insecticides. Also termed: stem and bulb nematode.

diuresis (medical) An unusually or abnormally large output of urine.

diuretics (pharmacology) Drugs used to reduce fluid in the body by increasing the excretion of water and mineral salts by the kidneys, so increasing urine production. Diuretics are divided into a number of distint classes in relation to their specific actions and uses: osmotic diuretics, loop diuretics, potassium-sparing diuretics, carbonic anhyrase inhibitors, aldosterone antagonists and thiazide and thiazide-like diuretics.

diurnal (physiology/pharmacology) Pertaining to a day, in the sense of occurring every 24 hours.

divalent (chemistry) Capable of combining with two atoms of hydrogen or their equivalent. Also termed: bivalent.

diverging lens (optics) Lens that spreads out a beam of light passing through it, often a concave lens.

dividend (mathematics) Number to be divided by another one (the divisor). The result of the division is the quotient.

division (1, 2 biology; 3 mathematics) 1 In biological classification, one of the major groups into which the plant kingdom is divided. The members of the group, although often quite different in form and structure, share certain common features, *e.g.* bryophytes include the mosses and liverworts. Divisions are divided into classes, often with an intermediate subdivision. The equivalent of a division in the animal kingdom is a phylum. 2 In biology, the formation of a pair of daughter cells from a parent cell. *See* cell division. 3 In mathematics, the inverse of multiplication, in which a dividend is divided by a divisor to give a quotient.

divisor (mathematics) Number that is divided into another one (the dividend). The result of the division is the quotient.

D-lines (physics) Pair of characteristic lines in the yellow region of the spectrum of sodium, used as standards in spectroscopy.

dm (measurement) An abbreviation of 'd'eci'm'etre (*USA*: decimeter).

DMed. (education) An abbreviation of 'D'octor of 'Med'icine.

dmg. (general terminology) An abbreviation of 'd'a'm'a'g'e.

DMLT (education) An abbreviation of 'D'iploma in 'M'edical 'L'aboratory 'T'echnology.

DMPB (education) An abbreviation of 'D'iploma in 'M'edical 'P'athology and 'B'acteriology.

DMR (education) An abbreviation of 'D'iploma in 'M'edical 'R'adiology; 'D'iploma in 'M'edical 'R'adiodiagnostics.

DMS (1 computing; 2 education; 3 chemistry) 1 An abbreviation of 'D'ata 'M'anagement 'S'ystem. 2 An abbreviation of 'D'octor of 'M'edical 'S'cience. 3 An abbreviation of 'D'ocumentation of 'M'olecular 'S'pectroscopy.

DMSO (chemistry) An abbreviation of 'D'i'M'ethyl 'S'ulph'O'xide (*USA*: dimethyl sulfoxide).

DMT (biochemistry) An abbreviation of N,N-'D'i'M'ethyl'T'ryptamine.

DNA (molecular biology) An abbreviation of 'D'eoxyribo'N'ucleic 'A'cid.

DNA hybridization (molecular biology) Technique in which DNA from one species is induced to undergo base pairing with DNA or RNA from another species to produce a hybrid DNA (a process known as annealing).

DOA (1 medical; 2 chemistry) 1 An abbreviation of 'D'ead 'O'n 'A'rrival. 2 An abbreviation of 'D'issolved 'O'xygen 'A'nalyser.

dob (statistics) An abbreviation of 'd'ate 'o'f 'b'irth.

DOCA (biochemistry) An abbreviation of 'D'es'O'xy'C'orticosterone 'A'cetate.

dod (1 statistics; 2 medical) 1 An abbreviation of 'd'ate 'o'f 'd'eath. 2 An abbreviation of 'd'ied 'o'f 'd'isease.

dodecanoic acid (chemistry) An alternative term for lauric acid.

dodecylbenzene (chemistry) $C_6H_5(CH_2)_{11}CH_3$ Hydrocarbon of the benzene family, important in the manufacture of detergents.

doenca da sono (parasitology) [Spanish] trypanosomiasis.

dog flea (veterinary parasitology) *See Ctenocephalides canis.*

dog tapeworm (veterinary parasitology) *See Dipylidium caninum.*

dog tick (veterinary parasitology) Varies with the country: American dog tick, *see Dermacentor variabilis*; Australian dog tick, *see Ixodes holocyclus*; British dog tick, *see Ixodes canisuga*; brown dod tick, *see Rhipicephalus sanguineus*; yellow dog tick, *see Haemaphysalis leachi leachi.*

dolich(o)- (terminology) A word element. [German] Prefix denoting *long*.

Dolichodorus (plant parasitology) A genus of nematodes in the order Tylenchida and family Dolichodoridae. They are of large size with adults 1.5–3 mm long with a head framework strongly sclerotized, long stylet that is 50–160 μm in length with its cone much longer than the shaft, and a body almost straight when killed by heat. *See Dolichodorus heterocephalus.*

Dolichodorus heterocephalus (plant parasitology) A species of nematodes in the genus *Dolichodorus* whose hosts cover a narrow range including celery, sweet corn, water chestnut, bean, tomato, pepper, cranberry, cabbage, balsam, and carnation, distributed over the Eastern USA, especially Florida, and also in South Africa. These nematodes are ectoparasite, preferring wet locations with high soil moisture such as swamps, marshes and the edges of lakes and streams. Males and females are both present, and the first larval moult (*USA*: molt) occurs in the egg, second stage larvae hatching after 14–17 days at 20–23°C. Males, females, and all larval stages feed, individuals sometimes remaining at one location on the root for up to seven days, although the nematode does not survive in fallowed pots for longer than three months. Feeding causes brownish cortical lesions and enlarged cortical nuclei. Root elongation ceases and secondary roots are attacked, resulting in stubby-root symptoms with some terminal galling ocurring. Control measures includes the use of preplant nematicides. Also termed: awl nematode.

DOM (biochemistry) An abbreviation of 2,5-'D'imeth'O'xy-4-'M'ethylamphetamine.

DOMA (biochemistry) An abbreviation of 3,4-'D'ihydr'O'xy'M'andelic 'A'cid.

domestic (epidemiology) Pertaining to an environment managed by humans.

dominant (genetics) In a heterozygous organism, describing the gene that prevents the expression of a recessive allele in a pair of homologous chromosomes. Thus the phenotype of an organism with a combination of dominant and recessive genes is similar to that with two dominant alleles.

donor (1 chemistry; 2 physics) 1 Atom that donates both electrons to form a co-ordinate bond. 2 An element that donates electrons to form an n-type semiconductor; *e.g.* antimony or arsenic may be donor elements for germanium or silicon.

DOPAC (biochemistry) An abbreviation of 3,4-'D'ihydr'O'xy'P'henylacetic 'AC'id.

dopamine (biochemistry) Precursor in the synthesis of adrenaline (*USA*: epinephrine) and noradrenaline (*USA*: norepinephrine) in animals. It is found in highest concentration in the corpus striatum of the brain, where it functions as a neurotransmitter.

DOPEG (biochemistry) An abbreviation of 3,4-'D'ihydr'O'xy'P'h'E'nyl'G'lycol.

DOPET (biochemistry) 3,4-'D'ihydr'O'xy'P'henyl 'ET'hanol.

dormancy (physiology) Period of minimal metabolic activity of an organism or reproductive body. It is a means of surviving a period of adverse environmental conditions, *e.g.* cold or drought. Examples of some dormant structures are spores, cysts and perennating organs of plants. Environmental factors such as day length and temperature control both the onset and ending of dormancy. Dormancy may also be prompted and terminated by the action of hormones, *e.g.* abscinic acid and gibberellins, respectively.

dors(o)- (terminology) A word element. [Latin] Prefix denoting the *back*, *the dorsal aspect*.

dorsal (anatomy) Describing the upper surface of an organism. In vertebrates this is the surface nearest to the backbone. In plants a dorsal surface is considered to be one facing away from the main stem or root.

DOS (computing) An abbreviation of 'D'isk 'O'perating 'S'ystem.

dosimeter (radiology) Instrument that measures the dose of radiation received by an individual or an area.

double bond (chemistry) Covalent bond that is formed by sharing two pairs of electrons between two atoms.

double decomposition (chemistry) Reaction between two dissolved ionic substances (usually salts) in which the reactants 'change partners' to form a new soluble salt and an insoluble one, which is precipitated, *e.g.* solutions of sodium chloride (NaCl) and silver nitrate ($AgNO_3$) react to form a solution of sodium nitrate ($NaNO_3$) and a precipitate of silver chloride (AgCl).

double recessive (genetics) Homozygote condition in which two recessive alleles of a particular gene are at the same locus on a pair of homologous chromosomes, so that the recessive form of the gene is expressed in the phenotype.

dourine (veterinary parasitology) A sexually transmitted trypanosomiasis of horses caused by *Trypanosoma equiperdum*, characterized by inflammation of the external genitalia, oedema (*USA*: edema) of the ventral abdominal wall, muscle weakness and incoordination. Severe loss of condition may be followed by emaciation to the point where euthanasia is necessary. Urticaria-like plaques called dollar spots occur on the skin in some forms of the disease.

down time (computing) Time periods when a computer is not available for use, usually because of hardware or software malfunction.

DP (computing) An abbreviation of 'D'ata 'P'rocessing.

DPath. (education) An abbreviation of 'D'iploma in 'Path'ology.

DPG (biochemistry) An abbreviation of 2,3-'D'i'P'hospho'G'lycerate.

DPH (education) An abbreviation of 'D'epartment/'D'iploma/'D'octor of 'P'ublic 'H'ealth.

DPhil. (education) An abbreviation of 'D'octor of 'Phil'osophy.

DPL (biochemistry) An abbreviation of 'D'i'P'almitoyl 'L'ecithin.

DPN (biochemistry) An abbreviation of 'D'i'P'hosphopyridine 'N'ucleotide.

DPPC (biochemistry) An abbreviation of 'D'i'P'almitoyl'P'hosphatidyl'C'holine.

Dr (education) An abbreviation of 'D'octo'r'.

Dr Med. (education) An abbreviation of 'D'octo'r' of 'Med'icine.

Dr PH (education) An abbreviation of 'D'octor of 'P'ublic 'H'ealth.

dracunculiasis (parasitology) Infection and infestation by nematodes of the genus *Dracunculus*. Also termed: dracunculosis.

Dracunculus (parasitology) A genus of spiruroid nematode parasites in the family Dracunculidae.

Dracunculus alii (veterinary parasitology) A species of spiruroid nematode parasites in the genus *Dracunculus* that may be found in reptiles.

Dracunculus dahomensis (veterinary parasitology) A species of spiruroid nematode parasites in the genus *Dracunculus* that may be found in reptiles.

Dracunculus globocephalus (veterinary parasitology) A species of spiruroid nematode parasites in the genus *Dracunculus* that may be found in reptiles.

Dracunculus fuelliborni (veterinary parasitology) A species of spiruroid nematode parasites in the genus *Dracunculus* that may be found in opossum.

Dracunculus insignis (veterinary parasitology) A species of spiruroid nematode parasites in the genus *Dracunculus* that may be found infesting dogs and wild carnivores. They cause cutaneous lesions and ulcers, sometimes internal lesions, *e.g.* in heart and vertebral column. Also termed: dragon; fiery dragon; guinea worm.

Dracunculus lutrae (veterinary parasitology) A species of spiruroid nematode parasites in the genus *Dracunculus* that may be found in otter.

Dracunculus medinensis (human/veterinary parasitology) A species of thread-like spiruroid nematode parasites in the genus *Dracunculus* that is frequently found in the subcutaneous and intermuscular tissues of humans, dogs, and sometimes horses and cattle where they cause cutaneous nodules and subsequently ulcers. They are widely distributed in North America, Africa, the Near East, East Indies and India.

Dracunculus ophidensis (veterinary parasitology) A species of spiruroid nematode parasites in the genus *Dracunculus* that may be found in reptiles.

Draschia (veterinary parasitology) A genus of the worm family of Habronematidae. Species includes *Draschia megastoma* (synonym: *Habronema megastoma*) that may be found in intramural nodules in the stomach wall of horses.

DRCPath (education) An abbreviation of 'D'iploma of the 'R'oyal 'C'ollege of 'Path'ologists.

dried fruit mite itch (parasitology) *See Carpoglyphus*.

droppings (parasitology) A term commonly applied to faeces (*USA*: feces), particularly from birds.

dry cell (physics) Electrolytic cell containing no free liquid electrolyte. A moist paste of ammonium chloride (NH_4Cl) often acts as the electrolyte. Dry cells are used in batteries for torches, portable radios, etc.

dry ice (chemistry) Solid carbon dioxide.

DS (education) An abbreviation of 'D'octor of 'S'cience. Also termed DSc.

DSc. (education) *See* DS.

DSc.- (PH) (education) An abbreviation of 'D'octor of 'Sc'ience – 'P'ublic 'H'ealth.

DSc.- Tech. (education) An abbreviation of 'D'octor of 'Sc'ience -'Tech'nical.

DTCH (education) An abbreviation of 'D'iploma in 'T'ropical 'C'hild 'H'ealth.

DTech. (education) An abbreviation of 'D'octor of 'Tech'nology.

DTH (education) An abbreviation of 'D'iploma in 'T'ropical 'H'ygiene.

DTM (1 immunology; 2 education) 1 An abbreviaion of 'D'elayed 'T'ype 'H'ypersensitivity. 2 An abbreviation of 'D'iploma in 'T'ropical 'M'edicine.

DTM&H (education) An abbreviation of 'D'iploma in 'T'ropical 'M'edicine and 'H'ygiene.

DTP (computing) An abbreviation of 'D'esk 'T'op 'P'ublishing.

DTPH (education) An abbreviation of 'D'iploma in 'T'ropical 'P'ublic 'H'ealth.

Duboscqia coptotermi (insect parasitology) A protozoal species of microsporidia that are parasites to the termites *Coptotermes*, *Reticulitermes* spp., and *Termes*. They may be found in the midgut and body fat with infected cells being hypertrophied.

Duboscqia legeri (insect parasitology) A protozoal species of microsporidia that are parasites to the termites *Coptotermes*, *Reticulitermes* spp., and *Termes*. They may be found in the midgut and body fat with infected cells being hypertrophied. Generally infections are light.

duck louse (avian parasitology) *See Anaticola crassicornis* (slender duck louse), *Trinoton querquedulae* (large duck louse).

ductless gland (anatomy) An alternative term for endocrine gland

Duncan applicator (control measures) A device used for the control of ticks in game animals; as the animal licks bait feed in the centre (*USA*: center) of the device it receives a dose of pour-on insecticide.

dung (parasitology) Faeces (*USA*: feces). Also termed: manure, droppings.

duodecimal system (mathematics) Number system that uses the base 12, which requires two additional numbers in addition to 1 to 9 (to represent ten and eleven), usually called A and B.

duodenal (anatomy) Of or pertaining to the duodenum.

duodenum (anatomy) First section of the small intestine which is mainly secretory in function, producing digestive enzymes. It also receives pancreatic juice from the pancreas and bile from the gall bladder.

Durikainema macropti (veterinary parasitology) A nematode parasite that may be found in the venous system of some kangaroo species.

dust mite (parasitology) *See* housedust mite.

dusting powders (control measures) A popular form of applying external parasiticides onto skin generally.

dv (medical) An abbreviation of 'd'ouble 'v'ision.

DVM (education) An abbreviation of 'D'octor of 'V'eterinary 'M'edicine.

dwarf tapeworm (parasitology) *See Hymenolepis nana.*

Dy (chemistry) The chemical symbol for dysprosium.

dye (chemistry) Any of various coloured (*USA*: colored) substances containing auxochromes and are thus capable of colouring (*USA*: coloring) substances to which they are applied; used for staining and coloring (*USA*: coloring), as test reagents, and as therapeutic agents.

dynamic isomerism (chemistry) An alternative term for tautomerism.

dyne (measurement) Force that gives an object of mass 1 gram an acceleration of 1 cm s^{-2}. The SI unit of force is the newton, equal to 10^5 dynes.

dynein (parasitological physiology) A protein from the microtubules of cilia and flagella which functions as an ATP-splitting enzyme and is essential to the motility of cilia and flagella.

dysentery (medical) Any of a number of disorders marked by inflammation of the intestine, especially of the colon, with abdominal pain, tenesmus, and frequent stools often containing blood and mucus. There are a variety of causative agents including protozoa, and parasitic worms.

See also coccidiosis; trichuriasis; *Entamoeba histolytica.*

dysprosium (Dy) (chemistry) Silvery metallic element in Group IIIB of the periodic table (one of the lanthanides) used to make magnets and nuclear reactor control rods. At. no.66; r.a.m. 162.50.

E

e (1 mathematics; 2 physics) 1 Fundamental mathematical constant and the base of natural (Napierian) logarithms. It is an irrational number, the limiting value of the series $(1 + 1/n)^n$ as n tends to infinity, approximately equal to 2.71828. It is the constant in an exponential function or series. 2 An abbreviation of 'e'lectron.

E (biochemistry) A symbol for glutamic acid.

E cells (physiology) A symbol for expiratory neurons.

e.g. (literary terminology) An abbreviation of *'e'xempli 'g'ratia.* [Latin, meaning, 'for example'.]

E$_1$ (biochemistry) A symbol for estrone.

E$_2$ (biochemistry) A symbol for estradiol.

EACA (biochemistry) An abbreviation of 'E'psilon-'A'mino'C'aproic 'A'cid.

ear (anatomy) One of a pair of hearing and balance sensory organs situated on each side of the head of vertebrates. In mammals it consists of three parts: the outer, middle and inner ear.

ear canker (parasitology) A lay term applied generally to otitis externa but sometimes specifically to that caused by ear mites.

ear drum (anatomy) Membrane at the inner end of the auditory canal of the outer ear which transmits sound vibrations to the ear ossicles of the middle ear. Also termed: tympanum.

ear mange (parasitology); Also termed auricular mange. *See* otodectic mange, psoroptic mange. *Otodectes cynotis*; *Raillietia*.

ear mites (parasitology) *See Psoroptes cuniculi*; *Raillietia auris*; *Raillietia caprae*.

ear ossicle (anatomy) Small bone found in the middle ear of vertebrates. In mammals there are three, which transmit sound waves from the ear drum to the hearing sensory cells in the cochlea of the inner ear. The three mammalian ossicles are the mailcus, incus and stapes. Amphibians, reptiles and birds have only one ear ossicle, the columella auris.

ear tick (parasitology) *See Otobius megnini.*

earsore (veterinary parasitology) Dermatitis around the ears of water buffalo caused by *Stephanofilaria zaheeri*.

earth (physics) In electric circuits, a connection to a piece of metal that is in turn linked to the Earth. It has the effect of preventing any earthed apparatus from retaining an electric charge.

earthworm (parasitology) The common oligochete worm of the genera *Lumbricus*, *Allobophora*, *Eisenia* etc.; they act as intermediate hosts for a number of internal parasites of livestock.

earworm (veterinary parasitology) Infestation of the ears of cattle by *Rhabditis bovis*, often complicated by blowfly infestation.

East African sleeping sickness (human/veterinary parasitology) A disease of humans caused by *Trypanosoma rhodesiense*. The parasite is infectious for many animal species which act as reservoirs for humans. The disease is fatal in humans if it is not treated. Keratitis and encephalitis occur in goats and sheep, facial paralysis and emaciation in horses.

ec (literary terminology) An abbreviation of *'e'xempli 'c'ausa.* [Latin, meaning, 'for example'.]

ecdysis (parasitological physiology) A term used in relation to arthropods. Shedding of larval skin during transformation from one stage to another.

ECF (histology) An abbreviation of 'E'xtra'C'ellular 'F'luid.

eCG (biochemistry) An abbreviation of 'e'quine 'C'horionic 'G'onadotropin.

Echidnophaga (parasitology) A genus of fleas that remain attached to the host for long periods.

Echidnophaga gallinaeea (human/avian parasitology) A species of fleas in the genus *Echidnophaga* that causes insect worry and blood loss in poultry. They do not transmit disease in animals but may transmit endemic murine typhus to humans. Also termed stickfast flea; sticktight flea.

Echidnophaga myrmecobii (veterinary parasitology) A species of fleas in the genus *Echidnophaga* that may be found in rabbits.

Echidnophaga perilis (veterinary parasitology) A species of flea whose principal hosts are rabbits.

Echinochasmus (veterinary parasitology) A genus of flukes of the family Echinostomatidae. Species includes *Echinochasmus* perfoliatus that may be found in the intestines of carnivores.

echinococcosis (human/veterinary parasitology) An infection of humans and animals, usually of the liver or lungs, caused by the larval stage (hydatid cysts) of tapeworms of the genus *Echinococcus*, marked by the development of expanding cysts.

See also hydatid disease.

Echinococcus (parasitology) A genus of small tapeworms of the family Taeniidae.

echinococcus cyst (parasitology) Hydatid cyst.

Echinococcus granulosus (veterinary parasitology) A species of tapeworms in the genus *Echinococcus* that are parasitic in dogs and wolves and occasionally in cats. Their larvae may develop in ungulates and macropods, forming hydatid cysts in the liver, lungs, kidneys and other organs.

Echinococcus multilocularis (human/veterinary parasitology) A species of tapeworms in the genus *Echinococcus* whose adult stage usually parasitizes the fox, dog and cat. It resembles *Echinococcus granulosus* but the larvae form alveolar or multilocular rather than unilocular cysts and occur principally in rodents but can infect humans.

Echinococcus oligarthus (veterinary parasitology) A species of tapeworms in the genus *Echinococcus* that may be found in wild cats with larval stages in rodents.

Echinococcus vogeli (human/veterinary parasitology) A species of tapeworms in the genus *Echinococcus* that may be found in domestic and wild dogs with intermediate stages in rodents and humans.

Echinolaelaps echidninus (veterinary parasitology) A mite of the Gamasidae family. The spiny rat mite, transmitter of *Hepatozoon* spp., the protozoan blood parasite. Also termed: lelapid mite.

Echinoparyphium (parasitology) A genus of fluke in the family Echinostomatidae.

Echinoparyphium paraulum (parasitology) *See Echinostoma revolutum.*

Echinoparyphium recurvatum (avian parasitology) A species of flukes in the genus *Echinoparyphium* which may be found in the small intestine of doves, pigeons and domesticated birds, and that can cause emaciation and anaemia (*USA*: anemia).

Echinophthirius (veterinary parasitology) A genus of lice that may be found on pinnipeds i.e. seals, walrus and sea lions.

Echinophthirius horridus (veterinary parasitology) A genus of lice that may cause pruritus and self-injury.

Echinorhynchus salmonis (fish parasitology) An acanthocephalan parasite of freshwater and marine fish that may be found in the intestine of salmonids.

Echinostoma (parasitology) A genus of flukes in the family Echinostomatidae. Species includes *Echinostoma aphylactum*, *Echinostoma caproni*, *Echinostoma hortense*, *Echinostoma jassyenese*, *Echinostoma lindoensis* and *Echinostoma suinum*.

Echinostoma iliocanum (human/veterinary parasitology) A species of flukes in the genus *Echinostoma* which may be found in the intestine of dogs, rodents and humans and that may cause enteritis.

Echinostoma revolutum (human/avian parasitology) (synonym: *Echinostoma paraulum*) A species of flukes in the genus *Echinostoma* which may be found in the rectum and caecum (*USA*: ceca) of birds and in humans. Severe infestations may cause enteritis.

Echinostomatidae (avian parasitology) A family of flukes (digenetic trematodes) that may be found in birds.

Echinuria (avian parasitology) A genus of spiruroid worms of the family Acuaridae. Species includes *Echinuria uncinata* that infests the upper alimentary canal of birds, causing caseous nodules in the wall of the oesophagus (*USA*: esophagus).

ECoG (physiology) An abbreviation of 'E'lectro-'Co'rtico'G'ram.

ecological niche (biology) Position that a particular species occupies within an ecosystem. The term both describes the function of a species in terms of interactions with other species, *e.g.* feeding behaviour (*USA*: behavior), and defines the physical boundaries of the environment occupied by the species, *e.g.* bats are said to occupy an airborne niche.

ecology (biology) Study of interaction of organisms between themselves and with their physical environment.

ecosystem (biology) Natural unit that contains living and non-living components (*e.g.* a community and its environment) interacting and exchanging materials, and generally balanced as a stable system; *e.g.* grassland or rainforest.

ect(o)- (terminology) A word element. [German] Prefix denoting *external, outside.*

ectoderm (histology) Outermost germ layer of the embryo of a metazoan which develops into tissues of the epidermis, *e.g.* skin, hair, sense organs. Also termed: epiblast.

-ectomy (terminology) A word element. [German] Suffix denoting *excision, surgical removal.*

ectoparasite (parasitology) Parasite that lives on the outside of its host; *e.g.* flea.

ectoparasitism (parasitology) The state in which the ectoparasite is living on the surface of the host's body.

ectophyte (plant parasitology) A plant parasite living on the surface of the host's body.

ectoplasm (cytology) Non-granulated jelly-like outer layer of cytoplasm that is located below the plasma membrane. It is characteristic of most amoeboid animal cells, in which at its boundary with plasmasol it aids cytoplasmic streaming, and thus movement; *e.g.* in amoeboid protozoa and leucocytes.

ectozoon (parasitology) Ectoparasite.

ectro- (terminology) A word element. [German] Prefix denoting *miscarriage*, *congenital absence*.

ED (pharmacology) An abbreviation of 'E'ffective 'D'ose.

Ed. in Ch. (literary teminology) An abbreviation of 'Ed'itor in 'Ch'ief.

Edbarellus (insect parasitology) *See Calaenostanus*.

eddy current (physics) Electric current within a conductor caused by electromagnetic induction. Such currents result in losses of energy in electrical machines (*e.g.* a transformer, in which they are overcome by laminating the core), but are utilized in induction heating and some braking systems.

Edison accumulator (physics) An alternative term for nickel-iron accumulator.

EDRF (physiology) An abbreviation of 'E'ndothelium-'D'erived 'R'elaxing 'F'actor.

EDTA (biochemistry) An abbreviation of 'E'thylene'D'iamine'T'etracetic 'A'cid. It is a white crystalline organic compound, used generally as its sodium salt as an analytical reagent.

EEG (physiology) An abbreviation of 'E'lectro-'E'ncephalo'G'ram.

EET (biochemistry) An abbreviation of 'E'poxy-'E'icosa'T'etraenoic acid.

EFF (parasitology) An abbreviation of 'E'lokomin 'F'luke 'F'ever.

effective resistance (physics) Total alternating current resistance of a conductor of electricity.

effector (physiology) Tissue or organ that responds to a nervous stimulus (*e.g.* endocrine gland, muscle).

efferent (anatomy) Leading away from, as applied to vessels, fibres (*USA*: fibers) and ducts leading from organs.

See also afferent.

effervescence (chemistry) Evolution of bubbles of a gas from a liquid.

efflorescence (chemistry) Property of certain crystalline salts that lose water of crystallization on exposure to air and become powdery; *e.g.* sodium carbonate decahydrate (washing soda), $Na_2CO_3 . 10H_2O$. It is the opposite of deliquescence.

effort (physics) In a simple machine (*e.g.* lever, pulley) the force that is applied to move a load. The ratio of the load to the effort is the mechanical advantage (force ratio).

effusion (physics) Passage of gases under pressure through small holes.

eflornithine (pharmacological) An antiprotozoal agent, that may be used as the hydrochloride in treatment of trypanosomiasis and *pneumocystis carinii* pneumonia.

EGA (computing) An abbreviation of 'E'nhanced 'G'raphics 'A'dapter.

EGF (histology) An abbreviation of 'E'pidermal 'G'rowth 'F'actor.

egg (1, 2, 3, 4 parasitology) 1 An ovum; a female gamete. 2 An oocyte. 3 A female reproductive cell at any stage before fertilization and its derivatives after fertilization and even after some development. 4 Helminth egg.

egg capsule (parasitological anatomy) A term used in relation to cestodes. A membranous structure containing eggs of a tapeworm, in the absence of a uterus.

egg count (parasitology) Counting of helminth eggs as an estimate of the parasite status in the animal or group. Flotation techniques and special counting chambers are used. The results are expressed as eggs per gram of faeces (*USA*: feces) or e.p.g.

egg membrane (histology) Thin protective membrane that surrounds the fertilized ovum of animals. It is secreted by the oocyte and the follicle cells.

See also chorion.

EHF (physics) An abbreviation of 'E'xtremely 'H'igh 'F'requency.

Eickwortius termes (insect parasitology) Mites in the Mesostigmata that are parasitic on the termites *Macrotermes michaelseni*.

Eimeria (avian parasitology) A genus of protozoan parasites in the family Eimeriidae. There are many species, mostly in birds and herbivores, and they are the principal cause of coccidiosis. *See Cystoisospora*; *Isospora*; *Tyzzeria*; *Wenyonella*.

Eimeria abramovi (avian parasitology) A species of protozoan parasites in the genus *Eimeria* that may be found in wild duck and geese.

Eimeria acervulina (avian parasitology) A species of protozoan parasites in the genus *Eimeria* that may be found in domestic poultry and quail.

Eimeria adenoeides (avian parasitology) A species of protozoan parasites in the genus *Eimeria* that may be found in turkeys.

Eimeria ahsata (veterinary parasitology) A species of protozoan parasites in the genus *Eimeria* that may be found in sheep and goats.

Eimeria alabamensis (veterinary parasitology) A species of protozoan parasites in the genus *Eimeria* that may be found in cattle.

Eimeria alijeva (veterinary parasitology) A species of protozoan parasites in the genus *Eimeria* that may be found in goats.

Eimeria alpacae (veterinary parasitology) A species of protozoan parasites in the genus *Eimeria* that may be found in llamas and alpacas.

Eimeria anatis (avian parasitology) A species of protozoan parasites in the genus *Eimeria* that may be found in mallard and domestic ducks.

Eimeria ankarensis (veterinary parasitology) A species of protozoan parasites in the genus *Eimeria* that may be found in water buffalo.

Eimeria anseris (avian parasitology) A species of protozoan parasites in the genus *Eimeria* that may be found in domestic and wild geese.

Eimeria apsheronica (veterinary parasitology) A species of protozoan parasites in the genus *Eimeria* that may be found in goats.

Eimeria arkhari (veterinary parasitology) A species of protozoan parasites in the genus *Eimeria* that may be found in sheep and goats.

Eimeria arloingi (veterinary parasitology) A species of protozoan parasites in the genus *Eimeria* that may be found in sheep and goats.

Eimeria auburnensis (veterinary parasitology) A species of protozoan parasites in the genus *Eimeria* that may be found in cattle.

Eimeria augusta (avian parasitology) A species of protozoan parasites in the genus *Eimeria* that may be found in grouse.

Eimeria aurata (fish parasitology) A species of protozoan parasites in the genus *Eimeria* that may be found in fish.

Eimeria azerbaidschanica (veterinary parasitology) A species of protozoan parasites in the genus *Eimeria* that may be found in water buffalo.

Eimeria bactriani (veterinary parasitology) A species of protozoan parasites in the genus *Eimeria* that may be found in one- and two-humped camels.

Eimeria bakuensis (veterinary parasitology) A species of protozoan parasites in the genus *Eimeria* that may be found in sheep.

Eimeria bareillyi (veterinary parasitology) A species of protozoan parasites in the genus *Eimeria* that may be found in water buffalo.

Eimeria batyakhi (avian parasitology) A species of protozoan parasites in the genus *Eimeria* that may be found in domestic ducks.

Eimeria bombayanis (veterinary parasitology) A species of protozoan parasites in the genus *Eimeria* that may be found in zebu cattle.

Eimeria bonasae (avian parasitology) A species of protozoan parasites in the genus *Eimeria* that may be found in grouse.

Eimeria boschadis (avian parasitology) A species of protozoan parasites in the genus *Eimeria* that may be found in wild ducks and geese.

Eimeria bovis (veterinary parasitology) A species of protozoan parasites in the genus *Eimeria* that may be found in cattle, zebu and water buffalo.

Eimeria brantae (avian parasitology) A species of protozoan parasites in the genus *Eimeria* that may be found in wild ducks and geese.

Eimeria brasiliensis (veterinary parasitology) A species of protozoan parasites in the genus *Eimeria* that may be found in cattle, zebu and water buffalo.

Eimeria brinkmanni (veterinary parasitology) A species of protozoan parasites in the genus *Eimeria* that may be found in rock ptarmigans.

Eimeria brunetti (avian parasitology) A species of protozoan parasites in the genus *Eimeria* that may be found in domestic poultry.

Eimeria bucephalae (avian parasitology) A species of protozoan parasites in the genus *Eimeria* that may be found in wild ducks and geese.

Eimeria bukidnonensis (veterinary parasitology) A species of protozoan parasites in the genus *Eimeria* that may be found in cattle, zebu and buffalo.

Eimeria cameli (veterinary parasitology) A species of protozoan parasites in the genus *Eimeria* that may be found in one- and two-humped camels.

Eimeria canadensis (veterinary parasitology) A species of protozoan parasites in the genus *Eimeria* that may be found in domestic cattle, zebu, bison and water buffalo.

Eimeria canis (veterinary parasitology) A species of protozoan parasites in the genus *Eimeria* that may be found in dogs and cats.

Eimeria caprina (veterinary parasitology) A species of protozoan parasites in the genus *Eimeria* that may be found in goats.

Eimeria caprovina (veterinary parasitology) A species of protozoan parasites in the genus *Eimeria* that may be found in goats.

Eimeria carinii (veterinary parasitology) A species of protozoan parasites in the genus *Eimeria* that may be found in rats.

Eimeria carpelli (fish parasitology) A species of protozoan parasites in the genus *Eimeria* that may be found in fish.

Eimeria caviae (veterinary parasitology) A species of protozoan parasites in the genus *Eimeria* that may be found in guinea pigs.

Eimeria cerdonis (veterinary parasitology) A species of protozoan parasites in the genus *Eimeria* that may be found in pigs.

Eimeria christenseni (veterinary parasitology) A species of protozoan parasites in the genus *Eimeria* that may be found in domestic goats.

Eimeria christianseni (avian parasitology) A species of protozoan parasites in the genus *Eimeria* that may be found in mute swans.

Eimeria clarkei (avian parasitology) A species of protozoan parasites in the genus *Eimeria* that may be found in lesser snow geese.

Eimeria coecicola (veterinary parasitology) A species of protozoan parasites in the genus *Eimeria* that may be found in rabbits.

Eimeria colchici (avian parasitology) A species of protozoan parasites in the genus *Eimeria* that may be found in pheasants.

Eimeria columbae (avian parasitology) A species of protozoan parasites in the genus *Eimeria* that may be found in pigeons.

Eimeria columbarum (avian parasitology) A species of protozoan parasites in the genus *Eimeria* that may be found in rock doves.

Eimeria coturnicus (avian parasitology) A species of protozoan parasites in the genus *Eimeria* that may be found in quails.

Eimeria crandallis (veterinary parasitology) A species of protozoan parasites in the genus *Eimeria* that may be found in domestic sheep, and small wild ruminants.

Eimeria cylindrica (veterinary parasitology) A species of protozoan parasites in the genus *Eimeria* that may be found in domestic cattle, zebu and water buffalo.

Eimeria cyprini (fish parasitology) A species of protozoan parasites in the genus *Eimeria* that may be found in fish.

Eimeria danailovi (avian parasitology) A species of protozoan parasites in the genus *Eimeria* that may be found in mallards.

Eimeria danielle (veterinary parasitology) A species of protozoan parasites in the genus *Eimeria* that may be found in domestic sheep.

Eimeria debliecki (veterinary parasitology) A species of protozoan parasites in the genus *Eimeria* that may be found in pigs.

Eimeria dispersa (avian parasitology) A species of protozoan parasites in the genus *Eimeria* that may be found in turkeys.

Eimeria dolichotis (veterinary parasitology) A species of protozoan parasites in the genus *Eimeria* that may be found in Patagonian cavey.

Eimeria dromedarii (veterinary parasitology) A species of protozoan parasites in the genus *Eimeria* that may be found in one- and two- humped camels.

Eimeria ellipsoidalis (veterinary parasitology) A species of protozoan parasites in the genus *Eimeria* that may be found in domestic cattle, zebu, European bison and water buffalo.

Eimeria elongata (veterinary parasitology) A species of protozoan parasites in the genus *Eimeria* that may be found in domestic rabbits.

Eimeria exigua (veterinary parasitology) A species of protozoan parasites in the genus *Eimeria* that may be found in rabbits and Greenland hares.

Eimeria falciformis (veterinary parasitology) A species of protozoan parasites in the genus *Eimeria* that may be found in mice.

Eimeria fanthami (avian parasitology) A species of protozoan parasites in the genus *Eimeria* that may be found in rock ptarmigans.

Eimeria farri (avian parasitology) A species of protozoan parasites in the genus *Eimeria* that may be found in white fronted geese.

Eimeria faurei (veterinary parasitology) A species of protozoan parasites in the genus *Eimeria* that may be found in sheep and small wild ruminants.

Eimeria ferrisi (veterinary parasitology) A species of protozoan parasites in the genus *Eimeria* that may be found in mice.

Eimeria fulva (avian parasitology) A species of protozoan parasites in the genus *Eimeria* that may be found in wild geese.

Eimeria gallopavonis (avian parasitology) A species of protozoan parasites in the genus *Eimeria* that may be found in turkeys.

Eimeria gilruthi (veterinary parasitology) A species of protozoan parasites in the genus *Eimeria* that may be found in sheep and goats.

Eimeria gokaki (veterinary parasitology) A species of protozoan parasites in the genus *Eimeria* that may be found in buffalo.

Eimeria gonzalei (veterinary parasitology) A species of protozoan parasites in the genus *Eimeria* that may be found in sheep.

Eimeria gorakhpuri (avian parasitology) A species of protozoan parasites in the genus *Eimeria* that may be found in guinea fowl.

Eimeria granulosa (veterinary parasitology) A species of protozoan parasites in the genus *Eimeria* that may be found in domestic and wild sheep.

Eimeria grenieri (avian parasitology) A species of protozoan parasites in the genus *Eimeria* that may be found in guinea fowl.

Eimeria guevarai (veterinary parasitology) A species of protozoan parasites in the genus *Eimeria* that may be found in pigs.

Eimeria hagani (avian parasitology) A species of protozoan parasites in the genus *Eimeria* that may be found in poultry.

Eimeria hasei (veterinary parasitology) A species of protozoan parasites in the genus *Eimeria* that may be found in rats.

Eimeria hawkinsi (veterinary parasitology) A species of protozoan parasites in the genus *Eimeria* that may be found in sheep and goats.

Eimeria hermani (avian parasitology) A species of protozoan parasites in the genus *Eimeria* that may be found in wild geese.

Eimeria hindlei (veterinary parasitology) A species of protozoan parasites in the genus *Eimeria* that may be found in mice.

Eimeria hirci (veterinary parasitology) A species of protozoan parasites in the genus *Eimeria* that may be found in goats.

Eimeria illinoisensis (veterinary parasitology) A species of protozoan parasites in the genus *Eimeria* that may be found in cattle.

Eimeria innocua (avian parasitology) A species of protozoan parasites in the genus *Eimeria* that may be found in turkeys.

Eimeria intestinalis (veterinary parasitology) A species of protozoan parasites in the genus *Eimeria* that may be found in rabbits.

Eimeria intricata (veterinary parasitology) A species of protozoan parasites in the genus *Eimeria* that may be found in sheep and wild small ruminants.

Eimeria irresidua (veterinary parasitology) A species of protozoan parasites in the genus *Eimeria* that may be found in rabbits and jackrabbits.

Eimeria jolchijevi (veterinary parasitology) A species of protozoan parasites in the genus *Eimeria* that may be found in goats.

Eimeria keilini (veterinary parasitology) A species of protozoan parasites in the genus *Eimeria* that may be found in mice.

Eimeria kocharli (veterinary parasitology) A species of protozoan parasites in the genus *Eimeria* that may be found in goats.

Eimeria kofoidi (avian parasitology) A species of protozoan parasites in the genus *Eimeria* that may be found in partridges.

Eimeria koganae (avian parasitology) A species of protozoan parasites in the genus *Eimeria* that may be found in wild ducks and geese.

Eimeria kosti (veterinary parasitology) A species of protozoan parasites in the genus *Eimeria* that may be found in cattle.

Eimeria kotlani (avian parasitology) A species of protozoan parasites in the genus *Eimeria* that may be found in domestic geese.

Eimeria krijgsmanni (veterinary parasitology) A species of protozoan parasites in the genus *Eimeria* that may be found in mice.

Eimeria labbeana (avian parasitology) A species of protozoan parasites in the genus *Eimeria* that may be found in pigeons.

Eimeria lagopodi (avian parasitology) A species of protozoan parasites in the genus *Eimeria* that may be found in ptarmigans.

Eimeria lamae (veterinary parasitology) A species of protozoan parasites in the genus *Eimeria* that may be found in alpacas.

Eimeria langeroni (avian parasitology) A species of protozoan parasites in the genus *Eimeria* that may be found in pheasants.

Eimeria leuckarti (veterinary parasitology) A species of protozoan parasites in the genus *Eimeria* that may be found in horses.

Eimeria lyruri (avian parasitology) A species of protozoan parasites in the genus *Eimeria* that may be found in partridges.

Eimeria macusaniensis (veterinary parasitology) A species of protozoan parasites in the genus *Eimeria* that may be found in llamas and alpacas.

Eimeria magna (veterinary parasitology) A species of protozoan parasites in the genus *Eimeria* that may be found in rabbits and hares.

Eimeria magnalabia (avian parasitology) A species of protozoan parasites in the genus *Eimeria* that may be found in wild geese.

Eimeria mandali (avian parasitology) A species of protozoan parasites in the genus *Eimeria* that may be found in peafowl.

Eimeria marsica (veterinary parasitology) A species of protozoan parasites in the genus *Eimeria* that may be found in sheep.

Eimeria matsubayashii (veterinary parasitology) A species of protozoan parasites in the genus *Eimeria* that may be found in domestic rabbits.

Eimeria maxima (avian parasitology) A species of protozoan parasites in the genus *Eimeria* that may be found in poultry.

Eimeria mayurai (avian parasitology) A species of protozoan parasites in the genus *Eimeria* that may be found in peafowl.

Eimeria media (veterinary parasitology) A species of protozoan parasites in the genus *Eimeria* that may be found in domestic and wild rabbits.

Eimeria megalostromata (avian parasitology) A species of protozoan parasites in the genus *Eimeria* that may be found in pheasants.

Eimeria meleagridis (avian parasitology) A species of protozoan parasites in the genus *Eimeria* that may be found in turkeys.

Eimeria meleagrimitis (avian parasitology) A species of protozoan parasites in the genus *Eimeria* that may be found in domestic turkeys.

Eimeria mieschultzi (veterinary parasitology) A species of protozoan parasites in the genus *Eimeria* that may be found in rats.

Eimeria mitis (avian parasitology) A species of protozoan parasites in the genus *Eimeria* that may be found in poultry.

Eimeria mivati (avian parasitology) A species of protozoan parasites in the genus *Eimeria* that may be found in domestic fowl.

Eimeria miyairii (veterinary parasitology) A species of protozoan parasites in the genus *Eimeria* that may be found in rats.

Eimeria mundaragi (veterinary parasitology) A species of protozoan parasites in the genus *Eimeria* that may be found in cattle and zebu.

Eimeria musculi (veterinary parasitology) A species of protozoan parasites in the genus *Eimeria* that may be found in mice.

Eimeria nadsoni (avian parasitology) A species of protozoan parasites in the genus *Eimeria* that may be found in grouse.

Eimeria nagpurensis (veterinary parasitology) A species of protozoan parasites in the genus *Eimeria* that may be found in rabbits.

Eimeria necatrix (avian parasitology) A species of protozoan parasites in the genus *Eimeria* that may be found in domestic fowl.

Eimeria neodebliecki (veterinary parasitology) A species of protozoan parasites in the genus *Eimeria* that may be found in domestic and wild pigs.

Eimeria neoleporis (veterinary parasitology) A species of protozoan parasites in the genus *Eimeria* that may be found in rabbits.

Eimeria ninakohlyakimovae (veterinary parasitology) A species of protozoan parasites in the genus *Eimeria* that may be found in sheep, goat and small wild ruminants.

Eimeria nocens (avian parasitology) A species of protozoan parasites in the genus *Eimeria* that may be found in domestic and wild geese.

Eimeria nochti (veterinary parasitology) A species of protozoan parasites in the genus *Eimeria* that may be found in rats.

Eimeria norvegicus (veterinary parasitology) A species of protozoan parasites in the genus *Eimeria* that may be found in rats.

Eimeria numida (avian parasitology) A species of protozoan parasites in the genus *Eimeria* that may be found in guinea fowl.

Eimeria ovina (veterinary parasitology) A species of protozoan parasites in the genus *Eimeria* that may be found in domestic sheep and small wild ruminants.

Eimeria ovoidalis (veterinary parasitology) A species of protozoan parasites in the genus *Eimeria* that may be found in buffalo.

Eimeria pacifica (avian parasitology) A species of protozoan parasites in the genus *Eimeria* that may be found in pheasants.

Eimeria pallida (veterinary parasitology) A species of protozoan parasites in the genus *Eimeria* that may be found in domestic sheep and goats.

Eimeria parva (veterinary parasitology) A species of protozoan parasites in the genus *Eimeria* that may be found in domestic sheep, goats and small wild ruminants.

Eimeria parvula (avian parasitology) A species of protozoan parasites in the genus *Eimeria* that may be found in geese.

Eimeria pavonina (avian parasitology) A species of protozoan parasites in the genus *Eimeria* that may be found in peafowl.

Eimeria pavonis (avian parasitology) A species of protozoan parasites in the genus *Eimeria* that may be found in peafowl.

Eimeria pelleryi (veterinary parasitology) A species of protozoan parasites in the genus *Eimeria* that may be found in bactrian camels.

Eimeria pellita (veterinary parasitology) A species of protozoan parasites in the genus *Eimeria* that may be found in cattle.

Eimeria perforans (veterinary parasitology) A species of protozoan parasites in the genus *Eimeria* that may be found in rabbits and hares.

Eimeria perminuta (veterinary parasitology) A species of protozoan parasites in the genus *Eimeria* that may be found in pigs.

Eimeria peruviana (veterinary parasitology) A species of protozoan parasites in the genus *Eimeria* that may be found in llamas.

Eimeria phasiani (avian parasitology) A species of protozoan parasites in the genus *Eimeria* that may be found in pheasants.

Eimeria piriformis (veterinary parasitology) A species of protozoan parasites in the genus *Eimeria* that may be found in domestic rabbits.

Eimeria polita (veterinary parasitology) A species of protozoan parasites in the genus *Eimeria* that may be found in pigs.

Eimeria porci (veterinary parasitology) A species of protozoan parasites in the genus *Eimeria* that may be found in pigs.

Eimeria praecox (avian parasitology) A species of protozoan parasites in the genus *Eimeria* that may be found in poultry.

Eimeria procera (avian parasitology) A species of protozoan parasites in the genus *Eimeria* that may be found in partridges.

Eimeria punctata (veterinary parasitology) A species of protozoan parasites in the genus *Eimeria* that may be found in sheep.

Eimeria punoensis (veterinary parasitology) A species of protozoan parasites in the genus *Eimeria* that may be found in alpacas.

Eimeria rajasthani (veterinary parasitology) A species of protozoan parasites in the genus *Eimeria* that may be found in dromedaries.

Eimeria ratti (veterinary parasitology) A species of protozoan parasites in the genus *Eimeria* that may be found in rats.

Eimeria saitamae (avian parasitology) A species of protozoan parasites in the genus *Eimeria* that may be found in ducks.

Eimeria scabra (veterinary parasitology) A species of protozoan parasites in the genus *Eimeria* that may be found in domestic and wild pigs.

Eimeria schueffneri (veterinary parasitology) A species of protozoan parasites in the genus *Eimeria* that may be found in mice.

Eimeria scrofae (veterinary parasitology) A species of protozoan parasites in the genus *Eimeria* that may be found in pigs.

Eimeria separata (veterinary parasitology) A species of protozoan parasites in the genus *Eimeria* that may be found in rats.

Eimeria solipedum (veterinary parasitology) A species of protozoan parasites in the genus *Eimeria* that may be found in horses.

Eimeria somateriae (avian parasitology) A species of protozoan parasites in the genus *Eimeria* that may be found in wild ducks.

Eimeria spinosa (veterinary parasitology) A species of protozoan parasites in the genus *Eimeria* that may be found in pigs.

Eimeria stiedae (veterinary parasitology) A species of protozoan parasites in the genus *Eimeria* that may be found in rabbits and hares.

Eimeria stigmosa (avian parasitology) A species of protozoan parasites in the genus *Eimeria* that may be found in domestic geese.

Eimeria striata (avian parasitology) A species of protozoan parasites in the genus *Eimeria* that may be found in wild geese.

Eimeria subepithelialis (fish parasitology) A species of protozoan parasites in the genus *Eimeria* that may be found in carp.

Eimeria subrotunda (avian parasitology) A species of protozoan parasites in the genus *Eimeria* that may be found in turkeys.

Eimeria subspherica (veterinary parasitology) A species of protozoan parasites in the genus *Eimeria* that may be found in cattle, zebu and water buffalo.

Eimeria suis (veterinary parasitology) A species of protozoan parasites in the genus *Eimeria* that may be found in domestic pigs.

Eimeria tenella (avian parasitology) A species of protozoan parasites in the genus *Eimeria* that may be found in domestic poultry.

Eimeria tetricis (avian parasitology) A species of protozoan parasites in the genus *Eimeria* that may be found in grouse.

Eimeria thianethi (veterinary parasitology) A species of protozoan parasites in the genus *Eimeria* that may be found in buffalo.

Eimeria tropicalis (avian parasitology) A species of protozoan parasites in the genus *Eimeria* that may be found in pigeons.

Eimeria truttae (fish parasitology) A species of protozoan parasites in the genus *Eimeria* that may be found in salmon.

Eimeria uniungulati (veterinary parasitology) A species of protozoan parasites in the genus *Eimeria* that may be found in horses and mules.

Eimeria weybridgensis (veterinary parasitology) A species of protozoan parasites in the genus *Eimeria* that may be found in sheep.

Eimeria wyomingensis (veterinary parasitology) A species of protozoan parasites in the genus *Eimeria* that may be found in cattle, zebu and water buffalo.

Eimeria zuernii (veterinary parasitology) A species of protozoan parasites in the genus *Eimeria* that may be found in cattle, zebu and water buffalo.

Eimeria schachdagica (avian parasitology) A species of protozoan parasites in the genus *Eimeria* that may be found in ducks.

Eimeria truncata (avian parasitology) A species of protozoan parasites in the genus *Eimeria* that may be found in domestic and wild geese.

Einstein equation (physics) Equation deduced from Einstein's special theory of relativity: $E = mc^2$, where E is energy, m is mass and c is the velocity of light. It is named after the German-born American physicist Albert Einstein (1879–1955).

einsteinium (Es) (chemistry) Radioactive metallic element in Group IIIB of the periodic table (one of the lathanides). It has several isotopes, with half-lives of up to 2 years. At. no. 99; r.a.m. 254 (most stable isotope).

EJP (physiology) An abbreviation of 'E'xcitatory 'J'unction 'P'otential.

Elaeophora (parasitology) A genus of filariid nematodes in the family Onchocercidae.

Elaeophora bohmi (veterinary parasitology) A species of filariid nematodes in the genus *Elaeophora* that may be found in the arteries and veins of the extremities of horses. Also termed: *Onchocerca bohmi*.

Elaeophora poeli (veterinary parasitology) A species of filariid nematodes in the genus *Elaeophora* that may be found in the aorta of cattle and other ruminants.

Elaeophora sagittus (veterinary parasitology) A species of filariid nematodes in the genus *Elaeophora* that may be found in bovine hearts. Also termed: *Cordophilus sagittus*.

Elaeophora schneideri (veterinary parasitology) A species of filariid nematodes in the genus *Elaeophora* that may be found in arteries of deer, elk and sheep, and which causes elaeophoriasis.

elaeophorial dermatitis (parasitology) *See* elaeophoriasis.

elaeophoriasis (parasitology) Dermatitis caused by vascular lesions of infestation with microfilariae of *Elaeophora schneideri*. The parasite can also cause rhinitis, keratoconjunctivitis and stomatitis in sheep infected by bites of horse flies. This and other species may also cause sporadic cases of disease due to vascular obstruction.

elaeophorosis (parasitology) *See* elaeophoriasis.

elaphostrongyliasis (parasitology) Infestation with *Paraelaphostrongylus tenuis*. *See* neurofilariasis.

Elaphostrongylus (parasitology) A genus of metastrongylid nematodes of the family Protostrongylidae.

Elaphostrongylus cervi (veterinary parasitology) A species of nematodes in the genus *Elaphostrongylus* which may be found in connective tissue and central nervous system of deer and that causes cerebrospinal nematodiasis.

Elaphostrongylus panticola (veterinary parasitology) A species of nematodes in the genus *Elaphostrongylus* that may be found in the brain of deer.

Elaphostrongylus rangiferi (veterinary parasitology) A species of nematodes in the genus *Elaphostrongylus* which may be found in the central nervous system and muscles of reindeer.

See also cerebrospinal nematodiasis.

Elaphostrongylus tenuis (parasitology) *Paraelaphostrongylus tenuis*. *See* neurofilariasis.

elastic Van Gieson stain (evg) (histology) A histological technique for staining elastic and connective tissue.

ele(o)- (terminology) A word element. [German] Prefix denoting *oil*.

electric charge (physics) Excess or deficiency of electrons in an object, giving rise to an overall positive or negative electric charge respectively.

electric constant (physics) An alternative term for permittivity of free space.

electric current (physics) Drift of electrons through a conductor in the same direction, usually because there is a potential difference across it, measured in amperes.

electric displacement (physics) Electric charge per unit area, in coulombs per square metre ($C\ m^{-2}$). Also termed: electric flux density.

electric field (physics) Region surrounding an electric charge in which a charged particle is subjected to a force.

electric flux (physics) Lines of force that make up an electric field.

electric motor (physics) Device that converts electrical energy into mechanical energy. A simple electric motor consists of a current-carrying coil that rotates in the magnetic field between the poles of a permanent magnet.

electric polarization (physics) Difference between the displacement of charge and the electric field strength in a dielectric.

electrical condenser (physics) An alternative term for capacitor.

electrical conductivity (measurement) Measure of the ability of a substance to conduct electricity, the reciprocal of resistivity. It is measured in $ohms^{-1}\ m^{-1}$.

electrical line of force (physics) Line radiating from an electric field.

electricity (physics) Branch of science that is concerned with all phenomena caused by static or dynamic electric charges.

electro- (terminology) A word element. [German] Prefix denoting *electricity*.

electrocardiogram (ECG) (physiology) Record of the electrical activity of the heart produced by an electrocardiograph machine. Also termed: cardiogram.

electrocardiograph (physiology) Machine that uses electrodes taped to the body to produce electrocardiograms.

electrochemical series (chemistry) List of metals arranged in order of their electrode potentials. A metal will displace from their salts metals lower down in the series. Also termed: electromotive series.

electrochemistry (chemistry) Branch of science that is concerned with the study of electrical chemical energy, such as the effects of electric current on chemicals (particularly electrolytes) and the generation of electricity by chemical action (as in an electrolytic cell).

electrode (1, 2 chemistry) 1 Conducting plate (anode or cathode) that collects or emits electrons from an electrolyte during electrolysis. 2 Conducting plate in an electrolytic cell (battery), discharge tube or vacuum tube.

electrode potential (chemistry) Potential developed by a substance in equilibrium with a solution of its ions.

electrodeposition (chemistry) Deposition of a substance from an electrolyte on to an electrode, as in electroplating.

electrodialysis (chemistry) Removal of salts from a solution (often a colloid) by placing the solution between two semipermeable membranes, outside which are electrodes in pure solvent.

electrodynamics (physics) Study of moving electric charges, especially in electric or magnetic fields, which has important applications in the design of generators and motors.

electroencephalogram (EEG) (physiology) Record of the electrical activity of the brain produced by an electroencephalograph machine.

electroencephalograph (physiology) Machine that uses electrodes taped to the skull to produce electroencephalograms.

electroimmunoassay (methods) *See* immunoelectrophoresis.

electroimmunodiffusion (methods) *See* immunoelectrophoresis.

electrokinetic potential (physics) An alternative term for zeta potential.

electrokinetics (physics) Branch of science concerned with the study of electric charges in motion.

electrolysis (chemistry) Conduction of electricity between two electrodes, through a solution of a substance (or a substance in its molten state) containing ions and accompanied by chemical changes at the electrodes.

electrolyte (chemistry) Substance that in its molten state or in solution can conduct an electric current.

electrolytic capacitor (physics) Electrolytic cell in which a thin film of nonconducting substance has been deposited on one of the electrodes by an electric current.

electrolytic cell (physics) Apparatus that consists of electrodes immersed in an electrolyte.

electrolytic dissociation (chemistry) Partial or complete reversible decomposition of a substance in solution or the molten state into electrically charged ions.

electrolytic rectifier (physics) Rectifier that consists of two electrodes and an electrolyte in which the current flows in one direction only.

electrolytic separation (chemistry) Method of separating metals from a solution by varying the applied potential according to the electrode potentials of the metals.

electromagnet (physics) Temporary magnet consisting of a current-carrying coil of wire wound on a ferromagnetic core. It is the basis of many items of electrical equipment, *e.g.* electric bells, solenoids and lifting magnets.

electromagnetic induction (physics) Electromotive force (emf) produced in a conductor when it is moved in a magnetic field. It is the working principle of an electrical generator (*e.g.* dynamo). It can give rise to a back emf and eddy currents.

electromagnetic interaction (physics) Interaction between electrically charged elementary particles.

electromagnetic radiation (physics) Energy that results from moving electric charges and travels in association with electric and magnetic fields, *e.g.* radio waves, heat rays, light and X-rays, which form part of the electromagnetic spectrum.

electromagnetic spectrum (physics) Range of frequencies over which electromagnetic radiation is propagated. In order of increasing frequency (decreasing wavelength) it consist of radio waves, microwaves, infrared radiation, visible light, ultraviolet radiation, X-rays and gamma-rays.

electromagnetic wave (physics) Wave formed by electric and magnetic fields, *i.e.* of electromagnetic radiation. Such waves do not require a medium in which to propagate, and will travel in a vacuum.

electromagnetism (physics) Combination of an electric field and a magnetic field, their interaction with stationary or moving electric charges, and their study and application. It therefore applies to light and other forms of electromagnetic radiation, as well as to devices such as electromagnets, electric motors and generators.

electromotive force (emf) (physics) Potential difference of a source of electric current, such as an electrolytic cell (battery) or generator. Often it can be measured only at equilibrium (when there is no current flow). An alternative term: voltage.

electromotive series (chemistry) An alternative term for electrochemical series.

electron (physics) Fundamental negatively-charged subatomic particle (radius 2.81777×10^{-15} m; rest mass 9.10908×10^{-31} kg; charge 1.602102×10^{-19} coulombs). Every neutral atom has as many orbiting electrons as there are protons in its nucleus. A flow of electrons constitutes an electric current.

electron affinity (physics) Energy liberated when an electron is acquired by a neutral atom.

electron capture (1, 2 physics) 1 Formation of a negative ion through the capture of an electron by a substance. 2 Transformation of a proton into a neutron in the nucleus of an atom (accompanied by the emission of X-rays) through the capture of an orbital electron, so converting the element into another with an atomic number one less.

electron charge (e) (measurement) Fundamental physical constant, equal to 1.602102×10^{-19} coulombs.

electron configuration (physics) *See* configuration.

electron density (physics) Density of electric charge.

electron diffraction (physics) Method of determining the arrangement of the atoms in a solid, and hence its crystal structure, by the diffraction of a beam of electrons.

electron donor (chemistry) An alternative term for reducing agent.

electron gun (physics) Electrode assembly for producing a narrow beam of electrons, as used, *e.g.* in cathode-ray tubes.

electron lens (physics) Arrangement of electrodes or of permanent magnets or electromagnets used to focus or divert beams of electrons in the same way as an optical lens modifies a beam of light, as in, *e.g.* an electron microscope.

electron microscope (equipment) A microscope using an electron beam of very short wavelength as the source of illumination. It has a resolving power of 2 nm, which is 100 times greater than with the light microscope. Types include the transmission electron microscope and the scanning electron microscope.

electron multiplier (physics) An alternative term for photomultiplier.

electron octet (chemistry) *See* octet.

electron optics (physics) Study of the control of free electrons by curved electric and magnet fields, particularly the use of such fields to focus and deflect beams of electrons.

electron probe microanalysis (EPM) (physics) Quantitative analysis of small amounts of substances by focusing a beam of electrons on to a point on the surface of the sample so that characteristic X-ray intensities are produced.

electron radius (r_e) (measurement) Fundamental physical constant, which is equal to 2.81777×10^{-15} m.

electron rest mass (m_e) (measurement) Fundamental physical constant, equal to 9.10908×10^{-31} kg.

electron transport (physiology) Process found mainly in aerobic respiration and photosynthesis that provides a source of energy in the form of atp. Hydrogen atoms are used in this system and taken up by a hydrogen carrier, *e.g.* fad; the electrons of the hydrogen pass along a chain of carriers which are in turn reduced and oxidized. This is coupled to the formation of ATP. The hydrogen atoms together with oxygen eventually form water.

electron tube (physics) An alternative term for a valve.

electron volt (eV) (measurement) A unit of energy equal to the energy acquired by an electron in being accelerated through a potential difference of 1 volt; equal to 1.602×10^{-19} joule.

electron-deficient compound (chemistry) Compound in which there are insufficient electrons to form two-electron covalent bonds between all the adjacent atoms (*e.g.* boranes).

electronegativity (chemistry) Power of an atom in a molecule to attract electrons. For elements arranged in the periodic table, it increases up a group and across a period.

electronics (physics) Branch of science concerned with the study of electricity in a vacuum, in gases and in semiconductors.

electron-spin resonance (ESR) (physics) Branch of microwave spectroscopy in which radiation of measurable frequency and wavelength is used to supply energy to protons.

electrophile (chemistry) Electron-deficient ion or molecule that attacks molecules of high electron density.

See also nucleophile.

electrophilic addition (chemistry) Chemical reaction that involves the addition of a molecule to an unsaturated organic compound across a double or triple bond.

electrophilic reagent (chemistry) Reagent that attacks molecules of high electron density.

electrophilic substitution (chemistry) Reaction that involves the substitution of an atom or group of atoms in an organic compound. An electrophile is the attacking substituent.

electrophoresis (physics) Movement of charged colloid particles in a solution placed in an electric field.

electrophorus (physics) Device for producing charges by electrostatic induction; an electrostatic generator.

electropositive (physics) Tending to form positive ions; having a deficiency of electrons.

electrostatic field (physics) Electric field associated with stationary electric charges.

electrostatic units (ESU) (measurement) System of electrical units based on the force exerted between two electric charges.

electrostatics (physics) Branch of electricity concerned with the study of electrical charges at rest.

electrostriction (physics) Change in the dimensions of a dielectric that is caused by the reorientation of molecules when an electric field is applied.

electrovalent bond (chemistry) An alternative term for ionic bond.

element (1 chemistry; 2 mathematics; 3 physics) 1 Substance consisting of similar atoms of the same atomic number. It cannot be decomposed by chemical action to a simpler substance. Also termed: chemical element. *See also* isotope; periodic table. 2 In mathematics, one of the members of a set. 3 In physics, one of several lenses in a compound lens, or one of several components in an electrical circuit.

elementary particle (physics) Subatomic particle not known to be made up of simpler particles.

elephant throat bot fly (veterinary parasitology) *See Pharyngobolus.*

elevation of boiling point (physics) Rise in the boiling point of a liquid caused by dissolving a substance in the liquid.

ELF (physics) An abbreviation of 'E'xtremely 'L'ow 'F'requency.

Elokomin fluke fever (fish/veterinary parasitology) A disease of Canidae, ferrets, bears and raccoons caused by a rickettsia-like agent, possibly a strain of *Neorickettsia helminthoeca*, transmitted by a fluke, *Naophyetus salmincola.* Fish are the intermediate host and dogs are infected by eating the fish. Clinical signs are similar to, but milder, than those of salmon poisoning.

eluate (chemistry) Solution obtained from elution.

eluent (chemistry) Solvent used for elution, the mobile phase.

elution (chemistry) Removal of an absorbed substance by washing the absorbent with a solvent (eluent). The technique is used in some forms of chromatography.

EM (microscopy) An abbreviation of 'E'lectron 'M'icroscope.

em (physics) An abbreviation of 'e'lectro'm'agnetic.

e-mail (computing) An abbreviation of 'e'lectronic 'mail'.

EMBO (society) An abbreviation of 'E'uropean 'M'olecular 'B'iology 'O'rganisation.

embryo (biology) Organism formed after cleavage of the zygote before birth. A maturing embryo is often termed a foetus (*USA*: fetus).

embryogenesis (medical) The development of the embryo.

embryol. (medical) An abbreviation of 'embryol'ogy.

embryology (biology) Study of embryos, their formation and development.

emer. (general terminology) An abbreviation of 'emer'gency.

emergence (parasitological physiology) A term used in relation to arthropods. The appearance of winged insect from the pupal stage.

emergency room (ER) (medical) The part of a hospital that provides immediate care.

emesis (medical) The act of vomiting.

emetics (medical) Drugs which cause vomiting.

emetine (pharmacology) An alkaloid derived from ipecac or produced synthetically, whose hydrochloride salt is used as an antiamoebic (*USA*: antiamebic).

emf (physics) An abbreviation of 'e'lectro'm'otive 'f'orce.

EMG (physiology) An abbreviation of 'E'lectro-'M'yo'G'ram.

emission spectrum (physics) Spectrum obtained when the light from a luminous source undergoes dispersion and is observed directly.

EMK (physics) An abbreviation of *'E'lektro-'M'otorische 'K'raft.* [German, meaning 'electromotive force'.]

empirical formula (chemistry) Chemical formula that shows the simplest ratio between atoms of a molecule, *e.g.* glucose, molecular formula $C_6H_{12}O_6$, and (ethanoic) acetic acid, $C_2H_4O_2$; both have the same empirical formula, CH_2O.

See also molecular formula; structural formula.

EMSA (society) An abbreviation of 'E'lectron 'M'icroscope 'S'ociety of 'A'merica.

EMTD (toxicology) An abbreviation of 'E'stimated 'M'aximum 'T'olerated 'D'ose.

emulsion (chemistry) Colloidal suspension of one liquid dispersed in another.

enamel (histology) White protective calcified outer coating of the crown of the tooth of a vertebrate. It is produced by epidermal cells, the ameloblasts, and it consists almost entirely of calcium salts bound together by keratin fibres (*USA*: keratin fibers).

enatiomer (chemistry) Molecule that is a mirror image of another, and which cannot be superimposed on it. Both molecules are optically active, but differ in the direction in which they rotate the plane of polarized light.

encephal(o)- (terminology) A word element. [German] Prefix denoting *brain.*

encephalin (biochemistry) One of two peptides which are natural analgesics, produced in the brain and released after injury. The encephalins have properties similar to morphine. Also termed: endorphin; enkephalin.

Encephalitozoon (parasitology) A genus of protozoa in the class Microsporea.

Encephalitozoon cuniculi (human/veterinary parasitology) A species of protozoa in the genus *Encephalitozoon* that may be found in rodents, dogs, primates and humans. They have a similar physical appearance and pathogenetic effect to toxoplasmosis causing encephalitozoonosis.

encephalitozoonosis (veterinary parasitology) A disease caused by *Encephalitozoon cuniculi* in rabbits that is characterized clinically by paralysis and lesions in the brain.

encystation (parasitological physiology) A term used in relation to amoebae (*USA*: amebae). The process of formation of the cyst from the trophozoite.

encysted (biology) Enclosed in a sac, bladder or cyst.

end point (chemistry) Point at which a chemical reaction is complete, such as the end of a titration.

See also volumetric analysis.

end(o)- (terminology) A word element. [German] Prefix denoting *within, inward.*

Endamoeba (parasitology) A genus of amoebae (*USA*: amebae) that are parasitic in the intestines of invertebrates. They differ from *Entamoeba* spp. because they lack a central endosome.

endectocides (pharmacology) Systemically administered parasiticides, which includes ivermectin and milbemycin.

endemic (epidemiology) Describing a disease that continually occurs among people or animals in a particular region.

See also epidemic; pandemic.

endemic haemoptysis (parasitology) *See* schistosomiasis.

endocrine gland (anatomy) Ductless organ or discrete group of cells that synthesize hormones and secrete them directly into the bloodstream. Such glands include the pituitary, pineal, thyroid, parathyroid and adrenal glands, the gonads and placenta (in mammals), islets of langerhans (in the pancreas) and parts of the alimentary canal. Their function is parallel to the nervous system, that of regulation of responses in animals. Also termed: ductless gland.

endocrinology (histology/biochemistry) Study of structure and function of endocrine glands and the roles of their hormones as the chemical messengers of the body.

endodyogeny (parasitological physiology) A term used in relation to *Toxoplasma* and *Sarcocystis*. The asexual form of division in which two daughter cells form within a mother cell. Both endozoites and cystozoites multiply by this method.

endogenous (biology) Produced or originating within an organism.

endogenous budding (parasitological physiology) A term used in relation to cestodes. Inward development from the germinal layer of a hydatid resulting in the formation of a daughter cyst or a brood capsule.

Endolimax (human/veterinary/avian parasitology) A genus of amoebas (*USA*: amebas) that may be found in the colon of humans, other mammals, birds, amphibians and cockroaches, but which appear not to be pathogenic.

Endolimax caviae (veterinary parasitology) A species of amoebas (*USA*: amebas) in the genus *Endolimax* that may be found in guinea pigs.

Endolimax nana (human/veterinary parasitology) A species of amoebas (*USA*: amebas) in the genus *Endolimax* that may be found in humans and monkeys.

Endolimax ratti (veterinary parasitology) A species of amoebas (*USA*: amebas) in the genus *Endolimax* that may be found in rats.

endolymph (anatomy) Fluid that fills the cavity of the middle, inner and semicircular canals of the ear.

endoparasite (parasitology) Parasite that lives inside the body of its host, *e.g.* fluke, malaria parasite, tapeworm.

endophyte (plant parasitology) A parasitic plant living within its host's body.

endoplasm (cytology) Central portion of cytoplasm, surrounded by the ectoplasm and containing organelles. Also termed: plasmasol.

endoplasmic reticulum (ER) (cytology) Structure that occurs in most eukaryotic cells in the form of a flattened membrane-bound sac of cell organelles, continuous with the outer nuclear membrane. When covered with ribosomes it is termed rough ER, in their absence smooth ER. Its main function is the synthesis of proteins and their transport within or to the outside of the cell. In liver cells ER is involved in detoxification processes and in lipid and cholesterol metabolism. In association with the Golgi apparatus, ER is involved in lysosome production.

endopolygeny (parasitological physiology) Asexual reproduction in protozoa in which new progeny are produced by budding within the parent cell.

endorphin (biochemistry) One of a group of peptides that are produced by the pituitary gland and which act as painkillers in the body.

endoscope (optics) Tubular optical device, perhaps using fibre optics (*USA*: fiber optics), that is inserted into a natural orifice or a surgical incision to study organs and tissues inside the body.

endospore (1, 2 biology) 1 Innermost layer of the wall of a spore. 2 Tough asexual spore that is formed by some bacteria to resist adverse conditions.

endothelium (histology) Tissue formed from a single layer of cells found lining spaces and tubes within the body, *e.g.* lining the heart in vertebrates.

See also epithelium.

endothermic (chemistry) Describing a process in which heat is taken in; *e.g.* in many chemical reactions.

See also exothermic.

endozoites (parasitological physiology) A term used in relation to *Toxoplasma* and *Sarcocystis.* Stages found in the macrophages and other host cells. As these multiply rapidly they are also known as tachyzoites. Formerly they were known as trophozoites or the proliferative forms. These stages measure between 4 to 7 by 2 to 4um. The shape is generally crescentic.

energy (physics) Capacity for doing work, measured in joules. Energy takes various forms: *e.g.* kinetic energy, potential energy, electrical energy, chemical energy, heat, light and sound. All forms of energy can be regarded as being aspects of kinetic or potential energy; *e.g.* heat energy in a substance is the kinetic energy of that substance's molecules.

energy level (physics) The energy of electrons in an atom is not continuously variable, but has a discrete set of values, *i.e.* energy levels. At any instant the energy of a given electron can correspond to only one of these levels.

enflurane (pharmacology) An inhalant general anaesthetic drug, which is similar to halothane. It is often used along with nitrous oxide-oxygen mixtures for the induction and maintenance of anaesthesia during major surgery.

enol (chemistry) Organic compound that contains the group C = CH(OH); the alcoholic form of a ketone.

enrichment (biology) Isolation of a particular type of organism by enhancing its growth over other organisms in a mixed population.

ENT (medical) An abbreviation of 'E'ar, 'N'ose and 'T'hroat.

Ent. (parasitology) An abbreviation of 'Ent'omology.

Entamoeba (parasitology) A genus of amoebas (*USA*: amebas) parasitic in the intestines of vertebrates. Member of the family Endamoebidae.

Entamoeba bovis (veterinary parasitology) A species of amoebas (*USA*: amebas) in the genus *Entamoeba* which may be found in cattle but that is considered to be nonpathogenic.

Entamoeba bubalis (parasitology) A species of amoebas (*USA*: amebas) in the genus *Entamoeba* that has a single nucleus in the cysts in its trophozoite.

Entamoeba canibuccalis (human/veterinary parasitology) (synonym: *Entamoeba gingivalis*) A species of amoebas (*USA*: amebas) in the genus *Entamoeba* that may be found in the mouth of cats, dogs, humans and primates, but which is considered nonpathogenic.

Entamoeba caviae (veterinary parasitology) A species of amoebas (*USA*: amebas) in the genus *Entamoeba* that may be found in guinea pigs, but which is considered nonpathogenic.

Entamoeba coli (human parasitology) A species of amoebas (*USA*: amebas) in the genus *Entamoeba* that may be found in the intestinal tract of humans. They are considered to be nonpathogenic but may be confused with the pathogenic species *Entamoeba histolytica.*

Entamoeba cuniculi (veterinary parasitology) A species of amoebas (*USA*: amebas) in the genus *Entamoeba* that may be found in the large bowel of rabbits.

Entamoeba equi (veterinary parasitology) A species of amoebas (*USA*: amebas) in the genus *Entamoeba* that may be found in horses.

Entamoeba equibuccalis (veterinary parasitology) A species of amoebas (*USA*: amebas) in the genus *Entamoeba* that may be found in the mouths of horses, but is considered to be nonpathogenic.

Entamoeba gedoelsti (veterinary parasitology) A species of amoebas (*USA*: amebas) in the genus *Entamoeba* that may be found in the large intestine of horses, but is considered to be nonpathogenic.

Entamoeba hartmanni (human/veterinary parasitology) A species of amoebas (*USA*: amebas) in the genus *Entamoeba* that may be found in the large intestine of humans and the colons of dogs, but is considered to be nonpathogenic.

Entamoeba histolytica (human/veterinary parasitology) A species of amoebas (*USA*: amebas) in the genus *Entamoeba* that causes amoebic dysentery (*USA*: amebic dysentery) and abscess of the liver in humans, and may also be found in monkeys, dogs, cats, rats and pigs.

Entamoeba invadens (veterinary parasitology) A species of amoebas (*USA*: amebas) in the genus *Entamoeba* that is the cause of entamoebiasis in reptiles. *See* amoebiasis.

Entamoeba moshkovskii (parasitology) A species of amoebas (*USA*: amebas) in the genus *Entamoeba* which may be found in sewage and that resembles *Entamoeba histolytica.*

Entamoeba muris (veterinary parasitology) A species of amoebas (*USA*: amebas) in the genus *Entamoeba* that may be found in the large intestine of rats and mice, but is considered to be nonpathogenic.

Entamoeba ovis (veterinary parasitology) A species of amoebas (*USA*: amebas) in the genus *Entamoeba* that may be found in sheep.

Entamoeba ranarum (veterinary parasitology) A species of amoebas (*USA*: amebas) in the genus *Entamoeba* that may be found in tadpoles.

Entamoeba suigingivalis (veterinary parasitology) A species of amoebas (*USA*: amebas) in the genus *Entamoeba* that may be found in the mouths of pigs.

Entamoeba suis (veterinary parasitology) A species of amoebas (*USA*: amebas) in the genus *Entamoeba* that may be found in swine.

Entamoeba wenyoni (parasitology) A species of amoebas (*USA*: amebas) in the genus *Entamoeba* that has eight-nucleated cysts in the trophozoite.

entamoebiasis (human/veterinary parasitology) Infection by *Entamoeba* spp. which occurs in most animal species but clinical illness is evident only in humans in the form of amoebic dysentery (*USA*: amebic dysentery) caused by *Entamoeba histolytica*.

enter(o)- (terminology) A word element. [German] Prefix denoting *intestine*.

enterobiasis (human/veterinary parasitology) Infection with nematodes of the genus *Enterobius*, especially *Enterobius vermicularis*. A disease of humans that also occurs in primates, that causes perianal irritation and aggressive behaviour (*USA*: behavior).

Enterobius (human/veterinary parasitology) A genus of nematodes of the family Oxyuridae. Species includes *Enterobius vermicularis*, the human pinworm, that causes enterobiasis.

Enteromonas (parasitology) A genus of the family Monocercomonadidae of protozoa, none of which appears to cause pathogenic effects.

Enteromonas caviae (veterinary parasitology) A species of protozoa in the genus *Enteromonas* that may be found in the caecum (*USA*: cecum) of guinea pigs.

Enteromonas hominis (human/veterinary parasitology) A species of protozoa in the genus *Enteromonas* that may be found in the caecum (*USA*: cecum) of humans, primates and rodents.

Enteromonas suis (veterinary parasitology) A species of protozoa in the genus *Enteromonas* that may be found in the caecum (*USA*: cecum) of pigs.

enterozoon (parasitology) An animal parasite in the intestines.

enthalpy (*H*) (physics) Amount of heat energy a substance possesses, measurable in terms of the heat change that accompanies a chemical reaction carried out at constant pressure. In any system, $H = U + pV$, where U is the internal energy, p the pressure and V the volume.

ento- (terminology) A word element. [German] Prefix denoting *within*, *inner*.

entomology (insect parasitology) That branch of biology concerned with the study of insects.

Entonyssus (veterinary parasitology) An entonyssid mite of snakes.

Entophionyssus (veterinary parasitology) An entonyssid mite of snakes.

entozoon (parasitology) An internal animal parasite.

entropy (S) (physics) In thermodynamics, quantity that is a measure of a system's disorder, or the unavailability of its energy to do work. In a reversible process the change in entropy is equal to the amount of energy adsorbed divided by the absolute temperature at which it is taken up.

envelope (mathematics) Curve that touches every one of a whole family of curves.

environment (biology) All the conditions in which an organism lives, including the amount of light, temperature, water supply and presence of other (competing) organisms.

enzootic (epidemiology) Peculiar to or present constantly in a location.

See also endemic.

enzyme (biochemistry) Protein that acts as a catalyst for the chemical reactions that occur in living systems. Without such a catalyst most of the reactions of metabolism would not occur under the conditions that prevail. Most enzymes are specific to a particular substrate (and therefore a particular reaction) and act by activating the substrate and binding to it.

Eoacanthocephala (fish parasitology) A class in the phylum Acanthocephala. The Eoacanthocephala are mainly parasites of fish but also occur in amphibians and reptiles. The class consists of two orders: Gyracanthocephala with one family, and Neoechinorhynchida with three families. There are about twenty five genera including: *Acanthogyrus*, *Microsentis*, *Neoechinorhynchus*, *Octospinifer*, *Pallisentis*, *Paulisentis*, *Quadrigyrus* and *Tenuisentis*. *See* Acanthocephala.

eosinophilic gastritis (veterinary parasitology) Diffuse infiltration or discrete nodules of eosinophils in the stomach wall occur rarely in dogs and may be immune-mediated, due to parasites.

eosinophilic ulcerative colitis (human/veterinary parasitology) A disease that occurs in humans and dogs, either as a primary disease or as part of an eosinophilic gastroenteritis. Characterized histologically by eosinophilic infiltration of the lamina propria and submucosa, it may be caused by parasites.

EP (physiology/biochemistry) An abbreviation of 'E'ndogenous 'P'yrogen.

Eperythrozoon (parasitology) A genus of microorganisms of the family Anaplasmataceae in the order Rickettsiales. They are small prokaryotic forms occurring on the surface of erythrocytes and may also be found free in the plasma.

Eperythrozoon coccoides (veterinary parasitology) A species of microorganisms in the genus *Eperythrozoon* that may be found in mice and may cause anaemia (*USA*: anemia).

Eperythrozoon felis (parasitology) *See Haemobartonella felis.*

Eperythrozoon ovis (veterinary parasitology) A species of microorganisms in the genus *Eperythrozoon* which may be found in sheep, often as an inapparent infection but they may cause anaemia (*USA*: anemia) and jaundice in an acute episode, or chronic poor condition.

Eperythrozoon parvum (veterinary parasitology) A species of microorganisms in the genus *Eperythrozoon* which may be found in pigs, but that are considered to be usually nonpathogenic.

Eperythrozoon suis (veterinary parasitology) A species of microorganisms in the genus *Eperythrozoon* that may be found in pigs, causing icteroanaemia (*USA*: icteroanemia) and may cause heavy mortality in piglets.

Eperythrozoon teganodes (veterinary parasitology) A species of microorganisms in the genus *Eperythrozoon* that may be found in cattle, but that are considered to be usually nonpathogenic. They are found only in the plasma and not attached to erythrocytes.

Eperythrozoon wenyoni (veterinary parasitology) A species of microorganisms in the genus *Eperythrozoon* which may be found in cattle and that may cause fever and anaemia (*USA*: anemia) but are usually nonpathogenic.

eperythrozoonosis (parasitology) Infection with *Eperythrozoon* spp. The infection is mostly innocuous but in times of stress, *e.g.* in the presence of another disease, it may cause an acute anaemia (*USA*: anemia) with fever. The infection is spread by insects and the disease may be seasonal as a result. In many cases the disease is a subacute one with illthrift as the main presenting sign.

Ephestia cautella (insect parasitology) *See Achroia grisella.*

Ephestia kühniella (insect parasitology) Arthropods in the Lepidoptera that is a parasite of the bees *Bombus fervidus*, feeding mainly on provisions but possibly on the brood.

epi- (terminology) A word element. [German] Prefix denoting *upon*.

epibiotic fouling (fish/veterinary parasitology) Fouling of external surfaces of mainly shellfish but sometimes finfish, with living organisms, principally protozoa of the genera *Zoothanium*, *Epistilis* and *Vorticella*.

epiblast (biology) An alternative term for ectoderm.

Epicauta vittata (veterinary parasitology) An insect which may infest hay and cause colic and frequent urination in horses. They contain the vesicant substance cantharidin. Also termed: blister beetle.

epidemic (epidemiology) Describing a disease that, for a limited time, affects many people or animals in a particular region.

See also endemic; pandemic.

epidemiologist (epidemiology) An expert in epidemiology.

epidemiology (epidemiology) Study of diseases as they affect the population, including their incidence and prevention.

epidermal nibbles (parasitology) Focal areas of epidermal oedema (*USA*: edema), eosinophils and necrosis, which are suggestive of ectoparasite injury to the skin.

epidermis (histology) Layer of cells at the surface of a plant or animal. In plants and some invertebrates, it forms a single protective layer, often overlaid by a cuticle which is impermeable to water. In vertebrates, it forms the skin and is composed of several layers of cells, the outermost becoming keratinized. *See* keratinization.

Epidermoptes (human/avian parasitology) A genus of parasitic mites of the family Epidermoptidae. Species includes *Epidermoptes bifurcatus* (*Epidermoptes bifurcata*) and *Epidermoptes bilobatus*; which may be found on the skin of birds, and that are capable of causing allergy in humans. They may coexist with fungal infections.

epidermotropic (parasitology) Predilection for epidermis.

epididymis (anatomy) Long coiled tube in the testes of some vertebrates through which sperm from the seminiferous tubules pass, before going into the vas deferens and to the exterior.

epiglottis (anatomy) Valve-like flap of cartilage in mammals that closes the opening into the larynx, the glottis, during swallowing.

epimastigote (parasitological physiology) A developmental stage in trypanosomes. The undulating membrane is shortened and the axoneme and the kinetoplast are anterior to the nucleus. In the previous stage they are near the tail and behind the nucleus. This stage occurs usually in the arthropod host. Also termed: crithidial stage.

epimerite (parasitological anatomy) An organelle of certain protozoa by which they attach themselves to epithelial cells.

***Epimetagea (Austeucharis)* sp.** (insect parasitology) Hymenoptera in the Eucharitidae that may be found in larvae and pupa, overwintering in cocoons, of *Myrmecia* spp.

epinephrine (biochemistry) An alternative term for adrenaline.

epipharynx (parasitological anatomy) A term used in relation to arthropods. An organ attached to the inner part of labrum.

epiphysis (1, 2 anatomy) 1 Growing end of a bone, at which cartilage is converted to solid bone. 2 An alternative term for pineal gland.

Epistylis (fish parasitology) A genus of ciliated protozoan parasites in the phylum Ciliophora that may be found commonly on the skin of freshwater fish, but which are considered to be usually nonpathogenic.

epithelium (histology) Animal lining tissue of varying complexity, whose main function is protective. It may be specialized for lining a particular organ, *e.g.* squamous epithelium lines capillaries and is permeable to molecules in solution, glandular epithelium contains cells that are secretory.

epithermal neutron (physics) Neutron that has energy of between 10^{-2} and 10^{2} electron volts (eV); a neutron having energy greater than that associated with thermal agitation.

epizoic (parasitology) Pertaining to or caused by an epizoon.

epizoon (parasitology) Plural: epizoa [German] An external animal parasite.

epizootic (epidemiology) A disease which attacks many subjects in a region at the same time but is only occasionally present in the population; when it occurs it is widely diffused and rapidly spreading. The rarely used equivalent of epidemic in veterinary medicine.

EPN (chemistry) An abbreviation of 'E'thyl '*P*''N'itrophenyl benzenethiophosphanate

Epomidiostomum (avian parasitology) A genus of nematode worms in the family Amidostomidae. The worms cause hemorrhage and necrosis in the proventriculus and the stomach and gizzard in geese, ducks and swans. Species includes *Epomidiostomum crispinum*, *Epomidiostomum skrjabini*, *Epomidiostomum uncinatum*, and *Epomidiostomum vogelsangi*.

epoxide (chemistry) Organic compound whose molecules include a three-membered oxygen ring (a cyclic ether).

epoxy resin (chemistry) Synthetic polymeric thermosetting resin with epoxide groups (*e.g.* polyethers). Such resins are used in surface coatings and as adhesives and electrical insulators. Also termed: epoxide resin.

EPP (1 haematology; 2 physiology) 1 An abbreviation of 'E'rythrocyte 'P'roto'P'orphyrin. 2 An abbreviation of 'e'nd 'p'late 'p'otential.

epsiprantel (pharmacology) A tapeworm anthelmintic.

Epsom salt (pharmacology) $MgSO_4.7H$ White crystalline salt, used in mineral waters, as a laxative. Also termed: epsomite; magnesium sulphate (*USA*: magnesium sulfate).

EPSP (physiology) An abbreviation for 'e'xcitatory 'p'ost's'ynaptic 'p'otential.

eq (chemistry) An abbreviation of 'eq'uivalent(s).

equation of state (physics) Any formula that connects the volume, pressure and temperature of a given system, *e.g.* van der Waals' equation.

equilibrium (1, 2 physics) 1 An object is in equilibrium when the forces acting on it are such that there is no tendency for the object to move. 2 State in which no change occurs in a system if no change occurs in the surrounding environment (*e.g.* chemical equilibrium).

equilibrium constant (K_c) (chemistry) Concentration of the products of a chemical reaction divided by the concentration of the reactants, in accordance with the chemical equation, at a given temperature.

equilibrium spinal ataxia (veterinary parasitology) *See* equine protozoal myeloencephalitis.

equimolecular mixture (chemistry) Mixture of substances in equal molecular proportions.

equine exfoliative eosinophilic dermatitis (veterinary parasitology) A condition of the skin characterized by infiltration of eosinophils and granulomatous inflammation with ulcerative stomatitis and wasting, that is suspected of being a hypersensitivity reaction to *Strongylus equinus* larvae.

equine leech (veterinary parasitology) *See* swamp cancer.

equine protozoal encephalomyelitis (veterinary parasitology) Inflammation of both the brain and spinal cord in horses which was originally ascribed to *Toxoplasma* spp. infection, but is now regarded as infection by *Sarcocystis neurona*. Also termed: equine protozoal myeloencephalitis.

equine seasonal conjunctivitis (veterinary parasitology) Irritation caused by flies, *Musca domestica*, or release of *Habronema* larvae. Also termed: summer conjunctivitis.

equivalence point (chemistry) Theoretical end point of a titration.

See also volumetric analysis.

equivalent fraction (mathematics) One of two or more fractions that represent the same number; *e.g.* the fractions 2/3, 4/6, 14/21 and 40/60 are equivalent.

equivalent proportions (chemistry) When two elements both form chemical compounds with a third element, a compound of the first two contains them in the relative proportions they have in compounds with the third one, *e.g.* carbon combines with hydrogen to form methane, CH_4, in which the ratio of carbon to hydrogen is 12:4; oxygen also combines with hydrogen to form water,

H_2O, in which the ratio of oxygen to hydrogen is 16:2. Carbon and oxygen form the compound carbon monoxide, CO, in which (and in accordance with the law) the ratio of carbon to oxygen is 12:16. Also termed: law of reciprocal proportions.

equivalent weight (chemistry) Number of parts by mass of an element that can combine with or displace one part by mass of hydrogen.

Er (chemistry) The chemical symbol for erbium.

ER (medical) An abbreviation of 'E'mergency 'R'oom.

erbium (Er) (chemistry) Metallic element in Group IIIB of the periodic table (one of the lanthanides), used in making lasers for medical applications. Its pink oxide is used as a pigment in ceramics. At. no. 68; r.a.m. 167.26.

Ereynetes (insect parasitology) Mites in the Prostigmata that may be parasitic on the ants, *Atta* and *Eciton*, and army ants.

Ereynetidae (insect parasitology) Mites in the Prostigmata that may be parasitic on the ants, *Atta*, *Eciton*, and army ants.

erg (measurement) Energy transferred when a force of 1 dyne moves through 1 cm, equivalent to 10^{-7} joules.

ERG (physiology) An abbreviation of 'E'lectro'R'etino'G'ram.

Ergasilus (parasitology) A miniature crustacean of the subclass Copepoda.

Ergasilus sieboldi (fish parasitology) A species of miniature crustacean in the genus *Ergasilus* that may be found as parasites on the gills of many freshwater fish causing impairment of the respiration, growth and sexual maturation. Heavy infestations can kill the host fish.

ergosterol (biochemistry) White crystalline sterol. It occurs in animal fat and in some micro-organisms. In animals it is converted to vitamin D_2 by ultraviolet radiation.

Eristalis (parasitology) A genus of flies in the family Syrphidae whose maggots breed in drains. Also termed: filth flies; rat-tailed maggots.

erogenous (biology) Originating outside an organism, organ or cell. The term may refer to such things as substances (*e.g.* nutrients) or stimuli (*e.g.* light).

ERPF (physiology) An abbreviation of 'E'ffective 'R'enal 'P'lasma 'F'low.

erythr(o)- (terminology) A word element. [German] Prefix denoting *red*, *erythrocyte*.

erythrocyte (haematology) Red blood cell. It contains haemoglobin (*USA*: hemaglobin) and carries oxygen around the body. In mammals, erythrocytes have no nuclei.

See also leucocyte.

Es (chemistry) The chemical symbol for einsteinium.

ES (society) An abbreviation of 'E'ntomological 'S'ociety.

ESA (society) An abbreviation of 'E'cological/'E'ntomological/'S'ociety of 'A'merica.

ESC (society) An abbreviation of 'E'ntomological 'S'ociety of 'C'anada.

-esis (terminology) A word element. [German] Suffix denoting *state*, *condition*.

esl (statistics) An abbreviation of 'e'xpected 's'ignificance 'l'evel.

esn (general terminology) An abbreviation of 'es'se'n'tial. Also termed esntl.

ESNZ (society) An abbreviation of 'E'ntomological 'S'ociety of 'N'ew 'Z'ealand.

eso- (terminology) A word element. [German] Prefix denoting *within*.

esophagostomiasis (parasitology) *See* oesophagostomiasis.

esophagostomosis (parasitology) Infestation with *Oesophagostomum* spp. worms, which causes necrotic nodules in the wall of the intestine. The resulting clinical syndrome includes poor condition and the passage of soft droppings containing more than normal amounts of mucus. *See* oesophagostomosis

esp. (general terminology) An abbreviation of 'esp'ecially.

espundia (parasitology) Leishmaniasis, particularly that caused by *Leishmania braziliensis*.

ESR (1 haematology; 2 physics) 1 An abbreviation of 'E'rythrocyte 'S'edimentation 'R'ate. 2 An abbreviation of 'E'lectron 'S'pin 'R'esonance.

essential amino acid (biochemistry) Any amino acid that cannot be manufactured in some vertebrates, including humans. These acids must therefore be obtained from the diet. They are as follows: arginine, histidine, isoleucine, leucine, lysine, methionine, phenylaianine, threonine, tryptophan and valine.

essential fatty acid (biochemistry) Any fatty acid that is required in the diet of mammals because it cannot be synthesized. They include linoleic acid and y-linolenic acid, obtained from plant sources.

essential oil (chemistry) Volatile oil with a pleasant odour (*USA*: odor), obtained from various plants. Such oils are widely used in perfumery.

est. (general terminology) An abbreviation of 'est'imated.

ester (chemistry) Compound formed when the hydrogen atom of the hydroxy group in an oxygen-containing acid is replaced by an alkyl group. Most important esters are derived from carboxylic acids.

esterification (chemistry) Formation of an ester, generally by reaction between an acid and an alcohol.

estimated (est.) (general terminology) An approximation.

estn (general terminology) An abbreviation of 'est'imatio'n'.

et al. (literary terminology) An abbreviation of '*et*' '*al'ii*. [Latin, meaning 'and others'.]

etc. (literary terminology) An abbreviation of '*et*' '*c'etera*. [Latin, meaning 'and so forth'.]

ethanal (chemistry) An alternative term for acetaldehyde.

ethanal trimer (chemistry) An alternative term for paraldehyde.

ethane (chemistry) C_2H_6 Gaseous alkane which occurs with methane in natural gas.

ethanedioic acid (chemistry) An alternative term for oxalic acid.

ethanoate (chemistry) An alternative term for acetate.

ethanoic acid (chemistry) An alternative term for acetic acid.

ethanol (chemistry) C_2H_5OH Colourless (*USA*: colorless) liquid alcohol. It is the active constituent of alcoholic drinks (in which it is produced by fermentation); it is also used as a fuel and in the preparation of esters, ethers and other organic compounds. Also termed: ethyl alcohol; alcohol.

ethanoyl chloride (chemistry) An alternative term for acetyl chloride.

ethene (chemistry) $CH_2 = CH_2$ Colourless (*USA*: colorless) gas with a sweetish smell, important in chemical synthesis. Also termed: ethylene.

ether (chemistry) Member of a group of organic compounds that have the general formula ROR', where R and R' are alkyl groups. The commonest, diethyl oxide (diethyl ether, or simply ether), $(C_2H_5)O_2$, is a useful volatile solvent formerly used as an anaesthetic (*USA*: anesthetic).

ethology (biology) Scientific study of animal behaviour (*USA*: behavior) in the wild.

ethyl (chemistry) C_2H_5 - Common alkyl group derived from ethane.

ethyl acetate (chemistry) An alternative term for ethyl ethanoate.

ethyl alcohol (chemistry) An alternative term for ethanol.

ethyl carbamate (chemistry) An alternative term urethane.

ethyl ethanoate (chemistry) $CH_3COOC_2H_5$ Colourless (*USA*: colorless) liquid ester with a fruity smell, produced by the reaction between ethanol (ethyl alcohol) and (ethanoic) acetic acid, used as a solvent and in medicine. Also termed: ethyl acetate.

ethyl *p*-nitrophenyl benzenethiophosphanate (EPN) (control measures) A nonsystemic organophosphorus acaricide.

ethylbenzene (chemistry) An alternative term for styrene.

ethylene (chemistry) An alternative term for ethene.

ethylene dibromide (chemistry) An alternative term for dibromoethane.

ethylene glycol (chemistry) An alternative term for ethanediol.

ethylene tetrachloride (chemistry) An alternative term for tetrachloroethene.

ethyne (chemistry) An alternative term for acetylene.

ETP (physiology) An abbreviation of 'E'lectron 'T'ransport 'P'article.

Eu (chemistry) The chemical symbol for europium.

eu- (terminology) A word element. [German] Prefix denoting *normal*, *good*, *well*, *easy*.

EUA (medical) An abbreviation of 'E'xamination 'U'nder 'A'naesthetic. (*USA*: examination under anesthetic).

Eubothrium (fish parasitology) A genus of tapeworms of the family Amphicotylidae that may be found in wild and captive salmonids.

Eucaryotae (biology) A kingdom of organisms that includes higher plants and animals, fungi, protozoa and most algae, except blue-green algae, all of which are made up of eucaryotic cells.

Eucestoda (fish parasitology) A subclass of flatworms in the class Cestoidea and phylum Platyhelminthes. Despite taxonomic controversies, twelve orders of fish tapeworms may be recognised as belonging to the subclass Eucestoda, nine of which are restricted to fish. Eucestodes are usually elongate, ribbon-like and segmented (strobilate), but in the Spathebothriidea, external segmentation is absent, although there are many sets of genitalia whilst in some eucestodes of the order Pseudophyllidea, an apparent prolific external segmentation far exceeds the actual number of internal sets of reproductive organs.

Eucharis ascendens (insect parasitology) Hymenoptera in the Eucharitidae that, similarly to *Eucharis bedeli*, *Eucharis myrmeciae*, *Eucharis punctata* and *Eucharis scutellaris*, may be found in larvae and pupa of *Cataglyphis*, *Formica*, *Messor*, *Myrmecia*, and *Myrmecocystus*.

Eucharis bedeli (insect parasitology) *See Eucharis ascendens*.

Eucharis myrmeciae (insect parasitology) *See Eucharis ascendens*.

Eucharis punctata (insect parasitology) *See Eucharis ascendens*.

Eucharis scutellaris (insect parasitology) *See Eucharis ascendens.*

Eucharomorpha (insect parasitology) Hymenoptera in the Eucharitidae that, similarly to *Isomerala, Kapala floridana, Kterminalis*, and *Obeza floridana*, may be found in larvae and pupa of Pheidole, Ectatomma Pogonomyrmex, Odontomachus, Camponotus and Pachycondyla, females dispersing only a few metres (*USA*: meters) away from the host nest.

Eucoleus aerophilus (parasitology) *Capillaria aerophila. See Capillaria.*

Eucotylidae (avian parasitology) A family of flukes (digenetic trematodes) that may be found in the kidneys of birds which includes the genera of *Tanaisia* spp. and *Eucotyle* spp., neither of which appears to have significant pathogenicity.

Eudiplogaster histophorus (insect parasitology) A species of nematode in the Diplogasteridae that are parasites to the ants *Lasius* spp., that may be found in the head glands.

eugenics (biology) Theory and practice of improving the human race through genetic principles. This can range from the generally discredited idea of selective breeding programmes to counselling of parents who may be carriers of harmful genes.

Euhopllopsyllus glacialis lynx (veterinary parasitology) A species of flea that parasitizes wild felines.

eukaryote (cytology) Cell with a certain level of complexity. Eukaryotes have a nucleus separated from the cytoplasm by a nuclear membrane. Genetic material is carried on chromosomes consisting of DNA associated with protein. The cell contains membrane-bounded organelles, *e.g.* mitochrondria and chloroplasts. All organisms are eukaryotic except for bacteria and cyanophytes, which are prokaryotes. Also termed: eucaryote.

eukaryotic (cytology) Describing or relating to a eukaryote.

eukaryotic cells (cytology) *See* cell.

eukaryotic transcription (molecular biology) *See* deoxyribonucleic acid.

Eulaelaps (parasitology) A genus of lelaptid mites in the family Gamasidae.

Eulaelaps stabularis (parasitology) A species of mite in the genus *Eulaelaps* which is a parasite of small mammals found in poultry houses, feed stores, etc., where they may cause irritation to workers. The species is a vector for *Francisella tularensis*.

Eumellitiphis (insect parasitology) *See Calaenostanus.*

European chick flea (parasitology) *Ceratophyllus gallinae.*

European harvest mite (parasitology) *See Trombicula autumnalis.*

European rabbit flea (veterinary parasitology) *See Spilopsyllus cuniculi.*

europium (Eu) (chemistry) Silvery-white metallic element in Group IIIB of the periodic table (one of the lanthanides), used in nuclear reactor control rods. At. no. 63; r.a.m. 151.96.

eury- (terminology) A word element. [German] Prefix denoting wide, broad.

Euryhelmis (parasitology) A genus of intestinal flukes (digenetic trematodes) in the family Heterophyidae.

Euryhelmis monorchis (veterinary parasitology) A species in the genus *Euryhelmis* which may be found in mink.

Euryhelmis squamula (veterinary parasitology) A species in the genus *Euryhelmis* which may be found in wild carnivora. It may cause fatal haemorrhagic (*USA*: hemorrhagic) enteritis in mink.

Eurytrema (parasitology) A genus of flukes (digenetic trematodes) in the family Dicrocoeliidae.

Eurytrema brumpti (parasitology) *Concinnum brumpti.*

Eurytrema fastosum (veterinary parasitology) A species in the genus *Eurytrema* which may be found in the pancreatic and biliara ducts of carnivores.

Eurytrema ovis (veterinary parasitology) A species in the genus *Eurytrema* which may be found in the perirectal fat of sheep.

Eurytrema pancreaticum (veterinary/insect parasitology) A species in the genus *Eurytrema* which may be found in the pancreatic ducts of sheep and cattle and that are parasites to the ants *Technomyrmex detorquens.*

Eurytrema procyonis (parasitology) *Concinnum procyonis.*

Eurytrema coelomaticum (veterinary parasitology) A species in the genus *Eurytrema* which may be found in the pancreatic ducts of sheep and cattle.

Euschongastia (parasitology) A genus of mites in the family Trombiculidae.

Euschongastia latchmani (veterinary parasitology) A species in the genus *Euschongastia* that may cause trombiculidiasis in cats.

Eustachian tube (anatomy) Channel that connects the middle ear with the pharynx at the back of the throat in mammals and some other vertebrates. It ensures that the air pressure on each side of the ear drum is equal. It is named after the Italian anatomist Bartolomeo Eustachio (1520–74). (*USA*: eustachian tube.)

Eustrongylides (avian/fish parasitology) A genus of nematodes in the family Dioctophymatidae which may be found in birds and wild and cultured fish.

Eustrongylides papillosus (avian parasitology) A species in the genus *Eustrongylides* which may be found in the oesophagus (*USA*: esophagus) and proventriculus of ducks and geese, that can cause extensive nodule formation.

Eustrongylides tubifex (avian parasitology) A species in the genus *Eustrongylides* which may be found in the intestine of anatine birds.

eutectic mixture (chemistry) Mixture of substances in such proportions that no other mixture of the same substances has a lower freezing point.

Eutrombicula (parasitology) A subgenus of mites in the genus *Trombicula* of the family Trombiculidae.

See also chigger.

Eutrombicula alfreddugesi (parasitology) The common chigger of the USA which may cause dermatitis in most species. Also termed *Trombicula alfreddugesi*.

Eutrombicula sarcina (veterinary parasitology) A species in the genus *Eutrombicula* that may cause leg itch and black soil itch of sheep, manifested by dermatitis on the lower limbs.

Euvarroa (insect parasitology) A species of mites in the Mesostigmata, that are parasitic to the bees *Apinae*.

eV (measurement) An abbreviation of 'e'lectron 'V'olt.

evaporation (physics) Process by which a liquid changes to its vapour. It can occur (slowly) at a temperature below the boiling point, but is faster if the liquid is heated and fastest when the liquid is boiling.

even-odd nucleus (chemistry) Atomic nucleus with an even number of protons and an odd number of neutrons.

event (statistics) The outcome of a random experiment.

evolution (biology) Successive altering of species through time. Evolutionary theory states that the origin of all species is through evolution, and thus they are related by descent.

See also Darwinism; natural selection.

ex. (literary terminology) An abbreviation of 'ex'ample. Also termed *e.g.*

exch. (chemistry) An abbreviation of 'exch'ange.

excitation (physics) Addition of energy to a system, such as an atom or nucleus, causing it to transfer from its ground state to one of higher energy.

excitation energy (physics) Energy required for excitation.

excited state (physics) Energy state of an atom or molecule that is higher than the ground state, resulting from excitation.

exclusion chromatography (methods) Chromatography in which the stationary phase is a gel having a closely controlled pore size. Molecules are separated based on molecular size and shape, smaller molecules being temporarily retained in the pores.

exclusion principle (physics) An alternative term for the Pauli exclusion principle.

excreta (physiology) Excretion products; waste material excreted or eliminated from the body, including faeces (*USA*: feces), urine and sweat. Mucus and carbon dioxide also can be considered excreta. The organs of excretion are the intestinal tract, kidneys, lungs and skin.

excrete (physiology) To throw off or eliminate, as waste matter, by a normal discharge.

excretion (physiology) Removal of waste products of metabolism, carried out by elimination from the body or storage in insoluble form. Products of protein metabolism are the main substances liberated. The chief organs of excretion in vertebrates are the kidneys.

excystation (parasitological physiology) Escape from a cyst or envelope, as in that stage in the life cycle of parasites occurring after the cystic form has been swallowed by the host.

exflagellation (parasitological physiology) The protrusion or formation of flagelliform microgametes from a microgametocyte in some sporozoa.

exo- (terminology) A word element. [German] Prefix denoting *outside of*, *outward*.

exocrine gland (anatomy) Gland that discharges secretions into ducts, *e.g.* salivary glands.

See also endocrine gland.

exoerythrocytic (parasitological physiology) Occurring or situated outside the red blood cells (erythrocytes), a term applied to a stage in the development of protozoan parasites that takes place in cells other than erythrocytes.

exogamy (1 parasitological physiology; 2 biology) 1 Protozoan fertilization by union of elements that are not derived from the same cell. 2 Outbreeding.

See also inbreeding.

exogenous (biology) Originating outside or caused by factors outside the organism.

exogenous budding (parasitological physiology) A term used in relation to cestodes. Outward or external development from the germinal layer of a larval cestode.

exogenous infection (parasitology) Infection caused by organisms not normally present in the body but which have gained entrance from the environment.

exoskeleton (parasitological anatomy) An external hard framework, as a crustacean's shell, that supports and protects the soft tissues of lower animals, derived from the ectoderm. In vertebrates the term is sometimes applied to structures produced by the epidermis, as hair, claws, hoofs, teeth, etc.

exosmosis (physics) Osmosis or diffusion from within outward.

exothermic (chemistry) Describing a process in which heat is evolved.

exotic disease (medical) A disease that does not occur in the subject country. Said of infectious diseases that may be introduced.

exp. (1-7 general terminology) An abbreviation of 1 'exp'eriment, 2 'exp'ires, 3 'exp'onential, 4 'exp'eriment, 5 'exp'erimental, 6 'exp'iration, 7 'exp'ire.

expansion of gas (physics) Increase in volume of an ideal gas is at the rate of 112 of its volume at 0°C for each degree rise in temperature. *See* Charles's Law.

See also kinetic theory.

expectation of life (medical) An epidemiological expression of the probability of dying between one age and the next. Based on the human cohort life table which describes the actual mortality experience of a group of animals which were all born at the same time.

expected (statistics) The expectation as predicted by the relevant formula or model.

expected frequency (statistics) The expected number of occurrences.

expected value (statistics) An estimate of the value of a population parameter, which would be achieved by sampling an infinite number of times.

experiment (parasitology) A study involving a comparison group in which the investigator intentionally alters one or more risk factors in order to discover or demonstrate some fact or general truth.

experimental (biology) Emanating from or pertaining to experiment.

experimental animals (parasitology) Animals kept expressly for the purposes of conducting experiments on them. Also termed: laboratory animals.

experimental epidemiology (epidemiology) Prospective population experiments designed to test epidemiological hypotheses, and usually attempt to relate the postulated cause to the observed effect. Trials of new anthelmintics are an example.

experimental model (parasitology) Experiment carried out using a model of a real system which contains some of the risk factors which apply in the real state; the model is a simplification of real life.

experimental pathology (parasitology) The study of artificially induced pathological processes.

experimental study (parasitology) A study in which all of the risk factors are under the direct control of the investigator.

expire (physiology) 1 To breathe out. 2 To die.

explant culture (parasitology) A small piece of tissue such as trachea or gut maintained in culture.

exploratory biopsy (medical) A combination of exploratory surgery to determine size and location of a lesion and the taking of a biopsy.

exploratory operation (medical) Incision into the body for determination of the cause of otherwise unexplainable symptoms.

exponent (mathematics) Number that indicates the power to which a quantity (the base) is to be raised, usually written as a superior number or symbol after the quantity, *e.g.* in 2^3, 2^x, the exponents are 3 and x.

exponential growth (biology) Growth that occurs, *e.g.* in cultures of micro-organisms, in which a population of cells increases in numbers logarithmically.

exponential series (mathematics) Mathematical series of functions of x that converges to e^x.

export (biology) To transport, secrete or excrete protein out of the cell.

exsheathing fluid (parasitological physiology) Fluid secreted by strongylid larvae preparatory to shedding the current larval stage sheath.

exsheathment (parasitological physiology) Shedding of the retained sheath of the second larval stage of strongyle nematodes before the parasitic stage of its existence can begin.

extermination (parasitology) Mass killing of animals or other pests. The term implies complete destruction of the species or other group.

external (biology) Situated or occurring on the outside, toward or near the outside; lateral.

external environment (biology) Environment outside the animal; includes ambient temperature, wind chill factor, feed and water supply.

external parasites (parasitology) *See* under individual listings for insects, insect larvae and helminths.

extra- (terminology) A word element. [Latin] Prefix denoting *outside, beyond the scope of, in addition.*

extracellular (histology) External to a cell; in a multicellular organism, extracellular tissue may still be within the organism.

See also intracellular.

extracellular constituents (biochemistry) All of the constituents of the body outside the cells; include water, electrolytes, protein, glucose, enzymes and hormones.

extracellular fluid (biochemistry) All of the body fluid lying outside the cells which includes intravascular fluid or plasma and the interstitial fluid. That part of the extracellular fluid that is in special cavities which have special characteristics, *e.g.* synovial fluid, urine, aqueous humor of eye, are termed transcellular fluids.

extracellular matrix (cytology) The network of proteins and carbohydrates that surround a cell or fill the intercellular spaces.

extracorporeal (biology) Situated or occurring outside the body.

extrafusal fibres (histology) The large contractile fibres (*USA*: fibers) contained in muscle; their movement causes muscle contraction. They consist of longitudinally arranged filaments of two kinds: myosin and actin. (*USA*: extrafusal fibers).

extrapolation (statistics) Estimation of a value outside the range of those already known, usually by graphical methods.

extrinsic (biology) Of external origin.

extrinsic incubation period (parasitology) The period between infection of the arthropod insect vector and the vector's ability to infect the next vertebrate host.

exudate (biochemistry/cytology) A fluid with a high content of protein and cellular debris which has escaped from blood vessels and has been deposited in tissues or on tissue surfaces, usually as a result of inflammation. It may be septic or nonseptic.

exudation (medical) 1 The escape of fluid, cell, or cellular debris from blood vessels and deposition in or on the tissue. 2 Exudate.

exudative (medical) Of or pertaining to a process of exudation.

eye (anatomy) Organ for detecting light.

eyeball (anatomy) The ball or globe of the eye.

eye fluke (fish parasitology) *Diplostomum* spp. larvae that may be found in the eyes of freshwater fish.

eyepiece (microsopy) The lens or system of lenses of a microscope nearest the user's eye, serving to further magnify the image produced by the objective.

eye tooth (anatomy) An alternative term for canine tooth.

eye worm (parasitology) *See Thelazia*; *Onchocerca.*

F

F (1 measurement; 2 general terminolgy; 3 chemistry; 4 biochemistry) An abbreviation of 1 'F'ahrenheit. Also termed Fahr. 2 'F'emale. 3 The chemical symbol for fluorine. 4 A symbol for phenylalanine.

f (microscopy) An abbreviation of 'f'ocal length.

F distribution (statistics) The theoretical distribution of the f ratio. It is the distribution of the ratio of the variances of two independent samples drawn from normal populations. It is used when testing the hypothesis that the samples yielding the F ratio have been drawn from the same population. It plays a part in numerous tests including analysis of variance and those used in regression analysis.

F ratio (statistics) The ratio of the variances of two independent samples from normal populations. It is used in analysis of variance. In testing data from psychological experiments by analysis of variance, F_1 stands for any effect in which subjects are the source of error variance, F_2 for any effect in which experimental materials are the source of error variance, and F' for any effect in which two or more sources contribute to the error variance. *See* F distribution.

F test (statistics) A statistical test based on the F ratio and F distribution. It is used in anova and in testing whether two variances are equal.

F_1 (1 genetics; 2 statistics) 1 The offspring of any two parents. *Compare* F_2. 2 *See* F ratio.

F_2 (1 genetics; 2 statistics) The offspring of two F_1 parents (*i.e.* the third generation). 2 *See* F ratio.

fa (chemistry) An abbreviation of 'f'atty 'a'cid/'f'olic 'a'cid.

FAAAS (society) An abbreviation of 'F'ellow of the 'A'merican 'A'ssociation for the 'A'dvancement of 'S'cience.

face fly (parasitology) *See Musca autumnalis.*

face louse (parasitology) *See Linognathus ovillus.*

FACG (education) An abbreviation of 'F'ellow of the 'A'merican 'C'ollege of 'G'astroenterology.

faci(o)- (terminology) A word element. [Latin] Prefix denoting face.

facial (biology) Of or pertaining to the face.

facilitated diffusion (biochemistry) Mode of transport through a membrane that involves carrier molecules in the membrane, which eases the transport of a specific substance but does not involve the use of energy; *e.g.* the uptake of glucose by erythrocytes (red blood cells). The transport system can become saturated with the transported substance, in contrast to simple diffusion.

facs. (general terminology) An abbreviation of 'facs'imile. Also termed fs, and facsim.

facsim. (general terminology) *See* facs.

facsimile machine (equipment) In data communications, transmission of written or pictorial matter over a conventional telephone line by one page at a time.

factor (1 general; 2,3 statistics; 4 mathematics) 1 Anything having a causal influence on something, usually where there are several different causal influences. 2 A discrete variable used to classify data (*e.g.* an independent variable) *Compare* level. 3 Any of the intervening variables discovered through factor analysis that accounts for a significant proportion of the variance of the data. 4 A number that divides into another number without a remainder.

factor analysis (statistics) A generic term for techniques whose objective is to discover whether the correlations or covariances between a set of observed variables can be accounted for in terms of their relationships to a small number of unobservable or intervening variables, *i.e.* the factors. Essentially factor analysis postulates a linear relation between the observed variables and the underlying factors, which may be estimated by a variety of methods. At its simplest, factor analysis is a method for data reduction, reducing a large number of intercorrelated variables to a smaller number of intervening variables which account for as much of the variance as possible.

factor axes (statistics) In factor analysis the axes representing the relations of the factors to one another; they may be orthogonal or oblique.

factor design (epidemiology) A design for an experiment in which all levels of each controlled variable is included with all levels of the others.

factor loading (statistics) In factor analysis the correlation between a test item and a factor, *i.e.* the extent to which the score on an item is determined by a factor.

factor matrix (statistics) A table showing the factor loading of each test item with each factor.

factorial (mathematics) Product of all the whole numbers from a given whole number n down to 1, written as $n!$; *e.g.* the factorial of 6 (written 6!) is $6 \times 5 \times 4 \times 3 \times 2 \times 1 = 720$.

factorial design (statistics) An experimental design in which each value (level) of each independent variable (factor) is combined with each value (level) of every other independent variable (factor); *e.g.* if there are two factors (A and B) each with two levels (1 and 2) the following combinations would be used: A1B1, A1B2, A2Bl, A2B2.

factorial validity (statistics) The extent to which scores on different tests purporting to measure the same thing are correlated.

facultative anaerobe (biology) Micro-organism that grows under either anaerobic or aerobic conditions.

facultative parasite (parasitology) A parasite that may be parasitic upon another organism but can exist independently.

FAD (1 medical; chemistry) 1 'F'lea 'A'llergy 'D'ermatitis. 2 Abbreviation of 'F'lavin 'A'denine 'D'inucleotide.

faecal (biology) Pertaining to or of the nature of faeces (*USA*: feces). (*USA*: fecal).

faecal egg count (parasitology) *See* egg count.

faecal examination for worms (parasitology) Adult worms in anthelmintic trials may be collected in sieves after sampling of faecal (*USA*: fecal) output. Faecal (*USA*: fecal) larvae may be counted by special techniques such as the Baermann technique.

faeces (medical) The waste products from the intestines, a third of which consists of dead or living bacteria. The remainder being made up of food which has not been digested, mostly cellulose and fibre (*USA*: fiber), dead cells from the lining of the intestine, water, mucus, and the whole coloured (*USA*: colored) by bile pigments. (*USA*: feces).

faex (biology) [Latin] *See* faeces.

Fahr. (measurement) An abbreviation of 'fahr'enheit.

Fahrenheit (F) (measurement) A unit of degree of temperature. The freezing point of water is 32°F; the boiling point of water is 212°F under standard atmospheric pressure. Named after the German physicist G. D. Fahrenheit (1686–1736).

Fahrenheit thermometer (equipment) A thermometer employing the Fahrenheit scale. The abbreviation 100°F should be read as 'one hundred degrees Fahrenheit'.

failure (terminology) Inability to perform or to function properly.

Fajans' rules (chemistry) Set of rules that state when ionic bonds are likely in a chemical compound (as opposed to covalent bonds). They were named after the Polish chemist Kasimir Fajans (1887–1975).

Falculifer (parasitology) A genus of the family of mites Demoglyphidae.

Falculifer clornutus (avian parasitology) A species in the genus *Falculifer* that is a feather mite of pigeons.

Falculifer rostratus (avian parasitology) A species in the genus *Falculifer* that is a feather mite of pigeons.

Fallopian tube (anatomy) Tube that in female mammals conducts ova (eggs) from an ovary to the uterus (womb) by ciliary action. Fertilization can occur when sperm meet eggs in the tube. It was named after the Italian anatomist Gabriel Fallopius (1523–62). Also termed: oviduct.

false hydatid (parasitology) *See Cysticercus tenuicollis.*

false negative (epidemiology) Pertaining to the situation when the result of a test in a patient is negative when the disease or condition which is the subject of the search is present. *Contrast* false positive.

false positive (epidemiology) A positive test result caused by a disease or condition other than the disease for which the test is designed, or in the absence of disease altogether. *Contrast* false negative.

false positive reactor (epidemiology) A person or animal that does not have the subject disease but reacts positively to a test for it.

familial (epidemiology) Occurring in or affecting members of a closely related group of animals at a frequency more than would be expected by chance.

family (biology) In biological classification, one of the groups into which an order is divided, and which is itself divided into genera. *See* taxonomy.

famphur (control measures) An organophosphorus insecticide used in cattle, sheep and goats. Also termed famophos.

FAMS (education) An abbreviation of 'F'ellow of the 'I'ndian 'A'cademy of 'M'edical 'S'ciences.

Fannia (parasitology) A genus of flies in the family Muscidae that pupate in faeces (*USA*: feces).

Fannia australis (veterinary parasitology) A species of fly in the genus *Fannia* that may cause tertiary blowfly strike.

Fannia benjamini (veterinary parasitology) A species of fly in the genus *Fannia* that may cause insect worry.

Fannia canicularis (veterinary parasitology) A species of fly in the genus *Fannia* that may cause urogenital myiasis.

Fannia scalaris (veterinary parasitology) A species of fly in the genus *Fannia* that may cause urogenital myiasis.

FANZAAS (education) An abbreviation of 'F'ellow of 'A'ustralian and 'N'ew 'Z'ealand 'A'ssociation for 'A'dvancement of 'S'cience.

FAO (govening body) An abbreviation of 'F'ood and 'A'griculture 'O'rganization (United Nations).

FAPHA (education) An abbreviation of 'F'ellow of the 'A'merican 'P'ublic 'H'ealth 'A'ssociation.

FAPT (education) An abbreviation of 'F'ellow of the 'A'ssociation of 'P'hotographic 'T'echnicians.

farad (F) (measurement) SI unit of electrical capacitance, defined as the capacitance that, when charged by a potential difference of 1 volt, carries a charge of 1 coulomb. It was named after the British scientist Michael Faraday (1791–1867).

faraday (F) (measurement) Unit of electric charge, equal to the quantity of charge that during electrolysis liberates one gram equivalent weight of an element. It has the value 9.6487×10^4 coulombs per gram-equivalent.

See also farad; faraday constant.

Faraday constant (F) (physics) Fundamental physical constant, the electric charge carried by one mole of singly-charged ions or electrons, equal to 9.6487×10^4 coulombs per mole. It is the product of the Avogadro constant and the electron charge.

Faraday effect (physics) Rotation of the plane of vibration of a beam of polarized light passing through a substance such as glass, in the direction of an applied magnetic field.

Faraday's laws of electrolysis (1, 2 chemistry) 1 The amount of chemical decomposition that takes place during electrolysis is proportional to the electric current passed. 2 The amounts of substances liberated during electrolysis are proportional to their chemical equivalent weights.

Faraday's laws of electromagnetic induction (1, 2 physics) 1 An induced electromotive force is established in an electric circuit whenever the magnetic field linking that circuit changes. 2 The magnitude of the induced electromotive force in any circuit is proportional to the rate of change of the magnetic flux linking the circuit.

FAS (society) 1 An abbreviation of 'F'ederation of 'A'merican 'S'cientists.

fascia (histology) Fibrous tissue organised to form sheets which lie just under the skin (superficial fascia), and round muscles (deep fascia). The superficial fascia contains fat and the nerves and blood vessels running to the skin. The deep fascia forms dense fibrous sheaths for the muscles and compartments in which groups of muscles lie together.

fascicle (histology) A slender bundle of nerve fibres (*USA*: nerve fibers). Also termed fasciculus.

Fasciola (parasitology) A genus of flukes in the family Fasciolidae.

Fasciola gigantica (veterinary parasitology) A species of fluke in the genus *Fasciola* that may be found in the liver of domestic livestock and is the common liver fluke of Africa. It is similar anatomically to *Fasciola hepatica* but considerably larger, and may cause fascioliasis.

Fasciola hepatica (human/veterinary parasitology) A species of fluke in the genus *Fasciola* which may be found in the liver parenchyma and bile ducts of humans, sheep, cattle and most other domesticated species and that may cause fascioliasis. In humans and horses, which are not common hosts, the flukes may be in lungs or other unusual sites and they are often in the lungs of cattle.

Fasciola jacksoni (veterinary parasitology) A species of fluke in the genus *Fasciola* which may be found in elephants and that may cause a disease similar to ovine fascioliasis.

fascioliasis (parasitology) The disease caused by infestation with *Fasciola*. Also termed distomatosis, distomiasis.

fascioliasis-ostertagiasis complex (veterinary parasitology) A disease of cattle in which the diarrhoea (*USA*: diarrhea) of ostertagiasis is accompanied by the anaemia (*USA*: anemia) of fascioliasis.

fasciolicidal (pharmacology) Lethal for *Fasciola* spp.

fasciolicidal drugs (pharmacology) Drugs that are lethal for *Fasciola* spp including drugs effective against mature flukes, such as halogenated hydrocarbons, *e.g.* carbon tetrachloride, the substituted salicylanilides, *e.g.* bromsalans, closantel, rafoxanide, oxyclozanide, the substituted phenols, *e.g.* nitroxynil, and the benzimidazoles, *e.g.* albendazole, and drugs affecting immature flukes, *e.g.* diamphenethide and trichlorbendazole.

fasciolicide (control measures) Fasciolicidal.

Fascioloides (parasitology) A genus of flukes of the family Fasciolidae.

Fascioloides magna (veterinary parasitology) A species of flukes in the genus *Fascioloides* that may be found in the liver and rarely the lungs of cattle, sheep, deer, pigs and horses and in many wild ruminants in North America and Europe. Sheep are the only species which appear to suffer ill effects because of damage to the liver by the unrestricted migration of the flukes in this species. In the other host species the flukes are enclosed in cysts.

fasciolopsiasis (human parasitology) Infection with *Fasciolopsis* spp. Principally a disease of humans manifested by intestinal inflammation and ulceration.

Fasciolopsis (parasitology) A genus of trematodes of the family Fasciolidae.

Fasciolopsis buski (human/veterinary parasitology) A species of flukes in the genus *Fasciolopsis*, considered the largest of the intestinal flukes. They may be found in the small intestines of humans and pigs throughout Asia, and may cause fasciolopsiasis.

FASEB (society) An abbreviation of 'F'ederation of 'A'merican 'S'ocieties for 'E'xperimental 'B'iology.

fast (general terminology/microscopy) 1 Immovable or unchangeable; resistant to the action of a specific drug, stain or destaining agent. 2 Quick.

fasting (general terminology) A length of time without food and water. Base levels are often obtained after the patient has fasted where the intake of food may influence the level of the factor under investigation, as is the case for instance with blood glucose levels.

fat body (parasitological cytology) A term used in relation to arthropods. Group of cells which are widely distributed in the body and act as a food store. They contain protein, fats and glycogen.

fatal (terminology) Causing death; deadly; mortal; fatal.

fatality rate (epidemiology) The number of deaths caused by a specific circumstance or disease, expressed as the absolute or relative number among individuals encountering the circumstance or having the disease.

fate map (genetics) A map of the cells or tissues in the fertilized egg or early embryo that shows where they will end up in the adult organism. The data can be collected by staining specific cells in the embryo with a non-diffusible dye; artificial chimaeras can also be used.

fats and oils (biochemistry) Naturally occurring esters (of glycerol and fatty acids) that are used as energy-storage compounds. They are hydrocarbons and members of a larger class of naturally occurring compounds called lipids.

fat-soluble (chemistry) Said of substances that occur naturally in fats and are soluble in fat solvents but not in water.

fatty acid (chemistry) Monobasic carboxylic acid, an essential constituent of fats and oils. The simplest fatty acids are formic (methanoic) acid, HCOOH, and (ethanoic) acetic acid, CH_3COOH.

fatty degeneration (histology) Disease of tissue caused by poisoning or lack of oxygen, in which droplets of fat form within cells.

fauces (anatomy) The opening between the mouth and the throat, bounded above by the soft palate, below by the tongue, and on each side by the tonsils. The two folds of mucous membrane containing muscle fibres (*USA*: muscle fibers) before and behind the tonsils are called the pillars of the fauces.

fax (general terminology) Facsimile transmission.

FBP (physics) An abbreviation of 'F'inal 'B'oiling 'P'oint.

FCA (parasitology) An abbreviation of 'F'reund's 'C'omplete 'A'djuvant.

FCAP (education) An abbreviation of 'F'ellow of the 'C'ollege of 'A'merican 'P'athologists.

FCST (governing body) An abbreviation of 'F'ederal 'C'ouncil for 'S'cience and 'T'echnology.

FD (microscopy) An abbreviation of 'F'ocal 'D'istance.

FDA (governing body) An abbreviation of 'F'ood and 'D'rug 'A'dministration.

Fe (chemistry) The chemical symbol for iron. The symbol is derived from Latin, 'ferrum'.

feather mites (avian parasitology) Mites that live on and in feathers, often in enormous numbers but have apparently little pathogenicity. They include the genera of *Analges* and *Megninia* of the family Analgesidae and the genus *Dermoglyphus* of the family Dermoglyphidae. Other miscellaneous genera include *Syringophilus*, *Falculifer*, *Freyana*, *Pterolichus*, and *Pteronyssus*.

febantel (pharmacology) An imidazothiazole anthelmintic used in sheep, cattle and horses that has a wide spectrum of efficiency against nematodes and that is safe for use in all stages of pregnancy but not to be used within a week of bromsalans.

febrifacient (medical) Producing fever.

febrile (medical) Pertaining to fever; feverish.

feedback inhibition (genetics) End-product inhibition. A genetic engineering control system in which the product, *e.g.* an enzyme, inhibits further production of itself.

feeder cells (immunology) Cells whose function is to maintain the growth of other cells, usually by secreting growth factors.

Fehling's solution (biochemistry) Test reagent consisting of two parts: a solution of copper(II) sulphate (*USA*: copper(II) sulfate), and a solution of potassium sodium tartrate and sodium hydroxide. When the two solutions are mixed, an alkaline solution of a soluble copper(II) complex is formed. In the presence of an aldehyde or reducing sugar, a pink-red precipitate of copper(I) oxide forms. It was named after the German chemist Hermann Fehling (1812–85).

See also Fehling's test.

Fehling's test (biochemistry) Test for an aldehyde group or reducing sugar, indicated by the formation of copper(I) oxide as a pink-red precipitate with Fehling's solution.

FEL (toxicology) An abbreviation of 'F'rank 'E'ffect 'L'evel.

FELASA (society) An abbreviation of 'F'ederation of 'E'uropean 'L'aboratory 'A'nimal 'S'cience 'A'ssociation.

Felicola subrostratus (veterinary parasitology) A biting louse of the superfamily Ischnocera that may be found on cats.

feline miliary dermatitis (veterinary parasitology) A papular, crusting skin disease located predominantly on the back, with varying degrees of pruritus. Ectoparasites are among the many possible causes. Also termed scabby cat disease.

feline pneumonitis (veterinary parasitology) Infection by *Chlamydia psittaci* that causes a chronic, often recurrent, conjunctivitis and infrequently lower respiratory disease.

feline scabies (veterinary parasitology) *See Notoedres cati.*

fem. (general terminolgy) An abbreviation of 'fem'ale/ 'fem'inine.

femur (anatomy) The thigh bone.

fenbendazole (pharmacology) A broad-spectrum, benzimidazole anthelmintic that may be used against nematodes in sheep, cattle and horses, but that can be acutely poisonous in cattle if administered with or within a few days of bromsalans having been administered as a treatment for liver fluke.

fenchlorphos (pharmacology) An organophosphorus insecticide suitable for use as a systemic treatment for warble fly larvae.

fenoxycarb (control measures) A carbamate insecticide, with ovicidal and larvicidal activity against fleas.

fenvalerate (control measures) A synthetic pyrethroid insecticide used in ear tags for the control of head fly and ear ticks on cattle and topically on many species for control of fleas, flies, tick and lice.

ferment (chemistry) 1 To undergo fermentation. 2 Any substance that causes fermentation.

fermentation (chemistry/biology) Energy-producing breakdown of organic compounds by micro-organisms (in the absence of oxygen); *e.g.* the breakdown of sugar by yeasts into ethanol, carbon dioxide and organic acids. Fermentation is a type of anaerobic respiration.

fermium (Fm) (chemistry) Radioactive element in Group IIIB of the periodic table (one of the actinides). It has several isotopes, with half-lives of up to 95 days. At. no. 100; r.a.m. (most stable isotope) 257.

-ferous (terminology) A word element. [Latin] Suffix denoting *bearing, producing.*

ferric (chemistry) Trivalent iron. Also termed: iron(III); iron(3+).

ferricyanide (chemistry) $[Fe(CN)_6]^{3-}$ Very stable complex ion of iron(III). A solution of the potassium salt gives a deep blue precipitate (Prussian blue) in the presence of iron(II) (ferrous) ions. Also termed: hexacyanoferrate(III).

ferrimagnetism (physics) Property of certain compounds in which the magnetic moments of neighbouring ions align in anti-parallel fashion.

ferrite (chemistry) Non-conducting ceramic material that exhibits ferrimagnetism; general formula MFe_2O_4, where M is a divalent metal of the transition elements. Ferrites are used to make powerful magnets in radars and other high-frequency electronic apparatus, such as computer memories.

ferrocene (chemistry) $C_{10}H_{10}Fe$ Orange orgaometallic compound, whose molecules consist of an iron atom 'sandwiched' between two molecules of cyclopentadiene. Also termed: dicyclopentadienyliron.

ferrocyanide (chemistry) $[Fe(CN)_6]^{4-}$ Very stable complex ion of iron(II). Also termed: hexacyanoferrate(II).

ferromagnetism (physics) Property of certain substances that in a magnetizing field have induced magnetism, which persists when the field is removed and they become permanent magnets. Examples include iron, cobalt and their alloys.

ferrous (chemistry) Bivalent iron. Also termed: iron(II); iron(2+).

fertilization (biology) Fusion of specialized sex cells or gametes which are haploid to form a single cell, a diploid zygote. It occurs in sexual reproduction; *e.g.* in vertebrates the ovum (female gamete) is fertilized by the sperm (male gamete).

fervescence (medical) Increase of fever or body temperature.

FES (education) An abbreviation of 'F'ellow of the 'E'ntomological 'S'ociety.

festoons (parasitological anatomy) A term used in relation to arthropods. Notches seen at the posterior part of the body of hard ticks.

Festuca (plant parasitology) Pasture grasses which are members of the family Poaceae, and that are toxic when infested with grass nematodes. Species includes *Festuca rubra* var.*commutata* (Chewing's fescue) that with *Anguina agrostis* nematodes and *Clavibacter toxicus* causes tunicaminyluracil poisoning.

Feulgen reaction for DNA (histology) Histological staining technique for nucleic acids. Feulgen reagent which contains fuchsin and sulphuric acid (*USA*: sulfuric acid) is named after the German physiologist and chemist, R. Feulgen (1884–1955).

FEV (physiology) An abbreviation of 'F'orced 'E'xpiratory 'V'olume in first second of forced expiration after maximum inspiration.

fever (physiology) A condition in which the body temperature is above normal, *i.e.* 37°C. Body temperature is controlled by a centre (*USA*: center) in the brain which maintains a balance between heat production and its loss from the surface of the body. Infection by bacteria or viruses excite the macrophage cells of the reticulo-endothelial system to produce pyrogens, substances which make the temperature rise. A great number of conditions can produce fever, and when the cause of the fever is not known, it is referred to as pyrexia of unknown (PUO).

FFA (biochemistry) An abbreviation of unesterified 'F'ree 'F'atty 'A'cid.

FGF (physiology) An abbreviation of 'F'ibroblast 'G'rowth 'F'actor.

FHR (medical) An abbreviation of 'F'oetal 'H'eart 'R'ate. (*USA*: Fetal Heart Rate).

fi (general terminology) An abbreviation of 'f'or 'i'nstance.

FIBiol (education) An abbreviation of 'F'ellow of the 'I'nstitute of 'Biol'ogy.

FIBMS (education) An abbreviation of 'F'ellow of the 'I'nstitute of 'B'io'M'edical 'S'ciences.

fibr(o)- (terminology) A word element. [Latin] Prefix denoting *fibre* (*USA*: fiber), *fibrous*.

fibre (anatomy) Any long thread-like structure in the body. (*USA*: fiber).

fibre optics (optics) Branch of optics that uses bundles of pure glass fibres (*USA*: glass fibers) within straight or curved 'pipes' or cables, along which light travels as it is internally reflected. A modulated light signal in such a cable can carry much more data, *e.g.* computer data, than a wire of similar dimensions. Fibre-optic cables are also used for making endoscopes. Also termed: optical fibres (*USA*: optical fibers).

fibrescope (medical) An endoscope using glass or plastic fibres (*USA*: fibers) to carry images and light, which means that the instrument is flexible and more readily introduced into the body cavity that is being examined than the older endoscopes which were rigid. (*USA*: fiberscope).

fibrillation (medical) When muscle fibres (*USA*: muscle fibers) contract independently and spontaneously.

fibrin (haematology) A protein substance which forms the framework of blood clots. It is precipitated from the blood by the action of the enzyme thrombin or fibrinogen, which creates a blood clot in a wound where blood is exposed to air.

fibrinogen (haematology) Soluble plasma protein found in blood which, after triggering of chemical factors (*e.g.* caused by a wound), is converted to fibrin as part of the blood-clotting mechanism.

fibro-adenoma (histology) A benign tumour (*USA*: tumor) which is commonly found in the breast, consisting of glandular and fibrous tissue.

fibroblast (cytology) Long flat cell found in connective tissue which secretes protein; *e.g.* collagen and elastic fibres (*USA*: elastic fibers).

fibroid (histology) A common benign tumour (*USA*: benign tumor) consisting of muscle and fibrous tissue enclosed within a capsule found in the uterus, usually in women over thirty years of age. There is often more than one, and symptoms may include pain, heavy menstrual periods, and the size of the tumour (*USA*: tumor) may interfere with the bladder and cause difficulty in passing urine.

fibroma (histology) A benign tumour (*USA*: benign tumor) composed of fibrous tissue, usually small and unimportant.

fibropruritic nodule (veterinary parasitology) Multiple, small cutaneous nodules located mainly over back that may be seen in dogs with chronic flea bite hypersensitivity.

fibrosarcoma (histology) A malignant tumour (*USA*: malignant tumor) of fibrous tissue which grows relatively slowly, often in muscles near the surface of the body. It invades neighbouring tissues but is slow to spread to other parts; if it does so, it may metastasise to the lungs.

fibrosis (histology) Formation of fibrous tissue or scar tissue usually in repair or replacement of cellular elements destroyed by injury, infection or deficient blood supply.

fibrous (histology) Composed of or containing fibres (*USA*: fibers).

fibrous tissue (histology) Scar tissue. A simple, strong structural or repair tissue comprising of two proteins, collagen and elastin, laid down by cells called fibroblasts.

fibula (anatomy) Leg bone, located below the knee and outside the tibia (shin bone) in the hind-limb of a tetrapod vertebrate.

fiche (general terminology) An abbreviation of micro'fiche'.

ficin (pharmacology) A highly active, crystallizable proteinase from the sap of fig trees, which catalyzes the hydrolysis of many proteins at acid (4.1) pH and digestion of some living worms, *e.g.* whipworms. Ficin is used as a protein digestant and to enhance the agglutination of red blood cells by IgG.

field (1 physics; 2 optics; 3 computing) 1 In physics, region in which one object exerts a force on another object; *e.g.* electric field, gravitational field, magnetic field. 2 In optics, area that is visible through an optical instrument. 3 In computing, specific part of a record, or a group of characters that make up one piece of information.

field-emission microscope (microscopy) Microscope used for the observation of the positions of atoms in a surface.

fiery dragon (parasitology) Guinea worm. *See Dracunculus insignis.*

fig. (literary terminology) An abbreviation of 'fig'ure.

figure (fig.) (literary terminology) Referring to a chart or table within a text.

FIH (education) An abbreviation of 'F'ellow of the 'I'nstitute of 'H'ygiene.

filament (physics) In electrical apparatus, a fine wire of high resistance which is heated by passing an electric current directly through it. Filaments are used in electric fires and incandescent lamps, and as heaters in thermionic valves.

Filaria (parasitology) A genus of filarid nematodes in the superfamily Filarioididea. Species includes *Filaria haemorrhagica* (synonym: *Parafilaria multi-papillosa*), and *Filaria* taxidea.

filaria (veterinary parasitology) Plural: filariae [Latin] A nematode worm of the superfamily Filarioididea.

Filaria taxidea (veterinary parasitology) A species of filarid nematodes in the genus *Filaria* that is reputed to cause filarial dermatitis in the badger.

filarial (parasitology) Pertaining to or emanating from filariae.

filarial dermatitis (parasitology) *See* cutaneous stephanofilarosis.

filarial dermatosis (parasitology) *See* elaeophoriasis.

filarial uveitis (parasitology) *See* onchocerciasis.

filariasis (medical/parasitology) Infestation with *Filaria*, a genus of parasitic thread-like worms, found mainly in the tropics and subtropics. The adults of *Filaria bancrofti* and *Filaria malayi* live in lymphatics, connective tissues or mesentery, where they may cause obstruction, but the embryos migrate to the blood stream. Completion of the life cycle is dependent upon passage through a mosquito.

filaricide (control measures) A substance that destroys filariae.

filarid (parasitology) Any filaria.

filariform (parasitology) Resembling filariae; threadlike.

Filarinema (veterinary parasitology) A genus of trichostrongylid nematodes that may be found in free-living kangaroos.

Filaroides (veterinary parasitology) A genus of worms in the family Filaroididae and the superfamily Metastrongyloidea that are parasites of the respiratory tract of mammals.

Filaroides bronchialis (veterinary parasitology) A species of parasites in the genus *Filaroides* that may be found in mink and polecats.

Filaroides cebus (veterinary parasitology) A species of parasites in the genus *Filaroides* that may be found in the lungs of capuchin and squirrel monkeys.

Filaroides gordius (veterinary parasitology) A species of parasites in the genus *Filaroides* that may be found in the lungs of capuchin and squirrel monkeys.

Filaroides hirthi (veterinary parasitology) A species of parasites in the genus *Filaroides* that may be found in the lungs of dogs.

Filaroides martis (veterinary parasitology) A species of parasites in the genus *Filaroides* that may be found in the lungs and blood vessels of mink and other Mustelidae.

Filaroides milksi (veterinary parasitology) A species of parasites in the genus *Filaroides* that may be found in the lungs of dogs.

Filaroides osleri (parasitology) *See Oslerus osleri.*

Filaroides pilbarensis (veterinary parasitology) A species of parasites in the genus *Filaroides* that may be found in Australian marsupials.

Filaroides rostratus (veterinary parasitology) A species of parasites in the genus *Filaroides* that may be found in cats where they may cause tracheobron-chitis.

file (computing) A group of data in a computer storage system that can be dealt with as a single unit. It can be retrieved and stored, deleted, or modified en bloc.

file extension (computing) Optional three letter suffix attached, after a period, at the end of a computer file name and used as a means of categorizing the file as, *e.g.* a data file, or a programme (*USA*: program).

file name (computing) The name given to each file by the user to enable the programme (*USA*: program) being used to access, open and then save the file.

filial generation (genteics) Any generation following the parental generation.

Filicollis (parasitology) A genus of small cylindrical worms (acanthocephalans) in the family Polymorphi-dae and the order Palaeacanthocephala.

Filicollis anatis (avian parasitology) A species of small cylindrical worms in the genus *Filicollis* which may be found in the intestine of wild and domestic aquatic birds, and that causes emaciation and some deaths due to peritonitis as a result of perforation of the bowel wall. Also termed: thorny-headed worm.

film (1 physics; 2 photography) 1 Thin layer of one substance on the surface of another substance, *e.g.* oil floating on water. Thin films can sometimes diffract light and produce rainbow colours (*USA*: colors). 2 Plastic strip carrying a light-sensitive emulsion that is used in photography.

filter (1 physics; 2 optics) 1 Porous material through which a liquid is passed to remove suspended matter. 2 In optics, light-absorbing semi-transparent material that passes only certain wavelengths (colours [*USA*: colors]). 3 In electronics, device that passes only certain a.c. frequencies.

filter pump (laboratory equipment) Simple vacuum pump in which a jet of water draws air molecules from the system. It can produce only low pressures and is commonly used to increase the speed of filtration by drawing through the filtrate.

filth fly (parasitology) *See Eristalis.*

filtrate (chemistry) Liquid obtained after filtration.

filtration (chemistry) Method of separating a suspended solid from a liquid by passing the mixture through a porous medium, *e.g.* filter paper or glass wool, through which only the liquid passes.

Fimbriaria (parasitology) A genus of tapeworms that belong to the family Fimbriariidae.

Fimbriaria fasciolaris (avian parasitology) A species of tapeworms in the genus *Fimbriaria* that may be found in the small intestine of chickens and anserine birds.

final host (parasitology) *See* definitive host.

fine chemical (chemistry) Chemical produced in pure form and in small quantities.

fine structure (physics) Splitting of certain lines in a line spectrum into a number of further discrete lines, which are observable only when high resolution is employed.

firing (physiology) The initiation of a nerve impulse.

firmware (computing) Computer software integral to the control of a peripheral device and stored in Read Only Memory (ROM).

first-stage (parasitological physiology) Said of larva; the first of several larval stages.

Fischoederius (veterinary parasitology) A genus of digenetic trematodes in the family Paramphistomatidae. Species includes *Fischoederius cobboldi* and *Fischoederius elongatus* which may be found in the rumen of cattle and other bovids.

fish kill (fish parasitology) Mass death of many fish, usually in a restricted area.

fish tapeworm (fish parasitology) *See Diphyllobothrium latum.*

Fisher's exact probability test (statistics) A test of independence in two-by-two contingency tables giving exact probability levels for all sample sizes.

Fisher's z-test (statistics) A test to determine the significance of a correlation coefficient employing fisher's z-transformation.

Fisher's z-transformation (statistics) A transformation of the product-moment correlation which yields an approximately normal distribution. It is useful in constructing tests of the significance of correlations and their differences.

fission (1 physics; 2 biology) Splitting. 1 In atomic physics, disintegration of an atom into two parts, usually with the release of energy and one or more neutrons. 2 In biology, division of a cell or single-celled organism into two. *See* meiosis; mitosis.

fissiparous (biology) Propagated by fission.

fissure (anatomy) A groove in a surface, particularly a comparatively large groove in the surface of the cortex. The term sulcus is used of smaller grooves in the cortex, but in practice the terms are often interchangeable.

fistula (medical) An opening either into the interior of the body or in an internal organ, such as can be produced by a wound, or by surgery (*e.g.* a tube implanted through the wall of the stomach). The term is used both for the opening in the tissue and for any tube that is passed through it.

fixation of nitrogen (biology) Part of the nitrogen cycle that involves the conversion and eventual incorporation of atmospheric nitrogen into compounds that contain nitrogen. Nitrogen fixation in nature is carried out by nitrifying soil bacteria or blue-green algae (Cyanophyta) in the sea. Soil bacteria may exist symbiotically (*see* symbiosis) in the root nodules of leguminous plants or they may be free-living. Small amounts of nitrogen are also fixed, as nitric oxide, by the action of lightning.

fixative (histology) A substance such as formalin used in preserving a histological or pathological specimen so as to maintain the normal structure of its constituent elements.

fixed pitch font (computing) Every letter is allocated the same horizontal space. Also termed: monospaced font.

fixed point (physics) Standard temperature chosen to define a temperature scale or at which properties are measured, *e.g.* the ice point (0°C) and the steam point (100°C). Also termed: fixed temperature.

fixing, photographic (photography) Process for removing unexposed silver halides after development of a photographic emulsion. It involves dissolving the silver salts by immersing the developed film in a fixing bath consisting of a solution of sodium or ammonium thiosulphate (*USA*: ammonium thiosulfate).

fl.rt. (general terminology) An abbreviation of 'fl'ow 'r'a't'e.

flagella (biology) *See* flagellum.

flagellar (biology) Of or pertaining to a flagellum.

flagellate (1, 2 biology; 3 parasitology) 1 Any microorganism having flagella. 2 Having flagella. 3 Any protozoon of the subphylum Mastigophora.

flagelliform (biology) Shaped like a flagellum.

flagellosis (parasitology) Infection with flagellate protozoa.

flagellum (parasitological anatomy) Long hair-like organelle whose beating movement causes locomotion or the movement of fluid over a cell. Flagella are present in most motile gametes and unicellular plants or animals (*e.g.* protozoa), in which they occur singly or in small clusters. In some multicellular organisms (*e.g.* sponges and hydra) they are used for circulation of water containing food and respiratory gases.

See also cilium.

flame cell (parasitological cytology) The excretory cell in cestodes and trematodes, the number and arrangement of which is used as a basis for identification. The cell has a tuft of cilia, whose beating resembles the flickering of a flame. The flame cells open into a collecting tubule. Also known as solenocyte.

flame test (chemistry) Qualitative chemical test in which an element in a substance is identified by the characteristic colour (*USA*: color) it imparts to a Bunsen burner flame.

flash point (chemistry) Lowest temperature at which a substance or a mixture gives off sufficient vapour to produce a flash on the application of a flame.

flask (equipment) A laboratory vessel, usually of glass and with a constricted neck.

flatulence (biology) Excessive formation of gases in the stomach or intestine, released through the anus.

flatulent (biology) Characterized by flatulence; distended with gas.

flatus (biology) 1 Gas or air in the gastrointestinal tract. 2 Gas or air expelled through the anus.

flatworm (parasitology) Any worm of the phylum Platyhelminthes, *i.e.* the flukes or trematodes, and the cestodes or tapeworms of domestic animals.

flav(o)- (terminology) A word element. [Latin] Prefix denoting *yellow*.

flavin adenine dinucleotide (FAD) (chemistry) Coenzyme that functions in the oxidation-reduction reactions of enzymes, *e.g.* the oxidative degradation of pyruvate, fatty acids and amino acids, and in electron transport. Also termed: flavine.

flavonoid (chemistry) Aromatic, oxygen-containing heterocyclic organic compound. Many natural pigments are flavonoids.

flavoprotein (chemistry) Member of a group of conjugated proteins in which the prosthetic group constitutes a derivative of riboflavin (*e.g.* FAD or FMN). Flavoprotein dehydrogenases (enzymes) are involved in the electron transport chain of aerobic respiration.

flea (parasitology) A small, wingless, blood-sucking insect. Many fleas are ectoparasites and may act as disease carriers. They are members of the order Siphonaptera. *See Ctenocephalides felis*; *Ctenocephalides canis*; *Archaeopsylla erinacei*; *Spilopsyllus cuniculi*; *Leptopsylla segnis*; *Ceratophyllus fasciatus*; *Xenopsylla cheopis*; *Pulex irritans*; *Tunga penetrans*; *Ceratophyllus gallinae*; *Ceratophyllus columbae*; *Ceratophyllus garei*; *Ceratophyllus niger*; *Dasypsyllus gallinulae*; *Echidnophaga gallinacea*; *Echidnophaga perilis*; *Echidnophaga myrmecobii*; *Vermipsylla ioffi*; *Vermipsylla perplexa*; *Vermipsylla alacurt*; *Vermipsylla dorcadia*.

flea allergy dermatitis (parasitology) The inflammatory lesions and self-trauma caused by a hypersensitivity to flea bites. Secondary infection is common.

flea bite hypersensitivity (parasitology) *See* flea allergy dermatitis.

flea collar (control measures) A collar or tag impregnated with insecticide, hung around the animal's neck. There is a slow release of the active ingredient, either as a vapor or powder, to kill ectoparasites on the body.

flea collar dermatitis (veterinary parasitology) A contact dermatitis in dogs and cats caused by the insecticide-impregnated polyvinyl chloride collars marketed for flea control. Although the initial and most severe skin reaction occurs where direct contact is made, surrounding skin may also become involved.

flea dip (control measures) Any external parasiticide applied to dogs as a rinse; dipping is not a practical form of application in most companion animals.

fleece fly strike (parasitology) *See* cutaneous myiasis.

fleece worm (parasitology) *See* cutaneous myiasis.

flesh (anatomy) The soft muscular tissue of the animal body.

flesh flies (parasitology) Large blowflies of the subfamily Sarcophaginae and the genera *Sarcophaga* and *Wohlfahrtia*.

flight range (parasitology/epidemiology) The distance that an insect or bird is capable of flying. An important factor in the spread of disease by vectors or mechanical spreaders.

flocculation (chemistry) Coagulation of a finely divided precipitate into larger particles.

flood plain staggers (plant/veterinary parasitology) Australian (north-western New South Wales) syndrome caused by tunicaminyluracils produced in seedhead galls on *Agrostis avenacea*, the galls produced by *Anguina funesta* (grass nematodes) and infected by *Clavibacter toxicus*. Clinically the disease is characterized by convulsions precipitated by driving and often death during the convulsion. Hypermetric ataxia is characteristic in less severe cases. Cattle are most affected, sheep and horses less frequently.

floppy disk (computing) Flexible, portable magnetic disk that provides data and program storage for microcomputers. The disk may be enclosed in a flexible or a rigid casing. Also termed: diskette.

See also hard disk.

flotation, principle of (physics) A floating object displaces its own weight of fluid (liquid or gas). An object floats if its weight equals the upthrust on it.

flour mite (parasitology) A mite capable of causing dermatitis. *See Tyroglyphus*; *Tyrophagus*; *Glycyphagus*; *Suidasia*; *Nesbitti*; *Carpoglyphus*.

fluff louse (parasitology) *See Goniocotes gallinae*.

fluid (physics) Form of matter that can flow; thus both gases and liquids are fluids. Fluids can offer no permanent resistance to changes of shape. Resistance to flow is manifest as viscosity.

fluid balance (physiology) On average the loss of fluid through urine, between 1200 ml and 1500 ml a day, and from perspiration and in the breath, up to 1000 ml, requires to be balanced by the intake of about two or three litres of fluid a day. Additional loss through vomiting or diarrhoea (*USA*: diarrhea), which also results in a loss of essential electrolytes, requires additional intake of fluid and electrolytes. *See* dehydration.

fluid ounce (measurement) Measure of the volume of liquids. 1 fl oz = 28.41 cm^3.

fluid retention (medical) A condition in which urine discharge from the urinary tract is drastically reduced. The most common cause is impaired kidney function. Fluid retention can have important consequences on both water and electrolyte balance and is treated symptomatically during maintenance therapy.

fluidity (physics) Property of flowing easily; the opposite of viscosity.

fluidrachm (measurements) Fluid dram.

fluke (human/avian/fish/veterinary parasitology) A helminth organism of the class Trematoda in the phylum Platyhelminthes, characterized by a body that is usually flat and often leaflike. Trematodes can infect the blood, liver, intestines and lungs. Some, such as *Chlonorchis sinensis*, the liver fluke, parasitises humans and causes disease. All of the species that infect domestic animals are in the subclass Digenea, that is they are digenetic flukes. Members of the subclass Monogenea occur on the gills or scales of fish.

flukicide (control measures) A substance that destroys flukes.

fluor. (1 physics; 2 chemistry) 1 An abbreviation of 'fluor'escent. 2 'fluor'ide.

fluorescein (chemistry) $C_{20}H_{12}O_5$ A dark orange-red dye which dissolves in alkalis (*USA*: alkalies) to give a green fluorescent solution. It is used as a chemical marker and for dyeing textiles. Also termed: resorcinolphthalein.

fluorescence (chemistry) Emission of radiation (generally visible light) after absorption of radiation of another wavelength (usually ultraviolet or near-ultraviolet) or electrons; it ceases when the stimulating source is removed.

See also luminescence; phosphorescence.

fluorescence microscope (microscopy) A microscope used for the examination of specimens stained with fluorochromes or fluorochrome complexes, *e.g.* a fluorescein-labeled antibody, which fluoresces in ultraviolet light.

fluorescent (chemistry) Having the quality of fluorescence.

fluorescent lamp (laboratory equipment) Mercury-vapour discharge lamp that uses phosphors to produce light by fluorescence.

fluoridation (chemistry) Addition of inorganic fluorides to drinking water to combat dental decay.

fluoride (chemistry) Compound containing fluorine; salt of hydrofluoric acid (HF).

fluorination (chemistry) Replacement of atoms, usually hydrogen, in an organic compound by fluorine.

fluorine (F) (chemistry) Gaseous nonmetallic element in Group VIIA of the periodic table (the halogens). A pale green-yellow poisonous gas, it is highly reactive and the most electronegative element, occurring in fluoride minerals such as fluorspar. It is used, as the gaseous uranium(VI) fluoride (UF_6), in the separation by diffusion of uranium isotopes and in making fluorocarbons. Inorganic fluorides are added to water supplies to combat tooth decay. At. no. 9; r.a.m. 18.9984.

fluorite (chemistry) An alternative term for fluorspar.

fluorocarbon (chemistry) Very stable organic compound in which some or all of the hydrogen atoms have been replaced by fluorine. Fluorocarbons are used as solvents, aerosol propellants and refrigerants. Their use is being limited because they have been implicated in damage to the ozone layer of the atmosphere.

See also freon.

fluorochrome (methods) A fluorescent compound, as a dye, used to mark protein with a fluorescent label.

See also fluorescein.

fluoroscope (radiology) Fluorescent screen that allows direct observation of X-ray images, often connected to a camera. It is used in medicine (radiography) and industrial X-ray applications.

fluorspar (chemistry) CaF_2 Naturally occurring calcium fluoride, used as a flux in glass and as a component of certain cements. Also termed: fluorite.

fly (parasitology) Members of the order Diptera. *See* black fly; blowfly; bot fly; *Chrysops*; *Cnephia*; *Cordylobia*; *Cuterebra*; *Eristalis*; flesh flies; *Haematobia*; *Hippobosca*; *Hydrotoea*; *Hypoderma*; *Musca*; *Phormia*; sandfly; screw-worm; *Simulium*; *Stomoxys calcitrans*; *Tabanus*; *Torsalo* grub; tsetse; warbles.

fly dermatitis (parasitology) Biting flies will inflict skin damage on the face and particularly ear tips of outdoor dogs, causing bleeding, dried crusts and moderate irritation that sometimes leads to the development of auricular haematomas (*USA*: hematomas). They may also be a common problem in zoo bears.

fly strike (parasitology) Cutaneous myiasis.

fly worry (veterinary parasitology) All fly infestations cause worry to their host animals. Heavy infestations with black flies in horses and buffalo flies in cattle may cause deaths from worry, blood loss, interference with grazing and intercurrent disease.

See also fly dermatitis.

flyblown (parasitology) Infested with fly maggots, usually blowfly larvae.

Fm (chemistry) The chemical symbol for fermium.

FMN (biochemistry) An abbreviation of 'F'lavin 'M'ono'N'ucleotide.

fn. (literary terminology) An abbreviation of 'f'oot'n'ote.

f-number (photography) Method of denoting the diameter of a lens aperture in a camera; for a simple lens it is the focal length divided by the diameter of the aperture. The smaller the f-number, the larger the aperture. In the usual sequence *f*22, f11, *f*8, *f*5.6, *f*4 etc., each aperture has twice the area (and hence admits twice the amount of light) as the preceding one in the series.

foam (physics) *See* froth.

focal (epidemiology) 1 Limited to a small area or volume. 2 Pertaining to or emanating from focus.

focal disease (epidemiology) A localized disease.

focal distance (FD) (microscopy) Distance from the optical centre (*USA*: center) of a lens to the point where light rays from a distant object converge. Also known as focal length.

focal length (optics) Distance from the centre (*USA*: center) of a lens or curved mirror to its focus. Also termed: focal distance.

focal liver disease (parasitology) Widely disseminated micro-abscesses or abscesses, migration paths of helminth larvae.

focal plane (microscopy) A plane perpendicular to the optical axis of a lens and passing through the focal point.

focal point (microscopy) The point on the optical axis of a lens at which parallel rays converge.

focus (1, 2 microscopy) 1 The point or plane at which light rays, or other electromagnetic rays, converge after passing through an optical system. An image on this plane is said to be 'in focus' and will have the minimum amount of blur. 2 An alternative term for focal length.

focusing (microsopy) Adjusting a lens or optical system so that light rays from points at a given distance converge to form an image in a particular plane.

foetal (biology) Of or pertaining to a foetus (*USA*: fetus) or to the period of its development. (*USA*: fetal).

foetal stage (medical) The stage of prenatal existence that follows the embryonic stage; it lasts from about the eighth week after conception until birth. (*USA*: fetal stage)

foetus (biology) The unborn mammalian embryo from the time that it has the recognisable features of its final form, *i.e.* from the end of the second month after conception. (pl. foetuses). (*USA*: fetus. pl. fetuses). *See* embryogenesis.

foliar nematode (plant parasitology) *See Aphelenchoides fragariae.*

folic acid (haematology/biochemistry) A water soluble vitamin of the B complex which acts as a coenzyme in various processes involving the metabolism of purines and pyrimidines and which is essential for the formation of normal red blood cells. Insufficient folic acid in the diet, available in green leafy vegetables and yeast, or impaired absorption because of intestinal disease leads to deficiency of this vitamin. Deficiency may result in a reduction of red blood cells and those that are present may be larger than they should be, a state called megaloblastic anaemia (*USA*: megaloblastic anemia).

follicle (histology) A very small secreting gland, or cyst.

follicle stimulating hormone (FSH) (biochemistry) Member of a group of hormones that are secreted by the anterior lobe of the pituitary gland in vertebrates. It stimulates the growth and maturation of ovarian follicles and the growth only of oocytes, which are matured under the action of luteinizing hormone. In males, FSH stimulates sperm formation in the testes.

follicular mange (parasitology) *See* demodectic mange.

follicular phase (physiology) The phase of the oestrous cycle (*USA*: estrous cycle) in which the ovum matures in women, roughly the first 14 days of the cycle.

follow-up study (epidemiology) A scholarly examination carried out to find out whether there has been change in the situation since the initial study.

fomes (epidemiology) Plural: fomites [Latin] An inanimate object or material on which disease-producing agents may be conveyed, *e.g.* faeces (*USA*: feces), bedding, etc.

font (computing) The size and style of a type face. Also termed: fount.

Fontana stain (histology) Histological staining technique for melanin.

food (biology) Materials taken into the body by mouth which provide nourishment in the form of energy or in the building of tissues.

Food and Drug Administration (FDA) (governing body) A U.S. government agency within the Department of Health and Human Services that monitors the purity and safety of food, cosmetics, and drugs; truth in packaging and labelling information; and sanitary practices in restaurants and other food-handling establishments.

food contaminants (parasitology) Contaminants such as parasites and toxic residues, that may be found in food.

foodborne (epidemiology) A term used when referring to an infection or other damaging agent that is transmitted via the animal or human food chain.

food vacuole (parasitological cytology) A term used in relation to amoebae (*USA*: amebae). A membrane bound vesicle in the cytoplasm formed around an ingested food particle.

foot (measurement) Unit of length in the Imperial system, corresponding to 12 inches or 1/3 of a yard, and equal to 30.48 cm.

foot louse (parasitology) *See Linognathus pedalis.*

foot maggot (parasitology) The larvae of the fly *Booponus intonsus.*

foot mange (parasitology) *See Chorioptes.*

foot-pound (measurement) Work done when a mass of 1 pound is lifted 1 foot against the force of gravity, equivalent to 1.3558 joules. Also termed: foot-pound-force.

footprinting (molecular biology) A technique for finding DNA sequences that have a protein bound to them which protects them from being cut by endonucleases. Two samples of the whole DNA molecule are prepared, one with the protein and one without. Both are digested with a restriction endonuclease and the resulting fragments are run on an electrophoretic gel. When the two gels are compared, the bands that have resisted cutting by the enzyme, because of being protected by the protein, will be distinguishable because they do not match the bands on the other gel. In this way, DNA sequences that bind proteins, such as regulatory sequences, promoters, and so on, can be identified.

for. (1 general terminology; 2 medical) 1 An abbreviation of 'for'eign. 2 An abbreviation of 'for'ensic.

forage mites (parasitology) *See Trombicula.*

Forania (insect parasitology) Mites in the Prostigmata that are parasitic on various ants.

force (physics) Influence that can make a stationary object move, or a moving object change speed or direction, *i.e.* that changes the object's momentum. It is equal to mass multiplied by acceleration. The SI unit of force is the newton.

force ratio (physics) An alternative term for mechanical advantage.

forebrain (anatomy) Largest and topmost portion of the vertebrate brain that comprises the cerebral hemispheres and the basal nucleus.

forensic medicine (medical) That part of medicine concerned with the law.

forest fly (parasitology) *Hydrotoea irritans. See Hydrotoea.*

forest tsetses (parasitology) *See Glossina.*

formal saline (histology) A 10% solution of formal saline is used as a fixative preparation.

formaldehyde (chemistry) HCHO Colourless (*USA*: colorless) pungent organic gas, an aldehyde, which is readily soluble in water. It is used as a disinfectant and in the manufacture of plastics. Also termed: methanal.

formaldehyde-malachite green (control measures) A treatment for external parasites in aquarium fish.

formalin (histology) A 40% solution of the gas formaldehyde, used for fixing and preserving tissues, and in dilution as an antiseptic.

formalin ether sedimentation of faeces (parasitology methods) A method of flotation of *Giardia* cysts.

formamidines (pharmacology) A group of acaricidal compounds, that may be used as plant sprays and topically on animals.

See also amitraz.

format (computing) 1 The initial preparation of a disk for use in a computer. 2 The layout or the laying out of a wordprocessing document.

formic acid (chemistry) HCOOH Simplest carboxylic acid, made commercially by the catalytic combination of carbon monoxide and superheated steam. It occurs naturally in the stings of certain insects such as ants, bees and wasps. Also termed: methanoic acid.

Formica fusca (parasitology) An ant, second intermediate host for *Dicrocoelium dendriticum.*

formol (chemistry) Formaldehyde solution.

formula (1 mathematics; 2 chemistry) 1 Mathematical expression that shows the relationship between various quantities; *e.g.* the formula for the area of a circle of radius r is $A = \pi r^2$. 2 Chemical composition of a substance indicated by the symbols of each element present in it and subscripts that show the number of each type of atom involved; *e.g.* the formula for water is H_2O, that for potassium dichromate is $K_2Cr_2O_7.2H_2O$. Also termed: chemical formula.

Fortin barometer (physics) Mercury barometer, used for the accurate measurement of atmospheric pressure. It was named after the French physicist Jean Fortin (1750–1831).

FORTRAN (computing) An abbreviation of 'FOR'mula 'TRAN'slation. A high-level computer programming language designed for mathematical and scientific use.

forward mutation (genetics) A change from the normal, or wild-type, version of an allele to a mutant form. The vast majority of mutations are of this type. *Contrast* back mutation.

fossa (anatomy) A depressed or hollow area.

Fouchet stain (histology) Histological staining technique for bile pigments.

Fourier analysis (mathematics) Mathematical method of expressing a complex function that represents a wave as a series of simpler sine waves. It was named after the French mathematical physicist Jean Fourier (1768–1830).

fovea (histology) Part of the retina of the eye that has a concentration of cones (but no rods), which comes into play when acute vision is required. Also termed: yellow spot.

fowl tick (avian parasitology) *See Argas persicus*.

fox mange (veterinary parasitology) *See* sarcoptic mange.

fp (1 general terminology; 2, 3 physics; 4 literary terminology) 1 An abbreviation of 'f'lame'p'roof. 2 An abbreviation of 'f'lash-'p'oint. 3 An abbreviation of 'f'reezing-'p'oint. 4 An abbreviation of 'f'rontis'p'iece.

FPAS (education) An abbreviation of 'F'ellow of the 'P'akistan 'A'cademy of 'S'ciences.

Fr (chemistry) The chemical symbol for francium.

fraction (mathematics) Part of a whole, represented mathematically by a pair of numbers. The upper numerator is written above the lower denominator and separated from it by a horizontal or diagonal line; *e.g.* 1/2, 2/3, 7/10. Fractions of a hundred may be given as percentages; *e.g.* 65/100 = 65 per cent (sometimes written 65%). In decimal fractions, the denominators are powers of 10 (10, 100, 1,000, ...), usually written using a decimal point and place values as for whole numbers. Thus 7/10 = 0.7, and 27/1,000 = 0.027.

See also common denominator.

fractional crystallization (chemistry) Separation of mixtures of substances by the repeated crystallization of a solution, each time at a lower temperature.

fractional distillation (chemistry) Separation of a number of liquids with different boiling points by distillation and collecting separately the liquids that come off at different temperatures. Also termed: fractionation.

fractionating column (chemistry) Long vertical tube containing bubble-caps, sieve plates, or various irregular packing materials, used for industrial fractional distillation.

fractionation (chemistry) An alternative term for fractional distillation.

frameshift mutation (genetics) A mutation in which one or more base-pairs are inserted into or deleted from a gene, so that the reading frame of every codon after that point is shifted along. *E.g.* if the original code was UCU-CAA-AGG-UUA, and the mutation put an extra U at the beginning, the message would be read as UUC-UCA-AAG and so on.

francium (Fr) (chemistry) Radioactive metallic element in Group IA of the periodic table (the alkali metals), made by proton bombardment of thorium. It has several isotopes with half-lives of up to 22 min. At. no. 87; r.a.m. 223 (most stable isotope).

frank effect level (FEL) (toxicology) A level of exposure to a chemical or mixture that provides an acute, unequivocal deleterious effect.

FRCPath. (education) An abbreviation of 'F'ellow of the 'R'oyal 'C'ollege of 'Path'ologists.

free electron (physics) Electron free to move from one atom or molecule to another under the influence of an electric field. Movement of free electrons enables a conductor to carry an electric current.

free energy (physics) Measure of the ability of a system to perform work.

free radical (chemistry) Intermediate and highly reactive molecule that has an unpaired electron and so easily forms a chemical bond.

freeze drying (chemistry) Method of drying heat-sensitive substances such as blood plasma or food by freezing them below 0°C and then removing the frozen water by volatilization in a vacuum.

freezing (physics) Solidification of a liquid that occurs when it is cooled sufficiently (to below its freezing point).

freezing microtome (equipment) A microtome used for cutting frozen tissues.

freezing mixture (chemistry) Mixture of two substances that absorbs heat and can be used to produce a temperature below 0°C.

See also eutectic mixture.

freezing point (physics) Temperature at which a liquid solidifies. Also termed: solidification point.

Frenkelia (avian parasitology) A genus of protozoa in the family Sarcocystidae. Cysts are found in the brain and spinal cord of birds of prey, the intermediate hosts being the prey.

Frenkelia clethrionomyobuteonis (avian/veterinary parasitology) A species of protozoa in the genus *Frenkelia* whose hosts are buzzards and voles.

FREnt.S (education) An abbreviation of 'F'ellow of the 'R'oyal 'Ent'omological 'S'ociety.

Freon (chemistry) Trade name for certain fluorocarbons and chlorofluorocarbons derived from methane and ethane. They are used as refrigerants.

Freon 112 (pharmacology) A flukicide in sheep. *See* Tetrachlorodifluorethane.

freq. (general terminology) An abbreviation of 'freq'uency.

frequency (freq.) (1 physics; 2 statistics) 1 Rate of recurrence of wave, *i.e.* number of cycles, oscillations or vibrations in unit time, usually one second. The frequency of a wave is inversely proportional to the wavelength. The SI unit of frequency is the hertz (which corresponds to 1 cycle per second). 2 The number of times a given phenomenon occurs.

frequency curve (statistics) A curve in which the frequency of occurrence of each of a set of consecutive values, or of values grouped in intervals, is plotted on the vertical axis and the values on the horizontal axis.

frequency distribution (statistics) The frequency with which the members of a sample or of a population take one of a set of consecutive values, or values grouped in intervals, or with which they are distributed among different categories.

FRES (education) An abbreviation of 'F'ellow of the 'R'oyal 'E'ntomological 'S'ociety of London.

Freyana (parasitology) A genus of mites in the family Dermoglyphidae.

Freyana chanayi (avian parasitology) A species of mites in the genus *Freyana* that may be found on the feathers of turkeys.

Friedman test (statistics) A non-parametric test applicable to data in which there are more than two matched samples. It is based on ranking the scores within each sample. *Compare* wilcoxon test.

fringed tapeworm (parasitology) *See Thysanosoma actiniodes.*

FRIPHH (education) An abbreviation of 'F'ellow of the 'R'oyal 'I'nstitute of 'P'ublic 'H'ealth and 'H'ygiene.

FRMS (education) An abbreviation of 'F'ellow of the 'R'oyal 'M'icroscopical 'S'ociety.

frontal (anatomy) Towards the front of the body or of an organ.

froth (physics) Collection of fairly stable small bubbles in a liquid produced by shaking, aeration or addition of a foaming agent (*e.g.* a detergent). Also termed: foam.

frozen section (histology) A biopsy which is taken, frozen and cut into sections for the pathologist to examine under the microscope and provide a rapid answer as to whether the tissue is malignant or not.

FRS (education) An abbreviation of 'F'ellow of the 'R'oyal 'S'ociety.

FRSSI (education) An abbreviation of 'F'ellow of the 'R'oyal 'S'tatistical 'S'ociety of 'I'reland.

FRSSS (education) An abbreviation of 'F'ellow of the 'R'oyal 'S'tatistical 'S'ociety of 'S'cotland.

FRSTM & H (education) An abbreviation of 'F'ellow of the 'R'oyal 'S'ociety of 'T'ropical 'M'edicine and 'H'ygiene.

fructose (chemistry) $C_6H_{12}O_6$ Fruit sugar, a monosaccharide carbohydrate (hexose) found in sweet fruits and honey. Also termed: laevulose.

fry (fish parasitology) Newly hatched fish. Also termed larvae.

fuchsin (chemistry) Any of several red to purple dyes.

-fugal (terminology) A word element. [Latin] Suffix denoting *driving away, fleeing from, repelling.*

fulminant (medical) Severe.

fumarate (chemistry) A salt of fumaric acid.

fumigation (control measures) The process of burning or volatilising substances in order to produce vapours which destroy infective organisms and vermin. It is now rarely employed.

function (1 general terminology; 2 mathematics; 3 computing) 1 The natural use to which anything can be put, its role or purpose. 2 Mathematical expression that involves one or more variables. 3 Any operation performed by a program, a use similar to the term's mathematical meaning.

functional group (chemistry) Atom or group of atoms that cause a chemical compound to behave in a particular way; *e.g.* the functional group in alcohols is the -OH (hydroxyl) group.

fundamental units (measurements) Units of length, mass and time that form the basis of most systems of units. *See* SI units.

fur mite (parasitology) *Lynxacarus radovsky. See Lynxacarus.*

furanose (chemistry) Any of a group of monosaccharide sugars (pentoses) whose molecules have a five-membered heterocyclic ring of four carbon atoms and one oxygen atom.

fuse (physics) Device used for protecting against an excess electric current passing through a circuit. It consists of a piece of metal, connected into the circuit, that heats and melts (thereby breaking the circuit) when the current exceeds a certain value.

fusiform (biology) Spindle shaped.

fusiform cell (cytology) A type of stellate cell in the cortex that has two short vertical projections which divide into many long processes running upwards or downwards from the cell body.

fusiform layer (histology/cytology) The innermost layer of nerve cells in the cerebral cortex, containing mostly fusiform cells.

fusion (1 cytology; 2 physics) 1 The intimate mixing of cells to produce a hybrid cell. 2 Act of melting or joining together.

fusion gene (genetics) A gene resulting from the accidental mixing of parts of two different genes. This may happen as the result of a small deletion occurring between two genes, leaving their ends joined up. Fusion genes can also be made *in vitro* using recombinant DNA technology.

G

G (1 measurement; 2 computing; 3 physics; 4 biochemistry) 1 An abbreviation of 'G'iga-. 2 An abbreviation of 'G'igabyte. 3 An abbreviation of 'G'ravitational constant. 4 A symbol for glycine.

g (measurement) An abbreviation of 'g'ram. Also termed gm.

g/l (measurement) grams per litre.

G6PD (biochemistry) An abbreviation of 'G'lucose '6'-'P'hosphate 'D'ehydrogenase.

Ga (chemistry) The chemical symbol for gallium.

GABA (biochemistry) An abbreviation for 'G'amma-'A'mino'B'utyric 'A'cid.

GABA-T (biochemistry) An abbreviation of 'GABA' 'T'ransaminase.

GAD (biochemistry) An abbreviation of 'G'lut'A'mate 'D'ecarboxylase.

gad fly (parasitology) *See Hypoderma.*

gadolinium (Gd) (chemistry) Silvery-white metallic element in Group IIIB of the periodic table (one of the lanthanides), which becomes strongly magnetic at low temperatures. At. no. 64; r.a.m. 157.25.

Gaigeria (parasitology) A genus of hookworms in the family Ancylostomatidae.

Gaigeria pachyscelis (veterinary parasitology) A species of hookworms in the genus *Gaigeria* which may occur in the duodenum of the goats and sheep, that is a voracious blood-sucker; an infestation of as few as twenty four can be fatal.

gain (physics) The degree to which a signal is amplified (positive gain) or reduced (negative gain) when it passes through a system or from one system to another.

galact(o)- (terminology) A word element. [German] Prefix denoting *milk*.

galactosaemic (biochemistry) A genetically determined disease in which the liver degenerates from early infancy owing to the inability to metabolize the sugar galactose; there is also serious mental deterioration. If the condition is diagnosed at birth, the baby can be put on a galactose-free (*i.e.* nonmilk) diet, in which case there are no ill effects. This diet can be relaxed after the age of five. The genetic cause is an autosomal recessive, and about 1 in 50,000 babies is affected. Heterozygotes (*i.e.* carriers) among the siblings of an affected child can be identified by biochemical tests. (*USA*: galactosemic).

galactose (chemistry) $C_6H_{12}O_6$ Monosaccharide sugar that occurs in milk and in certain gums and seaweeds as the polysaccharide galactan.

Galba (parasitology) The host snail for the intermediate stages of *Fasciola hepatica* in North America.

gall (1 biochemistry; 2 plant parasitology; 3 pharmacology) 1 The bile. 2 An excrescence on a plant, *e.g.* on the seedheads of *Lolium rididum*, caused by plant nematodes. 3 An extract of galls. Used in medicine as a bitter.

gall bladder (anatomy) Storage organ in some vertebrates that, stimulated by hormones, releases bile (along the bile duct) to the duodenum during digestion.

gall sickness (parasitology) *See* anaplasmosis.

Galleria melonella (insect parasitology) Arthropods in the Lepidoptera that is a parasite of the bees *Apis mellifera* and possibly *Bombus*, and which may kill small colonies representing a serious threat.

gallium (Ga) (chemistry) Blue-grey metallic element in Group IIIa of the periodic table, used in low-melting point alloys and thermometers. At. no. 31; r.a.m. 69.72.

gallon (measurement) Unit of liquid and dry capacity, equal to four quarts or eight pints. One Imperial gallon is the volume occupied by ten pounds of distilled water. One Imperial gallon = 4.54609 litres. One US gallon = 5/6 Imperial gallons, or 3.7854 litres.

gallstone (biochemistry) Accretion, usually of cholesterol or calcium salts, that occurs in the gall bladder or its ducts. Also termed: biliary calculus.

gallyas technique (histology) stain for neurofibrillary changes in CNS.

GALT (histology) An abbreviation of 'G'ut 'A'ssociated 'L'ymphoid 'T'issue.

Galumna (parasitology) Mites which act as intermediate hosts for *Moniezia* spp. tapeworms.

Galumna acutifrons (insect parasitology) Mites in the Orbibatidia (Cryptostigmata) that may be found in the nests of the ants *Pheidole vasliti acolhua*.

galvanic (physics) Pertaining to direct current, as opposed to alternating current.

galvanic cell (physics) An alternative term for a voltaic cell.

galvanic skin response (GSR) (physiology) A decrease in the electrical resistance of the skin, occurring in a state of arousal, whether pleasant or unpleasant; it is produced by activity of the sweat glands. The resistance change is often measured in order to detect arousal.

galvanometer (physics) Device that detects or measures small electric currents passing through it.

Gambian sleeping sickness (parasitology) *See Trypanosoma gambiense.*

Gambusia (control measures) A small, approximately one-inch long, pale fish which eats mosquito larvae and may be used in their control.

gamete (cytology) Specialized sex cell (*e.g.* an ovum or a sperm), which is usually haploid (containing half the normal number of chromsomes). Gametes are customarily produced by a male and a female, and they combine at fertilization to form a zygote that develops into a new organism (with the normal, diploid, chromosome number). Also termed: germ cell.

See also parthenogenesis.

gametogony (parasitological physiology) A term used in relation to *Toxoplasma* and *Sarcocystis*. The sexual form of division which involves the formation of micro and macrogametocytes.

gamma globulin (immunology) An alternative term for immunoglobulin.

gamma radiation (radiology) Penetrating form of electromagnetic radiation of shorter wavelength than X-rays, produced, *e.g.* during the decay of certain radio-isotopes. It is used to sterilize food, to make industrial radiographs and in medicine (radiotherapy).

gamma rays (radiology) High-energy photons that make up gamma radiation.

gamma-aminobutyric acid (GABA) (biochemistry) An amino acid that is an inhibitory neurotransmitter; it is found in the cortex, basal ganglia, and elsewhere in the central nervous system.

Gammarus (parasitology) Water crustaceans that are intermediate hosts for *Tetrameres* spp. worms.

gangli(o)- (terminology) A word element. [German] Prefix denoting *ganglion*.

ganglion (1, 2, medical) 1 A collection of nerve cells in the course of a nerve or a network of nerve fibres (*USA*: nerve fibers) outside the central nervous system. *Compare* nucleus. 2 A cystic swelling found in relation to a tendon.

ganglion cell (cytology) A cell with a large roughly spherical body. They are found, *e.g.* as the third-order retinal cells (whose axons enter the optic nerve) and as the first-order auditory cells, as well as in other sensory systems.

gangrene (medical) Death of a part or tissue due to failure of blood supply, disease, or injury.

Ganjam virus (veterinary parasitology) A member of the family *Bunyaviridae*, genus *Nairovirus*, that causes a tickborne virus infection of sheep and goats in India.

gap junction (physiology) A synapse at which the gap between the presynaptic and postsynaptic cells is only about 2 nm wide and is bridged by channels of membranous particles. Ions and other small molecules can pass through it from one cell to the other. Gap junctions are the basis of the electrical synapse.

gapeworm (parasitology) *See Syngamus trachea.*

gas (chemistry) Form (phase) of matter in which the atoms and molecules move randomly with high speeds, occupy all the space available, and are comparatively far apart; a vapour. A liquid heated above its boiling point changes into a gas.

gas chromatography (biochemistry) Method of analysing mixtures of substances. The sample is volatilized and then introduced into a column containing the stationary phase (a solid or a non-volatile liquid on an inert support), and an inert carrier gas (*e.g.* argon) is passed through the column. Components of the mixture are removed from the column by the carrier gas at different rates. A detector measures the conductivity of the gas leaving the column, which is recorded on a chart as a series of peaks corresponding to each of the components. The chart is calibrated by passing samples of known composition through the machine.

gas constant (R or R_0) (physics) Constant in the gas equation, value 8.31434 j mol^{-1} K^{-1}. Also termed: universal molar gas constant.

gas equation (physics) For n moles of a gas, $pV = nRT$, where p = pressure, V = volume, n = number of moles, R = the gas constant and T = absolute temperature.

gas exchange (physiology) Part of respiration in which organisms exchange gases (carbon dioxide and oxygen) with their environment: air for terrestrial plants and animals, water for aquatic ones. It may involve the use of lungs (mammals, birds, adult amphibians and reptiles), gills (larval amphibians, fish and other aquatic animals), spiracles (insects and other terrestrial arthropods) or stomata (green plants). In other organisms (*e.g.* aquatic plants, fungi) gas exchange takes place directly between cells and the environment.

gas laws (physics) Relationships between pressure, volume and temperature of a gas. The combination of Boyle's, Charles's and Gay-Lussac's laws is the gas equation.

gas thermometer (physics) Thermometer based on the variation in pressure or volume of a gas. Also termed: constant-volume gas thermometer.

gas-liquid chromatography (physics/chemistry) (GLC) Type of gas chromatography in which the column contains a non-volatile liquid on an inert support.

gasterophiliasis (parasitology) *See* gasterophilosis.

gasterophilosis (parasitology) The disease caused by infestation with *Gasterophilus* spp. Sporadic cases of abscess formation, even rupture of the stomach wall with local peritonitis, may occur but the infestation is generally considered of low pathogenicity. Also termed: gasterophiliasis.

Gasterophilus (human/veterinary parasitology) A genus of flies of the family Gasterophilidae, the horse bot flies, the larvae of which develop in the gastrointestinal tract of horses and may sometimes infect humans.

Gasterophilus equi (parasitology) *See Gasterophilus intestinalis.*

Gasterophilus haemorrhoidalis (veterinary parasitology) A species of horse bot flies in the genus *Gasterophilus* whose eggs are laid around the mouth and on the cheeks. The larvae are found in the mucosa of the tongue and establish finally in the stomach.

Gasterophilus inermis (veterinary parasitology) A species of horse bot flies in the genus *Gasterophilus* whose eggs are laid around the mouth and on the cheeks. The larvae are found in the cheek mucosa and settle finally in the rectum.

Gasterophilus intestinalis (veterinary parasitology) (synonym: *Gasterophilus equi*) A species of horse bot flies in the genus *Gasterophilus* whose eggs are laid near the front fetlocks and up the legs as far as the shoulder. The larvae are found in the mucosa of the tongue and subsequently at the gastric cardia.

Gasterophilus nasalis (veterinary parasitology) (synonym: *Gasterophilus veterinus*) A species of horse bot flies in the genus *Gasterophilus* whose larvae are laid in the intermandibular space. The larvae are found at the pylorus and in the duodenum.

Gasterophilus nigricornis (veterinary parasitology) A species of flies in the genus *Gasterophilus* that is an uncommon horse bot fly.

Gasterophilus pecorum (veterinary parasitology) A species of horse bot flies in the genus *Gasterophilus* whose eggs are laid on plants. The larvae are found in the mucosa of the cheeks and eventually at the gastric pylorus.

Gasterophilus veterinus (parasitology) *See Gasterophilus nasalis.*

Gasterostomata (fish parasitology) An order of flatworms in subclass Digenea, class Trematoda and the phylum Platyhelminthes. Members of the Gasterostomata are distinguished from all other Digenea because the mouth is in the middle of the body and leads, by way of a pharynx, into a small sacciform intestine. A ventral sucker is lacking. Adults are restricted to fish and usually are found only in carnivorous fish; metacercariae are common.

gastr(o)- (terminology) A word element. [German] Prefix denoting *stomach*.

gastric (medical) Pertaining to the stomach or digestion.

gastric fluid (biochemistry) *See* gastric juice.

gastric habronemiasis (parasitology) Large granulomatous masses in the gastric mucosa caused by invasion by *Draschia megastoma* larvae. They are usually clinically silent unless perforation of the gastric wall occurs. The larvae of *Habronema majus* and *Habronema muscae* cause mild gastritis but without the formation of tumours (*USA*: tumors).

gastric juice (biochemistry) Fluid secreted by glands in the stomach wall during digestion; it contains two principal enzymes, pepsin and rennin, and hydrochloric acid.

gastric lavage (medical) The washing of the stomach usually with saline solutions via a lavage tube. It is used as a rapid method to remove stomach contents in poisoning cases, particularly in cases in which emesis is not recommended.

gastric ulcer (medical) *See* Peptic ulcer.

gastritis (medical) Inflammation of the stomach.

gastrocolic (anatomy) Pertaining to or communicating with the stomach and colon.

Gastrodiscoides (parasitology) A genus of intestinal flukes (digenetic trematodes) in the family Paramphistomatidae.

Gastrodiscoides hominis (human/veterinary parasitology) (synonym: *Gastrodiscus hominis*) A species of flukes in the genus *Gastrodiscoides* that is a natural parasite of the colon of the pig but that may also be found in the human caecum (*USA*: cecum).

Gastrodiscus (parasitology) A genus of intestinal flukes (digenetic trematodes) in the family Paramphistomatidae.

Gastrodiscus aegyptiacus (veterinary parasitology) A species of flukes in the genus *Gastrodiscus* that may be found in the intestines of horse, pig and wart hog.

Gastrodiscus hominis (parasitology) *See Gastrodiscoides hominis.*

Gastrodiscus secundus (veterinary parasitology) A species of flukes in the genus *Gastrodiscus* that may be found in the colon of horse and elephant.

gastroduodenal (anatomy) Pertaining to the stomach and duodenum.

gastroduodenitis (medical) Inflammation of the stomach and duodenum.

gastroduodenoscopy (medical) Endoscopic examination of the stomach and duodenum.

gastrodynia (medical) Pain in the stomach.

gastroenteralgia (medical) Pain in the stomach and intestines.

gastroenteric (anatomy) Pertaining to the stomach and intestines.

gastroenteritis (medical) Inflammation of the lining of the stomach and intestine. The clinical manifestations are vomiting and diarrhoea (*USA*: diarrhea).

See also gastritis.

gastroenterologist (medical) A physician specializing in gastroenterology.

gastroenterology (medical) The study of the stomach and intestine and their diseases.

gastrointestinal (anatomy) Pertaining to the digestive tract. In humans, this consists of the buccal cavity, oesophagus (*USA*: esophagus), stomach, small intestine (duodenum, jejunum and ileum), large intestine (colon and rectum), with various associated glands and accessory structures.

gastrointestinal myiasis (parasitology) *See* gastric habronemiasis; gasterophilosis.

gastrointestinal tract (anatomy) The stomach and intestines in continuity.

gastrojejunal (anatomy) Pertaining to the stomach and jejunum.

gastrology (medical) Study of the stomach and its diseases.

Gastrophilus (parasitology) *See Gasterophilus*.

gastroscope (medical) An endoscope for viewing the interior of the stomach. No anaesthetic (*USA*: anesthetic) is needed for the examination, which is perhaps uncomfortable but safe.

gastrostomy (medical) An operation on the stomach in which an opening is made between the stomach and the overlying abdominal wall, used in cases in which the oesophagus (*USA*: esophagus) is blocked or the patient is unable to swallow. A self-retaining catheter or tube is introduced into the opening through which the patient is fed.

Gastrothylax (parasitology) A genus of stomach flukes (digenetic trematodes) in the family Paramphistomatidae.

Gastrothylax crumenifer (veterinary parasitology) A species of elongated circular flukes in the genus *Gastrothylax* that may be found in the rumen and reticulum in sheep, cattle, zebu and buffalo.

gate (computing) An electronic circuit (switch) that produces a single output signal from two or more input signals. Also termed: logic element.

gauze (equipment) A light, open-meshed fabric of muslin or similar material.

Gay-Lussac's law of volume (physics) When gases react, their volumes are in a simple ratio to each other and to the volume of products, at the same temperature and pressure. It was named after the French chemist and physicist Joseph Gay-Lussac (1778–1850).

G-CSF (physiology) An abbreviation of 'G'ranulocyte 'C'olony-'S'timulating 'F'actor.

Gd (chemistry) The chemical symbol for gadolinium.

Ge (chemistry) The chemical symbol for germanium.

Gedoelstia (parasitology) A genus of flies of the family Oestridae.

Gedoelstia cristata (veterinary parasitology) A species of flies in the genus *Gedoelstia* that may be found in wild ruminants, whose larvae are deposited in the eye and pass via a vein to the cardiovascular system. Larvae migrate up the trachea to the nasal cavity. Aberrant infection also occurs in domestic ruminants.

Gedoelstia hassleri (veterinary parasitology) A species of flies in the genus *Gedoelstia* that may be found in hartebeeste and wildebeeste, whose larvae are deposited in the conjunctival sac and migrate to the nasal cavity via blood vessels, meninges and subdural space, without causing apparent clinical illness. Aberrant infection occurs in domestic ruminants causing severe ocular and neural disease. Also termed: ophthalmomyiasis, uitpeuloog, gedoelstial myiasis.

Gedoelstial myiasis (parasitology) Infestation with the larvae of the fly *Gedoelstia* spp.

Geiger counter (radiology) Instrument for detecting atomic and sub-atomic particles (*e.g.* alpha particles and beta particles), used for radioactivity measurements. It was named after the German physicist Hans Geiger (1882–1945). Also termed: Geiger-Müller counter.

See also counter.

Geiger-Nuttal law (physics) Empirical law for calculating the distance that an alpha particle can travel once it is emitted from a radioactive substance.

gel filtration (chemistry) Type of chromatography in which compounds are separated according to their molecular size. Molecules of the components of a mixture penetrate the surface of an inert porous material in proportion to their size. Also termed: gel permeation.

gel filtration chromatography (biochemistry) *See* chromatography; gel filtration.

gel. (chemistry) An abbreviation of 'gel'atin.

gelatin (chemistry) A colourless (*USA*: colorless) and transparent substance made from the collagen of the connective tissue of animals. It has the consistency of jelly. *See* colloid.

-gen (terminology) A word element. [German] Suffix denoting *an agent that produces*.

gen. (1, 2 classification; 3 genetics; 4 anatomy) 1 An abbreviation of 'gen'eric. 2 An abbreviation of 'gen'us. 3 An abbreviation of 'gen'etic. 4 An abbreviation of 'gen'ital.

gene (genetics) The basic unit of heredity. A gene is a sequence of DNA, occupying its own place, or locus, on a chromosome. Most genes are structural, *i.e.* they code for a particular protein; these genes are divided into housekeeping and luxury genes. Other genes code for the RNA molecules that are necessary for protein synthesis, and others again provide recognition markers for the polymerase enzymes involved in gene regulation. A gene may exist in two or more alternative forms, called alleles. Genes do not change except in the very rare event of mutation; there are many genes in the human genome that have been passed unchanged down the generations for millions of years, since before we became humans. *See* mutation.

gene amplification (genetics) The process by which extra copies of a gene that is needed temporarily are made. This can occur naturally, as in *Drosophila* females who multiply the genes that code for the proteins that compose the membranes of their eggs: all the eggs are matured within about five hours, and to achieve this, the relevant genes are multiplied about ten times. Gene amplification can also be artificially induced, and has been observed in mice, who respond to a sub-lethal dose of cadmium by multiplying copies of the genes coding for the protein metallothioncin that neutralizes the poison.

gene cloning (genetics) The technique of taking a complete gene from an organism's genome and inserting it, via a plasmid, into a host bacterial cell. To make sure that the gene will be expressed by the bacterial cell, regulatory regions as well as the coding sequence of the gene must be included. This technique, perhaps the most basic one in genetic engineering, can be used to produce large amounts of the product of any gene that has been isolated. Human genes have been cloned in *Escherichia coli* to produce once scarce proteins for treatment of people whose own gene is defective; these proteins include growth hormone and insulin.

gene divergence (genetics) The process by which a sequence of related genes arises from an original single gene that has been duplicated a number of times and then undergone divergent changes in the different copies.

gene dosage (genetics) The number of times that a gene is represented in an organism's cells, or total genome. It can range from as low as one, if there is a single-copy gene on the X chromosome in animals, to many thousands for a highly duplicated gene.

gene expression (genetics) The production of RNA and cellular proteins.

gene flow (genetics) The movement of genes from one population into a neighbouring one. It may happen by interbreeding or migration or both.

gene frequency (genetics) The frequency with which a particular allele occurs at a given locus. It is expressed as a decimal: if there are only two alleles, the gene frequencies could be, *e.g.* 0.1 and 0.9 (the total must add up to 1). Evolution is really a matter of changing gene frequencies: the old allele gets rarer as the new one comes in, and finally the new one may become fixed, so that there is a complete change in the genetic constitution of the population in respect of that locus. Taken over a sufficient number of loci (which could perhaps be rather few), this gives rise to a new species. Changes in gene frequency are brought about by natural selection, gene flow, founder effect and genetic drift.

gene library (genetics) A collection of the genes from a species, maintained as gene clones. It is different from a DNA library in that the items stored are whole genes, not DNA fragments.

gene pool (genetics) The total number of alleles in an interbreeding population.

gene product (genetics) The molecule that is made from the information in a single gene. In the majority of cases it is a protein or polypeptide, but some genes code for a sequence of RNA. *See* protein synthesis.

gene regulation (genetics) The various systems by which genes that are not constitutive (*i.e.* in use all the time) are switched on and off. Gene regulation is the means by which cells specialise into the very many different cell types found in even quite simple organisms, and by which organisms respond moment by moment to changes in their external and internal environment. Most of what is known about regulation has been discovered in prokaryotes, and eukaryote gene regulation is still hardly understood. The key feature in prokaryotes is the operon, a sequence of DNA in which the regulatory genes and the several structural genes (*e.g.* for a metabolic process) are in a line, one next to the other. Some operons are inducible, that is they are not transcribed until they are switched on by the presence of some molecule within the cell. Others are repressible, *i.e.* they are normally transcribed until they are switched off by some molecule within the cell.

gene splicing (genetics) *See* recombinant DNA.

gene switch (genetics) The moment when a developing organism changes over from one gene (or set of genes) to another related one. A classic example is the change in humans from foetal haemoglobin (*USA*: fetal hemaglobin) to the adult equivalent. The series of 'choices' made by gene switching are irreversible.

genera (biology) Plural of genus.

general formula (chemistry) Expression representing the common chemical formula of a group of compounds *e.g.* C_nH_{2n+2} is the general formula for an alkane. A series of compounds of the same general formula constitute an homologous series.

generally regarded as safe (GRAS) (toxigenicity) Term used by the Food and Drug Administration applied to substances that are not known to cause harm when used as directed.

generation time (1 genetics; 2 biology) 1 The average time in a species between the birth of an individual and its production of offspring. 2 Time within a population of cells that it takes for them to undergo division to form pairs of daughter cells; *i.e.* the doubling time.

generic (classification) The official name of a drug is the generic name, as opposed to the brand name given it by the manufacturer.

genet (genetics) An organism raised clonally from a single zygote.

See also clone; ortet.

genetic code (genetics) A code which consists of three nucleotide bases occurring in a DNA or RNA molecule. It determines which amino acid is incorporated into a protein at a given point in the sequence. The code thus contains instructions for all enzymes produced in the cell, and consequently the characteristics of the organism.

genetic death (genetics) The death of a gene, in the sense that the individual who carried it failed to reproduce. This may have been because the individual itself died as a result of carrying the gene, but not necessarily. From the genetical point of view, it is the fact that fertility was reduced to zero that counts.

genetic distance (genetics) A measure of the amount of relatedness of two individuals or populations, in terms of the probability of their possessing the same alleles at a locus or loci.

genetic drift (genetics) Random fluctuation in the frequency of particular genes in a population. It becomes an important effect in small populations when drift can cause the loss of an allele from the gene pool even though it was a useful one.

genetic engineering (genetics) Manipulation of genetic material such as DNA for practical use, *e.g.* the introduction of foreign genes into micro-organisms for the production of a useful protein (such as human insulin). The DNA so produced is called recombinant DNA, and the term 'recombinant DNA technology' is often used. The technique is also used in the study of genetic material.

genetic load (genetics) The hidden burden on a population of deleterious recessive mutations. While these alleles are in the heterozygous state, they do no harm, but once they meet in the homozygous state, they damage the individual and therefore the population.

genetic map (genetics) A diagram that shows the position on a chromosome of the various genes that are known to be on it. A method has been devised using the frequency of crossing-over between two genes as a measure of how far apart they are on a chromosome: two genes exactly next to each other will very seldom have a cross-over between them, but genes at opposite ends of a chromosome will be separated by any cross-over along the length of the chromosome. The basic technique of mapping is therefore to make large numbers of crosses involving two or more loci at which there are distinguishable variant alleles, and to record the number of times that recombination of the original types has taken place. The frequency of the crossovers is expressed as a percentage, and one percentage point is known as a map unit. When three or more loci are mapped in this way, the map units add up as they should. *E.g.* if locus A and locus B have a 7 per cent cross-over frequency between them, and locus B and locus C have 10 per cent, loci A and C have crossovers in 17 per cent of instances. Other methods of locating genes on chromosomes include deletion mapping, and restriction mapping.

genetic variance (genetics) The proportion of the total variance of a trait that is due to genes. It may have several components, due to the additive effects of separate genes, or dominance, or interaction. The rest of the variance is environmentally produced. The relative contribution of genetic factors to variance is not a fixed feature of any trait. *See* heritability.

genetics (genetics) Study of inheritable characteristics of organisms; *i.e.* heredity.

-genic (terminology) A word element. [German] Suffix denoting *giving rise to, causing.*

genito- (terminology) A word element. [Latin] Prefix relating to the organs of reproduction.

genito-urinary (GU) (anatomy) Pertaining to both the reproductive structures and the urinary tract.

genocopy (genetics) A genetically determined character that mimics the appearance of another genetically determined character, though caused by an allele at a different locus.

genome (genetics) The total amount of genetic information that an organism possesses; the sum of its genes. In eukaryotes, this means the entire set of genes in a haploid set of chromosomes. The genome of a virus may be made up of either DNA (single or double stranded) or RNA; it can be as small as 3500 base-pairs and may contain as few as three genes. The genome of a prokaryote (bacteria, etc.) is contained in a double-stranded DNA molecule (not strictly a chromosome), and contains from about 750,000 to 4,500,000 base-pairs, comprising up to about 1500 genes This molecule is extremely compact. The eukaryotes have even larger genomes, and in general, the higher up the evolutionary scale, the larger the genome (but with exceptions: some species of amphibians and fish have genomes nearly 40 times larger than the average for mammals). The size of the genome bears no simple relationship to the number of chromosomes: closely related species may have either a large number of small chromosomes or vice versa, for roughly the same-sized genome. Humans have a genome of about 2900 million basepairs of DNA, which, if it were all in one DNA molecule, would be 1 metre long. (This is the haploid total; each of our body cells has twice this amount.) It would be physically possible for there to be as many as one million genes encoded in that much DNA, but in common with other eukaryotes, humans have many instances of repetitive sequences of DNA that do not seem to have any meaning, and the actual number of working genes is probably much lower. Many genes are also present in multiple copies.

genotype (genetics) Genetic constitution of an organism, *i.e.* the characteristics specified by its alleles. It is the outward appearance of the organism, as opposed to the way its genes are expressed (which is the phenotype).

-genous (terminology) A word element. [German] Suffix denoting *arising or resulting from, produced by.*

gentian violet (chemistry) An aniline dye used by microscopists as a stain. As crystal violet it may be used on unbroken skin as an antiseptic. At one time administered orally for the treatment of pinworm and liver fluke infections in humans. Also termed: crystal violet; methylrosaniline chloride.

gentianophilic (microscopy) Staining readily with gentian violet.

gentianophobic (microscopy) Not staining with gentian violet.

genus (gen.) (taxonomy) In biological classification, one of the groups into which a family is divided, and which is itself divided into species.

See also binomial nomenclature.

geo- (terminology) A word element. [German] Prefix denoting *the earth, the soil.*

geographic (epidemiology) Pertaining to geography.

geometric isomerism (chemistry) Form of stereoisomerism that results from there being no free rotation about a bond between two atoms. Groups attached to each atom may be on the same side of the bond (the *cis*-isomer) or on opposite sides (the *trans*-isomer). Also termed: cistrans isomerism.

geometric mean (statistics) A measure of central tendency, arrived at by taking the nth root of the product of all scores, where n is the number of scores. *Contrast* arithmetic mean.

geometric progression (statistics) Sequence of numbers in which the ratio of any two successive terms is constant (the common ratio *r*); *e.g.* for the geometric progression, 2, 8, 32, 128, the common ratio is 4. For the general series $a + ar + ar^2 + ar^3 \ldots + ar^{n-1}$, the sum of the series to *n* terms is $a(r^n - 1)/(r - 1)$.

geometric series (statistics) Series whose terms form a geometric progression.

germ (biology) Imprecise term for a micro-organism that can cause disease; a pathogen. Germs include bacteria; protozoa; viruses etc.

germ cell (genetics) *See* gamete.

germ layer (histology) Layer of cells present in an embryo. In triploblastic organisms the layers consist of ectoderm, mesoderm and endoderm, which each give rise to particular tissues.

germanium (Ge) (chemistry) Grey-white semimetallic element in Group IVA of the periodic table, which occurs in some silver ores. It is an important semiconductor, used for making solid-state diodes and transistors. Its oxide is used in optical instruments and infra-red cameras. At. no. 32; r.a.m. 72.59.

germinal stage (physiology) The first stage of embryonic development, during which there is little cellular differentiation; it lasts about two weeks in the human embryo.

gero(nto)- (terminology) A word element. [German] Prefix denoting *old age, the aged.*

gerontology (physiology) The study of aging.

gestation (physiology) The process of carrying the embryo and fetus from conception to birth. The duration of pregnancy.

Getah virus (veterinary parasitology) A member of the family Togaviridae, genus *Alphavirus*, that is transmitted by mosquitoes and that causes a highly infectious disease of horses.

GeV (measurement) An abbreviation of 'G'iga-'e'lectron-'V'olt, which is 10^9 electron volts.

GFR (physiology) An abbreviation of 'G'lomerular 'F'iltration 'R'ate.

gg (immunology) An abbreviation of 'g'amma 'g'lobulin.

GI (anatomy) An abbreviation of 'G'astro'I'ntestinal.

giant kidney worm (parasitology) *See Dioctophyme renala.*

giant liver fluke (parasitology) *See Fasciola gigantica; Fascioloides magna.*

giant pyramidal cell (cytology) *See* betz cells.

giant schizonts (parasitological physiology) Schizonts that are up to 300 µm diameter in sheep and goats, as part of the life cycle of *Eimeria* spp.

Giardia (parasitology) A genus of flagellate protozoa parasitic in the intestines of most animals where they are capable of causing protracted, intermittent diarrhoea (*USA*: diarrhea) suggestive of malabsorption, sometimes dysentery, but many infections may also be innocuous.

Giardia bovis (veterinary parasitology) A species of flagellate protozoa in the genus *Giardia* that may be found in cattle.

Giardia canis (veterinary parasitology) A species of flagellate protozoa in the genus *Giardia* that may be found in dogs.

Giardia caprae (veterinary parasitology) A species of flagellate protozoa in the genus *Giardia* that may be found in goats.

Giardia cati (veterinary parasitology) A species of flagellate protozoa in the genus *Giardia* that may be found in cats.

Giardia caviae (veterinary parasitology) A species of flagellate protozoa in the genus *Giardia* that may be found in guinea pigs.

Giardia chinchillae (veterinary parasitology) A species of flagellate protozoa in the genus *Giardia* that may be found in chinchillas.

Giardia duodeualis (veterinary parasitology) A species of flagellate protozoa in the genus *Giardia* that may be found in rabbits.

Giardia equi (veterinary parasitology) A species of flagellate protozoa in the genus *Giardia* that may be found in horses.

Giardia felis (veterinary parasitology) A species of flagellate protozoa in the genus *Giardia* that may be found in cats.

Giardia intestinalis (parasitology) *See Giardia lamblia.*

Giardia lamblia (human/avian/veterinary parasitology) A species of flagellate protozoa in the genus *Giardia* that may be found in found in humans, pigs, budgerigars, and monkeys.

Giardia muris (veterinary parasitology) A species of flagellate protozoa in the genus *Giardia* that may be found in mice and rats.

giardiasis (human veterinary parasitology) Infection with *Giardia* protozoa, a parasite with flagellae found in the small intestine. It produces cysts, which are passed in the faeces (*USA*: feces) and may be spread by contaminated food and water. In humans it causes acute or chronic diarrhoea (*USA*: chronic diarrhea), and is a cause of traveller's diarrhoea (*USA*: traveller's diarrhea) in some cases. It is also liable to infect those with immune deficiency syndromes. Infection in dogs and cats is common, with subclinical to severe disease resulting. There may be profuse watery diarrhoea (*USA*: diarrhea) with borborygmus. Chronic small bowel diarrhoea (*USA*: diarrhea) can also result, with weight loss and signs of malabsorption. The condition is treated with metronidazole or tinidazole; cases of cyst carriers who have no symptoms may also be treated with these drugs.

GIBiol. (education) An abbreviation of 'G'raduate of the 'I'nstitute of 'Biol'ogy.

Giemsa for helicobacter pylori (histology) A histological staining technique used to make *helicobacter pylori* visible under the microscope.

Giemsa stain (histology) A histological staining technique used to make parasites visible under the microscope.

giga- (G) (measurement) A prefix meaning 1,000,000,000 [one billion], used with units of measurement.

gigabyte (G) (computing) A computer term denoting one billion bytes or 1,000 megabytes of information or storage space.

gigaelectron volt (GeV) (measurements) One thousand million electron volts (10^9 eV).

Gigantobilharzia (avian parasitology) A genus of blood flukes in the family Schistosomatidae that may be found in birds, which inhabit the blood vessels of their hosts.

Gigantocotyle (parasitology) A genus of flukes (digenetic trematodes) in the family Paramphistomatidae.

Gigantocotyle explanatum (veterinary parasitology) A species of flukes in the genus *Gigantocotyle* that may be found in the duodenum, bile ducts and gall bladder of cattle and buffalo.

gill (1 fish anatomy; 2 insect anatomy) 1 External breathing apparatus of fish which is very susceptible to a wide range of diseases. 2 A term used in relation to arthropods. A specialised respiratory organ seen in aquatic insects.

gill parasites (fish parasitology) The following external parasites are commonly found on the gills, and elsewhere on the skin, in aquarium fish: the monogenetic flukes *Gyrodactylus elegans*, *Diplozoon barbi*, and *Diplozoon paradoxum*, and the crustacean *Ergasilus sieboldi*. Freshly caught seahorses may carry the crustacean *Argulus* spp. Pond fish may carry the anchor worm *Lernaea* spp. These are all visible with the naked eye and can be removed manually.

Gimenez stain (histology) A histological staining technique used to make chlamydia visible under the microscope.

gingivo- (terminology) A word element. [Latin] Prefix denoting *gingival*.

GIP (physiology/biochemistry) An abbreviation of 'G'astric 'I'nhibitory 'P'eptide.

gizzard (avian parasitology) The muscular stomach of the bird, separated from the more cranial proventriculus or glandular stomach by a constriction. Also termed ventriculus.

gizzard strongyles (avian parasitology) *See Amidostomum anseris.*

gizzard worms (avian parasitology) *See Habronema incertum.*

Gla (biochemistry) A symbol for gammacarboxyglutamic acid.

glabrous (biology) Smooth.

glacial (chemistry) Describing a compound of ice-like crystalline form, especially that of the solid form of a liquid, *e.g.* glacial (ethanoic) acetic acid.

glacial acetic acid (chemistry) Crystalline form of (ethanoic) acetic acid below its freezing point.

glands (histology) A specialised group of cells which secrete or excrete substances which are not the same as those needed for their own metabolism, and which act on other tissues and in places other than the gland itself. Glands are 'exocrine', when the secretion is removed through a duct, or 'endocrine', when it passes into the bloodstream. The word is also used to mean a lymph node.

glass (chemistry) Hard brittle amorphous mixture of the silicates of sodium and calcium, or of potassium and calcium. In some particularly strong or heat-resistant forms of glass, boron replaces some of the atoms of silicon. Glass is usually transparent or translucent.

glass electrode (laboratory equipment) Glass membrane electrode used to measure hydrogen ion concentration or pH.

glass wool (chemistry) Material that consists of fine glass fibres (*USA*: glass fibers), used in filters, as a thermal insulator and for making fibreglass (*USA*: fiberglass).

Glauber's salt (chemistry) An alternative term for sodium sulphate (*USA*: sodium sulfate), named after the German physician Johann Glauber (1603–68).

GLC (chemistry) An abbreviation of 'G'as-'L'iquid 'C'hromatography. *See* chromatography.

glia (medical) An abbreviation of neuro'glia'.

glioma (histology) A tumour (*USA*: tumor) of the brain or spinal cord arising from neuroglial tissue.

Gliricola (parasitology) A genus of biting lice of the suborder Amblycera.

Gliricola pistoi (veterinary parasitology) A species of biting lice in the genus *Gliricola* that may be found on New World primates.

Gliricola porcelli (veterinary parasitology) A species of biting lice in the genus *Gliricola* that may be found on guinea pigs and rodents.

Gln (biochemistry) An abbreviation of 'Gl'utami'n'e.

global search (computing) A technique of searching, in a single pass, a number of files to find a specified character string.

globidiosis (parasitology) A name formerly given to a disease thought to be caused by *Globidium*. Abomasal and cutaneous forms were described, which are now classified as *Eimeria* and *Besnoitia* respectively.

Globidium (parasitology) A poorly defined genus of protozoa, now classified as *Eimeria*.

Globidium gilruthi (veterinary parasitology) A species of protozoa in the genus *Globidium* that may be found in goats and sheep.

Globidium leuckarti (veterinary parasitology) A species of protozoa in the genus *Globidium*. Infestation with this parasite may cause diarrhoea (*USA*: diarrhea), but is rarely pathogenic in young horses. Also termed: *Eimeria leuckarti*.

globins (biochemistry) A group of proteins involved in the transport of oxygen in blood, muscles, etc. *See* haemoglobin.

Globocephaloides (parasitology) A genus of nematodes in the Trichostrongyloid family Herpetostrongyloidae.

Globocephaloides trifidospicularis (veterinary parasitology) A species of nematodes in the genus *Globocephaloides* that is a blood-sucking parasite in the small intestine of the grey (*USA*: gray) kangaroo, *Macropus giganteus*.

Globocephalus (veterinary parasitology) A genus of blood-sucking nematodes in the subfamily Ancylostominae that occur in the intestine of pigs and probably cause anaemia (*USA*: anemia) if the infestation is heavy. Species includes *Globocephalus longemucronatus*, *Globocephalus samoensis*, *Globocephalus urosubulatus*, *Globocephalus versteri*.

globular nematode (parasitology) *See Tetrameres americana*.

globulin (chemistry) Water-insoluble protein that is soluble in aqueous solutions of certain salts. Globulins generally contain glycin and are coagulated by heat; *e.g.* immunoglobulin.

glomerulonephritis (medical) A disease of the kidneys affecting the glomeruli, formerly known as Bright's disease. *See* kidney.

glomerulus (anatomy) Ball of capillaries located in the Bowman's capsule of the kidney.

gloss. (literary terminology) An abbreviation of 'gloss'ary.

gloss(o)- (terminology) A word element. [German] Prefix denoting *tongue*.

Glossina (human/veterinary parasitology) A genus of biting flies in the family Glossinidae. It includes the tsetse flies, which serve as vectors of trypanosomes causing various forms of trypanosomiasis in humans and animals. Species of tsetse flies include *Glossina brevipalpis*, *Glossina fusca* (forest tsetse), *Glossina longipalpis*, *Glossina morsitans* (savannah tsetse) *Glossina pallidipes*, *Glossina palpalis* (riverine tsetse), *Glossina tachinoides*.

glossitis (medical) Inflammation of the tongue.

glottis (anatomy) Front opening of the larynx, through which air passes from the pharynx to the trachea (windpipe) of vertebrates.

glove box (laboratory equipment) Closed box that has gloves fixed into holes in the walls, and in which operations involving hazardous substances, such as radioactive materials or toxic chemicals, may be carried out safely. Also termed: dry box.

GLP (quality standards) An abbreviation of 'G'ood 'L'aboratory 'P'ractice.

GLP-1, -2 (biochemistry) An abbreviation of 'G'lucagon'L'ike 'P'olypeptide-1, -2.

Glu (biochemistry) An abbreviation of 'Glu'tamic acid.

gluc(o)- (terminology) A word element. [German] Prefix denoting *sweetness*, *glucose*.

glucagon (biochemistry) In animals, a polypeptide hormone that (like insulin) is synthesized and secreted by the islets of langerhans in the pancreas. Produced in response to low blood pressure, it stimulates glycogen breakdown in the liver, with release of glucose into the bloodstream. Its action is therefore opposite to that of insulin (which reduces blood glucose levels).

glucocorticoids (biochemistry) Steroid hormones secreted by the adrenal cortex that promote the conversion of fats and proteins to glycogen and glucose.

glucoreceptor (biochemistry) A hypothetical receptor sensitive to the level of glucose in the blood, thought to exist in the hypothalamus.

glucose (chemistry) $C_6H_{12}O_6$ Monosaccharide carbohydrate, a soluble colourless (*USA*: colorless) crystalline sugar (hexose). It is the substance into which all higher carbohydrates are converted in the body; it is the main source of energy when it is broken down into water and carbon dioxide. It is absorbed without having to be digested. Also termed: dextrose; grape sugar.

glucose-6-phosphate dehydrogenase deficiency (G6PD deficiency) (genetics/biochemistry/parasitology) A hereditary lack of an enzyme involved in the metabolism of carbohydrates. Inability to produce this enzyme is inherited as an X-linked recessive. Under normal circumstances, people with this gene show no ill effects, but when they eat broad beans (also known as fava beans), they suffer a severe bout of haemolytic anaemia (*USA*: hemolytic anemia) (anaemia caused by the destruction of red blood cells). The gene causing G6PD deficiency is found at frequencies of up to 30 per cent in countries around the Mediterranean (Sardinia, Greece, Israel), in parts of Africa (especially the Congo Basin) and in India, and it seems likely that its distribution is connected with that of malaria, because the trait gives some resistance to malarial infection (as do the genes for sickle-cell anaemia [*USA*: sickle cell anemia] and thalassaemia [*USA*: thalassemia]).

glucoside (chemistry) *See* glycoside.

Glugea (fish parasitology) A genus of protozoa in the class Microsporea which may be found in the tissues of fish where they cause large cysts, called glugea-cysts or xenomas, which result in deformity or intestinal obstruction. Species includes *Glugea anomala*, *Glugea hertwigi*, and *Glugea stephani*.

glutamic acid (biochemistry) An amino acid that is the precursor of GABA. It is thought to be a neurotransmitter.

gluten (chemistry) Protein that occurs in cereals, particularly wheat flour. People with coeliac disease cannot tolerate gluten, and have to cat a gluten-free diet.

Gly (biochemistry) An abbreviation of 'Gly'cine.

glyceride (chemistry) Ester of glycerol with an organic acid. The most important glycerides are fats and oils.

glycerol (chemistry) $HOCH_2CH(OH)CH_2OH$ Colourless (*USA*: colorless) sweet syrupy liquid, a trihydric alcohol that occurs as a constituent of fats and oils (from which it is obtained). It is used in foodstuffs, medicines and in the preparation of alkyd resins and nitroglycerine (glyceryl trinitrate). Also termed: glycerin; glycerine; propan-1,2,3-triol.

glycine (biochemistry) $CH_2(NH_2)CO_2H$ Simplest amino acid, found in many proteins and certain animal excretions. It is a precursor in the biological synthesis of purines, porphyrins and creatine. It is also a component of glutathione and the bile salt glycocholate. It acts as a neurotransmitter at inhibitory nerve synopses in vertebrates. Also termed: aminoacetic acid; aminoethanoic acid.

glyco- (terminology) A word element. [German] Prefix denoting sweetness, glucose.

glycobiarsol (pharmacology) An organic arsenic derivative with anthelmintic and antiprotozoal activity. Used in the treatment of whipworm (*Trichuris vulpis*) infections in dogs. Also termed: bismuth glycollylarsanilate.

glycogen (biochemistry) A polysaccharide stored in the liver, which when required can be broken down into glucose for release into the blood stream under the control of adrenaline (*USA*: epinephrine). Amylase enzymes convert it to glucose, for use in metabolism.

glycogen vacuole (parasitological physiology) A term used in relation to amoebae (*USA*: amebae). A glycogen reserve which stains deeply with iodine and is found mainly in the cysts.

glycol (chemistry) An alternative term for any diol or, specifically, ethylene glycol (ethanediol).

glycol ethers (chemistry) A class of aliphatic compounds containing both a hydroxyl group (-OH) and an ether (-C-O-C-) linkage (*e.g.* ethylene glycol monomethyl ether, HOCH,CH,OCH,). Widely used as solvents, they are miscible with water and organic solvents. Glycol ethers are not toxic acutely, but toxicity of some members of the series has been noted in both humans and animals.

glycolipid (biochemistry) Member of a family of compounds that contain a sugar linked to fatty acids. Glycolipids are present in higher plants and neural tissue in animals. Also termed: glycosylacylglycerols; glycosyldiacylglycerols.

glycolysis (biochemistry) Conversion ofglucose to lactic or pyruvic acid with the release of energy in the form of adenosine triphosphate (ATP). In animals it may occur during short bursts of muscular activity.

glycoproteins (biochemistry) Proteins with carbohydrate groups attached at specific locations. They include blood glycoproteins, some hormones and enzymes.

glycoside (chemistry) Compound formed from a monosaccharide in which an alcoholic or phenolic group replaces the first hydroxyl group. If the monosaccharide is glucose, it is termed a glucoside.

glycosuria (biochemistry) An excess of glucose in the urine.

glycosylase (biochemistry) An enzyme which recognizes and removes physically or chemically modified bases (*e.g.* alkyl purines) from the sugar phosphate DNA chain leaving behind a hole (an abasic site).

glycosylation (biochemistry) The attachment of a carbohydrate molecule to another molecule such as a protein.

Glycyphagus (parasitology) A genus of mites in the family Acaridae.

Glycyphagus destructor (parasitology) A species of mites in the genus *Glycyphagus* which is a cause of hay itch.

Glycyphagus domesticus (human parasitology) A species of mites in the genus *Glycyphagus* which is very common as a parasite in stored food products and as a cause of dermatitis (grocer's itch) in humans.

Glyphidomastax (insect parasitology) Mites in the Prostigmata that may be parasitic on the ants, *Atta* and *Eciton*, and army ants.

gm (measurement) An abbreviation of 'g'ra'm'. Also termed gramme; g.

gm mol; g.mol. (measurement) gram-molecule, molecular weight in grams.

GM tube (radiology/physics) An abbreviation of 'G'eiger 'M'üller tube.

GMAG (governing body) An abbreviation of 'G'enetic 'M'anipulation 'A'dvisory 'G'roup.

GM-CSF (physiology) An abbreviation of 'G'ranulocyte-'M'acrophage 'C'olony-'S'timulating 'F'actor.

gnat (parasitology) *See Simulium*; *Cnephia*.

gnath(o)- (terminology) A word element. [German] Prefix denoting jaw.

Gnathostoma (parasitology) A genus of spiruroid nematodes of the family Gnathostomatidae.

Gnathostoma doloresi (veterinary parasitology) A species of spiruroid nematodes of the genus *Gnathostoma*, which may be found in domestic and wild pigs.

Gnathostoma hispidum (veterinary parasitology) A species of spiruroid nematodes of the genus *Gnathostoma*. Adults may be found in the stomach of pigs and cause gastritis, while larvae migrate through the liver causing hepatitis.

Gnathostoma nipponicum (veterinary parasitology) A species of spiruroid nematodes of the genus *Gnathostoma* which causes granuloma in the oesophagus (*USA*: esophagus) of weasels.

Gnathostoma spinigerum (human/veterinary parasitology) A species of spiruroid nematodes of the genus *Gnathostoma* which may be found in the stomach of cats, dogs, minks, polecats and wild carnivora and occassionally as a parasite under the skin of humans. The parasite causes damage to the liver and other organs while migrating, and establishes large cyst-like structures containing worms in the stomach wall.

gnathostomiasis (fish/veterinary parasitology) Infestation with the nematodes *Gnathostoma* spp. contracted by eating the intermediate host, *e.g.* undercooked fish infected with the larvae of *Gnathostoma spinigerum* which results in gastritis and cysts in the stomach wall caused by the adults, and hepatitis caused by the larvae.

goat louse (veterinary parasitology) *Damalinia caprae*, biting louse; *Linognathus stenopsis*, sucking louse.

goblet cell (cytology) Mucus-secreting cell in mucous membranes.

gold (Au) (chemistry) Soft yellow metallic element in Group IB of the periodic table. It occurs as the free metal (native) in lodes and placer (alluvial) deposits. Gold and many of its compounds are used in human medicine and occasionally in veterinary medicine, as well as in electroplating electronic circuits and components. At. no. 79; r.a.m. 196.9665.

gold dust (fish parasitology) A disease of aquarium fish caused by the flagellate protozoon *Oodinium limnecicum*. Affected fish develop a varnished look caused by a very heavy infestation of the protozoa on the skin and die within a few days.

gold standard (quality control) The ultimate standard to which all undertakings aspire.

Golgi apparatus (cytology) Organale that occurs in most eukaryotic cells as stacks of flattened membrane-bounded sacs. It is involved in the formation of zymogen granules, synthesis and transport of secretory polysaccharides (*e.g.* cellulose in cell plate or secondary cell wall formation) and formation of mucus in goblet cells; assembly of glycoproteins; packing of hormones in nerve cells that carry out neurosecretion; formation of lysosomes.3 and probably production of the plasma membrane. The synaptic vesicles and their contents are manufactured in the Golgi apparatus. It was named after the Italian histologist Camillo Golgi (1843–1926). Also termed: Golgi body; Golgi complex.

Golgi cell (cytology) A large interneuron found in the cerebellum. They are activated by mossy fibres (*USA*: mossy fibers), climbing fibres (*USA*: climbing fibers), and parallel fibres (*USA*: parallel fibers), and they inhibit the granule cells.

Golgi stain (histology) A histological staining technique used to make nerve cells visible under the microscope. It is a silver stain which percolates to all parts of a living nerve cell.

gomori trichrome (histology) Connective tissue stain for muscle biopsy.

gonad (anatomy) Reproductive organ of an animal, *e.g.* ovary or testis, in which ova (eggs) and sperm are formed respectively. Gonads may also function as endocrine glands, secreting sex hormones.

gonadal (anatomy) Pertaining to the gonads.

See also gonad.

gonadotrophic hormone (biochemistry) The anterior pituitary hormones that promote growth or activity of the gonads, controlling the initiation of puberty, the menstrual cycle and lactation in females and sperm-formation in males. It is produced by the pituitary gland. Also termed: gonadotrophin.

gonadotrophin (biochemistry) *See* gonadotrophic hormone.

gonadotrophin releasing hormone (GNRH) (physiology/biochemistry) An hypothalamic hormone that acts on the pituitary gland to release the gonadotrophins, i.e. luteinizing hormone, LH and follicle-stimulating hormone, FSH.

gonads (anatomy) The organs that produce spermatozoa and ova, *i.e.* the testis and ovary.

Gonderi (parasitology) A genus of protozoa related to *Theileria*.

See also Gonderia mutans.

Gonderia mutans (parasitology) An alternative term for *Theileria mutans*.

Gonglyonema (parasitology) A genus of spiruroid nematodes in the superfamily Spiruroidea.

Gonglyonema crami (avian parasitology) A species of spiruroid nematodes in the genus *Gonglyonema* which may be found in the crop of domestic fowls.

Gonglyonema ingluvicola (avian parasitology) A species of spiruroid nematodes in the genus *Gonglyonema* which may be found in the crop of domestic fowls.

Gonglyonema minimum (veterinary parasitology) A species of spiruroid nematodes in the genus *Gonglyonema* which may be found in the oesophagus (*USA*: esophagus) of primates; may cause anaemia (*USA*: anemia), vomiting, gastritis and enteritis.

Gonglyonema monnigi (veterinary parasitology) A species of spiruroid nematodes in the genus *Gonglyonema* which may be found in the rumen of sheep and goats.

Gonglyonema neoplasticum (veterinary parasitology) A species of spiruroid nematodes in the genus *Gonglyonema* which may be found in the tongue, oesophagus (*USA*: esophagus) and stomach of laboratory rats and mice but causes little epithelial reaction.

Gonglyonema pulchrum (human/veterinary parasitology) (synonym: *Gonglyonema scutatum*) A species of spiruroid nematodes in the genus *Gonglyonema* which may be found in the oesophageal (*USA*: esophageal) mucusa of ruminants and occasionally in humans.

Gonglyonema sumani (avian parasitology) A species of spiruroid nematodes in the genus *Gonglyonema* which may be found in the crop of domestic fowl.

Gonglyonema verrucosum (veterinary parasitology) A species of spiruroid nematodes in the genus *Gonglyonema* which may be found in the rumen of ruminants.

Goniocotes (avian parasitology) A genus of lice in the superfamily Ischnocera which may infest the feathers of domestic fowl and guinea fowl.

Goniocotes gallinae (avian parasitology) (synonym: *Goniocotes hologaster, Goniodes hologaster*) A species of lice of the genus Goniocotes which may be found in the feathers of domestic fowl. Also termed fluff lice.

Goniocotes hologaster (parasitology) *See Goniocotes* gallinae.

Goniocotes maculatus (avian parasitology) A species of lice of the genus Goniocotes which may be found in the feathers of guinea fowl.

Goniodes (parasitology) A genus of biting lice of the superfamily Ischnocera.

Goniodes dissimilis (avian parasitology) A species of biting lice of the genus *Goniodes* which may be found on domestic fowl. Also termed brown chicken louse.

Goniodes gigas (avian parasitology) (synonym: *Goniocotes gigas*) Also termed large chicken louse.

Goniodes hologaster (parasitology) (synonym: *Goniocotes gallinae*) *See Goniocotes gallinae.*

Goniodes meleagridis (parasitology) *See Chelopistes.*

gono- (terminology) A word element. [German] Prefix denoting *seed, semen.*

Gonometa (plant parasitology) An African moth which infests *Acacia erioloba* and *Acacia mellifera* producing very many cocoons of silk fibre (*USA*: fiber) which cause ruminal impaction in cattle browsing on the tree. Also termed Molopo moth.

good laboratory practices (GLP) (quality standards) Regulations governing good laboratory are in effect in both the USA and OECD cover all phases of toxicity testing, including facilities, personnel training, data gathering, laboratory inspections, animal health and welfare, chemical analysis and sample preparation. Good laboratory practices as laid down by the appropriate regulatory agency.

goodness of fit (statistics) The extent to which data points match theoretical expectations.

goose body louse (avian parasitology) *See Trinoton anserinum.*

goose septicaemia (avian parasitology) A fatal septicaemia (*USA*: septicemia) caused by *Borrelia* (*Treponema*) *anserina* and transmitted by the tick *Argas persicus*. (*USA*: goose septicemia).

Gordon & Sweets stain (histology) A histological staining technique used to make reticulin visible under the microscope.

GP (1, 2 medical) 1 An abbreviation of 'G'eneral 'P'aralysis. 2 An abbreviation of 'G'eneral 'P'ractitioner.

gpl (measurement) An abbreviation of 'g'rams 'p'er 'l'itre.

gr. (measurement) An abbreviation of 'gr'ain; approximately 65 mg.

Graafian follicle (histology) Fluid-filled ball of cells in the mammalian ovary inside which an oocyte develops. It matures periodically and then bursts at the surface of the ovary (at ovulation) to release an ovum (egg). The follicle then temporarily becomes a solid body, the corpus luteum. It was named after the Dutch anatomist Regnier de Graaf (1641–73). Also termed: ovarian follicle.

gradient (1, 2 mathematics) 1 Amount of inclination of a line (or a curve at a particular point) to the horizontal; its slope. 2 Rate of rise or fall of a variable quantity such as temperature or pressure.

graduated (general terminology) Marked by a succession of lines, steps or degrees.

Grafia (insect parasitology) A species of mites in the Mesostigmata, that are parasitic to the bees Meliponinae.

graft (medical) Transplantation of an organ or tissue from one organism into the body of another or from one position on or in the body to another. A homograft is a graft where the recipient and the donor are of the same species; a heterograft is one carried out between two different species.

Graham's law (physics) Velocity of diffusion of a gas is inversely proportional to the square root of its density. It was named after the British chemist Thomas Graham (1805–69).

Grahamella (veterinary parasitology) Small intra-erythrocytic rickettsiae parasitizing small mammals and turtles, which are transmitted by blood-sucking ectoparasites, especially fleas. Related to *Bartonella* spp.

grain (1, 2 measurement) 1 Small weight in apothecaries' and troy weight (in which it equals 1/5760 lb), and in avoirdupois weight (in which it equals 1/7000 lb). It is equivalent to 0.0648 grams. 2 One-quarter of a (metric) carat, equal to 0.050 grams.

grain fumigants (control measures) Substances used to fumigate silos full of grain to kill insect pests. Use of these agents other than as recommended by the makers may lead to poisoning.

grain itch mite dermatitis (veterinary parasitology) A transient, superficial dermatitis, mostly about the head in horses although it may be all over the body in pigs, that it is caused by *Pediculoides ventricosus* or *Tyroglyphus*.

grain itch mites (parasitology) *See Pediculoides ventricosus.*

grain rash (parasitology) Grain itch mite dermatitis.

gram (measurement) Unit of mass in the metric system, and defined ass the mass of one cubic millimetre (USA: millimeter) of water at 4°C [39°F]. One gram is equal to 1/1000 kg. Also termed: gramme. *See* SI units.

-gram (terminology) A word element. [German] Suffix denoting *written, recorded.*

gram molecule (measurement) Molecular weight of a substance in grams; one mole.

Gram/Twort stain (histology) A histological staining technique used to make bacteria visible under the microscope.

Gram's stain (bacteriology) A staining technique used to make bacteria visible under the microscope. The staining method is to use crystal violet, then iodine solution, then ethanol or ethanol-acetone, and then counter-stain. The bacteria which are stained purple are called Gram-positive, those which have been decolourised (*USA*: decolorized) and have taken up the counter-stain Gram-negative. The different uptake of stain is due to differences in cell-wall structure. This method of staining separates bacteria into two large groups, Gram-positive and Gram-negative. Named after the Danish physician, H. Gram (1853–1938).

Grammocephalus (parasitology) A genus of hookworms in the subfamily Bunostominae.

Grammocephalus clathratus (veterinary parasitology) A species of hookworms in the genus *Grammocephalus* which may be found in the bile ducts of the African elephant.

Grammocephalus hybridatus (veterinary parasitology) A species of hookworms in the genus *Grammocephalus* which may be found in the bile ducts of the Indian elephant.

Grammocephalus intermedius (veterinary parasitology) A species of hookworms in the genus *Grammocephalus* which may be found in the bile ducts of the rhinoceros.

Grammocephalus varedatus (veterinary parasitology) A species of hookworms in the genus *Grammocephalus* which may be found in the bile ducts of Indian elephants.

granular layer (1, 2 histology) 1 Either of two layers of nerve cells in the cerebral cortex; numbered from the surface inwards, the external granular layer is the second layer and the internal granular layer is the fourth layer; the name derives from the large number of golgi type 11 cells or granule cells in these layers. 2 The innermost of the three layers of the cerebellum, which contains granule cells and receives an input from the mossy fibres (*USA*: mossy fibers).

granulation tissue (histology) The tissue formed by fibroblasts and endothelial cells which grows over a raw surface in the process of healing.

granule (physics) A small particle or grain.

granule cell (cytology) A small stellate cell; the name 'granule' is used because these cells look granular under a microscope. The expression 'granule cell' is preferred to 'stellate cell' for the granule cells in the granular layers of the cortex and cerebellum.

granuloma (histology) Aggregation and proliferation of macrophages to form small (usually microscopic) nodules.

grape sugar (chemistry) An alternative term for glucose.

grapevine fanleaf virus (plant parasitology) *See Xiphinema index.*

graph (statistics) Mathematical diagram that shows how one quantity varies in relation to another.

See also bar chart; histogram; pie chart.

Graphidium (parasitology) A genus of netnatodes in the family Trichostrongylidae.

Graphidium strigosum (veterinary parasitology) A species of nematodes in the genus *Graphidium* that may be found in the stomach and small intestines of rabbits and hares, and which may cause digestive upset and poor body condition.

-graphy (terminology) A word element [German] Suffix denoting *writing or recording, a method of recording.*

GRAS (toxicology) An abbreviation of 'G'enerally 'R'ecognised 'A's 'S'afe.

grass nematode (plant parasitology) *Anguina lolii. See Anguina.*

grass-seed nematode poisoning (parasitology) *See Anguina.*

grass-seed nematode (plant parasitology) The grass seed nematode *Anguina lolii* infests Wimmera ryegrass and causes a fatal poisoning in animals eating the grass. Also termed *Anguina fenesta*; *Anguina agrostis.*

See also Lolium rigidum.

gravimetric analysis (chemistry) Quantitative chemical analysis made ultimately by weighing substances.

gravitation (physics) An alternative term for the force of gravity.

gravity weight (physics) Tendency toward the centre (*USA*: center) of the earth.

gray (Gy) (chemistry) Amount of absorbed radiation dose in Si units, equal to supplying 1 joule of energy per kg. It is equivalent to 100 rad (the unit it superseded).

Greco Latin square (statistics) A latin square in which there are two conditions, each with the same number of values, and in which each combination of values appears once in each row and once in each column.

green bottle fly (parasitology) *See Lucilia.*

green vitriol (chemistry) Old name for iron(II) sulphate (*USA*: iron(II) sulfate) (ferrous sulphate [*USA*: ferrous sulfate]).

grey crab disease (parasitology) Infection of blue crabs by *Paramoeba perniciosa* causing grey (*USA*: gray) translucency of central aspect of the body and limbs. (*USA*: gray crab disease).

grey fever (parasitology) Q fever. (*USA*: gray fever).

grey matter (anatomy) Any region of the central nervous system containing many cell bodies and their processes, and having unmyelinated axons. All parts of a cell look grey except for a myelinated axon. The expression usually refers to the layers of the cerebral cortex, but grey matter also exists in the cerebellar cortex and spinal cord.

grey meat fly (parasitology) (*USA*: gray meat fly). *Sarcophaga carnaria. See Sarcophaga.*

Grimelius stain (histology) A histological staining technique used to make argyrophil cells visible under the microscope.

grocer's itch (parasitology) *See Glycyphagus domesticus.*

gross weight (gr wt) (measurement) The entire weight of a product, including its container. Found on packaging and labelling.

ground itch (parasitology) The pruritic eruption caused by the entrance into the skin of the hookworm larvae.

See also bunostomiasis; ancylostomiasis; uncinariasis.

ground state (physics) Lowest energy state of an atom or molecule, from which it can be raised to a higher energy state by excitation.

See also energy level.

group (chemistry) Column, or vertical row, of elements in the periodic table (horizontal rows are periods). The group number (I to VIII and 0) indicates the number of electrons in the atom's outermost shell.

grouping (statistics) Assigning data to class intervals.

grouping error (statistics) Any statistical error introduced by the way in which data are grouped, *e.g.* by grouping in such a way that the distribution of scores within a group is not normal.

grouse disease (avian parasitology) *See Trichostrongylus tenuis.*

growth (genetics) Increase in size of an organism, either by an increase in cell size or, much more usually, by an increase in cell number. Most organisms have a species-typical growth rate and final size, both of which are, to some extent, genetically controlled, as is shown by the fact that they can be changed by artificial selection, but are also much influenced by environmental factors. Some fish and reptiles grow all their lives.

Grp (general terminology) An abbreviation of 'Gr'ou'p'.

GRP (physiology/biochemistry) An abbreviation of 'G'astrin-'R'eleasing 'P'olypeptide.

grub (insect parasitology) Maggot-like or caterpillar-like insect larva, that is said of beetles and flies, but most properly of beetles.

grubby gullets (parasitology) Gullet's damaged by migrating *Hypoderma* spp.

GTP (biochemistry) An abbreviation of 'G'uanosine 'T'ri'P'hosphate.

GU (medical) An abbreviation of 'G'enito-'U'rinary.

guanine (molecular biology) $C_5H_5N_5O$ Colourless (*USA*: colorless) crystalline organic base (a purine derivative) that occurs in DNA.

gubernaculum (parasitological anatomy) A term used in relation to nematodes. A protuberance on the wall of cloaca. It apparently guides the spicule during copulation.

guinea fowl feather louse (avian parasitology) *Goniodes numidae. See Goniodes.*

guinea worm (parasitology) *See Dracunculus medinensis.*

Gulf Coast tick (parasitology) *See Amblyomma maculatum.*

gullet (anatomy) *See* oesophagus.

gullet worm (parasitology) *See Capillaria*; *Echinuria*; *Gonglyonema.*

Gurleya spraguei (insect parasitology) A protozoal species of microsporidia that are parasites to the termites *Macrotermes estherae.* Maximum infections occur at the end of winter months.

Gurltia (veterinary parasitology) A genus of worms in the family Angiostrongylidae that may be found in the veins of the leptomeninges of Felidae.

Gurltia paralysans (veterinary parasitology) A species of worms in the genus *Gurltia* that may be found in the thigh veins of cats, and that may cause paralysis.

gust (measurement) A unit used in scales of taste; one gust is produced by a 1 per cent sucrose solution.

gustation (physiology) The sense of taste.

gustometer (measurement) A device for measuring taste thresholds. It consists of a U-tube with a small hole at the bottom; the tube is placed upright on a particular part of the tongue, and a solution is poured through it: the tongue is thus exposed to a constant amount of solution.

gut (anatomy) An alternative term for alimentary canal.

Gyalocephalus (parasitology) A genus of nematodes in the subfamily Cyathostominae.

Gyalocephalus capitatus (veterinary parasitology) A species of nematodes in the genus *Gyalocephalus* that is a rare finding in the large intestine of horses, and that may contribute to the clinical signs of cyathostomiasis.

gyn. (1, 2 medical) An abbreviation of 1 'gyn'aecology (*USA*: gynecology). Also termed gynae.; gynacol. 2 'gyn'aecologist (*USA*: gynecologist).

gynaecologist (medical) A medical doctor specialising in gynaecology (*USA*: gynecology). (*USA*: gynecologist).

gynaecology (gyn.; gynac.; gynacol) (medical) That branch of medicine dealing with the diseases of women, especially concerning the reproductive system of women. (*USA*: gynecology).

gyn(e)- (terminology) A word element. [German] Prefix denoting *woman.*

gyno- (terminology) A word element. [German] Prefix denoting *woman.*

gypsum (chemistry) $CaSO_4.2H_2O$ Very soft calcium mineral, a form of calcium sulphate (*USA*: calcium sulfate), used in making cement and plasters. Also termed: calcium sulphate dihydrate (*USA*: calcium sulfate dihydrate).

Gyrocotylidea (fish parasitology) An order of flatworms in subclass Cestodaria, class Cestoidea and the phylum Platyhelminthes. Gyrocotylideans are restricted to the intestine of holocephalans. These unusual Platyhelminthes are now generally given a status independent of monogeneans and cestodes, although they have affinities with both.

Gyrodactylida (fish parasitology) An order of flatworms in class Monogenea and the phylum Platyhelminthes. A large order, with one family, of viviparous monogeneans found on marine, freshwater and brackish water teleosts. Examples includes *Macrogyrodactylus* and *Gyrodactylus.*

Gyrodactylus (parasitology) *See Dactylogyrus.*

Gyrodactylus elegans (fish parasitology) A species of monogenetic nematode in the genus *Gyrodactylus* that is a common parasite of fish and found attached to their skin.

Gyrodactylus salavis (fish parasitology) A species of nematode in the genus *Gyrodactylus* that is an important pathogen of wild Atlantic salmon in

Scandinavia, that is spread by hatchery-reared salmonids.

gyromagnetic ratio (physics) Ratio of the magnetic moment of an atom or nucleus to its angular momentum.

Gyropus ovalis (veterinary parasitology) A species of louse parasite in the genus *Gyropus* of the superfamily Amblycera, that may be found on guinea pigs and other rodents.

Gyrostigma (parasitology) A genus of stomach flukes.

Gyrostigma conjugens (veterinary parasitology) A species of flukes in the genus Gyrostigma that may be found in the forestomachs of the black rhinoceros.

Gyrostigma pavesii (veterinary parasitology) A species of flukes in the genus Gyrostigma that may be found in the forestomachs of the white rhinoceros.

gyrus (anatomy) One of the convolutions on the surface of the brain.

H

H (1 chemistry; 2 biochemistry) 1 The chemical symbol for hydrogen. 2 An abbreviation of 'H'istidine.

h (1 measurement; 2 physics) 1 An abbreviation of 'h'our. 2 Symbol for Planck's constant.

HA (biochemistry) General symbol for an acid.

habitat (epidemiology) The environment inhabited by a specific organism or animal.

Habronema (parasitology) A genus of nematodes in the order Spirurida.

Habronema incertum (avian parasitology) A species of nematodes in the genus *Habronema* that may be found in the proventriculus and gizzard of companion birds causing chronic digestive upset and diarrhoea (*USA*: diarrhea). They may also cause sudden death. Also termed: *Spiroptera incerta*.

Habronema khalili (veterinary parasitology) A species of nematodes in the genus *Habronema* that may be found in the large intestine of elephants and rhinoceroses.

Habronema majus (veterinary parasitology) (synonym: *Habronema microstoma*) A species of nematodes in the genus *Habronema* that, similarly to *Habronema muscae*, are parasites in the stomach of horses.

Habronema megastoma (parasitology) *Draschia megastoma*. *See Draschia*.

Habronema microstoma (parasitology) *See Habronema majus*.

Habronema muscae (veterinary parasitology) A species of nematodes in the genus *Habronema* that, similarly to *Habronema majus* (synonym: *Habronema microstoma*), are parasites in the stomach of horses.

habronemiasis (veterinary parasitology) A disease of horses caused by the nematodes *Habronema muscae*, *Habronema majus* (*Habronema microstoma*) and *Draschia megastoma*.

HACCP (quality control) An abbreviation of 'H'azard 'A'nalysis 'C'ritical 'C'ontrol 'P'oints.

haem (chemistry) Iron-containing group of atoms attached to a polypeptide chain; *e.g.* in haemoglobin (*USA*: hemaglobin) and myoglobin. (*USA*: heme).

haem- (terminology) A word element. [German] Prefix denoting *blood*.

haem. (1, 2 haematology) An abbreviation of 1'haem'oglobin (*USA*: hemoglobin); 2 'haem'orrhage (*USA*: hemorrhage). (*USA*: hem).

Haemadipsa (parasitology) A genus of leeches in the class Hirudinea. Small, 1 to 1.5 inches, but heavy infestations can cause anaemia (*USA*: anemia).

haemagglutination (haematology/immunology) Agglutination of red blood cells caused by certain antibodies, virus particles, or high molecular weight polysaccharides. (*USA*: hemagglutination).

haemangioma (histology) A tumour (*USA*: tumor) composed of blood vessels. (*USA*: hemangioma).

Haemaphysalis (parasitology) A large genus of small ticks in the family Ixodidae.

Haemaphysalis bancrofti (veterinary parasitology) A species of small ticks of the genus *Haemaphysalis* that may be found on cattle and marsupials. Also termed wallaby tick.

Haemaphysalis bispinosa (veterinary parasitology) A species of small ticks of the genus *Haemaphysalis* that may be found on cattle.

Haemaphysalis chordeilis (avian parasitology) (synonym: *Haemaphysalis* cinnabarina, *Haemaphysalis* punctata) A species of small ticks of the genus *Haemaphysalis* that may be found on birds.

Haemaphysalis cinnabarina punctata (avian/veterinary parasitology) A species of small ticks of the genus *Haemaphysalis* that may be found on most mammals and on birds. They have been implicated in the transmission of *Babesia bigemina*, *Babesia motasi*, *Anaplasma centrale*, and *Anaplasma marginale*.

Haemaphysalis humerosa (parasitology) A species of small ticks of the genus *Haemaphysalis* that may transmit Q fever.

Haemaphysalis inermis (parasitology) A species of small ticks of the genus *Haemaphysalis* that has a widespread distribution.

Haemaphysalis leachi leachi (veterinary parasitology) A species of small ticks of the genus *Haemaphysalis* that occurs on domestic carnivora and rodents and which may transmit canine piroplasmosis, *Rickettsia conori* and *Coxiella burnetii*. Also termed yellow dog tick.

Haemaphysalis leachi mushami (veterinary parasitology) A species of small ticks of the genus *Haemaphysalis* that occurs on small carnivora.

Haemaphysalis leporispalustris (avian/veterinary parasitology) A species of small ticks of the genus *Haemaphysalis* that may be found on rabbits, other small mammals and birds, and which may transmit

Coxiella burnetii, and *Francisella tularensis*. Also termed rabbit tick.

Haemaphysalis longicornis (human/veterinary parasitology) A three-host tick that may be found on most mammals including humans which may transmit *Theileria* spp. and *Coxiella burnetii*.

Haemaphysalis otophila (veterinary parasitology) A species of small ticks of the genus *Haemaphysalis* that may be found on rodents and which may transmit *Francisella tularensis*.

Haemaphysalis parmata (veterinary parasitology) A species of small ticks of the genus *Haemaphysalis* that may be found on carnivora and antelope.

Haemaphysalis punctata (parasitology) *See Haemaphysalis cinnabarina punctata*.

haemat(o)- (terminology) A word element. [German] Prefix denoting *blood*.

Haematobia (parasitology) A genus of small, grey (*USA*: gray), blood-sucking flies. Also termed *Lyperosia* spp., *Haematobosca* spp.

Haematobia atripalpis (parasitology) A species of blood-sucking flies of the genus *Haematobia* which is a vector for *Parafilaria bovicola*.

Haematobia exigua (veterinary parasitology) (synonym: Siphona spp.) A species of blood-sucking flies of the genus *Haematobia* which is a parasite of buffalo and cattle and that may transmit *Trypanosoma evansi* and *Habronema majus*. Also termed buffalo fly.

Haematobia irritans (veterinary parasitology) A species of blood-sucking flies of the genus *Haematobia* which may be found on cattle, and occasionally on other mammals. They create skin lesions around horns, and along the neck and back, which are often invaded by screw-worms, and may transmit *Stephanofilaria stilesi*. Also termed horn fly.

Haematobia minuta (veterinary parasitology) A species of blood-sucking flies of the genus *Haematobia* which is a parasite of buffalo and cattle and that may transmit *Trypanosoma evansi* and *Habronema majus*.

Haematobia stimulans (veterinary parasitology) A species of blood-sucking flies of the genus *Haematobia* which may be a pest of cattle.

Haematobosca (parasitology) *See Haematobia*.

haematochezia (medical) Blood in the faeces (*USA*: feces). (*USA*: hematochezia).

haematogenous (medical) Disseminated by the bloodstream. (*USA*: hematogenous).

haematology (haematology) The branch of medicine which deals with the blood and its disorders. (*USA*: hematology).

Haematomyzus (parasitology) An exotic, rare lice of aberrant constitution that is the only genus in the family Rhynchophthirina.

Haematomyzus elephantis (veterinary parasitology) A species of lice of the genus *Haematomyzus* which may be found on elephants.

Haematomyzus hopkinsi (veterinary parasitology) A species of lice of the genus *Haematomyzus* which may be found on wart hogs.

haematophagous (parasitological physiology) Subsisting on blood, *e.g.* haematophagous (*USA*: hematophagous) flies. (*USA*: hematophagous).

Haematopinus (parasitology) A genus of sucking lice in the order Phthiraptera.

Haematopinus asini (veterinary parasitology) A species of sucking lice of the genus *Haematopinus* which may be found on horses.

Haematopinus bufali (veterinary parasitology) A species of sucking lice of the genus *Haematopinus* which may be found on buffalo.

Haematopinus eurysternus (veterinary parasitology) A species of sucking lice of the genus *Haematopinus* which may be found on cattle. Also termed shortnosed cattle lice.

Haematopinus quadripertusus (veterinary parasitology) A species of sucking lice of the genus *Haematopinus* which may be found on cattle.

Haematopinus suis A species of sucking lice of the genus *Haematopinus* which may be found on pigs.

Haematopinus trichechi (veterinary parasitology) A species of sucking lice of the genus *Haematopinus* which may be found on walrus.

Haematopinus tuberculatus (veterinary parasitology) A species of sucking lice of the genus *Haematopinus* which may be found on cattle, camels, yaks and buffalo.

haematopoietic stem cells (cytology) (*USA*: hematopoietic stem cells). *See* stem cells.

Haematopota (veterinary parasitology) A genus of large, blood-sucking flies predatory on vertebrates, of the family Tabanidae, which causes significant insect worry to horses. They are a) mechanical transmitters of anaplasmosis, anthrax, equine infectious anaemia (*USA*: anemia), and *Typanosoma evansi*, b) cyclical transmitters of *Trypanosoma theilri*. Species include *Haematopota pluvialis*.

Haematosiphon (human/veterinary parasitology) A genus of blood-sucking, true bugs of the family Cimicidae which includes, *e.g.* the bed bug of humans.

Haematosiphon indora (avian parasitology) A species of blood-sucking true bugs of the genus *Haematosiphon*, which is an important parasite of domestic poultry in Central America.

Haematosiphon nodorus (avian parasitology) A species of blood-sucking true bugs of the genus *Haematosiphon*, which is the most important species of bird bugs, which infests poultry, turkeys and some wild birds.

Haematoxenus (parasitology) A genus of nonpathogenic protozoan parasites in the family Theileriidae.

Haematoxenus separatus (veterinary parasitology) A species of protozoan parasites of the genus *Haematoxenus* that may be found in sheep which is transmitted by *Rhipicephalus evertsi*.

Haematoxenus veliferus (veterinary parasitology) A species of protozoan parasites of the genus *Haematoxenus* that may be found in the bloodstream of cattle which is transmitted by *Amblyomma variegatum*.

haematoxylin (histology) An acid colouring (*USA*: coloring) matter from the heartwood. A histological stain used to make cell nuclei visible under the microscope. (*USA*: hematoxylin).

See also hematoxylin and eosin stain.

haematoxylin & eosin (H&E) stain (histology/ haematology) A mixture of haematoxylin (*USA*: hematoxylin) in distilled water and an aqueous eosin solution; a stain used routinely for examination of tissues. (*USA*: hematoxylin & eosin stain).

haematozoon (parasitology) Any animal microorganism living in the blood; refers usually to protozoa. (*USA*: hematozoon).

haematuria (medical) The discharge of blood in the urine. The urine may be slightly blood tinged, grossly bloody, or a smoky brown colour (*USA*: color). (*USA*: hematuria).

haemo- (terminology) A word element. [German] Prefix denoting *blood*.

Haemobartonella (veterinary parasitology) A genus of parasitic microorganisms in the family Anaplasmataceae that parasitize the cell surface of erythrocytes, although only the infection in cats appears to result in a naturally occurring disease.

Haemobartonella bovis (veterinary parasitology) A species of parasitic microorganisms in the genus *Haemobartonella* which may be found in cattle.

Haemobartonella canis (veterinary parasitology) A species of parasitic microorganisms in the genus *Haemobartonella* which may be found in dogs but is nonpathogenic unless the dog is splenectomized.

Haemobartonella felis (veterinary parasitology) A species of parasitic microorganisms in the genus *Haemobartonella* which may be found in cats and in animals whose resistance is reduced, and that causes feline infectious anaemia (*USA*: anemia). Also termed *Eperythrozoon felis*.

Haemobartonella muris (veterinary parasitology) A species of parasitic microorganisms in the genus *Haemobartonella* which may be found in rats, that is transmitted by lice. Their presence is usually asymptomatic, but they are capable of causing a fatal anaemia (*USA*: anemia) in animals subjected to splenectomy or that are immunosuppressed.

***Haemobartonella* tyzzeri** (veterinary parasitology) A species of parasitic microorganisms in the genus *Haemobartonella* which may be found in guinea pigs.

haemocoele (parasitological anatomy) The body cavity of arthropods.

haemocyanin (haematology) Blood pigment that contains copper as its prosthetic group for the transport of oxygen. It is confined to lower animals, *e.g.* molluscs. (*USA*: hemocyanin).

haemocyte (haematology) A blood cell. (*USA*: hemocyte).

haemocytometer (haematology) Apparatus for counting blood cells. (*USA*: hemocytometer).

haemodiagnosis (medical) Diagnosis by examination of the blood. (*USA*: hemodiagnosis).

Haemodipsus (parasitology) A genus of lice in the family Hoplopleuridae.

Haemodipsus ventricosus (veterinary parasitology) A species of lice in the genus *Haemodipsus* which may be found on field rabbits.

Haemogamasus (parasitology) A genus of mites in the family Gamasidae.

Haemogamasus pontiger (veterinary parasitology) A species of mites in the genus *Haemogamasus* which may be found in bedding and on rodents and insectivorous animals.

haemoglobin (Hb) (medical) The oxygen carrying pigment of the red blood cells (erythrocytes). It is a conjugated protein containing four haem groups (*USA*: heme groups) and globin. A molecule of haemoglobin contains four globin polypeptide chains designated alpha, beta, gamma and delta. In the adult, Haemoglobin A predominates (alpha2, beta2). (*USA*: hemoglobin).

haemoglobinuria (haematology) The appearance of haemoglobin (*USA*: hemaglobin) in the urine. (*USA*: hemoglobinuria).

Haemogregarina (fish parasitology) A genus of protozoa parasitic on fish, which includes coccidia that cause hepatic necrosis, castration and occlusion of the air bladder.

Haemogregarina sachae (fish parasitology) A species of protozoa in the genus *Haemogregarina* which participates in the production of a leech vector transmitted lymphoma in cultured turbot.

Haemolaelaps (parasitology) A genus of mites of the family Gamasidae.

Haemolaelaps casalis (avian parasitology) A species of mite in the genus *Haemolaelaps* which may infest birds and their litter.

haemolymph (haematology) 1 Blood and lymph. 2 The bloodlike fluid of invertebrates having open blood-vascular systems. (*USA*: hemolymph).

haemolysis (haematology) Breakdown of red blood cells (erythrocytes) which results in the release of haemoglobin (*USA*: hemaglobin) from the red cells and its appearance in plasma. (*USA*: hemolysis).

haemolytic anaemia (medical) Anaemia (*USA*: anemia) caused by the increased destruction of erythrocytes which may occur in the vascular system, or due to phagocytosis by the monocyte-macrophage system. It may result from protozoan infections such as babesiosis. (*USA*: hemolytic anemia).

Haemonchus (parasitology) A genus of blood-sucking nematodes of the family Trichostrongylidae.

Haemonchus bedfordi (veterinary parasitology) A species of blood-sucking nematodes of the genus *Haemonchus* which may be found in African buffalo and gazelles.

Haemonchus contortus (veterinary parasitology) A species of blood-sucking nematodes of the genus *Haemonchus* which may be found in the abomasum of most ruminants and which is the cause of serious losses in sheep from haemonchosis. There have been attempts to subdivide the species, *e.g.* *Haemonchus* contortus cayugensis, but the differences between the subspecies have not been substantiated.

Haemonchus dinniki (veterinary parasitology) A species of blood-sucking nematodes of the genus *Haemonchus* which may be found in gazelles.

Haemonchus krugeri (veterinary parasitology) A species of blood-sucking nematodes of the genus *Haemonchus* which may be found in impalas (*Aepyceros melampus*).

Haemonchus lawrenci (veterinary parasitology) A species of blood-sucking nematodes of the genus *Haemonchus* which may be found in duikers (small antelopes).

Haemonchus longistipes (veterinary parasitology) A species of blood-sucking nematodes of the genus *Haemonchus* which may be found in camels.

Haemonchus mitchelli (veterinary parasitology) A species of blood-sucking nematodes of the genus *Haemonchus* which may be found in gazelles, elands, and oryx.

Haemonchus placei (veterinary parasitology) A species of blood-sucking nematodes of the genus *Haemonchus* which may be found in cattle and that can cause hemonchosis in that species.

Haemonchus similis (veterinary parasitology) A species of blood-sucking nematodes of the genus *Haemonchus* which may be found in cattle and deer.

Haemonchus vegliai (veterinary parasitology) A species of blood-sucking nematodes of the genus *Haemonchus* which may be found in oryx and antelope.

haemopathology (haematology) The study of diseases of the blood. (*USA*: hemopathology).

haemopathy (haematology) Any disease of the blood. (*USA*: hemopathy).

haemophagous (parasitology) Feeding on blood. (*USA*: hemophagous).

haemophil (parasitology) 1 Thriving on blood. 2 A microorganism that grows best in media containing haemoglobin (*USA*: hemoglobin). (*USA*: hemophil).

Haemoproteus (avian parasitology) A genus of protozoan blood parasites of the family Plasmodiidae which are the cause of avian 'malaria'.

Haemoproteus antigonis (avian parasitology) A species of protozoan blood parasites of the genus *Haemoproteus* which may be found in cranes.

Haemoproteus canachites (avian parasitology) A species of protozoan blood parasites of the genus *Haemoproteus* which may be found in grouse, and that is transmitted by *Culicoides* spp.

Haemoproteus columbae (avian parasitology) A species of protozoan blood parasites of the genus *Haemoproteus* which may be found in pigeons, doves and some wild birds, and that is transmitted by hippoboscid flies. The species may cause anaemia (*USA*: anemia) in baby pigeons.

Haemoproteus danilewski (avian parasitology) A species of protozoan blood parasites of the genus *Haemoproteus* which may be found in wild birds.

Haemoproteus lophortyx (avian parasitology) A species of protozoan blood parasites of the genus *Haemoproteus* which may be found in quail, which may cause anaemia (*USA*: anemia) and is transmitted by biting flies.

Haemoproteus meleagridis (avian parasitology) A species of protozoan blood parasites of the genus *Haemoproteus* which may be found in turkeys.

Haemoproteus nettionis (avian parasitology) A species of protozoan blood parasites of the genus *Haemoproteus* which may be found in ducks, geese and swans.

Haemoproteus sacharovi (avian parasitology) A species of protozoan blood parasites of the genus *Haemoproteus* which may be found in doves, and that is possibly transmitted by *Culicoides* spp.

haemoptysis (medical) The spitting of blood or blood-stained sputum. (*USA*: hemoptysis).

haemorrhage (medical) Bleeding, which may be internal or external, arterial, venous or capillary. In arterial bleeding, the blood is bright red, and may spurt out with the beat of the heart; venous blood is darker and escapes in a steady stream; capillary blood oozes from the damaged tissues. (*USA*: hemorrhage).

haemorrhagic (medical) Pertaining to or characterized by haemorrhage (*USA*: hemorrhage). (*USA*: hemorrhagic).

haemorrhagic disease (medical) An undifferentiated disease manifested by unprovoked haemorrhage (*USA*: hemorrhage) and caused by any one of a number of factors. (*USA*: haemorrhagic disease).

haemorrhagic filariasis (parasitology) *See Parafilaria multipapillosa*. (*USA*: hemorrhagic filariasis).

haemostasis (medical) The process of stopping blood flow. (*USA*: hemostasis).

haemostat (1 medical; 2 chemistry) 1 A small surgical clamp for constricting blood vessels. 2 An antihemorrhagic agent. (*USA*: hemostat).

haemostatic (pharmacology) 1 Checking blood flow. 2 A substance that checks the flow of blood. (*USA*: hemostatic).

hafnium (Hf) (chemistry) Silvery metallic element in Group IVB of the periodic table (a transition element), used to make control rods for nuclear reactors. At. no. 72; r.a.m. 178.49.

Haga-Yamagchi methenamine silver stain (histology) A histological staining technique used to make senile plaques visible under the microscope.

hahnium (Ha) (chemistry) Element no. 105 (a post-actinide). It is a radioactive metal with short-lived isotopes, made in very small quantities by bombardment of an actinide with atoms of an element such as carbon or oxygen.

hair (1, 2 anatomy) 1 Derivative of ectoderm composed of insoluble proteins or keratins. Its role in mammals includes assisting regulation of body temperature. 2 In plants, any of various outgrowths from the epidermis, *e.g.* root hairs, which absorb water. Also termed: trichome.

hairworms (parasitology) *See Capillaria.*

Halarachne (veterinary parasitology) A genus of mites in the family Halarachnidae which may be found in the nostrils of sea lions and seals.

half-cell (physics) Half of an electrolytic cell, consisting of an electrode immersed in an electrolyte.

half-life (1, 2 measurement/radiology) 1 Time taken for something whose decay is exponential to reduce to half its value. 2 More specifically, time taken for half the nuclei of a radioactive substance to decay spontaneously. The half-life of some unstable substances is only a few seconds or less, whereas for other substances it may be thousands of years; *e.g.* lawrencium has a half-life of 8 seconds, and the isotope plutonium-239 has a half-life of 24,400 years.

Halicephalobus (parasitology) *See Micronema deletrix.*

Halictacarus (insect parasitology) A species of mites in the Astigmata, that, similarly to *Histiostoma* and *Konglyphus*, are parasitic to the bees Halictinae and Bombinae.

halide (chemistry) Binary compound containing a halogen: a fluoride, chloride, bromide or iodide.

Haliotrema (fish parasitology) A genus of flukes (monogenetic trematodes) that are parasitic on fish.

halite (chemistry) Naturally occurring form of sodium chloride. Also termed: common salt; rock salt.

Hall effect (physics) Production of a voltage in a semiconductor or metal carrying an electric current in a strong transverse magnetic field. It was named after the American physicist Edwin Hall (1855–1938).

haloalkane (chemistry) An alternative term for halogenoalkane.

haloform (chemistry) Organic compound of the type CHX_3, where X is a halogen (chlorine, bromine or iodine); *e.g.* chloroform (trichloromethane), iodoform (triiodomethane). The compounds are prepared by the action of the halogen on heating with ethanol in the presence of sodium hydroxide.

halofuginone (pharmacology) A coccidiostat derived from ornamental hydrangeas that may be used in poultry.

halogen (chemistry) Element in Group VIIA of the Perodic Table: fluorine, chlorine, bromine, iodine or astatine.

halogenation (chemistry) Chemical reaction that involves the addition of a halogen to a substance.

halogenoalkane (chemistry) Halogen derivative of an alkane, general formula $C_nH_{2n+1}X$, where X is a halogen (fluorine, chlorine, bromine or iodine). Also termed: haloalkane; monohalogenoalkane.

halophilic (biology) Exhibiting a preference for an environment containing salt (*e.g.* sea-water). The term is usually applied to bacteria.

halothane (pharmacology) An inhaled general anaesthetic drug, which is used for induction and maintenance of anaesthesia during surgery.

haloxon (pharmacology) An organophosphorus anthelmintic once used against nematodes of the abomasum and small intestine in ruminants.

halquinol (pharmacology) A topical hydroxyquinoline antibiotic and amoebicide (*USA*: amebicide).

haltere (parasitological anatomy) Club shaped structure in place of the second pair of wings, seen in Diptera.

ham beetle (parasitology) The 0.5-inch long larvae of the beetle which infest hams. Also termed: *Dermestes lardarius.*

ham fly (parasitology) The larvae or skippers, so called because of their habit of leaping long distances, of the fly *Piophila casei* which invade cured ham.

Hammondia (parasitology) A genus of protozoa in the family Eimeriidae.

Hammondia hammondi (veterinary parasitology) A species of protozoa in the genus *Hammondia* that may be found in the intestine of cats without apparent pathogenicity. The oocysts, shed in faeces (*USA*: feces), are identical in appearence to those of *Toxoplasma gondii*.

Hammondia heydorni (veterinary parasitology) A species of protozoa in the genus *Hammondia* that may be found in dogs without apparent pathogenicity.

Hannemania (veterinary parasitology) Chigger mites that may infest amphibians, especially frogs, causing red spots and vesicles.

Haplobothriidea (fish parasitology) An order of flatworms in subclass Eucestoda, class Cestoidea and the phylum Platyhelminthes. *Haplobothrium globuliforme* and *Haplobothrium bistrobilae* are the only representatives of this order. They are found in the intestine of the archaic bony fish, holostean *Amia calva*. The scolex has four unarmed tentacles.

haploid (cytology) Having half the number of chromosomes of the organism in the cell nucleus. The haploid state is found in gametes (resulting from meiosis), the gametophyte generation and spores. It occurs after meiosis or reduction division.

haplosporidiosis (fish parasitology) Protozoan infection by the genus *Haplosporidium* of haemocytes (*USA*: hemocytes) and organs of oysters and clams in the USA.

Haplosporidium (veterinary parasitology) A genus of parasitic protozoa in the order Balanosporida found in segmented worms and leeches (annelids).

Haplosporidium nelsoni (fish parasitology) A species of parasitic protozoa in the genus *Haplosporidium* that is the cause of multinucleate sphere unknown (MSX) disease in the American oyster.

haptor (parasitological anatomy) Posterior disk of a monogenetic trematode.

hard copy (computing) Printed version of a disk file, programme (*USA*: program) output etc. that people can read (*e.g.* a print-out in plain language).

hard disk (computing) Rigid magnetic disk that provides data and program storage for computers, including microcomputers. Hard disks can hold a high density of data. Also termed: diskette.

See also floppy disk.

hard tick (parasitology) Ticks of the family Ixodidae and members of *Ixodes*, *Boophilus*, *Margaropus*, *Hyalomma*, *Rhipicephalus*, *Haemaphysalis*, *Aponomma*, *Dermacentor*, *Amblyomma*, and *Rhipicentor* spp. They have a hard chitinous shield on the dorsal surface of the body, on the entire back of the male but only the anterior portion of the female.

hardness of water (chemistry) Property of water that prevents it forming a lather with soap because of the presence of dissolved compounds of calcium or magnesium.

See also soft water

hardware (computing) Electronic, electrical, magnetic and mechanical parts that make up a computer system.

See also software.

harmonic progression (mathematics) Series of numbers whose reciprocals have a constant difference (form an arithmetic progression), *e.g.* 1/1 + 1/2 + 1/3+ 1/4.

Harteria gallinarum (insect parasitology) A species of nematode in the Spiuridae that are parasites to the termites *Hodotermes pretoriensis*.

Hartertia (parasitology) A genus of nematodes of the family Spirocercidae.

Hartertia gallinarum (avian parasitology) A species of nematodes in the genus *Hartertia* that may be found in the intestine of fowls and wild bustards. They can cause emaciation, weakness and diarrhoea (*USA*: diarrhea) and may transmit *Histomonas meleagridis*.

Hartmannella (parasitology) A protozoan parasite, free-living and thought not to be pathogenic, although they were at one time listed as being so.

harvest (plant parasitology) Pertaining to or emanating from grain or cereal crops or the harvesting of them.

harvest mites (human/veterinary parasitology) Pests of grain and hay where they are predators on arthropods. The larvae ordinarily parasitize rodents but can infest other animals including humans. The infections are self-limiting but can cause dermatitis of the face and the lower limbs. The lesions are itchy, small scabs, which cause rubbing and stamping of the feet. In pigs the lesions are distributed over most of the body. Also termed chigger mites; grain mites; *Pyemotes ventricosus*; *Neotrombicula autumnalis*; *Eutrombicula alfreddugesi*; *Eutrombicula splendens*; *Eutrombicula batatas*; *Lepotrombidium* spp.; *Schoengastia* spp.

Haversian canal (anatomy) Any of the numerous channels that occur in bone tissue, containing blood vessels and nerves. An organic matrix is laid down in layers encircling each Haversian canal. It was named after the English physician Clopton Havers (? –1702).

hazard (general terminology) A risk.

hb (general terminology) An abbreviation of 'h'uman 'b'eing.

Hb (haematology) An abbreviation of 'H'aemo'g'lobin (*USA*: hemoglobin).

HBE (physiology) An abbreviation of 'H'is 'B'undle 'E'lectrogram.

Hb-meter (equipment) An abbreviation of 'H'aemoglo'b'ino'meter'. (*USA*: hemoglobinometer).

HbO_2 (physiology/biochemistry) A symbol for Oxyhemoglobin.

HCC (biochemistry) An abbreviation of 25-'H'ydroxy'C'hole'C'alciferol.

Hct (haematology) An abbreviation of 'H'aemato'c'ri't' (*USA*: Hematocrit).

HDL (biochemistry) An abbreviation of 'H'igh-'D'ensity 'L'ipoprotein.

He (chemistry) The chemical symbol for helium.

H & E (histology/haematology) An abbreviation of 'h'aematoxylin & 'e'osin.

head (computing) In a tape recorder, video recorder, record player or computer input/output device, an electromagnetic component that can read, erase or write signals off or onto tapes and disks.

head bot (parasitology) *Cephenemyia*.

head crash (computing) A serious equipment malfunction in computer disk drive causing loss of data and damage to the disk.

head fly (parasitology) *Hydrotoea irritans*.

head grub (parasitology) *Oestrus ovis*.

head louse (parasitology) *Cyclotogaster heterographus*.

head mange (parasitology) Notoedric mange.

Health and Safety at Work Act (legislation) Important UK legislation of 1974, resulting from the recommendations of the Robens committee in July 1972, that covers all aspects of safety at the workplace, including the potential exposure of workers to toxic and infectious substances.

Health and Safety Commission (governing body) A commission, consisting of employers' and union representatives, that was established under the UK Health and Safety at Work Act. The Commission, which is advised by various Advisory Committees on health hazards in the workplace from toxic chemicals and infectious agents, and related hazards to the public, has overall powers to propose health and safety regulations and to approve codes of practice. It is directly responsible to the Secretary of State for Employment.

See also Health and Safety at Work Act.

Health and Safety Executive (governing body) The body that enforces the statutory duties laid down in the UK Health and Safety at Work Act. The Executive is responsible to the Health and Safety Commission.

heart (anatomy) Muscular organ that occurs in vertebrates which pumps blood into a system of arteries. Blood returns to the heart from the tissues via veins, and is passed to the lungs to become reoxygenated. The mammalian heart is divided into four chambers, the right and left atria (auricles) and right and left ventricles, so that deoxygenated and oxygenated blood remain separate.

heartbeat (physiology) Alternate contraction and relaxation of the heart, corresponding to dyastole and systole.

heartwater (veterinary parasitology) A disease of cattle, sheep, goats and wild ungulates caused by *Cowdria ruminantium*, transmitted by the *Amblyomma* spp. ticks.

heartworm (parasitology) The common name for *Dirofilaria immitis*.

heartworm dermatitis (veterinary parasitology) Cutaneous dirofilariasis; a variety of skin lesions have been seen in dogs infested by *Dirofilaria immitis*, including hypersensitivity reactions, pyogranulomas and seborrheic dermatitis.

heartworm disease (veterinary parasitology) The syndrome of pulmonary artery disease with hypertension, heart failure (primarily cor pulmonale), and occasionally liver failure and interstitial, tubular and glomerular renal lesions caused by infestation by *Dirofilaria immitis*. Many species may be infected, but dogs are most commonly affected by chronic cough, weight loss and ultimately congestive heart failure. Infestation by the parasite, and the disease, can be prevented with appropriate prophylactic chemotherapy. Also termed dirofilariasis.

See also occult heartworm infection.

heat (physics) Form of energy, the energy of motion (kinetic energy) possessed by the atoms or molecules of all substances at temperatures above absolute zero.

heat capacity (physics) Quantity of heat required to produce unit rise of temperature in an object.

heat of activation (physics) Difference between the values of the thermodynamic functions for the activated complex and the reactants for a chemical reaction (all the substances being in their standard states).

heat of atomization (measurement) Amount of heat that is required to convert 1 mole of an element into the gaseous state.

heat of combustion (measurement) Heat change that accompanies the complete combustion of 1 mole of a substance.

heat of formation (measurement) Heat change that occurs when 1 mole of a compound is formed from its elements, in their normal states.

heat of neutralization (measurement) Heat change that occurs when 1 mole of aqueous hydrogen ions are neutralized by a base in a diluted solution.

heat of reaction (measurement) Heat change that occurs when the molar quantities (lowest possible multiples of 1 mole) of reactants as stated in a chemical equation react together.

heat of solution (measurement) Heat change that occurs when 1 mole of a substance is dissolved in so much water that further dilution with water produces no further heat change.

hecto- (terminology) A word element. [French] Prefix denoting *hundred.*

hedgehog fleas (veterinary parasitology) *See flea.*

hedgehog tick (veterinary parasitology) *See Ixodes hexagonus.*

heel bug (parasitology) *See Trombicula autumnalis.*

heel fly (parasitology) *See Hypoderma.*

height (ht or hgt) (measurement) The linear measurement of an object from top to bottom.

Heinz bodies (haematology) Dark-staining bodies found in erythrocytes that lie on the inner surface.

Heisenberg uncertainty principle (physics) The precise position and momentum of an electron cannot be determined simultaneously. It was named after the German physicist Werner Heisenberg (1901–1976).

Hela cells (cytology) A line of human cancer cells that has been maintained in culture since 1951 and has been much used in cancer research. It is named after the patient from whom the cells were taken, Henrietta Lacks, who had cancer of the cervix. *See* tisuue culture.

heli(o)- (terminology) A word element. [German] Prefix denoting *sun.*

Helictometra giardi (parasitology) *See Thysaniezia giardi.*

helium (He) (chemistry) Gaseous element in Group 0 of the periodic table (the rare gases), which occurs in some natural gas deposits. It is inert and noninflammable, used to fill airships and balloons (in preference to inflammable hydrogen) and in helium-oxygen 'air' mixtures for divers (in preference to the nitrogen-oxygen mixture of real air, which can cause the bends). It is also used in gas lasers. Liquid helium is employed as a coolant in cryogenics. At. no. 2; r.a.m. 4.0026.

helix (molecular biology) A spiral shape. The shape of the thread of a bolt or screw. It occurs in nature as the three-dimensional structure of DNA, which is a double helix. The alpha helix, described by Linus Pauling, is the secondary structure of many proteins, *i.e.* they are twisted threads that are subsequently folded into a particular three-dimensional shape.

helmet cell (veterinary parasitology) *See* schistocyte.

helminth (parasitology) A parasitic worm; a nematode, cestode or trematode.

helminth hypersensitivity (parasitology) A hypersensitivity state that may occur following an invasion with helminths of, for example, the lungs.

helminthagogue (pharmacology/control measures) Anthelmintic; vermifuge; a substance that expels worms or intestinal animal parasites.

helminthemesis (parasitology) The vomiting of worms.

helminthiasis (parasitology) A disease caused by an helmintic infection which may take any one of a number of forms: 1) Worms living virtually free in the gut lumen, *e.g. Moniezia*, that are relatively nonpathogenic. 2) Those that suck blood that may cause anaemia (*USA*: anemia). 3) Those causing mucosal damage, *e.g. Ostertagia* spp., that are followed by inappetence, malabsorption, a protein-losing enteropathy. 4) A group that burrows into the gastrointestinal wall, *e.g. Oesophagostomum* spp., that cause physical disturbances to gut function. Those that migrate through other tissues, for example *Strongylus vulgaris*, that cause clinical signs related to that migration. Also termed: helminthosis.

helminthology (parasitology) The scientific study of parasitic worms.

helminthoma (parasitology) A tumour (*USA*: tumor) caused by a parasitic worm.

helminthosis (parasitology) *See* helminthiasis.

helmintic (parasitology) Pertaining to or emanating from helminths.

helmintic immunity (immunology/parasitology) Antibodies produced by the host to antigens in the worm's exsheathing fluid, and enzymes produced by the worm, that are thought to damage the worm and encourage its expulsion.

hematozoa (parasitology) Plural of hematozoon. (*USA*: hematozoa).

hemi- (terminology) A word element. [German] Prefix denoting *half.*

hemianaesthesia (medical) Loss of sensation on one side of the body; it is caused by unilateral damage to the somaesthetic system. (*USA*: hemianesthesia).

hemianopia (medical) Loss of vision in the right or left half of the visual field of one or both eyes. It is caused by unilateral damage to the visual cortex, optic radiations, or optic chiasm. Also termed hemiopia.

Hemicycliophora (plant parasitology) A genus of nematodes in the order Tylenchida and family Criconematidae. They are sexually dimorphic with females being 0.6–2mm long, rather stout, anterior end usually bluntly rounded, posterior end tapered to cylindrical with a bluntly rounded or cone-shaped terminus. Males are much thinner than females, juveniles are similar to females and lack the extra cuticular scales of juveniles in Criconemella. *See Hemicycliophora arenaria.*

Hemicycliophora arenaria (plant parasitology) A species of nematodes in the genus *Hemicycliophora* whose hosts cover a narrow range, including citrus, tomato, beans, celery, squash, pepper, and Tokay grape; also *Hymenoclea salsola* (cheesebush) and coyote melon (*Cucurbita palmata*) on virgin desert soil. They are migratory ectoparasites feeding near the root tip. The life cycle from egg to egg requires around 15 to 19 days at 28 to 30°C. Females can lay approximately one egg per hour in water. Egg hatch occurs in three to five days. During feeding, the stylet is inserted two to three cells deep, through and between cells. The nematode has a feeding tube, an adhesive polysaccharide tube, which becomes firmly attached to the root, and the nematode has to writhe and twist to detach. Nematodes may appear as a fringe around the root tip. Nurse cells increase in volume, and walls thicken; some cells are multinucleate. Cells collapse when depleted and are pushed to the surface by meristem activity, thus providing a continuous supply of new food cells for the nematode. Nematode infestation causes root-tip swelling and stunted root/plant growth. Root galls are formed by an increase in cell divisions, hyperplasia, giving rise to enlarged cortex. Control measures include the use of preplant nematicides and postplant nematicides on perennials. The nematode is sensitive to reduced aeration, and mortality is associated with irrigation. Also termed: sheath nematode.

hemihydrate (chemistry) Compound containing one water molecule for every two molecules of the compound; *e.g.* $CaSO_4.1/2H_2O$.

hemiopia (medical) *See* hemianopia.

hemiplegia (medical) Paralysis of one side of the body; it is caused by unilateral damage to part of the motor system, usually caused by a stroke.

Hemiptera (parasitology) An order of arthropods (class Insecta) which includes some 30,000 species, known as the true bugs. They are characterized by having mouthparts adapted to piercing or sucking, and usually having two pairs of wings.

hemoflagellate (parasitology) Any flagellate protozoan parasitic in the blood.

See also Trypanosoma.

hemoglobin (haematology) *See* haemoglobin.

hemomelasma ilei (parasitology) Small, up to 1 by 2 inches, slightly elevated, red-brown plaques, under the serosa of the terminal small intestine, that is caused by migrating larvae of *Strongylus* spp. in horses.

hemonchosis (veterinary parasitology) Infestation with the abomasal worms *Haemonchus contortus* and *Haemonchus placei*. The disease in sheep, goats and cattle is characterized by acute anaemia (*USA*: anemia), anasarca and death due to anaemic (*USA*: anemic) anoxia.

See also Haemonchus.

hemoparasite (veterinary parasitology) An animal parasite, specifically a hemoflagellate or filarid worm, living in the blood of a vertebrate, *e.g. Babesia* spp., *Eperythrozoon* spp., *Trypanosoma* spp.

hemophagous (veterinary parasitology) *See* haemophagous.

Henneguya (fish parasitology) A genus of protozoa causing disease in free-living fish. Genus of class Myxosporea, the myxosporids.

Henneguya exilis (fish parasitology) A species of protozoa in the genus *Henneguya* that causes cysts in the gills of channel catfish and death due to suffocation.

Henneguya zschokkei (fish parasitology) A species of protozoa in the genus *Henneguya* that causes boil disease in salmonid fishes.

henry (H) (measurement) SI unit of electrical inductance, defined as the inductance that produces an induced electromotive force of 1 volt for a current change of 1 ampere per second. It was named after the American physicist Joseph Henry (1797–1878).

Henry's law (physics) Weight of gas dissolved by a liquid is proportional to the gas pressure. It was named after the British chemist William Henry (1774–1836).

heparin (haematology) A substance found in many tissues of the body, but mostly in the liver, which prevents the clotting of blood by interfering with the formation of thrombin from prothrombin. It is used in medicine to prevent clotting, and acts quickly; but the action is shortlived, and it has to be given by injection, either into the veins or under the skin. It is particularly useful in cardiac surgery and renal dialysis. If haemorrhage (*USA*: hemorrhage) should be provoked, the antidote is protamine sulphate (*USA*: protamine sulfate). *See* coagulation of the blood.

hepat(o)- (terminology) A word element. [German] Prefix denoting *liver*.

hepatic (anatomy) Emanating from or pertaining to the liver.

hepatic cell (cytology) Hepatocyte.

hepatic coccidiosis (veterinary parasitology) Infection of the liver by coccidia, an apparently infrequent condition, to date only having been recorded in an isolated incident in a calf.

hepatic distomatosis (parasitology) Infection of the liver with flukes, *e.g. Fasciola hepatica, Fascioloides magna, Metorchus conjunctus.*

See also fascioliasis.

hepatic fascioliasis (parasitology) Infection of the liver with *Fasciola* species resulting in acute or chronic hepatic insufficiency. Acute fascioliasis in sheep is characterized by sudden death due to blood loss from the liver. Chronic fascioliasis in sheep and cattle causes anaemia (*USA*: anemia), weight loss, submandibular oedema (*USA*: edema) and mucosal pallor. The liver is badly damaged due to fibrosis and cholangitis.

hepatic portal system (physiology) In vertebrates, system of blood capillaries into which dissolved foods (except for fatty acids and glycerol) pass from the intestinal lining for transport to the liver.

hepatic(o)- (terminology) A word element. [German] Prefix denoting *hepatic duct.*

hepatitis (medical) Inflammation of the liver which may be toxic or infectious in origin; characterized by signs due to diffuse injury to the liver.

hepatitis cysticercosa (parasitology) Damage to the liver caused by migrating cestode larvae, including oncospheres and post-oncopheral stages, cysticerca in much the same way as migrating fluke larvae cause damage.

hepatocystic (anatomy) Pertaining to the liver and gall bladder.

Hepatocystis (parasitology) A genus of intra-erythrocytic protozoon parasites in the family Plasmodiidae.

Hepatocystis kochi (veterinary parasitology) A species of intra-erythrocytic protozoon parasites in the genus *Hepatocystis* that is the common malarial parasite of many primates. They rarely cause disease but may cause white to grey (*USA*: gray) nodular foci on the liver.

hepatocyte (cytology) A liver cell.

hepatoma (histology) A malignant tumor of the liver.

hepatomegaly (medical) Enlargement of the liver.

hepatopathy (medical) Any disease of the liver.

hepatosplenomegaly (medical) Enlargement of the liver and spleen.

hepatotoxicants (toxicology) Chemicals that cause adverse effects on the liver.

See also hepatotoxicity.

hepatotoxicity (toxicology) Adverse effects on the liver. The liver is particularly susceptible to chemical injury because of its anatomical relationship to the most important portal of entry, the gastrointestinal tract, and its high concentration of xenobiotic metabolizing enyzmes. Many of these enzymes, particularly the cytochrome P450-dependent monooxygenase system, metabolize xenobiotics to produce reactive intermediates that can react with endogenous macromolecules such as proteins and DNA to produce adverse effects.

Hepatozoon (parasitology) A genus of protozoan parasites in the family Haemogregarinidae.

Hepatozoon canis (veterinary parasitology) A species of protozoan parasites in the genus *Hepatozoon* that may be found in dogs and other canines and in cats, in circulating leukocytes and bone marrow cells. Transmitted by the tick *Rhipicephalus sanguineus*, they may cause intermittent fever, anaemia (*USA*: anemia), loss of weight, splenomegaly and posterior paralysis, although many infected animals are clinically normal.

Hepatozoon cuniculi (veterinary parasitology) A species of protozoan parasites in the genus *Hepatozoon* that may be found in rabbits.

Hepatozoon griseisciuri (veterinary parasitology) A species of protozoan parasites in the genus *Hepatozoon* that may be found in grey (*USA*: gray) squirrels.

Hepatozoon muris (veterinary parasitology) A species of protozoan parasites in the genus *Hepatozoon* that may be found in rats and that is transmitted by rat mites.

Hepatozoon musculi (veterinary parasitology) A species of protozoan parasites in the genus *Hepatozoon* that may be found in white mice.

hepatozoonosis (parasitology) Infection with the protozoa *Hepatozoon.*

hepta- (terminology) A word element. [German] Prefix denoting *seven.*

heptane (chemistry) C_7H_{16} Liquid alkane hydrocarbon, the seventh member of the methane series, present in petrol.

heptavalent (chemistry) With a valency of seven. Also termed: septivalent.

herbage (plant parasitology) Standing growing nonwoody plants, mostly annuals and especially grasses.

herbage larval counts (plant parasitology) Harvesting of a representative sample of herbage, washing it and sieving off larvae for the purpose of assessing levels of contamination of pasture with infective larvae of strongylid nematodes.

heredity (genetics) The process by which characteristics are genetically transmitted from one generation to the next. *See* genetics.

heritability (genetics) The proportion of the variability of the phenotype of a trait that is attributable to genetic factors. Heritability is not a fixed property of the genes concerned but depends in each situation on how much variation there is both in genotypes and in the environment.

hermaphrodite (genetics) An animal that has the reproductive organs of both sexes. Animals that are normally hermaphroditic include flatworms, leeches, land snails, some fish and some flies. They are capable of self-fertilization, unless they are sequential hermaphrodites (*i.e.* they have first one set of sex organs, then the other). Hermaphrodites do not have sex chromosomes. In humans, hermaphroditism is a rare abnormality. The person has both ovarian and testicular tissue, which may occur as (a) one ovary plus one testis, (b) two ovotestes or (c) one ovotestis plus either an ovary or a testis; the external genitalia are ambiguous. The condition is the result of a single ovum being fertilized simultaneously by an X and a Y sperm, so that the person is a chimaera of XX and XY cells. *See* sex determination.

Herpetomonas (parasitology) A genus of protozoa in the family Trypanosomatidae that may be found in invertebrates.

Herpetosoma (parasitology) A subgenus of the genus *Trypanosoma*, containing nonpathogenic species of *Trypanosoma lewisi.*

herring worm (fish parasitology) *See Anisakis.*

hertz (Hz) (measurement) SI unit of frequency. 1 Hz = 1 cycle per second. It was named after the German physicist Heinrich Hertz (1858–94).

Hess's law (chemistry) Total energy change resulting from a chemical reaction is dependent only on the initial and final states, and is independent of the reaction route. It was named after the Austrian-born American physicist Victor Hess (1883–1964).

heter(o)- (terminology) A word element. [German] Prefix denoting *other, dissimilar.*

Heterakis (parasitology) A genus of nematodes in the family Heterakidae.

Heterakis beramporia (avian parasitology) A species of nematodes in the genus *Heterakis* that may be found in nodules in the caecal (*USA*: cecal) wall of chickens.

Heterakis brevispiculum (avian parasitology) A species of nematodes in the genus *Heterakis* that is a parasite of chickens and guinea fowl.

Heterakis dispar (avian parasitology) A species of nematodes in the genus *Heterakis* that may be found in geese and ducks.

Heterakis gallinae (avian parasitology) (synonym: - *Heterakis gallinarum*) A species of nematodes in the genus *Heterakis* that *Heterakis vesicularis* and *Heterakis papillosa* may be found in most species of domestic birds and some wild species, and that is a vector for *Histomonas meleagridis.*

Heterakis indica (avian parasitology) A species of nematodes in the genus *Heterakis* that may be found in chickens.

Heterakis isolonche (avian parasitology) A species of nematodes in the genus *Heterakis* that may be found in pheasants, quail and other gallinaceous birds which causes nodular typhlitis, diarrhoea (*USA*: diarrhea), emaciation and death.

Heterakis linganensis (avian parasitology) A species of nematodes in the genus *Heterakis* that may be found in chickens.

Heterakis meleagris (avian parasitology) A species of nematodes in the genus *Heterakis* that may be found in turkeys.

Heterakis papillosa (avian parasitology) A species of nematodes in the genus *Heterakis* that, similarly to *Heterakis vesicularis and Heterakis gallinae* (synonym: *Heterakis gallinarum*), may be found in most species of domestic birds and some wild species, and that is a vector for *Histomonas meleagridis.*

Heterakis pavonis (avian parasitology) A species of nematodes in the genus *Heterakis* that may be found in peafowls and pheasants.

Heterakis spumosa (veterinary parasitology) A species of nematodes in the genus *Heterakis* that may be found in the caecum (*USA*: cecum) of rats.

Heterakis vesicularis (avian parasitology) A species of nematodes in the genus *Heterakis* that *Heterakis papillosa* and *Heterakis gallinae* (synonym: *Heterakis gallinarum*) may be found in most species of domestic birds and some wild species, and that is a vector for *Histomonas meleagridis.*

heterarchy (computing) The characteristic of a computer program that does not process data merely through successively higher levels, but can call many different routines to help in the processing of the data at each level, including routines from higher levels.

hetero- (biology) Prefix denoting other or different.

See also homo-.

Heterobilharzia (parasitology) A genus of blood flukes (digenetic trematodes) in the family Schistosomatidae.

Heterobilharzia americanum (veterinary parasitology) A species of blood flukes (digenetic trematodes) in the genus *Heterobilharzia* that may be found in raccoons, bobcats, dogs and other mammals.

heterochromatic flicker photometer (physics) *See* photometer.

heterochromatin (molecular biology) Chromatin, the substance of which chromosomes are made, that is in a highly condensed state because the DNA, not being actively transcribed, can be packed densely around the histones. Examples are the DNA in the centromeres and in Barr bodies. Because of its more condensed state, heterochromatin is more intensely coloured (*USA*: colored) when stained than ordinary chromatin (euchromatin).

heterochrony (measurment) A difference in the rate or timing of events.

heterocyclic (chemistry) Describing a cyclic organic compound that contains atoms other than carbon in the ring, *e.g.* pyridine.

See also homocyclic.

heterocytotropic (biology) Having an affinity for cells from different species.

Heterodera (plant parasitology) A genus of nematodes in the order Tylenchida and family Heteroderidae. Sexually dimorphic, the mature females are obese, lemon-shaped, 0.5–1 mm long, with a distinct neck. The cuticle is thick, whitish at first but gradually turns brown to black to form a cyst protecting the many eggs retained inside. The males are elongated, approximately 1 mm long, vermiform, whose posterior region becomes twisted through 90–180 degrees when killed by heat. Species includes the alfalfa root nematode, *Heterodera goettingiana*, the almond cyst nematode, *Heterodera amygdali* and the amu-darya nematode, *Heterodera oxiana*. *See Heterodera schachtii.*

Heterodera schachtii (plant parasitology) A species of nematodes in the genus *Heterodera* whose hosts cover a narrow range including sugar beet, cabbage, cauliflower, brussels sprouts, mustard, radish, spinach, chard, and which are distributed widely in all sugar beet growing areas of the world. They are sedentary endoparasites whose second-stage juveniles that hatch from eggs are the infective stage. Adult females, which are white or light yellow in colour (*USA*: color), retain most of the eggs within their bodies and some are laid into a gelatinous matrix. Eggs laid in the gelatinous matrix hatch and the second-stage juveniles invade new roots. When the female nematodes are fully mature, or when feeding is interrupted, the body wall undergoes a tanning process and becomes brown in colour (*USA*:color). At this stage the dead female body containing eggs (about 300–500) is referred to as a 'cyst'. Eggs in the 'cyst' may remain viable for long periods of time and may not hatch readily unless stimulated by root secretions, usually of host plants. Eggs within cysts remain viable for many years. Males are required for reproduction. The optimum temperature for growth and reproduction is 21–27°C and development takes about 17 days. Infested crops are typically stunted, have yellow foliage and wilt in warm weather. The white females and brown cysts attached to the roots are frequently visible to the unaided eye. Control measures include crop rotation, as generally three to four years growing of non-hosts reduces soil populations to a low enough level to allow the growing of a susceptible crop for one year. Also termed: sugar beet cyst nematode.

Heterodoxus (parasitology) A genus of lice in the superfamily Amblycera.

Heterodoxus longitarsus (veterinary parasitology) A species of lice in the genus *Heterodoxus* that may be found on kangaroos.

Heterodoxus macropus (veterinary parasitology) A species of lice in the genus *Heterodoxus* that may be found on kangaroos and wallabies.

Heterodoxus spiniger (veterinary parasitology) A species of lice in the genus *Heterodoxus* that parasitizes dogs.

heteroduplex (molecular biology) A nucleic acid molecule that consists of two strands of different genetic origins. It can be either a double-stranded DNA molecule or a DNA-RNA combination. Where the sequences are homologous, the two strands will link; non-homologous sequences can then be seen under the electron microscope looking like bubbles. Examination of a heteroduplex made from DNA strands from wild-type and mutant-type DNA will show where there are deletions or insertions; this is called 'heteroduplex mapping'.

heteroecious (parasitology) Requiring different hosts in different stages of development; a characteristic of certain parasites.

heterogametic (genetics) Describing an organism that produces two kinds of gametes, each possessing a different sex chromosome. These are usually produced by the male; *e.g.* in human males half the sperms contain an X-chromosome and half a Y-chromosome. The opposite is homogametic.

heterogeneous nuclear RNA (genetics) The RNA molecule that reads the DNA code in eukaryotes. *See* messenger RNA.

heterogeneous Relating to more than one phase, *e.g.* describing a chemical reaction that involves one or more solids in addition to a gas or a liquid phase.

See also homogeneous.

heterograft (medical) *See* graft.

heterokaryon (genetics) A cell or individual that contains two nuclei, each of different genetic origin. It can happen naturally when cells of two different fungi fuse, and it can be artificially induced by mixing the cells of higher organisms *in vitro*.

heterolytic fission (chemistry) Breaking of a two-electron covalent bond to give two fragments, with one fragment retaining both electrons. Also termed: heterolytic cleavage.

See also homolytic fission.

Heterophyes (avian/veterinary parasitology) A genus of intestinal flukes (digenetic trematodes) of the family Heterophyidae. Other genera of this family infest birds, *e.g. Cryptocotyle* spp.

Heterophyes heterophyes (human/veterinary parasitology) A species of flukes in the genus *Heterophyes* that may be found in the intestine of dogs, cats, foxes and humans, but that is regarded as being of low pathogenicity.

heterophyiasis (veterinary parasitology) Infection with *Heterophyes* spp. in dogs and cats, which may cause a generally mild, but occasionally severe haemorrhagic diarrhoea (*USA*: hemorrhagic diarrhea).

heterosis (genetics) The technical term for the phenomenon more generally known as hybrid vigour.

heterosome (genetics) The X or Y chromosome, each of which, unlike the other chromosomes, can pair with a different chromosome. Compare autosome.

heterotroph (biology) Organism that requires an organic carbon source.

heterotrophism (physiology) Mode of nutrition exhibited by most animals, fungi, bacteria and some flowering plants. It involves the intake of organic substances from the environment due to the inability of the organism to synthesize them from inorganic materials.

heteroxenous (parasitology) Requiring more than one host to complete the life cycle.

heteroxenous apicomplexan protozoa (parasitology) Protozoa requiring more than one host to complete the life cycle, such as *Besnoitia*, *Frenkelia*, *Hammondia*, *Sarcocystis* and *Toxoplasma*.

heterozygote (genetics) An organism having two different alleles at the same locus on a pair of chromosomes. The term 'double heterozygote' refers to an individual who, at two given loci, has unlike alleles. *Contrast* homozygote.

heterozygous (genetics) Describing an organism that possesses two dissimilar alleles in a pair of chromosomes. A dominant allele can be expressed in the heterozygous or homozygous state, but a recessive allele can be expressed only in the homozygous state.

heterozygous advantage (genetics) The situation where an organism that is heterozygous at a given locus has greater genetic fitness than either of the two homozygotes. It is one way in which two different alleles can be maintained within a population at stable frequencies: neither is better on its own, but the two in combination are the most successful genotype, and therefore neither allele is lost from the population. The classic example of this is sickle-cell anaemia (*USA*: sickle-cell anemia) in humans, in which homozygotes for the mutant allele have severe anaemia (*USA*: anemia), homozygotes for the normal allele are liable to get malaria, but heterozygotes are neither anaemic (*USA*: anemic) nor subject to malaria. In a more general sense, organisms that are heterozygous at many loci are at an advantage over more homozygous ones; this is known as hybrid vigour. *See* polymorphism.

heuristic (philosophy) Describing an approach to problem-solving that is based on trial and error rather than theory. The term is sometimes used to describe computer programs that can 'learn' from their mistakes.

hex(a)- (terminology) A word element. [German] Prefix denoting *six*.

hexacanth embryo (parasitological physiology) The infective stage (larvae, oncosphere) in a cestode egg after fertilization takes place.

hexachloroethane (pharmacology) An anthelmintic used in the treatment of fascioliasis.

hexachloroparaxylene (pharmacology) A chlorinated derivative of benzene used as a fasciolicide that is effective only against adult flukes. Also termed chloxyle.

hexachlorophene (pharmacology/control measures) A compound used as a tenicide in poultry, a fasciolicide in cattle, and as a germicidal agent in soaps and dermatological preparations. Also termed: hexachlorophane.

hexadecimal (computing) Describing a number system based on 16 (its digits are 1–9, A, B, C, D, E, F), commonly used in digital computers.

Hexamastix (parasitology) A genus of flagellated protozoa in the family Monocercomonadidae.

Hexamastix caviae (veterinary parasitology) A species of flagellated protozoa in the genus *Hexamastix* that, similarly to *Hexamastix robustus*, may be found in the caecum (*USA*: cecum) of guinea pigs.

Hexamastix muris (veterinary parasitology) A species of flagellated protozoa in the genus *Hexamastix* that may be found in the caecum (*USA*: cecum) of rats and other rodents.

Hexamastix robustus (veterinary parasitology) A species of flagellated protozoa in the genus *Hexamastix* that, similarly to *Hexamastix caviae*, may be found in the caecum (*USA*: cecum) of guinea pigs.

Hexamermis (insect parasitology) A genus of nematode in the Mermithidae that are parasites to the ants *Pheidole* and *Polyrhachis*.

hexamine (chemistry) $C_6H_{12}N_4$ Organic compound made by condensing methanal (formaldehyde) with ammonia, used as a camping fuel and antiseptic drug. It can be nitrated to make the high explosive cyclonite. Also termed: hexamethylenetetramine.

Hexamita (parasitology) A genus of flagellated protozoa in the family Hexamitidae.

Hexamita columbae (avian parasitology) A species of flagellated protozoa in the genus *Hexamita* that may be found in the small intestine of pigeons and that may cause enteritis.

Hexamita meleagridis (avian parasitology) A species of flagellated protozoa in the genus *Hexamita* that is a major parasite of turkeys in which it produces an enteritis, which may cause serious losses with a mortality rate of up to 80%. They may also cause disease in other species of birds.

Hexamita muris (veterinary parasitology) A species of flagellated protozoa in the genus *Hexamita* that may be found in the intestines of rodents and that causes enteritis.

Hexamita pitheci (veterinary parasitology) A species of flagellated protozoa in the genus *Hexamita* that may be found in the large intestine of rhesus and other monkeys.

Hexamita salmonis (fish parasitology) A species of flagellated protozoa in the genus *Hexamita* that may be found in the intestine of salmonids and may cause acute enteritis or may become systemic and cause death. *Hexamita*-like organisms can be associated with hole-in-the-head disease in aquarium fish, especially *Symphysodon discus* and *Artromotus ocellatus*.

Hexamita truttae (fish parasitology) A species of flagellated protozoa in the genus *Hexamita* that may be found in the intestines of fish, and that causes enteritis, peritonitis, cholecystitis in debilitated fish.

hexamitiasis (avian parasitology) An infectious catarrhal enteritis of turkey poults caused by the protozoan *Hexamita meleagridis*.

hexamitosis (fish parasitology) *See Hexamita salmonis.*

hexane (chemistry) C_6H_{14} Colourless (*USA*: colorless) liquid alkane, used as a solvent.

hexanedioic acid (chemistry) An alternative term for adipic acid.

hexose (chemistry) Monosaccharide carbohydrate (sugar) that contains six carbon atoms and has the general formula $C_6H_{12}O_6$; *e.g.* glucose and fructose.

hexylresorcinol (pharmacology) A superseded anthelmintic for intestinal roundworms and trematodes.

Hf (chemistry) The chemical symbol for hafnium.

hf (physics) An abbreviation of 'h'igh 'f'requency.

HFG (physics) An abbreviation of 'H'igh 'F'requency 'G'as.

Hg (chemistry) The chemical symbol for mercury. The symbol is derived from Latin hydragyrum, meaning mercury.

hg (measurement) An abbreviation of 'h'ecto'g'ram.

HGH (biochemistry) An abbreviation of 'H'uman 'G'rowth 'H'ormone.

hgt (measurement) An abbreviation of 'h'ei'g'h't'. Also termed ht.

hi temp (measurement) An abbreviation of 'hi'gh 'temp'erature.

5-HIAA (biochemistry) An abbreviation of 5-'H'ydroxy'I'ndole'A'cetic 'A'cid.

hidr(o)- (terminology) A word element. [German] Prefix denoting *sweat*.

high blood pressure (medical) *See* hypertension.

high frequency (HF) (1, 2 physics) 1 A radio frequency with a range of 3 to 30 megahertz. 2 Rapidly alternating electric current or wave.

high iron diamine stain (histology) A histological staining technique used to make sulphomucins (*USA*: sulfomucins) visible under the microscope.

high voltage (HV) (physics) Marked on wires carrying a high load of electricity to indicate danger of electrocution.

highest common factor (HCF) (mathematics) Of two or more numbers is the largest number that divides exactly into each of them; *e.g.* the HCF of 21, 35 and 63 is 7.

Highmans congo red stain (histology) A histological staining technique used to make amyloid visible under the microscope.

high-pass filter (physics) A filter that transmits from a waveform only the frequencies higher than a specified frequency.

high-risk behavior (epidemiology) A term used to describe certain activities that increase the risk of disease exposure.

high-risk groups (epidemiology) Those groups of people that show a behavioral risk for exposure to a disease or condition.

HIOMT (biochemistry) An abbreviation of 'H'ydroxy'I'ndole-'O'-'M'ethyl'T'ransferase.

H-ion (chemistry) An abbreviation of 'H'ydrogen 'ion'.

hip joint (anatomy) The ball-and-socket joint between the ballshaped head of the femur, the thigh-bone, and the acetabulum, a cup-shaped hollow in the side of the pelvis.

Hippobosca (veterinary parasitology) A genus of flies in the family Hippoboscidae that cause fly worry for horses and cattle. Also termed horse louse fly.

Hippobosca camelina (veterinary parasitology) A species of flies in the genus *Hippobosca* that attacks camels and horses.

Hippobosca capensis (veterinary parasitology) A species of flies in the genus *Hippobosca* that attacks dogs.

Hippobosca equina (avian/veterinary parasitology) A species of flies in the genus *Hippobosca* that, similarly to *Hippobosca rufipes* and *Hippobosca maculata*, attacks cattle and horses usually, and that transmits *Trypanosoma theileri* to cattle and *Haemoproteus* spp. to birds.

Hippobosca maculata (avian/veterinary parasitology) A species of flies in the genus *Hippobosca* that, similarly to *Hippobosca equina* and *Hippobosca rufipes*, attacks cattle and horses usually, and that transmits *Trypanosoma theileri* to cattle and *Haemoproteus* spp. to birds.

Hippobosca rufipes (avian/veterinary parasitology) A species of flies in the genus *Hippobosca* that, similarly to *Hippobosca equina* and *Hippobosca maculata*, attacks cattle and horses usually, and that transmits *Trypanosoma theileri* to cattle and *Haemoproteus* spp. to birds.

hippoboscid fly (parasitology) *See Hippobosca; Lipoptena; Pseudolynchia; Melophagus.*

hippocampus (histology) Structure of nerve tissue within the vertebrate brain. It is implicated in emotions and also possibly in memory, since bilateral removal in humans produces severe and permanent anterograde deficits in memory.

Hirudinea (biology) Class of annelids consisting of the leeches, some of which are blood-sucking.

His. (biochemistry) An abbreviation of 'His'tidine.

histamine (biochemistry) Organic compound that is released from cells in connective tissue during an allergic reaction. It causes dilation of capillaries and constriction of bronchi. *See* allergy.

Histidine (His.) (chemistry) $(C_3H_3N_2)CH_2CH(NH_2)COOH$ Crystalline soluble solid, an Optically active basic essential amino acid. Also termed: 2-amino-3-imidazolylpropanoic acid.

Histiocephalus (avian parasitology) Spiruroid worms found in the gizzard of free-living birds, that is rarely pathogenic.

hist(io)(o)- (terminology) A word element. [German] Prefix denoting *tissue.*

Histiostoma (insect parasitology) *See Halictacarus.*

Histiostoma formosana (insect parasitology) Mites in the Mesostigmata that, similarly to *Laelaptonyssus* spp. and *Termitacarius cuneiformis*, are phoretic on the heads of workers and soldiers of the termites *Amitermes*, *Coptotermes* and other various termites.

histocompatibility (histology) *See* T cells.

histocompatibility complex (genetics) The system whereby the body recognizes as foreign any tissue other than its own. It is medically important in that differences between donor and recipient lead to the rejection of *transplants*. Each of the individual's own cells has on its surface a number of specific proteins, the histocompatibility antigens (also known as HLA, human leucocyte antigens), of which there are six different types, each with dozens of allelic variants. The probability of any two people other than identical twins having exactly the same combination is correspondingly remote.

histogram (statistics) A chart containing adjacent parallel bars, usually vertical, whose lengths represent the frequency distribution on a continous dimension.

histology (histology) The study of the structure of tissue, particularly at the microscopic level.

histomonad (parasitology) A member of the genus *Histomonas.*

Histomonas (parasitology) A genus of amoeboid (*USA*: ameboid), single flagellate protozoa in the family Monocercomonadidae.

Histomonas meleagridis (avian parasitology) A species of amoeboid (*USA*: ameboid), single flagellate protozoa in the genus *Histomonas* that may be found in many species of birds. It is the cause of histomoniasis in turkeys and is transmitted by the nematode *Heterakis gallinarum*. Chickens are resistant carriers.

Histomonas wenrichi (parasitology) *See Parahistomonas wenrichi.*

histomoniasis (avian parasitology) A very widespread disease of gallinaceous birds, especially turkeys, caused by the protozoan parasite *Histomonas meleagridis*. It is characterized by ulceration of the caecal (*USA*: cecal) mucosa and necrotic foci in the liver.

histone (biochemistry) One of a group of small proteins with a large proportion of basic amino acids, *e.g.* arginine or lycine. Histones are found in combination with nucleic acid in the chromatin of eukaryotic cells.

Hisudo (parasitology) A genus of leeches in the class Hirudinea.

Hisudo medicinalis (parasitology) A species of leech in the genus *Hisudo* that is up to 5 inches long and 1 inch in diameter, that may attach to animals. It may be used medicinally.

hit (computing) The detection of a signal when it is there.

HIV (virology) An abbreviation of 'H'uman 'I'mmunodeficiency 'V'irus.

hives (immunology) Nettle rash or urticaria.

HIV-positive (virology) Presence of the human immunodeficiency virus in the body.

HLA (haematology) An abbreviation of 'H'uman 'L'eucocyte 'A'ntigen.

HLA system (immunology) An abbreviation of 'H'uman 'L'eucocyte 'A'ntigen system.

HMG-CoA reductase (biochemistry) An abbreviation of 3-'H'ydroxy-3-'M'ethyl'G'lutaryl 'Co'enzyme 'A' 'reductase'.

HMO (medical) An abbreviation of 'H'ealth 'M'aintenance 'O'rganisation.

HNC (education) An abbreviation of 'H'igher 'N'ational 'C'ertificate.

HND (education) An abbreviation of 'H'igher 'N'ational 'D'iploma.

hndbk (general terminology) An abbreviation of 'h'a'nd' 'b'oo'k'.

hnRNA (genetics) An abbreviation of 'h'eterogeneous 'n'uclear 'RNA'.

Ho (chemistry) The chemical symbol for holmium.

Hoareosporidium (parasitology) A genus of coccidia in the famiy Eimiriidae.

Hoareosporidium pellerdyi (veterinary parasitology) A species of coccidia in the genus *Hoareosporidium* that may be found in the tissues of dogs, and that causes mild inflammation.

HOC (chemistry) An abbreviation of 'H'eavy 'O'rganic 'C'hemical.

Hohorstiella lata (avian parasitology) A species of lice in the genus *Hohorstiella* that may be found on pigeons.

hol(o)- (terminology) A word element. [German] Prefix denoting *entire, whole.*

holandric (genetics) Occurring only in males, *i.e.* a gene that is on the Y chromosome.

hollow viscus (anatomy) Tubular as distinct from solid, *e.g.* intestine.

holmium (Ho) (chemistry) Silvery metallic element in Group IIIB of the periodic table (one of the lanthanides). At. no. 67; r.a.m. 164.9303.

holoenzyme (biochemistry) Enzyme that forms from the combination of a coenzyme and an apoenzyme. The former determines the nature and the latter the specificity of a reaction.

Holomenopon (avian parasitology) Bird lice of the superfamily Amblycera.

Holomenopon leucoxanthum (avian parasitology) A species of lice in the genus *Holomenopon* that parasitizes ducks causing soiling and loss of waterproofing of feathers that may lead to the bird becoming chilled and to die of pneumonia.

holomyarian (parasitological anatomy) Arrangement of the somatic musculature in a group of trematodes, including *Trichuris* spp., which have their body cavity completey surrounded by longitudinal muscle cells.

holophytic (parasitological physiology) One of the three forms of nutrition in Mastigophora protozoa. It comprises the synthesis of simple carbohydrates from carbon dioxide and water by chlorophyll contained in chloroplasts.

holoptic (insect anatomy) Said of eyes which are very close together as in some insects.

holozoic (parasitological physiology) One of the forms of nutrition in Mastigophora protozoa, comprising the capture and assimilation of organic materials in the environment.

homatropine (pharmacology) An alkaloid derived from atropine, used to dilate the pupil of the eye.

homeo- (terminology) A word element. [German] Prefix denoting *similar, same, unchanging.*

homeo. (medical) An abbreviation of 'homeo'pathic.

homeostasis (physiology) The tendency of a physiological system to maintain itself in balance, whatever the changes in the internal or external environment. From the Greek 'homoio', meaning 'equal', and 'stasis', meaning 'state'. Homeostasis is an evolved mechanism, operated by the switching off and on of many genes. An example is maintenance of body temperature or the balance of salts in the blood.

homeotic mutation (genetics) A mutation that replaces a normal structure with another structure that is also normal but is in the wrong place.

homidium (1 pharmacology; 2 methods) 1 A trypanocide that is used as the bromide or chloride salt. Also termed: ethidium. 2 A dye used for staining DNA fragments after electrophoretic separation in agarose gels. The molecule intercalates across the double helix and is detected by UV illumination.

homo- (1 terminology; 2 chemistry) 1 A word element. [German] Prefix denoting *same, similar.* 2 Chemical prefix indicating addition of one CH_2 group to the main compound.

homo sapiens (genetics) The species to which all human beings alive belong *Homo sapiens* has 22 pairs of autosomal chromosomes plus two sex chromosomes (XX or XY); the amount of DNA in each human somatic cell is about 6 picograms, and if fully extended would be about 1.9 metres long. The number of genes is estimated at somewhere in the region of 100,000.

homocyclic (chemistry) Describing a chemical compound that contains one or more closed rings comprising carbon and hydrogen atoms only, *e.g.* benzene. Also termed: carbocyclic.

See also heterocyclic.

homoeopathy (medical) System of alternative medicine that treats disorders by introducing substances into the body that provoke similar symptoms and so encourage the body's own defences. Medical opinion on the value of homoeopathy is divided. Also termed: homeopathy.

homogametic (genetics) Describing an organism with homologous sex chromosomes (*e.g.* XX). This sex can only produce one type of gamete, with a single X chromosome, and therefore does not determine the sex of the offspring. In humans the female is the homogametic sex. *See* heterogametic; Barr body.

homogeneity of variance (statistics) Similarity of variance, *i.e.* lack of a significant difference in variance. Most parametric tests are only valid if the samples being compared are from populations with the same variance.

homogeneous (1 biology; 2 chemistry; 3 mathematics) 1 In biology, describing similar structures found in different species that are thought to have originated from a common ancestor. 2 In chemistry, relating to a single phase (*e.g.* describing a chemical reaction in which all the reactants are solids, or liquids or gases); describing a system of uniform composition. 3 In mathematics, describing a polynomial whose terms all have the same degree; *e.g.* $2x^3 - 3x^2y + xy^2 + 5y^3$ is homogeneous (of degree 3).

See also heterogeneous.

homograft (medical) A graft taken from another individual of the same species. Also called an allograft.

homoiothermic (physiology) Describing animals that maintain a more-or-less constant body temperature (*e.g.* mammals, birds). Also termed: warm-blooded.

homologous (biology) Describing things with common origin, but not necessarily the same appearance or function (*e.g.* the arms of a human and the wings of a bat).

homologous chromosomes (genetics) the members of a pair of chromosomes that carry the same genes but not always the same alleles for a given character as the other member of the pair. All the autosomes come in homologous pairs. In the case of the sex chromosomes, the X and the Y are not homologous as they do not carry the same gene sequence, but two X chromosomes are homologous with each other. *See* meiosis.

homologous series (chemistry) Family of organic chemical compounds with the same general formula; *e.g.* alkanes, alkenes and alkynes.

homologue (1 chemistry; 2 biology) 1 In chemistry, member of a homologous series. 2 In biology, a homologous chromosome.

homolytic fission (chemistry) Breaking of a two-electron covalent bond in such a way that each fragment retains one electron of the bond. Also termed: homolytic cleavage.

See also heterolytic fission.

homozygote (genetics) An organism having two identical alleles at the same site on a pair of chromosomes.

homozygous (genetics) Having the same two alleles for a given gene. Two organisms that are homozygous for the same alleles breed true for the character in question, thus producing progeny which are homozygous and identical to the parent with respect to that gene.

hookworm (parasitology) Nematodes of the genera *Ancylostoma*, *Bunostomum*, *Gaigeria*, *Necator* and *Uncinaria*.

hookworm dermatitis (veterinary parasitology) Penetration of the skin by third-stage larvae of the hookworm species causes an inflammatory reaction. It can occur in many species, but is seen particularly in dogs on skin that comes in contact with the ground.

See also cutaneous larva migrans.

hookworm disease (parasitology) *See* bunostomiasis, ancylostomiasis, uncinariasis.

hookworm infestation (parasitology) Ancylostomiasis. Parasitisation by one of the roundworms (nematodes) *Ancylostoma duodenale* or *Necator americanus*.

hoose (parasitology) Lungworm disease.

Hoplodontophorus (veterinary parasitology) A genus of nematodes found in hyrax.

Hoplopleura (veterinary parasitology) A genus of sucking lice in the family Hoplopleuridae. Species includes *Hoplopleura acanthopus*, *Hoplopleura captiosa*, *Hoplopleura pacifica* that are sucking lice of rodents.

horizontal transmission (genetics) The transmission of DNA by viral infection into cells. Normal genetic transmission is vertical transmission. *See* transduction.

hormone (biochemistry) A chemical released by an endocrine gland and carried round in the bloodstream to alter the behaviour (*USA*: behavior) of other specific target cells or tissues. Steroid hormones are small lipid molecules that enter the target cells; they include the sex hormones. Other hormones are polypeptides (chains of amino acids shorter than proteins) which bind on to the outside of the target cell; they include growth hormone and insulin. Many hormones can be synthesised to provide treatment for patients with a deficiency; some can now also be produced by gene cloning.

horn fly (parasitology) *See Haematobia irritans*.

horse fly (veterinary parasitology) *See Tabanus*.

horse louse fly (veterinary parasitology) *See Hippobosca*.

horseradish peroxidase (biochemistry) An enzyme that is taken up by nerve cells and that flows in a retrograde direction from axon terminals to the cell body and dendrites; after being made radioactive, it can be used to identify the location of cell bodies whose axonal terminals are elsewhere.

horse-sick (parasitology) Said of a field that has had horses grazing on it for too long or at too heavy a stocking rate resulting in pasture that is too rank and clumpy and with a population of *Strongyle* spp. and other helminth larvae that is likely to be high.

hosp. (medical) An abbreviation of 'hosp'ital.

host (1 parasitology; 2 computing) 1 An animal or plant that harbours (*USA*: harbors) and provides sustenance for another organism (the parasite). Includes paratenic, intermediate, etc. 2 The computer that one dials up when operating from a remote terminal.

host determinants (parasitology) Characteristics in the host which determine its susceptibility to a disease.

host risk factors (epidemiology) Epidemiological factors contributing to the development of a disease and which are contributed by the host.

host specificity (parasitology) The characteristic of a parasite that renders it capable of infecting only one or more specific hosts.

host variable (parasitology) *See* host determinants.

host-parasite reaction (parasitology) The inflamm-atory reaction that sometimes occurs around a parasite in tissues, *e.g.* a warble fly larva in the oesophageal (*USA*: esophageal) wall.

host-parasite relationship (parasitology) A relationship that may be at any one of a series of classified levels in two groups, those of disease and symbiosis. In the disease category there are velogenic, mesogenic and lentigenic.

hot spot (molecular biology) A site in a DNA molecule where mutations or recombinations occur with much higher frequency than elsewhere.

house fly (parasitology) *See Musca domestica.*

house-dust mite (parasitology) *See Allodermanyssus, Dermatophagoides, Cheyletus eruditus, Tyrophagus farinae.*

housekeeping (computing) Terminology for tidying up storages, culling defunct files and records, backing-up, etc.

housekeeping genes (genetics) A gene that has a product essential for the normal metabolic requirements of any cell. These genes are active in most or all cells, in contrast to the cell-specific luxury genes. The parts of the chromosome where they are situated show up as light-staining interbands.

housemouse mite (veterinary parasitology) *See Allodermanyssus.*

Houttuynia (parasitology) A genus of tapeworms in the family Davaineidae.

Houttuynia struthionis (avian parasitology) A species of tapeworms in the genus *Houttuynia* that may be found in the intestine of the ostrich and rhea, and that causes unthriftiness and diarrhoea (*USA*: diarrhea) in chicks.

hPL (physiology/biochemistry) An abbreviation of 'h'uman 'P'lacental 'L'actogen.

HPLC (biochemistry) An abbreviation of 'H'igh 'P'erformance 'L'iquid 'C'hromatography.

HPR (parasitology) An abbreviation of 'H'ost 'P'arasite 'R'eaction.

HPRT (immunology) An abbreviation of 'H'ypoxanthine 'P'hospho 'R'ibosyl 'T'ransferase.

HSC (governing body) An abbreviation of 'H'ealth and 'S'afety 'C'ommission.

HS-CoA (physiology/biochemistry) a symbol for reduced coenzyme A.

HSS (society) An abbreviation of 'H'istory of 'S'cience 'S'ociety.

ht (measurement) An abbreviation of 'h'eigh't'. Also termed: hgt.

H-T, HT (measurement) An abbreviation of 'H'igh-'T'emperature.

human equivalent dose (pharmacology/toxicology) A dose which, when administered to humans, produces an effect equal to that produced by a given dose in another animal.

human leucocyte antigens (HLA) (haematology) Protein markers of self used in histocompatibility testing. It plays a most important part in determining whether transplanted tissues will be rejected by the recipient, and some HLA types also correlate with certain auto-immune diseases. *E.g.* ankylosing spondylitis is 120 times more likely to occur in people with allele B27.

humerus (anatomy) Upper bone in the forelimb of a tetrapod vertebrate; in human beings, the upper arm bone.

humi. (physics) An abbreviation of 'humi'dity.

humidity (physics) Measure of the amount of water vapour in a gas, *e.g.* the air, usually expressed as a percentage. Relative humidity is the amount of vapour divided by the maximum amount of vapour the gas will hold (at a particular temperature).

humoral immunity (immunology) The production of antibodies for defense against infection or disease.

hundredweight (cwt) (measurement) Unit of mass equal to 112 lb (100 lb in the United States). 20 cwt = ton.

HV (physics) An abbreviation of 'H'igh 'V'oltage.

HVA (biochemistry) An abbreviation of 'H'omo-'V'anillic 'A'cid.

hyal(o)- (terminology) A word element. [German] Prefix denoting *glassy*.

Hyalomma (parasitology) A genus of ticks in the family Ixodidae. The several stages of these ticks transmit *Babesia caballi, Babesia equi, Theileria parva, Theileria annulata, Theileria dispar, Coxiella burnetii, Rickettsia bovis* and *Rickettsia conori.* There are significant differences of opinion about the nomenclature of these ticks. A further subdivision creating new genera of *Hyalommina* and *Hyalommosta* is proposed.

See also sweating sicknes.

Hyalommina (parasitology) Suggested new genus, subdivision of *Hyalomma* spp.

Hyalommosta (parasitology) Suggested new genus, subdivision of *Hyalomma* spp.

hyaluronidase (biochemistry) A naturally occurring enzyme which has the property of breaking down hyaluronic acid.

Hybomitra (parasitology) A genus of the fly family Tabanidae that cause significant insect worry and blood loss and are capable of transmitting diseases such as anaplasmosis. Also termed horse fly.

hybrid (genetics) The offspring of parents that are genetically unalike, particularly of parents of different species or of different varieties within a species.

hybrid DNA (molecular biology) DNA that is artificially made *in vitro* by mixing two strands of DNA (or one of DNA and one of RNA) of different genetic origins. It is not the same as recombinant DNA.

hybrid vigour (genetics) The phenomenon in which the hybrid is more vigorous than either of its parental strains resulting from an increase in genetical variation.

hybridization (1, 2 genetics) 1 Crossing of animals or plants to produce a hybrid. 2 Combination of atomic orbitals to produce hybrid orbitals.

hybridoma (immunology/genetics) An artificially produced hybrid cell, combining a normal lymphocyte and a myeloma (cancerous) lymph cell, and used in the preparation of monoclonal antibodies.

hydatid (parasitological physiology) 1 A hydatid cyst 2 Any cyst-like structure.

hydatid cyst (parasitological physiology) The larval stage of the tapeworm *Echinococcus granulosus* or *Echinococcus multilocularis*, containing daughter cysts, each of which, if fertile, will have many proto-scoleces; it is the cause of hydatid disease. Also termed *Echinococcus* cyst; hydatid.

hydatid disease (human/veterinary parasitology) An infection in humans, sheep, cattle, pigs and horses, and occasionally in many other mammal species. The infection is usually of the liver or lungs, caused by larval forms (hydatid cysts) of tapeworms of the genus *Echinococcus*, and characterized by the development of expanding cysts. In the infection caused by *Echinococcus granulosus*, single or multiple cysts that are unilocular in character are formed, and in that caused by *Echinococcus multilocularis*, the host's tissues are invaded and destroyed as the cyst or cysts enlarge by peripheral budding. Also termed echinococcosis.

hydatid sand (parasitological physiology) Free protoscoleces lying inside a hydatid.

hydatid tapeworm (parasitology) *Echinococcus granulosus*.

hydatidiform (parasitology) Resembling a hydatid.

hydatidocele (parasitology) A tumour (*USA*: tumor) of the scrotum containing hydatids.

hydatidoma (parasitology) A tumour (*USA*: tumor) containing hydatids.

hydatidosis (parasitology) Hydatid disease.

hydatiduria (parasitology) Excretion of hydatid cysts in the urine.

hydatigera (parasitology) *See Taenia.*

hydr(o)- (terminology) A word element. [German] Prefix denoting *hydrogen*, *water*.

hydrargyrum (chemistry) Mercury.

hydrate (chemistry) Chemical compound that contains water of crystallization.

hydrated (1, 2 chemistry) 1 Describing a substance after treatment with water. 2 Describing a compound that contains chemically bonded water, a hydrate.

hydrated lime (chemistry) An alternative term for calcium hydroxide.

hydration (chemistry) Attachment of water to the particles (particularly ions) of a solute during the dissolving process.

hydrazine (chemistry) (NH2NH2) Colourless (*USA*: colorless) liquid, a powerful reducing agent used in organic synthesis and as a rocket fuel.

hydrazone (chemistry) Member of a family of organic compounds that contain the group -C= NH_2. Hydrazones are formed by the action of hydrazine on an aldehyde or ketone, and are used in identifying them.

hydride (chemistry) Compound formed between hydrogen and another element (*e.g.* calcium hydride, CaH_2).

hydro- (biology) Prefix denoting water.

hydrobromic acid (chemistry) (HBr) Colourless (*USA*: colorless) acidic aqueous solution of hydrogen bromide; its salts are bromides.

hydrocarbon (chemistry) Organic compound that contains only carbon and hydrogen. The chief naturally occurring hydrocarbons are bitumen, coal, methane, natural gas and petroleum. Most of these, and hydrocarbons derived from them, are used as fuels. The aliphatic compounds form three homologous series: alkanes, alkenes and alkynes. Aromatic hydrocarbons are cyclic compounds. *See* aromatic compounds.

hydrocephalus (medical) An abnormal amount of cerebrospinal fluid within the ventricles of the brain, eventually leading to malformation of brain tissue.

hydrochloric acid (chemistry) HCl Colourless (*USA*: colorless) acidic aqueous solution of hydrogen chloride, a strong acid that dissolves most metals with the release of hydrogen; its salts are chlorides. It is contained in normal stomach juices in a diluted form.

hydrocortisone (pharmacology) A cortisone molecule with a hydroxyl group at carbon atom 11, which administered by intravenous injection, can be used to treat inflammation. Also termed: cortisol.

hydrocyanic acid (chemistry) (HCN) Very poisonous solution of hydrogen cyanide in water, its salts are cyanides. Also termed: prussic acid.

hydrofluoric acid (chemistry) (HF) Colourless (*USA*: colorless) corrosive aqueous solution of hydrogen fluoride; its salts are fluorides. It is used for etching glass (and must be stored in plastic bottles).

hydrogen (H) (chemistry) Gaseous element usually given its own place at the beginning of the periodic table, but sometimes assigned to Group IA, Colourless

(*USA*: colorless), odourless (*USA*: odorless) and highly inflammable, it is the lightest gas known and occurs abundantly in combination in water (H_2O), coal and petroleum (mainly as hydrocarbons) and living things (mainly as carbohydrates). It is also the major constituent of the Sun and other stars, and is the most abundant element in the Universe. Hydrogen is made commercially by electrolysis of aqueous solutions, by cracking of petroleum, by the Bosch process or as synthetic gas. In the laboratory, hydrogen is generally prepared by the action of a dilute acid on zinc. It has many uses: for hydrogenating (solidifying) oils, to make ammonia by the Haber process, as a fuel (particularly in rocketry), in organic synthesis, and as a moderator for nuclear reactors. In addition to the common form (sometimes called protium, r.a.m. 1.00797) there are two other isotopes: deuterium or heavy hydrogen (r.a.m. 2.01410) and the radioactive tritium (r.a.m. 3.0221). At. no. 1; r.a.m. (of the naturally occurring mixture of isotopes) 1.0080.

hydrogen bond (chemistry) Strong chemical bond that holds together some molecules that contain hydrogen, *e.g.* water molecules, which become associated as a result. A hydrogen atom bonded to an electronegative atom interacts with a (non-bonding) lone pair of electrons on another electronegative atom.

hydrogen bromide (chemistry) (HBr) Pale yellow gas which dissolves in water to form hydrobromic acid.

hydrogen chloride (chemistry) (HCl) Colourless (*USA*: colorless) gas which dissolves readily in water to form hydrochloric acid. It is made by treating a chloride with concentrated sulphuric acid (*USA*: sulfuric acid) or produced as a by-product of electrolytic processes involving chlorides.

hydrogen cyanide (chemistry) (HCN) Colourless (*USA*: colorless) poisonous gas, which dissolves in water to form hydrocyanic acid (prussic acid). It has a characteristic smell of bitter almonds.

hydrogen electrode (physics) Half-cell that consists of hydrogen gas bubbling around a platinum electrode, covered in platinum black (very finely divided platinum). It is immersed in a molar acid solution and used for determining standard electrode potentials. Also termed: hydrogen half-cell.

hydrogen fluoride (HF) (chemistry) Colourless (*USA*: colorless) fuming liquid, which is extremely corrosive and dissolves in water to form hydrofluoric acid.

hydrogen half-cell (physics) An alternative term for hydrogen electrode.

hydrogen halide (chemistry) Compound of hydrogen and a halogen; *e.g.* hydrogen fluoride, HF, hydrogen iodide, HI.

hydrogen ion (chemistry) H^+ Positively-charged hydrogen atom; a proton. A characteristic of an acid is the production of hydrogen ions, which in aqueous solution are hydrated to hydroxonium ions, H_3O^+. Hydrogen ion concentration is a measure of acidity, usually expressed on the pH scale.

hydrogen peroxide (chemistry) H_2O_2 Colourless (*USA*: colorless) syrupy liquid with strong oxidizing powers, soluble in water in all proportions. Dilute solutions are used as an oxidizing agent, disinfectant and bleach; in concentrated form it is employed as a rocket fuel.

hydrogen spectrum (physics) Spectrum produced when an electric discharge is passed through hydrogen gas. The hydrogen molecules dissociate and the atoms emit light at a series of characteristic frequencies.

hydrogen sulphate (chemistry) Acidic salt containing the ion HSO_4^-. Also termed: bisulphate (*USA*: bisulfate). (*USA*: hydrogen sulfate).

hydrogen sulphide (chemistry) H_2S Colourless (*USA*: colorless) poisonous gas with a characteristic smell (when impure) of bad eggs. It is formed by rotting sulphur (*USA*: sulfur) containing organic matter and the action of acids on sulphides (*USA*: sulfides).

hydrogen sulphite (chemistry) Acidic salt containing the ion HSO_3^-. Also termed: bisulphite (*USA*: bisulfite). (*USA*: hydrogen sulfite).

hydrogenation (chemistry) Method of chemical synthesis by adding hydrogen to a substance. It forms the basis of many important industrial processes, such as the conversion of liquid oils to solid fats.

hydrogencarbonate (chemistry) Acidic salt containing the ion HCO_3^-. Also termed: bicarbonate.

hydrolysis (biochemistry) Cleavage of a molecule by the addition of water, with a hydroxyl group (-OH) from the water taking part in the reaction; *e.g.* esters hydrolyse to form alcohols and acids.

hydrolytic enzymes (biochemistry) Any enzyme capable of splitting a molecule into components by inserting water. Hydrolytic enzymes include lysosomal acid hydrolases and a number of the enzymes in protein, carbohydrate, lipid and nucleic acid catabolism.

hydrometer (physics) Instrument for measuring the density of a liquid. It consists of a weighted glass bulb with a long graduated stem, which floats in the liquid being tested.

hydrophilic (biochemistry) Possessing an affinity for water. Refers to chemicals that are water-soluble or to the regions of chemicals that are polar and therefore attracted to water. Hydrophilic compounds do not diffuse easily through membranes.

hydrophobic (chemistry) Water-repellent; having no attraction for water.

hydrophobic binding (biochemistry) When two non-polar groups come together they exclude the water between them, and this mutual exclusion of water results in a hydrophobic interaction. In the aggregate they present the least possible disruption of interactions among polar water molecules, and thus can lead to stable complexes. Some authorities consider this a special case involving van der Waals' forces. The minimization of thermodynamically unfavorable contact of a polar grouping with water molecules provides the major stabilizing effect in hydrophobic interactions.

hydrops fetalis (medical) Accumulation of fluid in the body cavities and subcutaneous tissues of the fetus.

hydrosol (chemistry) Aqueous solution of a colloid.

Hydrotoea (parasitology) A genus of head flies in the family Muscidae that cause insect worry in all animal species.

hydrous (chemistry) Containing water.

hydroxide (chemistry) Compound of a metal that contains the hydroxyl group (-OH) or the hydroxide ion (OH^-). Many metal hydroxides are bases.

hydroxonium ion (chemistry) Hydrated hydrogen ion, H_3O^+.

hydroxybenzene (chemistry) An alternative term for phenol.

hydroxybenzoic acid (chemistry) An alternative term for salicyllic acid.

hydroxyl group (chemistry) (-OH) Group containing oxygen and hydrogen, characteristics of alcohols and some hydroxides.

hydroxypropionic acid (chemistry) An alternative term for lactic acid.

hydroxystilbamidine isethionate (pharmacology) A systemic antiprotozoal drug.

hygro- (terminology) A word element. [German] Prefix denoting *moisture*.

hygrometer (physics) Instrument for measuring the humidity of air, the amount of water vapour in the atmosphere.

hygroscope (physics) Instrument for indicating the humidity of air.

hygroscopic (chemistry) Having the tendency to absorb moisture from the atmosphere.

Hyl (biochemistry) An abbreviation of 'Hy'droxy'l'ysine.

hymenolepiasis (parasitology) Infection due to species in the tapeworm genus of *Hymenolepis*.

Hymenolepis (parasitology) A genus of cyclophyllidean tapeworms of the family Hymenolepididae.

Hymenolepis cantaniana (avian parasitology) A species of tapeworms in the genus *Hymenolepis* that may be found in the small intestine of chickens and other birds.

Hymenolepis carioca (avian parasitology) A species of tapeworms in the genus *Hymenolepis* that may be found in fowls.

Hymenolepis diminuta (human/veterinary parasitology) A species of tapeworms in the genus *Hymenolepis* that may be found in wild rodents and in humans.

Hymenolepis lanceolata (avian parasitology) A species of tapeworms in the genus *Hymenolepis* that may be found in ducks and geese.

Hymenolepis microstoma (veterinary parasitology) A species of tapeworms in the genus *Hymenolepis* that may be found in the duodenum, gall bladder and bile ducts of rodents.

Hymenolepis nana (human/veterinary parasitology) A species of tapeworms in the genus *Hymenolepis* that may be found in primates, rodents and humans. Also termed: dwarf tapeworm.

Hymenoptera (insect parasitology) An order of the class Insecta which includes the ants, wasps, hornets, bees, fireants and sawflies, that are characterized by two pairs of shiny, membranous wings.

Hymenoptera sting (parasitology) A cause of injury and sometimes serious toxic and hypersensitivity reactions, particularly in dogs, which may be local or systemic.

hyostrongylosis (veterinary parasitology) Disease of pigs caused by infestation with the worm *Hyostrongylus rubidus* which is characterized by loss of condition, anaemia (*USA*: anemia) and diarrhoea (*USA*: diarrhea).

Hyostrongylus (parasitology) A genus of gastric nematodes in the family Trichostrongylidae.

Hyostrongylus rubidus (veterinary parasitology) A species of nematodes in the genus *Hyostrongylus* that causes hyostrongylosis in pigs. The worms are found in the stomach of the pig.

Hyp (biochemistry) An abbreviation of 4-'Hy'droxy'p'roline.

hyp- (terminology) A word element. [German] Prefix denoting *abnormally decreased*, *deficient*, *beneath*, *under*.

hyper- (terminology) A word element. [German] Prefix denoting *abnormally increased*, *excessive*.

hyperaemia (haematology) Congestion of a part with blood. (*USA*: hyperemia).

hyperchlorhydria (biochemistry) Excess of hydrochloric acid in the stomach.

hyperparasite (parasitology) A parasite that preys on a parasite.

hyperphoshataemia (haematology/biochemistry) Elevated levels of phosphates in the blood. (*USA*: hyperphoshatemia).

hypersensitivity (immunology) Over-reaction by the immune system. It involves the mounting of a large-scale immunological defence against antigens that are in low concentration and are possibly not harmful to the organism in that low concentration though they would be at a higher one. *See* allergy.

hypertension (medical) High arterial blood pressure.

hypertext (computing) Information which contains both formatting instructions and links to other documents.

hypertext markup (computing) Language set of codes embedded in a hypertext document which give instructions to the browser about the formatting of the document. Also termed HTML.

hypertext transfer protocol (computing) Provides the set of rules used by one computer to obtain an HTML (hypertext markup language) document from another.

hypertrophy (histology) Increased size of an organ due to enlargement of individual cells.

hypn(o)- (terminology) A word element. [German] Prefix denoting *sleep, hypnosis.*

hypo (chemistry) Popular name for sodium thiosulphate (*USA*: sodium thiosulfate).

hypo- (terminology) A word element. [German] Prefix denoting *abrnormally decreased, deficient, beneath, under.*

Hypoaspis (insect parasitology) A species of mites in the Mesostigmata, that are parasitic and cleptoparasitic on the bees *Aphis*, Meliponinae and Bombinae.

hypobiosis (parasitological physiology) Arrested stage of development, as of nematode larvae in the gut mucosa of the definitive host.

hypocalcaemia (biochemistry) A low level of calcium in the blood.

hypochlorite (chemistry) Salt of hypochlorous acid (containing the ion ClO^-). Hypochlorites are used as bleaches and disinfectants.

hypochlorous acid (chemistry) (HOCl) Weak liquid acid stable only in solution, used as an oxidizing agent and bleach. Its salts are hypochlorites. Also termed: chloric(I) acid.

Hypochthoniidae (insect parasitology) Mites in the Orbibatidia (Cryptostigmata) that may be found on the ants *Eciton* but that are possible commensals.

Hypoderaerum (parasitology) A genus of the family Echinostomatidae of flukes.

Hypoderaerum conoideum (avian parasitology) A species of flukes that may be found in the terminal portion of the small intestine of pigeon, fowl and most aquatic birds. It causes localized enteritis in ducks.

Hypoderma (parasitology) A genus of flies of the family Oestridae, whose larvae invade tissues causing damage to tissue, and to the skin as they emerge through it.

Hypoderma actaeon (parasitology) A species of flies in the genus *Hypoderma*. A central European fly.

Hypoderma aeratum (veterinary parasitology) A species of flies in the genus *Hypoderma* that may be found in sheep and goats.

Hypoderma bovis (veterinary parasitology) Hairy flies about 0.7 inch long, which similarly to *Hypoderma lineatum* (synonym: *Hypoderma* lineata) parasitize cattle, bison and rarely horses. The larvae cause warbles under the skin and an emission puncture to mar the hide of the animal.

Hypoderma capreola (veterinary parasitology) A species of flies in the genus *Hypoderma* that may be found in roe deer.

Hypoderma crossi (veterinary parasitology) A species of flies in the genus *Hypoderma* that may be found in goats and sheep.

Hypoderma diana (veterinary parasitology) A species of flies in the genus *Hypoderma* that may be found in red deer and roe deer.

Hypoderma lineatum (veterinary parasitology) (synonym: *Hypoderma* lineata) Hairy flies about 0.7 inch long, which similarly to *Hypoderma bovis* parasitize cattle, bison and rarely horses. The larvae cause warbles under the skin and an emission puncture to mar the hide of the animal.

Hypoderma moschiferi (veterinary parasitology) A species of flies in the genus *Hypoderma* that may be found in musk deer.

Hypoderma silenus (veterinary parasitology) A species of flies in the genus *Hypoderma* that attacks horses and goats.

hypodermiasis (human/veterinary parasitology) A creeping eruption of the skin in humans and cattle caused by the larvae of *Hypoderma* spp.

hypodermis (parasitological anatomy) A term used in relation to nematodes. Epidermis. A layer between the cuticle and the somatic musculature.

hypoglycaemia (biochemistry) A low level of glucose in the blood and symptoms are likely to appear if the level falls below 2.5 mmol/1. By far the most common cause of this is failure in a diabetic to cover the dose of insulin or similar drug with food, although there are a number of rare conditions that can lower the blood sugar. (*USA*: hypoglycemia).

hypokalaemia (biochemistry) Abnormally low level of potassium in the blood. (*USA*: hypokalemia)

hyponatraemia (biochemistry) Abnormally low level of sodium in the blood. (*USA*: hyponatremia).

hypopharynx (parasitological anatomy) A term used in relation to arthropods. Tongue-shaped structure which lies between the labrum and the labium. It is connected with the salivary duct.

hypophysis (anatomy) An alternative term for pituitary.

hypoplasia (histology) Underdevelopment of a tissue or organ.

hypostome (insect anatomy) An appendage on the ventral aspect of the oral opening of some insects and arachnids.

hyposulphuric acid (chemistry) An alternative term for dithionic acid. (*USA*: hyposulfuric acid).

hypotension (medical) Abnormally low blood pressure.

hypoth. (biology) An abbreviation of 'hypoth'esis.

hypothalamic-releasing factors (HTRF) (biochemistry) Hormones produced by nerve cells in the hypothalamus that selectively promote the release of five pituitary hormones adrenocorticotropin, luteinizing hormone, thyrotropin, follicle-stimulating hormone, and growth hormone.

hypothesis (philosophy) Provisional scientific explanation that is unproved, but either thought of as probably true or used as a basis for further investigation.

hypotonic (biochemistry) Having a lower concentration of a solute, used of one solution with reference to another. *Compare* hypertonic.

hypotrophy (medical) *See* atrophy.

hyps(o)- (terminology) A word element. [German] Prefix denoting *height*.

Hyptiasmus (parasitology) A genus of flukes (digenetic trematodes) in the family Cyclocoelidae.

Hyptiasmus tumidus (avian parasitology) A species of flukes (digenetic trematodes) in the genus *Hyptiasmus* that causes rhinitis and sinusitis in ducks and geese.

hyster(o)- (terminology) A word element. [German] Prefix denoting *uterus*, *hysteria*.

hysterosalpingogram (radiology) X-ray examination to outline the cavity of the uterus and lumina of the fallopian tubes by the injection of radio-opaque dye.

hysteroscopy (medical) Inspection of the uterine cavity through a fibreoptic (*USA*: fiberoptic) scope.

Hystrichis (avian parasitology) A genus of nematodes of the family Dioctophymatidae that may be found in aquatic birds.

Hystrichis tricolor (avian parasitology) A species of nematodes in the genus *Hystrichis* that causes nodules and some destruction of tissue in the proventriculus of ducks.

Hz (measurement) An abbreviation of 'H'ert'z'.

I

I (1 measurement; 2 biochemistry) 1 An abbreviation for 'I'ntensity. 2 A symbol for 'I'soleucine.

I (chemistry) Chemical symbol for iodine.

I Biol (society) Institute of Biology.

I cells (physiology) A symbol for Inspiratory neurons.

i.e. (literary terminology) An abbreviation of 'i'd 'e'st. Latin, meaning 'that is'. Used to clarify or restate what has been said or written.

-ia (terminology) A word element. 1 Suffix denoting certain classes in animal taxonomy. 2 Suffix denoting a disease or pathological condition.

IAMS (society) An abbreviation of 'I'nternational 'A'ssociation of 'M'icrobiological 'S'ocieties/'S'tudies.

IAP (education) An abbreviation of 'I'nternational 'A'cademy of 'P'athology.

-iasis (terminology) A word element. [German] Suffix denoting condition, state.

IATA (association) An abbreviation of 'I'nternational 'A'ir 'T'ransport 'A'ssociation.

iatr(o)- (terminology) A word element. [German] Prefix denoting medicine, physician.

-iatric (terminology) Suffix denoting a specified form of medical treatment or care.

iatrogenic (medical) Caused by medical treatment or mistreatment.

IBID (literary terminology) Latin, meaning, 'in the same place'. Refers to the source most recently cited. Found in footnotes and text of dissertations, theses, papers, and the like.

ibotenic acid (pharmacology) The insecticidal agent in the mushroom *Amanita muscaria*.

ibp (physics) An abbreviation for 'i'nitial 'b'oiling 'p'oint.

ICAM (cytology) An abbreviation of 'I'nter'C'ellular 'A'dhesion 'M'olecule.

ICD (taxonomy) An abbreviation for 'I'nternational 'C'lassification of 'D'iseases (WHO).

ice (chemistry) Water in its solid state (*i.e.* below its freezing point, 0°C). It is less dense than water, because hydrogen bonds give its crystals an open structure, and it therefore floats on water. This also means that water expands on freezing.

ice point (measurement) Freezing point of water, 0°C, used as a fixed point on temperature scales.

Iceland spar (chemistry) Very pure transparent form of calcite (calcium carbonate), noted for the property of double refraction.

Ich (fish parasitology) *See Ichthyophthirius multifiliis* and *Cryptocaryon irritans*.

Ichthyoboda (fish parasitology) A genus of small, flagellate protozoan parasites of the skin of freshwater and marine fish which cause a grey (*USA*: gray) discolouration (*USA*: discoloration) of the skin and often respiratory distress.

ichthyoid (biology) Fishlike.

ichthyology (biology) The study of fishes.

ichthyophagous (biology) Eating or subsisting on fish.

Ichthyophthirius (fish parasitology) A genus of protozoan parasites in the phylum Ciliophora.

Ichthyophthirius multifiliis (fish parasitology) A species of protozoan parasites in the genus *Ichthyophthirius* that causes white spot disease or Ich in freshwater fish in aquariums and hatcheries which is characterized by white nodules on the skin. Nodules may coalesce and cause sloughing of the skin and the gills may also be affected leading to respiratory distress. Many affected fish die.

Ichthyosporidium (fish parasitology) A genus of obligate, intracellular, protozoan parasites in the class Microsporea.

Ichthyosporidium giganteum (fish parasitology) A species of obligate, intracellular, protozoan parasites of the genus *Ichthyosporidium* which may be found in the connective tissue of the body wall of fish, causing large ventral swellings which are full of cysts.

ICSH (biochemistry) An abbreviation for 'I'nterstitial 'C'ell-'S'timulating 'H'ormone.

Ictalurus (fish/veterinary parasitology) A genus of catfish located in the USA that are intermediate hosts for parasites of animals.

Ictalurus melas (fish/veterinary parasitology) The black bullhead catfish, paratenic host for *Dioctophyme renale*.

Ictalurus nebulosa (fish/veterinary parasitology) The brown bullhead catfish, the paratenic host for *Dioctophyme renale*.

ICU (medical) An abbreviation of 'I'ntensive 'C'are 'U'nit.

id (measurement) An abbreviation of 'i'nner 'd'iameter.

-id (terminology) [German] Suffix denoting *resembling*, or *having the shape of*.

ID (parasitology) An abbreviation of 'I'nfective 'D'ose.

Id (parasitology) An abbreviation of 'Id'iotype.

ID$_{50}$ (parasitology) A term used to denote the median infective dose; the dose that will infect 50% of the experimental group.

idio- (terminology) A word element. [German] Prefix denoting self, peculiar to a substance or organism.

Idiogenes (avian parasitology) A genus of tapeworms of bustards.

idiogram (genetics) A formal representation of all the chromosomes in an individual's karyotype, usually consisting of photographed chromosomes from a single cell-division, which have been cut out and arranged in order.

idiopathic (medical) A term applied to diseases when their cause is unknown, or of spontaneous origin.

idiopathic hyperesthesia syndrome (veterinary parasitology) Dogs and especially cats may show an increased sensitivity to being touched or handled, with intense chewing or licking over the sensitive area, which is commonly the back of one or more limbs. No dermatitis is present initially, but it does develop with continuing self-trauma. The cause is usually unknown, but tapeworm infestation amongst others have been suggested. Also termed acral lick dermatitis, psychogenic alopecia, feline hyperesthesia syndrome.

idiopathy (medical) An undiagnosed morbid state arising without known cause.

idiotypes (immunology) The unique and characteristic parts of an antibody's variable region, which can themselves serve as antigens.

IDL (biochemistry) An abbreviation of 'I'ntermediate-'D'ensity 'L'ipoprotein.

IF (immunology) An abbreviation of 'I'nter'F'eron.

IFA (immunology) An abbreviation of 'I'mmuno'F'luorescent 'A'ntibody.

IFEMS (society) An abbreviation of 'I'nternational 'F'ederation of 'E'lectron 'M'icroscope 'S'ocieties.

IFN (physiology/biochemistry) An abbreviation of 'I'nter'F'ero'N'.

Ig (immunology) An abbreviation of 'I'mmuno-'g'lobulin.

IgA (immunology) An abbreviation of 'I'mmuno-'g'lobulin 'A'. *See* immunoglobulin.

IgD (immunology) An abbreviation of 'I'mmuno-'g'lobulin 'D'. *See* immunoglobulin.

IgE (immunology) An abbreviation of 'I'mmuno-'g'lobulin 'E'. *See* immunoglobulin.

IgG (immunology) An abbreviation of 'I'mmuno-'g'lobulin 'G'. *See* immunoglobulin.

IgM (immunology) An abbreviation of 'I'mmuno-'g'lobulin 'M'. *See* immunoglobulin.

IGR (insect biology) An abbreviation of 'I'nsect 'G'rowth 'R'egulator.

IHA test (methods) An abbreviation of 'I'ndirect 'H'aem'A'gglutination (*USA*: 'H'em'A'gglutination) test.

^{123}I-IMP (biochemistry) An abbreviation of '^{123}I'-labeled 'I'odoa'MP'hetamine.

IIT (methods) An abbreviation of 'I'ndirect 'I'mmunofluorescence 'T'est.

IJP (physiology) An abbreviation of 'I'nhibitory 'J'unction 'P'otential.

IL (physiology/biochemistry) An abbreviation of 'I'nter'L'eukin.

Ile (biochemistry) An abbreviation of 'I'so'le'ucine.

ile(o)- (terminology) A word element. [Latin] Prefix denoting ileum.

ileac (anatomy) 1 Pertaining to the ileum. 2 Of the nature of ileus.

ileal (anatomy) Pertaining to the ileum.

ileitis (medical) Inflammation of the ileum or distal portion of the small intestine, manifested by chronic or intermittent diarrhoea (*USA*: diarrhea) and weight loss.

ileocaecal (anatomy) Pertaining to the ileum and caecum (*USA*: cecum). (*USA*: ileocecal).

ileocolitis (medical) Inflammation of the ileum and colon.

ileocolostomy (medical) Surgical anastomosis of the ileum to the colon.

ileocolotomy (medical) Incision of the ileum and colon.

ileum (anatomy) Last section of the small intestine continuous with the jejunum above and the colon below, where both digestion and absorption take place.

See also duodenum; jejunum.

illuminance (E) (physics) The amount of light falling per unit area on a surface as measured by photometry; the standard unit of measurement is the lux.

illuminometer (physics) A device for measuring luminance by matching a comparison field of known brightness to the standard.

image (optics) Point from which rays of light entering the eye appear to have originated. A real image, *e.g.* one formed by a converging lens, can be focused on a screen; a virtual image, *e.g.* one formed in a plane mirror, can be seen only by the eyes of the observer and has no physical existence.

image converter (physics) Electron tube for converting infra-red or other invisible images into visible images.

imago (insect parasitology) Plural: imagoes, imagines [Latin] The adult or definitive form of an insect.

IMI (medical) An abbreviation of 'I'ntra'M'uscular 'I'njection.

imidazole (biochemistry) $C_3H_4N_2$ Aromatic heterocyclic compound whose ring contains three carbon atoms and two nitrogen atoms. Also termed: glyoxaline.

imidazole amines (biochemistry) A class of biogenic amines containing an imidazole ring, *e.g.* histamine.

imide (chemistry) Organic compound derived from an acid anhydride, general formula R-CONHCO-R', where R and R' are organic radicals. Also termed: imido compound.

imido compound (chemistry) An alternative term for imide.

imidocarb (pharmacology) An antiprotozoal agent, used as the hydrochloride or dipropionate, in the treatment of babesiosis and ehrlichiosis.

imido-urea (chemistry) An alternative term for guanidine.

imine (chemistry) Secondary amine, an organic compound derived from ammonia, general formula RNHR', where R and R' are organic radicals. Also termed: imino compound.

imino compound (chemistry) An alternative term for imine.

IML (physiology) An abbreviation of 'I'nter'M'edio'L'ateral gray column.

immature (parasitology) Unripe or not fully developed.

immediate (general terminology) Direct, precipitating or primary; having no intermediate stage or mechanism.

immediate hypersensitivity (immunology) Antibody-mediated hypersensitivity, i.e. Types I, II and III, characterized by a response that appears within minutes to hours, resulting either from a release of histamine and other mediators of hypersensitivity from IgE-sensitized mast cells, causing increased vascular permeability, oedema (*USA*:edema) and smooth muscle contraction (type I), from antibody-mediated lysis of red blood cells (type II), or from immune complex mediated pathology (type III).

immediate-type hypersensitivity reaction (immunology) *See* immediate hypersensitivity.

immersion (microscopy) The use of the microscope with the object and object glass both covered with a liquid.

immersion objective (microscopy) A lens designed to have its tip and the coverglass over the specimen connected by a liquid instead of air.

immiscible (chemistry) Describing two or more liquids that will not mix (when shaken together they separate into layers); *e.g.* oil and water.

immun. (immunology) An abbreviation of 'immun'ity/'immun'isation/'immun'ology.

immune (immunology) Being highly resistant to a disease because of the formation of humoral antibodies or the development of immunologically competent cells, or both, or as a result of some other mechanism, such as interferon activities in viral infections.

immune adherence (immunology) The binding of antibody-antigen-complement complexes to complement receptors found on red blood cells.

immune complex (immunology) A cluster of interlocking antigens and antibodies. *See* antibody-antigen complex.

immune complex disease (immunology) Disease induced by the deposition of or association with antigen-antibody-complement complexes in the microvasculature of tissues. Fixation of complement component C3 by the complexes initiates inflammation.

See also hypersensitivity.

immune complex reaction (immunology) Type III hypersensitivity.

immune cytolysis (immunology) Cell lysis produced by antibody with the participation of complement.

immune deficiency disease (medical) One in which animals have inadequate immune responses and so are more susceptible to infectious disease.

immune haemolysis (immunology) *See* immune-mediated haemolytic anaemia. (*USA*: immune hemolysis).

immune interferon (immunology) *See* interferon.

immune-mediated (immunology) Caused by an unspecified immune reaction.

immune modulator (immunology) *See* immunomodulation.

immune reaction (immunology) *See* immune response.

immune reaction fever (immunology) Aseptic fever occurring in anaphylaxis, angio-oedema (*USA*: angioedema).

immune response (immunology) The reaction of the immune system to invasion by a foreign substance (antigen). It involves the production of specific antibody molecules, that may be present in body fluids or carried by lymphocytes which combine with the antigen to form an antigen-antibody complex, cell-mediated immunity, development of hypersensitivity, or as immunological tolerance. Also termed immune reaction.

See also immunity.

immune status (immunology) The state of the body's natural defense to diseases. It is influenced by heredity, age, past illness history, diet and physical and mental health. It includes production of circulating and local antibodies and their mechanism of action.

immune system (immunology) The system by which the body overcomes infections and other invasions by foreign bodies. 'Natural', or 'non-specific', immunity works by phagocytosis and by the action of the protein interferon. 'Specific immunity', *i.e.* a response to one particular infective agent (all of which, whether viruses, bacteria, parasites, or non-living particles are known as antigens in this context), is most highly developed in vertebrates, including humans. It is of two kinds: 'humoral immunity', in which antibody molecules circulate in the lymph and blood; and 'cell-mediated' immunity, in which lymph cells directly bind to the antigens. In humoral immunity, the arrival of the antigen stimulates the appropriate line of B lymphocytes in the bone marrow; these cells, and their descendants in the lymph nodes and spleen, produce the specific antibody molecules which bind on to the antigens, either making them incapable of infective action or holding them in lumps so that they can be easily engulfed and dissolved by the large macrophage cells (this is facilitated by a group of proteins collectively known as complement, activated by the antibody-antigen complex). Cell-mediated immunity is similar, with the antigen binding being done by T lymphocytes which originate in the thymus gland. The humoral immunity produced by circulating antibody is most effective against infections by bacteria and viruses while they are outside the cells; cell-mediated immunity is better against viruses inside cells, parasites, cancer cells and foreign tissue. *See* immunisation.

immune tolerance (immunology) *See* immunological tolerance.

immunisation (immunology) The technique of artificially creating active immunity by injecting the patient with non-virulent antigens that resemble the disease-causing pathogens. A specific antibody or immunoglobulin is formed in response and may persist in the body, preventing further infection by the same organism. Some diseases have a more lasting immunity than others: the acquired immunity to measles and mumps is generally lifelong, but in some cases (*e.g.* flu) it is very shortlived. This is probably because of changes in the virus responsible, which affect its antigenic properties.The term is also used for giving passive immunity by the injection of antiserum. Immunisation can also mean any experimental process in which animals are injected with some substance to which they acquire immunity.

immunity (immunology) Protection by an organism against infection. Defence may be divided into passive and active mechanisms. Passive processes prevent the entry of foreign invasion, *e.g.* skin, mucous membranes. Active mechanisms include phagocytosis by leucocytes and the immune response in animals. Plants can have immunity, *e.g.* by means of phytoalexins.

immunize (immunology) To render immune.

immunoabsorbent (immunology) A preparation of antigen attached to a solid support or antigen in an insoluble form, which absorbs homologous antibodies from a mixture of immunoglobulins. *See* immunosorbent; immunoassay.

immunoassay (immunology) Any of a number of assays based on the binding of antibody to antigen. Often the antibody is linked to a marker such as a fluorescent molecule, a radioactive molecule or an enzyme. Immunoassays include radial immunodiffusion (RID) and immunoelectrophoresis (IEF), both of which utilise antibody incorporated into a uniform gel; radioimmunoassay (RIA) where the antibody is linked to a radioactive tracer; and enzyme-linked immunosorbent assay (ELISA) where the antibody is linked to an enzyme.

immunoaugmenting (immunology) To enhance immune response in a nonantigen specific manner by stimulating macrophage and reticuloendothelial function.

immunobiology (biology) That branch of biology dealing with immunological effects on such phenomena as infectious disease, growth and development, recognition phenomena, hypersensitivity, heredity, aging, cancer and transplantation.

immunoblastic (immunology) Pertaining to or involving the stem cells of lymphoid tissue.

immunoblot (methods) *See* Western blot.

immunochemistry (biology) The study of the chemical basis of immune phenomena and their interactions.

immunocompetent (immunology) Capable of developing an immune response

immunocomplex (immunology) *See* immune complex.

immunocompromised (medical) Having reduced immune responsiveness as a result of inherited defects or infection, particularly by retroviruses and herpesviruses or by administration of immunosuppressive drugs, by irradiation, by malnutrition, and by certain disease processes, *e.g.* cancer.

immunoconglutinin (immunology) Antibody formed against complement components that are part of an antibody-antigen complex, especially C3.

immunocyte (cytology) Any cell of the lymphoid series which can react with antigen to produce antibody or to participate in cell-mediated immunity. Also termed: immunologically competent cell.

immunocytoadherence (immunology) The aggregation of red blood cells to form rosettes around lymphocytes with surface immunoglobulins.

immunocytochemical staining (methods) The use of antibody to detect and stain particular cell types, *e.g.* immunoperoxidase staining.

immunocytochemistry (methods) The application of immunochemical techniques to cytochemistry; includes direct and indirect fluorescent antibody techniques.

immunodeficiency (immunology) A deficiency in the immune system, either mediated by antibody or T lymphocytes, or both.

immunodepression (immunology) An absence or deficient supply of the components of either humoral or cellular immunity, or both.

immunodiagnostic (immunology) Pertaining to diagnosis by immune reactions.

immunodiffusion (methods) The diffusion of antigen and antibody from separate wells, usually cut in agar, such that precipitation lines form in the agar between the wells.

immunodiffusion tests (methods) Techniques utilizing immunodiffusion, such as double immunodiffusion (Ouchterloney technique), single immunodiffusion (Oudin technique), and radial immunodiffusion.

immunodominance (immunology) The property of an antigenic determinant that causes it to be responsible for the major immune response in a host.

immunoelectrophoresis (methods) Separation, usually in an agar gel, of complex mixtures of antigens which then combine, following immunodiffusion, with antibody to form precipitation lines for each separated antigen.

immunofluorescence technique (immunology) A method of locating targets by linking them to antibody tagged with fluorescent dyes so that they can be made visible under a fluorescent-microscope.

immunofluorescent (chemistry) Having the characteristic of immunofluorescence.

immunofluorescent antibody test (methods) *See* fluorescence microscope.

immunofluorescent microscopy (microscopy) *See* fluorescence microscope.

immunogen (immunology) A substance that elicits an immune response.

immunogenic (immunology) Producing immunity; evoking an immune response.

immunogenicity (immunology) The ability of a substance to provoke an immune response or the degree to which it provokes a response.

immunoglobulin (biochemistry) Large globular proteins found in the body fluids. The basic unit of immunoglobulin structure consists of four polypeptide chains, two identical light chains and two identical heavy chains, which are disulphide bonded (*USA*: disulfide bonded) to form two identical antigen-binding (variable) regions. Humans possess five distinct classes of immunoglobulin (IgG, IgA, IgM, IgD, IgE) distinguishable by differences in the carboxy-terminal portion of the heavy chain. Immunoglobulin is a generic term. Immunoglobulins known to bind specifically with a particular antigen are referred to as antibodies.

See also antibody.

immunoglobulin deficiency (immunology) *See* immunodeficiency.

immunoglobulin genes (immunology) Genes that code for the light and heavy chains of immunoglobulins.

immunoglobulin superfamily (immunology) Immunoglobulins and a number of other proteins including T cell receptor, major histocompatibility molecules, T cell accessory proteins and some adhesion molecules that are structurally related to immunoglobulins.

immunogold (methods) A labelling technique for the detection and localization of particular antigens in a specimen.

immunohistochemical (methods) Denoting the application of antibody-antigen interactions to histochemical techniques, as in the use of immunofluorescence.

immunoincompetent (immunology) Lacking the ability or capacity to develop an immune response to antigenic challenge.

immunol. (immunology) An abbreviation of 'immunol'ogy.

immunological (immunology) Emanating from or pertaining to immunology.

immunological domains (biology) The structure of cell receptors is conveniently considered in terms of three functional domains: transmembrane, ligand binding and immunological; the latter contain primary antigenic regions.

immunological injury (medical) *See* immunopathy.

immunological reactions (immunology) *See* immune response, immunity.

immunological status (immunology) A reference to the immunocompetence of a host.

immunological tolerance (immunology) Specific nonreactivity of the immune system to a particular antigen, which is capable under other conditions of inducing an immune response.

immunologist (immunology) A specialist in immunology.

immunology (immunol.) (immunology) The scientific study of immunity, antigens and antibodies and their role in immune mechanisms and infection.

immunomodulating (immunology) Having the capacity for and used for immunomodulation.

immunomodulation (immunology) Adjustment of the immune response to a desired level, as in immunopotentiation, immunosuppression, or induction of immunological tolerance.

immunomodulator (immunology) Agents that alter the immune response by suppression or enhancement.

immunopathogenesis (immunology) The process of development of a disease in which an immune response or the products of an immune reaction are involved.

immunopathological (immunology) Emanating from or pertaining to immunopathogenesis.

immunopathology (medical) 1 That branch of biomedical science concerned with immune reactions associated with disease, whether the reactions are beneficial, without effect, or harmful. 2 The structural and functional manifestations of the immune response involved in a disease.

immunopathy (immunology) Any abnormal immune response.

immunoperoxidase staining (methods) A technique of histological staining that provides morphological details and immunological identification. Analogous to immunofluorescence techniques, but uses peroxidase conjugated to immunoglobulin instead of fluorescent dyes.

immunophysiology (immunology) The physiology of immunological processes.

immunopotency (immunology) The capacity of an individual antigenic determinant of an antigen molecule to initiate an immune response.

immunopotentiation (immunology) Accentuation of the response to an immunogen by administration of another substance, *e.g.* adjuvant.

immunoprecipitation (immunology) The precipitation resulting from interaction of antibody and soluble antigen.

immunoprophylaxis (immunology) Prevention of disease by the administration of vaccines or hyperimmune sera.

immunoradiometry (methods) The use of radiolabeled antibody, in the place of radiolabeled antigen, in radioimmunoassay techniques.

immunoreactant (immunology) A substance exhibiting immunoreactivity.

immunoreactive (immunology) Exhibiting immunoreactivity.

immunoregulation (immunology) The control of specific immune responses and interactions between B and T lymphocytes and macrophages.

immunoresponsiveness (immunology) The capacity to react immunologically.

immunosorbent (immunology) The binding of antigen or antibody to a solid support, usually a plastic surface at the bottom of a well in a microtitre plate or on to a latex particle.

immunostimulants (immunology) Agents that boost the natural immune response. Therapeutic immunostimulants have been used in cancer treatments to restore or augment the antitumor response.

immunostimulation (immunology) Stimulation of an immune response, *e.g.* by use of vaccine.

immunosuppressant (immunology) immunosuppressive.

immunosuppression (immunology) The state of an altered immune system that may lead to impaired immune function. Drugs such as cytotoxic drugs that inhibit cell division are powerful immunosuppressants and depression of immune function can lead to increased susceptibility to bacterial, viral and parasitic infections, and possibly increased incidence of neoplasm.

immunosuppressive (pharmocology) Describing a drug that suppresses the immune response, given to recipients of transplanted organs to minimize that chance of rejection.

immunosuppressor (immunology) A cell or soluble mediator that suppresses immune responses. Immunosuppressors may function to regulate immune responses, preventing excessive, potentially tissue-damaging responses.

immunosurveillance (immunology) The monitoring function of the immune system whereby it recognizes and reacts against aberrant cells arising within the body.

immunotolerance (immunology) *See* immunological tolerance.

immunotoxin (immunology) An antitoxin.

immunotransfusion (immunology) Transfusion of blood from a donor previously rendered immune to the disease affecting the recipient.

IMP (chemistry) An abbreviation of 'I'nosine 5-'M'ono'P'hosphate.

impact printer (computing) A printer in which a printhead strikes a carbon or cloth ribbon to produce an imprint. It may be a 'line' printer printing an entire line at a time or a 'character' printer printing one character at a time across the page, usually bidirectionally. The printhead comes in a variety of forms presenting print of differing legibility.

Imparipes (insect parasitology) A species of mites in the Prostigmata, that similarly to *Leptus* are parasitic to the bees *Apinae*, *Halictinae* and Bombinae, and are also parasitic on the ants, *Atta*, *Eciton*, and army ants.

impedance (*Z*) (physics) Property of an electric circuit or circuit component that opposes the passage of a current. For direct current (d.c.) it is equal to the resistance (*R*). For alternating current (a.c.) the reactance (*X*)also has an effect, such that $Z^2 = R^2 + X^2$, or $Z = R + iX$, where $i^2 = -1$.

imperial system (measurement) Comprehensive system of weights and measures (feet and inches, avoirdupois weights, pints and gallons, etc.) that was formerly used throughout the British Empire. Si units have replaced the imperial system for scientific measurement.

impermeable (physics) Describing a substance that will not allow a fluid (gas or liquid) to pass through it (*e.g.* granite is impermeable to water).

implantation (physiology) Process in which a fertilized ovum (egg) or embryo becomes attached to the lining of the uterus (womb) of a mammal. It is the beginning of pregnancy.

implicit function (mathematics) Variable *x* is an implicit function of *y* when *x* and *y* are connected by a function that is not explicit (*i.e.* in which *x* is not directly expressed in terms of *y*).

impression smear (methods) A smear made by pressing a laboratory glass slide against tissue.

improper fraction (mathematics) Fraction whose upper part (numerator) is larger than the lower part (denominator). *e.g.* 7/3, 11/4, 3/25. It is always greater than 1, as can be seen by converting it to a mixed number (the previous examples become 2 1/3, 2 3/4, 1 7/25).

impulse (1 biology; 2 physics) 1 In biology, transmission of a message along a nerve fibre (*USA*: nerve fiber). The nerve impulse is an electrical phenomenon which results in depolarization of the nerve membrane. This action potential lasts for a millisecond before the resting potential is restored. *See also* all-or-none response. 2 In physics, when two objects collide, over the period of impact there is a large reactionary force between them whose time integral is the impulse of the force (equal to either object's change of momentum).

IMR (statistics) An abbreviation of 'I'nfant 'M'ortality 'R'ate.

IMS (chemistry) An abbreviation of 'I'ndustrial 'M'ethylated 'S'pirit.

in situ (biology) In the normal or natural location.

in utero (biology) Within the uterus.

in vitro (biology) From the Latin, meaning 'in glass'. A term for processes that happen in a controlled environment outside of a living organism. It is, strictly speaking, an adverb but is often used adjectivally. *Contrast in vivo.*

in vivo (biology) From the Latin, meaning 'in a living thing'. A term used for processes that happen in a living organism. It does not imply that the process described is as it occurs in natural conditions, because the word is applied to what happens in experiments; it is used solely in contrast to what happens *in vitro*. It is strictly speaking an adverb, but is often used adjectivally. *Contrast in vitro.*

inactivated (biology) Rendered inactive; the activity is destroyed.

inactivation (biology) The destruction of activity, by the action of heat, chemical or other agent.

inapparent (general terminology) Not clearly seen.

inapparent infection (medical) Infection without clinical signs.

inbreeding (biology) Reproduction between closely related organisms of a species.

incandescence (physics) Light emission that results from the high temperature of a substance; *e.g.* the filament in an electric lamp is incandescent.

inch (measurement) Unit of length equal to 1/12 of a foot; 1 in = 25.4 mm.

incid. (general terminology) An abbreviation of 'incid'ental.

incidence (epidemiology) The rate of occurrence of new cases of a disorder over a given period of time, usually expressed as the number of new cases per 100,000 members of the population.

incidence data (epidemiology) Data related to the occurrence of specific disease incidents.

incidence reporting schemes (epidemiology) Prospective gathering of epidemiological data on incidence of nominated diseases.

incident (epidemiology) Impinging upon.

incidental parasite (parasitology) An accidental parasite.

incineration (chemistry) The act of burning to ashes.

incipient (biology) Beginning to exist; coming into existence.

incised wound (medical) A wound caused by a cutting instrument.

incision (medical) 1 A cut or a wound made by a sharp instrument. 2 The act of cutting.

incisional biopsy (histology) Biopsy of a selected portion of a lesion.

incisor (anatomy) Chisel-shaped cutting tooth of mammals located at the front of the upper and lower jaws; human beings have eight incisors. They grow continually in rodents, which use them for gnawing.

inclusion (cytology) 1 The act of enclosing or the condition of being enclosed. 2 Anything that is enclosed; a cell inclusion.

incoagulability (chemistry) The state of being incapable of coagulation.

incognito mange (veterinary parasitology) A form of sarcoptic mange in well-groomed dogs in which lesions are virtually absent, but there is still intense pruritus.

incompatibility (biology) The quality of being incompatible.

incompatible (biology) Not suitable for combination, simultaneous administration, or transplantation; mutually repellent.

incompetent (biology) Not able to function properly.

incomplete dominance (genetics) Condition in which neither of a pair of alleles completely masks the presence of the other phenotypically.

incomplete metamorphosis (parasitological physiology) A term used in relation to arthropods. Hemimetabolous. The morphological difference between adults and immature stages are mainly of size and sexual maturity. Each stage, therefore, resembles the other morphologically.

incr. (general terminology) An abbreviation of 'incr'ease.

incrustation (medical) 1 The formation of a crust. 2 A crust, scab or scale.

incubate (biology) 1 To subject to or to undergo incubation. 2 Material that has undergone incubation.

incubation (biology) 1 The provision of proper conditions for growth and development, *e.g.* as for tissue cultures. 2 Development of a disease by multiplication of an infectious agent within the host. 3 The development of the embryo in the eggs of oviparous animals. 4 The maintenance of an artificial environment for a neonate, especially a premature one.

incubation period (1 medical; 2 biology) 1 The time elapsing between infection with the organisms of a disease and the appearance of symptoms, e.g., with malaria, the interval between the entry of sporozoites and the first clinical manifestation. In any given disease it is relatively constant. People who have been exposed to an infectious disease and may be incubating it and who may become infectious, are known as contacts. 2 The time period that agar plates, tissue cultures, etc., are held in incubators before being examined.

incubator (equipment) An apparatus for maintaining optimal conditions (temperature, humidity, etc.) for growth and development, especially one used for cultures.

incus (anatomy) One of the ear ossicles. An alternative term: anvil.

indandione (control measures) Any of a group of synthetic anticoagulants derived from 1,3-indandione, including pindone and diphacinone, which impair the hepatic synthesis of the vitamin K-dependent coagulation factors (prothrombin, factors VII, IX and X), which may be used in human medicine as an anticoagulant and as a rodenticide in a manner similar to that of warfarin.

independence (statistics) Two variables are independent if the expected distribution of one variable is the same for each and every value of the other. Independence implies zero correlation, but the reverse is not necessarily true.

independent assortment (genetics) The second of Mendel's laws, which states that genes are transmitted independently from parents to offspring and assort freely. Thus, there is an equal chance of any particular gene being transmitted to the gametes. It does not apply to genes that exhibit linkage.

independent variable (mathematics) If y is a function of x, *i.e.* $y = f(x)$, x is the independent variable of the function (and y is the dependent one).

index (mathematics) An exponent or power; *e.g.* in the terms 5^4 and $x^{-1/2}$, 4 and 1/2 are indices. Multiplication is achieved by adding indices (*e.g.* $x^3 \times x^2 = x^{3+2} = x^5$); division by subtracting indices (*e.g.* $x^3 \mathbf{B} x^2 = x^{3-2} = x$).

index case (epidemiology) The first case recorded in an outbreak.

Index Medicus (medical) A monthly publication of the national library of medicine, in which the world's leading biomedical literature is indexed by author and subject.

Index Veterinarius (veterinary) A periodic listing of all publications in the veterinary literature by the Commonwealth Agricultural Bureaux, United Kingdom.

indicator (1 chemistry; 2 biology) 1 In chemistry, substance that changes colour (*USA*: color) to indicate the end of a chemical reaction or the pH of a solution; *e.g.* litmus. Indicators are commonly used in titrations in volumetric analysis. 2 In biology, organism that survives only in certain environments; its presence gives information about the environment *e.g.* the presence in water of certain bacteria that normally live in faeces (*USA*: feces) indicates that the water is polluted with sewage.

indigenous flora (biology) Normal or resident flora.

indigo (chemistry) $C_{16}H_{10}N_2O_2$ Blue organic dye, a derivative of indole, that occurs as a glucoside in plants of the genus Indigofera.

indirect (biology) Completed through an intermediate animal or function.

indirect cause (biology) All causes other than the direct cause.

indirect fluorescent antibody (methods) *See* fluorescence microscope.

indirect peroxidase technique (histology) Immunohistochemical technique for visualising material from frozen section.

indirect-transmission (epidemiology) The transmission of a disease to a susceptible person by means of vectors or by airborne route.

indium (In) (chemistry) Silvery-white metallic element of Group IIIA of the periodic table, used in making mirrors and semiconductors. At. no. 49; r.a.m. 114.82.

Indofilaria (parasitology) A genus of filaroid nematodes in the family Filariidae.

Indofilaria patabiramani (veterinary parasitology) A species of filaroid nematodes in the genus *Indofilaria* that causes dermatitis in elephants.

indole (biochemistry) C_8H_7N Colourless (*USA*: colorless) organic solid, a heterocyclic aromatic compound consisting of fused benzene and pyrrole rings, normally found in faeces (*USA*: feces) which is in part responsible for the characteristic smell. Also termed: benzpyrrole.

indole amines (biochemistry) A class of biogenic amine formed by the combination of indole with an amine group. They are neurotransmitters and include serotonin.

induced magnetism (physics) Creation of a magnet by aligning the magnetic domains in a ferromagnetic substance by placing it in the magnetic field of a permanent magnet or electromagnet.

induced radioactivity (physics) An alternative term for artificial radioactivity.

inducer (molecular biology) A molecule that binds to a repressor to turn on the transcription of an inducible operon. The inducer is usually a small molecule, smaller than the repressor, a protein, but able to alter the latter's configuration so that it dissociates from the operator region and allows transcription to proceed. The inducer is often the substrate of the enzyme whose production is being induced. *See* gene regulation.

inducible enzyme (molecular biology) An enzyme which is produced only when its substrate is present to act as inducer. The opposite is a repressible enzyme. *See* operon; gene regulation.

inductance (1 physics; 2 measurement) 1 Property of a current-carrying electric circuit or circuit component that causes it to form a magnetic field and store magnetic energy. 2 Measurement of electromagnetic induction.

induction (1, 2, 3 physics) Magnetization or electrification produced in an object. 1 Electromagnetic induction is the production of an electric current in a conductor by means of a varying magnetic field near it. 2 Magnetic induction is the production of a magnetic field in an unmagnetized metal by a nearby magnetic field. 3 Electrostatic induction is the production of an electric charge on an object by a charged object brought near it.

induction coil (physics) Type of transformer for producing high-voltage alternating current from a low-voltage source. Induction coils can be used to produce a high voltage pulse, *e.g.* for firing spark plugs in a petrol engine.

induction heating (physics) Heating effect that arises from the electric current induced in a conducting material by an alternating magnetic field.

inductor (1 chemistry; 2 physics) 1 Substance that accelerates a chemical reaction between two other substances by reacting rapidly with one of them. 2 Any component of an electrical circuit that possesses significant inductance. Also termed: choke; coil.

industrial organophosphorus compound (control measures) *See* organophosphorus compound.

Inermicapsifer (parasitology) A genus of tapeworms in the family Anoplocephalidae.

Inermicapsifer cubensis (veterinary/human parasitology) A species of the genus *Inermicapsifer* which may be found in rodents and hyracoids and occasionally in humans.

Inermicapsifer madagascariensis (veterinary/human parasitology) A species of the genus *Inermicapsifer* which may be found in rodents and hyracoids and occasionally in humans.

inert (chemistry) Chemically nonreactive, *e.g.* gold is inert in air at normal temperatures.

inert gas (chemistry) Member of Group 0 of the periodic table; the unreactive elements helium, neon, argon, krypton, xenon and radon. Also termed: noble gas; rare gas.

inertia (physics) Resistance offered by an object to a change in its state of rest or motion. Inertia is a property of the mass of an object.

infarction (medical) When an artery is suddenly blocked by thrombosis or by an embolus and there is no alternative circulation to keep the tissues nourished.

infection (parasitology) Invasion by and multiplication of micro-organisms in body tissue which may or may not result in overt disease. The organisms concerned are called pathogens, and include protozoa, and worms.

infection control (parasitology) The utilization of procedures and techniques in the surveillance, investigation and compilation of statistical data in order to reduce the spread of infection/infestation, particularly nosocomial infections.

infection rate (parasitology) Percentage of the population from which a specific pathogen is isolated.

infectious (parasitology) Caused by or capable of being communicated by infection/infestation.

infectious cyclic thrombocytopenia (veterinary parasitology) A condition in which recurring cycles of parasitemia and reduced numbers of thrombocytes in the peripheral blood are seen in dogs infected with *Ehrlichia platys*. Clinical signs are rarely observed, but coinfection may potentiate clinical disease caused by *Ehrlichia canis* which is transmitted by the brown dog tick *Rhipicephalus sanguineus*.

infectious disease (parasitology) A disease caused by small living organisms, including protozoa and metazoan parasites.

infectious myxomatosis of rabbits (veterinary parasitology) A highly infectious, mosquito and rabbit flea or contact transmitted, generalized disease caused by myxoma virus which is very similar to the rabbit fibroma virus.

infectious sinusitis (avian parasitology) A contagious disease of turkeys caused by *Mycoplasma gallisepticum*; the same infection also causes chronic respiratory disease of chickens.

infective endocarditis (medical) Infection of the lining of the heart, which may be acute or subacute.

inferential statistics (statistics) The branch of statistics that enables one to assess the validity of a conclusion drawn from the data, *e.g.* whether two means are significantly different from one another, or whether there is a significant correlation between two variables.

inferior (anatomy) Situated below, or directed downward; the lower surface of a structure, or the lower of two (or more) similar structures. In bipeds, it is usually synonymous with caudal, in quadripeds with ventral.

infestation (parasitology) The presence on or in the body of parasites mites and fleas, or other multicellular organisms, such as ticks; sometimes used to denote parasitic invasion of the organs and tissues, as by helminths.

infinity (mathematics) Quantity that is larger than any quantified concept. It may be considered as the reciprocal of zero.

infl. (general terminology) Abbreviation of 'infl'ammable.

inflammatory response (immunology) Redness, warmth and swelling in response to injury or infection; the result of increased blood flow and a gathering of immune cells and secretions. It involves dilation of blood vessels, migration of leucocytes to the site of injury, and movement of fluid and plasma proteins into the inflamed tissue.

inflection (optics) Point on a curve where it changes from being concave to convex (or convex to concave).

info. (general terminology) information

information retrieval (computing) Science of storing and accessing data, which may use microfilm, microfiche, magnetic tape and computer storage devices.

informed consent (biology) Agreement to take part in an experiment or medical trial, or to receive a medical procedure, with sufficient knowledge of the protocol and risks to reach a rational decision.

infra- (terminology) A word element. [Latin] Prefix denoting *beneath*.

infradian rhythm (biology) Any biological rhythm with cycles longer than a day, *e.g.* the menstrual cycle.

infra-red radiation (physics) Electromagnetic radiation in the wavelength range from 0.75μm to 1 mm approximately; between the visible and microwave regions of the electromagnetic spectrum. It is emitted by all objects at temperatures above absolute zero, as heat (thermal) radiation.

infrasound (physics) Sound waves with a frequency below the threshold of human hearing, *i.e.* less than about 20 Hz.

infusoria (veterinary parasitology) Protozoa of the class Infusoria; microscopic, aquatic, with vibratile cilia, which may be found in ruminant forestomachs.

inheritance (genetics) The acquisition of a characteristic through genetic factors or the characteristics so acquired.

inhibited larval development (parasitology) *See* hypobiosis.

inhibition (biology) The prevention of life or deactivation of a process.

inj (1, 2 medical) 1 An abbreviation of 'inj'ection. 2 An abbreviation of 'inj'ury.

inner diameter (i.d.) (measurement) The distance from the centre (*USA*: center) of the opening of a tube to the inside edge of the material the tube is made from.

inner ear (anatomy) Fluid-filled part of the ear that contains both the organs that convert sound to nerve impulses and the organs of balance.

innocent (histology) A term sometimes used of benign as opposed to malignant tumours (*USA*: maglignant tumors).

inoculation (medical) Accidental or intentional introduction of foreign matter into a living organism or culture medium. Used of the injection of vaccines to prevent disease. *See* vaccine.

inorganic chemistry (chemistry) Study of non-carbon based compounds and their reactions.

inorganic compound (chemistry) Compound that does not contain carbon, with the exception of carbon's oxides, metallic carbides, carbonates and hydrogencarbonates.

input (computing) Information entered into a computer.

input device (computing) Part of a computer that feeds in with data and program instructions. The many types of input devices include a keyboard, punched card reader, paper tape reader, optical character recognition, light pen (with a VDU) and various types of devices equipped with a read head to input magnetically recorded data (*e.g.* on magnetic disk, tape or drum).

input file (computing) The file to be processed by a programme (*USA*: program).

insect bites and stings (insect parasitology) Injuries caused by the mouth parts and venom of insects and of certain related creatures, known as arachnids i.e. spiders, scorpions, and ticks, but popularly classified with insects. Bites and stings can be the cause of much discomfort and a local infection can develop from scratching. Some insects establish themselves on the skin as parasites, others inject poison, and still others transmit disease.

insect (biology) Any individual of the class Insecta.

insect growth regulator (IGR) (control measures) Any of a number of substances found naturally in insects which regulate morphogenesis and reproduction. Synthetic chemicals with similar activity are used topically and in the environment to control ectoparasites, particularly fleas, as a larvicide and ovicide. Also termed: juvenoids.

See also methoprene; fenoxycarb.

insect larva (insect parasitology) The second stage in the standard insect life cycle, the maggot or caterpillar.

insect pupa (insect parasitology) The third stage in the insect life cycle. An inert, dormant stage from which the adult emerges.

insect vector (insect parasitology) Insects may carry infection mechanically on feet or mouthparts, by passage through the digestive tract but without the insect being infected, or by becoming an intermediate host with some part of the parasite's life cycle taking place in insect tissues.

insect worry (veterinary parasitology) Swarms of biting insects may cause sufficient worry to interfere with grazing and may cause the animals to lose weight.

Insecta (biology) A class of arthropods whose members are characterized by division of the body into three distinct regions: head, thorax, and abdomen, and that have three pairs of legs.

insecticide (control measures) Substance used to kill insects of agricultural or public health importance. There are two main types: those that are eaten (with food) or inhaled by insects, and those that kill by contact. Important classes of insecticides include: chlorinated hydrocarbons (including DDT analogs, chlorinated alicyclic compounds, cyclodienes and chlorinated terpenes); organophosphates; carbamates; thiocyanates; dinitrophenols; botanicals (including pyrethroids, rotenoids and nicotinoids); juvenile hormone analogs; growth regulators; and inorganics (including arsenicals and fluorides). Non-biodegradable insecticides may persist for a long time and become concentrated in food chains, where they have a damaging ecological effect.

insertion (molecular biology) The addition of one or more base-pairs into a DNA sequence. The effect of a small insertion may be a frameshift mutation, but a larger insertion can upset gene regulation around it, either by activating genes that should be inactive or vice versa. *See* insertional inactivation.

insertional inactivation (genetics) A technique used in genetic engineering to prevent expression of a gene by inserting a foreign DNA sequence adjacent to it or in its coding region.

insoluble (chemistry) Describing a substance that does not dissolve in a given solvent; not capable of forming a solution.

Inst. (education) An abbreviation of 'Inst'itute/ 'Inst'itution.

insulation (physics) Layer of material (an insulator) used to prevent the flow of electricity or heat.

insulator (physics) Substance that is a poor conductor of electricity or heat; a non-conductor. Most non-metallic elements (except carbon) and polymers are good insulators. The presence of entrapped air (as in a foam plastic or woollen garment) increases the effectiveness of a thermal insulator.

integer (mathematics) Whole number; it may be positive or negative.

integral (1, 2 mathematics) 1 Value that results from the process of integration. 2 Describing a whole number value.

integration (mathematics) In calculus, the process of summation of the series of infinitely small quantities that make up the difference between two values of a given function. It is the inverse of differentiation and is used in the solution of such problems as finding the area enclosed by a given curve or the volume enclosed by a given surface.

intensity (physics) Power of sound, light or other waveform (*e.g.* the loudness of a sound or the brightness of light), determined by the amplitude of the wave.

Intensive Care Unit (ICU) (medical) The section of a hospital where critically ill patients are monitored around the clock.

inter- (terminology) A word element. [Latin] Prefix denoting *between*.

interaction (physics) In atomic physics, exchange of energy between a particle and a second one or electromagnetic radiation.

interaction variance (statistics) The proportion of the total variance caused by the interaction of the independent variables.

interactive (biology) Characterized by interaction.

intercellular (biology) Between cells; *e.g.* intercellular fluid surrounds cells, maintaining a constant internal environment.

See also intracellular.

intercostal muscle (histology) Any of the muscles between a mammal's ribs, which are important in breathing movements.

intercross (genetics) A cross between two individuals or strains that are themselves hybrid. *See* F2.

intercurrent (medical) Term applied to a second disease or infection occurring during the course of the original disease.

interdigital dermatitis (veterinary parasitology) Inflammatory, usually moist, skin disease between the toes in dogs and sometimes cats, that is often associated amongst others with hookworm penetration.

interdigital pyoderma (veterinary parasitology) Infection of the interdigital skin in dogs that may be associated with *Demodex canis* infestation.

intereceptor (physiology) A receptor located in or on an internal organ.

interface (1 physics; 2 computing) 1 Boundary of contact (the common surface) of two adjacent phases, either or both of which may be solid, liquid or gaseous. 2 The point at which two systems, or two parts of one system, interact.

interference (physics) Interaction between two or more waves of the same frequency emitted from coherent sources. The waves may reinforce each other or tend to cancel each other; the resultant wave is the algebraic sum of the component waves. The phenomenon occurs with electromagnetic waves and sound.

interference microscope (microscopy) A microscope similar to the phase contrast microscope but that delivers a three-dimensional image. Also termed Nomarski interference phase microscope.

interferon (IF) (immunology) A protein produced by the body to counteract a viral infection by slowing or stopping the virus from replicating. Its action is not specific to any particular group of viruses. Genes for human interferons have been cloned so the proteins can be produced in bulk.

interkrometer (physics) Instrument that makes use of interference for measuring such things as wavelengths and other small distances.

intermediate (1, 2 chemistry) 1 In industrial chemistry, compound to be subjected to further chemical treatment to produce finished products such as dyes and pharmaceuticals. 2 Short-lived species in a complex chemical reaction.

intermediate compound (chemistry) Compound of two or more metals that are present in definite proportion, although they frequently do not follow normal valence rules.

intermediate host (parasitology) Required host in the life cycle in which essential larval development must occur before a parasite is infective to its definitive host or to additional intermediate hosts.

intermolecular force (physics) Force that binds one molecule to another. Intermolecular forces are much weaker than the bonding forces holding together the atoms of a molecule.

See also van der Waals' force.

internal carotid artery arteritis (veterinary parasitology) Parasitic arteritis of the external carotid artery as it courses around the edge of the guttural pouch may lead to copious nasal bleeding in horses when the artery ruptures.

internal conversion (physics) Efrect on the nucleus of an atom produced by a gamma-ray photon emerging from it and giving up its energy on meeting an electron of the same atom.

internal energy (physics) Total quantity of energy in a substance, the sum of its kinetic energy and potential energy.

internal friction *See* viscosity.

internal parasitic mites (avian parasitology) Poultry cutaneous mites in a number of families are found in the trachea, air sacs and subcutaneous tissues. Examples include *Cytodites*, *Epidermoptes*, *Laminosioptes*, *Pneumonyssus*, *Rivoltsia* and *Sternastoma* spp.

internal resistance (physics) Electrical resistance in a circuit of the source of current, *e.g.* a cell.

International Air Transport Association (IATA) (association) An association which sets the rules for air transport, including those concerning air transport of animals.

international candle (measurement) Former unit of luminous intensity.

International Classification of Diseases (ICD) (governing body) A classification, made by the World Health Organization, of diseases including mental disorders.

interphase (cytology) State of cells when not undergoing division. Preparation for division (mitosis) is carried out during this phase, including replication of DNA and cell constituents.

interpolation (statistics) Inferring a value or values for points on a dimension from the obtained values of points lying to each side. *Compare* extrapolation.

interpose (genetics) The stage on the cell cycle between divisions. This is the phase during which the chromosomes are active in synthesising proteins.

interpreter (computing) A program that translates another program written in a high-level language into machine code and executes each instruction before translating the next. An interpreter is slower in execution than a compiler and the user's program must be translated afresh every time it is run.

interquartile range (IQR) (statistics) In a frequency distribution, the difference between the upper and lower quartiles; hence it contains the middle 50 per cent of cases.

intersegmental tracts (anatomy) Short tracts connecting different segments of the spinal cord.

interstitial atom (chemistry) Atom that is in a position other than a normal lattice place.

interstitial cells (cytology) In mammals, cells present in the male and female gonads. In males they are found between the testis tubules and in females in the ovarian follicle. When stimulated by luteinizing hormone, they produce androgens in males and oestrogen (*USA*: estrogen) in females.

interstitial cell-stimulating hormone (ICSH) (biochemistry) A gonadotropin from the adenohypophysis that stimulates the testicular interstitial cells to produce androgens. It is equivalent to luteinizing hormone (LH) in the female. Its secretion is stimulated by hypothalamic gonadotropin releasing hormone (GnRH).

interstitial compound (chemistry) Chemical compound formed by penetration of non-metallic atoms of small diameter between the atoms of a metal (usually a transition element).

interstitial fluid (histology) The extracellular fluid bathing the cells in must tissues, excluding the fluid within the lymph and blood vessels.

interval scale (measurement) A scale having equal intervals, but no zero point *e.g.* Centigrade.

intervening variable (statistics) A variable operating within a theoretical model that mediates the effects of the independent variables on the dependent variables. There may be several intervening variables all interacting with one another. The value of an intervening variable is usually affected by more than one independent variable and it may in turn affect several dependent variables.

intestinal (anatomy) Pertaining to the intestine.

intestinal eosinophilic granuloma (parasitology) *See Angiostrongylus costaricensis.*

intestinal juice (biochemistry) The liquid secretion of glands in the intestinal lining.

intestinal parasitism (parasitology) Infestation of the intestinal lumen and wall by nematodes, cestodes and immature trematodes.

intestine (anatomy) The alimentary canal after it leaves the stomach. It is divided into a small and a large intestine; the small intestine is the longer and is continuous with the stomach at the gastro-duodenal junction. The small intestine is concerned with the further digestion of food and the absorption into the bloodstream of already digested amino acids and monosaccharides. The large intestine is mainly concerned with the absorption of water from semi-solid indigestible remains, which form the faeces (*USA*: feces).

intestinum (anatomy) Plural: *intestina* [Latin] denoting *intestine.*

intestinum crassum (anatomy) Large intestine.

intestinum tenue (anatomy) Small intestine.

intima (cytology) The innermost coat of a blood vessel. Also termed: tunica intima.

intimal (anatomy) Pertaining to or emanating from vascular intima.

intimal tracks (veterinary parasitology) Aggregations of cellular debris, fibrin and inflammatory cells, especially eosinophils, in the form of tortuous tracks in the intima of arteries in horses which are caused by migrating larvae of *Strongylus vulgaris.*

intra- (terminology) A word element. [Latin] Prefix denoting *inside of, within.*

intra vitam (terminology) [Latin] Denoting *during life.*

intra-abdominal (biology) Within the abdomen.

intra-aortic (biology) Within the aorta.

intra-arterial (biology) Within an artery.

intra-articular (biology) Within a joint.

intracanalicular (biology) Within a canaliculus.

intracapsular (biology) Within a capsule.

intracardiac (biology) Within the heart.

intracartilaginous (biology) Within a cartilage.

intracellular (cytology) Occurring within the boundary of a cell or cells.

See also intercellular.

intracellular fluid (cytology) The fluid contained within the cell membrane.

intracervical (biology) Within the canal of the cervix uteri.

intrachromosomal recombination (genetics) Exchange of sequences of DNA between sister chromatids (*i.e.* two halves of the same chromosome) during meiosis. Because the sister chromatids are identical, there is no genetic effect, unlike in the normal process of recombination, when non-sister chromatids from homologous chromosomes exchange genetic material.

intracisternal (biology) Within a subarachnoid cistern.

intracornual (biology) Within the horns of the uterus.

intracranial (biology) Inside the skull.

intracutaneous (biology) Within the substance of the skin.

intracystic (biology) Within a bladder or a cyst.

intradermal (biology) In the skin.

intradermal test (methods) *See* skin test.

intraductal (biology) Within a duct.

intradural (biology) Within or beneath the dura mater.

intraepidermal (biology) Within the epidermis.

intraepidermal lymphocytes (immunology) Cellular components of the cutaneous immune system.

intraepithelial (biology) Within epithelial cells.

intraepithelial lymphocytes (immunology) Cellular components of the mucosal immune system.

intrafusal (anatomy) Pertaining to the striated fibres (*USA*:fibers) within a muscle spindle.

intrahepatic (biology) Within the liver.

intralobar (biology) Within a lobe.

intralocular (biology) Within the loculi of a structure.

intraluminal (biology) Within the lumen of a tubular structure.

intramammary (biology) Within or into the mammary gland.

intramuscular (biology) Inside a muscle.

intraocular (biology) Within the eye.

intraoral (biology) Within the mouth.

intraorbital (biology) Within the orbit.

intrapleural (biology) Within the pleura.

intraspinal (biology) Within the spinal column.

intrathecal (biology) Within a sheath; through the theca of the spinal cord into the subarachnoid space.

intrathoracic (biology) Within the thorax.

intratracheal (biology) Endotracheal.

intratympanic (biology) Within the tympanic cavity.

intrauterine (biology) Within the uterus.

intravascular (biology) Within a vessel or vessels.

intravenous (IV) (medical) Directly into a vein.

intraventricular (biology) Within a ventricle.

intravital (medical) Occurring during life.

intrinsic (biology) Situated entirely within, or pertaining exclusively to, a part.

intrinsic host determinants (epidemiology) Characteristics peculiar to the host that affect the spread and occurrence of a disease.

intubation (medical) The introduction of a tube into the body, used mostly of the process of inserting a tube into the trachea to maintain an airway during the administration of an anaesthetic (*USA*: anesthetic), or for the purpose of artificial respiration by a breathing machine.

invasive (parasitology) 1 Having the quality of invasiveness. 2 Involving puncture or incision of the skin or insertion of an instrument or foreign material into the body; said of diagnostic techniques.

invasiveness (parasitology) 1 The ability of microorganisms to enter the body and spread in the tissues. 2 The ability to infiltrate and actively destroy surrounding tissue.

inverse function (mathematics) Mathematical function of such a nature in respect to another operation, relation, etc. that the starting point of one function is the conclusion of the other, and vice versa. A function that is opposite in effect or nature, *e.g.* inverse hyperbolic functions and inverse trigonometrical functions.

inverse square law (physics) The law that the intensity of a wave, *e.g.* light or acoustic waves, travelling in a homogenous medium, decreases in proportion to the square of the distance from the source. For instance, in optics, the quantity of light from a given source on a surface of definite area is inversely proportional to the square of the distance between the source and the surface.

inversion (1 genetics; 2 chemistry) 1 A mutation consisting of the reversal of the order of the genes in part of a chromosome. 2 Splitting of dextrorotatory higher sugars (*e.g.* sucrose) into equivalent amounts of laevorotatory lower sugars (*e.g.* fructose and glucose).

invert sugar (chemistry) Natural disaccharide sugar that consists of a mixture of glucose and fructose, found in many fruits.

invertebrate (biology) Animal that does not possess a backbone.

See also vertebrate.

inverted-U curve (statistics) A curve that looks like a letter U upside-down, starting low, reaching a peak and then declining; when performance on a task is plotted against arousal level or drive strength, such curves are frequently obtained.

involuntary muscle (histology) Muscle not under conscious control, located in internal organs and tissues, *e.g.* in the alimentary canal and blood vessels. Also termed: smooth muscle.

See also voluntary muscle.

IOAT (society) An abbreviation of 'I'nternational 'O'rganisation 'A'gainst 'T'rachoma.

IoB (society) An abbreviation of 'I'nstitute 'o'f 'B'iology.

Iodamoeba (parasitology) A genus of amoebae (*USA*: amebae) in the family Endamoebidae.

Iodamoeba buetschlii (human/veterinary parasitology) A parasitic species of the genus *Iodamoeba* that may cause intestinal ulceration in humans. May also be found in pigs.

Iodamoeba suis (veterinary parasitology) A species of the genus *Iodamoeba* that may found in pigs.

iodide (chemistry) Compound of iodine and another element; salt of hydriodic acid (HI).

iodinated (chemistry) A compound into which iodine has been incorporated.

iodine (I) (chemistry) Non-metallic element in Group VIIA of the periodic table (the halogens), extracted from Chile saltpetre (in which it occurs as an iodate impurity of sodium nitrate) and certain seaweeds. It forms purpleblack crystals that sublime on heating to produce a violet vapour. It is essential for the secretion of the thyroid hormones thyroxine and tri-iodothyronine, usually present in sufficient quantities in normal diets. Iodine and its organic compounds are used in medicine; silver iodide is used in photography. At. no. 53; r.a.m. 126.9044.

iodine number (chemistry) Number that indicates the amount of iodine taken up by a substance, *e.g.* by fats or oils; it gives a measure of the number of unsaturated bonds present. Also termed: iodine value; iodine absorption.

iodine solution (strong) (chemistry) Contains 5% free iodine and 10% potassium iodide in an aqueous solution.

iodine solution (chemistry) Contains 2% free iodine and 2.4% sodium iodide in an aqueous solution.

iodoform (chemistry) An alternative term for triiodomethane.

iodoquinol (pharmacology) An amoebicide (*USA*: amebicide) and antibiotic, that may be used in the treatment of intestinal amoebiasis (*USA*: amebiasis) in dogs.

ion (chemistry) Atom or molecule that has positive or negative electric charge because of the loss or gain of one or more electrons. Many inorganic compounds dissociate into ions when they dissolve in water. Ions are the electric current-carriers in electrolysis and in discharge tubes.

ion channels (physiology/pharmacology) Channels or pores composed of proteins that allows exchange of ions or water between the cytoplasm and the extracellular fluid. Ion channels are vital in the control of cell volume and the activity of electrically excitable cells. Conformation changes within the protein channel lead to opening or closing of the pore. Some drugs bind to sites on the extracellular surface of the ion channel to influence channel opening or closing, other drugs influence channel opening following interaction with a G-protein coupled receptor.

ion exchange (chemistry) Reaction in which ions of a solution are retained by oppositely-charged groups covalently bonded to a solid support, such as zeolite or a synthetic resin. The process is used in water softeners, desalination plants and for isotope separation.

ion pair (chemistry) Two charged fragments that result from simultaneous ionization of two unchanged ones, a positive and a negative ion.

ion pump (physics) High-vacuum pump for removing a gas from a system by ionizing its atoms or molecules and adsorbing the resulting ions on a surface.

ionic (chemistry) Pertaining to an ion or ions.

ionic bond (physics) Electrostatic attraction that occurs between two oppositely charged ions, *e.g.* proteins binding with metal ions. The degree of binding varies with the chemical nature of each compound and the net charge. Dissociation of ionic bonds usually occurs readily, but some members of the transition group of metals exhibit high association constants (*i.e.* low Kd values) and exchange is slow. Also termed: electrovalent bond.

See also covalent bond.

ionic product (chemistry) Product (in moles per litre) of the concentrations of the ions in a liquid or solution, *e.g.* in sodium chloride solution, the ionic product of sodium chloride is given by $[Na^+]$ $[Cl^-]$. In a pure liquid, it results from the dissociation of molecules in the liquid.

ionization (chemistry/physics) Formation of ions. It is generally achieved by chemical or electrical processes, or by dissociation of ionic compounds in solution, although at extremely high temperatures (such as those in stars) heat can cause ionization.

ionization chamber (physics) Apparatus consisting of a gas-filled container with a pair of high-voltage electrodes. It is used to study the ionization of gases or ionizing radiation.

ionization potential (chemistry) Electron bonding energy, the energy required to remove an electron from a neutral atom.

ionize (chemistry) To convert into ions.

ionizing radiation (radiology) Radiation of sufficiently high energy to produce ions in the medium through which it passes, *e.g.* high-energy (electrons, protons) or short-wave radiation (UV, X-rays).

IPA (chemistry) An abbreviation of 'I'so'P'ropyl 'A'lcohol, also termed isopropanol.

Ipiduropoda (insect parasitology) *See Dinychus.*

ipsi- (terminology) A word element. [Latin] Prefix denoting *same, self.*

IPSP (physiology) An abbreviation of 'I'nhibitory 'P'ostS'ynaptic 'P'otential.

irid(o)- (terminology) A word element. [German] Prefix denoting *iris of the eye.*

iridium (Ir) (chemistry) Steel-grey metallic element in Group VIII of the periodic table (a transition element). It is used (with platinum or osmium) in hard alloys for bearings, surgical tools and crucibles. At. no. 77; r.a.m. 192.22.

iris (anatomy) Pigmented part of the human eye that controls the amount of light entering the eye.

See also iris diaphragm.

iris diaphragm (photography) Adjustable aperture in a camera or incorporated in a lens to control the amount of light passing through the lens.

iritis (medical) inflammation of the iris. The iris, ciliary body and choroid make up the uveal tract, and they have the same blood supply, so that infection can easily spread from one to the other. There are a number of causes, including toxoplasmosis.

iron (Fe) (chemistry) Silver-grey magnetic metallic element in Group VIII of the periodic table (a transition element). It is the fourth most abundant element in the Earth's crust and probably forms much of the core. It is also the most widely used metal, particularly (alloyed with carbon and other elements) in steel. It occurs in various ores, chief of which are haematite (*USA*: hematite), limonite and magnesite, which are refined in a blast furnace to produce pig iron. Inorganic iron compounds are used as pigments; the blood pigment haemoglobin (*USA*: hemaglobin) is an organic iron compound. At. no. 26; r.a.m. 55.847.

iron(II) (chemistry) An alternative term for ferrous.

iron(II) sulphate (chemistry) $FeSO_4.7H_2O$. Green crystalline compound, used in making inks, in printing and as a wood preservative. A white monohydrate is also known. Also termed: ferrous sulphate (*USA*: ferrous sulfate). (*USA*: iron(II) sulfate)

iron(III) (chemistry) An alternative term for ferric.

iron(III) chloride (chemistry) $FeCl_3.6H_2O$ Brown crystalline compound, used as a catalyst, a mordant and for etching copper in the manufacture of printed circuits. Also termed: ferric chloride.

iron(III) oxide (chemistry) Fe_2O_3. Red insoluble compound, the principal constituent of haematite (*USA*: hematite). It is used as a pigment, catalyst and polishing compound. Also termed: ferric oxide.

irradiance (physics) The power of light falling on an area of a surface. It can be measured in watts per square metre.

irradiation (1 physics; 2 radiology) 1 Radiant energy per unit of intercepting area. 2 Treatment by exposure to radiation of any kind, but usually refers to treatment by ionising radiation. *See* radiation.

irrational number (mathematics) Real number that cannot be expressed as a fraction, *i.e.* as the ratio of two integers; *e.g.* e, π, $\sqrt{2}$.

irreversible reaction (chemistry) Chemical reaction that takes place in one direction only, therefore proceeding to completion.

irrigant (medical) Fluid used to irrigate cavities.

irrigation (medical) 1 The washing-out of a cavity or wound by a stream of water or other liquid. 2 Artificial watering of agricultural crops and pasture by flood, furrow, drip or sprinkler system. This has a significant implication for animals both nutritionally and in terms of health because of the great change in the soil microclimate and the reaction of this on parasite larvae and fungi.

irritability (physiology) Ability to respond to a stimulus, evident in all living material.

irritants (toxicology) Any non-corrosive substance that, on immediate, prolonged or repeated contact with normal living tissue produces a local inflammatory reaction.

Isakis migrans (insect parasitology) A species of nematode in the Diplogasteridae that are parasites to the termites *Reticulitermes* spp.

ISCB (society) An abbreviation of 'I'nternational 'S'ociety for 'C'ell 'B'iology.

ischaemia (medical) Inadequate blood supply to a part of the body, caused by spasm or disease of the blood vessels or failure of the general circulation; if it is prolonged and severe, the tissue dies. (*USA*: ischemia).

ischi(o)- (terminology) A word element. [German] Prefix denoting *ischium.*

ischiorectal abscess (medical) An abscess occurring between the rectum and the ischium, part of the pelvis.

ISCP (society) An abbreviation of 'I'nternational 'S'ociety of 'C'linical 'P'athology.

ISGE (society) An abbreviation of 'I'nternational 'S'ociety of 'G'astro'E'nterology.

-ism (terminology) A word element. Suffix meaning a *state*, *process* or *condition*.

iso- (terminology) A word element. [German] Prefix denoting *equal*, *alike*, *same*.

isobar (1, 2 physics) 1 Curve that relates to qualities measured at the same pressure. 2 One of a set of atomic nuclei having the same total of protons and neutrons (*i.e.* the same nucleon number or mass number) but different numbers of protons and therefore different identities.

isocellular (cytology) Made up of identical cells.

isochromatic (physics) Of the same colour (*USA*: color) throughout.

isochromatophil (microscopy) Staining equally with the same stain.

isochromosome (genetics) an abnormal chromosome in which two identical arms have become joined at the centromere because the centromere divided transversely and not longitudinally during cell division.

isochrony (physics) The property of having the same time interval.

Isocyamus delpini (veterinary parasitology) An ectoparasitic copepod that parasitizes the skin of whales. Also termed whale lice.

isocyanide (chemistry) Organic compound of general formula RNC, where R is an organic radical. Also termed: isonitrile; carbylamine.

isoelectric focusing (biochemistry) A method used to separate mixtures containing proteins of different pI. The migration of ampholytes (*e.g.* proteins) occurs through a pH gradient under an applied electric field. Molecules possessing an electric charge migrate towards a region in which they are isoelectric.

See also isoelectric point.

isoelectric point (biochemistry) The Hydrogen ion concentration (pH) at which a species (amino acid, protein or colloid) is electrically neutral.

See also isoelectric focusing.

isoenzymes (biochemistry) A set of enzymes all carrying out the same chemical reaction. Several isoenzymes may be present within the same organism, as they are coded for by genes at different loci. Also termed isozymes.

isogamy (biology) Sexual fission of gametes that are similar in structure and size, such as occurs in some protozoa.

See also anisogamy.

isolation (1 medical; 2 biology) 1 The separation of an infective patient from others, so that the infecting organism is not spread. 2 The growth and separation of micro-organisms.

isoleucine (biochemistry) Crystalline amino acid; it is a constituent of proteins, and essential in the diet of human beings.

isomer (chemistry) Substance that exhibits chemical isomerism.

Isomerala (insect parasitology) *See Eucharomorpha.*

isomerism (1 chemistry; 2 physics) 1 In chemistry, the existence of substances that have the same molecular composition (and therefore the same chemical formula), but different structures. *See also* optical isomerism. 2 In physics, the existence of atomic nuclei with the same atomic numbers and the same mass numbers, but different energy states.

isonitrile (chemistry) An alternative term for isocyanide.

isopod (fish parasitology) Member of the order Isopoda, suborders of which include Flabellifera that are aquatic parasites in marine fish.

isopotential (physics) Lines connecting points of equal electrical potential.

isoprecipitin (immunology) An isoantibody that acts as a precipitin.

isopropanol (chemistry) $(CH_3)_2CHOH$ One of the two isomers of propanol. Also termed: isopropyl alcohol.

isopropyl (chemistry) Denotes the 1-methylethyl group, $-CH(CH_3)_2$.

isopropyl alcohol (chemistry) An alternative term for isopropanol.

Isoptera (insect parasitology) The insect order to which the termites belong. The name Isoptera (iso = equal; ptera = wing) refers to the adult primary reproductives which possess two pairs of equal length wings. Termites are most closely related to cockroaches (Blattaria). There are approximately 2,753 validly named termite species in 285 genera in the world. The vast majority of termite species occur in the tropics.

Isospora (veterinary/avian parasitology) A genus of apicomplexan parasites in the family Eimeriidae found mainly in dogs and cats. These protozoa develop intracellularly in the cells of the intestinal epithelium and some of them are associated with attacks of diarrhoea (*USA*: diarrhea). Includes *Isospora almaataensis* (pigs), *Isospora bahiensis* (dogs), *Isospora bigemina* (*Cystoisospora burrowsi*), *Isospora burrowsi* (*Cystoisospora burrowsi*), *Isospora butonis* (raptor birds), *Isospora* canis (*Cystoisospora canis*), *Isospora felis* (*Cystoisospora* felis), *Isospora heydoni* (*Cystoisospora heydoni*), *Isospora neorivolta* (dog), *Isospora ohioensis* (*Cystoisospora ohioensis*), *Isospora ratti* (rats), *Isospora rivolta* (*Cystoisospora rivolta*), *Isospora suis* (pigs), *Isospora wallacei* (*Besnoitia wallacei*).

isothermal (physics) Having the same temperature.

isothermal process (physics) Process that occurs at a constant or uniform temperature; *e.g.* the compression of a gas under constant temperature conditions.

isotonic (chemistry) Solutions which have the same osmotic pressure. Such solutions will not bring about diffusion one into the other through an intervening membrane.

isotope (physics) A chemical element which has the same atomic number as another, but a different atomic mass. It has the same number of protons in the nucleus, but a different number of neutrons. Radioactive isotopes change into another element over the course of time, the change being accompanied by the emission of electromagnetic radiations.

isotopic number (chemistry) Difference between the number of neutrons in an isotope and the number of protons. Also termed: neutron excess.

isotopic weight (chemistry) Atomic weight of an isotope. Also termed: isotopic mass.

isotropic (chemistry) Describing a substance whose physical properties are the same in all directions (*e.g.* most liquids).

isotropy (chemistry) The quality or condition of being isotropic.

isotype (immunology) The antigenic variability between related proteins, *e.g.* between immunoglobulin classes of a single species.

isotypic determinant (biology) A determinant that occurs in all individuals of the same species.

isotypical (biology) Of the same kind.

isozymes (biochemistry) *See* isoenzymes.

Israeli turkey encephalomyelitis (avian parasitology) Nonsuppurative encephalomyelitis of turkeys caused by a flavivirus and carried by insects, probably mosquitoes. Manifested by a progressive paralysis.

issue (medical) A discharge of pus, blood, or other matter; a suppurating lesion emitting such a discharge.

-ist (terminology) A suffix meaning one who specializes in.

Isthmiophora (parasitology) A genus of intestinal flukes in the family Echinostomatidae.

Isthmiophora melis (veterinary parasitology) A species of the genus *Isthmiophora* which may be found in the intestine of cat, fox and many other small wild mammals, including mink. It is considered non-pathogenic except in mink, in which it causes a severe haemorrhagic (*USA*: hemorrhagic) enteritis.

isthmus (parasitological anatomy) A term that may be used in relation to nematodes. Middle portion of the oesophagus.

ISV (quality standards) An abbreviation of 'I'nternational 'S'cientific 'V'ocabulary.

itch (medical) A skin disease attended with itching.

itch mite (parasitology) *See Psorergates ovis.*

itchiness (medical) Pruritus.

itching (medical) Pruritus; an unpleasant cutaneous sensation, provoking the desire to scratch or rub the skin.

itchy leg (parasitology) *See Chorioptes.*

item (statistics) A single observation.

iterate (computing) A term meaning to repeat automatically, but under control of a programme (*USA*: program); a series of processing steps until a predefined criterion is reached.

iteration (methods) Automatic repetition of one cycle.

-itis (terminology) Plural: -itides. A word element [German]. Suffix denoting *inflammation.*

ITP (biochemistry) An abbreviation of 'I'nosine 'T'ri'P'hosphate.

IU (measurement) An abbreviation of 'I'nternational 'U'nit(s)

IUBS (society) An abbreviation of 'I'nternational 'U'nion of 'B'iological 'S'ciences.

IV (medical) An abbreviation of 'I'ntra'V'enous.

-ive (terminology) A suffix meaning *pertaining to.*

ivermectin (pharmacology/control measures) An avermectin with broad activity against many helminths and arthropods. A broad-spectrum anthelmintic, acaricide and insecticide, used orally, subcutaneously and as a pour-on.

IVT (medical) An abbreviation of 'I'ntra'V'enous 'T'ransfusion.

Ixodes (parasitology) A genus of hard-bodied ticks in the family Ixodidae. Some species are vectors of disease.

Ixodes angustus (veterinary parasitology) A species of hard-bodied ticks of the genus *Ixodes* which may be found on dogs.

Ixodes canisuga (veterinary parasitology) A species of hard-bodied tick of the genus *Ixodes* which may be found on dogs, and also on foxes and occasionally other species in Europe.

Ixodes cookei (parasitology) A species of hard-bodied tick of the genus *Ixodes* which may be found on most species.

Ixodes cornuatus (veterinary parasitology) A species of hard-bodied tick of the genus *Ixodes* which may be found on dogs and other species in Australia and which may cause paralysis.

Ixodes dammini (parasitology) A species of hard-bodied tick of the genus *Ixodes*. A three-host tick and important transmitter of *Borrelia burgdorferi* in the USA.

Ixodes hexagonus (veterinary parasitology) A species of hard-bodied tick of the genus *Ixodes* which may be found on hedgehogs, and also on dogs and other species in Europe.

Ixodes holocyclus (veterinary parasitology) A species of hard-bodied tick of the genus *Ixodes* which may be found on bandicoots in Australia, but also may be found on other species. Transmits *Coxiella burnetii* and causes tick paralysis by a toxin secreted by its salivary glands. It also produces a cardiovascular component which causes intense vasoconstriction, high blood pressure and death.

Ixodes kingi (veterinary parasitology) A species of hard-bodied tick of the genus *Ixodes*. The rotund tick of dogs.

Ixodes loricatus (veterinary parasitology) A species of hard-bodied tick of the genus *Ixodes* which may be found as a very rare infestation in New World primates.

Ixodes muris (veterinary parasitology) A species of hard-bodied tick of the genus *Ixodes* which may be found on mice and also dogs.

Ixodes ornithorhynchi (veterinary parasitology) A species of hard-bodied tick of the genus *Ixodes* which may be found on platypus.

Ixodes pacificus (parasitology) A species of hard-bodied tick of the genus *Ixodes*. The California black-legged tick, which may be found on most species.

Ixodes persulcatus (parasitology) A species of hard-bodied tick of the genus *Ixodes* which transmits *Babesia* spp.

Ixodes pilosus (parasitology) A species of hard-bodied tick of the genus *Ixodes*. The bush, sour-veld or russet tick found on most species, but which does not cause paralysis.

Ixodes ricinus (human/avian/veterinary parasitology) A species of hard-bodied tick of the genus *Ixodes* which may be found on many species of mammals and birds in Europe. The castor-bean tick, transmits *Babesia divergens*, *Babesia bovis*, *Anaplasma*, tick pyemia, *Coxiella burnetii*, several human encephalitides and also causes paralysis.

Ixodes rubicundus (parasitology) A species of hard-bodied tick of the genus *Ixodes* which may infest most species but not cats, horses or birds. Infestation may cause paralysis.

Ixodes rugosus (veterinary parasitology) A species of hard-bodied tick of the genus *Ixodes* which may be found on dogs.

Ixodes scapularis (parasitology) A species of hard-bodied tick of the genus *Ixodes* which may be found on most species. The shoulder or black-legged tick, which may transmit anaplasmosis and tularemia.

Ixodes sculptus (veterinary parasitology) A species of hard-bodied tick of the genus *Ixodes* which may be found on dogs.

Ixodes texanus (veterinary parasitology) A species of hard-bodied tick of the genus *Ixodes* which may be found on dogs.

ixodiasis (parasitology) Any disease or lesion due to tick bites; infestation with ticks.

ixodic (parasitology) Pertaining to, or caused by, ticks.

ixodid (parasitology) A tick of the family Ixodidae.

Ixodidae (parasitology) A family of ticks comprising the hard-bodied ticks.

J

J (measurement) An abbreviation of 'J'oule.

J exon (immunology) *See* J gene.

J gene (immunology) A short sequence of DNA coding for part of the hypervariable region of immunoglobulin light or heavy chains near to the site of joining to the constant region. Also termed J exon.

J region (immunology) An abbreviation of 'J'oining 'region'. The part of an immunoglobulin molecule that lies between the constant and the variable regions.

J/deg. (measurement) An abbreviation of 'J'oule per 'deg'ree.

jail fever (medical) *See* typhus.

jaundice (medical) A condition characterised by a yellow appearance of the whites of the eyes, the skin and mucous membranes. This discolouration (*USA*: discoloration) is due to the presence of excess bilirubin in the blood. Jaundice is a common symptom of disorders of the gall bladder or liver.

jaws (1, 2 anatomy) 1 In vertebrates, bony structure enclosing the mouth, often furnished with teeth for grasping prey and/or chewing food, consisting of an upper maxilla and lower mandible. 2 In invertebrates, grasping structure surrounding the mouth.

jejunectomy (medical) Excision of part of the jejunum.

jejunitis (medical) Inflamation of the jejunum.

jejunocecostomy (medical) Anastomosis of the jejunum to the caecum (*USA*: cecum).

jejunoileal (anatomy) Pertaining to the jejunum and ileum; connecting the proximal jejunum with the distal ileum.

jejunoileitis (medical) Inflammation of the jejunum and ileum.

jejunoileum (anatomy) The jejunum and the ileum considered as a single organ.

jejunotomy (medical) Incision of the jejunum.

jejunum (anatomy) Part of the intestine of mammals, located between the duodenum and the ileum, the main function of which is absorption.

Jembrana disease (veterinary parasitology) A highly fatal, infectious, generalized disease of cattle probably caused by an Ehrlichia carried by the tick *Boophilus microplus*. The disease is characterized by diarrhoea (*USA*: diarrhea), haemorrhages (*USA*: hemorrhages), rarely erosions in the mucosae, anaemia (*USA*: anemia) and lymphadenopathy, and a high fever.

Jerne plaque assay (immunology) A method of quantitating antibody-producing cells.

jetting (control measures) Application of insecticide to sheep by use of a high-pressure spraying machine. The jets at the head of the handheld appliance are combed through the wool so that the jetted fluid penetrates to the skin.

JG cells (histology) An abbreviation of 'J'uxta'G'lomerular cells.

jigger flea (parasitology) *See Tunga penetrans*.

jigger (parasitology) A flea *Tunga penetrans* which burrows into the skin of the feet, causing intense irritation and ulceration. Also termed chigoe; jigger flea. *See Tunga penetrans*.

jnl (literary terminology) An abbreviation of 'j'our'n'a'l'. Also termed jl/jour.

joint (anatomy) Point of articulation of limbs or bones. The bones are connected to each other by connective tissue ligaments, and are well lubricated. Common types of joints include ball-and-socket joints (*e.g.* the hip joint), hinge joints (*e.g.* the elbow) and sliding joints (*e.g.* between vertebrae).

joint probability (statistics) The probability that two or more events will occur together.

Jonckheere test (statistics) A non-parametric planned comparison trend test of the hypothesis that the values of a variable of different samples are ordered in a specific sequence.

Jones methenamine silver stain (histology) A histological staining technique used to make basement membrane and mesangium visible under the microscope.

joule (J) (measurement) The SI unit of energy, work and quantity of heat. Equal to the work done when one newton of force moves an object one metre. Named after James Prescott Joule (1818–89).

joule's law (1, 2 physics) 1 Internal energy of a given mass of gas is dependent only on its temperature and is independent of its pressure and volume. 2 If an electric current I flows through a resistance R for a time t, the heat produced Q, in joules, is given by $Q = I^2Rt$.

Joule-Thompson effect (physics) When a gas is allowed to undergo adiabatic expansion through a porous plug, the temperature of the gas usually drops. This results from the work done in breaking the intermolecular forces in the gas, and is a deviation from joule's law. The effect is important in the liquefaction of gases by cooling. Also termed: Joule-Kelvin effect.

jour. (literary terminology) An abbreviation of 'jour'nal.

journal (jour.) (literary terminology) A scientific/ medical etc. newspaper or periodical.

Joyeuxiella (parasitology) A genus of tapeworms of dogs and cats of the family Dipylidiidae.

jugular vein (anatomy) One of a pair of veins draining the brain and joining with the subclavian veins before discharging into the anterior vena cava.

juice (biology) Any fluid from animal or plant tissue.

jumping genes (genetics) Genes that move within chromosomes. They influence the regulation of gene activity, probably by physically removing a gene from its promoter. *See* operon.

junctionopathy (medical) A disorder of the neuromuscular junction, *e.g.* tick paralysis.

junk DNA (genetics) Sequences of DNA in the genome that have no apparent genetic function. Also known as selfish DNA.

jurisprudence (medical) Medical jurisprudence, or forensic medicine, concerned with any aspect of medicine which relates to the law.

juv. (general terminology) An abbreviation of 'juv'enile.

juvenile (juv.) (general terminology) 1 A minor; an animal that has not reached the age of majority. 2 A cell, tissue, disease or organism intermediate between the immature and mature forms.

juvenoids (insect parasitology) *See* insect growth regulators.

juxta- (terminology) A word element. [Latin] Prefix denoting *situated near, adjoining*.

K

K (1 chemistry; 2 measurement; 3 biochemistry) 1 Chemical symbol for potassium, derived from Latin: Kalium. 2 An abbreviation of 'K'ilodalton/'K'elvin/ 'K'ilobyte/'K'ilogram. 3 A symbol for lysine.

k bar (measurement) An abbreviation of 'k'ilo'bar'.

K cells (immunology) An abbreviation of 'K'iller cells.

K complex (physiology) An isolated slow wave occurring in the electroencephalogram, usually during stage II sleep.

kala azar (human parasitology) one of the forms of Leishmaniasis, caused by *Leishmania* organisms and spread to humans by the bite of sandflies. Kala azar occurs mainly in the Mediterranean area and in India, but is spreading westward into Europe. There is fever, anaemia (*USA*: anemia), lymph node swelling, enlargement of the spleen and liver, damage to the bone marrow, malnutrition and loss of immune capacity (immunosuppression). The condition can be diagnosed by the elisa test and is treated with drugs containing antimony. Also known as visceral leishmaniasis.

Kalicephalus (veterinary parasitology) A genus of hookworm of the family Diaphanocephalidae, found in snakes.

kalium (chemistry) [Latin] *potassium* (symbol K).

kallikreins (biochemistry) Proteases that release the peptides bradykinin and lysylbradykinin from the precursor proteins, high-molecular-weight kininogen and low-molecular-weight kininogen. There are two kallikreins: plasma kallikrein, which circulates in an inactive form, and which with high-molecular-weight kininogen also catalyse the activation of factor XII; and tissue kallikrein, which is located primarily on the apical membranes of cells concerned with transcellular electrolyte transport.

kamala (pharmacology) An anticestodal agent derived from the plant *Mallotus philippinensis*; now replaced by more effective and safer compounds.

kaolin (chemistry) Aluminium silicate, or china clay. It is used in the treatment of mild diarrhoea (*USA*: diarrhea). Also termed china clay.

kaolinite (chemistry) Hydrated aluminium silicate mineral.

Kapala floridana (insect parasitology) *See Eucharomorpha.*

kappa (terminology) The tenth letter of the Greek alphabet.

kappa chain (immunology) One of the two light chains in immunoglobulin. The other is the lambda chain.

karyapsis (cytology) *See* karyogamy.

karyo- (terminology) A word element. [German] Prefix denoting *nucleus*. From the Greek *karuon*, meaning 'a nut'.

karyocyte (cytology) A nucleated cell.

karyogamy (cytology) The coming together and fusing of the nuclei of gametes.

karyogenesis (cytology) The formation of a cell nucleus.

karyogram (cytology) A graphic representation of a karyotype.

karyokinesis (cytology) *See* mitosis.

karyolymph (cytology) The fluid in the nucleus of a cell.

karyolysis (cytology) Destruction of a cell nucleus.

karyomegaly (cytology) An increase in the size of the nuclei of the cells of a tissue.

karyomorphism (cytology) The shape of a cell nucleus.

karyon (cytology) The cell nucleus.

karyophage (parasitology) A protozoon that phagocytizes the nucleus of the cell it infects.

karyoplasm (cytology) The protoplasm of the cell nucleus.

karyopyknosis (cytology) Shrinkage of a cell nucleus, with condensation of the chromatin.

karyorrhexis (cytology) Rupture of the cell nucleus in which the chromatin disintegrates.

karyosome (cytology) A spherical mass of aggregated chromatin material in a resting (interphase) nucleus.

karyotheca (anatomy) The nuclear membrane.

karyotype (genetics) The chromosomal constitution of an individual as seen in the nucleus of a somatic cell. It provides information about the species or strain, because a karyotype is characteristic to the cell of a particular organism.The chromosomes are classified by their size and centromere position, and for convenience are usually shown in a specific order in an idiogram.

karyotyping (genetics) Preparation of a karyotype.

Kastenbaum-Bowman test (statistics) A test of statistical significance used in calculation of mutation frequencies.

kat (measurement) An abbreviation of 'kat'al.

kat(a)- (terminology) A word element. [German] Prefix denoting *down, against.*

katal (kat) (measurement) A unit of enzyme activity in the SI system. 1 international unit is equal to 16.6 nanokatal.

katathermometer (equipment) A thermometer with a wet bulb and a dry bulb, for detecting cooling rates.

Katayama syndrome (medical /parasitology) A set of allergic phenomena associated with the penetration of the skin and invasion of the body by the larval stage (cercariae) of schistosomiasis. The effects include local itching, urticaria, fever, headache, muscle aches, abdominal pain, cough, patchy pneumonia and enlargement of the spleen.

kb (measurement) An abbreviation of 'k'ilo'b'ase.

kc (measurement) An abbreviation of 'k'ilo'c'ycle.

Kcal (measurement) An abbreviation of 'K'ilo 'cal'orie. Also termed k.cal.

Kda (measurement) An abbreviation of 'K'ilo'da'lton.

K_e (physiology/biochemistry) A symbol for exchangeable body potassium.

keloid (histology) An overgrowth of scar tissue at the site of a cut or burn. The scar, instead of disappearing, spreads and sends out offshoots like claws which pucker the surrounding skin.

kelvin (K) (measurement) SI unit of thermodynamic temperature with 0K being absolute zero [-273.15°C or -459.7°F]. One Kelvin degree is equal to one Celsius degree. Water freezes at 273.15K, 32°F, and 0°C. Water boils at 373.15K, 212°F, and 100°C. Named after the British physicist Lord Kelvin (William Thomson) (1824–1907).

See also Kelvin temperature.

Kelvin effect (physics) An alternative term for the Thomson effect.

Kelvin temperature (measurement) Scale of temperature that originates at absolute zero, with the triple point of water defined as 273.16K. The freezing point of water (on which the Celsius scale is based) is 273.15K. Also termed: Kelvin thermodynamic scale of temperature.

Kendall's coefficient of concordance (W) (statistics) A measure of the extent to which two or more rank orderings agree with one another. Complete agreement between the rankings gives W = 1; lack of agreement gives W = 0 (or nearly 0).

Kendall's tau (T) (statistics) A coefficient of rank correlation between two sets of scores based on the number of inversions of ranks in one ranking compared with the other.

keno- (terminology) A word element. [German] Prefix denoting *empty.*

Kenya typhus (human/veterinary parasitology) *See* Boutonneuse fever.

kerat(o)- (terminology) A word element. [German] Prefix denoting *horny tissue, cornea.*

kerat-, kerato- (histology) Combining form denoting cornea or horny keratosis. From the Greek 'keras', meaning horn.

keratin (biochemistry) A fibrous protein containing sulphur (*USA*: sulfur), the substance of which horn, hair, the outer layer of the skin, and the nails are composed. It is also part of the structure of the enamel of the teeth.

keratinization (histology) Replacement of the cytoplasm of cells in the epidermis by keratin, thus resulting in hardening of skin. Also termed: cornification.

keratitis (medical) Inflammation of the outer lens of the eye, the cornea.

keratoconjunctivitis (medical) Inflammation of the cornea and conjunctiva.

keratometer (equipment) An instrument for measuring the radius of curvature of the cornea.

keratoscope (equipment) An instrument for examining the cornea that makes it possible to detect irregularities of curvature.

ketene (chemistry) Unstable organic compound of general formula R_2CCO, where R is an organic radical. Ketenes react with other unsaturated compounds to form 4-membered rings.

keto- (terminology) A word element denoting *ketone group.*

keto acids (chemistry) Compounds containing both of the groups CO (carbonyl) and COOH (carboxyl).

ketoacidosis (biochemistry) Acidosis resulting from the accumulation of ketone bodies in the blood.

keto-enol tautomerism (chemistry) Existence of a chemical compound in two double-bonded structural forms, keto and enol, which are in equilibrium. The keto changes to the enol by the migration of a hydrogen atom to form a hydroxyl group with the ketone oxygen; the position of the double bond also changes.

ketogenesis (biochemistry) The formation of acid ketone bodies, as in uncontrolled diabetes, starvation or as a result of a diet with a very high fat content.

ketogenic (biochemistry) (of an amino acid) Giving rise to ketone bodies.

ketonaemia (biochemistry) Raised ketones in the blood. Low levels are normal. (*USA*: ketonemia).

ketone (chemistry) Member of a family of organic compounds of general formula RCOR', where R and R' are organic radicals and =CO is a carbonyl group. Ketones may be made in various ways, such as the oxidation of a secondary alcohol; *e.g.* oxidation of isopropanol gives acetone (propanone) $(CH_3)_2CO$.

ketosis (biochemistry) Poisoning caused by an accumulation of ketones (acetoacetate, D-0-hydroxy-butyrate, acetone) in the blood.

ketotic (biochemistry) Pertaining to ketosis.

Kety method (physiology) Method for measuring cerebral blood flow. The average cerebral blood flow in young adults is 54mL/100g/min. The Kety method gives no information about regional differences in blood flow.

keV (measurement) An abbreviation of 'k'ilo-'e'lectron 'V'olt, a unit of particle energy equivalent to 10^3 electron volts.

key signs (medical) The important signs on which a diagnosis can be based. Equivalent to keywords in literature searches.

keyboard (computing) Computer input device which a human operator uses to type in data as alphanumeric characters. It consists of a standard keyboard, usually with additional function keys.

Keyes punch (methods) A skin biopsy instrument similar to and used like a cork borer by pushing down on the palm-held handle and rotating the instrument so that its sharp, circular cutting edge cuts through the skin.

keystroke (computing) One stroke of a key on a computer keyboard.

keyword (computing) The significant word or words in a piece of information; often used to accurately store information and by which it can be retrieved.

Kf (physiology/biochemistry) A symbol for glomerular ultrafiltration coefficient.

kg (measurement) An abbreviation of 'k'ilo'g'ram.

kg cal (measurement) An abbreviation of 'k'ilo'g'ram 'cal'oric.

kg cum (measurement) An abbreviation of 'k'ilo'g'rams per 'cu'bic 'm'etre.

kg f (measurement) An abbreviation of 'k'ilo'g'ram 'f'orce.

kg m (measurement) An abbreviation of 'k'ilo'g'ram 'm'etre.

kH (measurement) An abbreviation of 'k'ilo'H'ertz. Also termed kHz.

Khalilia (parasitology) A genus of strongyles in the family Strongylidae.

Khalilia buta (veterinary parasitology) A species of strongyles in the genus *Khalilia* that, similar to *Khalilia pileata* and *Khalilia sameera*, may be found in the large intestine of elephants.

Khalilia pileata (veterinary parasitology) A species of strongyles in the genus *Khalilia* that similar to *Khalilia buta* and *Khalilia sameera* may be found in the large intestine of elephants.

Khalilia sameera (veterinary parasitology) A species of strongyles in the genus *Khalilia* that similar to *Khalilia buta* and *Khalilia pileata* may be found in the large intestine of elephants.

Khawia (parasitology) A genus of tapeworms in the family Caryophyllaeidae.

***Khawia* sinensis** (fish parasitology) A species of tapeworm in the genus *Khawia* that causes haemorrhagic enteritis (*USA*: hemorrhagic enteritis) in carp.

kHz (veterinary parasitology) An abbreviation of 'k'ilo'H'ert'z'.

kidney (anatomy) One of a pair of excretory organs that filter waste products (particularly nitrogenous waste) from the blood and concentrate them in urine. Kidneys also have an important function in the regulation of the balance of water and salts in the body. The main processes of the kidney occur in a large number of tubular structures called nephrons. Water and waste products pass from the kidneys to the bladder via the ureters.

kidney cysts (histology) Small, fluid-filled cavities in the kidneys that are common and benign and usually cause neither symptoms nor danger.

kidney failure (medical) The stage in kidney disease in which neither organ is capable of excreting body waste products fast enough to prevent their accumulation in the blood.

kill (fish parasitology) Term used to describe heavy mortalities in wild fish or fauna, especially those in the wild, or at extensive range, and usually where the deaths are unexpected and generally not easily explained.

killed vaccine (immunology) *See* vaccine.

killer cells (immunology) A subclass of large, granular lymphocytes. These are important elements in the immune system and are the final effectors in the process by which damaged, infected or malignant cells are recognised and destroyed.

kilo- (terminology) A word element. [German] Metric prefix meaning a thousand times ($\times 10^3$).

kilobase (molecular biology) A unit of length of DNA or RNA of 1000 bases, used in describing genes or shorter sequences. When measuring double stranded DNA, it is actually base-pairs that are counted, so that the unit is comparable with counting bases in a single-stranded molecule.

kilobyte (K) (computing) A computer term denoting 1,024 bytes of information or storage space.

kilocalorie (Kcal) (measurement) 1,000 small calories [cal] or one large calorie [Cal]. This is a scientific unit of energy required to burn off consumed food.

kiloelectron volt (keV) (measurement) One thousand electron volts (10^3 eV).

kilogram (kg) (measurement) SI unit of mass, equal to 1,000 grams. 1 kg 2.2046 lb.

kilohertz (kHz) (measurement) A unit of frequency equal to 1,000 hertz, or 1,000 cycles per second.

kilojoule (kj) (measurement) Unit of energy equal to 1,000 joules.

kilometre (km) (measurement) Unit of length equal to 1,000 metres. 1 km = 0.62137 miles. (*USA*: kilometer).

kilopascal (measurement) One thousand (10^3) pascals; the metric unit of pressure; equal to 7 pounds per square inch; abbreviated kPa.

kilowatt (kw) (measurement) A unit of electrical power equal to 1,000 watts.

kilowatt-hour (kwh or kwhr) (measurement) A unit of measure equal to 1,000 watts of power over the period of an hour. Also termed: unit.

Kiluluma (veterinary parasitology) A genus of nematode parasite of the family Strongylidae, which may be found in the intestine of rhinoceros.

kinaesthesis (physiology) Process by which sensory cells in muscles and organs relay information concerning the relative position of the limbs and the general orientation of an organism in space; a type of biological feedback.

kinase (chemistry) Enzyme that causes phosphorylation by ATP.

kine- (terminology) A word element. [German] Prefix denoting *movement*.

kinesi(o)- (terminology) A word element. [German] Prefix denoting *movement*.

kinesiology (biology) Scientific study of movement of body parts.

kinesis (biology) Simplest kind of orientation behaviour (*USA*: behavior) that occurs in response to a stimulus (*e.g.* the concentration of a nutrient or an irritant). The speed of an animal's random motion increases until the stimulus reduces.

See also taxis.

-kinesis (terminology) A word element. [German]. Suffix denoting *movement*, *motion*.

kinetic energy (physics) Energy possessed by an object because of its motion, equal to $1/2mv^2$, where m = mass and v = velocity. The kinetic energy of the particles that make up any sample of matter determine its heat energy and therefore its temperature (except at absolute zero, when both are equal to zero). *See* kinetic theory.

kinetic theory (physics) Theory that accounts for the properties of substances in terms of the movement of their component particles (atoms or molecules). The theory is most important in describing the behaviour (*USA*: behavior) of gases (when it is referred to as the kinetic theory of gases). An ideal gas is assumed to be made of perfectly elastic particles that collide only occasionally with each other. Thus, *e.g.* the pressure exerted by a gas on its container is then the result of gas particles colliding with the walls of the container.

kinetics (chemistry) Study of the rates at which chemical reactions take place.

kinetochore (cytology) Point of attachment of the spindle in a cell.

kinetogenic (biology) Causing or producing movement.

kinetoplast (parasitological anatomy) An accessory body found in many protozoa, primarily the Mastigophora. It stains with nuclear dyes, contains DNA and replicates independently.

kingdom (biology) Highest rank in the classification of living organisms, which encompasses phyla (for animals) and divisions (for plants). Criteria determining members of a kingdom are broad, and consequently members are very diverse. Traditionally, there were two kingdoms: the plants and the animals. However, fungi, protists and prokaryotes are now often placed in kingdoms of their own.

kininase I (biochemistry) An enzyme involved in the cardiovasular regulatory mechanism. A carboxypeptidase that metabolizes the kinins bradykinin and lysylbradykinin to inactive fragments by removing the C-terminal Arg.

kininase II (biochemistry) An enzyme involved in the cardiovasular regulatory mechanism. A ipeptidylcarboxypeptidase that inactivates bradykinin and lysylbradykinin by removing Phe-Arg from the C terminal. Kininase II is the same enzyme as angiotensin-converting enzyme, which removes His-Leu from the C-terminal end of angiotensin I.

kininogen (immunology) An α_2-globulin of plasma that is a precursor of the kinins.

kinins (biochemistry) Vasodilator hormones. Two related vasodilator peptides called kinins are found in the body. One is the nanapeptide brabykinin and the other is the decapeptide lysylbradykinin, also known as kalidin.

Kinyoun's carbol fuschsin stain (methods) An acid-fast stain useful for demonstration of coccidial oocysts.

Kirchhoff's laws (1, 2 physics) Extensions of Ohm's law that are used in the analysis of complex electric circuits. 1 The sum of the currents flowing at any junction is zero. 2 Around any closed path, the sum of the emfs equals the sum of the products of the currents and impedances.

Kiricephalus (parasitology) A genus of pentastomid parasites of colubrid snakes.

Kisenyi sheep disease (veterinary parasitology) A virus disease of the nervous system of sheep in Africa transmitted by the tick *Rhipicephalus appendiculatus*.

kissing bug (parasitology) The reduviid bug that transmits chagas' disease. It is so called because its nocturnal bite is barely felt. *See Triatoma.*

kJ (measurement) An abbreviation of 'k'ilo'j'oule.

klino-taxis (biology) Movement of an animal in response to light.

Klossiella boae (veterinary parasitology) A species of coccidians in the genus *Klossiella* which may be found in the intestine of the boa constrictor. It may cause anorexia, restlessness, haemorrhagic enteritis (*USA*: hemorrhagic enteritis) and intussusception.

Klossiella equi (veterinary parasitology) A species of coccidians in the genus *Klossiella* which may be found in the kidneys of horses, donkeys, zebra and burro.

Klossiella (parasitology) A genus of coccidians in the family Eimeriidae.

Klossiella cobayae (veterinary parasitology) A species of coccidians in the genus *Klossiella* which may be found in guinea pigs.

Klossiella muris (veterinary parasitology) A species of coccidians in the genus *Klossiella* which may be found in mice where it causes renal coccidiosis.

km (measurement) An abbreviation of 'k'ilo'm'eter.

knife (equipment) A single-bladed cutting instrument other than a scalpel and usually designed for a special purpose.

knowledge base (computing) The part of an expert system in which the expert knowledge is stored.

Koch's postulates (bacteriology) A set of criteria to be obeyed before it is established that a particular organism causes a particular disease. The organism must be present in every case and must be isolated, cultured and identified; it must produce the disease when a pure culture is given to susceptible animals; and it must he recoverable from the diseased animal. Named after the German bacteriologist, R. Koch (1843–1910).

KOH (chemistry) The chemical formula of potassium hydroxide.

KOH-cleared specimens (microscopy) Specimens of external parasites cleared of epithelial debris by potassium hydroxide solution.

koilo- (terminology) A word element. [German] Prefix denoting *hollowed, concave*.

koilocytosis (medical) Ballooning degeneration.

Konglyphus (insect parasitology) *See Halictacarus.*

kPa (measurement) An abbreviation of 'k'ilo'pa'scal.

Kr (chemistry) The chemical symbol for krypton.

Krebs cycle (biochemistry) A cyclical sequence of 10 biochemical reactions, brought-about by mitochondrial enzymes, that involves the oxidation of a molecule of acetyl-CoA, to two molecules of carbon dioxide and water. It forms the second stage of aerobic respiration, in which pyruvate or lactic acid produced by glycolysis is oxidized to carbon dioxide and water, thus producing a large amount of energy in the form of ATP molecules. It was named after the German-born British biochemist Hans Krebs (1900–82). Also termed: citric acid cycle; tricarboxylic acid cycle.

Kruskal-Wallace test (statistics) A non-parametric rank test of the hypothesis that two or more independent samples have been drawn from the same population.

krypton (Kr) (chemistry) Gaseous nonmetallic element of Group 0 of the periodic table (the rare gases), which occurs in trace quantities in air (from which it is extracted). It is used in gas-filled lamps and discharge tubes. At. no. 36; r.a.m. 83.80.

Kterminalis (insect parasitology) *See Eucharomorpha.*

kumri (parasitology) *See* cerebrospinal nematodiasis.

kunka (parasitology) Cheese-like granules found in granulomatous lesions of habronemiasis.

Kupffer cells (histology) Cells that line the fine blood sinuses (capillaries) of the liver and act as scavengers to remove bacteria and other foreign material. Named after the German anatomist, K. W. von Kupffer (1829–1902).

kurchatovium (chemistry) An alternative term for the element rutherfordium.

Kuzinia (insect parasitology) (Formerly *Tyrophagus*). A species of mites in the Astigmata, that are phoretic and cleptoparasitic to the bees *Apis* and Bombinae.

kV (measurement) An abbreviation of 'k'ilo'V'olt.

kVA (measurement) An abbreviation of 'k'ilo'V'olt-'A'mpere.

kw (measurement) An abbreviation of 'k'ilo'w'att.

Kyasanur forest disease (human parasitology) An arbovirus haemorrhagic fever (*USA*: hemorrhagic fever) that occurs in Mysore State, India, in the villages around the Kyasanur forest. The disease is caused by a virus of the same group as that causing Japanese B encephalitis and the infection is transmitted by tick bite.

L

l (biochemistry) A symbol for geometric isomer of D form of chemical compound.

L (biochemistry) A symbol for leucine.

l (measurement) An abbreviation of 'l'itre (*USA, liter*).

La (chemistry) The chemical symbol for lanthanum.

I_a antigen (immunology) Histocompatibility antigens found primarily on B lymphocytes but also on some macrophages, T lymphocytes and skin.

lab. (general terminology) An abbreviation of 'lab'oratory.

label (chemistry) An isotope (radioactive or stable) that replaces a stable atom in a compound. The course of a chemical or biochemical reaction or physical process can be followed by tracing the radioactivity using a counter or, in the case of stable isotopes, a mass spectrometer.

labelling (methods) To attach a label. (*USA*: labeling).

labellum (parasitological anatomy) Mouthparts of insects that carry tubes for the passage of aspirated fluids.

labial palpi (parasitological anatomy) A term used in relation to arthropods. A pair of appendages arising from each side of the labium.

labiate (biology) Having lips or structures resembling lips.

labile (1 parasitology; 2 chemistry) 1 Gliding; moving from point to point over the surface; unstable; fluctuating. 2 Liable to change; usually applied with respect to particular conditions, *e.g.* heat-labile.

labio- (terminology) A word element. [Latin] Prefix denoting *lip*.

Labiostrongylus (parasitology) The largest of the common nematodes in the stomach of macropods.

labium (parasitological anatomy) A term used in relation to arthropods. The lower lip which forms the floor of the mouth.

laboratory (lab) (general terminology) A room for scientific experiments and demonstrations.

laboratory rat (parasitology) *See* Sprague-Dawley; Wistar.

labour (physiology) Parturition, the process of giving birth. When pregnancy in the human has lasted for more or less 280 days the contractions of the uterus known as labour (*USA*: labor) pains begin. Labour is divided into three stages: dilation of the cervix, the neck of the womb; delivery of the child; and expulsion of the placenta, the after birth. (*USA*: labor).

labrum (parasitological anatomy) Plural: *labra* [Latin] An edge, rim or lip.

labyrinth (anatomy) Part of the inner ear of vertebrates, containing the cochlea and vestibular apparatus. Also known as the inner ear.

labyrinthitis (medical) Inflammation of the inner ear, which causes vertigo, nausea and vomiting.

lacrimal gland (anatomy) Gland that produces tears. Fluid is continuously secreted to protect and moisten the cornea; it also contains the bactericidal enzyme lysozyme. Also termed: lachrimal gland.

lact(o)- (terminology) A word element. [Latin] Prefix denoting *milk*.

lactate (1 chemistry; 2 biology) 1 Salt or ester of lactic acid (2-hydroxypropanoic acid). 2 To produce milk.

lactation (physiology/biochemistry) Production of milk. May be physiological (after pregnancy) or pathological (galactorrhoea).

lacteal (physiology) Lymph vessel of the villi in the intestine of vertebrates. Fat passes into the lacteals as an emulsion of globules to be circulated in the lymphatic system.

lactic acid (biochemistry) $CH_3CH(OH)COOH$ Colourless (*USA*: colorless) liquid organic acid. A mixture of (+)-lactic acid (dextrorotatory) and (-)-lactic acid (laevorotatory) is produced by bacterial action on the sugar lactose in milk during souring. The (+)-form is produced in animals when anaerobic respiration takes place in muscles because of an insufficient oxygen supply during vigorous activity. Also termed: 2-hydroxypropanoic acid.

lactic dehydrogenase (LDH) (biochemistry) An enzyme that catalyses the reduction of pyruvate to lactate. The enzyme is found in all cells capable of glycolysis. LDH exists as several different isozymes, and different tissues have either different isozymes or different sets of isozymes. Release of LDH into the blood is a sign of tissue damage and can occur under many circumstances. Identification of the particular isozyme can be used to identify the organ involved.

lactoflavin (biochemistry) An alternative term for riboflavin.

lactose (chemistry) $C_{12}H_{22}O_{11}$ White crystalline disaccharide sugar that occurs in milk, formed from the union of glucose and galactose. It is a reducing sugar. Also termed: milk-sugar.

Laelapidae (insect/avian parasitology) Mites in the Mesostigmata that are parasites to the ants *Solenopsis* spp., and *Formica*, and that are found as occasional infestations on chickens and pigeons.

***Laelaptonyssus* spp.** (insect parasitology) *See Histiostoma formosana.*

Laelaspis brevichelis (insect parasitology) A species of mites in the Mesostigmata that, similarly to *Laelaspis dubitatus* and *Laelaspis equitans*, are phoretic on adults of the ants Aphaenogaster, Crematogaster, and Tetramorium.

Laelaspis dubitatus (insect parasitology) *See Laelaspis brevichelis.*

Laelaspis equitans (insect parasitology) *See Laelaspis brevichelis.*

Laelaspoides (insect parasitology) A species of mites in the Mesostigmata, that are cleptoparasitic on the bees Halictinae.

laevorotatory (chemistry) Describing a compound with optical activity that causes the plane of polarized light to rotate in an anti-clockwise direction.

laevulose (chemistry) Fruit sugar, also called fructose. A monosaccharide carbohydrate.

lag (biology) 1 The time elapsing between application of a stimulus and the resulting reaction. 2 The early period after inoculation of bacteria into a culture medium, in which the growth or cell division is slow.

Lagochilascaris (parasitology) A genus of nematodes in the family Ascarididae.

Lagochilascaris minor (veterinary parasitology) A species of nematodes in the genus *Lagochilascaris* that may be found in wild felines and didelphoids (American opossums).

Lamarckism (genetics) Discredited thoery proposed by the French natuarlist Jean-Baptiste Lamarck (1744–1829) that evolutionary change could be achieved by the transmission of acquired characteristics from parents to offspring. The theory was superseded by Darwinism.

lambda (1 terminology; 2 genetics) 1 The eleventh letter of the Greek alphabet. 2 One of the most intensively studied of the bacteriophages. It attacks *Escherichia coli*. The genome is double-stranded DNA, but with single-stranded, mutually complementary 'tails' that join up to make a circular DNA after entry into the host cell.

lambda chain (immunology) One of the two light chains of immunoglobulins; the other is the kappa chain.

lambda particle (physics) Type of elementary particle with no electric charge.

lambda point (chemistry) Temperature at which liquid helium (helium I) becomes the superfluid known as helium II.

Lambert's law (physics) Equal fractions of incident light radiation are absorbed by successive layers of equal thickness of the light-absorbing substance. It was named after the German mathematician and physicist J Lambert (1728–77).

Lamblia (parasitology) *See Giardia.*

lambliasis (parasitology) *See* giardiasis.

laminar flow (physics) Streamlined, or non-turbulent, flow in a gas or liquid.

Laminosioptes (parasitology) A genus of mites in the family Laminosioptidae.

Laminosioptes cysticola (avian parasitology) A species of mites in the genus *Laminosioptes* that cause small, flat, oval nodules in the subcutaneous tissue of fowls. Also termed cyst mites.

Lamor precession (physics) Orbital motion of an electron about the nucleus of an atom when it is subjected to a small magnetic field. The electron processes about the direction of the magnetic field. Named after the British physicist J Larmor (1857–1942).

lance (equipment) 1 Lancet. 2 To cut or incise with a lancet.

lanceolate fluke (parasitology) *See Opisthorchis tenuicollis.*

lancet (equipment) A small, pointed, two-edged surgical knife.

lancet fluke (parasitology) *See Dicrocoelium dendriticum.*

landscape epidemiology (epidemiology) Epidemiology of a disease in relation to the entire ecosystem under study.

Lankesterella (avian parasitology) A genus of protozooan parasites of lymphocytes and monocytes of birds.

lanolin (chemistry) Yellowish sticky substance obtained from the grease that occurs naturally in wool. It is used in cosmetics, as an ointment and in treating leather. Also termed: lanoline; wool fat.

lanthanide (chemistry) Member of the Group IIIB elements of atomic number 57 to 71. The properties of these metals are very similar, and consequently they are difficult to separate. Also termed: lanthanoid; rare-earth element.

lanthanum (La) (chemistry) Silver-white metallic element in Group IIB of the periodic table, the parent element of the lanthanide series. It is used in making lighter flints. At. no. 57; r.a.m. 138.9055.

laparo- (terminology) A word element. [German]. Prefix denoting *loin* or *flank, abdomen.*

laparoscopy (medical) Inspection of intra-abdominal structures through fibreoptic (*USA*: fiberoptic) scope. Very widely used in gynaecology (*USA*: gynecology).

lapine canker (veterinary parasitology) Inflammation of the ears of rabbits caused by the mites *Psoroptes communis* or *Chorioptes cuniculi* in which the ear canal is filled with an accumulation of serum and sebaceous material.

LaPlace law (physics) An alternative term for Ampere's law.

larder beetle (parasitology) *See Dermestes lardarius.*

large (biology) Dimensionally big.

large chicken louse (parasitology) *See Goniodes gigas.*

large intestine (anatomy) *See* colon; intestine.

large liver fluke (parasitology) *See Fascioloides magna; Fasciola gigantica.*

large roundworm (parasitology) *Ascaris suum. See Ascaris.*

large stomach worm (parasitology) *See Haemonchus.*

large strongyle (parasitology) *See Craterostomum, Oesophagodontus, Strongylus asini, Strongylus edenatus, Strongylus equinus, Strongylus vulgaris* and *Triodontophorus* spp.

larva (parasitological physiology) An independent, immature stage in a life cycle in which the stage is unlike the parent and must undergo changes in form and size to reach the adult stage. There may be one or several, three is common, larval stages in the one life cycle. In fish larvae are also called fry.

larva currens (parasitology) A variant of larva migrans caused by *Strongyloides stercoralis*, in which the linear progress of the lesions is much more rapid.

larval (parasitology) 1 Pertaining to larvae. 2 Larvate.

larval migrans (parasitology) *See* cutaneous larva migrans; visceral larva migrans.

larvate (medical) Masked; concealed: said of a disease or of a clinical sign of a disease.

larvicide (control measures) A substance that kills insect larvae.

laryngitis (medical) Inflammation of the larynx.

larynx (anatomy) Region of trachea (windpipe) that usually houses the vocal cords (composed of membrane folds that vibrate to produce sounds).

laser (physics) An abbreviation of 'l'ight 'a'mplification by 's'timulated 'e'mission of 'r'adiation.

Lasioacarus (insect parasitology) A species of mites in the Astigmata, that are cleptoparasitic to the bees *Apis mellifera.*

***Lasioseius* sp.** (insect parasitology) Mites in the Mesostigmata that are parasites to the ants *Atta sexdens.*

lassitude (medical) Weakness; exhaustion.

latency period (parasitology) The period of time between the application of an agent to a living organism and a demonstrable effect of such application. In relation to malaria, the duration between the primary attack of malaria and the relapse. There are no parasites in the circulation during this period.

latent (medical) Dormant or concealed; not manifest; potential.

latent heat (physics) Heat energy that is needed to produce a change of state during the melting (solid-to-liquid change) or vaporisation (*USA,vaporization*) (liquid-to vapour/gas change) of a substance; it causes no rise in temperature. This heat energy is released when the substance reverts to its former state (by freezing/ solidifying or condensing/liquefying).

latent trait (genetics) Any trait that is not expressed in an organism's phenotype, but that can be passed on through its genes to descendants.

lateral (anatomy) Lying towards the side; away from the mid-line. *Contrast* medial.

lateral dorsal nucleus (cytology) A nucleus that projects to the cingulate gyrus.

lateral horn (histology) Grey matter lying on either side of the spinal cord between the dorsal and ventral horns, and forming part of the autonomic system.

latex (chemistry) Milky fluid produced in some plants after damage, containing sugars, proteins and alkaloids. It is used in manufacture, *e.g.* of rubber. A suspension of synthetic rubber is also called latex.

latin square design (statistics) An experimental design whose aim is to remove experimental error due to variation arising from two sources. The design is identified with the rows and columns of a square. The number of conditions arising from each source and the number of treatments are the same, and each treatment occurs once in each column and row.

LATS (physiology/biochemistry) An abbreviation of 'L'ong-'A'cting 'T'hyroid 'S'timulator.

lattice theory (immunology) The interaction of multivalent antigen with multivalent antibody will, at optimum proportions of each (zone of equivalence), result in the formation of a lattice and a precipitate. In the zone of equivalence all antibody and antigen are bound in the lattice. With an excess of either antigen or antibody, the lattice is incomplete or disassembled so that the amount of precipitate is reduced; the basis for precipitin tests.

laughing gas (chemistry) An alternative term for dinitrogen oxide.

Laurer's canal (parasitological anatomy) A term used in relation to trematodes. A tubular structure connecting the oviduct with the exterior. Also termed: copulatory canal.

lauric acid (chemistry) $CH_3(CH_2)_{10}COOH$ White crystalline carboxylic acid, used in making soaps and detergents. Also termed: dodecanoic acid; dodecylic acid.

lavage (medical) 1 Irrigation or washing out of an organ or cavity, as of the stomach or intestine. 2 To wash out, or irrigate.

law (science) Simple statement or mathematical expression for the generalisation (*USA, generalization*) of results relating to a particular phenomenon or known facts.

lawrencium (Lr) (chemistry) Lr Radioactive element in Group IIIB of the periodic table (the last of the actinides). At. no. 103; r.a.m. 257 (most stable isotope).

laxative (pharmacology) A medicine that loosens the bowel contents and encourages evacuation. A laxative with a mild or gentle effect on the bowels is also known as an aperient; one with a strong effect is referred to as a cathartic or a purgative.

LCAT (physiology/biochemistry) 'L'ecithin-'C'holesterol 'A'cyl'T'ransferase.

LCD (physics) An abbreviation of 'l'iquid-'c'rystal 'd'isplay.

LCM (mathematics) An abbreviation of 'l'owest 'c'ommon 'm'ultiple.

LCR (physiology) An abbreviation of 'L'ocus 'C'ontrol 'R'egion.

ld (toxicology) An abbreviation of 'l'ethal 'd'ose.

LD5O (toxicology) A term used to denote the dose of a substance that will kill 50 per cent of the organisms receiving it. LD50 tests are used on laboratory animals for testing the toxicity of drugs, etc. designed for human use.

LDH (physiology/biochemistry) An abbreviation of 'L'actate 'D'e'H'ydrogenase.

LDL (biochemistry) An abbreviation of 'L'ow-'D'ensity 'L'ipoprotein.

Le Chatelier's principle (physics) If a change occurs in one of the factors (such as temperature or pressure) under which a system is in equilibrium, the system will tend to adjust itself so as to counteract the effect of that change. It was named after the French physicist H Le Chatelier (1850–1936). Also termed: Le Chatelier-Braun principle.

leaching (chemistry) Washing out of a soluble material from a solid by a suitable liquid.

lead (Pb) (chemistry) Silver-blue poisonous metallic element in Group IVA of the periodic table, obtained mainly from its sulphide (*USA*: sulfide) ore galena. Various isotopes of lead are final elements in a radioactive decay series. The metal is used in building, as shielding against ionizing radiation, as electrodes in accumulators and in various alloys (such as solder, metals for bearings and type metal). Its inorganic compounds are employed as pigments; tetraethyllead is used as an anti-knock agent in petrol. At. no. 82; r.a.m. 207.19.

lead arsenate (control measures) A substance which has been used as an insecticidal spray in orchards.

lead dioxide (chemistry) An alternative term for lead(IV) oxide.

lead equivalent (radiology) Factor that compares any form of shielding against radioactivity to the thickness of lead that would provide the same measure of protection.

lead monoxide (chemistry) An alternative term for lead(II) oxide.

lead tetraethyl(IV) (chemistry) An alternative term for tetraethyllead.

lead(II) oxide (chemistry) PbO Yellow crystalline substance, used in the manufacture of glass. Also termed: lead monoxide; litharge.

lead(IV) (chemistry) An alternative term for plumbic.

lead(IV) oxide (chemistry) PbO_2 Brown amorphous solid, a strong oxidising (*USA, oxidizing*) agent, used in lead-acid accumulators. Also termed: lead dioxide; lead peroxide.

lead(II) (chemistry) An alternative term for plumbous.

leader sequence (molecular biology) The part of a molecule of messenger RNA between the 5' end and the start of the coding sequence. It contains sequences that are concerned with binding to the ribosome.

leaders (histology) The popular name for tendons.

least squares (statistics) A method of regression analysis. The line on a graph that best summarizes the relationship between two variables is the one that ensures that there is the least value of the sum of the squares of the deviation between the fitted curve and each of the original data points.

Lecanicephalidea (fish parasitology) An order of flatworms in subclass Eucestoda, class Cestoidea and the phylum Platyhelminthes. Lecanicephalideans are tapeworms of the intestine of elasmobranchs in which the scolex is divided horizontally into anterior and posterior regions: the scolex may have four small suckers plus hooks and/or tentacles. There are four distinct families.

lecithin (biochemistry) Type of phospholipid, a glyceride in which one organic acid residue is replaced by a group containing phosphoric acid and the base choline. It is a major component of a cell membrane and found in large amounts in the brain and nerves as well as semen and the yolk of eggs. Also termed: phosphatidyl choline.

lecitho- (terminology) A word element. [German] Prefix denoting *the yolk of an egg, ovum.*

lectins (biochemistry) A general term for proteins or glycoproteins of non-immune origin that have multiple highly specific carbohydrate-binding sites.

LED (physics) An abbreviation of 'l'ight-'e'mitting 'd'iode.

Lee biopsy needle (equipment) A needle designed for cutting and aspiration of tissue for biopsy, particularly of aerated lung tissue.

leech (parasitology) Any of the annelids of the class Hirudinea, especially *Hirudo medicinalis*; some species are bloodsuckers, and used for drawing blood.

LEED (physics) An abbreviation of 'l'ow-'e'nergy 'e'lectron 'd'iffraction.

leg mange (parasitology) *See* chorioptic mange.

leg-itch mite (veterinary parasitology) A mite of sheep. *See Eutrombicula sarcina.*

***Leidyana* sp.** (insect parasitology) A protozoal species of gregarinida that are parasites to the bees *Apis mellifera*. They may be found attached to epithelium of the midgut.

Leiperia (veterinary parasitology) A genus of parasites in the class Pentastomida which may be found in crocodiles.

Leipernema (parasitology) A genus of worms of the family Kathlaniidae.

Leipernema galebi (veterinary parasitology) A species of worm in the genus *Leipernema* which may be found in the intestine of Indian elephants.

Leipernema leiperi (veterinary parasitology) A species of worm in the genus *Leipernema* which may be found in the intestine of the African elephant and hippopotamus.

Leishman stain (histology) A histological staining technique used to make blood cells and parasites visible under the microscope.

Leishman-Donovan bodies (parasitology) Round or oval bodies found in reticuloendothelial cells, especially those of the spleen and liver, in visceral leishmaniasis; they are amastigote intracellular stages of *Leishmania donovani*. The term is also used to designate similar forms of *Leishmania tropica* found in macrophages in lesions of cutaneous leishmaniasis.

Leishmania (parasitology) A genus of protozoan parasites transmitted by sandflies, which also act as intermediate hosts.

Leishmania adleri (veterinary parasitology) A species of protozoan parasites in the genus *Leishmania* which may be found in lizards and other mammals.

Leishmania aethiopica (veterinary parasitology) A species of protozoan parasites in the genus *Leishmania* whose reservoir hosts are hyraxes.

Leishmania brasiliensis brasiliensis (human/veterinary parasitology) A species of protozoan parasites in the genus *Leishmania* which causes mucocutaneous leishmaniasis in humans, and whose reservoir hosts are forest rodents.

Leishmania brasiliensis guyanensis (human/veterinary parasitology) A species of protozoan parasites in the genus *Leishmania* which may infect dogs and humans, the disease in the latter being the cutaneous form in most cases.

Leishmania brasiliensis panamensis (veterinary parasitology) A species of protozoan parasites in the genus *Leishmania* whose reservoirs are sloths, kinkajous and many other forest animals.

Leishmania chagasi (human/veterinary parasitology) A species of protozoan parasites in the genus *Leishmania* which causes visceral leishmaniasis in humans and dogs.

Leishmania donovani (human/veterinary parasitology) A species of protozoan parasites in the genus *Leishmania* which causes visceral leishmaniasis in humans and in carnivores.

Leishmania enriettii (veterinary parasitology) A species of protozoan parasites in the genus *Leishmania* which causes cutaneous leishmaniasis in guinea pigs.

Leishmania infantum (human/veterinary parasitology) A species of protozoan parasites in the genus *Leishmania* which causes visceral leishmaniasis in dogs and other carnivores. In humans it is children who are most commonly affected.

Leishmania major (human/veterinary parasitology) A species of protozoan parasites in the genus *Leishmania* whose reservoir hosts are dogs and bush mammals. In humans this is the cause of oriental sore, the important cutaneous form of the disease.

Leishmania mexicana amazonensis (human/veterinary parasitology) A species of protozoan parasites in the genus *Leishmania* which causes cutaneous leishmaniasis in humans. Rodents and bush animals are reservoir hosts.

Leishmania mexicana mexicana (human/veterinary parasitology) A species of protozoan parasites in the genus *Leishmania* that causes cutaneous leishmaniasis in humans and whose reservoir hosts are rodents.

Leishmania mexicana pifanoi (human parasitology) A species of protozoan parasites in the genus *Leishmania* which causes chronic cutaneous leishmaniasis in humans.

Leishmania peruviana (human/veterinary parasitology) A species of protozoan parasites in the genus *Leishmania* which causes cutaneous leishmaniasis in humans and probably infests dogs.

Leishmania tropica (human/veterinary parasitology) A species of protozoan parasites in the genus

Leishmania which causes cutaneous leishmaniasis in humans and dogs.

leishmaniasis (human/veterinary parasitology) Infection with a protozoon of the genus *Leishmania*, which occurs in humans, domestic Canidae and

rodents, is transmitted by sandflies, and is an important zoonosis throughout the world, except Australia. Three forms occur: (l) Visceral (kala-azar, black fever, dumdum fever), caused by *Leishmania donovani*; (2) Cutaneous (Oriental sore, Bagdad boil, Aleppo button), caused by *Leishmania tropica*; and (3) Mucocutaneous, caused by *Leishmania brasiliensis*. Named after the English bacteriologist, Sir W. Leishman (1865–1926).

leishmanoid (parasitology) Like leishmaniasis.

lelapid mite (parasitology) *See Echinolaelaps echidninus.*

lens (optics) Any transparent substance with two opposite surfaces that refract light, most often used of a disc having a spherical curvature on one or both sides, which through refraction can be used to form an image, to magnify an image, etc. By analogy also a current-carrying coil that focuses a beam of electrons (as in an electron microscope). The mammalian eye contains a single lens with convex faces, lying behind the cornea. By changing its convexity, it brings the image of objects at different distances into correct focus on the retina. The refractive power of the human lens is about 20 diopters. *See* accommodation.

lens fluke (parasitology) *See Diplostomum spathaceum.*

lentigenic (parasitology) A host-parasite relationship in which the host shows little or no disease, the relationship persists for a very long time, and the host often survives or dies of intercurrent disease.

Lenz's law (physics) When a wire moves in a magnetic field, the electric current induced in the wire generates a magnetic field that tends to oppose the motion. It was named after the Russian physicist H Lenz (1804–65).

***Lepeophtherius* salmonis** (fish parasitology) The salmon louse.

Lepotrombidium (parasitology) *See* harvest mites.

lepto- (terminology) A word element. [German] Prefix denoting *slender, delicate.*

leptokurtic (statistics) Of a frequency distribution, having a sharper peak than a reference distribution such as the normal curve.

Leptomonas (parasitology) A genus of the protozoan family Trypanosomatidae, the members of which are found in invertebrates.

Leptomonas apis (insect parasitology) A protozoal species of flagellates that are commonly found parasites to the bees *Apis mellifera* but with no apparent harmful effects.

Leptopsylla segnis (veterinary parasitology) A flea found on house mice, rats and field rodents.

Leptopsylla segnis (veterinary parasitology) A species of flea whose principal hosts are house mice, rats, and wild rodents.

leptotene (genetics) One of the stages in meiosis. One of the two processes of cell division.

Leptotrombidium (human parasitology) Genus of the family Trombiculidae. The mites act as vectors of the human scrub typhus agent.

Leptus (insect parasitology) *See Imparipes.*

Lernaea (fish parasitology) A genus of crustaceans in the class Crustacea which attach themselves to the skin of freshwater fish causing ulceration. Affected fish swim sluggishly and grow poorly, and the mortality may be heavy. *Lernaea ciprinacea* is the common species. Also termed: anchor worms.

LES (physiology) An abbreviation of 'L'ower 'E'sophageal 'S'phincter.

lesion (medical) A region of damage to an organ, whether produced deliberately or by pathology.

lesion nematode (plant parasitology) *See Pratylenchus penetrans.*

lesser liver fluke (parasitology) *See Dicrocoelium dendriticum.*

lethal mutation (genetics) Any mutation that has such a severe effect on the organism as to cause early death. From the genetic point of view, a mutation is considered 'lethal' if the death occurs at any time before the individual reaches reproductive age. Common usage in relation to humans, however, includes only those mutations that cause death before birth or in early infancy. Lethal mutations can be dominant, recessive or X-linked, and there are some known as 'semi-lethals'. Lethal mutations are quite common in humans.

Leu. (biochemistry) An abbreviation of 'Leu'cine.

leuc(o)- (terminology) A word element. [German] Prefix denoting white, leucocyte.

leucine (Leu.) (biochemistry) $(CH_3)_2CHCH_2CH(NH_2)COOH$ Colourless (*USA*: colorless) crystalline amino acid; a constituent of many proteins. Also termed: 2-amino-4-methylpentanoic acid.

leucine-amino peptidase stain (histology) An enzyme histochemical staining technique.

leuco- (general terminology) Prefix denoting white; *e.g.* leucocyte.

Leucochloridiomorpha (parasitology) A genus of intestinal flukes (digenetic trematodes).

Leucochloridiomorpha constantiae (veterinary/avian parasitology) A species of intestinal flukes (digenetic trematodes) in the genus *Leucochloridiomorpha* that may be found in the intestine of raccoons and the bursa of fabricius in ducks.

leucocyte (haematology) Any kind of white blood cell. They are classified as polymorphonuclear leucocytes or polymorphs, lymphocytes and monocytes. Granulocytes are further divided into neutrophils, basophils or eosinophils, according to their staining characteristics. In health they number approximately 4 to 11 $\times$ 10^9 per litre in the blood, 60–70% of which are neutrophils. Their number is raised (leucocytosis) in inflammatory conditions and abnormally low (leucopenia) in other conditions, such as poisoning of the bone marrow by drugs. (*USA*: leukocyte).

leucocytosis (haematology) Elevated white blood cell count. (*USA*: leukocytosis).

Leucocytozoon (avian parasitology) A protozoan parasite of avian erythrocytes transmitted by the 'blackfly' *Simulium* spp. Infection causes the disease leucocytozoonosis.

Leucocytozoon andrewsi (avian parasitology) *See Leucocytozoon caulleryi.*

Leucocytozoon bonasae (avian parasitology) A species of protozoan parasite which may be found in grouse and ptarmigan.

Leucocytozoon caulleryi (avian parasitology) (synonym: *Leucocytozoon* andrewsi) A species of protozoan parasite which may be found in chickens, and that causes leucocytozoonosis.

Leucocytozoon grusi (avian parasitology) A species of protozoan parasite which may be found in cranes.

Leucocytozoon mansoni (avian parasitology) A species of protozoan parasite which may be found in grouse.

Leucocytozoon marchouxi (avian parasitology) A species of protozoan parasite which may be found in doves and pigeons.

Leucocytozoon sabrazesi (avian parasitology) A species of protozoan parasite which may be found in fowl.

Leucocytozoon sakharoffi (avian parasitology) A species of protozoan parasite which may be found in crows and other wild birds.

Leucocytozoon schoutedeni (avian parasitology) A species of protozoan parasite which may be found in chickens only in east Africa.

Leucocytozoon simondi (avian parasitology) (synonym: *Leucocytozoon anseris*; *Leucocytozoon anatis*) A species of protozoan parasite which may be found in ducks and geese.

Leucocytozoon smithi (avian parasitology) A species of protozoan parasite which may be found in turkeys.

leucocytozoonosis (avian parasitology) An acute disease in young birds caused by infection with *Leucocytozoon* spp. and characterized by emaciation, weakness and debility. Death is usual within a few days and the mortality rate may be very high. In adult birds the course is longer and the signs may include dyspnea.

leucopenia (haematology) Low white blood cell count. (*USA*: leukopenia).

leucorrhoea (medical) An abnormal white discharge from the vagina. (*USA*: leukorrhoea).

leukocytozoonosis (parasitology) *See* leucocytozoonosis.

levamisole (pharmacology) A broad-spectrum anthelmintic of proven efficiency against gastrointestinal and lung worms that can be administered orally, by injection and by pour-on in cattle. It has no effect on tapeworms or liver fluke; it has a secondary effect in enhancing depressed immune responsiveness by stimulation of T lymphocytes and phagocytosis.

level (statistics) A sub-classification of a factor. The level of a factor corresponds to the value of an independent or intervening variable. A 'fixed level' is any one level of a factor where all possible levels are investigated. A 'random level' is a level chosen to be representative of the factor levels where only some levels are investigated.

levelling effect (statistics) The tendency after many trials for scores to cluster closely around the mean.

levo- (terminology) A word element. [Latin] Prefix denoting *left*.

Lewis acid and base (chemistry) Concept of acids and bases in which an acid is defined as a substance capable of accepting a pair of electrons, whereas a base is able to donate a pair of electrons to a bond. It was named after the American chemist G Lewis (1875–1946).

LFO (physics) An abbreviation of 'L'ow 'F'requency 'O'scillator.

LH (biochemistry) An abbreviation of 'L'uteinising 'H'ormone.

LHRF (biochemistry) An abbreviation of 'L'uteinising 'H'ormone 'R'eleasing 'F'actor.

LHRH (physiology/biochemistry) An abbreviation of 'L'uteinizing 'H'ormone-'R'eleasing 'H'ormone.

Li (chemistry) The chemical symbol for lithium.

libido (physiology) The sexual impulse.

Libyostrongylus (parasitology) A genus of nematodes in the family Trichostrongylidae.

Libyostrongylus douglassii (avian parasitology) A species of nematodes in the genus *Libyostrongylus* that causes proventriculitis in ostriches. Young birds are emaciated, poorly grown, anaemic (*USA*: anemic) and death losses may be heavy.

lice (parasitology) Plural of louse.

licked beef (veterinary parasitology) Diffuse, often haemorrhagic (*USA*: hemorrhagic), jelly-like oedema (*USA*:edema) along the back of a beef carcass, caused by warble fly larvae.

lien(o)- (terminology) A word element. [Latin] Prefix denoting *spleen*.

life cycle (biology) Progressive sequence of changes that an organism undergoes from fertilization to death. In the course of the cycle a new generation is usually produced. Reproduction may be sexual or asexual, both meiosis and mitosis may occur.

ligament (histology) Tough elastic connective tissue that hold the bones together and stabilise the joints.

ligand (1 chemistry; 2 biochemistry) 1 Any molecule or ion that has at least one electron pair that donates its electrons to a metal ion or other electron acceptor, often forming a co-ordinate bond. 2 Any molecule that interacts with or binds to a receptor that has an affinity for it.

ligase (biochemistry) Enzyme that repairs damage to the strands that make up DNA, widely used in recombination techniques to seal the joins between DNA sequences.

ligation (molecular biology) The joining together of separate sequences of DNA by DNA ligase enzymes. It occurs during replication, DNA repair, and is also widely used in genetic engineering to stick DNA molecules together *in vitro*.

ligature (medical) A piece of material, *e.g.* thread or catgut, used to tie off blood vessels or other structures.

light (physics) Visible part of the electromagnetic spectrum, of wavelengths between about 400 and 760 nanometres.

light intensity (physics) The radiant energy emitted per unit time. The SI unit of measurement is the watt.

light meter (physics) Instrument for measuring levels of illumination (*e.g.* in photography), usually by means of a photocell.

light microscope (microscopy) Used for examining unstained or stained particles or the cellular structure of tissues that have been cut into sections and stained. It has a resolving power of 0.2 µm. Modern light microscopes have an eyepiece and objective lenses which provide magnification, and a condenser beneath the stage which gathers and focuses light on the object being examined.

light-emitting diode (LED) (physics) Semiconducting diode that gives off light, used to form letters and numbers on a display panel.

lignocaine (pharmacology) A commonly used local anaesthetic (*USA*: anesthetic).

Ligula (parasitology) A genus of tapeworms in the family Diphyllobothriidae.

Ligula intestinalis (fish/avian/veterinary parasitology) A species of tapeworms in the genus *Ligula*, the plerocercoids of which are found in the body cavity of freshwater fish, where they cause infertility and loss of condition. The adults are found in the alimentary canal of piscivorous birds.

likelihood (statistics) Probability.

likelihood ratio (statistics) The ratio of the likelihood of obtaining an observed set of data under one hypothesis to the likelihood of obtaining the same data under another. It is widely used in significance tests.

lime (chemistry) General term for quicklime (calcium oxide, CaO), slaked lime and hydrated lime (both calcium hydroxide, $Ca(OH)_2$). They are obtained from limestone.

lime water (chemistry) Solution of calcium hydroxide ($Ca(OH)_2$) in water, used as a test for carbon dioxide (which turns lime water milky when bubbled through it due to the precipitation of calcium carbonate; after prolonged bubbling, the solution goes clear again due to the formation of soluble calcium hydrogencarbonate).

Limey disease (avian parasitology) Renal coccidiosis in birds.

limit (mathematics) Value to which a sequence or series tends as more and more terms are included.

limiting friction (physics) Maximum value of a frictional force.

limiting step (chemistry) An alternative term for rate-determining step.

Limnatis (parasitology) A genus of leeches in the class Hirudinea, that live in water and infest animals that pass through or drink the water.

Limnatis africana (human/veterinary parasitology) A species of leeches in the genus *Limnatis* that may be found in the nasal cavity, vagina and urethra of humans, dogs and monkeys.

Limnatis nilotica (parasitology) A species of leeches in the genus *Limnatis* that may be found in the pharynx and nasal cavity in all species, and which causes anaemia (*USA*: anemia), local oedema (*USA*: edema)and obstruction.

lindane (pharmacology/control measures) The gamma isomer of benzene hexachloride used as a topical pediculicide and scabicide. Also termed: gamma benzene hexachloride.

line breeding (genetics) The technique of breeding from animals as closely related as possible.

linea alba (histology) A whitish line running down the middle of the muscular wall of the abdomen, formed by the junction of the flat tendons of the external oblique, internal oblique and transverse muscles after they have split to enclose the two longitudinal rectus muscles.

lineage (genetics) A sequence of species through evolutionary time, from the ancestral species to its present-day descendants.

linear (1 statistics; 2 mathematics) 1 Pertaining to a straight line or plane. 2 Describing an equation of the first order. *See* linear equation.

linear absorption coefficient (physics) Measure of a medium's ability to absorb a beam of radiation passing through it, but not to scatter or diffuse it.

linear accelerator (physics) Apparatus for accelerating charged particles.

linear attenuation coefficient (physics) Measure of a medium's ability to diffuse and absorb a beam of radiation passing through it.

linear energy transfer (LET) (physics) Linear rate of energy dispersion of separate particles of radiation when they penetrate an absorbing medium.

linear equation (mathematics) Algebraic equation of the first order. In co-ordinate geometry, the equation of a straight line, general formula $y = mx + c$, where m is the gradient (slope) of the line and c (a constant) is its intercept with the *y-axis*.

linear function (1 statistics; 2 general terminology) 1 A sum of weighted variables; they can be graphically expressed as a line or plane. 2 Any function describing the relationship of output to input in a linear system.

linear molecule (chemistry) Molecule whose atoms are arranged in a line.

linear regression (statistics) Fitting a straight line to the data points produced when one variable is plotted against another.

linear system (statistics) A system in which the output varies in direct proportion to the input, and in which if there is more than one input, the response is proportional to the weighted sum of the inputs.

linear transformation (statistics) A transformation of scores (*e.g.* test scores) that changes their mean and standard deviation, but retains other aspects of the original distribution. *Compare* normalization.

Lineweaver-Burk plot (statistics) One of several ways of linearising enzyme- or receptor-binding data, based on derivations of the law of mass action.

Linguatula (parasitology) A genus of parasites in the class Pentastomida.

Linguatula serrata (veterinary parasitology) A species of parasite in the genus *Linguatula* whose adults occur in the nasal cavities of canines and larvae in mesenteric lymph nodes of the horse, goat, sheep and rabbit. They causes sneezing and a bloody nasal discharge, and the larval stages may be confused with tuberculosis.

linguatulosis (parasitology) Disease caused by infection with *Linguatula*. Also termed: linguatuliasis.

linguo- (terminology) A word element. [Latin] Prefix denoting *tongue*.

linkage (genetics) Occurrence of two genes on the same chromosome. Genes that are close together are likely to be inherited together; genes that are further apart may become separated during crossing over.

linkage group (genetics) A group of genes that can be observed to be linked together rather than to assort independently. In practice, a linkage group means a group of genes all on the same chromosome.

Linnaean system (biology) System that classifies and names all organisms according to scientific principles. Each species has two names; the first indicates the organism's general type (genus), the second names the unique species. It was named after the Swedish botanist C Linne (Linnaeus) (1707–78). Also termed: binomial classification.

Linognathus (veterinary parasitology) A genus of sucking lice of the family Linognathidae, mostly parasitic on ungulates.

Linognathus africanus (veterinary parasitology) A species of sucking lice in the genus *Linognathus* that may be found on goats.

Linognathus ovillus (veterinary parasitology) A species of sucking lice in the genus *Linognathus* that may be found on sheep. Also termed: blue louse.

Linognathus pedalis (veterinary parasitology) A species of sucking lice in the genus *Linognathus* that may be found on hairless parts of limbs of sheep. Also termed: foot louse.

Linognathus setosus (veterinary parasitology) A species of sucking lice in the genus *Linognathus* that may be found on foxes and dogs.

Linognathus stenopis (veterinary parasitology) A species of sucking lice in the genus *Linognathus* that may be found on goats.

Linognathus vituli (veterinary parasitology) A species of sucking lice in the genus *Linognathus* that may be found on cattle. Also termed: longnosed cattle louse.

lip(o)- (terminology) A word element. [German] Prefix denoting *fat*, *lipid*.

lipaemia (biochemistry) Abnormally high lipid level in the blood. (*USA*: lipemia)

lipase (biochemistry) In vertebrates, an enzyme in intestinal juice and pancreatic juice that catalyses the hydrolysis of fats to glycerol and fatty acids.

Lipeurus (parasitology) A genus of lice in the suborder Ischnocera.

Lipeurus caponis (avian parasitology) The wing louse of fowl and pheasants.

lipid (biochemistry) Member of a group of naturally occurring fatty or oily compounds that share the property of being soluble in organic solvents, but sparingly soluble in water; they are an important part of the cell membrane. Also, all lipids yield monocarboxylic acids on hydrolysis.

lipolyte (cytology) Lipid-containing cell. Also termed: fat cell.

lipoma (histology) A harmless tumour (*USA*: tumor) composed of fat cells often occurring just below the skin.

lipoprotein (biochemistry) A molecule consisting of a protein complexed with a lipid. They exist in a wide range of forms. Some have a transport function within cells, and others have roles in the metabolism of cholesterol, particularly in the context of some forms of heart disease.

Lipoptena (parasitology) A parasitic genus of flies in the family Hippoboscidae.

Lipoptena caprina (veterinary parasitology) A species of parasitic flies in the genus *Lipoptena* that is a ked of goats.

Lipoptena cervi (veterinary parasitology) A species of parasitic flies in the genus *Lipoptena* that is a parasite of deer, wild boar and badger.

liposoluble (chemistry) Soluble in fats.

liposome (cytology) Droplet of fat in the cytoplasm of a cell, particularly that of an egg.

liq. (general terminology) An abbreviation of 'liq'uid.

liquefaction of gases (chemistry) All gases can be liquefied by a combination of cooling and compression. The greater the pressure, the less the gas needs to be cooled, but there is for each gas a certain critical temperature below which it must be cooled before it can be liquefied.

liquid (liq.) (chemistry) Fluid that, without changing its volume, takes the shape of its container. According to the kinetic theory, the molecules in a liquid are not bound together as rigidly as those in a solid but neither are they as free to move as those of a gas. It is therefore a phase that is intermediate between a solid and a gas.

liquid crystal (chemistry) Compound that is liquid at room temperature and atmospheric pressure but shows characteristics normally expected only from solid crystalline substances. Large groups of its molecules maintain their mobility but nevertheless also retain a form of structural relationship. Some liquid crystals change colour (*USA*: color) according to the temperature.

liquid-crystal display (LCD) (physics) A device displaying letters and numbers on objects such as control panels. These images are produced when an electrical field causes a capsule of transparent liquid crystal to become opaque.

liquid-liquid extraction (chemistry) An alternative term for solvent extraction.

Listrophorus (parasitology) A genus of mites in the family Listrophoridae.

Listrophorus gibbus (veterinary parasitology) A species of mites in the genus *Listrophorus* that may cause a mange-like condition in rabbits.

liter (measurement) *See* litre.

litharge (chemistry) An alternative term for lead(II) oxide.

lithium (Li) (chemistry) Silver-white metallic element in Group IA of the periodic table (the alkali metals), the solid with the least density. Its compounds are used in lubricants, ceramics, drugs and the plastics industry. Lithium, when given as a salt, partially replaces sodium in body tissues, thus affecting the permeability of membranes. It is used to reduce mania, and on a long-term basis, to alleviate manic-depressive illness. At. no. 3; r.a.m. 6.941.

lithium aluminium hydride (chemistry) $LiAlH_4$ Powerful reducing agent, used in organic chemistry. Also termed: lithium tetrahydridoaluminate(III).

litmus (chemistry) Dye made from *Rocella tinctoria* and other lichens, used as an indicator to distinguish acids from alkalis (*USA*: alkalies). Neutral litmus solution or litmus paper is naturally violet-blue; acids turn it red, alkalis (*USA*: alkalies) turn it blue.

Litobothridea (fish parasitology) An order of flatworms in subclass Eucestoda, class Cestoidea and the phylum Platyhelminthes.

litre (l) (measurement) A basic unit of capacity in the metric system, determined by volume of distilled water at 4^0 Celsius. (*USA, liter*). 1 litre = 1.7598 pints.

live vaccine (immunology) A vaccine prepared from live, usually attenuated, microorganisms.

liver (anatomy) In vertebrates, a large organ in the abdomen, the main function of which is to regulate the chemical composition of the blood by removing surplus carbohydrates and amino acids, converting the former into glycogen for storage and the latter into urea for excretion. Its glandular secretion is known as bile, and it is secreted into the small intestine via the common bile duct. In those species having a gall bladder, including humans, the common bile duct originates at the point at which the hepatic duct, draining the intra-hepatic bile passages, comes together with the cystic duct that connects the gall bladder to the common duct. Bile salts function in the digestion of fats in the small intestine. The liver is supplied with blood via the hepatic artery and the hepatic portal vein, and drains into the inferior vena cava via hepatic veins. The liver receives blood, via the hepatic portal vein, from the small intestine, thus receiving the products of digestion and absorption of food. It forms glycogen from glucose, proteins from amino acids and glycerides from fatty acids. It also carries out many reactions of intermediary metabolism. It is responsible for the excretion of bilirubin and biliverdin, products formed by degradation of haem (*USA*: heme).

liver fluke (human/fish/veterinary parasitology) *Clonorchis, Fasciola and Opisthorcis* are genera of flukes which infect the liver of humans. They live in the biliary tract, and their intermediate hosts are snails. *Clonorchis and Opisthorcis* find a further secondary host in freshwater fish, and the infection occurs where people eat raw fish, particularly in the Far East. Flukes may infect fish-eating mammals other than man, such as cats and dogs. *Fasciola*

is the sheep liver fluke, and the snail infects vegetation and water; in sheep and cattle raising countries the infection spreads to humans through infected water. *See* fluke.

liver fluke disease (parasitology) *See* hepatic fascioliasis.

liver function (physiology) Summation of the functions of the liver.

liver function tests (methods) Biochemical tests capable of demonstrating that the liver's functions are, or are not, at full capacity.

lizard poisoning (veterinary parasitology) Infection of lizards with the fluke *Platynosomum fastosum*.

LL (general terminology) An abbreviation of 'L'ower 'L'imb.

LMA (physics) An abbreviation of 'L'ow 'M'oisture 'A'vidity.

LMRSH (education) An abbreviation of 'L'icentiate 'M'ember of the 'R'oyal 'S'ociety for the 'P'romotion of 'H'ealth.

Loa (parasitology) A genus of onchocercid worms in the superfamily Filarioidea.

Loa loa (human/veterinary parasitology) A filarial worm of West Africa, which infests the subcutaneous tissues in humans and primates, causes transient swellings (Calabar swellings) and intense itching, and is sometimes seen crossing the eye beneath the conjunctiva. It is transmitted by *Chrysops* spp, and treated with diethylcarbamazine citrate. *See* filariasis.

loaiasis (parasitology) Infection with nematodes of the genus *Loa*; loiasis.

lobar (bilogy) Pertaining to a lobe.

lobe (anatomy) A well-defined part of an organ, by virtue of their shape, by partitions of connective tissue, or by fissures in the organ. The brain, lungs, liver and thyroid gland are, for example, made up of lobes.

loc. cit. (literary terminology) An abbreviation of 'loc'o 'cit'ato. Latin, meaning 'in the place cited'. Used in papers and dissertations to cite a work previously referred to.

local (general terminology) Restricted to or pertaining to one spot or part; not general.

local anaesthesia (medical) Anaesthesia of a localised area of the body, as opposed to general anaesthesia, when the whole body is rendered insensitive to pain (*USA*: local anesthesia).

lochia (physiology) The normal discharge from the womb after childbirth. It may last for one or two weeks.

locus (1 mathematics; 2 genetics) 1 Path traced by a moving point, *i.e.* a line that can be drawn through adjacent positions of a point, each position of that point satisfying a particular set of conditions. 2 Position of a gene on a chromosome. (pl. 'loci').

Locustacarus (insect parasitology) (synonym: *Bombacarus*) A species of mites in the Prostigmata, that are parasitic to the bees Bombinae, puncturing the trachea and sucking the haemolymph (*USA*: hemolymph).

lodestone (chemistry) Fe_3O_4 Naturally occurring magnetic oxide of iron. Also termed: loadstone; magnetite.

LOEL (toxicology) An abbreviation of 'L'owest 'O'bserved 'E'ffect 'L'evel.

log (mathematics) An abbreviation of 'log'arithm to base 10.

logarithm (mathematics) Number related to an ordinary number in such a way that addition or subtraction of logarithms corresponds to multiplication or division of ordinary numbers. Logarithms are given to a particular base. The logarithm of a number to a given base is the power to which the base must be raised to give the number, *i.e.* if y is a number and x is the base, $y = x^n$ where n is the logarithm of y to the base x; *e.g.* the logarithm of 100 to the base 10 is 2, because 10^2 is 100.

See also characteristic; common logarithm; natural logarithm.

logarithmic curve (mathematics) A curve governed by an equation of the form $y = a.\log x$.

logarithmic phase (biology) Period of maximal growth rate of a micro-organism in a culture medium.

logarithmic scale (measurement) Non-linear scale of measurement. For common logarithms (to the base 10), an increase of one unit represents a tenfold increase in the quantity measured.

logic (1 mathematics; 2 computing) 1 The use of methods from mathematics and formal logic to analyse the underlying principles on which mathematical systems are based. 2 In electronic data-processing systems, the principles that define the interactions of data in the form of physical entities.

logo- (terminology) A word element. [German] Prefix denoting *words*, *speech*.

-logy (terminology) A word element. [German] Suffix denoting *science*, *treatise*, *sum of knowledge in a particular subject*.

Lolium (plant parasitology) Grass genus of the family Poaceae; toxic when infested with the grass nematode *Anguina lolii*. Species includes *Lolium* *multiflorum* (Italian rye grass), *Lolium perenne* (perennial rye), *Lolium rididum* (Wimmera rye grass) and *Lolium temulentum* (darnel).

Lolium rigidum (plant parasitology) An annual pasture grass whose seeds may be parasitized by a gall-inducing grass nematode *Anguina lolii*, and if it is accompanied by a *Clavibacter toxicus* which produces tunicaminyluracils (corynetoxins), it is very poisonous. Also termed Wimmera rye grass.

lone pair of electrons (chemistry) Pair of unshared electrons of opposite spin (in the same orbital) that under suitable conditions can form a co-ordinate bond *e.g.* the nitrogen atom in ammonia has a lone pair of electrons; the oxygen in water has two lone pairs.

lone star tick (parasitology) *See Amblyomma americanum.*

Longidorus (plant parasitology) A genus of nematodes in the order Dorylaimida and family Longidoridae. They are very long, being 2–11 mm in length, slender nematodes with a very long, 44–180 µm, needle-like mouth spear (odonstyle) plus a rather weak basal extension (odontophore). Males have massive, banana-shaped spicules without a gubernaculum but with small lateral accessory pieces, and prominent copulatory muscles in the posterior region cause the tail end to curl ventrally when killed by heat. *See Longidorus africanus.*

Longidorus africanus (plant parasitology) A species of nematodes in the genus *Longidorus* whose hosts cover a broad range including sorghum, snap bean, lima bean, sugarbeet, barley, bermuda grass, corn, wheat, cotton, okra, lettuce, cucumber, eggplant, tomato, grape; fair-poor - oat, sunflower, alfalfa, pea, carrot, cantaloupe, crookneck squash, zucchini, watrmelon, pepper, spinach, spearmint, onion, broccoli, raddish, cabbage, and cauliflower, over a wide distribution including Zimbabwe, Israel, South Africa, and California in the USA. They are migratory ectoparasites, all stages of which are found outside of roots in the soil, that feed at root tips. Males are rare and reproduction is probably by parthenogenesis. First stage juvenile hatches from the egg and develops to the adult form in seven to nine weeks at 28°C. They can survive in fallow soil at least 3 months at 25°C. Galls are formed at root tips, and infestation is devastating to seedling root systems. Control measures include preplant nematicides, timing of planting of fall crops to avoid temperature conditions favourable (*USA*: favorable) to the nematode and planting when soil temperatures are below approximately 22°C. Also termed: needle nematode.

Longistrongylus (veterinary parasitology) A genus of worms in the family Trichostrongylidae. Species includes *Longistrongylus albifrontis* and *Longistrongylus meyeri* which may be found in the abomasum of antelope, gazelle and African buffalo.

longitudinal study (statistics) Any study of a group over a long period of time, usually years. It avoids the problems that arise in retrospective studies and in cross-sectional studies, but cannot take into account the effects of repeated testing.

longitudinal veins (parasitological anatomy) A term used in relation to arthropods. A series of six veins following the subcosta. These are numbered in sequence.

long-necked bladder worm (parasitology) *See Cysticercus tenuicollis.*

long-nosed louse (parasitology) *See Linognathus vituli.*

Lophodispus (insect parasitology) Mites in the Prostigmata that are phoretic on the heads of the ants *Lasius niger.*

Lophophyton (avian parasitology) A genus of fungus that causes dermatitis in fowl which usually accompanies infestation with the mite *Epidermoptes bilobatus.*

loptotene (genetics) Stage of prophase in the first cell division in meiosis. At this stage, chromosomes can be seen to carry chromomeres.

louse (human/veterinary parasitology) Plural: lice; a general name for various species specific parasitic insects, the true lice, which infest mammals and belong to the order Phthiraptera. This is divided into two suborders, Mallophaga, the biting lice, and Anoplura, the sucking lice. They are greyish (*USA*: grayish), wingless, dorsoventrally flattened, and vary in length from about 1.5 to 4 mm. They stimulate rubbing, scratching and restlessness, causing damage to fleece and loss of production in animals. Heavy infestations with sucking lice may cause serious anaemia (*USA*: anemia). Louse infestation is also called pediculosis. Only two of the many species of lice breed on man, *Pediculus* and *Pthirus*. *Pediculus humanus capitis* lives on the head, and is common in schools, where it is spread by direct contact and by combs and hairbrushes. *Pediculus humanus humanus* is the body louse, bigger than the head louse and capable of spreading typhus, relapsing fever and trench fever. It lives in the clothes rather than on the body. *Pthirus pubis* is the crab louse, found in the pubic hair and usually spread by sexual intercourse. The term louse is also used loosely with respect to other external parasites.

louse fly (parasitology) *See Hippobosca.*

low birthweight (physiology) Birthweight less than 2.5 kg. Babies either preterm (two thirds) or small-for-dates (one third).

lowest common denominator (mathematics) Number that is the lowest common multiple of all the denominators of a set of fractions, necessary in order to add or subtract the fractions; *e.g.* the lowest common denominator of 1/2, 1/3 and 1/4 is 12 (because these fractions can be expressed as 6/12, 4/12 and 3/12, and thereby added).

lowest common multiple (LCM) (mathematics) Smallest number that all the members of a group of numbers will divide into; *e.g.* the lowest common multiple of 2, 3, 4 and 5 is 60.

low-pass filter (physics) A filter that transmits from a waveform only frequencies below a given frequency.

LOX (physics) An abbreviation of 'L'iquid 'OX'ygen.

Lp (general terminolgy) An abbreviation of 'L'im'p'.

lp (medical) An abbreviation of 'l'atent 'p'eriod.

Lr (chemistry) The chemical symbol for lawrencium.

LRCVS (education) An abbreviation of 'L'icentiate of the 'R'oyal 'C'ollege of 'V'eterinary 'S'urgeons.

LRP (physiology/biochemistry) An abbreviation of 'L'DL 'R'eceptor-related 'P'rotein.

LSHTM (education) An abbreviation of 'L'ondon 'S'chool of 'H'ygiene and 'T'ropical 'M'edicine.

LTH (1 biochemistry; 2 education) 1 An abbreviation for 'L'uteo'T'ropic 'H'ormone. 2 An abbreviation of 'L'icentiate in 'T'ropical 'M'edicine.

LTP (physiology) An abbreviation of 'L'ong-'T'erm 'P'otentiation.

Lu (chemistry) The chemical symbol for lutetium.

lubricant (physics) Any substance used to reduce friction between surfaces in contact; *e.g.* oil, graphite, molybdenum disulphide (*USA*: molybdenum bisulfide), silcone grease.

luciferase (biochemistry) Enzyme that initiates the oxidation of luciferin.

luciferin (biochemistry) Substance that occurs in the light-producing organ of some animals, *e.g.* firefly. When oxidized (through the action of luciferase) it produces bioluminescence.

Lucilia (veterinary parasitology) A genus of blowflies in the family Calliphoridae. Species includes *Lucilia caesar* and *Lucilia illustris*, flies with bright metallic colors which are called copper-bottle or green-bottle flies, and *Lucilia cuprina and Lucilia sericata* which are important causes of blowfly strike in sheep.

See also cutaneous myiasis.

lufenuron (control measures) A benzoylphenyl urea insecticide which acts as an insect development inhibitor, blocking normal synthesis and deposition of chitin, that may be used to control fleas.

Lugol's iodine (chemistry) A solution of iodine 5% and potassium iodide 10% in water.

lum. (physics) An abbreviation of 'lum'inous.

lumb(o)- (terminology) A word element. [Latin] Prefix denoting *loin.*

lumbar (anatomy) Pertaining to the lower back, a region that comes between the sacral and thoracic regions.

lumbar paralysis (parasitology) Paraplegia generally and specifically due to cerebrospinal nematodiasis.

lumbar puncture (medical) The introduction of a hollow needle into the spinal canal in order to draw off a specimen of cerebrospinal fluid for laboratory examination, or to introduce drugs, spinal anaesthetics (*USA*: spinal anesthetics) or radio-opaque substances for X-ray investigations. The puncture is made between the third and fourth or fourth and fifth lumbar vertebrae, where the point of the needle cannot harm the spinal cord, for it ends at the level of the second spinal vertebra. The puncture is usually made under local anaesthesia (*USA*: local anesthesia).

lumbar tap (medical) *See* lumbar puncture.

lumbricoid (parasitology) Resembling the earthworm; designating the ascaris, or intestinal roundworms.

Lumbriculus variegatus (parasitology) The intermediate host of *Dioctophyme renale*. Also termed: mudworm.

lumbricus (parasitology) Plural: lumbrici [Latin] ascaris.

lumen (1 physics; 2 biology) 1 The SI unit of luminous flux, equal to the amount of light emitted by source of 1 candela through unit solid angle. 2 The space enclosed by a duct, vessel or tubular organ.

luminance (L) (physics) The photometric intensity of the light emitted or reflected by a surface per unit area. The standard unit is candela per square metre.

luminescence (physics) Emission of light by a substance without any appreciable rise in temperature. *See* bioluminescence; fluorescence; incandescence; phosphorescence.

luminosity (physics) The apparent brightness of a surface or light source. It tends to be used to mean the apparent intensity of a light as derived from the luminous efficiency function.

luminous flux (physics) The total amount of light per unit time emitted from a source, as measured by photometry; the standard unit of measurement is the lumen. *See* photometry.

luminous intensity (physics) The luminous flux emitted per unit solid angle (steradian) by a light source; the standard unit of measurement is the candela. *See* photometry.

lunar caustic (chemistry) Silver nitrate.

Lund's fly (parasitology) *See Cordylobia rodhaini.*

lung (anatomy) The paired or single respiratory organ located in the thorax. Its surface contains a large area of thinly folded, moist epithelium membrane so that it occupies little volume. This membrane is richly supplied by blood capillaries which allow for efficient and easy gaseous exchange. Air enters and leaves lungs through the bronchus, which branches into bronchioles ending in clusters of alveoli, where the main gaseous exchange takes place. The respiratory system consists of the lungs, air passages, the muscles that control breathing and the pleural cavities.

lung fluke (parasitology) A term used for flukes that may be found in the lung but also includes *Dasymetra*, *Stomatrema* etc., which infest the mouths of reptiles and can be found in the lungs. *See Paragonimus.*

lung mites (parasitology) *Pneumonyssus*; *Halarachne*; *Orthohalarachne*.

lungworm (parasitology) Any parasitic worm that invades the lungs, but generally refers to worms that preferentially invade the lungs, come to maturity there and either lay their eggs or produce viable larvae there causing chronic cough and respiratory distress.

lungworm disease (parasitology) A disease caused by species-specific lungworms. The worms involved include various members in the genus *Angiostrongulus*, *Dictyocaulus*, *Filaroides*, *Metastrongylus*, *Muellerius* and *Protostrongylus*, and the clinical picture is one of persistent cough and respiratory distress.

lupoid leishmaniasis (parasitology) *See* cutaneous leishmaniasis.

luteal phase (physiology) The stage of the menstrual cycle that starts with ovulation and ends, unless fertilization occurs, with the onset of menstruation; during this stage, progesterone is secreted by the corpus luteum.

Lutetium (Lu) (chemistry) Metallic element in Group IIIB of the periodic table (one of the lanthanides). The irradiated metal is a beta-particle emitter, used in catalytic processes. At. no. 71; r.a.m. 174.97.

Lutzomyia (parasitology) A genus of sandflies, *e.g. Lutzomyia trapidoi*, that is capable of transmitting *Leishmania* spp.

lux (measurement) SI unit of illumination, equal to one lumen per square metre. Also termed: metre-candle.

See also photometry.

luxol fast blue/cresyl violet stain (histology) A histological staining technique used to make myelin, nissl substance visible under the microscope.

luxury gene (genetics) A term rarely used nowadays. A gene that codes for a product specific to one particular type of cell, in contrast to housekeeping genes. An example of a luxury gene would be the gene for insulin, which is active only in certain cells in the pancreas, though present in all cells. Luxury genes are switched off in all cells except the ones where their function is needed. While they are inactive, the DNA of luxury genes is highly condensed as heterochromatin, and it shows up as a dark-staining band.

lv (physics) An abbreviation of 'l'ow 'v'oltage.

LVS (education) An abbreviation of 'L'icentiate in 'V'eterinary 'S'cience.

lye (chemistry) Solution of strong caustic alkali (*e.g.* potassium hydroxide, sodium hydroxide).

Lyme disease (medical) First described as the result of cases occurring in Lyme in the USA in 1977, the disease is caused by the micro-organism *Borrelia bergdorferi*, and spread by ticks, mosquitoes and biting flies. The disease may continue, some months after the original infection, to involve the joints, especially the knee; the heart, with irregular pulse, shortness of breath and pain in the chest; and the central and peripheral nervous system, with encephalitis and neuritis. Similar illnesses have occurred in Europe where the tick *Ixodes ricinus* is found to carry the disease, possibly from dogs.

Lymnaea (parasitology) A genus of snails, some of which act as intermediate hosts of *Fasciola hepatica*. The snails involved include *Lymnaea truncatula*, *Lymnaea tomentosa*, *Lymnaea columella*, and *Lymnaea viridis*.

lymph (histology) The fluid found in the lymph vessels; it is clear and slightly yellow, containing lymph cells and, if it derives from intestinal vessels, particles of fat. It is similar in salt concentration to plasma, but possesses a lower protein concentration. It originates in the tissue spaces, being derived from the fluid which filters through the walls of the capillary blood vessels. It may contain particles as big as bacteria if it is draining from an infected area.

lymph node (histology) Flat, oval structure made of lymphoid tissue that lies in the lymphatic vessels and occurs in clusters in the neck, armpit or groin. Its main function is the manufacture of antibodies and leueocytes. Lymph nodes also act as a defence barrier against the spread of infection by filtering out foreign bodies and bacteria, thus preventing their entry into the bloodstream. Also termed: lymph gland. Also known as lymph glands.

lymph(o)- (terminology) A word element. [Latin] Prefix denoting *lymph*, *lymphoid tissue*, *lymphatics*, *lymphocytes*.

lymphadenitis (histology) Inflammation of lymph glands, or nodes.

lymphadenopathy (medical) Disease of the lymph nodes.

lymphatic system (histology) Tissue fluid, consisting mainly of fluid forced out of the capillaries by the pressure of circulating blood, is known as lymph, and is in part recirculated by a system known as the lymphatic system. The lymphatic system arises as capillaries in the tissues that come together to form an interconnecting network of progressively larger vessels, similar to veins in structure. Flow is maintained largely by contraction of body muscles with valves preventing back flow; the system empties into the vena cava via the thoracic duct. As lymph moves through the system it passes through lymph nodes (lymph glands is not correct usage). Lymph nodes contain T and B cells derived from the bone marrow stem cells and remove foreign particles by phagocytosis. The spleen and the thymus are both largely composed of lymphoid tissue.

See also spleen; thymus.

lymphocytes (haematology) The primary nucleated cells of the lymphatic system. Resting lymphocytes are small, densely staining and have little cytoplasm. Activated lymphocytes enlarge and have increased cytoplasm. Lymphocytes comprise 20–80% of the nucleated blood cells and more than 99% of the cells in the lymphatic system. Lymphocytes originate from stem cells in the bone marrow and migrate to secondary lymphoid organs for further maturation. The two major classes of lymphocytes, which are morphologically indistinguishable, are the B cells, the effector cells of the humoral immune response, and T cells, the effector cells of the cellular immune response. Lymphocytes circulate through the blood stream and lymphatic vessels, passing through the spleen and lymph nodes, which have filtered and retained antigen for presentation to the B and T cells. They enter the lymph nodes from the bloodstream via high endothelial venules (HEV) in the node, percolate through the lymph node and exit via the efferent lymphatics that drain into the venous system at the thoracic duct. The spleen is primarily a blood-filtering organ, and lymphocytes enter and exit via the capillaries.

lymphoid tissue (histology) Tissue found in dense aggregations in lymph nodes, tonsils, thymus and spleen. It produces lymphocytes and macrophagocytic cells, which ingest bacteria and other foreign bodies. Also termed: lymphatic tissue.

lymphokines (biochemistry) Biologically active molecules produced by lymphocytes that have diverse effects on many different cell types. They can be produced in response to stimulation by antigen, cell contact or other lymphokines. The majority of lymphokines have been defined only functionally, although the structures of a few, such as interleukin are known. Most lymphokines have actions on several different target cells.

lymphoma (medical) A number of conditions that involve swelling of the lymph nodes which is not caused by inflammation, metastatic malignant disease or Hodgkin's disease, have been called lymphomas. They are rare, and the diagnosis is made on microscopical examination of lymph nodes removed at biopsy.

Lynxacarus (parasitology) A genus of mites. Species include the cat fur mite *Lynxacarus radovsky*, that may cause pruritus and exudative, crusted skin lesions.

lyophilic (chemistry) Possessing an affinity for liquids.

lyophilized (chemistry) Freeze dried.

lyophobic (chemistry) Liquid-repellent, having no attraction for liquids.

Lys (biochemistry) An abbreviation of 'Lys'ine.

lyse (biology) 1 To cause or produce disintegration of a compound, substance or cell. 2 To undergo lysis.

lysine (chemistry) $H_2N(CH_2)_4CH(NH_2)COOH$ Essential amino acid that occurs in proteins and is responsible for their base-neutralizing powers because of its two NH_2 groups. Also termed: diaminocaproic acid.

lysis (biology) The break-up of a cell after the rupture of its cell wall. This can be the result of attack by chemicals (*e.g.* detergents) or infection (as happens when bacteria are invaded by bacteriophages), or the cell may simply dissolve itself (autolysis).

-lysis (terminology) A word element. [German] Suffix denoting *dissolution*.

lysogeny (genetics) The colonisation of a host cell by a virus, which gets itself replicated by the host's genetic machinery.

lysosome (cytology) Membrane-bound organelle of eukaryotes that contains a range of digestive enzymes, such as proteases, phosphatases, lipases and nucleases. The functions of lysosomes include contributing enzymes to white blood cells during phagocytosis and the destruction of cells and tissue during normal development. Lysosomes may be produced directly from the endoplasmic reticulum or by budding of the Golgi apparatus.

lysozyme (biochemistry) Enzyme in saliva, egg white, tears and mucus. It catalyses the destruction of bacterial cell walls by hydrolysis of their mucopeptides, and thus has a bactericidal effect.

lyze (biology) *See* lyse.

M

M (1 general terminolgy; 2 biochemistry) 1 An abbreviation of 'M'ale. 2 An abbreviation of 'M'ethionine.

m (measurement) An abbreviation of 'm'etre (*USA meter*).

M cells (histology) An abbreviation of 'M'icrofold 'cells'.

m pt (physics) An abbreviation of 'm'elting 'p'oint.

m.k.s. unit (measurement) Metre-kilogram-second unit, a metric unit used in science in preference to c.g.s. units and now superseded by si units. Also termed: MKS unit; Giorgi unit (after the Italian physicist Giovanni Giorgi (1871–1950), who devised it).

M.Sc. (education) - (Med.) Master of Science in Medicine; -(Med.Sc.) Master of Science in Medical Science; -(N) Master of Science in Nursing; -(Nutr.) Master of Science in Nutrition; (Pharm.) Master of Science in Pharmacy; - (V. of Sc.) Master of Veterinary Science

mA (physics) An abbreviation of 'm'illi'A'mpere.

MA test (methods) An abbreviation of 'M'icroscopic 'A'gglutination test.

mAb (immunology) An abbreviation of 'm'onoclonal 'A'nti'b'ody.

mabp (physiology) An abbreviation of 'm'ean 'a'rterial 'b'lood 'p'ressure.

Macchiavello stain (parasitology) A basic fuscin solution for staining chlamydial elementary bodies.

Macdonaldius (veterinary parasitology) A genus of filaroid worms found in the blood vessels of colubrid and viperine snakes. In aberrant host snakes they cause cutaneous ulceration.

maceration (histology) The softening of a solid by fluid. In medicine, the softening and damaging of the tissues by water.

Macheth illuminometer (physics) A device for measuring luminance; an observer views a surface through it and adjusts a light of known intensity to appear the same brightness as the light from the surface.

machine code (computing) The basic set of instructions for a computer, which are in binary code and can be directly implemented by the central processor. They are limited in both number and scope. These instructions are used to construct the more complex and versatile instructions of high-level languages, which must be translated back into machine code when a higher-level program is run.

machine language (computing) *See* machine code.

macr(o)- (terminology) A word element. [German] Prefix denoting large, long.

Macracanthorhynchus (parasitology) A genus of large acanthocephalans in the family Oligacanthorhynchidae.

Macracanthorhynchus catalinum (veterinary parasitology) A species in the genus *Macracanthorhynchus* which may be found in the small intestine of dogs, wolves, badgers and foxes.

Macracanthorhynchus hirudinaceus (veterinary parasitology) A species in the genus *Macracanthorhynchus* which may be found in the small intestine of domestic and wild pigs, where they may cause granulomatous lesions in the intestinal wall, sometimes perforation and peritonitis.

Macracanthorhynchus ingens (veterinary parasitology) A species in the genus *Macracanthorhynchus* which may be found in the small intestine of wild mammals including skunk, mink, raccoon and mole.

macro (computing) A procedure or operator built from a sequence of other procedures or operators.

Macrocheles (insect parasitology) A species of mites in the Mesostigmata, that similarly to *Melichares* are parasitic on the bees *Aphis*, Apinae and Bombinae.

Macrocheles dibamos (insect parasitology) Mites in the Mesostigmata that, similarly to *Macrocheles rettenmeyeri*, are parasites to the ants *Eciton* spp., feeding on body fluid.

Macrocheles rettenmeyeri (insect parasitology) *See Macrocheles dibamos.*

Macrochelidae (insect parasitology) Mites in the Mesostigmata that, similarly to Uropodidae, Trachyuropoda, Urobovella, Urodiscella and Urozercon, are phoretic on the heads of workers of the termites Odontotermes, Glyptotermes, and other termite genera.

macrocyclic (chemistry) Describing a chemical compound whose molecules have a large ring structure.

macrocyclic lactones (pharmacology) Potent nematicidal and insecticidal compounds derived from *Streptomyces* spp.

macrogametocyte (parasitology) 1 The female gametocyte of ampicomplexan protozoa which matures into a macrogamete. 2 A cell that produces macrogametes.

macromolecule (chemistry) Very large molecule containing hundreds or thousands of atoms; *e.g.* natural polymers such as cellulose, rubber and starch, and synthetic ones, including plastics.

macronutrient (biology) Food substance needed in fairly large amounts by living organisms, which may be an inorganic element (*e.g.* phosphorus or potassium in plants) or an organic compound (*e.g.* amino acids and carbohydrates in animals).

macrophages (cytology) Large cells present in connective tissue and in the walls of blood vessels. They can be found in the pleural and peritoneal cavities, the lungs (alveolar macrophages), the liver (Kupffer cells), connective tissue (histiocytes), the lymph nodes and other tissues. Macrophages are highly phagocytic and serve in the first line of defence against micro-organisms. They form part of the reticulo-endothelial system.

macroscopic (biology) Pertaining to the state of being of large size; visible to the unaided eye.

macroscopy (parasitology) Examination with the unaided eye.

macule (histology) A flat discoloured (*USA*: discolored) spot in the skin, as distinct from a papule, which is a raised spot.

Maculinea teleius (insect parasitology) Arthropods in the Lepidoptera that are a parasitic to the ants *Myrmica laevinodis*, attacking the larvae, and which finds the host by following ant trails.

maduramicin (pharmacology) A polyether ionophore used in poultry feed as a coccidiostat.

mag (physics) An abbreviation of 'mag'net/'mag'netic/'mag'netism.

maggot (parasitological physiology) The soft-bodied larva of an insect, especially one living in decaying flesh or tissue debris.

magnesia (chemistry) Magnesium oxide, MgO, particularly a form that has been processed and purified. It is used as an antacid.

magnesium (chemistry) A metallic element used in medicine in the form of its salts: magnesium carbonate, hydroxide and trisilicate are used as antacids in the treatment of peptic ulcers and gastritis. Magnesium hydroxide mixture (milk of magnesium) has a slight laxative action and magnesium sulphate (*USA*: magnesium sulfate) (Epsom salts) a more powerful one. The element is a constituent of chlorophyll, and so an essential part of life; it is essential for human beings, playing a part in the functioning of nerves and muscles, and in energy metabolism. A normal diet supplies the requirements.

magnesium carbonate (chemistry) $MgCO_3$ White crystalline compound, soluble in acids and insoluble in water and alcohol, which occurs naturally as magnesite and dolomite. It is used, often as the basic carbonate, as an antacid.

magnesium chloride (chemistry) $MgCl_2$ White crystalline compound obtained from sea-water and the mineral carnallite, used as a source of magnesium. The hexahydrate is hygroscopic and used as a moisturizer for cotton in spinning.

magnesium hydroxide (pharmacology) $Mg(OH)_2$ White crystalline compound, used as an antacid.

magnesium (Mg) (chemistry) Reactive silver-white metallic element in Group IIA of the periodic table (the alkaline earths). It burns in air with a brilliant white light, and is used in flares and lightweight alloys. It is the metal atom in chlorophyll and an important trace element in plants and animals. At. no. 12; r.a.m. 24.305.

magnesium oxide (chemistry) MgO White crystalline compound insoluble in water and alcohol, made by heating magnesium carbonate. It is used as a refractory and antacid. Also termed: magnesia.

magnesium silicate *See* talc.

magnesium sulphate *See* Epsom salt.

magnet (physics) Object possessing the property of magnetism, either permanently (a permanent magnet, made of a ferromagnetic material) or temporarily under the influence of another magnet or the magnetic field associated with an electric current (an electromagnet).

See also paramagnetism.

magnetic amplifier (physics) Transducer so arranged that a small controlling direct current input can produce large changes in coupled alternating current circuits.

magnetic circuit (physics) Completely closed path described by a given set of lines of magnetic flux.

magnetic core (computing) Computer storage device consisting of a ferromagnetic ring wound with wires; a current flowing in the wires polarizes the core, which can therefore adopt one of two states (making it a bistable device).

magnetic disk (computing) Device for direct-access storage and retrieval of data, used in computers and similar systems. It consists of a rotatable flexible or rigid plastic disc (*i.e.* a floppy or hard disk) coated on one or both surfaces with magnetic material, such as iron oxide. Data is stored or retrieved through one or more read/write heads. Also termed: magnetic disc.

magnetic domain (physics) Group of atoms with aligned magnetic moments that occur in a ferromagnetic material. There are many randomly oriented domains in a permanent magnet.

magnetic drum (computing) Computer storage device consisting of a rotatable drum coated with magnetic material, such as iron oxide. Data is stored or retrieved through one or more read/write heads.

magnetic field (physics) Field of force in the space around the magnetic poles of a magnet.

magnetic field strength (physics) An alternative term for magnetic intensity.

magnetic flux (physics) Measure of the total size of a magnetic field, defined as the scalar product of the flux density and the area. Its SI unit is the weber.

magnetic flux density (physics) Product of magnetic intensity and permeability. Its SI unit is the tesla (formerly weber m^{-2}).

magnetic induction (physics) In a magnetic material, magnetization induced in it, *e.g.* by placing it in the electromagnetic field of a current-carrying coil or by stroking it with a permanent magnet.

magnetic intensity (physics) Magnitude of a magnetic field. Its SI unit is the ampere m^{-2}. Also termed: magnetic field strength; magnetizing force.

magnetic lens (microscopy) Arrangement of electromagnets used to focus a beam of charged particles (*e.g.* electrons in an electron microscope).

magnetic resonance imaging (MRI) (physics) *See* Nuclear magnetic resonance imaging.

magnetic tape (computing) Flexible tape coated with magnetic material and used for the bulk storage of computer data. Suffers from a lack of immediate access compared to disks but can provide a large volume of removable data storage space.

magnetic tape drive (computing) The device that reads data onto and off magnetic tape from and to a computer.

magnetism (physics) Presence of magnetic properties in materials. Diamagnetism is a weak effect common to all substances and results from the orbital motion of electrons. In certain substances this is masked by a stronger effect, paramagnetism, due to electron spin. Some paramagnetic materials such as iron also display ferromagnetism, and are permanently magnetic.

magnetization (physics) Difference between the ratio of the magnetic induction to the pemicability and the magnetic intensity; its SI unit is the ampere m^{-1}. It represents departure from randomness of magnetic domains.

magnetochemistry (chemistry) Study of the magnetic properties of chemicals.

magnetohydrodynamics (MHD) (physics) Branch of physics that deals with the behaviour (*USA*: behavior) of a conducting fluid under the influence of a magnetic flux.

magnetometer (physics) Instrument for measuring the strength of a magnetic field, used, *e.g.* in prospecting for minerals.

magneton (measurement) Unit for the magnetic moment of an electron. Also termed: Bohr-magneton.

magnification (microscopy) 1 Apparent increase in size, as under the microscope. 2 The process of making something appear larger, as by use of lenses. 3 The ratio of apparent (image) size to real size. 4 Radiological magnification; a factor of object to film distance.

mainframe computer (computing) *See* computer.

maintenance level (physiology) The level of growth at which further physical development ceases.

major histocompatibility complex (genetics) The group of genes that control the HLA (human leucocyte antigen) system.

See also immune system.

mal (medical) Sickness, disease.

malaccol (control measures) An insecticidal substance present in the roots of *Derris* spp. plants.

malachite green (fish control measures/parasitology) A green dye used to stain bacteria. Also used, with great caution, as a treatment of cutaneous mycosis in aquarium fish.

malaise (medical) A vague feeling of general discomfort.

malaria (parasitology) An infectious disease in which the red blood cells are attacked by the protozoan parasite Plasmodium, which is transmitted from one infected person to another by *Anopheles* mosquitoes. Malaria is common in tropical areas, and also used to be common around the Mediterranean until mosquito numbers were reduced, partly through loss of their habitat as marshes were drained and partly by insecticidal campaigns. In those areas still affected by malaria, prevention is better than cure, and synthetic quinine-like drugs are effective in deterring the *Plasmodium* parasite. Malaria, which causes high fever that often recurs for years after the original infection, is a dangerous disease, so it is not surprising that in areas where it is (or was) endemic, genes that confer protection against it have been selected for and are common in the population. Unfortunately, in most cases these genes 'buy' protection against malaria at the cost of giving people who are homozygous for the gene another disease: the two best examples are sickle-cell anaemia (*USA*: sickle-cell anemia) and thalassaemia (*USA*: thalassemia), and another is glucose-6-phosphate dehydrogenase deficiency (but one antigen in the Duffy blood group series seems to give protection without deleterious effects).

malate (chemistry) A salt of malic acid.

malathion (pharmacology/control measures) One of the least toxic and most widely used organophosphorus insecticides in companion animals. Toxicity when it occurs is usually due to gross over concentration of the compound in the topical preparation used.

See also organophosphorus compound.

malic acid (chemistry) $COOHCH_2CH(OH)COOH$ Colourless (*USA*: colorless) crystalline carboxylic acid with an agreeable sour taste resembling that of apples, found in unripe fruit. It is used as a flavouring agent.

malignant tertian malaria (medical) Malaria caused by *Plasmodium falciparum*.

malleus (anatomy) A small hammer-shaped bone in the middle ear, one of the ossicles. It is set in motion by the ear drum and transmits sound vibrations to the incus.

Mallophaga (avian/veterinary parasitology) A suborder of insects comprising the biting or bird lice, which is also found on mammals.

See also louse.

Mallory cell (physics) An alternative term for mercury cell.

malnutrition (biochemistry) A condition arising from deficiency in the diet or deficiency in the absorption or metabolism of food.

Malpighamoeba mellificae (insect parasitology) A protozoal species of amoeba (*USA*:ameba) that are parasites to the bees *Apis mellifera*. They are associated with bee virus X and *Nosema apis* but have few apparent effects.

Malpighian body (anatomy) Part of the mammalian kidney, Bowman's capsule and glomerulus. Its function is to filter blood. It was named after the Italian biologist Marcello Malpighi (1628–1694).

malt (immunology) An abbreviation of 'M'ucosal-'A'ssociated 'L'ymphoid 'T'issue.

maltose (chemistry) $C_{12}H_{22}O_{11}$ Common disaccharide sugar, composed of two molecules of glucose. It is found in starch and glycogen.

maltotriose (biochemistry) A trisaccharide product of amylase digestion of starch.

mamm(o)- (terminology) A word element. [Latin] Prefix denoting breast, mammary gland.

mammary gland (histology) The gland in the female breast that secretes milk.

Mammomonogamus (veterinary parasitology) A genus of nematodes in the family Syngamidae which are parasites of the nasal sinuses and trachea of mammals.

Mammomonogamus auris (veterinary parasitology) A species of nematodes in the genus *Mammomonogamus* which may be found in the puma.

Mammomonogamus ierei (veterinary parasitology) A species of nematodes in the genus *Mammomonogamus* which may be found in cats, and that may cause chronic nasopharyngitis.

Mammomonogamus indicus (veterinary parasitology) A species of nematodes in the genus *Mammomonogamus* which may be found in the pharynx of the Indian elephant.

Mammomonogamus laryngeus (human/veterinary parasitology) A species of nematodes in the genus *Mammomonogamus* which may be found in the larynx of cattle, goats, water buffalo and deer, and occasionally, in humans.

Mammomonogamus loxodontus (veterinary parasitology) A species of nematodes in the genus *Mammomonogamus* which may be found in the trachea of the African elephant.

Mammomonogamus mcgaughei (veterinary parasitology) A species of nematodes in the genus *Mammomonogamus* which may be found in cats.

Mammomonogamus nasicola (veterinary parasitology) A species of nematodes in the genus *Mammomonogamus* which may be found in the nasal cavities of cattle, sheep, goats and deer.

man. (general terminology) An abbreviation of 'man'ual. A guidebook explaining the use of a product or the operations of an organisation.

Mancini technique (methods) *See* radial immunodiffusion.

mandibles (parasitological anatomy) A term used in relation to arthropods. A pair of upper jaws.

manganese (Mn) (chemistry) Metallic element in group VIIA of the periodic table used mainly for making special alloy steels and as a deoxidizing agent. It is also an essential trace element for plants and animals. At. no. 25; r.a.m. 54.9380.

manganese dioxide (chemistry) An alternative term for manganese (IV) oxide.

manganese(IV) oxide (chemistry) MnO_2 Black amorphous compound, used as oxidizing agent, catalyst and depolarizing agent in dry batteries. Also termed: manganese dioxide.

manganic (chemistry) An alternative term for manganese(III) in manganese compounds.

manganous (chemistry) An alternative term for manganese(II) in manganese compounds.

mange (parasitology) A skin disease of domestic animals, caused by a number of genera of mites and described under those headings. *See* chorioptic mange, demodectic mange, notoedric mange, otodectic mange, psoroptic mange and sarcoptic mange.

Mann Whitney U Test (statistics) A non-parametric test of whether two independent samples come from the same population. It is equivalent to Wilcoxon's test. The test compares each value in one sample with each value in the other but does not require ranking, as in Wilcoxon's algorithm. It is a non-parametric version of a t-test for independent samples.

mannitol (chemistry) $HO.CH_2(CHOH)_4CH_2OH$ Soluble hexahydric alcohol that occurs in many plants, and is used in medicine as a diuretic.

mannose (biochemistry) An aldohexose, a monosaccharide produced from mannitol by oxidation.

manometer (methods) An instrument for ascertaining the pressure of liquids or gases.

manometer (physics) Device for measuring fluid pressure.

manoptoscope (measurement) A device for measuring eye dominance.

Mansonella (parasitology) A genus of nematode parasites of the superfamily Filaroidea which may be transmitted by *Culicoides* and *Simulium* spp.

Mansonella ozzardi (human parasitology) A species of nematode parasite of the genus *Mansonella* which may be found in the mesentery and visceral fat of humans in Central and South America, and possibly also in domestic animals.

Mansonia (veterinary parasitology) A genus of mosquitoes in the family Culicidae and vectors of Dirofilaria immitis of dogs and the virus of Rift Valley fever in all species. Now termed *Taeniorhynchus* spp.

Manson's eyeworm (parasitology) *See Oxyspirura.*

mantissa (mathematics) Fractional part of a logarithm (the other part is the characteristic); *e.g.* in the logarithm 2.3010, 3010 is the mantissa.

MAO (biochemistry) An abbreviation for 'M'ono'A'mine 'O'xidase.

MAOI (biochemistry) An abbreviation for 'M'ono'A'mine 'O'xidase 'I'nhibitors.

map unit (genetics) A measure of the distance between two loci on the same chromosome. Loci are 1 map unit apart if cross-over occurs between them in 1 per cent of cases. A map unit has the formal name of centimorgan.

marasmus (medical) Progressive wasting away, particularly in infants.

march fly (parasitology) *See Tabanus.*

Marchi stain (histology) A histological stain that makes myelinated nerve fibres (*USA*: nerve fibers) visible.

Margaropus (parasitology) A genus of ticks in the family Ixodidae, similar physically to *Boophilus* spp.

Margaropus reidi (veterinary parasitology) A species of ticks in the genus *Margaropus* which may be found on giraffes.

Margaropus winthemi (veterinary parasitology) A species of ticks in the genus *Margaropus* which may be found on horses, occassionally cattle. Also termed Argentine tick.

marine (biology) Of or pertaining to the sea.

marine biologist (biology) Specialist in the biology of marine life.

marker (terminology) A visual or electronic signal that permits identification and therefore sorting of individual items from a group.

marker genes (genetics) Genes with a known location on a chromosome and an obvious phenotype which are used as reference points when mapping other genes.

mark-sensing (computing) A method of entering data into a computer without needing to transcribe it and which may be performed by marking the collected data form at predetermined positions.

marrow (histology) The marrow of the bones in an adult is red or yellow. Red marrow is found in the skull, ribs, pelvis, breastbone, the bodies of the vertebrae and the ends of the long bones. It is actively engaged in making the cells of the blood. Yellow marrow is full of fat; in early life, it is active and red, but by the end of the period of growth it has become inactive. In times of crisis it is still capable of regaining activity, and it then becomes red again.

Marseilles fever (parasitology) Boutonneuse fever.

Marshallagia (parasitology) A genus of intestinal worms in the family Trichostrongylidae, which are not known to have significant pathogenicity.

Marshallagia dentispicularis (veterinary parasitology) A species of intestinal worms in the genus *Marshallagia* which may be found in sheep.

Marshallagia marshalli (veterinary parasitology) A species of intestinal worms in the genus *Marshallagia* which may be found in the abomasum of amongst others sheep, goats, antelopes, bighorn sheep.

Marshallagia mongolica (veterinary parasitology) A species of intestinal worms in the genus *Marshallagia* which may be found in the abomasum of sheep, goats and camels.

Marshallagia orientalis (veterinary parasitology) A species of intestinal worms in the genus *Marshallagia* which may be found in hillgoats (*Capra sibirica*).

Marshallagia schikhobalovi (veterinary parasitology) A species of intestinal worms in the genus *Marshallagia* which may be found in sheep.

Marteilia (fish parasitology) Parasitic protozoa in the order Occlusosporidia. Species includes *Marteilia refringens* in European oysters, and *Marteilia sydnei* in Sydney rock oysters.

marteiliosis (veterinary parasitology) A disease of the digestive gland of molluscs, caused by the protozoan parasite *Marteilia* spp.

Martindale (pharmacology) The Extra Pharmacopoeia; published in 30 editions over a period of 110 years by the Royal Pharmaceutical Society of Great Britain; contains over 5000 monographs on substances used in pharmacy and medicine including those used to treat parasitic diseases.

Martinotti cells (cytology) Spindle-shaped cells with cell bodies in the internal granular layer of the cortex, and with axons extending to the cell bodies of the outer pyramidal layer. *See* cortical layers.

MAS (education) An abbreviation of 'M'aster of 'A'pplied 'S'cience. Also termed MASc.

mAs (measurement) An abbreviation of 'm'illi'A'mpere 's'econds.

maser (physics) An abbreviation of 'm'icrowave 'a'mplification by 's'timulated 'e'mission of 'r'adiation. A precise energy-emitting device that amplifies electromagnetic waves into the microwave spectrum.

mass (physics) Quantity of matter in an object, and a measure of the extent to which it resists acceleration if acted on by a force (*i.e.* its inertia). The SI unit of mass is the kilogram.

mass decrement *See* mass defect.

mass defect (1, 2 chemistry) 1 Difference between the mass of an atomic nucleus and the masses of the particles that make it up, equivalent to the binding energy of the nucleus (expressed in mass units). 2 Mass of an isotope minus its mass number. Also termed: mass decrement; mass excess.

mass number (A) (chemistry) Total number of protons and neutrons in an atomic nucleus. Also termed: nucleon number.

See also isotope.

mass spectrograph (physics) Vacuum system in which positive rays of charged atoms (ions) are passed through electric and magnetic fields so as to separate them in order of their charge-to-mass ratios on a photographic plate. It measures relative atomic masses of isotopes with precision.

mass spectrometer (physics) Mass spectrograph that uses electrical methods rather than photographic ones to detect charged particles.

mass spectrum (physics) Indication of the distribution in mass, or in mass-to-charge ratio, of ionized atoms or molecules produced by a mass spectrograph.

mass-energy equation (physics) Deduction from Einstein's special theory of relativity that energy has mass; $E = mc^2$, where E is the energy, m is the amount of mass, and c the speed of light.

Masson/Fontana stain (histology) A histological staining technique used to make melanin visible under the microscope.

massons trichome (histology) A histological staining technique used to make connective tissue visible under the microscope.

mast cells (cytology) Large cells found in many places in the body, particularly in connective tissues and the mucosal surfaces. They are damaged, or activated, by an antibody-antigen reaction, they release histamine and other substances which increase the passage of fluid from the blood vessels, lower the blood pressure, increase secretion from the mucous membranes, and contract smooth muscles. They also attract white blood cells. Massive activation of the mast cells results in anaphylactic shock.

masterate (education) A master's degree.

mastic (microscopy) A resin derived from the tree *Pisiacia lentiscus*, used in microscopy.

masticatory nerve (histology) An alternative term for mandibular nerve.

Mastigophora (parasitology) A subphylum of protozoa, including all those that have one or more flagella throughout most of their life cycle and a simple, centrally located nucleus; many are parasitic in both invertebrates and vertebrates. Included are the Kinetoplastida (*e.g. Trypanosoma*), Diplomonadina (*e.g. Hexamita*) and Trichomonadida (*e.g. Trichomonas*).

mastigote (parasitology) Any member of the subphylum Mastigophora.

masto- (terminology) A word element. [German] Prefix denoting mammary gland.

mastocyte (cytology) A mast cell.

mastoiditis (medical) Inflammation of the air cells in the mastoid process of the temporal bone, once not uncommon in association with middle ear disease, but since the advent of antibiotics rarely seen.

Mastophorus muris (veterinary parasitology) A species of worms in the genus *Mastophorus* which may be found in the stomach of a wide range of rodents.

Mastophorus (parasitology) A genus of worms in the family Spirocercidae.

MAT (methods) An abbreviation of 'M'icroscopic 'A'gglutination 'T'est.

mat. (medical) An abbreviation of 'mat'ernity.

matched groups (statistics) Groups assigned to different experimental conditions but not differing in other ways, *e.g.* in IQ or age, that might be relevant to the outcome of the experiment. The expression is sometimes used of the matched pairs method, which is a special case.

matched pairs method (statistics) Obtaining two matched groups by matching each member of one group against a member of the other group by reference to qualities other than those under investigation.

matched sample (1 statistics; 2 general terminology) A sample chosen in such a way that each member of it is the same on some characteristic or characteristics (other than those being investigated) as a member of another sample to be compared with it; *e.g.* the same subjects may each be run under two conditions. In this case, a paired comparison test can be run on the differences in the effects of the two conditions. 2 A sample chosen in such a way that its members are on average the same on some category as those of another sample, but are not necessarily matched pair by pair.

matching (statistics) Comparison for the purpose of selecting objects having similar or identical characteristics.

mathematics (mathematics) Branch of science concerned with the study of numbers, quantities and space.

Mathevotaenia (veterinary parasitology) A genus of cestodes in the family Linstowiidae. Species includes *Mathevotaenia oklahomensis*, *Mathevotaenia pedunculata*, and *Mathevotaenia wallacei* which may be found in raccoons and skunks.

mating system (genetics) The system within a population by which mates are chosen. At one extreme are parthenogenesis and self-fertilization; at the other is random mating. There may be systematic inbreeding or outbreeding, or assortative mating in respect of one or more traits. Some mating systems involve equal numbers of both sexes; in others, members of one sex have multiple mates. All mating systems have an effect on the genetic constitution of the succeeding generations: under random mating, it stays the same; inbreeding and positive assortative mating increase homozygosity; outbreeding and negative assortative mating increase heterozygosity.

matrix (1 biology; 2 mathematics) 1 Extracellular substance that embeds and connects cells, *e.g.* connective tissue. 2 Square or rectangular array of elements, *e.g.* numbers.

matter (1 physics; 2 medical) 1 Physical material having form and weight under ordinary conditions of gravity. 2 Pus.

Mattesia geminata (insect parasitology) A protozoal species of gregarinida that are parasites to the ants *Solenopsis geminata*, *Solenopsis invicta* and *Solenopsis richteri*. These are unusual in that they infect the hypodermis.

maturation (1 biology; 2 medical) 1 The stage or process of attaining maximal development. In biology, a process of cell division during which the number of chromosomes in the germ cell is reduced to one-half the number characteristic of the species. 2 The formation of pus.

Maurer's dots (parasitology) A term used in relation to malaria. Coarse and irregular stippling seen on *Plasmodium falciparum* infected red cells in Romanowsky stained films.

maw worm (parasitology) *See Oxyuris equi.*

max. (measurement) An abbreviation of 'max'imum. The highest amount possible.

Maxam-Gilbert method (methods) A method used in DNA sequencing. Four samples of end-labeled DNA restriction fragments are chemically cleaved at different specific nucleotides, the resulting subfragments separated by gel electrophoresis, and the labeled fragments are detected by autoradiography. The sequence of the original end-labeled restriction fragment can be determined directly from parallel electropherograms of the four samples.

maxilla (anatomy) The bone of the upper jaw, which also takes part in the formation of the orbit or eye-socket, the nose and the hard palate. It contains the maxillary antrum or sinus, a hollow cavity which communicates with the nose and is liable to become inflamed in sinusitis.

maximum (biology) The greatest quantity, effect or value possible or achieved under given circumstances.

maximum containment laboratory (parasitology) A laboratory designed and equipped to provide the highest level of security in the handling of infectious agents that are serious pathogens for humans and animals.

maximum individual risk (MIR) (toxicology) Increased risk for an individual exposed to the highest measured or predicted concentration of a toxicant.

maximum likelihood principle (statistics) The principle that we should estimate the value of a population parameter as the value which maximizes the likelihood of the obtained data.

maxwell (measurement) C.g.s. unit of magnetic flux, the SI unit being the weber. 1 maxwell = 10^{-8} weber. It was named after the British physicist James Clerk Maxwell (1831–79).

May Grunwald-Giemsa (histology) Histological stain for air dried cytology preparations.

Mazzotti reaction (human parasitology) A skin reaction that occurs in humans when the microfilariae of *Onchocerca volvulus* in cutaneous sites are killed by the administration of diethylcarbamazine.

MB (education) An abbreviation of *'M'edicinae 'B'accalaureus.* [Latin, meaning 'Bachelor of Medicine'.]

mb (measurement) An abbreviation of 'm'illi'b'ar.

mbc (physiology) An abbreviation of 'm'aximum 'b'reathing 'c'apacity.

mbp (physiology) An abbreviation of 'm'ean 'b'lood 'p'ressure.

mbr (general terminology) An abbreviation of 'm'em'b'e'r'.

mc (1, 2 measurement) 1 An abbreviation of 'm'ega-'c'ycle. 2 An abbreviation of 'm'illi'c'uries, a measurement for radium.

mcg (measurement) An abbreviation of 'm'i'c'ro'g'ram. (1/1,000,000 of a gram). Scientifically expressed as µg.

mcg (measurement) An abbreviation of 'm'i'c'ro'g'ram.

McMaster technique (methods) A rapid, simple, quantitative technique for counting parasite eggs in ruminant faeces (*USA*: feces), based on flotation on concentrated salt solution (specific gravity of 1.1 to 1.3) in a counting chamber.

Mcnemar's test (statistics) A non-parametric test for the significance of the difference between two proportions in matched samples, usually used when the same set of subjects are tested at two different times.

MCPA (education) An abbreviation of 'M'ember of the 'C'ollege of 'P'athologists of 'A'ustralia.

M-CSF (physiology) An abbreviation of 'M'acrophage 'C'olony-'S'timulating 'F'actor.

MD (1, 2 education) 1 An abbreviation of 'M'edical 'D'epartment. 2 An abbreviation of 'M'edicinae 'D'octor. [Latin, meaning, 'Doctor of Medicine'. The graduate degree awarded to a student who has successfully completed a predetermined course of study in medicine.]

Md (chemistry) The chemical symbol for Element 101, mendelevium.

mdr (physiology) An abbreviation of 'm'inimum 'd'aily 'r'equirement.

ME (1 forensics; 2 physiology) 1 An abbreviation of 'M'edical 'E'xaminer. 2 An abbreviation of 'M'etabolizable 'E'nergy.

mealworm (parasitology) *Alphitobius diaperinus*.

mean (statistics) The arithmetical average, *i.e.* the sum of all the values divided by the number of individuals.

See also median; mode; normal distribution.

mean (statistics) An average; a numerical value intermediate between two extremes. Also termed arithmetic mean.

mean average. *See* arithmetic mean; geometric mean.

mean corpuscular haemoglobin (haematology) A measure of the haemoglobin (*USA*: hemaglobin) content of red blood cells. (*USA*: mean corpuscular hemaglobin).

mean corpuscular volume (haematology) A measure of the volume of red blood cells.

mean deviation (statistics) Of a group of numbers, the sum of all deviations from the arithmetic mean divided by the quantity of numbers in the group.

mean deviation (statistics) The average value of a set of absolute deviations from the mean of a set of observations.

mean life (1 statistics; 2 physics) 1 Average time for which the unstable nucleus of a radioisotope exists before decaying. 2 Average time of survival of an elementary particle, ion, etc. in a given medium or of a charge carrier in a semiconductor.

mean square (statistics) The arithmetic mean of the squared deviations from the mean, *i.e.* the variance.

meatus (anatomy) Duct or channel between body parts, *e.g.* external auditory meatus, the opening to the ear.

mebendazole (pharmacology) A broad-spectrum anthelmintic efficient against all gastrointestinal nematodes, lungworms and *Moniezia* in ruminants. It is not effective against *Trichuris* spp. and has reduced activity against benzimidazole-resistant worm species. It is not very effective against larval forms of *Taenia* spp., not effective against *Draschia* or *Habronema* spp. in horses, nor against *Trichostrongulus axei*. It is effective against *Strongylus* spp., but not against migrating *Strongylus vulgaris* larvae nor against *Trichostrongulus axei*, Strongyloides spp. or *Anoplocephala perfoliata*. It is recommended for general use in horses and combined with metriphonate for use against bot fly larvae. It is useful in dogs but does not remove *Echinococcus granulosus* or *Dipylidium caninum*. The drug has very low toxicity but has caused severe acute hepatic necrosis in some dogs.

mechanical vector (parasitology) An arthropod vector that transmits the infective organisms from one host to another but is not essential to the life cycle of the parasite.

mechanoreceptor (physiology) A receptor sensitive to a mechanical force, *e.g.* pressure receptors, hair cells, muscle receptors.

Mecistocirrus (parasitology) A genus of worms in the family Trichostrongylidae.

Mecistocirrus digitatus (veterinary parasitology) A species of worms in the genus *Mecistocirrus* which may be found in the abomasum of domestic ruminants and buffalo, and in the stomach of pigs. In endemic areas it is an important parasite, causing effects similar to those of *Haemonchus* spp.

meconium (physiology) A dark green semi-fluid material consisting of bile, mucus, desquamated cells, and debris discharged from the infant's bowel: at birth or immediately afterwards.

MED (pharmacology) An abbreviation of 'M'inimal 'E'ffective 'D'ose.

med lab (general terminology) An abbreviation of 'med'ical 'lab'oratory.

Med Sc D (education) Doctor of Medical Science

Med Tech (education) An abbreviation of 'Med'ical 'Tech'nology.

med. (1 medical; 2 statistics) 1 An abbreviation of 'med'ical/'med'icine. 2 An abbreviation of 'med'ium/'med'ian.

Med.RC (education) An abbreviation of 'Med'ical 'R'esearch 'C'ouncil. Also termed MRC.

medial (anatomy) Toward the mid-line of the body or of an organ. *Contrast* lateral.

medial dorsal nucleus (cytology) An alternative term for dorsomedial nucleus.

medial forebrain bundle (MFB) (histology) A tract of axons passing through the lateral hypothalamus and running in both directions between the forebrain and brainstem. It has many connections with the limbic system and self-stimulation can readily be obtained from it.

medial geniculate nucleus (cytology) A nucleus in the thalamus relaying auditory information received from the inferior colliculus to the cortex.

medial hypothalamic area (histology) A region of the hypothalamus, stimulation of which tends to arouse the parasympathetic system.

medial lemniscus (histology) A nerve tract which is part of the lemniscus and which conveys somaesthetic information from the medulla to the thalamus.

medial plane (histology) The vertical plane that divides the body into two symmetrical halves.

medial rectus (histology) An extraocular muscle that pulls the eye inward towards the nose.

median (1 statistics; 2 anatomy) 1 The value in a distribution that comes halfway between the highest and the lowest. *See also* mean; mode; normal distribution. 2 An alternative term for medial.

median eminence (histology) A swelling at the base of the hypothalamus that releases hormones into the portal circulation of the anterior pituitary.

median test (statistics) A non-parametric test for the significance of the difference between two or more samples. The median for all samples combined is calculated; in each sample the number of values above and below the combined median is found and the results are evaluated by a chi square test.

mediastinum (anatomy) The space in the chest between the two lungs. It contains the heart and great vessels, the oesophagus (*USA*: esophagus), the lower end of the trachea, the thoracic duct, various nerves, the thymus gland and lymph nodes.

Medical Examiner (ME) (medical) A forensics specialist, coroner, or other official authorized to perform autopsies.

Medical Research Council (MRC) (governing body) A government organization that supports MRC research institutes in the UK, as well as providing funds for research projects carried out at universities.

Medicines Act (legislation) UK legislation, passed in 1968, that restricts the supply and manufacture of all medicines to license holders.

medina worm (parasitology) *See Dracunculus medinensis.*

Mediterranean coast fever (parasitology) *See Theileria annulata.*

medulla (1, 2 histology) 1 The inner part of a structure or organ or tissue (*e.g.* adrenal medulla). *See also* cortex. 2 An alternative term for medulla oblongata.

medulla oblongata (anatomy) The lower of the two divisions of the hindbrain; it is immediately above the spinal cord and below the pons. It contains many nerve tracts, and also the autonomic nuclei implicated in the control of breathing, heartbeat, and is concerned with the co-ordination of nerve impulses from hearing, touch and taste receptors.

medullary layer (histology) The innermost of the three layers of the cerebellar cortex.

medullary sheath (histology) An alternative term for myelin.

medullated nerve fibre *See* nerve fibre.

mega- (measurement) Metric prefix meaning million times; × 10^6 (*e.g.* megahertz).

mega- (terminology) A word element. [German] Prefix denoting *large*.

megabyte (computing) A computer term denoting a million bytes of storage space.

megacolon (histology) Abnormal enlargement of the colon, which may happen in consequence of prolonged use of certain laxatives such as senna and cascara. It may be congenital: in Hirschsprung's disease (aganglionic megacolon) there are no nerve ganglion cells in the wall of the rectum; the absence of nerve cells may extend upwards into the colon. The affected part of the gut cannot move, and therefore forms an obstruction, above which the colon becomes loaded with faeces (*USA*: feces) and grossly distended. The baby is constipated, its abdomen becomes distended, and it may vomit. The symptoms may take some time to develop, in some cases only becoming apparent after a year or longer. The treatment is surgical removal of the affected part of the bowel, with functional continuity being restored. The disease is named after the Danish physician, H. Hirschsprung (1830–1916).

megaelectron volt (MeV) (measurement) One million electron volts (10^6 eV).

megahertz (MHz) (measurement) Unit of frequency of one million hertz.

megal(o)- (terminology) A word element. [German] Prefix denoting *large*, abnormal *enlargement*.

meglumine antimonate (pharmacology) A pentavalent antimonial used as an antiprotozoal, and a preferred drug in the treatment of leishmaniasis.

Megninia (avian parasitology) A genus of feather mites in the family Analgesidae that are generally considered to be nonpathogenic but may cause depluming itch on rare occasions.

Megninia columbae (avian parasitology) A genus of feather mites of the genus *Megninia* which may be found on pigeons.

Megninia ginglymura (avian parasitology) A genus of feather mites of the genus *Megninia* which may be found on fowl.

Megninia phasiani (avian parasitology) A genus of feather mites of the genus *Megninia* which may be found on pheasants and peacocks.

Megninia velata (avian parasitology) A genus of feather mites of the genus *Megninia* which may be found on ducks.

Megniniacubitalis (avian parasitology) A genus of feather mites of the genus *Megninia* which may be found on fowl.

megohm (measurement) One million ohms.

megrim (medical) Another term for headache; migraine.

Mehlis' gland (parasitological anatomy) Part of the reproductive system of a trematode. A unicellular gland which encircles the ootype. Its function is not known.

meiosis (genetics) The central event in heredity. Type of cell division in which the number of chromosomes in the daughter cells is halved; thus they are in the haploid state. Two successive divisions occur in the process, giving four daughter cells. The first division takes place in four stages: prophase, metaphase, anaphase and telophase. The second division has three stages: metaphase, anaphase and telophase. In animals meiosis occurs in the formation of gametes, *e.g.* eggs and sperm. Also termed: reduction division. *See also* mitosis.

meiotic (genetics) Describing or referring to meiosis.

meiotic drive (genetics) Any mechanism that causes unequal proportions of the possible gametes to be formed at meiosis. One example is B chromosomes, which tend to be included preferentially in the ovum during meiosis in the female.

melaena (haematology) Black motions, caused by altered blood originating from haemorrhage (*USA*: hemorrhage) in the stomach or intestines, by iron taken as a medicine, or by some red wines.

melan(o)- (terminology) A word element. [German] Prefix denoting *black*, *melanin*.

mélangeur (equipment) [French] An instrument for drawing and diluting specimens of blood for examination.

melanin (biochemistry) The black or dark brown pigment that gives colour (*USA*: color) to skin, hair and eyes in animals, including humans. Melanin production in melanocytes is stimulated by ultraviolet radiation, either natural (*i.e.* from the sun) or artificial, and is catalysed by the enzyme tyrosinase, lack of which causes albinism.

melanism (biochemistry/histology) Occurrence in populations of dark-coloured (*USA*: colored) individuals having an excess of melanin in their tissues.

melanocyte (histology) A pigmented cell in the skin containing melanin.

melanocyte-stimulating hormone (MSH) (biochemistry) A peptide hormone produced by the intermediate lobe of the pituitary that increases production of melanin.

melarsomine (pharmacology) A trivalent arsenical with activity against adult *Dirofilaria immitis*.

melarsoprol (pharmacology) An antiprotozoal effective against *Trypanosoma* spp.

melatonin (biochemistry) A substance found in the Pineal gland that is implicated in circadian rhythms.

Melichares (insect parasitology) *See Macrocheles*.

melitose (chemistry) *See* raffinose.

melittin (biochemistry) One of the toxic peptides in bee sting.

Melőe cicatricosus (insect parasitology) *See Melőe* spp.

Melőe laevis (insect parasitology) *See Melőe* spp.

***Melőe* spp.** (insect parasitology) Arthropods in the Coleoptera including *Melőe cicatricosus*, *Melőe laevis* and *Melőe variegatus*, that are a parasitic to the bees *Apis mellifera*. Triungulins burrow into joints of the bee's abdomen and extracts haemolymph (*USA*: hemolymph).

Melőe variegatus (insect parasitology) *See Melőe* spp.

Meloidogyne (plant parasitology) A genus of nematodes in the order Tylenchida and family Heteroderidae. They are sexually dimorphic. The female is saccate to globose, 0.4–1.3 mm long with a body that is soft, pearl-white in colour (*USA*: color) and does not form a cyst, which is usually embedded in root tissue which is often swollen or galled. Males are vermiform, similar to *Heterodera*, but the lip region has a distinct head cap which includes a labial disc surrounded by lateral and medial lips. The head skeleton is usually weaker than *Heterodera* and the stylet less robust and shorter, 18–24 um long for many species. Infective second stage juveniles which are often free in the soil, are usually 0.3–9.5 mm long, but they are less robust than *Heterodera* juveniles, the stylet is delicate with small basal knobs, under 20 um long, and the head skeleton weak. The tail is conoid often ending in a narrow rounded terminus but tail length variable, 1.5–7 anal body widths, between species, it often ends in a clear hyaline region the extent of which can help to distinguish species. *See Meloidogyne incognita*.

Meloidogyne incognita (plant parasitology) A species of nematodes in the genus *Meloidogyne* whose hosts cover a very broad range with more than 700 hosts. These include most cultivated crops and ornamentals, and is distributed worldwide but more common in temperate, subtropical and tropical areas. The cosmopolitan distribution is the result of the movement of rooted plants in commerce and at the local level through movement of water, soil and equipment and rooted seedlings of crop plants and ornamentals. They are sedentary endoparasites, and are parthenogenic. Second-stage infective juveniles hatch from the eggs and these invade roots in the region of elongation near the root cap. They migrate between and through cells and position themselves with the head in the vascular tissues.

Cell damage occurs as a result of the migration and if several juveniles enter the root tip, cell division stops and there is no root elongation. As feeding continues several cells near the head begin to enlarge and become multinucleate. These are called giant cells and there are usually three to six associated with each nematode. The formation of giant cells and galls is the result of cell enlargement and of increased numbers of cells. These changes are induced by substances in salivary secretions introduced into cells and surrounding tissues during the feeding of the nematode. During this process, the xylem vessels become disrupted and the roots cannot function normally with respect to water and nutrients. During the process of gall formation, the nematodes undergo the second, third and fourth moults (*USA*: molts) to reach the adult stages. Mature females are saccate (pear-shaped) and lay eggs into a gelatinous matrix. This matrix may protrude from the surface of small roots or may be entirely within the gall. Eggs hatch in about 7 days. The entire life cycle is completed in 20–25 days at 70°F. Males are vermiform but not required in reproduction. Infestation results in reduced root systems and galling. There is poor top growth and the foliage is frequently chlorotic (yellow) because essential elements are not taken in and transported by the impaired root system. Severe infections cause wilting of the foliage and the plants require more frequent irrigations. Control measures includes preplant nematicides and postplant nematicides on some perennials. Some resistant or immune varieties and rootstocks are available in tomato, bean, sweet potato, peach, almond, walnut and grape. Hot water treatment of rooted grape cuttings 125°F for 5 minutes will give 100 percent control. Because of broad host range, crop rotation is frequently not feasible. Also termed: cotton root-knot nematode.

Melophagus (parasitology) A genus of insects in the family Hippoboscidae.

Melophagus ovinus (veterinary parasitology) A species of insects in the genus *Melophagus* that is a permanent ectoparasite, the wingless, brown, leathery ked of sheep. Heavy infestations cause anaemia (*USA*: anemia), loss of condition and wool damage.

melting point (physics) Temperature at which a solid begins to liquefy; a fixed (and therefore characteristic) temperature for a pure substance.

melting profile (genetics) A graph that shows how much DNA in a sample melts over time in an increasing temperature. DNA melts in the range 60–80°C (140–175°F). DNA with a preponderance of A-T base-pairs melts at lower temperatures than DNA with more G-Cs, so the melting profile gives a crude indication of the composition of the sample.

membrane (1 physics; 2 biology) 1 Any thin material, *e.g.* plastic film. 2 Structure that forms a dynamic interface, *e.g.* the outer membrane between a cell and its surroundings. The structure may be selective in allowing the passage of certain molecules through (*e.g.* a permeable membrane) or specific (*e.g.* in active transport). Membranes are widely distributed and very important in all organisms. *See* plasma membrane.

membranella (anatomy) A membrane formed of a fused row of cilia.

membrane potential (physics) The voltage difference between the inside and outside of a cell's membrane, particularly of a nerve membrane.

membranous (biology) Pertaining to or emanating from a membrane.

memo. (general terminology) An abbreviation of 'memo'randum.

memory (computing) Part of a computer that stores data and instructions (programs), usually referring to the immediate access store. *See also* random access memory (RAM); read-only memory (ROM).

memory cells (immunology) Memory, or the ability to mount a quantitatively and qualitatively different response upon secondary exposure to a specific antigen, is one of the hallmarks of the vertebrate immune response. In the humoral, or antibody, secondary response, antibody is produced more quickly, in larger quantities, for a longer period of time, and of different classes and affinities than in a primary response. The secondary response in cell-mediated immunity produces faster elimination of viral antigens, faster and more severe delayed-type hypersensitivity reactions and decreased graft rejection time. The ability to mount a secondary immune response is due to the generation of antigen-specific memory cells during the primary response. Memory B cells and probably memory T cells, as well as effector B cells (plasma cells) and effector T cells (cytotoxic T cells) are produced from naive B and T cells during the primary response to that antigen. The effector cells are generally short-lived, but a clone of long-lived antigen-primed memory cells is thought to survive and provide for the heightened secondary response by bypassing the early stages of clonal expansion.

See also immune system.

Menacanthus (parasitology) A genus of lice in the superfamily Amblycera.

Menacanthus cornutus (avian parasitology) A species of lice of the genus *Menacanthus* that resembles *Menacanthus stramineus*.

Menacanthus pallidulus (avian parasitology) A species of lice of the genus *Menacanthus* that is a body louse of chickens.

Menacanthus stramineus (avian parasitology) (synonym: *Eomenacanthus stramineus*) A species of lice of the genus *Menacanthus*. Yellow body louse of poultry.

Mendel's laws (genetics) Conclusions drawn from work on inheritance carried out by Gregor Mendel in breeding experiments. The first is the law of segregation: an inherited characteristic is controlled by a pair of factors (alleles), which separate and become incorporated into different gametes. The second is the law of independent assortment: the separated factors are independent of each other when gametes form.

Mendelevium (Md) (chemistry) Radioactive element in Group IIIB of the periodic table (one of the actinides); it has several isotopes, with half-lives of up to 54 days. At. no. 101; r.a.m. 258 (most stable isotope).

Mendelism (genetics) Study of inheritance, and therefore genetics. The principles of genetics put forward by the Austrian monk Gregor Mendel (1822–84), who postulated the existence of genes, both dominant and recessive, and the rules governing their transmission. *See* Mendel's laws.

See also heredity.

mening(o)- (terminology) A word element. [German] Prefix denoting meninges, membrane.

meningeal (anatomy) Pertaining to the meninges.

meningeal worm (parasitology) *See Paraelaphostrongylus tenuis.*

meninges (histology) The three layers of protective tissue that enclose the brain and spinal cord, which are, from outside to inside, the dura mater, arachnoid layer, and pia mater.

meningitis (medical) Inflammation of the meninges, membranes that cover the brain and spinal cord. The symptoms of meningitis are in general headache, fever, nausea, vomiting, backache, and dislike of light (photophobia). There is stiffness of the neck, and in children there may be convulsions.

meningoencephalomyelitis (medical) Inflammation of the meninges, brain and spinal cord.

Meningonema peruzzi (human/veterinary parasitology) A filarial parasite of the central nervous system of a number of African monkeys including talopin monkeys. Thought also to cause cerebral filariasis of humans.

meningopathy (medical) Any disease of the meninges.

meniscus (1 anatomy; 2 physics) 1 A semicircular or crescentic cartilage in a joint, usually used to refer to the semilunar cartilages in the knee. 2 Curved surface of a liquid where it is in contact with a solid. The effect is due to surface tension.

Menopon (avian parasitology) A genus of bird lice in the suborder Amblycera.

Menopon gallinae (avian parasitology) A species of bird lice in the genus *Menopon*. The 'shaft louse' off owls, ducks and pigeons.

Menopon pallidum (avian parasitology) A species of bird lice in the genus *Menopon*. A chicken louse found to harbour (*USA*: harbor) the virus of equine encephalomyelitis.

Menopon phaeostomum (avian parasitology) A species of bird lice in the genus *Menopon* which may be found on peacocks.

mensuration (measurement) Science of measurement.

menthol (chemistry) $C_{10}H_{19}OH$ White solid alicyclic compound. It is an alcohol with a minty smell and taste, used as a flavouring and in medicines.

menu (computing) A list of options which are displayed at each step in a computer programme (*USA*: program) and from which the computer user can choose.

menu bar (computing) Section at the top of a window on a computer screen which contains items available in pull-down menus.

menu-driven programme (computing) A programme (*USA*: program) which uses menus to advance to the next step by responding to a prompt from the user. (*USA*: menu-driven program).

mercaptan (chemistry) An alternative term for a thiol.

mercuric (chemistry) An alternative term for mercury(II).

mercurous (chemistry) An alternative term for mercury(I).

mercury arc (chemistry) Bright blue-green light obtained from an electric discharge through mercury vapour.

mercury cell (1, 2 physics) 1 Electrolytic cell that has a cathode made of mercury. *See* polarography. 2 Dry cell that has a mercury electrode. Also termed: Mallory cell.

mercury (Hg) (chemistry) Dense liquid metallic element in Group IIB of the periodic table (a transition element), used in lamps, batteries, switches and scientific instruments. It alloys with most metals to form amalgams. Its compounds are used in drugs, explosives and pigments. Also termed: quicksilver. At. no. 80; r.a.m. 200.59.

mercury(I) chloride (chemistry/control measures) HgCl White crystalline compound, used as an insecticide and in a mercury cell. Also termed: mercurous chloride; calomel.

mercury(II) chloride (chemistry) $HgCl_2$ Extremely poisonous white compound. Also termed: mercuric chloride; corrosive sublimate.

mercury(II) oxide (chemistry) HgO Red or yellow compound, slightly soluble in water, which reduces to metallic mercury on heating. Also termed: mercuric oxide.

mercury(II) sulphide (chemistry) HgS Red compound, which occurs naturally as cinnabar, used as a pigment (vermilion) and source of mercury. (*USA*: mercury(II) sulfide).

mercury-vapour discharge lamp (physics) Lamp that uses a mercury arc in a quartz tube; it produces ultraviolet radiation.

merge (computing) The combining of two or more files into one, according to a previously specified order.

Mermis (insect parasitology) A genus of nematode in the Mermithidae that are parasites to the ants *Aphaenogaster*, *Colobopsis*, *Lasius* spp., *Pheidole*, *Solenopsis* and *Teramorium*. Partly aberrant colour (*USA*: color) morphs (yellow ants) and occasionally morphological and behavioural (*USA*: behavioral) changes may develop.

Mermis albicans (*Hexamermis albicans*) (insect parasitology) A species of nematode in the Mermithidae that, similarly to *Mermis nigrescens* and *Mermis subnigriscens*, are parasites to the bees *Apis mellifera*. They may be found in workers, queens, and drones. It is possible that workers take up eggs with water from foliage.

Mermis nigrescens (insect parasitology) *See Mermis albicans.*

Mermis racovitzai (insect parasitology) A species of nematode in the Mermithidae that are parasites to the ants *Camponotus* and *Formica.*

Mermis subnigrescens (insect parasitology) A worm in the family Mermithidae which is parasitic in grasshoppers and earwigs. Also termed rainworm. *See Mermis albicans.*

mero- (terminology) A word element. [German] Prefix denoting: 1 part, 2 thigh.

meromyarian (parasitology) A nematode whose somatic musculature is composed of closely packed units of three or four flattened muscle cells. Includes those in the genus *Ancylostoma* and *Necator* spp.

merozoite (parasitological physiology) One of the organisms formed in certain protozoa by multiple asexual fission (schizogony) of a sporozoite within the body of the host. The basic structure of this stage resembles endozoites.

Merthiolate (chemistry) Trademark for an alcohol, acetone and water preparation of thimerosal.

Merthiolate-iodine-formalin perservative (methods) A preparation used for the preservation and staining of protozoa.

mes(o)- (terminology) A word element [German] Prefix denoting middle.

mesencephalon (anatomy) An alternative term for midbrain.

mesenteritis (medical) Inflammation of the mesentery.

mesenterium (anatomy) Mesentery.

mesenteron (medical) The midgut.

mesentery (anatomy) A membranous sheet attaching various organs to the body wall, especially the peritoneal fold attaching the intestine to the dorsal body wall.

meso- (biology) Prefix meaning middle.

Mesocestoides (human/veterinary parasitology) A genus of tapeworms in the family Mesocestoididae. Species include *Mesocestoides corti*, *Mesocestoides lineatus*, and *Mesocestoides variabilis* which may be found in the intestines in a range of carnivores including dogs, foxes, cats and humans. They may cause enteritis in humans, but are innocuous in other species. The worm has an unusual life cycle and has a stage of tissue invasion which may cause peritonitis.

mesocolon (anatomy) The mesentery that attaches the colon to the dorsal abdominal wall. It is called ascending, descending or transverse, according to the portion of the colon to which it attaches.

mesocolopexy (medical) Suspension or fixation of the colon.

mesocoloplication (medical) Plication of the mesocolon to limit its mobility.

mesoderm (histology) Tissue in an animal embryo that develops into tissues between the gut and ectoderm.

mesogenic (parasitology) A host-parasite relationship in which the parasite dominates but the host usually survives.

Mesogyna (parasitology) A genus of tapeworms in the family Mesocestoididae.

Mesogyna hepatica (veterinary parasitology) A species of tapeworms in the genus *Mesogyna* which may be found in the small bile ducts and blood vessels of the liver in a little known fox (*Vulpis macrotis*) in California, USA.

mesoileum (anatomy) The mesentery of the ileum.

mesojejunum (anatomy) The mesentery of the jejunum.

mesomerism (chemistry) Phenomenon in which a chemical compound can adopt two or more different structures by the alteration of (covalent) bonds, the atoms in the molecules remaining in the same relationship to each other. (*e.g.* Kekulé forms of benzene). Also termed: resonance.

See also tautomerism.

meson (physics) Member of a group of unstable elementary particles with masses intermediate between those of electrons and nucleons, and with positive, negative or zero charge. Mesons are emitted by nuclei that have been bombarded by high-energy electrons.

mesonotum (parasitological anatomy) A term used in relation to arthropods. The dorsal surface of the second thoracic segment.

messenger RNA (MRNA) (molecular biology) Ribonucleic acid that conveys instructions from DNA by copying the code of DNA in the cell nucleus and passing it out to the cytoplasm. It is translated into a polypeptide chain formed from amino acids which join in a sequence according to the instructions in the messenger RNA.

See also transcription.

Messocarus mirandus (insect parasitology) Mites in the Mesostigmata that, similarly to *Myrmozercon*, may be cleptoparasitic to various ants.

mesulfen (pharmacology/control measures) A sulfur-containing scabicide and antipruritic.

meta- (terminology) Word element. [German] Prefix denoting: 1 change, transformation, exchange, 2 after, next, 3 the 1,3- position in derivatives of benzene.

metabasis (medical) Change in the manifestations or course of a disease.

metabiosis (parasitology) The dependence of one organism upon another for its existence; commensalism.

metabolic (biology) Pertaining to internal metabolism.

metabolism (biochemistry) Biochemical reactions that occur in cells and are a characteristic of all living organisms. Metabolic reactions are initiated by enzymes and liberate energy in a usable form. Organic compounds may be broken down to simple constituents (catabolism) and used for other processes. Simple compounds may be built up to more complex ones (anabolism).

metabolite (biochemistry) Molecule participating in metabolism, which may be synthesized in an organism or taken in as food. Autotrophic organisms need only to take in inorganic metabolises; heterotrophs also need organic metabolises.

metacercaria (parasitological physiology) Plural: metacercariae. The encysted resting or maturing stage of a trematode parasite in the tissues of a second intermediate host or on vegetation.

metacestode (parasitological physiology) The larval stage of cestodes.

metacyclic trypanosome (parasitological physiology) Infective forms of trypanosomes which develop in the vector.

metacyst (parasitological physiology) A term used in relation to amoebae (*USA*: amebae). The trophozoite which emerges from the cyst.

Metagonimus (parasitology) A genus of digenetic trematodes in the family Heterophydiiae.

Metagonimus yokogawai (human/veterinary parasitology) A species of digenetic trematodes in the genus *Metagonimus* that may be found in the small intestine of dogs, cats, pigs, pelicans and humans. They may cause enteritis in humans but considered to be nonpathogenic in animals.

metal (chemistry) Any of a group of elements and their alloys with general properties of strength, hardness and the ability to conduct heat and electricity (because of the presence of free electrons). Most have high melting points and can be polished to a shiny finish. Metallic elements (about 80 per cent of the total) tend to form cations.

See also metalloid.

metallocene (chemistry) Member of a group of chemicals formed between a metal and an aromatic compound in which the oxidation state of the metal is zero; *e.g.* ferrocene.

metalloid (chemistry) Element with physical properties resembling those of metals and chemical properties typical of non-metals (*e.g.* arsenic, germanium, selenium). Many metalloids are used in semiconductors.

metallurgy (chemistry) Scientific study of metals.

metamorphosis (biology) Change of structure or shape, particularly, transition from one developmental stage to another, as from larva to adult form.

metaphase (genetics) Second stage of mitosis and meiosis, in which chromosomes are lined up along the equator of the nuclear spindle. Also termed: aster phase.

metaplasia (cytology) Transformation of one normal tissue type into another as a response to a disease or abnormal condition.

metastasis (histology) Process by which disease-bearing cells are transferred from one part of the body to another via the lymph and blood vessels; the term is usually applied to the spread of cancers. The term also applies to the newly diseased area arising from the process.

metastatic (chemistry) Describing electrons that leave an orbital shell, either entering another shell or being absorbed into the nucleus.

Metastrongylidae (parasitology) A family of nematode parasites, the adults of which invade the bronchi and lung parenchyma.

Metastrongyloidae (parasitology) A superfamily of nematodes whose members are found in the bronchi, parenchyma and blood vessels of the lungs; the lungworms.

Metastrongylus (parasitology) A genus of nematodes in the family Metastrongylidae.

Metastrongylus apri (parasitology) *See Metastrongylus elongatus.*

Metastrongylus brevivaginatus (parasitology) *See Metastrongylus pudendotectus.*

Metastrongylus elongatus (human/veterinary parasitology) (synonym: *Metastrongylus apri*) A species of nematodes in the genus *Metastrongylus* which may occur in the bronchi and bronchioles of pigs, occasionally in ruminants and very rarely in humans, but which has no significant pathogenic effect.

Metastrongylus madagascariensis (veterinary parasitology) A species of nematodes in the genus *Metastrongylus* which may be found in the bronchi of pigs, where it causes bronchitis, verminous pneumonia and loss of body weight.

Metastrongylus pudendotectus (veterinary parasitology) (synonym: *Metastrongylus brevivaginatus*) A species of nematodes in the genus *Metastrongylus* which may be found in the lungs of pigs.

Metastrongylus salmi (veterinary parasitology) A species of nematodes in the genus *Metastrongylus* which is a pig lungworm.

metatarsal (anatomy) One of the rod-shaped bones that forms the lower hind limb or part of the hind foot in tetrapods and the arch of the foot in human beings.

Metathelazia (parasitology) A genus of nematode worms of the family Pneumospiruridae, that may be found in the lungs of various species.

Metathelazia ascaroides (veterinary parasitology) A species of nematode in the genus *Metathelazia* that may be found in the langur monkey.

Metathelazia californica (veterinary parasitology) A species of nematode in the genus *Metathelazia* that may be found in various cats.

Metathelazia felis (veterinary parasitology) A species of nematode in the genus *Metathelazia* that may be found in *Felis pardalis*, etc.

Metathelazia multipapillata (veterinary parasitology) A species of nematode in the genus *Metathelazia* that may be found in hedgehogs.

metathorax (parasitological anatomy) A term used in relation to arthropods. Last or third thoracic segment.

Metazoa (biology) The division of the animal kingdom that includes the multicellular animals, *i.e.* all animals except the Protozoa.

metazoan (biology) Member of the zoological division of Metazoa.

metazoon (biology) Plural: metazoa [German] An individual organism of the Metazoa.

metazoonosis (parasitology) A zoonosis transmitted to vertebrate hosts by invertebrates. It depends on the invertebrate vectors or other intermediate hosts to complete the life cycle. Babesiosis and fascioliasis are examples.

-meter (terminology) A word element. [German] Suffix denoting *instrument for measuring*.

meter (measurement) *See* metre.

methanal (chemistry) HCHO An alternative term for formaldehyde.

methanoic acid (chemistry) HCOOH An alternative term for formic acid.

methanol (chemistry) CH_3OH Simplest primary alcohol, a poisonous liquid used as a solvent and added to ethanol to make methylated spirits. Also termed: methyl alcohol; wood spirit.

methionine (chemistry) $CH_3S(CH_2)_2CH(NH_2)COOH$ Sulphur-containing (*USA*: sulfur-containing) amino acid; a constituent of many proteins. Also termed: 2-amino-4-methylthiobutanoic acid.

methodology (methodology) The science dealing with principles of procedure in research and study.

methoprene (control measures) An insect growth regulator used to control ectoparasites on animals and in their environment. A common ingredient of household flea-bombs.

methyl alcohol (chemistry) An alternative term for methanol.

methyl chloroform (chemistry) An alternative term for 1,1,1-trichloroethane.

methyl cyanide (chemistry) An alternative term for acetonitrile.

methyl orange (chemistry) Orange dye used as a pH indicator.

methyl red (chemistry) Red dye used as a pH indicator.

methyl salicylate (pharmacology/microscopy) Methyl ester of salicyllic acid, used as a topical analgesic and as a clearing agent when mounting parasites. Also termed: oil of wintergreen; oil of sweet birch.

methylamine (chemistry) CH_3NH_2 Simplest primary amine, a gas smelling like ammonia, used in making herbicides.

methylaniline (chemistry) An alternative term for toluidine.

methylated spirit (chemistry) A mixture of methyl alcohol (4%) and ethyl alcohol.

methylation (chemistry) Chemical reaction in which a methyl group is added to a chemical compound.

methylbenzene (chemistry) An alternative term for toluene.

methylbutadiene (chemistry) An alternative term for isoprene.

methylene blue (chemistry) Blue dye used as a pH indicator.

methylphenol (chemistry) $CH_3C_6H_4OH$ Derivative of phenol in which one of the hydrogens of the benzene ring has been substituted by a methyl group. There are three isomers (ortho-, meta- and para-), depending on the positions of the substituents in the ring. Also termed: cresol.

Metoecus paradoxus (insect parasitology) Arthropods in the Coleoptera that, similarly to *Metoecus vespae*, are a parasitic to the wasps *Dolichovespula* spp., *Paravespula* spp. and *Vespula* spp., initially developing as endoparasites of wasp larvae, later becoming ectoparasitic.

Metoecus vespae (insect parasitology) *See Metoecus paradoxus*.

Metorchis (parasitology) A genus of flukes (digenetic trematodes) parasitic in the gall bladder and bile ducts, members of the family Opisthorchiidae.

Metorchis albidis (human/avian/veterinary parasitology) A species of flukes (digenetic trematodes) in the genus *Metorchis* that may be found in dogs, cats, foxes, grey (*USA*: gray) seals, birds and humans.

Metorchis conjunctus (veterinary parasitology) A species of flukes (digenetic trematodes) in the genus *Metorchis* that may be found in cats, dogs, mink, raccoons, and foxes.

metoxenous (parasitology) Requiring two hosts for the entire life cycle.

metre (m) (measurement) SI unit of length. 1 m = 39.37 inches. (*USA*: meter).

metre-candle (measurement) An alternative term for lux. (*USA*: meter-candle).

metric system (measurement) Decimal-based system of units. *See* SI units.

Metroliasthes (parasitology) A genus of tapeworms in the family Paruterinidae.

Metroliasthes lucida (avian parasitology) A species of tapeworms in the genus *Metroliasthes* that may be found in the small intestine of fowl and turkeys.

metronidazole (pharmacology) An antimicrobial compound effective against protozoa and anaerobic bacteria. Commonly used to treat trichomoniasis, amoebiasis (*USA*: amebiasis), giardiasis and balantidiasis.

-metry (terminology) A word element. [German] Suffix denoting *measurement*.

Mexican axolotl (parasitology) *Ambystoma mexicanum*. *See Ambystoma*.

MGP (physiology/biochemistry) An abbreviation of 'M'atrix 'G'la 'P'rotein.

MHC (1,2 physiology) 1 An abbreviation of 'M'ajor 'H'istocompatibility 'C'omplex; 2 An abbreviation of 'M'yosin 'H'eavy 'C'hain.

mho (measurement) Unit of conductance. Also termed: reciprocal ohm.

MHPG (biochemistry) An abbreviation of 3-'M'ethoxy-4-'H'ydroxy'P'heny'G'lycol.

micr(o)- (1 terminology; 2 measurement) 1 A word element. [German] Prefix denoting *small*. 2 Metric prefix meaning a millionth; $\times 10^{-6}$ (*e.g.* microfarad). It is sometimes represented by the Greek letter μ (*e.g.* μF).

microbalance (equipment) Balance capable of weighing very small masses (*e.g.* down to 10^{-5} mg).

microbe (biology) Imprecise term for any micro-organism, particularly a disease-causing bacterium.

microbiology (biology) The study of microorganisms, including bacteria and pathogenic protozoa, and the diseases they cause.

Microbothriidae (fish parasitology) A family of flatworms in class Monogenea and the phylum Platyhelminthes. Microbothriidae are parasites of the body surface of elasmobranchs and their affinities are uncertain. *Microbothrium apiculatum* is a representative of this family.

Microcotyle (fish parasitology) A monogenetic trematode of marine fish that attaches to the skin of the host.

microcyst (cytology) A cyst visible only under a microscope.

microfauna (biology) Microscopic animals, *e.g.* protozoa.

microfilaremia (parasitology) The presence of microfilariae in the circulating blood.

microfilaria (parasitology) The larva of worms in the superfamily Filarioidea. They are produced by adult worms residing in the bloodstream, tissues or body cavities, from where they can be ingested by biting insects. There they pass through a developmental stage and are transmitted to another permanent host when it is bitten by the insect. The microfilariae of some species are nocturnal and are therefore available for transmission only at night.

microfilarial (parasitology) Emanating from or pertaining to microfilariae.

microfilarial pityriasis (veterinary parasitology) A name given to a seasonal skin lesion on the backs of horses, characterized by hair loss and severe pruritus. Some of the lesions contain microfilariae, hence the name; but many of the cases appear to be allergic dermatitis.

See also dermatitis.

microfilariasis (medical/veterinary parasitology) Any disease caused by microfilariae. These include equine microfilarial pityriasis, heartworm in dogs, filarial dermatoses, cerebrospinal nematodiasis and miscellaneous localizations in individual organs, *e.g.* *Onchocerca* spp. invasions of the eye.

microflora (biology) Living microorganisms including protozoa that are so small that they can be seen only with a microscope and that maintain a more or less constant presence in a particular area, *e.g.* the pharynx or the rumen.

microgametocyte (parasitological physiology) In relation to malaria, the male gametocyte which produces a number of microgametes. In relation to *Toxoplasma* and *Sarcocystis*, the male stage of the parasite found in the epithelial cells of the intestine. The nucleus undergoes many divisions leading to the formation of microgametes.

micrometre (measurement) Small unit of length equal to 10^{-6} m, formerly called a micron. (*USA*: micrometer).

micron (μ) (measurement) Former name for the micrometre.

Micronema deletrix (veterinary parasitology) A saprophytic worm usually found in decaying organic matter but also found in lesions about the head and in internal organs of a horse, including the brain.

micronemes (parasitological cytology) A term used in relation to *Toxoplasma* and *Sarcocystis*. These are electron dense, small, cord-like structures located near the conoid. In cross-section they appear as oval or spherical bodies. Their function is not known. Also termed: taxonemes.

micronucleus (parasitological anatomy) 1 In ciliate protozoa, the kinetoplast, the smaller of two types of nucleus in each cell, which functions in sexual reproduction. 2 A small nucleus. 3 Nucleolus.

micronutrient (biochemistry) General term for any of the trace elements or vitamins.

micro-organism (biology) Organism that may be seen only with the aid of a microscope. Micro-organisms include microscopic bacteria, viruses and single-celled animals (*e.g.* protozoans).

microorganism (biology) A microscopic organism, including bacteria and protozoa.

micropathology (medical) 1 The sum of what is known about minute pathological change. 2 Pathology of diseases caused by microorganisms.

Microphallidae (avian parasitology) A family of intestinal digenetic trematodes that parasitize anserine and gallinaceous birds and may cause losses if the infestations are heavy.

microphotograph (photography) A photograph of small size.

micropipette (equipment) A pipette for handling small quantities of liquids up to 1 ml.

microprocessor (computing) The central processing unit of a computer.

microscope (microscopy) Instrument that produces magnified images of structures invisible to the naked eye. There are two major optical types: the simple microscope, consisting of one short focal-length convex lens giving a virtual image, and the compound microscope, consisting of two short focal-length convex lenses which combine to give high magnification. Highest magnifications are produced by an electron microscope.

microscopic (biology) Pertaining to something of extremely small size, which is visible only with the aid of a microscope.

microscopical (microscopy) Pertaining to a microscope or to microscopy.

microscopist (biology) A person skilled in using a microscope.

microscopy (biology) Examination with a microscope.

microsecond (measurement) One-millionth (10^{-6}) of a second; abbreviated μs or μsec.

microspectroscope (microscopy) A spectroscope and microscope combined.

Microsporidia (fish/insect/veterinary parasitology) An order of the subphylum Sporozoa characterized by having small spores and one polar capsule. Common occurrence is as parasites of arthropods and fish. The order includes the genus *Nosema*. In finfish Microsporidia may cause xenomas, while in crustaceans they may cause 'cotton flesh'.

***Microsporidium* sp.** (insect parasitology) A protozoal species of microsporidia that are parasites to the workers of the termites *Macrotermes championi*.

Microsporidium termitis (insect parasitology) A protozoal species of microsporidia that are parasites to the termites *Reticulitermes flavipes*.

microsyringe (measurement) A syringe fitted with a screw-threaded micrometer for accurate measurement of minute quantities.

Microtetramere (avian parasitology) A genus of spirurid nematode found in the proventriculus of birds.

Microthoracius (veterinary parasitology) A genus of sucking lice in the family Linognathidae. Species includes *Microthoracius cameli* which may be found on camels and dromedaries; and *Microthoracius mazzai*, *Microthoracius minor*, *Microthoracius praelongiceps* which may be found on the neck of the llama.

microtome (histology) Instrument for cutting thin slices (of the order of a few micrometres) of biological materials for microscopic examination.

microtomy (histology) The cutting of thin sections.

microtrauma (medical) A microscopic lesion or injury.

microtubule (histology) Minute cylindrical unbranched tubule composed of globular protein subunits found either singly or in groups in the cytoplasm of eukaryotic cells, in which it has the skeletal function of maintaining their shape. Microtubules are also associated with spindle formation, and hence are responsible for chromosomal movement during nuclear division.

microvolt (measurement) One-millionth (10^{-6}) of a volt; abbreviated μV.

microwave (physics) Electromagnetic radiation with a wavelength in the approximate range 1 mm to 0.3 m, *i.e.* between infra-red radiation and radio waves.

microwave spectroscope (physics) Study of atomic and/or molecular resonances in the microwave region.

mid-brain (anatomy) Part of the brain that connects the fore-brain to the hind-brain, concerned with processing visual information passed from the fore-brain. Fishes, amphibians and birds have a well developed mid-brain roof, the tectum, which forms the integration centre (*USA*: center) of their brain. Mammals have a less well developed mid-brain. Also termed: mesencephalon.

middle ear (anatomy) Air-filled part of the ear that is inside the ear drum and transmits sound waves from the outer ear to the inner ear. Also termed: tympanic cavity.

midges (parasitology) *See* Ceratopogonidae and *Culicoides*.

Mikoletzkya aerivora (insect parasitology) A species of nematode in the Diplogasteridae that are parasites to the termites *Reticulitermes* spp., that may be found in the head glands, the host becoming sluggish.

Mikrocytos (fish parasitology) Minute protozoan parasites infecting the haemocytes (*USA*: hemocytes) of oysters. *See* mikrocytosis.

mikrocytosis (fish parasitology) Disease caused by minute protozoans infecting the haemocytes (*USA*: hemocytes) of oysters. Includes *Mikrocytos macini* which causes Denman island disease in Pacific oysters, and *Mikrocytos roughlei* which causes winter mortality in Sydney rock oysters.

mil (1, 2 measurement) 1 A millilitre. 2 One-thousandth of an inch, equivalent to 0.0254 mm.

milbemycin (pharmacology) A macrolone lactone, derived from *Streptomyces cyanogeneus*, similar to the avermectins, and effective as an anthelmintic and parasiticide.

milbemycin oxime (pharmacology) A substance used for the prevention of heartworm in dogs and that has found a use in the treatment of generalized demodecosis.

mile (measurement) Unit of length equal to 1,760 yards or 5,280 feet. 1 mile = 1.60934 kilometres. A nautical mile is 6,080 feet (= 1.85318 km).

milk sugar (chemistry) An alternative term for lactose.

milli- (measurement) Metric prefix meaning a thousandth; $\times 10^{-3}$ (*e.g.* milligram).

milligram (mg) (measurement) Thousandth of a gram.

millilitre (ml) (measurement) Thousandth of a litre, equivalent to a cubic centimetre (cc or cm^3).

millimetre (mm) (measurement) Thousandth of a metre, equal to a tenth of a centimetre. 1 mm = 0.03937 inches. (*USA*: millimeter).

millimetre of mercury (mmHg) (measurement) Unit of pressure, equal to 1/760 atmospheres. (*USA*: millimeter of mercury).

Millipore filter (equipment) Trademark for cellulose acetate filters with pure sizes of 8 µm to 10 nm; such membranes are widely used for sterilizing liquid media.

min (measurement) An abbreviation of 'min'ute(s).

mineral acid (chemistry) Inorganic acid such as sulphuric (*USA*: sulfuric), hydrochloric or nitric acid.

mineral oil (chemistry) Hydrocarbon oil obtained from mineral sources or petroleum (as opposed to an animal oil or vegetable oil).

mineral salts (chemistry) Dissolved salts that occur in soil, derived from weathered rock and decomposed plants. They contain essential nutrients for plant growth, which are in turn utilized by herbivores (and carnivores that feed on them).

minute (1, 2 measurement) 1 Unit of time equal to 1/60 of an hour. 2 Unit of angular measure equal to 1/60 of a degree. Both types of minutes are made up of 60 seconds.

miracidium (parasitology) Plural: miracidia. [German] The free-swimming, ciliated larva of a trematode parasite which emerges from an egg and penetrates the body of a snail host.

MIS (physiology/biochemistry) An abbreviation of 'M'üllerian 'I'nhibiting 'S'ubstance.

miscarriage (medical) An alternative term for a spontaneous abortion.

miscible (chemistry) Describing two or more liquids that will mutually dissolve (mix) to form a single phase. They can be separated by fractional distillation.

MIT (biochemistry) An abbreviation of 'M'ono'I'odo'T'yrosine.

mite (parasitology) Any arthropod of the order Acarina except the ticks; they are characterized by minute size, usually transparent or semitransparent body, and other features distinguishing them from the ticks. They may be free living or parasitic on animals or plants, and may produce various irritations of the skin. *See* mange; chigger; harvest mites; *Psorergates ovis*; demodectic mange; otodectic mange; and many locality names, *e.g.* cat fur mites; ear mites; nasal mites; and other special titles, *e.g.* house-dust mites.

mite fever (parasitology) *See* scrub typhus.

miticide (control measures) A substance destructive to mites.

mitochondrion (cytology) Cell organelle in the cytoplasm of eukaryotic cells, concerned with aerobic respiration and hence energy production from the reduction of ATP to ADP. Its shape varies from spherical to cylindrical. Large concentrations of mitochondria are observed in areas of high energy consumption, such as muscle tissue. Also termed: chondriosome.

mitosis (genetics) The usual type of cell division in which the parent nucleus splits into two identical daughter nuclei, which contain the same number of chromosomes and identical genes to that of the parent nucleus. Also termed: karyokinesis.

See also meiosis.

mixed number (mathematics) Sum of a whole number (integer) and a fraction (*e.g.* 3 2/3, 12 15/16).

mixture (chemistry) Combination of two or more substances that do not react chemically and can be separated by physical methods (*e.g.* a solution).

Miyagawanella (parasitology) An obsolete name for a genus of organisms, the species of which are now assigned to the genus *Chlamydia.*

mode (1 statistics; 2 computing) Of a group of numbers, the number that occurs most often in the group. *See also* median. 2 A particular state of the system or programme (*USA*: program), *e.g.* edit or print, selected by the operator for a particular task.

modem (computing) Acronym of 'mo'dulator/ 'dem'odulator, a device for transmitting computer data over long distances (*e.g.* by telephone line).

molality (m) (chemistry) Concentration of a solution given as the number of moles of solute in a kilogram of solvent.

molar (1 chemistry; 2 anatomy) 1 Describing a quantity of a substance that is proportional to its molecular weight (a mole). *See* molarity. 2 Molar tooth.

molar concentration *See* molarity.

molar conductivity (physics) Electrical conductivity of an electrolyte with a concentration of 1 mole of solute per litre of solution. Expressed in siemens cm^2 mol^{-1}.

molar heat capacity (measurement) Heat required to increase the temperature of 1 mole of a substance by 1 kelvin. Expressed in joules K^{-1} mol^{-1}.

molar solution (chemistry) Solution that contains 1 mole of solute in 1 litre of solution.

molar volume (chemistry) Volume occupied by 1 mole of a substance under specified conditions.

molarity (M) (chemistry) Concentration of a solution given as the number of moles of solute in a litre of solution.

mole (mol) (chemistry) SI unit of amount of substance. In chemistry, it is the amount, of a substance in grams that corresponds to its molecular weight, or the amount that contains particles equal in number to the Avogadro constant. Also termed: gram-molecule.

molecular biology (molecular biology) Study of biological macromolecules (*e.g.* nucleic acids, proteins).

molecular formula (chemistry) Method of describing the composition of a molecule of a chemical compound, using the chemical symbols of the constituent elements with numerical suffixes that indicate the number of atoms of each element in the molecule *e.g.* H_2O and Na_2SO_4 are the molecular formulae of water and sodium sulphate (*USA*: sodium sulfate), respectively. The molecular formula gives no indication how the component atoms are arranged.

See also empirical formula; structural formula.

molecular orbital (chemistry) Region in space occupied by a pair of electrons that form a covalent bond in a molecule, formed by the overlap of two atomic orbitals.

molecular oxygen (chemistry) O_2 Diatomic molecular form of oxygen.

molecular sieve (chemistry) Method of separating substances by trapping (absorbing) the molecules of one within cavities of another, usually a natural or synthetic zeolite. Molecular sieves are used in ion exchange, desalination and as supports for catalysts.

molecular spectrum (physics) *See* spectrum.

molecular weight (chemistry) *See* relative molecular mass.

molecule (chemistry) Group of atoms held together in fixed proportions by chemical bonds; the fundamental unit of a chemical compound. The simplest molecules are diatomic molecules, consisting of two atoms (*e.g.* O_2, HCl); the most complex are biochemicals and macromolecules. The atoms may be joined by covalent bonds, dative bonds or bonds.

mollusc (parasitology) Members of the phylum Mollusca, which comprises about 50,000 species, and includes gastropods, which are intermediate hosts for animal flukes and lungworms.

molluscicide (control measures) A substance used for killing molluscs (mainly snails and slugs), *e.g.* copper sulphate (*USA*: sulfate), metaldehyde, methiocarb.

monad (1 biology; 2 chemistry; 3 genetics) 1 A single-celled protozoon or coccus. 2 A univalent radical or element. 3 In meiosis, one member of a tetrad.

monatomic (chemistry) Describing a molecule that contains only one atom (*e.g.* the rare gases).

Moniezia (veterinary parasitology) A genus of tapeworms in the family Anoplocephalidae. Species include *Moniezia benedini* and *Moniezia expansa* that are common inhabitants of the small intestine in young ruminants up to the age of 6 months. Massive infestations may cause diarrhoea (*USA*: diarrhea) and wasting.

Moniezia expensa (insect parasitology) A species of helminth in the Cestoda that are parasites to the ants *Solenopsis invicta.*

Moniliformis moniliformis (veterinary parasitology) A species of acanthocephalid (thorny headed) worms in the genus *Moniliformis* that may be found in the small intestine of rodents.

mono- (terminology) A word element. [German] Prefix denoting *one*, *single*, *limited to one part*, *combined with one atom*. (*e.g.* monobasic, monocotyledon).

monobasic acid (chemistry) Acid that on solvation produces one mole of hydroxonium ion (H_3O^+) per mole of acid; an acid with one replaceable hydrogen atom in its molecule (*e.g.* hydrochloric acid, HCl, and nitric acid, HNO_3). It cannot therefore form acid salts.

monocarboxylic acid (chemistry) Carboxylic acid with only one carboxylic group (*e.g.* (ethanoic) acetic acid, CH_3COOH).

Monocercomonas caviae (veterinary parasitology) A species of protozoa in the genus *Monocercomonas* that, similarly to *Monocercomonas minuta* and *Monocercomonas pistillum*, may be found in the caeca (*USA*: ceca), with no apparent pathogenic effect, of guinea pigs.

Monocercomonas cuniculi (veterinary parasitology) A species of protozoa in the genus *Monocercomonas* that may be found in the caeca (*USA*: ceca), with no apparent pathogenic effect, of rabbits.

Monocercomonas gallinarum (avian parasitology) A species of protozoa in the genus *Monocercomonas* that may be found in the caeca (*USA*: ceca), with no apparent pathogenic effect, of chicken.

Monocercomonas minuta (veterinary parasitology) A species of protozoa in the genus *Monocercomonas* that, similarly to *Monocercomonas caviae* and *Monocercomonas pistillum*, may be found in the caeca (*USA*: ceca), with no apparent pathogenic effect, of guinea pigs.

Monocercomonas pistillum (veterinary parasitology) A species of protozoa in the genus *Monocercomonas* that, similarly to *Monocercomonas minuta* and *Monocercomonas caviae*, may be found in the caeca (*USA*: ceca), with no apparent pathogenic effect, of guinea pigs.

Monocercomonas ruminantium (veterinary parasitology) A species of protozoa in the genus *Monocercomonas* that may be found in the rumen, with no apparent pathogenic effect, of cattle.

monochromatic light (physics) Light of a single wavelength.

monochromator (physics) *See* spectrometer.

monoclonal (immunology) Derived from a single parent clone.

monoclonal antibody (immunology) Antibody produced by a single-cell clone and hence consisting of a single amino acid sequence. Such cell clones are produced by the artificial fusion of cancerous and antibody-forming cells from the mouse spleen. The hybrid cells are grown in vitro as clones of cells, with each producing only a single type of antibody molecule.

Monocotylidae (fish parasitology) A family of flatworms in class Monogenea and the phylum Platyhelminthes. These are monogeneans of holocephalans and elasmobranchs. *Calicotyle* is an example of a typical monocotylid.

monocular (microscopy/biology) 1 Having but one eyepiece, as in a microscope. 2 Pertaining to one eye.

monocyte (haematology) The largest of the phagocytic leucocytes which are 10 to 12 micrometres and have a monogranulated cytoplasm with a large oval nucleus.

monodelphic (parasitological anatomy) Having a single set of reproductive systems in the female nematode.

monoecious (parasitological anatomy) A term used in relation to trematodes. Individuals containing gonads of both sexes, *i.e.* hermaphroditic.

Monogenea (fish parasitology) A class of flatworms in the phylum Platyhelminthes. There are five major monogenean groups, two families and three orders: the family Microbothriidae; order Dactylogyrida; order Gyrodactylida; family Monocotylidae; and order Polyopisthocotylida. *See* Microbothriidae; Dactylogyrida; Gyrodactylida; Monocotylidae; Polyopisthocotylida.

monogenean (parasitology) Pertaining to or emanating from Monogenea.

monohydrate (chemistry) Chemical compound (a hydrate) that contains 1 mole of water of crystallization in each of its molecules *e.g.* iron(II) sulphate (*USA*: iron(II) sulfate) forms a monohydrate, $FeSO_4.H_2O$.

monohydric (chemistry) Describing a chemical compound that has one hydroxyl group in each of its molecules (*e.g.* ethanol, C_2H_5OH, is a monohydric alcohol).

monomer (chemistry) Small molecule that can polymerize to form a larger molecule. *See* polymer.

monomorphic (parasitological physiology) When only a single morphological type occurs in one life cycle.

Monordotaenia taxidiensis (veterinary parasitology) A tapeworm of the family Taeniidae recorded in wild animals, such as skunks, mink, ermines, martens and badgers.

monosaccharide (chemistry) $C_nH_{2n}O_n$ Member of the simplest group of carbohydrates, which cannot be hydrolysed to any other smaller units; *e.g.* the sugars glucose, fructose.

monosodium glutamate (MSG) (chemistry) White crystalline solid, a sodium salt of the amino acid glutamic acid, made from soya bean protein and used as a flavour enhancer. Eating it can cause an allergic reaction in certain susceptible people.

monospaced font (computing) *See* fixed pitch font.

monosulfiram (control measures) An ectoparasiticide, used extensively against mites, but toxic if overused. Also termed: Tetmosol; sulfiram.

monovalent (chemistry) Having a valence of one. Also termed: univalent.

monoxenous (parasitology) Requiring only one host to complete the life cycle.

moose disease (veterinary parasitology) *See* neurofilariasis. Also termed: moose sickness.

moose tick (veterinary parasitology) *See Dermacentor albipictus*.

morantel (pharmacology) A general purpose anthelmintic for horses and cattle, used as the tartrate. Its mode of action is similar to levamisole and it can be used against worms that have benzimidazole resistance.

mordant (microscopy) 1 A substance capable of intensifying or deepening the reaction of a specimen to a stain. 2 To subject to the action of a mordant before staining.

Morellia (parasitology) A genus of flies in the family Muscidae. They are essentially flies of flowers and vegetation, but are attracted to sweat and mucus and can be a significant source of insect worry in summer. The common species are *Morellia aenescens*, *Morellia hortorum* and *Morellia simplex*. Also termed: sweat flies.

morpholine (chemistry) C_4H_9O Heterocyclic secondary amine, used as a solvent.

morphology (biology) Study of the origin, development and structures of organisms.

morphometry (measurement) The measurement of forms.

-morphous (terminology) A word element. [German] Suffix denoting *shape, form*.

mosaic (genetics) Organism derived from a single embryo that displays the characteristics of different genes in different parts of its body. *See* chimaera.

Moseley's law (physics) The X-ray spectrum of an element can be divided into several distinct line series: K, L, M and N. The law states that for certain elements the square root of the frequency f of the characteristic X-rays of one of these series is directly proportional to the element's atomic number Z. It was named after the British physicist Henry Moseley (1887–1915).

mosquito (insect parasitology) Blood-sucking insect of the genera *Aedes*, *Anopheles*, *Culex*, *Taeniorhynchus* (*Mansonia*) and *Psorophora*. Some species are concerned with the transmission of diseases, such as malaria, filarial nematodes, and avian malaria.

mosquito-bite dermatitis (parasitology) A condition in which pruritic papules and plaques develop on the face of cats due to hypersensitivity reactions to mosquito bites.

mother cell (biology) A cell that divides to form new, or daughter cells.

motile (biology) Describing an organism or structure that can move.

motility (biology) The ability to move spontaneously.

motor (biology) 1 Pertaining to motion. 2 A muscle, nerve or centre (*USA*: center) that effects movements.

motor neurone (cytology) Nerve cell that transmits impulses from the spinal cord or the brain to a muscle. Also termed: motor neuron; motor nerve.

mount (microscopy) To prepare specimens and slides for study.

mounting (microscopy) *See* mount.

mouse (computing) A peripheral device used to control cursor movement by means of a ball and a sensor.

mouse tick (veterinary parasitology) *See Ixodes muris*.

moxidectin (pharmacology) A macrolide derived from *Streptomyces cyaneogriseus noncyanogenus* with activity against nematodes and arthropods.

MPGF (physiology/biochemistry) An abbreviation of 'M'ajor 'P'ro'G'lucagon 'F'ragment.

MPP+ (biochemistry) An abbreviation of 1-'M'ethyl-4-'P'henyl'P'yridinium.

MPR (physiology/biochemistry) An abbreviation of 'M'annose-6-'P'hosphate 'R'eceptor.

MPTP (biochemistry) An abbreviation of 1-'M'ethyl-4-'P'henyl-1,2,5,6-'T'etrahydro'P'yridine.

MRI (physics) An abbreviation of 'M'agnetic 'R'esonance 'I'maging.

mRNA (physiology/biochemistry) An abbreviation of 'm'essenger 'RNA'.

MSH (physiology/biochemistry) An abbreviation of 'M'elanocyte-'S'timulating 'H'ormone.

MT (biochemistry) An abbreviation of 3-'M'ethyoxy'T'yramine.

mucin (biochemistry) Any of a number of glycoproteins that occur in mucus.

muck itch (parasitology) *See* sweet itch.

mucociliary (medical) Pertaining to a combination of cilia and mucus.

mucocutaneous (medical) Pertaining to mucous membrane and skin.

mucopus (medical) Mucus blended with pus.

mucous (cytology) Pertaining to or resembling mucus; secreting mucus.

mucous cell (cytology) A cell which secretes mucus.

mucous membrane (histology) Moist, mucus-lined epithelium which itself lines vertebrate internal cavities, including the alimentary, respiratory and reproductive tracts, which are continuous with the outer environment.

mucus (cytology) Slimy substance secreted by the goblet cells of mucous membrane. It lubricates and protects the epithelial layer on which it is secreted.

mud snail (parasitology) *Lymnaea truncatula. See Lymnaea.*

mudworm (parasitology) 1 Polychete worms, especially *Polydora websteri* which infect the shells of molluscs and that can cause significant losses. 2 *Lumbriculus variegatus.*

Muellerius (parasitology) A genus of nematodes in the family Protostrongylidae.

Muellerius capillaris (veterinary parasitology) A species of nematodes in the genus *Muellerius* that is the common lungworm of sheep and which may also occur in goats and chamois. In sheep the worm causes little harm, but in goats it may cause dyspnea and cough and occasionally severe interstitial pneumonia.

multi- (terminology) A word element. [Latin] Prefix denoting *many.*

multicausal disease (parasitology) 1 A number of causative agents are needed to combine to cause the disease. 2 The same disease can be caused by a number of different agents.

multicellular (biology) Describing plants and animals that have bodies consisting of many cells.

multi-CSF (physiology/biochemistry) An abbreviation of 'multi'potential 'C'olony-'S'timulating 'F'actor.

multifactorial (biology) Of or pertaining to, or arising through the action of, many factors.

multifactorial disease (medical) *See* multicausal disease.

multifactorial inheritance (genetics) Existence of more than two alleles for one gene.

multifocal (biology) Arising from or pertaining to many foci.

multilocular (biology) Having many compartments.

multilocular hydatid (parasitological physiology) Larval stage of *Echinococcus multilocularis* in which exogenous development occurs, resulting in infiltration of tissues. *See Echinococcus multilocularis.*

multimeter (physics) Instrument that can be used as a galvanometer, ammeter and voltmeter.

multiple bond (chemistry) Chemical bond that contains more electrons than a single bond (which contains 2 electrons); *e.g.* a double bond (4 electrons) or a triple bond (6 electrons).

multiple fission (parasitological physiology) A method of reproduction in protozoa. *See* schizogony.

murine respiratory mycoplasmosis (veterinary parasitology) A disease of mice caused by *Mycoplasma pulmonis* and characterized by dyspnea, nasal discharge, head tilt and incoordination. In most mice, infection occurs without clinical signs. Also termed: chronic respiratory disease of rats and mice; chronic murine pneumonia.

murine typhus (human/veterinary parasitology) A disease of rats caused by *Richettsia typhi*, transmitted by the rat flea *Xenopsylla cheopis* and the rat louse *Polyplax spinulosa,* that is an important disease of humans.

murrina (parasitology) [Spanish] *Trypanosoma evansi* infection.

Musca (insect parasitology) A genus of flies in the family Muscidae, many of which act as intermediate hosts for *Thelazia* spp worms.

Musca amica (insect parasitology) A species of flies in the genus *Musca* that, similarly to *Musca convexifrans* and *Musca larvipora*, are intermediate hosts for *Thelazia* spp worms.

Musca autumnalis (insect parasitology) A species of flies in the genus *Musca* that congregates on the face of cattle and may spread infectious keratoconjunctivitis. Also termed: face fly.

Musca bezzi (insect parasitology) A species of flies in the genus *Musca* that, similarly to *Musca fasciata, Musca lusoria, Musca pattoni, Musca vetustissima*, and *Musca vitripennis*, follow the blood-sucking fly species, cleaning up spilt blood.

Musca conducens (insect parasitology) A species of flies in the genus *Musca* that transmits *Stephanofilaria* spp.

Musca convexifrans (insect parasitology) A species of flies in the genus *Musca* that, similarly to *Musca amica* and *Musca larvipora*, are intermediate hosts for *Thelazia* spp worms.

Musca crassirostris (insect parasitology) A species of flies in the genus *Musca* that is able to draw blood.

Musca domestica (insect parasitology) A species of flies in the genus *Musca* that acts as a mechanical carrier of many bacterial, viral and protozoan diseases and is an intermediate host for the helminths *Habronema* and *Raillietina* spp. Also termed: house fly.

Musca fasciata (insect parasitology) A species of flies in the genus *Musca* that, similarly to *Musca bezzi, Musca lusoria, Musca pattoni, Musca vetustissima*, and *Musca vitripennis*, follow the blood-sucking fly species, cleaning up spilt blood.

Musca fergusoni (insect parasitology) A species of flies in the genus *Musca* that, similarly to *Musca hilli* and *Musca terraeregina*, are bush flies.

Musca hilli (insect parasitology) A species of flies in the genus *Musca* that, similarly to *Musca fergusoni* and *Musca terraeregina*, are bush flies.

Musca larvipora (insect parasitology) A species of flies in the genus *Musca* that, similarly to *Musca convexifrans* and *Musca amica*, are intermediate hosts for *Thelazia* spp worms.

Musca lusoria (insect parasitology) A species of flies in the genus *Musca* that, similarly to *Musca fasciata*, *Musca bezzi*, *Musca pattoni*, *Musca vetustissima*, and *Musca vitripennis*, follow the blood-sucking fly species, cleaning up spilt blood.

Musca pattoni (insect parasitology) A species of flies in the genus *Musca* that, similarly to *Musca fasciata*, *Musca lusoria*, *Musca bezzi*, *Musca vetustissima*, and *Musca vitripennis*, follow the blood-sucking fly species, cleaning up spilt blood.

Musca sorbens (insect parasitology) A species of flies in the genus *Musca* that are a complex of bush flies.

Musca terraeregina (insect parasitology) A species of flies in the genus *Musca* that, similarly to *Musca hilli* and *Musca fergusoni*, are bush flies.

Musca vetustissima (insect parasitology) The Australian bush fly. A species of flies in the genus *Musca* that is an intermediate host for *Habronema* spp. and similarly to *Musca fasciata*, *Musca lusoria*, *Musca pattoni*, *Musca bezzi*, and *Musca vitripennis* follow the blood-sucking fly species, cleaning up spilt blood.

Musca vitripennis (insect parasitology) A species of flies in the genus *Musca* that, similarly to *Musca fasciata*, *Musca lusoria*, *Musca pattoni*, *Musca vetustissima*, and *Musca bezzi*, follow the blood-sucking fly species, cleaning up spilt blood.

muscle (histology) Animal tissue that contracts (by means of muscle fibres [*USA*: muscle fibers]) to produce movement, tension and mechanical energy. *See* involuntary muscle; voluntary muscle.

muscular (anatomy) 1 Pertaining to a muscle. 2 Having well developed muscles.

muscular parasitic diseases (parasitology) Parasitic diseases involving muscle, including cysticercosis, hepatozoonosis, *Neosprum caninum* myositis, sarcocystosis, toxoplasmosis, and trichenellosis.

musculature (anatomy) The muscular system of the body, or the muscles of a particular region.

musculocutaneous (biology) Pertaining to muscle and skin.

musculomembranous (biology) Pertaining to muscle and membrane.

musculoskeletal (biology) Pertaining to muscle and skeleton.

musculotropic (parasitology) Exerting its principal effect upon muscle.

mutagen (toxicology) Chemical or physical agent that induces or increases the rate of mutation; *e.g.* ethyl methane sulphonate (*USA*: ethyl methan sulfonate), ultra-violet light, X-rays and gamma-rays.

mutant (molecular biology) Organism that arises by mutation.

mutarotation (chemistry) Change in the optical activity of a solution containing photo-active substances, such as sugars.

mutation (molecular biology) Alteration in the sequence of bases encoded by DNA, resulting in a permanent inheritable change in the gene and consequently the protein encoded. Mutations may occur in different ways and may be induced by a mutagen or occur spontaneously. A mutation can be detrimental, *e.g.* those thought to be involved in carcinogenesis (formation of cancer). However, some mutations can be advantageous, *e.g.* in evolution, where favourable characteristics may be endowed.

mutualism (biology) Relationship between two organisms from which each benefits (*e.g.* cellulose-digesting micro-organisms and animals, such as ruminants, whose gut they inhabit).

See also symbiosis.

MVV (physiology) An abbreviation of 'M'aximal 'V'oluntary 'V'entilation.

my(o)- (terminology) A word element. [German] Prefix denoting *muscle*.

myc(o)- (terminology) A word element. [German] Prefix denoting *fungus*.

Mycoplasma (biology) A genus of highly pleomorphic, aerobic or facultatively anaerobic bacteria that lack cell walls, including the pleuropneumonia-like organisms (PPLO).

mycoplasmal (medical) Emanating from or pertaining to infection with *Mycoplasma*.

mycoplasmosis (medical) Any disease caused by infection with *Mycoplasma* spp.

myel(o)- (terminology) A word element. [German] Prefix denoting marrow (often with specific reference to the spinal cord).

myelin sheath (cytology) Thin fatty layer of membranes, produced by Schwann cells, that covers the axon of most vertebrate neurones (nerve cells).

myelitis (medical) Inflammation of the spinal cord. Myelitis is a sigificant lesion in equine protozoal myeloencephalitis.

myeloid tissue (histology) Tissue usually present in bone marrow which produces red blood cells and other blood constituents.

myiasis (human/veterinary parasitology) Invasion of the body by the larvae of flies, characterized as cutaneous (subdermal tissue), gastrointestinal, nasopharyngeal, ocular or urinary, depending on the region invaded.

Mylabris phalerata (insect parasitology) A beetle containing cantharidin. Also termed Chinese blistering beetle, Chinese blister fly, mylabris.

Mylabris sidae (insect parasitology) *See Mylabris phalerata.*

myo- (terminology) A word element. [German] Prefix denoting muscle.

Myobia (parasitology) A genus of parasitic mites in the family Myobiidae.

Myobia musculi (veterinary parasitology) A species of parasitic mites in the genus *Myobia* that causes alopecia and dermatitis in laboratory mice.

myocardial (anatomy) Relating to myocardium.

myocarditis (medical) Inflammation of the muscular walls of the heart (the myocardium). The condition may result from, among other things, bacterial or viral infections or from trypanosomiasis or Chagas' disease.

myocardium (histology) Muscle tissue of the vertebrate heart.

Myocoptes (parasitology) A genus of mites in the family Listrophoridae.

Myocoptes musculinus (veterinary parasitology) A species of mites in the genus *Myocoptes* that may be found on the hair of guinea pigs and laboratory mice.

Myocoptes romboutsi (parasitology) *See Trichoecius romboutsi.*

myoglobin (biochemistry) In vertebrate muscle fibre (*USA*: muscle fiber), a haem (*USA*: heme) protein capable of binding with one atom of oxygen per molecule.

myology (histology) Study of muscles.

Myoptes musculinus (parasitology) *See Myocoptes musculinus.*

myosin (biochemistry) Fibrous protein which, with actin, makes up muscle. Movement of myosin fibres (*USA*: myosin fibers) between actin fibres (*USA*: actin fibers) causes muscle contraction.

myositis (medical) Inflammation of a voluntary muscle. Causes heat, swelling, pain and lameness if a limb is affected.

See also trichinous myositis.

Myriapoda (biology) Originally a class of arthropods, including the millipedes and centipedes now superseded by two classes: Diplopoda (millipedes) and Chilopoda (centipedes).

myring(o)- (terminology) A word element. [Latin] Prefix denoting tympanic membrane.

Myrmecodispus (insect parasitology) Mites in the Prostigmata that may be larval or egg parasites on the ants *Eciton*, *Neivamyrmex* and *Nomamyrmex*.

Myrmecolax **sp.** (insect parasitology) Arthropods in the Strepsiptera that are parasitic to the ants *Camponotus*, *Pseudomyrmex* and *Solenopsis*.

Myrmozercon (insect parasitology) (synonym: *Myrmonyssus*) *See Messocarus mirandus.*

myx(o)- (terminology) A word element. [German] Prefix denoting mucus, slime.

Myxobolus cerebralis (fish parasitology) A myxosporean parasite which may invade the cranial cartilages of juvenile rainbow trout, causing whirling disease.

myxomatosis (medical) 1 The development of multiple myxomas. 2 myxomatous degeneration.

Myxosoma (fish parasitology) A genus of myxozoan parasites in the class Myxosporea which may cause serious disease losses in free-living fishes.

Myxosoma cartilaginis (fish parasitology) A species of myxozoan parasites in the genus *Myxosoma* which may invade the cartilage in the head of fish without causing apparent deformity or nervous sign.

Myxosoma cerebralis (fish parasitology) A species of myxozoan fish parasites in the genus *Myxosoma* which causes whirling disease and twist disease characterized by tail chasing, black pigmentation of the tail and deformity, including sunken heads and twisted spines. The parasite invades and destroys skeletal discharge.

Myxosoma dujardini (fish parasitology) A species of myxozoan fish parasites in the genus *Myxosoma* which causes cysts in the gills, followed by dyspnea and death due to asphyxia.

myxosporean (fish parasitology) Cyst-producing parasite found in many organs of fish. Previously regarded as protozoans, now believed to be metazoans.

myxozoan (parasitology) Member of the phylum Myxozoa, a phylum of parasitic protozoa found chiefly in fish, amphibians and reptiles.

Myzorhynchus (parasitology) One of the mosquito vectors of the canine heartworm, *Dirofilaria immitis.*

myzorhyncus (fish parasitology) The apical glandular region of the scolex of a cestode belonging to the order Tetraphyllidea, parasitic in sharks.

N

N (biochemistry) 1 The chemical symbol for nitrogen. 2 A symbol for asparagine.

N gen (biology) An abbreviation of 'N'ew 'gen'us.

N terminal (biochemistry) A term for the end of peptide or protein having a free -NH2 group.

NA (1, 2 general terminology) 1 An abbreviation of 'N'ot 'A'pplicable. 2 An abbreviation of 'N'ot 'A'vailable.

Na (chemistry) The chemical symbol for sodium. The symbol is derived from Latin *natrium.*

Na⁺ (chemistry) Sodium ion.

NaCl (chemistry) Sodium chloride.

NAD⁺ (biochemistry) An abbreviation of 'N'ictinamide 'A'denine 'D'inucleotide. Co-enzyme form of the vitamin nicotinic acid, necessary in certain enzyme-catalysed oxidation reduction reactions in cells. Its reduced form is a precursor in the fixation of carbon dioxide in chloroplasts during photosynthesis.

NADP⁺ (biochemistry) An abbreviation of 'N'icotinamide 'A'denine 'D'inucleotide 'P'hosphate.

Na_e (physiology/biochemistry) A symbol for exchangeable body sodium.

Naegleria (parasitology) Genus of protozoa in the family Vahlkampfiidae.

Naegleria fowleri (human parasitology) A species of protozoa in the genus *Naegleria* that persists in thermally heated swimming pools and may cause fatal meningoencephalitis in humans.

nagana (parasitology) *See* trypanosomiasis.

nail (histology) Layer of keratin that grows on the upper surface of the fingers of human beings and other primates (except tree-shrews, which have claws).

Nairobi sheep disease (veterinary parasitology).An infectious disease of sheep transmitted by the tick Rhipicephalus appendiculatus, caused by an arbovirus in the genus Nairovirus.

naive (immunology) An individual that has not been exposed to a particular antigen.

naled (control measures) An organophosphorus insecticide.

nano- (measurement) A word element. [German]. Prefix meaning a thousand-millionth; $\times$ 10^{-9}; *e.g.* 1 nanosecond is 10^{-9} s.

nanogram (measurement) One-thousand-millionth (10^{-9}) gram.

nanometre (measurement) Thousand-millionth of a metre (*USA*: meter).; 10^{-9} m. It is the usual unit for wavelengths of light and interatomic bond lengths in chemistry. (*USA*: nanometer).

nanophyetiasis (parasitology) *See* salmon poisoning; *Nanophyetus salmincola.*

Nanophyetus (parasitology) A genus of digenetic trematodes in the family Nanophyetidae, which are parasites in the intestines of mammals.

Nanophyetus salmincola (human/avian/fish/veterinary parasitology) A species of digenetic trematodes in the genus *Nanophyetus* which may be found in the small intestines of cats, dogs and many small, wild mammals, and occasionally in piscivorous birds and humans. The second intermediate host is fish. The fluke carries the rickettsia that causes salmon poisoning and elokomin fluke fever in the terminal host.

nanosecond (nS) (measurement) One billionth of a second. Also termed nsec.

naphthalene (chemistry) $C_{10}H_8$ Solid aromatic hydrocarbon that consists of two fused benzene rings, insoluble in water but soluble in hot ethanol. It is a starting material in the manufacture of dyes.

naphthalophos (pharmacology) A species-specific anthelmintic against *Haemonchus contortus*, most useful in areas where benzimidazole resistance has developed in the worm population and where frequent treatment is necessary. It also has the advantage that it is used at a single dose level for sheep of all sizes. It is an organophosphate compound but is safe at the recommended dose rates. Higher dose rates are of moderate efficiency against most intestinal nematodes of sheep, but may be toxic. Also termed phthalophos: naftalofos.

Napierian logarithm (mathematics) Logarithm to the base *e* (e = 2.71828...), named after the Scottish mathematician John Napier (1550–1617). Also termed: natural logarithm.

narc. (biochemistry) An abbreviation of 'narc'otic.

narco- (terminology) A word element. [German] Prefix denoting *stupor, stuporous state.*

narcotic (pharmacology) Analgesic drug that, in addition to killing pain, causes loss of sensation or consciousness (*e.g.* morphine and other opiates).

narcotic antagonist (pharmacology) Any drug that reduces the effects of a narcotic.

nares (anatomy) The external openings of the nose; the nostrils.

NAS (education) An abbreviation of 'N'ational 'A'cademy of 'S'cience.

nasal (biology) Pertaining to the nose.

nasal acariasis (parasitology) Nasal infestation with arthropod parasites of the order Acarina. Characterized by mild nasal discharge and hyperaemia (*USA*: hyperemia), and occasionally severe rhinitis.

nasal bot fly (veterinary parasitology) A bot fly, an infestation of which in sheep, causes sneezing and constant nasal discharge. The presence of the flies in the flock also causes some insect worry. *See Oestrus ovis.*

nasal cavity (anatomy) Cavity located in the head of tetrapods, containing the olfactory sense organs.

nasal mites (parasitology) *See* nasal acariasis.

nasal myiasis (parasitology) *See Oestrus ovis* infestation.

nasal schistosomiasis (parasitology) Infection with the blood fluke *Schistosoma*, which is largely asymptomatic but can cause dyspnea, snoring and profuse nasal discharge.

nasal swab (parasitology) A cotton swab on a stick, passed up the nostril to obtain a sample of exudate and epithelial debris for microbiological or cellular examination.

naso- (terminology) A word element. [Latin]. Prefix denoting nose.

nasopharyngeal (anatomy) Pertaining to the part of the pharynx above the level of the soft palate.

National Science Foundation (NSF) (governing body) A US government agency supporting research and education programs to encourage interest in science.

natural abundance (chemistry) Relative proportion of the various isotopes in a naturally occurring sample of an element.

natural cytotoxic cells (immunology) *See* killer cells.

natural logarithm (mathematics) An alternative term for Napierian logarithm.

natural number (mathematics) One of the set of ordinary counting numbers (*e.g.* 1, 2, 3, 4, etc.).

natural selection (genetics) One of the conclusions drawn by the British naturalist Charles Darwin (1809–82) from the theory of evolution: certain organisms with particular characteristics are more likely to survive and hence pass on their characteristics to their offspring, *i.e.* survival of the fittest. Thus the characteristics of a population are controlled by this process.

Naumenko/Feigin stain (histology) Histological staining technique for astrocytes.

Nauta stain (histology) Histological staining technique. A silver stain that is taken up by degenerating axons, and hence identifies dead or degenerating neurons.

Nb (chemistry) The chemical symbol for niobium.

NB (general terminology) An abbreviation of *'N'ota 'B'ene*, Latin, meaning 'note well'.

n-butyl chloride (pharmacology) A largely superseded anthelmintic with activity against ascarids, hookworms and whipworms in dogs.

nearest neighbour method (epidemiology) A method for establishing the population density of a particular species of animal or plant. One of the units is identified and the distance to its nearest neighbour (*USA*: neighbor) of the same species measured. The number of units of that species in an area can then be established, the accuracy increasing with the number of measurements made. (*USA*: nearest neighbor method).

Necator (parasitology) A genus of hookworm in the subfamily Necatorinae.

Necator americanus (human/veterinary parasitology) A species of hookworm in the genus *Necator*. The common hookworm of humans, it may also be found in pigs and dogs.

Necator suillus (veterinary parasitology) A species of hookworm in the genus *Necator* that may be found in pigs.

necatoriasis (parasitology) Infection with organisms of the genus *Necator*. A hookworm disease manifested by anaemia (*USA*: anemia) and melaena (*USA*: melena).

neck (parasitological anatomy) The connecting tissues between the scolex and strobila of a tapeworm. This part is unsegmented.

necro- (terminology) A word element [German] Prefix denoting *death*.

needle nematode (plant parasitology) *See Longidorus africanus.*

NEFA (physiology/biochemistry) An abbreviation of 'Non'E'sterified free 'F'atty 'A'cid.

negative number (mathematics) Number less than zero.

negative staining (microscopy) A procedure visualizing specimens by either light or electron microscopy. In light microscopy, India ink, which blocks the transmission of light, may be used as a negative stain. In electron microscopy, electron-dense salts, such as sodium phosphotungstate, may be used in the examination of particles.

nematocerans (insect parasitology) Members of the suborder of insects Nematocera which includes the gnats, mosquitoes, midges, black flies and gall flies.

nematocide (pharmacology) 1 Destroying nematodes. 2 A substance that destroys nematodes, including phenothiazine, piperazine, benzimidazole, the imidazothiazoles, the tetrahydropyrimidines, organophosphorus compounds and a wide variety of miscellaneous compounds, including avermectins.

Nematoda (fish parasitology) A phylum of invertebrates comprising the roundworms. The nematodes are a large, extremely diverse and successful group, consisting of 256 families. The majority are free-living and occupy environments ranging from the sea depths to deserts and ice caps, while the parasitic nematodes exploit a great variety of hosts in both plant and animal kingdoms. The 125 families of zooparasitic nematodes are thought to have originated from freeliving soil-dwelling forms on land rather than in the sea. Fish nematodes are suggested to have been derived from terrestrial vertebrate nematodes, as nearly three quarters of the 17 nematode families found in fish are also found in land vertebrates. This hypothesis is supported by the absence of any superfamily, and the presence of only five families of nematodes unique to fish, suggesting that they are an acquired fauna. Nematodes are usually divided into two groups according to the presence or absence of, 1 special gland cells associated with the oesophagus known as stichocytes which, if arranged in a longitudinal row, are known as stichosome, 2 caudal papillae, 3 a reserve organ, called a trophosome, apparently syncytial, formed by the transformation of the oesophagus, 4. a pair of glandular sensory organs, called phasmids, situated laterally in the caudal region and opening to the surface by a slit or pore, 5. polar plugs to the eggs. Also termed: roundworms. *See* Adenophorea; Secernentea.

nematode (parasitology) Any individual organism of the class Nematoda.

nematode galls (plant parasitology) Hard, fibrous excrescences produced in the seedheads of grasses by chronic inflammation created by an invasion by larvae of grass seed nematodes, *e.g.* *Anguina* spp.

nematodiasis (parasitology) State of infestation with nematodes.

Nematodirella (parasitology) A genus of worms in the trichostrongyloid family Molineidae.

Nematodirella cameli (veterinary parasitology) A species of worms in the genus *Nematodirella* that may be found in camel and reindeer.

Nematodirella dromedarii (veterinary parasitology) A species of worms in the genus *Nematodirella* that may be found in dromedary and sheep.

Nematodirella longispiculata (veterinary parasitology) A species of worms in the genus *Nematodirella* that may be found in the small intestine of sheep, goats and other ruminants.

nematodiriasis (parasitology) Infestation in the small intestine with the nematode *Nematodirus* spp. Usually found in mixed infestations. Characterized clinically by persistent diarrhoea (*USA*: diarrhea) and wasting.

Nematodirus (parasitology) A genus of roundworms in the trichostrongyloid family Molineidae that are the cause of nematodiriasis.

Nematodirus abnormalis (veterinary parasitology) A species of roundworms in the genus *Nematodirus* that may be found in sheep, goats and camels.

Nematodirus andrewi (veterinary parasitology) A species of roundworms in the genus *Nematodirus* that may be found in sheep.

Nematodirus aspinosus (veterinary parasitology) A species of roundworms in the genus *Nematodirus* that may be found in rabbits and hares.

Nematodirus battus (veterinary parasitology) A species of roundworms in the genus *Nematodirus* that is a parasite of sheep, causing nematodiriasis.

Nematodirus filicollis (veterinary parasitology) A species of roundworms in the genus *Nematodirus* that may be found in sheep, cattle, goats and deer.

Nematodirus helvetianus (veterinary parasitology) A species of roundworms in the genus *Nematodirus* that may be found in cattle and occasionally sheep.

Nematodirus hsuei (veterinary parasitology) A species of roundworms in the genus *Nematodirus* that may be found in sheep.

Nematodirus lamae (veterinary parasitology) A species of roundworms in the genus *Nematodirus* that may be found in sheep.

Nematodirus leporis (veterinary parasitology) A species of roundworms in the genus *Nematodirus* that may be found in rabbits and hares.

Nematodirus odocoilei (veterinary parasitology) A species of roundworms in the genus *Nematodirus* that may be found in sheep.

Nematodirus oiratianus (veterinary parasitology) A species of roundworms in the genus *Nematodirus* that may be found in camels and wild ruminants.

Nematodirus spathiger (veterinary parasitology) A species of roundworms in the genus *Nematodirus* that may be found in the small intestine of sheep, cattle and other ruminants.

Nematodirus tarandi (veterinary parasitology) A species of roundworms in the genus *Nematodirus* that may be found in sheep.

neo- (general terminology) Word element. [German]. Prefix meaning *new.*

neoarsphenamine (pharmacology) An organic arsenical antiprotozoal agent which used to be used extensively, particularly in the treatment of swine dysentery and blackhead of turkeys, but is now replaced by more effective compounds.

Neoascaris vitulorum (parasitology) *See Toxocara vitulorum.*

Neobenedinia (fish parasitology) A genus of cutaneous, ocular and oral flukes of marine fish, in the order Capsalidae that are characterized by being transparent. Also termed: *Benedinia.*

Neoberlesia mexicana (insect parasitology) Mites in the family Mesostigmata that may be found in nests of the ants *Pheidole acolhua.*

Neocuterebra (parasitology) A genus of flies in the family Oestridae, the larvae of which parasitize animals.

Neocuterebra squamosa (veterinary parasitology) A species of flies in the genus *Neocuterebra* the larvae of which may be found in the African elephant.

Neodactylogyrus (fish parasitology) A genus of monogenetic flukes in the family Dactylogyridae, which parasitize marine and freshwater fish.

Neodiplostomum (parasitology) A genus of intestinal digenetic trematodes in the family Diplostomatidae.

Neodiplostomum multicellulata (avian parasitology) A species of monogenetic flukes in the genus *Neodiplostomum* which may be found in herons.

Neodiplostomum perlatum (avian parasitology) A species of monogenetic flukes in the genus *Neodiplostomum* which may be found in piscivorous birds.

Neodiplostomum tamarini (veterinary parasitology) A species of monogenetic flukes in the genus *Neodiplostomum* which may be found in New World primates.

neodymium (Nd) (chemistry) Metallic element in Group IIIB of the periodic table (one of the lanthanides), used in special glass for lasers. At.no. 60; r.a.m. 144.24.

Neogregarinidae sp. (insect parasitology) Protozoal species of gregarinida that, similarly to *Mattesia geminata*, are parasites to the ants *Solenopsis geminata*, *Solenopsis invicta* and *Solenopsis richteri*. These are unusual in that they infect the hypodermis.

neo-lamarckism (genetics) *See* Lamarckism.

neon fish disease (fish parasitology) *See Plistophora.*

neon (Ne) (chemistry) Gaseous nonmetallic element in Group 0 of the periodic table (the rare gases) which occurs in trace quantities in air (from which it is extracted). It is used in discharge tubes (for advertising signs) and indicator lamps. At. no. 10; r.a.m. 20.179.

neonatal (medical) Concerning the newborn.

Neonyssus (parasitology) A genus of mites in the family Rhinonyssidae.

Neonyssus columbae (avian parasitology) A species of nasal mites in the genus *Neonyssus* that, similarly to *Neonyssus melloi*, may be found in pigeons.

Neonyssus melloi (avian parasitology) A species of nasal mites in the genus *Neonyssus* that, similarly to *Neonyssus columbae*, may be found in pigeons.

neoplasm (histology) Tumour (*USA*: tumor) or group of cells with uncontrolled growth. It may be benign and localized, or if cells move from their normal position in the body and invade other organs the tumour is malignant.

Neoschongastia (parasitology) A genus of mites in the family Trombiculidae.

Neoschongastia americana (avian parasitology) A species of mites in the genus *Neoschongastia* which infest chicken, quail and turkey, and that may cause dermatitis.

Neospora (veterinary parasitology) A genus of protozoa that closely resembles *Toxoplasma.* The life cycle is unknown but transplacental transmission is known to occur; causes ascending paralysis in kittens, puppies, calves and abortion in cattle. Species includes *Neospora caninum.*

neosporosis (veterinary parasitology) Infection by the protozoa *Neospora caninum.* In dogs, there are clinical signs referable to the central nervous system and myositis. Young puppies are most commonly and severely affected, showing a characteristic ascending limb paralysis with hyperextension, cervical weakness and dysphagia. Infection is a significant cause of late abortion, and some defective live calves, in cattle in Australia and New Zealand. An infected carnivore is the likely source of infection.

Neostrongylus (parasitology) A genus of nematodes in the family Protostrongylidae.

Neostrongylus linearis (veterinary parasitology) A species of nematodes in the genus *Neostrongylus* that may be found in the lungs of sheep and goats.

Neotarsonemoides (insect parasitology) A species of mites in the Prostigmata, that are phoretic on social bees.

Neotrombicula autumnalis (parasitology) An acarine mite in the family Trombiculidae that causes dermatitis, usually on the head and extremities, in most animal species. Also termed: *Trombicula autumnalis.*

NEP (biochemistry) An abbreviation of 'N'eutral 'E'ndo'P'eptidase.

nephr(o)- (terminology) A word element. [German] Prefix pertaining to *kidney.*

nephritis (medical) Inflammation of the kidney.

nephron (histology) Functional filtering unit of the vertebrate kidney, consisting of Bowman's capsule and the glomerulus.

neptunium (Np) (chemistry) Radioactive element in group IIIB of the periodic table (one of the actinides); it has several isotopes. At. no. 93; r.a.m. 237 (most stable isotope).

nequinate (pharmacology) A quinolone cocciostat used in poultry.

nerve (anatomy) Structure that carries nervous impulses to and from the central nervous system, consisting of a bundle of nerve fibres (*USA*: nerve fibers), and often associated with blood vessels and connective tissue.

See also nerve fibre; neurone.

nerve cell (cytology) An alternative term for a neurone.

nerve cord (histology) Cord of nervous tissue in invertebrates that forms part of their central nervous system.

nerve fibre (histology) Extension of a nerve cell. Nerve fibres (*USA*: nerve fibers) may be surrounded by a myelin sheath (except at the nodes of Ranvier), as in many vertebrates; or they may be unmyelinated and bound by a plasma membrane. (*USA*: nerve fiber).

nerve impulse (physiology) Electrical signal conveyed by a nerve to carry information throughout the nervous system. External stimuli trigger nerve impulses in receptor cells, and travel along afferents towards the central nervous system. Alternatively the impulses are generated within the central nervous system and travel along efferents towards organs and tissues (*e.g.* along motor nerves to muscles).

nervous coccidiosis (veterinary parasitology) A condition presenting where a small number of calves during an outbreak of classical coccidiosis may develop severe nervous signs including hyperesthesia, nystagmus, tremor, orthotonus and convulsions and die within a few hours. There is no detectable lesion in the brain.

nervous system (anatomy) System that provides a rapid means of communication within an organism, enabling it to be aware of its surroundings and to react accordingly. In most animals it consists of a central nervous system (CNS) that integrates the sensory input from peripheral nerves which transmit stimuli from receptors (afrerents) to the CNS, allowing the appropriate response from the electors.

net weight (nt wt) (measurement) Used on packaging and labelling to show the weight of the product itself, not including the weight of the packaging the product is stored in.

nettle rash (immunology) *See* urticaria.

network (physics/computing) System of interconnected points and their connections; *e.g.* a grid of electricity supply lines or a set of interconnected terminals online to one or more computers.

neur(o)- (terminology) A word element. [German] Prefix denoting *nerve, neurology, neurological.*

neur. (medical) An abbreviation of 'neur'ological/ 'neur'ology. Also termed neurol.

neural (anatomy) Pertaining to nerve cells or neurons, or pertaining to the nervous system.

neurinoma (histology) a benign tumour (*USA*: benign tumor) arising from the sheath of a nerve. Also termed a neurofibroma.

neuro. (medical) An abbreviation of 'neuro'tic.

neuroanatomy (anatomy) The study of the structure of the nervous system, its constituents, and their connections.

neurochemistry (biochemistry) The study of the chemistry of neurons.

Neurofilaria cornelliensis (parasitology) *See Paraelaphostrongylus tenuis.*

neurofilariasis (veterinary parasitology) A disease caused by *Paralaphostrongylus tenuis*; principally a parasite of deer but also infests sheep in which the worms invade the spinal cord and cause a syndrome of lameness, incoordination and paralysis. Also termed: moose disease, moose sickness.

neuroglia (histology) Connective tissue between nerve cells (neurones) of the brain and spinal cord.

neurohumour (biochemistry) Any chemical substance specifically secreted by neurons, particularly substances involved in synaptic transmission such as neurotransmitters.

neurol. (medical) *See* neur.

neurology (medical) A medical specialty dealing with.the diagnosis and treatment of patients with damage to, or disorders of, the nervous system, and with the causation of the symptoms. It is not concerned with mental diseases that have no known pathological basis.

neuroma (histology) A tumour (*USA*: tumor) composed of nerve cells or nerve fibres (*USA*: nerve fibers); a tumour growing from a nerve.

neuromuscular junction (anatomy) The interface between a motor nerve fibre (*USA*: motor nerve fiber) and a skeletal muscle.

neuromuscular junction disease (parasitology) A disease affecting the interface between a motor nerve fibre (*USA*: motor nerve fiber) and a skeletal muscle, *e.g.* as occurs with tick paralysis.

neurone (cytology) Basic cell of the nervous system which transmits nerve impulses. Each cell body typically possesses a nucleus and fine processes: short dendrites and a long axon. The axon carries impulses to distant effector cells and other neurones. Neurones also make functional contacts over the surface of shorter, thread-like projections from the cell body (dendrites). Also termed: neuron; nerve cell.

neuropathology (medical) The branch of pathology that is concerned with diseases of the nervous system.

neuropharmacology (pharmacology) The study of the effects of drugs on the nervous system.

neurosecretory cell (biochemistry) A neuron that secretes a hormone into the extracellular fluid.

neurosurgery (medical) The medical speciality concerned with surgery on the nervous system, particularly on the brain.

neurotransmitter (biochemistry) Chemical released by neurone endings to either induce or inhibit transmission of nerve impulses across a synapse. Neurotransmitters are typically stored in small vesicles near the synapse and released in response to arrival of an impulse. There are more than 100 different types, *e.g.* acetylcholine, noradrenaline (*USA*: norepinephrine) and serotonin. Also termed: transmitter.

neurotrophic (biology) Having a selective affinity for nerve tissue.

neutral (1 physics; 2 chemistry) 1 Having neither positive nor negative electrical charge; *e.g.* a neutron is a neutral subatomic particle. 2 Describing a solution with pH equal to 7 (*i.e.* neither acidic nor alkaline).

neutral oxide (chemistry) Oxide that is neither an acidic oxide nor a basic oxide (*e.g.* dinitrogen oxide [N_2O], water [H_2O]).

neutralization (chemistry) Chemical reaction between an acid and a base in which both are used up; the products of the reaction are a salt and water. The completion of the reaction (end-point) can be detected by an indicator.

neutron (chemistry) Uncharged particle that is a constituent of the atomic nucleus, having a rest mass of 1.67482×10^{-27} kg (similar to that of a proton). Free neutrons are unstable and disintegrate by beta decay to a proton and an electron; outside the nucleus they have a mean life of about 12 minutes.

neutron excess (chemistry) An alternative term for isotopic number.

neutron flux (chemistry) Product of the number of free neutrons per unit volume and their mean speed. Also termed: neutron flux density.

neutron number (chemistry) Number of neutrons in an atomic nucleus, the difference between the nucleon number of an element and its atomic number.

neutrophil (haematology) A granular leucocyte important in the immune system because it ingests bacteria, whose protoplasm can be stained by neutral dyes. Neutrophils have a nucleus with 3 to 5 lobes connected.

New Forest fly (parasitology) *See Hippobosca equina.*

New Zealand cattle tick (veterinary parasitology) *See Haemaphysalis longicornis.*

newton (N) (measurement) SI unit of force, defined as the force that provides a mass of 1 kg with an acceleration of 1 m s^{-2}. It was named after the British mathematician and physicist Isaac Newton (1642–1727).

Newton per metre (N/m) (measurement/physics) A Unit of surface tension. (*USA*: Newton per meter).

Newton's formula (optics) For a lens, the distances p and q between two conjugate points and their respective foci (f) are related by $pq = f^2$.

Newton's rings (optics) Circular interference fringes formed in a thin gap between two reflective media, *e.g.* between a lens and a glass plate with which the lens is in contact. There is a central dark spot around which there are concentric dark rings.

nexin (anatomy) The connecting link between microtubules in cilia and flagella.

NGF (physiology/biochemistry) An abbreviation of 'N'erve 'G'rowth 'F'actor.

niacin (biochemistry) Vitamin B_3, the only one of the B vitamins that is synthesized by animal tissues. It is used by the body to manufacture the enzyme nad, and a deficiency causes the disease pellagra. Also termed: nicotinic acid.

NIAID (education) An abbreviation of 'N'ational 'I'nstitute of 'A'llergy and 'I'nfectious 'D'iseases.

NIC (chemistry) An abbreviation of 'NI'tro'C'ellulose.

nicarbazin (pharmacology) An efficient coccidiostat but causes mottling of the egg yolk and lowered egg production in laying hens. Excessive dosing causes incoordination, inanition and loss of weight.

niche (physiology) Status or way of life of an organism (or group of organisms) within an environment, which it cannot share indefinitely with another competing organism. Also termed: ecological niche.

nickel (Ni) (chemistry) Silver-yellow metallic element in Group VIII of the periodic table (a transition element). It is used in vacuum tubes, electroplating and as a catalyst, and its alloys (*e.g.* stainless steel, cupronickel, German silver, nickel silver) are used in making cutlery, hollow-ware and coinage. At. no. 28; r.a.m. 58.71.

nickel-iron accumulator (physics) Rechargable electrolytic cell (battery) with a positive electrode of nickel oxide and a negative electrode of iron, in a potassium hydroxide electrolyte. Also termed: Edison accumulator; NiFe cell.

niclofolan (pharmacology) A nitrosubstituted analog of hexachlorophane used as a fasciolicide and effective against mature liver flukes in sheep.

niclosamide (pharmacology) A nitrosalicylanilide anthelmintic effective against tapeworms in all species except *Echinococcus granulosus* and *Dipylidium caninum* in dogs. It also has some activity against paramphistomes in ruminants.

Nicol prism (optics) Pair of calcite crystals glued together and cut in such a way that they polarize a beam of light passing through them. It was named after the

British physicist William Nicol (1768–1851). *See* polarized light.

nicotine (control measures) Poisonous alkaloid, found in tobacco, which potentially binds to the receptor for the neurotransmitter acetylcholine. It is used as an insecticide.

nicotine sulphate (pharmacology/control measures) A substance which has been used as an anthelmintic but that is very poisonous. It has also been used as an insecticide and acaricide. It was once used against sheep scab and is still used against poultry lice. (*USA*: nicotine sulfate).

nicotinic acid (biochemistry) An alternative term for niacin.

NiFe cell (physics) An alternative term for nickel-iron accumulator.

NIH (institute) An abbreviation of 'N'ational 'I'nstitutes of 'H'ealth.

nimorazole (pharmacology) A 5-nitroimidazole derivative used in the treatment of intestinal protozoal infection.

niobium (Nb) (chemistry) Metallic element in Group VB of the periodic table (a transition element). Its alloys are used in high-temperature applications and superconductors. At. no. 41; r.a.m. 92.9064.

Nippostrongylus braziliensis (veterinary parasitology) A species of nematode in the genus *Nippostrongylus* that may be found in the intestine of rats.

Nippotaenidea (fish parasitology) An order of flatworms in subclass Eucestoda, class Cestoidea and the phylum Platyhelminthes.

nitrate (chemistry) Salt of nitric acid, containing the NO_3^- anion. Nitrates are employed as oxidizing agents and commonly used as fertilizers, but their misuse can give rise to pollution of water supplies.

nitration (chemistry) Chemical reaction in which a nitro group ($-NO_2$) is incorporated into a chemical structure, to make a nitro compound. Nitration of organic compounds is usually achieved using a mixture of concentrated nitric and sulphuric acids (*USA*: sulfuric acid) (known as nitrating mixture).

nitre (chemistry) Old term for potassium nitrate, KNO_3, also commonly known as saltpetre.

nitric acid (chemistry) HNO_3 Strong extremely corrosive mineral acid. It is manufactured commercially by the catalytic oxidation of ammonia to nitrogen monoxide (nitric oxide) and dissolving the latter in water. Its salts are nitrates. The main use of the acid is in making explosives and fertilizers.

nitric oxide (chemistry) An alternative term for nitrogen monoxide.

nitrification (plant physiology) Conversion of ammonia and nitrites to nitrates by the action of nitrifying bacteria. It is one of the important parts of the nitrogen cycle, because nitrogen cannot be taken up directly by plants except as nitrates.

nitrile (chemistry) Member of a group of organic compounds that contain the nitrite group (-CN). Also termed: cyanide.

nitrite (chemistry) Salt of nitrous acid, containing the NO_2^- anion. Nitrites are used to preserve meat and meat products.

nitro compound (chemistry) Organic compound in which a nitro group ($-NO_2$) is present in the basic molecular structure. Usually made by nitration, some nitro compounds are commercial explosives.

nitrobenzene (chemistry) $C_6H_5NO_2$ Aromatic liquid organic compound in which one of the hydrogen atoms in benzene has been replaced by a nitro group ($-NO_2$). It is used to make aniline and dyes.

nitrocellulose (chemistry) An alternative term for cellulose trinitrate.

nitrofurazone (pharmacology) A nitrofuran derivative used mostly for topical application, and orally for treatment of coccidiosis in poultry and in swine dysentery. Also termed Furacin.

nitrogen cycle (biochemistry) Circulation of nitrogen and its compounds in the environment. The main reservoirs of nitrogen are nitrates in the soil and the gas itself in the atmosphere (formed from nitrates by denitrification). Nitrates are also taken up by plants, which are eaten by animals, and after their death the nitrogen-containing proteins in plants and animals form ammonia, which nitrification converts back into nitrates, Some atmospheric nitrogen undergoes fixation by lightning or bacterial action, again leading to the eventual formation of nitrogen dioxide NO_2 .

nitrogen fixation (biology) *See* fixation of nitrogen.

nitrogen monoxide (chemistry) NO Colourless (*USA*: colorless) gas made commercially by the catalytic oxidation of ammonia and used for making nitric acid. It reacts with oxygen (*e.g.* in air) to form nitrogen dioxide. Also termed: nitric oxide.

nitrogen (N) (chemistry) Gaseous nonmetallic element in Group VA of the periodic table. It makes up about 80% of air by volume, and occurs in various minerals (particularly nitrates) and all living organisms. It is used as an inert filler in electrical devices and cables, and is an essential plant nutrient. At. no. 7; r.a.m. 14.0067.

nitrogen oxides (chemistry) Compounds containing nitrogen and oxygen in various ratios, including N_2O, NO, N_2O_3, NO_2, N_2O_4, N_2O_3, N_2O_5, NO_3 and N_2O_6. The most important are dinitrogen oxide, nitrogen dioxide and nitrogen monoxide.

nitroimidazoles (pharmacology) Antiprotozoal agents, the most common being metronidazole.

nitrophenide (pharmacology) A coccidiostat with little usefulness because of its toxicity, manifested by paralysis, at relatively low dose rates.

nitrophenol (control measures) A parasiticide which may be applied to buildings and fixtures.

nitroscanate (pharmacology) An anthelmintic effective against some cestodes which may be used in cats and dogs.

nitrous acid (chemistry) HNO_2 Weak, unstable mineral acid, made by treating a solution of one of its salts (nitrites) with an acid.

nitrous oxide (chemistry) An alternative term for dinitrogen oxide.

nitroxynil (pharmacology) An injectable fasciolicide with good efficiency against *Fasciola hepatica* in cattle and sheep and against Haemonchus *contortus* strains that are resistant to benzimidazoles. Also termed: nitroxinil.

nits (parasitology) The egg containers of lice. *See* louse.

NK (general terminology) An abbreviation of 'N'ot 'K'nown.

nm (measurement) An abbreviation for 'n'ano'm'etre.

NMDA (biochemistry) An abbreviation of 'N'-'M'ethyl-'D'-'A'spartate.

nmr (physics) Abbreviation of 'n'uclear 'm'agnetic 'r'esonance.

No (chemistry) The chemical symbol for nobelium.

nobelium (No) (chemistry) Radioactive element in Group IIIB of the periodic table (one of the actinides); it has various isotopes with half-lives of up to 3 min. At. no. 102; r.a.m. 255 (most stable isotope).

noble gas (chemistry) Any of the elements in group 0 of the periodic table: helium, neon, argon, krypton, xenon and radon. They have a complete set of outer electrons, which gives them great chemical stability (very few noble gas compounds are known); radon is radioactive. Also termed: inert gas; rare gas.

noble metal (chemistry) Highly unreactive metal, *e.g.* gold and platinum.

noci- (terminology) A word element. [Latin] Prefix denoting *harm, injury*.

nocturia (medical) Excessive urination at night.

nocturnal mite (parasitology) *See Dermanyssus gallinae*.

node (1 anatomy; 2 mathematics; 3 physics) 1 Thickening or junction of an anatomical structure, *e.g.* lymph node (gland), sinoatrial node, node of ranvier. 2 Meeting point of one or more arcs on a network. 3 Stationary point (*i.e.* point with zero amplitude) on a standing wave.

node of Ranvier (histology) One of several regular constrictions along the myelin sheath of a nerve fibre (*USA*: nerve fiber). It was named after the French histologist Louis-Antoine Ranvier (1835–1922).

nodular intestinal worm disease (parasitology) *See oesophagostomiasis*.

nodular lungworm (parasitology) *See Muellerius capillaris*.

nodular worm (parasitology) *See Oesophagostomum columbianum*.

nodular worm disease (parasitology) *See oesophago-stomiasis*.

non-homologous chromosomes (genetics) Any chromosomes that do not have an exactly matching pair. The main example is the sex chromosomes, in which the X and Y are non-homologous.

non-metal (chemistry) Substance that does not have the properties of a metal. Nonmetallic elements are usually gases (*e.g.* nitrogen, halogens, noble gases) or low-melting point solids, *e.g.* phosphorus, sulphur (*USA*: sulfur), iodine; bromine is exceptional in being liquid at ordinary temperatures. They have poor electrical and thermal conductivity, form acidic oxides, do not react with acids and tend to form covalent bonds. In ionic compounds they usually form anions.

non-Newtonian fluid (physics) Fluid that consists of two or more phases at the same time. The coefficient of viscosity is not a constant but is a function of the rate at which the fluid is sheared as well as of the relative concentration of the phases.

non-parametric (statistics) Describing significance tests that do not involve the assumptions of normality or homogeneity of variance. In such tests, no assumptions are made about the distribution or parameters of the population from which the observations were sampled. The chi-square test is an example of a commonly employed non-parametric tests. *Contrast* parametric.

nonscutate tick (parasitology) *See* soft ticks.

nonsense codon (genetics) One of the three codons that specify the end of a genetic message. These codons do not code for any amino acid but signal the termination of the protein being synthesised They are UAG ('amber'), UAA ('ochre') and UGA ('opal'). Also known as 'stop codon'. *See* genetic code; protein synthesis.

nonsense mutation (genetics) A mutation in which a codon that codes for an amino acid is changed to a codon that codes for 'stop', resulting in a shortened protein. The degree to which the protein is shortened depends on how far along the gene the nonsense mutation occurs. *See* genetic code; nonsense codon; protein synthesis.

non-specific urethritis (medical) One of the most commonly encountered sexually transmitted diseases which produces frequency of micturition, some pain and a discharge. It is in many cases caused by the organism *Chlamydia trachomatis*, the organism of trachoma, in others possibly by *Ureaplasma urealyticum;* the treatment is by tetracyclines or erythromycin.

nonsporulating (biology) Does not produce spores.

nonstoichiometric compound (chemistry) Chemical whose molecules do not contain small whole numbers of atoms.

See also stoichiometric compound.

non-viable (biology) Not capable of independent life.

noradrenaline (biochemistry) Hormone secreted by the medulla of the adrenal glands for the regulation of the cardiac muscle, glandular tissue and smooth muscles. It is also a neurotransmitter in the sympathetic nervous system, where it acts as a powerful vasoconstrictor on the vascular smooth muscles. In the brain, levels of noradrenaline (*USA*: norepinephrine) are related to normal mental function, *e.g.* lowered levels lead to mental depression. (*USA*: norepinephrine).

norepinephrine (biochemistry) An alternative term for noradrenaline.

norm(o)- (terminology) A word element. [Latin] Prefix denoting *normal, usual, conforming to the rule.*

norm. (general terminology) An abbreviation of 'norm'al.

normal (1 mathematics; 2 chemistry) 1 A plane or line that is perpendicular to another. At any point on a curve, the normal is perpendicular to the tangent to the curve at that point. 2 Describing a solution that contains 1 gram-equivalent of solute in 1 litre of solution. It is denoted by the symbol N and its multiples (thus 3N is a concentration of 3 times normal; N/10 or decinormal is a concentration of one-tenth normal).

normal distribution (statistics) The statistical distribution in which the frequency of occurrences of values decreases with their distance from the mean. When plotted as a graph, a normal distribution looks symmetrically bell-shaped. The mean (the average of all the individual values), the mode (the value that occurs most often) and the median (the half-way point between the highest and the lowest values) all coincide. The standard deviation is a measure of how steep-sided the 'bell' is. In biological data, many traits that show continuous variation conform to a normal distribution. Also termed Gaussian distribution.

normalization (statistics) Transforming data by mathematical operations to make them fit a preconceived pattern.

normalize (statistics) To convert a set of data by, for example, converting them to logarithms or reciprocals so that their previous non-normal distribution is converted to a normal one.

normoblast (haematology) Immature red blood cell which still has a nucleus. As the cell matures and enters the bloodstream, the nucleus breaks up, and the cell containing the remains of the nucleus is called reticulocyte. Finally, when it is mature, the red blood cell has no nucleus. Normoblasts are not seen in the circulation unless new red cells are being formed abnormally quickly.

North American cattle tick (veterinary parasitology) *See Boophilus annulatus.*

North American chigger (parasitology) *See Eutrombicula alfreddugesi.*

northern black bullhead (parasitology) A frog that is a paratenic host for the worm *Dioctophyma renale.*

Northern blotting (biochemistry) A similar procedure to Southern blotting, except that the nucleic acid being transferred is RNA and not DNA. Nucleic acids, previously resolved by agarose gel electrophoresis, are transferred to a nitrocellulose filter by capillary action. They are then bound to the sheet by heating and can be probed with radioactive labeled nucleic acids followed by autoradiography. Whereas Southern blotting was called after its inventor, Northern blotting got its name by analogy. Also termed Northern transfer.

See also Western blotting.

northern fowl mite (avian parasitology) *See Ornithonyssus sylviarum.*

Northern transfer (methods) *See* Northern blotting.

Norwegian scabies (parasitology) A variety of scabies seen in immunocompromised patients that is characterized by immense numbers of mites and marked scaling of the skin.

nos(o)- (terminology) A word element. [German] Prefix denoting disease.

nos. (general terminology) Numbers. The plural of no.

nose mite (veterinary parasitology) *Speleognathus australis.* A mite that occurs in wild ruminants and may cause bouts of sneezing.

See also nasal acariasis.

Nosema (parasitology) A genus of protozoa in the class Microsporea. All members of the genus are obligate intracellular parasites.

Nosema apis (insect parasitology) A protozoal species of microsporidia that, similarly to *Nosema cerana,* are parasites to the bees, *Apis mellifera* and *Apis cerana.* There is an association with black queen cell virus, bee Y virus, and with *Malpighamoeba.* Queens become sterilized and colony growth reduced, with lower honey yields.

Nosema bombi (insect parasitology) A protozoal species of microsporidia that are parasites to the bees *Bombus* spp. and *Apis florea,* and the wasps *Vespa germanica.* They can cross-infect among *Bombus* spp., and with workers dying quickly colonies develop badly.

Nosema branchialis (fish parasitology) A species of protozoa in the genus *Nosema* which are obligate intracellular parasites found in fish.

Nosema cerana (insect parasitology) A protozoal species of microsporidia that, similarly to *Nosema apis*, are parasites to the bees, *Apis mellifera* and *Apis cerana*. There is an association with black queen cell virus, bee Y virus, and with *Malpighamoeba*. Queens become sterilized and colony growth reduced, with lower honey yields.

Nosema cuniculi (parasitology) *See Encephalitozoon cuniculi.*

Nosema lophi (fish parasitology) A species of protozoa in the genus *Nosema* which are obligate intracellular parasites found in fish.

nosematosis (parasitology) *See* encephalitozoonosis.

nosocomial (parasitology) Pertaining to or originating in a hospital.

nosocomial infections (medical) Infections acquired during hospitalization or during attendance at any medical facility.

nosogeny (parasitology) The development of a disease; pathogenesis.

nosography (parasitology) A nomenclature of disease entities.

nosology (parasitology) Classification of patients into groups.

nosoparasite (parasitology) An organism found in a disease that it is able to modify, but not to produce.

nosopoietic (parasitology) Causing disease.

Nosopsyllus fasciatus (human/veterinary parasitology) A flea that infests mice and rats. It maintains plague in rats but is reluctant to bite humans and is rarely involved in cases in them.

nostril fly (parasitology) *See Oestrus ovis.*

not applicable (NA) (general terminology) Term used in tables and charts to indicate that the information is not pertinent.

not available (NA) (general terminology) Term used in tables and charts to indicate that the information is not available.

not(o)- (terminology) A word element. [German] Prefix denoting *the back.*

notifiable diseases (biology) Certain communicable diseases that are by law notifiable to the appropriate authority.

Notocotylus attenuatus (avian parasitology) A species of digenetic trematodes in the genus *Notocotylus* which may be found in the caeca (*USA*: ceca) and rectum of fowl, duck, goose and wild aquatic birds. They can cause erosion, diarrhoea (*USA*: diarrhea) and emaciation in young birds.

Notocotylus (parasitology) A genus of intestinal flukes (digenetic trematodes) in the family Notocotylidae.

Notocotylus impricatus (avian parasitology) A species of digenetic trematodes in the genus *Notocotylus* which may be found in domestic poultry.

Notocotylus thienemanni (avian parasitology) A species of digenetic trematodes in the genus *Notocotylus* which may be found in domestic and wild ducks.

Notoedres (parasitology) A genus of mange mites in the family Sarcoptidae.

Notoedres cati (veterinary parasitology) A species of mange mites in the genus *Notoedres* that causes notedric mange in cats and occasionally rabbits.

Notoedres douglassi (veterinary parasitology) A species of mange mites in the genus *Notoedres* that may be found in grey (*USA*: gray) squirrels, and that causes severe dermatitis in koalas and bandicoots in captivity.

Notoedres muris (veterinary parasitology) A species of mange mites in the genus *Notoedres* that causes ear mange in rats.

Notoedres oudemansi (veterinary parasitology) A species of mange mites in the genus *Notoedres* that may be found on rats.

notum (parasitological anatomy) A term used in relation to arthropods. The dorsal part of a segment.

Np (chemistry) The chemical symbol for neptunium.

NPN (physiology/biochemistry) An abbreviation of 'N'on'P'rotein 'N'itrogen.

npt (physics) An abbreviation of 'n'ormal 'p'ressure and 't'emperature.

nS (measurement) An abbreviation of 'n'ano's'econd.

NSF (governing body) An abbreviation of 'N'ational 'S'cience 'F'oundation.

nsp (biology) An abbreviation of 'n'ew 'sp'ecies.

NSS (chemistry) An abbreviation of 'N'ormal 'S'aline 'S'olution.

nt (physics) An abbreviation of 'n'ormal 't'emperature.

nt wt (physics) An abbreviation of 'n'e't' 'w'eigh't'.

NTP (physics) An abbreviation of 'N'ormal 'T'emperature and 'P'ressure.

nuclear barrier (chemistry) Region of high potential energy that a charged particle must pass through in order to enter or leave an atomic nucleus.

nuclear division (biology) *See* meiosis; mitosis.

nuclear envelope (cytology) Double membrane that surrounds the nucleus of eukaryotic cells.

nuclear force (chemistry) Strong force that operates during interactions between certain subatomic particles. It holds together the protons and neutrons in an atomic nucleus.

nuclear isomerism (chemistry) Property exhibited by nuclei with the same mass number and atomic number but different radioactive properties.

nuclear magnetic resonance (nmr) (physics) An effect observed when radio-frequency radiation is absorbed by matter. Nuclear magnetic resonance spectroscopy is used in chemistry for the study of molecular structure.

See also nuclear magnetic resonance imaging.

nuclear magnetic resonance imaging (medical/ physics) A technique of imaging by computer using a strong magnetic field and radio frequency signals to examine thin slices of the body. It has the advantage over computed tomography in that no X-rays are used, thus no biological harm is thought to be caused to the subject.

See also nuclear magnetic resonance

nuclear membrane (cytology) Membrane that encloses the nucleus of a cell.

nuclease (biochemistry) Type of enzyme that splits the 'chain' of the DNA molecule. Nucleases that act at specific sites are called restriction enzymes.

nucleic acid (molecular biology) Complex organic acid of high molecular weight consisting of chains of nucleotides. Nucleic acids commonly occur, conjugated with proteins, as nucleoproteins, and are found in cell nuclei and protoplasm. They are responsible for storing and transferring the genetic code. *See* deoxyribonucleic acid (DNA); RNA (ribonucleic acid).

nucleoid (cytology) In a prokaryotic cell, the DNA-containing region, similar to the nucleus of a eukaryotic cell but not bounded by a membrane.

nucleolus (cytology) Spherical body that occurs within nearly all nuclei of eukaryotic cells. It is associated with ribosome synthesis and is thus abundant in cells that make large quantities of protein. It contains protein DNA and much of the nuclear RNA.

nucleon (chemistry) Comparatively massive particle in an atomic nucleus; a proton or neutron.

nucleon number (chemistry) Total number of neutrons and protons in an atomic nucleus. *See* mass number.

nucleophile (chemistry) Electron-rich chemical reactant that is attracted by electron-deficient compounds. Examples include an anion such as chloride (Cl^-) or a compound with a lone pair of electrons such as ammonia (NH_3).

nucleophilic addition (chemistry) Chemical reaction in which a nucleophile adds onto an electrophile.

nucleophilic reagent (chemistry) Chemical reactant that contains electron-rich groups of atoms.

nucleophilic substitution (chemistry) Substitution reaction that involves a nucleophile.

nucleoprotein (molecular biology) Compound that is a combination of a nucleic acid (DNA, RNA) and a protein. *E.g.* in eukaryotic cells DNA is associated with histones and protamines; RNA in the cytoplasm is associated with protein in the form of the ribosomes.

nucleoside (molecular biology) Compound formed by partial hydrolysis of a nucleotide. It consists of a base, such as purine or pyrimidine, linked to a sugar, such as ribose or deoxyribose; *e.g.* adenosine, cytidine and uridine.

nucleotide (molecular biology) Compound that consists of a sugar (ribose or deoxyribose), base (purine, pyrimidine or pyridine) and phosphoric acid. These are the basic units from which nucleic acids are formed.

nucleus (1 chemistry/physics; 2 cytology) 1 The most massive, central part of the atom of an element, having a positive charge given by Ze, where Z is the atomic number of the element and e the charge on an electron. It is composed of chiefly protons and (except for hydrogen) neutrons, and is surrounded by orbiting electrons. *See also* isotope. 2 The largest cell organelle (about 20 micrometres in diameter), found in nearly all eukaryotic cells. It is spherical to oval, containing the genetic material DNA, and hence controlling all cell activities. A nucleus is absent from mature mammalian erythrocytes (red blood cells) and the mature sieve-tube elements of plants. It is surrounded by a double membrane that forms the nuclear envelope.

null (statistics) An absence of information, as contrasted with zero or blank or nil, about a value.

null method (measurement) Any measuring system that establishes an unknown value from other known values, when a particular instrument registers zero, *e.g.* a potentiometer. Also termed: zero method.

numeral (mathematics) Symbol that represents a number; *e.g.* Arabic numerals 1, 2, 3, 4, 5, etc; Roman numerals I, II, III, IV, V, etc.

numerator (mathematics) Top part of a fraction (the lower part being the denominator).

Numidicola antennatus (avian parasitology) A species of lice in the genus *Numidicola* which may be found on guinea fowl.

nutritional (parasitology) Pertaining to or emanating from nutrition.

nutritious (biology) Affording nourishment.

nutritive (biology) Pertaining to or promoting nutrition.

Nuttallia (parasitology) *See Babesia.*

nuttalliosis (parasitology) *See* babesiosis.

nyct(o)- (terminology) A word element. [German] Prefix denoting *night, darkness.*

nymph (parasitological physiology) A developmental stage in certain arthropods such as ticks, mites and lice, between the larval form and the adult, and resembling the latter in appearance.

O

ob- (terminology) A word element. [Latin] Prefix denoting *against*, *in front of*, *towards*.

Obeliscoides (parasitology) A genus of nematodes in the family Trichostrongylidae.

Obeliscoides cuniculi (veterinary parasitology) A species of nematodes in the genus *Obeliscoides* which may be found in the stomach of rabbits, and that in heavy infestations causes gastritis.

Obeza floridana (insect parasitology) *See Eucharomorpha.*

object (optics) With a mirror, lens or optical instrument, the source of light rays that form an image.

objective (microscopy) Lens of an optical system (*e.g.* microscope) that is nearest the object.

obligate (parasitology) Necessary, essential.

obligate aerobe (biology) Micro-organism that lives and grows freely in air and cannot grow under anaerobic conditions.

obligate parasite (parasitology) *See* obligatory parasite.

obligatory (biology) Unavoidable; something that is bound to occur.

obligatory parasite (parasitology) A parasite that is entirely dependent upon a host for its survival.

OBP (physiology/biochemistry) An abbreviation of 'O'dorant-'B'inding 'P'rotein.

observational epidemiology (epidemiology) Epidemiology that is based on clinical and field observations, not on experiments.

observed frequency (statistics) The actual frequency; as opposed to the expected frequency.

obtuse (mathematics) Describing an angle that is more than 90 degrees but less than 180 degrees.

occlusion (1 biology; 2 chemistry) 1 Closure of an opening (*e.g.* the way an animal's teeth meet when the mouth closes). 2 Absorption of a gas on a solid mass or on the surface of solid particles by forming a solid solution, by the formation of a chemical compound, or by the condensation of the gas on the surface of the solid.

occult (biology) Obscure or hidden from view.

occult heartworm infection (parasitology) Infection by *Dirofilaria immitis* in which circulating microfilariae cannot be detected in the peripheral blood by the usual test methods.

occurrence (epidemiology) Frequency of a disease without defining incidence or prevalence.

ocelli (parasitological anatomy) Simple eyes of insects.

ochre (chemistry) Mineral of clay and iron(III) oxide (Fe_2O_3), used as a light yellow to brown pigment.

OCR (computing) An abbreviation of 'O'ptical 'C'haracter 'R'ecognition.

octa- (terminology) A word element [German, Latin] Prefix denoting eight.

octadecanoic acid (chemistry) An alternative term for stearic acid.

octahedral compound (chemistry) Chemical compound whose molecules have a central atom joined to six atoms or groups located at the vertices of an octahedron.

octahydrate (chemistry) Chemical containing eight molecules of water of crystallization. *See* hydrate.

octanoic acid (chemistry) $C_7H_{15}COOH$ Colourless (*USA*: colorless) oily carboxylic acid, used in the manufacture of dyes and perfumes. Also termed: caprylic acid.

octet (chemistry) Stable group of eight electrons; the configuration of the outer electron shell of most rare gases, and the arrangement achieved by the atoms of other elements as a result of most cases of chemical combination between them. Also termed: electron octet.

ocul(o)- (terminology) A word element. [Latin] Prefix denoting *eye*.

ocular (biology/microscopy) 1 Pertaining to the eye. 2 Eyepiece of a microscope.

ocular filariasis (veterinary parasitology) The occurrence of filariae, particularly of *Dirofilaria immitis* in dogs, in the anterior chamber or vitreous body. *See* thelaziasis; onchocerciasis.

ocular larva migrans (parasitology) Infection of the eye with the larvae of the roundworm *Toxocara canis* or *Toxocara cati*, which may lodge in the choroid or retina or migrate to the vitreous; on the death of the larvae, a granulomatous inflammation occurs, the lesion varying from a translucent elevation of the retina to massive retinal detachment and pseudoglioma.

ocular myiasis (parasitology) *See* onchocerciasis, thelaziasis.

oculovascular myiasis (parasitology) Infestation by maggots of *Gedoelstia hassleri*, in which the eye is invaded by larvae per medium of the vascular system.

odd-even nucleus (chemistry) Atomic nucleus with an odd number of protons and an even number of neutrons.

odd-odd nucleus (chemistry) Atomic nucleus with an odd number of both protons and neutrons; it is usually unstable.

odont(o)- (terminology) A word element. [German] Prefix denoting *tooth*.

-odynia (terminology) A word element. [German] Suffix denoting pain.

Oeciacus vicarius (avian parasitology) The cliff swallow bug, possibly an overwintering vector for the western equine encephalomyelitis virus.

oedema (medical) A swelling due to the accumulation of excess tissue fluid. (*USA*: edema).

Oedemagena (parasitology) A genus of flies similar to *Hypoderma* spp.

Oedemagena tarandi (veterinary parasitology) A species of flies in the genus *Oedemagena* which may be found in the caribou, musk-ox and reindeer, and that causes significant damage to the skin, the lesions being conducive to blowfly strike. Also termed: reindeer warble fly.

oesophageal (anatomy) Of or pertaining to the oesophagus (*USA*: esophagus). (*USA*: esophageal)

oesophageal osteosarcoma (veterinary parasitology) A bone-producing malignant tumour (*USA*: tumor) which occurs in dogs in association with the parasite *Spirocerca lupi*. (*USA*: esophageal osteosarcoma).

Oesophagodontus (parasitology) A blood-sucking nematode genus in the family Strongylidae.

Oesophagodontus robustus (veterinary parasitology) A blood-sucking nematode in the genus *Oesophagodontus* which may be found in horses, and that causes strongylosis.

oesophagostomiasis (parasitology) Infestation with *Oesophagostomum* spp. that causes necrotic nodules in the wall of the intestine. The resulting clinical syndrome of affected animals includes poor condition and the passage of soft droppings containing more than normal amounts of mucus. (*USA*: esophagostomiasis). Also termed oesophagostomosis (*USA*: esophagostomosis).

oesophagostomosis (parasitology) *See* oesophagostomiasis.

Oesophagostomum (veterinary parasitology) A genus of roundworms in the family Chabertiidae. Found in the large intestine, they cause the important disease oesophagostomiasis (*USA*: esophagostomiasis) in sheep.

Oesophagostomum aculeatum (veterinary parasitology) (synonym: *Oesophagostomum apiosternum*) A species of roundworms in the genus *Oesophagostomum* that may be found in monkeys.

Oesophagostomum asperum (veterinary parasitology) (synonym: *Oesophagostomum indicum*) A species of roundworms in the genus *Oesophagostomum* that may be found in goats and sheep.

Oesophagostomum bifurcum (veterinary parasitology) A species of roundworms in the genus *Oesophagostomum* that may be found in monkeys.

Oesophagostomum brevicaudatum (veterinary parasitology) A species of roundworms in the genus *Oesophagostomum* that may be found in pigs.

Oesophagostomum columbianum (veterinary parasitology) A species of roundworms in the genus *Oesophagostomum* that may be found in sheep, goats, camels and wild antelopes.

Oesophagostomum dentatum (veterinary parasitology) A species of roundworms in the genus *Oesophagostomum* that may be found in pigs.

Oesophagostomum georgianum (veterinary parasitology) A species of roundworms in the genus *Oesophagostomum* that may be found in pigs.

Oesophagostomum granatensis (veterinary parasitology) A species of roundworms in the genus *Oesophagostomum* that may be found in pigs.

Oesophagostomum hsiungi (veterinary parasitology) A species of roundworms in the genus *Oesophagostomum* that may be found in pigs.

Oesophagostomum longicaudatum (veterinary parasitology) (synonym: *Oesophagostomum quadrispinulatum*) A species of roundworms in the genus *Oesophagostomum* that may be found in pigs.

Oesophagostomum maplestonei (veterinary parasitology) A species of roundworms in the genus *Oesophagostomum* that may be found in pigs.

Oesophagostomum multifoliatum (veterinary parasitology) A species of roundworms in the genus *Oesophagostomum* that may be found in goats and sheep.

Oesophagostomum okapi (veterinary parasitology) A species of roundworms in the genus *Oesophagostomum* that may be found in okapi.

Oesophagostomum radiatum (veterinary parasitology) A species of roundworms in the genus *Oesophagostomum* that may be found in water buffaloes and cattle.

Oesophagostomum rousseloti (veterinary parasitology) A species of roundworms in the genus *Oesophagostomum* that may be found in pigs.

Oesophagostomum staphanostomum (veterinary parasitology) A species of roundworms in the genus *Oesophagostomum* that may be found in monkeys.

Oesophagostomum venulosum (veterinary parasitology) A species of roundworms in the genus *Oesophagostomum* that may be found in camels, deer, goats and sheep.

Oesophagostomum walkeri (veterinary parasitology) A species of roundworms in the genus *Oesophagostomum* that may be found in elands.

oesophagus (anatomy) Muscular tube between the pharynx and stomach (the gullet), through which food passes by peristalsis. (*USA*: esophagus).

oestrid myiasis (parasitology) Infestation and invasion of tissues by larvae of *Oestrus* spp. and *Hypoderma* spp.

oestrogen (biochemistry) Female sex hormone, a member of a group of steroid hormones that act on the sex organs. The most important is oestradiol, which is responsible for the growth and activity of much of the female reproductive system. (*USA*: estrogen).

Oestromyia leporina (veterinary parasitology) A fly similar to *Hypoderma* spp. which parasitizes moles and muskrats.

Oestrus (parasitology) A genus of bot flies in the family Oestridae.

Oestrus ovis (human/veterinary parasitology) A widespread species of bot fly in the genus *Oestrus* that deposits its larvae in the nostrils of sheep and goats. Invasion of the nasal cavity causes irritation manifested by sneezing, nose rubbing, noisy breathing and nasal discharge. It may cause ocular myiasis in humans. Also termed sheep nasal bot fly.

OGF (physiology/biochemistry) An abbreviation of 'O'varian 'G'rowth 'F'actor.

Ogmocotyle (parasitology) A genus of intestinal flukes (digenetic trematodes) in the family Notocotylidae.

Ogmocotyle indica (veterinary parasitology) A species of intestinal flukes (digenetic trematodes) in the genus *Ogmocotyle* that is a parasite of sheep, goats and cattle, infecting stomach and small and large intestines, but that is thought to be nonpathogenic.

ohm Ω (measurement) SI unit of electrical resistance. It is the resistance of a conductor in which the current is 1 ampere when a potential difference of 1 volt is applied across it. It was named after the German physicist Georg Ohm (1787–1854).

Ohm's law (physics) Relationship stating that the voltage across a conductor is equal to the product of the current flowing through it and its resistance. It is written $V = IR$, where V is voltage, I current and R resistance.

oil of vitriol (chemistry) Old name for sulphuric acid (*USA*: sulfuric acid).

oil of wintergreen (chemistry) An alternative term for methyl salicylate.

-ol (chemistry) Chemical suffix that denotes an alcohol or phenol.

oleate (chemistry) Ester or salt of oleic acid.

olefin (chemistry) An alternative term for alkene.

oleic acid (chemistry) $C_{17}H_{33}COOH$ Unsaturated fatty acid that occurs in many fats and oils. It is a colourless (*USA*: colorless) liquid that turns yellow on exposure to air, and is used in varnishes.

oleo- (terminology) A word element. [Latin] Prefix denoting *oil.*

oleum (chemistry) $H_2S_2O_7$ Oily solution of sulphur trioxide (*USA*: sulfur trioxide) (SO_3) in concentrated sulphuric acid (*USA*: sulfuric acid). Also termed: disulphuric acid (*USA*: disulfuric acid); fuming sulphuric acid (*USA*: fuming sulfuric acid).

olfaction (physiology) Process of smelling. In vertebrates, the incoming nerve impulses from the olfactory sense organs are processed in the olfactory lobes of the brain.

olfactory (physiology) Concerning the sense of smell.

olifantvel (veterinary parasitology) The thickened, wrinkled, alopecic skin caused by *Besnoitia besnoiti* in cattle.

olig(o)- (terminology) A word element. [German] Prefix denoting *few, little, scanty.*

oligomer (chemistry) Polymer formed from the combination of a few monomer molecules.

Ollulanus tricuspis (veterinary parasitology) A species of roundworms in the genus *Ollulanus* that may be found in the stomach of cats, foxes, wild cats and pigs. They cause chronic gastritis and emaciation in pigs and vomiting and wasting in cats.

Ollulanus (parasitology) A genus of roundworms in the family Ollulanidae.

-oma (terminology) A word element. [German] Suffix denoting *tumour* (*USA*: tumor), *neoplasm.*

omphal(o)- (terminology) A word element. [German] Prefix denoting *umbilicus.*

Onchocerca (parasitology) A genus of nematode parasites in the family Onchocercidae. They are important as causes of onchocercosis. Their life cycles depend on the carriage of larval microfilariae by a variety of insects, chiefly mosquitoes, black flies and midges (families Simuliidae and Ceratopogonidae). The microfilariae are found in the lymph and connective tissue spaces of the skin. Also termed *Wehrdikmansia.*

Onchocerca armillata (veterinary parasitology) A species of nematode parasites in the genus *Onchocerca.* The microfilariae are found in the skin of the hump, withers, neck, dewlap and umbilicus and the adults in the aorta of cattle, buffaloes, sheep, goats and donkeys.

Onchocerca bohmi (parasitology) *See Elaeophora bohmi.*

Onchocerca cebei (veterinary parasitology) A species of nematode parasites in the genus *Onchocerca* that may be found in nodules on the brisket and the lateral aspects of the hindlimbs of buffalo. Also termed: *Onchocerca swetae.*

Onchocerca cervicalis (veterinary parasitology) A species of nematode parasites in the genus *Onchocerca* that may be found in the ligamentum nuchae of the horse and mule. The microfilariae are commonest in the area of the linea alba. Aetiologically (*USA*: etiologically) related to the occurrence of fistulous withers, poll evil, equine conjunctivitis and possibly equine recurrent ophthalmitis.

Onchocerca cervipedis (veterinary parasitology) A species of nematode parasites in the genus *Onchocerca* that may be found in the subcutaneous tissues of the neck and limbs of deer.

Onchocerca dukei (veterinary parasitology) A species of nematode parasites in the genus *Onchocerca.* Adults are found in nodules in subcutaneous and perimuscular sites in cattle, mostly in the thorax, abdomen, diaphragm and thighs.

Onchocerca flexuosa (veterinary parasitology) A species of nematode parasites in the genus *Onchocerca* that may be found in deer.

Onchocerca garmsi (veterinary parasitology) A species of nematode parasites in the genus *Onchocerca* that are parasites of deer.

Onchocerca gibsoni (veterinary parasitology) A species of nematode parasites in the genus *Onchocerca* that may be found in nodules under the skin of the brisket or the lateral aspect of the hindlimbs of *Bos taurus* and *Bos indicus* cattle.

Onchocerca gutturosa (veterinary parasitology) A species of nematode parasites in the genus *Onchocerca* that may be found in cattle and buffalo in the ligamentum nuchae, on the scapular cartilage and in the hip, shoulder and stifle areas.

Onchocerca lienalis (veterinary parasitology) A species of nematode parasites in the genus *Onchocerca* that may be found in the gastrosplenic ligament, on the capsule of the spleen and above the xiphisternum in cattle.

Onchocerca ochengi (veterinary parasitology) A species of nematode parasites in the genus *Onchocerca* that may be found in cattle, in nodules in cutaneous and subcutaneous sites on the udder, scrotum and flanks.

Onchocerca raillieti (veterinary parasitology) A species of nematode parasites in the genus *Onchocerca.* Adult worms are found in the ligamentum nuchae, in subcutaneous cysts on the penis and in perimuscular connective tissue of donkeys.

Onchocerca reticulata (veterinary parasitology) A species of nematode parasites in the genus *Onchocerca* that is a parasite of horses, mules and donkeys. Adult worms are found in the connective tissue of flexor tendons and suspensory ligament of the fetlock, mostly in the forelimb.

Onchocerca rugosicauda (veterinary parasitology) A species of nematode parasites in the genus *Onchocerca* that may be found in the subcutaneous fascia in the shoulders and back of roe deer.

Onchocerca sweetae (parasitology) *See Onchocerca cebei.*

Onchocerca synceri (veterinary parasitology) A species of nematode parasites in the genus *Onchocerca* that may be found in subcutaneous tissues in African buffaloes.

Onchocerca tarsicola (veterinary parasitology) A species of nematode parasites in the genus *Onchocerca* that is a parasite of deer.

Onchocerca tubingensis (veterinary parasitology) A species of nematode parasites in the genus *Onchocerca* that is a parasite of deer.

Onchocerca volvulus (human parasitology) A species of nematode parasites in the genus *Onchocerca* that causes dermatitis and ocular disease in humans.

onchocerciasis (parasitology) Infestation by the filarioid worm *Onchocerca* spp., which can cause a variety of diseases listed generally under the heading onchocercosis.

onchocercosis (human/veterinary parasitology) The diseases caused by infection with the nematode, *Onchocerca* spp. The presence of large numbers of microfilariae in the skin is associated with a severe, summer dermatitis also termed wahi, kasen in cattle and summer mange or allergic dermatitis in horses. An association is suspected between periodic ophthalmia in horses and onchocerciasis because of the finding of microfilariae in the eye and because *Onchocerca volvulus* is a known cause of blindness in humans. The adult worms cause little problem to the animals but the hides and carcasses are damaged and reduced in value. The relationship between the worms and the occurrence of fistulous withers, poll evil and tendonitis is unproven. The lesions in the aorta caused by *Onchocerca armillata* are often extensive but appear to cause little clinical disease.

See also sweet itch.

onchosphere (parasitology) *See* oncosphere.

Oncicola (parasitology) A genus of thorny-headed worms (acanthocephalans) in the family Macracanthorhynchidae.

Oncicola campanulatus (veterinary parasitology) A species of thorny-headed worms (acanthocephalans) in the genus *Oncicola* that may be found in the intestine of dogs.

Oncicola canis (veterinary parasitology) A species of thorny-headed worms (acanthocephalans) in the genus *Oncicola* that may be found in the intestine of dogs, coyotes, cats, lynx and bobcats. Chance infections can occur in young turkeys and cause cysts in the oesophageal (*USA*: esophageal wall).

onco- (terminology) A word element. [German] Prefix denoting *tumour* (*USA*: tumor), *swelling*, *mass*.

oncogenic (toxicology) Cancer-producing.

oncosphere (parasitological physiology) The larva of the tapeworm contained within the external embryonic envelope within the egg and armed with six hooks. Also termed: onchosphere.

on-line (computing) Describing part of a computer (*e.g.* an input device) that is linked directly to and under the control of the central processor.

onych(o)- (terminology) A word element. [German] Prefix denoting *the nails*.

oo- (terminology) A word element. [German] Prefix denoting *egg*, *ovum*.

Oochoristica (veterinary parasitology) A genus of tapeworms in the family Listowiidae, that is primarily a parasitie of reptiles.

Oochoristica megatoma (veterinary parasitology) A species of tapeworms in the genus *Oochoristica* that may be found in squirrel monkeys.

oocyst (parasitological physiology) The resistant stage of the life cycle of coccidial parasites. It contains a zygote and under appropriate conditions sporulates to become a mature infective oocyst. It may also remain infective for long periods in dry conditions.

oocyst patches (parasitology) *See* oocyst plaques.

oocyst plaques (veterinary parasitology) Raised patches up to 0.5 inch in diameter in small intestinal epithelium in goats and sheep caused by heavy infestation by coccidial oocysts.

Oodinium (fish parasitology) A protozoan parasite of fish affecting the skin and the gills. Also termed; Piscinoodinium; Amyloodinium; velvet disease.

Oodinium limneticum (fish parasitology) A species of protozoan parasite in the genus *Oodinium* that causes a dermatitis in fish which gives them a varnished appearance. It may kill the fish within a few days.

Oodinium pillularis (fish parasitology) A species of protozoan parasite in the genus *Oodinium* that is a skin parasite of fish, tadpoles, axolotls and newts, causing a greenish appearance to the skin, and kills by blocking the gills.

Oodinychus (insect parasitology) *See Dinychus*.

ookinete (parasitological physiology) A term used in relation to malaria. The motile stage of the zygote preceding the oocyst stage.

Oolaelaps oophilus (insect parasitology) Mites in the Mesostigmata that feed on salivary secretions deposited on eggs of the ants *Formica* spp.

oophor(o)- (terminology) A word element. [German] Prefix denoting *ovary*.

ootype (parasitological anatomy) A term used in relation to trematodes. The fertilising chamber where the ovum is fertilised by the spermatozoon.

Opalina (fish/veterinary parasitology) A common ciliated protozoon in the gut of fish and amphibians.

opaque (physics) Not allowing a wave motion (*e.g.* light, sound, X-rays) to pass; not transmitting light, not transparent.

operand (mathematics) Entity or quantity on which a mathematical operator acts.

operator (1 mathematics; 2 molecular biology) 1 Symbol or term that represents a mathematical operation to be carried out on a particular operand. 2 Region of DNA to which a molecule or repressor may bind to regulate the activity of a group of closely linked structural genes.

operculum (parasitological anatomy) A lid-like structure covering certain cestode and most trematode eggs.

operon (genetics) Groups of closely linked structural genes which are under control of an operator gene. The operator may be switched off by a repressor, produced by a regulator gene separate from the operon. Another substance, the effector, may inactivate the repressor.

Ophidascaris (veterinary parasitology) A genus of roundworms found in reptiles. Species includes *Ophidascaris morelia* in pythons and *Ophidascaris roberttsi*, and *Ophidascaris labiato-papillosa* in a variety of snakes.

Ophionyssus (parasitology) A genus of mites of the family Dermanyssidae.

Ophionyssus natricis (veterinary parasitology) A species of blood-sucking mite in the genus *Ophionyssus* which may be found on captive snakes. They may cause anaemia (*USA*: anemia) or transmit bacterial disease.

ophthalm(o)- (terminology) A word element. [German] Prefix denoting *eye*.

ophthalmic (medical) Pertaining to the eye.

ophthalmology (medical) The branch of medicine dealing with the eye and its diseases.

ophthalmomyiasis (parasitology) Infection of the conjunctival sac by fly larvae. *See Gedoelstia* and *Oestrus ovis*.

opisthorchiasis (parasitology) Infection and obstruction of the biliary tract by the liver flukes *Opisthorchis* spp. that is characterized by abdominal pain, diarrhoea (*USA*: diarrhea), jaundice and ascites.

Opisthorchis (human/avian/veterinary parasitology) A genus of flukes (digenetic trematodes) parasitic in the liver and biliary tract of various reptiles, birds and mammals, including humans, that causes opisthorchiasis.

Opisthorchis felineus (parasitology) *See Opisthorchis tenuicollis.*

Opisthorchis sinensis (parasitology) *See Clonorchis sinensis.*

Opisthorchis tenuicollis (human/veterinary parasitology) A species of flukes (digenetic trematodes) in the genus *Opisthorchis* that may be found in the bile ducts, rarely the pancreatic duct and intestine, of dogs, cats, foxes, pigs and the Cetacea and humans.

Opisthorchis viverrini (human/veterinary parasitology) A species of flukes (digenetic trematodes) in the genus *Opisthorchis* that may be found in domestic and wild cats, civets, dogs and humans.

Oplitis exopodi (insect parasitology) Mites in the Mesostigmata that, similarly to *Oplitis virgilinus*, may be phoretic on the ant *Atta* and *Solenopsis* spp.

Oplitis virgilinus (insect parasitology) *See Oplitis exopodi.*

optic (optics) Concerning the eye and vision.

optic nerve (histology) Cranial nerve of vertebrates that transmits stimuli from the eye to the brain.

optical activity (chemistry) Phenomenon exhibited by some chemical compounds which, when placed in the path of a beam of plane-polarized light, are capable of rotating the plane of polarization to the left (laevorotatory) or right (dextrorotatory). Also termed: optical rotation.

optical axis (optics) Line that passes through the optical centre (*USA*: optical center) and the centre of curvature (*USA*: center of curature) of a spherical lens or mirror. Also termed: principal axis.

optical centre (optics) Point at the centre of a lens through which a ray continues straight on and undeviated. (*USA*: optical center).

optical character reader (computing) Computer input device that 'reads' printed or written alphanumeric characters and feeds the information into a computer system.

optical character recognition (OCR) (computing) Technique that uses an optical character reader.

optical fibre (optics) *See* fibre optics.

optical glass (microscopy) Very pure glass free from streaks and bubbles, used for lenses, etc.

optical isomerism (chemistry) Property of chemical compounds with the same molecular structure, but different configurations. Because of their molecular asymmetry they exhibit optical activity.

optical prism (optics) Transparent solid with triangular ends and rectangular sides, with refracting surfaces at acute angles with each other.

optical rotation (chemistry) An alternative term for optical activity.

optically active (chemistry) Describing a substance that exhibits optical activity.

optics (optics) Branch of physics concerned with the study of light.

opto- (terminology) A word element. [German] Prefix denoting *visible*, *vision*, *sight*.

oral (anatomy) Concerning the mouth and, in some contexts, speech.

orange oxide (chemistry) An alternative term for uranium(VI) oxide.

Orasema aenea (insect parasitology) Hymenoptera in the Eucharitidae that, similarly to *Orasema assectator*, *Orasema coloradensis*, *Orasema costaricensis*, *Orasema crassa*, *Orasema minutissima*, *Orasema rapo*, *Orasema robertsoni*, *Orasema sixaolae*, *Orasema tolteca*, *Orasema viridis* and *Orasema wheeleri*, may be found in larvae and pupa of *Formica* spp., *Pheidole* spp., *Solenopsis* spp. and *Wasmannia*. They are ectoparasitic on pupae, and parasites mimic the host chemically.

Orasema assectator (insect parasitology) *See Orasema aenea.*

Orasema coloradensis (insect parasitology) *See Orasema aenea.*

Orasema costaricensis (insect parasitology) *See Orasema aenea.*

Orasema crassa (insect parasitology) *See Orasema aenea.*

Orasema minutissima (insect parasitology) *See Orasema aenea.*

Orasema rapo (insect parasitology) *See Orasema aenea.*

Orasema robertsoni (insect parasitology) *See Orasema aenea.*

Orasema sixaolae (insect parasitology) *See Orasema aenea.*

Orasema tolteca (insect parasitology) *See Orasema aenea.*

Orasema viridis (insect parasitology) *See Orasema aenea.*

Orasema wheeleri (insect parasitology) *See Orasema aenea.*

orbital (chemistry) Region around the nucleus of an atom in which there is high probability of finding an electron. *See* atomic orbital; molecular orbital.

orbital electron (chemistry) Electron that orbits the nucleus of an atom. Also termed: planetary electron. *See* atomic orbital; molecular orbital.

Orchopeas howardii (avian/veterinary parasitology) A flea in the genus *Orchopeas* that parasitizes squirrels but that may also be found on poultry.

order (taxonomy) In biological classification, one of the groups into which a class is divided, and which is itself divided into families; *e.g.* Lagomorpha (lagomorphs), Rodentia (rodents).

order of reaction (chemistry) Classification of chemical reactions based on the power to which the concentration of a component of the reaction is raised in the rate law. The overall order is the sum of the powers of the concentrations.

ordinal number (mathematics) Number that indicates the rank of a quantity; *e.g.* 1st, 2nd, 3rd, etc. (as opposed to ordinary counting or cardinal numbers: 1, 2, 3, etc.).

ordinate (mathematics) In co-ordinate geometry, the *y* co-ordinate of a point (distance to the *x-axis*). The other (*x*) co-ordinate is the abscissa.

organ (anatomy) Specialized structural and functional unit made up of various tissues, in turn formed of many cells, found in animals and plants; *e.g.* heart, kidney, leaf.

organ culture (histology) Maintenance of an organ in vitro (after removal from an organism) by the artificial creation of the bodily environment.

organ of corti (anatomy) Organ concerned with hearing, located in the cochlea of the ear. It was named after the Italian anatomist Alfonso Corti (1822–88).

organelle (histology) Discrete membrane-bound structure that performs a specific function within a eukaryotic cell; *e.g.* nucleus, mitochondrion, chloroplast, endoplasmic reticulum.

organic acid (chemistry) Organic compound that can give up protons to a base; *e.g.* carboxylic acids, phenol.

organic base (chemistry) Organic compound that can donate a pair of electrons to a bond; *e.g.* amines.

organic chemistry (chemistry) Study of organic compounds.

organic compound (chemistry) Compound of carbon, with the exception of its oxides and metallic carbonates and carbides. Other elements are involved in organic compounds, principally hydrogen and oxygen but also nitrogen, the halogens, sulphur (*USA*: sulfur) and nitrogen.

organometallic compound (chemistry) Chemical compound in which a metal is directly bound to carbon in an organic group.

organophosphate (chemistry) Term commonly used to describe organophosphorus compounds.

organophosphorus compound (pharmacology/ control measures) An organic ester of phosphoric or thiophosphoric acid; such compounds are powerful acetylcholinesterase inhibitors and are used as insecticides and anthelmintics.

organosilicon compound (chemistry) Chemical compound in which silicon is directly bound to carbon in an organic group.

organotropism (parasitology) The special affinity of chemical compounds or pathogenic agents for particular tissues or organs of the body.

oribatid mite (parasitology) Freeliving, nonparasitic mites which are members of the superfamily Oribatoidea, that are intermediate hosts to tapeworms found in grazing animals. Examples include *Moniezia*, *Anoplocephala*, *Paranoplocephala*, and *Avitellina* spp.

oriental avian eye fluke (parasitology) *See Philophthalmus gralli.*

oriental blood fluke (parasitology) *See Schistosoma japonicum.*

oriental liver fluke (parasitology) *See Clonorchis sinesis.*

oriental lung fluke (parasitology) *See Paragonimus westermani.*

oriental sore (parasitology) *See* leishmaniasis.

oriental theileriosis (parasitology) A mild form of East Coast fever caused by *Theileria orientalis.*

Orientobilharzia (parasitology) *See Ornithobilharzia.*

origin (mathematics) In co-ordinate geometry, the point where the *x*- and *y*-axes cross (and from which Cartesian co-ordinates are measured).

ormetoprim (pharmacology) A close analog of diaveridine and a folic acid antagonist used in potentiated sulphonamide (*USA*: sulfonamide) mixtures used is coccidiostats.

ornidazole (pharmacology) A nitroimidazole used to treat anaerobic enteric protozoa.

ornithic (biology) Of or pertaining to birds.

ornithine (chemistry) $NH_2(CH2)_3CH(NH_2)COOH$ Amino acid, involved in the formation of urea in animals. Also termed: 1,6-diaminovaleric acid.

Ornithobilharzia (parasitology) A genus of blood flukes (digenetic trematodes) in the family Schistosomatidae.

Ornithobilharzia bomfordi (veterinary parasitology) A species of blood flukes (digenetic trematodes) in the genus *Ornithobilharzia* that may be found in the mesenteric veins of zebu cattle.

Ornithobilharzia turkestanicum (veterinary parasitology) A species of blood flukes (digenetic trematodes) in the genus *Ornithobilharzia* that may be found in the mesenteric veins of most grazing herbivores and cats. They cause hepatic cirrhosis and nodules in the intestinal wall accompanied by loss of body weight in small ruminants.

Ornithobius mathisi (avian parasitology) A species of louse in the genus *Ornithobius* that may be found on duck and geese.

Ornithocoris (avian parasitology) Cimicid bug pest of poultry. Species includes *Ornithocoris toledoi* (also termed: Brazilian chicken bug), and *Ornithocoris pallidus*.

Ornithodorus (parasitology) A genus of soft-bodied ticks in the family Argasidae.

Ornithodorus coriaceus (veterinary parasitology) A species of soft-bodied ticks in the genus *Ornithodorus* which may be found on cattle and deer, that are a vector of epizootic bovine abortion.

Ornithodorus erraticus (parasitology) A species of soft-bodied ticks in the genus *Ornithodorus* which is a possible vector of African swine fever.

Ornithodorus guerneyi (human/veterinary parasitology) A species of soft-bodied ticks in the genus *Ornithodorus*. The kangaroo tick, a vector of *Coxiella burnetii*, the bite from which can cause severe local and systemic reactions in humans.

Ornithodorus lahorensis (parasitology) A species of soft-bodied ticks in the genus *Ornithodorus* which causes tick paralysis.

Ornithodorus moubata (human/avian/veterinary parasitology) A species of soft-bodied ticks in the genus *Ornithodorus* which is a parasite of mammals, including humans, birds and reptiles. It is a vector of African swine fever and Q fever and possibly of *Borrelia anserina* and *Aegyptianella pullorum* in fowls.

Ornithodorus moubata porcinus (veterinary parasitology) (synonym: *Ornithodorus porcinus*) A species of soft-bodied ticks in the genus *Ornithodorus* which infests wart hog burrows and that transmits African swine fever.

Ornithodorus puertoicensis (parasitology) A species of soft-bodied ticks in the genus *Ornithodorus* which is a possible transmitter of the African swine fever virus.

Ornithodorus savignyi (human/veterinary parasitology) A species of soft-bodied ticks in the genus *Ornithodorus* which may be found on most domestic animals and which may bite humans.

Ornithodorus turicata (parasitology) A species of soft-bodied ticks in the genus *Ornithodorus* which is a vector of Q fever, *Theileria* and *Anaplasma* spp., and a cause of tick paralysis.

Ornithofilaria (parasitology) A roundworm genus in the family Onchocercidae. Also termed: *Splendido filaria*.

Ornithofilaria fallisensis (avian parasitology) A species of roundworm in the genus *Ornithofilaria* that may be found in the subcutaneous tissues in ducks, including domesticated species.

Ornithonyssus (parasitology) A mite genus in the family Dermanyssidae.

Ornithonyssus bacoti (human/veterinary parasitology) (synonym: *Bdellonyssus/Liponyssus bacoti*) A species of mites in the genus *Ornithonyssus* that are parasitic in rats and humans, the intermediate host of the filarial nematode of rats, *Litomosoides sigmodontis*, vector of the cause of human plague, *Yersinia pestis*, murine typhus and Q fever. Also termed: tropical rat mite.

Ornithonyssus bursa (avian parasitology) A species of mites in the genus *Ornithonyssus* that parasitizes birds. Also termed tropical fowl mite.

Ornithonyssus sylviarum (avian parasitology) A species of mites in the genus *Ornithonyssus* that are a parasite of birds.

Ornithostrongylus (parasitology) A genus of worms in the superfamily Trichostrongyloidea.

Ornithostrongylus quadriradiatus (avian parasitology) A species of worms in the genus *Ornithostrongylus* that may be found in the crop, proventriculus and intestine of pigeons. They cause catarrhal and haemorrhagic (*USA*: hemorrhagic) enteritis and may cause heavy losses in young birds.

ortet (genetics) The original organism from which a clone is derived.

orth(o)- (chemistry) Word element. [German] Prefix that denotes a benzene compound with substituents in the 1,2 positions.

orth. (medical) An abbreviation of 'orth'opaedic (*USA*: orthopedic).

orthoarsenic acid (chemistry) *See* arsenic acid.

orthochromatic (microscopy) Staining normally.

Orthohalarachne (veterinary parasitology) Mesostigmatid mites of the family Halarachnidae. Species includes *Orthohalarachne* attenuata and *Orthohalarachne diminuta* that may be found in the nostrils, trachea and bronchi of seals and sea lions, causing sneezing and a nasal discharge.

orthopaedic (orth.) (medical) Relating to the medical science of treating injured bones and muscles (*USA*: orthopedic).

orthophosphoric acid (chemistry) An alternative term for phosphoric(V) acid.

Orthoptera (insect parasitology) An order of insects, some of which act as the intermediate host for some worms. Includes grasshoppers, crickets, and cockroaches.

Os (chemistry) The chemical symbol for osmium.

osche(o)- (terminology) A word element. [German] Prefix denoting *scrotum.*

oscillator (physics) Device or electronic circuit for producing an alternating current of a particular frequency, usually controlled by altering the value of a capacitor in the oscillator circuit.

oscillo- (terminology) A word element. [Latin] Prefix denoting *oscillation.*

oscilloscope (physics) An instrument incorporating a cathode-ray tube, time-base generators, triggers etc. for displaying a wide range of waveforms by electron beam.

-osis (terminology) A word element. [German] Suffix denoting *disease*, *morbid state*, abnormal *increase.*

Osleroides massinoi (parasitology) *See Vogeloides.*

Oslerus (parasitology) A genus of nematodes in the family Filaroididae and the superfamily Metastrongyloidea, that may be found in the respiratory tract of mammals.

Oslerus osleri (veterinary parasitology) A species of nematodes in the genus *Oslerus* that may be found in the trachea and bronchi of dogs.

osm (measurement) An abbreviation of 'osm'ole(s).

osmium (Os) (chemistry) Metallic element in Group VIII of the periodic table (a transition element). The densest element, it is used in hard alloys and as a catalyst. At. no. 76; r.a.m. 190.2.

osmium(iv) oxide (chemistry) OsO_4 Volatile crystalline solid with a characteristic penetrating odour (*USA*: odor) reminiscent of chlorine. It is used in the preparation of tissues for observation with an electron microscope. Its aqueous solutions are used as a catalyst in organic reactions. Also termed: osmium tetroxide.

osmophiles (biology) Microorganisms including protozoa that can withstand high osmotic pressures.

osmoregulation (physiology) Process that controls the amount of water and electrolyte (salts) concentration in an animal's body. In a saltwater animal, there is a tendency for water to pass out of the body by osmosis, which is prevented by osmoregulation by the kidneys. In freshwater animals, osmoregulation by the kidneys (or by contractile vacuoles in simple creatures) prevents water from passing into the animal by osmosis.

osmose (physics) To diffuse by osmosis.

osmosis (physics) Movement of a solvent from a dilute to a more concentrated solution across a semi-permeable (or differentially permeable) membrane, thus tending to equal the concentration of the solute on either side.

osmotic pressure (physics) The pressure required to prevent osmosis i.e. the flow of water through a semipermeable membrane to the side on which there is a greater concentration of a solute.

ossicles (anatomy) Small bones; usually used to refer to the small bones of the middle ear.

ossification (histology) Process by which bone is formed, especially the transformation of cartilage into bone.

OST (governing body) An abbreviation of 'O'ffice of 'S'cience and 'T'echnology.

oste(o)- (terminology) A word element. [German] Prefix denoting *bone.*

Ostertagia (veterinary parasitology) A genus of worms in the family Trichostrongylidae which are found in the abomasum, rarely the intestine, of ruminants, and that are thin and brown. They cause ostertagiasis. Also termed: brown stomach worms.

Ostertagia bisonis (veterinary parasitology) A species of worms in the genus *Ostertagia* that may be found in cattle and wild ruminants.

Ostertagia circumcincta (veterinary parasitology) A species of worms in the genus *Ostertagia* that may be found in goats and sheep.

Ostertagia crimensis (veterinary parasitology) (synonym: *Ostertagia leptospicularis*; *Ostertagia hamata*) A species of worms in the genus *Ostertagia* that may be found in springbok.

Ostertagia leptospicularis (veterinary parasitology) A species of worms in the genus *Ostertagia* that may be found in cercid deer and cattle.

Ostertagia lyrata (veterinary parasitology) (synonym: *Skrjabinagia lyrata*; *Grosspiculagia lyrata*) A species of worms in the genus *Ostertagia* that may be found in cattle.

Ostertagia orloffi (veterinary parasitology) A species of worms in the genus *Ostertagia* that may be found in Barbary sheep, cattle and deer.

Ostertagia ostertagi (veterinary parasitology) A species of worms in the genus *Ostertagia* that may be found in cattle, goats, horses and rarely sheep

Ostertagia podjapolskyi (veterinary parasitology) (synonym: *Grosspiculagia podjapolskyi*) A species of worms in the genus *Ostertagia* that may be found in cattle, sheep and moufflon.

Ostertagia trifurcata (veterinary parasitology) A species of worms in the genus *Ostertagia* that may be found in sheep, goats and cattle.

ostertagiasis (parasitology) *See* ostertagiosis.

ostertagiosis (veterinary parasitology) A disease of ruminants caused by invasion of the abomasum by *Osteragia* spp. Two forms occur. Type I is in lambs or calves in their first summer at pasture and is characterized by the presence of large numbers of adult worms in the abomasum, profuse watery diarrhoea (*USA*: diarrhea), depressed appetite and a high morbidity rate. Type II occurs in cattle in the late winter after that first summer and sometimes in adults. It is characterized by emergence of large numbers of inhibited larvae from the abomasal mucosa, by chronic diarrhoea (*USA*: diarrhea), emaciation, a high death rate and a greatly thickened and oedematous (*USA*: edematous) abomasal mucosa, subcutaneous oedema (*USA*: edema) and high plasma pepsinogen levels. The timing of the two forms varies between countries. Also termed: ostertagiasis.

Oswaldocruzia (veterinary parasitology) A trichostrongyloid nematode genus that may be found in amphibians and rarely in lizards.

ot(o)- (terminology) Word element. [German] Prefix denoting *ear*.

otitis (medical) Inflammation of the ear.

otoacariasis (parasitology) Infestation of the ear with mites, *e.g.* with *Otodectes cynotis*, *Raillietia auris*, *Raillietia caprae*, causing shaking of the head and scratching at the ear, and for the ear canal to contain a waxy exudate. Also termed: parasitic otitis.

Otobius (parasitology) A genus of soft-bodied ticks in the family Agasidae.

Otobius lagophilus (veterinary parasitology) A species of soft-bodied ticks in the genus *Otobius* that may be found on rabbits.

Otobius megnini (veterinary parasitology) A species of soft-bodied ticks in the genus *Otobius* that may be found in the ears of most mammals, but most commonly in dogs, sheep, horses and cattle. They cause irritation, head shaking and general debility. Mainly a problem in cattle and horses, but recorded from a wide range of hosts.

Otodectes (parasitology) Acarid mite genus in the family Psoroptidae.

Otodectes cynotis (veterinary parasitology) A species of mite in the genus *Otodectes*.The ear mite of dogs, foxes, cats, raccoons and ferrets and that causes otodectic mange.

otodectic mange (parasitology) The ear and skin disease caused by infestation with *Otodectes cynotis*. Signs include those of otitis externa, with thick, brownish-red crusts in the ear canal, and occasionally a pruritic dermititis.

otopathy (medical) Any disease of the ear.

otopharyngeal (biology) Pertaining to the ear and pharynx.

otorhinology (medical) The branch of medicine dealing with ear and nose.

otoscope (equipment) An instrument for inspecting the ear.

otoscopy (medical) Examination of the external acoustic meatus with an otoscope.

Otostrongylus (parasitology) A genus of lungworms in the family Crenosomatidae.

***Otostrongylus* circumlitus** (veterinary parasitology) A species of lungworms in the genus *Otostrongylus* that may be found in the lungs or heart of true seals, causing anorexia and coughing.

Ouchterloney technique (methods) *See* immunodiffusion tests.

ounce, apothecary (measurement) Unit of weight, 1/12 of a pound (apothecary). 1 oz ap. = 1.0971 oz (avoirdupois) = 31.103481 grams. Also termed: ounce, troy.

ounce, avoirdupois (measurement) Unit of weight, 1/16 of a pound. 1 oz = 28.349527 grams.

outer ear (anatomy) Part of the ear that transmits sound waves from external air to the ear drum.

outlier (statistics) An extremely high or low value lying beyond the range of the bulk of the data.

output (biology) The yield or total of anything produced by any functional system of the body.

output device (computing) Part of a computer that presents data in a form that can be used by a human operator; *e.g.* a printer, visual display unit (VDU), chart plotter, etc. A machine that writes data onto a portable magnetic medium (*e.g.* magnetic disk or tape) may also be considered to be an output device.

ova (biology) Plural of ovum.

oval window (histology) Membranous area at which the 'sole' of the stirrup (stapes) bone of the inner ear makes contact with the cochlea. Also termed: fenestra ovalis.

ovari(o)- (terminology) A word element. [Latin] Prefix denoting *ovary*.

ovarian follicle (histology) An alternative term for Graaflan follicle.

ovary (anatomy) Female reproductive organ. In vertebrates, there is a pair of ovaries, which produce the ova (eggs) and sex hormones.

ovary ossicle (anatomy) An alternative term for an ear ossicle.

overdose (pharmacology) 1 To administer an excessive dose. 2 An excessive dose.

overwrite (computing) To record new data over the top of existing data thus obliterating it.

ovi- (terminology) A word element. [Latin] Prefix denoting egg, ovum.

ovicide (pharmacology) A substance destructive to the ova of certain organisms, usually helminths and arthropods.

oviduct (anatomy) Tube that conducts released ova (eggs) from the ovaries after ovulation. Also termed: Fallopian tube.

oviferous (biology) Producing ova.

oviform (biology) Egg-shaped.

ovine (biology) Pertaining to, characteristic of, or derived from sheep.

ovipositor (parasitological anatomy) A term used in relation to arthropods. A tubular structure through which eggs are laid.

OVLT (physiology) An abbreviation of 'O'rganum 'V'asculosum of the 'L'amina 'T'erminalis.

ovo- (terminology) A word element. [Latin] Prefix denoting *egg, ovum.*

ovocytes (biology) Embryonic stages of ova.

ovoid (biology) Having the oval shape of an egg.

ovoid ovoplasm (cytology) The cytoplasm of all unfertilized ovum.

ovulation (physiology) In vertebrates discharge of an ovum (egg) from a mature Graafian follicle at the surface of an ovary. In mature human females, ovulation occurs from alternate ovaries at about every 28 days until the menopause occurs.

ovum (cytology) Unfertilized non-motile female gamete produced by the ovary. Also termed: egg cell; egg.

owl midge (parasitology) *See* sandfly.

oxalate (chemistry) Ester or salt of oxalic acid.

oxalic acid (chemistry) $(COOH)_2.2H_2O$ White crystalline poisonous dicarboxylic acid. It occurs in rhubarb, wood sorrel and other plants of the *Oxalis* genus. It is used in dyeing and volumetric analysis. Also termed: ethanedioic acid.

oxatyl (chemistry) An alternative term for carbonyl group.

oxfendazole (pharmacology) A benzimidazole anthelmintic used extensively in ruminants against a wide spectrum of worms. A special formulation is available for intraruminal injection. Also effective against equine strongyles.

oxibendazole (pharmacology) A benzimidazole anthelmintic similar to oxfendazole.

oxidase (biochemistry) Collective name for a group of enzymes that promote oxidation within plant and animal cells.

oxidation (chemistry) Process that involves the loss of electrons by a substance; the combination of a substance with oxygen. It may occur rapidly (as in combustion) or slowly (as in rusting and other forms of corrosion).

oxidation number (chemistry) Number of electrons that must be added to a cation or removed from an anion to produce a neutral atom. An oxidation number of zero is given to the elements themselves. In compounds, a positive oxidation number indicates that an element is in an oxidized state; the higher the oxidation number, the greater is the extent of oxidation. Conversely, a negative oxidation number shows that an element is in a reduced state.

oxidation-reduction reaction (chemistry) An alternative term for a redox reaction.

oxide (chemistry) Compound of oxygen and another element, usually made by direct combination or by heating a carbonate or hydroxide.

oxidize (chemistry) To cause to combine with oxygen or to remove hydrogen.

oxidized (chemistry) Having been modified by the process of oxidation.

oxidizing agent (chemistry) Substance that causes oxidation. Also termed: electron acceptor.

oxime (chemistry) Compound containing the group NH_2OH, derived by the condensation of an aldehyde or ketone with hydroxylamine (NH_2OH).

oxime (chemistry) Any of a series of compounds formed by action of hydroxylamine on an aldehyde or ketone.

oximeter (equipment) A device for measuring oxygen concentration.

oximetry (medical) Measurement of the oxygen content of arterial blood.

2-oxopropanoic acid (chemistry) An alternative term for pyruvic acid.

oxy- (terminology) A word element. [German] Prefix denoting *sharp, quick, sour, presence of oxygen in a compound.*

oxyacid (chemistry) Acid in which the acidic (*i.e.* replaceable) hydrogen atom is part of a hydroxyl group, *e.g.* organic carboxylic acids and phenols, and inorganic acids such as phosphoric acid and sulphuric acid (*USA*: sulfuric acid).

oxyclozanide (pharmacology) A useful treatment for adult liver fluke in dairy cattle. It has a short withholding period and is also effective against immature paramphistomes.

oxygen debt (physiology) Physiological condition that induces anaerobic respiration in an otherwise aerobic organism. It occurs during anoxia, caused *e.g.* by violent exercise.

oxygen (O) (chemistry) Gaseous nonmetallic element in Group VIA of the periodic table. A colourless (*USA*: colorless) odourless (*USA*: odorless) gas, it makes up about 20% of air by volume, from which it is extracted, and is essential for life. It is the most abundant element in the Earth's crust, occurring in all water and most rocks. It is used in welding, steel-making and as a rocket fuel. It has a triatomic allotrope, ozone (O_3). At. no. 8; r.a.m. 15.9994.

oxygenation (chemistry) Saturation with oxygen.

oxyhaemoglobin (haematology) Product of respiration formed by the combination of oxygen and haemoglobin (*USA*: hemaglobin). (*USA*: oxyhemaglobin).

Oxylipeurus (avian parasitology) A genus of lice that parasitize turkeys. Species includes *Oxylipeurus polytrapezius*, and *Oxylipeurus corpelentis*.

Oxyspirura (avian parasitology) A genus of nematodes in the family Thelaziidae. They are parasites of the eyes of birds and cause ophthalmitis, with ocular discharge, and scratching at the eyes. Species include *Oxyspirura mansoni* and *Oxyspirura parvorum* that may be found under the third eyelid of fowls, turkeys and peafowl, and *Oxyspirura petrowi* and many others that may be found in the eyes of wild birds.

oxyuriasis (veterinary parasitology) The disease caused by infestation with *Oxyuris* nematodes that is manifested by intense irritation of the perianal region causing rubbing and biting at the tail.

oxyuricide (pharmacology) A substance that kills oxyurids.

oxyurid (parasitology) An individual nematode of the family Oxyuridae.

Oxyurida (fish parasitology) An order of roundworms in subclass Secernentea, and the phylum Nematoda. Five genera within the order Oxyurida are found in fish: *Travmena*, *Icthyouris*, *Synodentisia*, *Laurotravassoxyuris* and *Cithariniella*. The first of these has an oesophagus divided into two equal parts, each dilated into a bulb at its posterior extremity.

Oxyuris (parasitology) A genus of nematodes in the family Oxyuridae. Occupants of the large intestine, the females have long, tapering tails.

Oxyuris equi (veterinary parasitology) A species of nematodes in the genus *Oxyuris* that may be found in the horse.

Oxyuris karamoja (veterinary parasitology) A species of nematodes in the genus *Oxyuris* that may be found in rhinoceros.

Oxyuris poculum (veterinary parasitology) A species of nematodes in the genus *Oxyuris* that may be found in the horse.

Oxyuris tenuicorda (veterinary parasitology) A species of nematodes in the genus *Oxyuris* that may be found in Burchell's zebra *Equus burchelli*.

Ozobranchus branchiatus (veterinary parasitology) A leech found on green sea turtles in association with cutaneous fibroepitheliomas.

ozone (chemistry) O_3 Allotrope of oxygen that contains three atoms in its molecule. It is formed from oxygen in the upper atmosphere by the action of ultraviolet light, where it also acts as a shield that prevents excess ultraviolet light reaching the Earth's surface. It is a powerful oxidizing agent, often used in organic chemistry.

P

p (1, 2 physics; 3 chemistry) 1 An abbreviation of 'p'article; 2 An abbreviation of 'p'roton; 3 An abbreviation of 'p'ara.

P (biochemistry) An abbreviation of 'P'roline.

P factor (physiology) An abbreviation of hypothetical 'p'ain producing substance produced in ischemic muscle.

P sac (anatomy) Pericardial cavity. The cavity within which the heart lies.

p tgt (chemistry) An abbreviation of 'p'rimary 't'ar'g'e't'.

P cells (1,2 physiology) An abbreviation of 'P'rincipal 'cells' in the renal tubules; 2 An abbreviation of 'P'acemaker 'cells' of sinoatrial and atrioventricular nodes.

P450 (genetics) A gene superfamily, consisting of large groups of genes that code for proteins (cytochromes) which metabolise drugs or other foreign chemicals. The P450 cytochromes are coded for by 20 different families of genes, ten of which are the same in all mammals. Any one mammalian species may have between 60 and 200 individual P450 genes, each making a unique cytochrome. The drug or chemical acts in many cases as the inducer for the relevant P450 gene. In humans, polymorphisms of P450 genes are responsible for many of the examples of patients' variable responses to clinical drugs.

P_{50} (physiology/biochemistry) An abbreviation of partial pressure of O_2 at which haemoglobin (*USA*: hemoglobin) is half-saturated with O_2.

Pa (1 chemistry; 2 measurement) 1 The chemical symbol for Element 91, protactinium. 2 An abbreviation of 'Pa'scal (SI unit of pressure).

PABA (chemistry) An abbreviation of 'P'ara-'A'mino'B'enzoic 'A'cid. An organic acid found in yeast. It absorbs ultraviolet light and is used in suntan lotions.

pachy- (terminology) A word element. [German] Prefix denoting *thick*.

pachymeningitis (histology) Inflammation of the dura mater of the brain. *See* meningitis.

pachytene (genetics) Stage in prophase or first division of meiosis, in which the paired chromosomes shorten and thicken, appearing as two chromatids.

Pacific Coast tick (parasitology) *See Dermacentor occidentalis*.

Pacinian corpuscle (histology) A small structure about 4 mm long found below the skin and in other parts of the body, which in section is like an oval onion. It is attached to a sensory nerve fibre (*USA*: nerve fiber), and is sensitive to pressure. Named after the Italian anatomist, F. Pacini (1812–83).

packing fraction (radiology) Difference between the actual mass of an isotope and the nearest whole number divided by the mass number.

paediatrics (medicine) The medical specialty concerned with childhood diseases and disorders. (*USA*: pediatrics).

PAF (physiology/haematology) An abbreviation of 'P'latelet-'A'ctivating 'F'actor.

PAH (biochemistry) An abbreviation of 'P'ara-'A'mino' 'H'ippuric acid.

Pajaroello tick (parasitology) *See Ornithodorus coriaceus*.

Palaeacanthocephala (fish/veterinary/avian parasitology) A class in the phylum Acanthocephala. The class is the largest and most diversified of acanthocephalans, paratizing fish, amphibia, reptiles, birds and mammals. There are two orders but members of only one, the Echinorhynchida with ten families, are found in fish. There are about fifty genera, including: *Acanthocephalus*, *Diplosentis*, *Heteracanthocephalus* and *Illiosentis*. *See* Acanthocephala.

palat(o)- (terminology) A word element. [Latin] Prefix denoting *palate*.

pale(o)- (terminology) A word element. [German] Prefix denoting *old*.

pali(n)- (terminology) A word element. [German] Prefix denoting *again, pathological repetition*.

palisade worms (parasitology) *See Strongylus*.

palladium (Pd) (chemistry) Silver-white metallic element in Group VIII of the periodic table (a transition element), used as a catalyst and in making jewellery. At. no. 46; r.a.m. 106.4.

palmate hairs (parasitological anatomy) Fan shaped hairs on the dorsal surface of Anopheline larva.

palmitic acid (chemistry) $C_{15}H_{31}COOH$ Long-chain carboxylic acid which occurs in oils and fats (*e.g.* palm oil) as its glyceryl ester, used in making soap.

PAM (physiology/haematology) An abbreviation of 'P'ulmonary 'A'lveolar 'M'acrophage.

PAMS (chemistry) An abbreviation of 'P'ara-'AM'ino 'S'alicylic acid.

pan- (terminology) A word element. [German] Prefix denoting all.

Panacur (pharmacology/control meaures) A proprietary name for fenbendazole.

pancreas (anatomy) Gland situated near the duodenum that has digestive and endocrine functions. The enzymes amylase, trypsin and lipase are released from it during digestion. Special groups of cells (the islets of langerhans) produce the hormones insulin and glueagon for the control of blood sugar levels.

pancreatic fluke (parasitology) *See Eurytrema.*

pancreatico- (terminology) A word element. [German] Prefix denoting *pancreatic duct.*

pancreato- (terminology) A word element. [German] Prefix denoting *pancreas.*

pandemic (epidemiology) Describing a disease that affects people or animals throughout the world.

See also endemic; epidemic.

Pangonia (parasitology) A genus of flies in the family Tabanidae, some of which feed on spilled blood, rather than sucking blood, and cause insect worry in horses and cattle. Also termed: deer fly.

pant(o)- (terminology) A word element. [German] Prefix denoting *all, the whole.*

pantothenic acid (biochemistry) Constituent of coenzyme A, a carrier of acyl groups in biochemical processes. It is required as a B vitamin by many organisms, including vertebrates and yeast.

papain (biochemistry) Proteolytic enzyme, which digests proteins, found in various fruits and used as a meat tenderizer.

paper chromatography (chemistry) Type of chromatography in which the mobile phase is liquid and the stationary phase is porous paper. Compounds are separated on the paper, and can then be identified.

papilla (parasitological anatomy) Plural: Papillae [Italian] A small, nipple-shaped projection or elevation. As in the case of alae, these may be cervical or caudal.

papillary (anatomy) Pertaining to a papilla; having the characteristics of a papilla.

papulation (parasitology) The formation of papules.

papule (parasitology) A small circumscribed, solid, elevated lesion of the skin.

para- (1, 2 chemistry; 3 terminology) 1 Prefix that denotes the form of diatomic molecule in which both nuclei have opposite spin directions. 2 Relating to the 1, 4 positions in the benzene ring. Also termed *p-*. 3 A word element. [German] Prefix denoting *beside,* beyond, *accessory to, apart from, against.*

Parabronema (veterinary parasitology) A genus of alimentary canal nematodes that may be found in wild animals.

Parabronema africum (veterinary parasitology) A species of alimentary canal nematodes in the genus *Parabronema* that, similarly to *Parabronema rhodesiense,* may be found in African elephants.

Parabronema rhinocerotis (veterinary parasitology) A species of alimentary canal nematodes in the genus *Parabronema* may be found in rhinoceros and elephant.

Parabronema rhodesiense (veterinary parasitology) A species of alimentary canal nematodes in the genus *Parabronema* that, similarly to *Parabronema africum,* may be found in African elephants.

paraclinical tests (methods) Laboratory tests in the area of parasitology, pathology, clinical pathology, immunology, microbiology and toxicology.

Paracooperia (parasitology) A genus of nematodes in the family Trichostrongylidae.

Paracooperia nodulosa (veterinary parasitology) (synonym: *Paracooperia matoffi*) A species of nematodes in the genus *Paracooperia* which may be found in the intestines of buffalo and other wild animals and that may cause nodules in the intestinal wall.

Paracoroptes allenopitheci (veterinary parasitology) A mange mite species in the family Psoroptidae that may be found in the ears and on the body of primates.

Parafasciolopsis (parasitology) A genus of flukes (digenetic trematodes) in the family Fasciolidae.

Parafasciolopsis fasciolaemorpha (veterinary parasitology) A species of flukes in the genus *Parafasciolopsis* that may be found in the digestive tract and gall bladder of the elk and deer.

paraffin embedding technique (methods) The most commonly used technique for the preparation of slides of tissue for light microscopic examination.

paraffin wax (chemistry) Mixture of solid paraffins (alkanes) which takes the form of a white translucent solid that melts below 80°C. It is used to make candles. Also termed: petroleum wax.

Parafilaria (parasitology) A genus of nematodes in the filarioid family Filariidae.

Parafilaria antipini (veterinary parasitology) A species of nematodes in the genus *Parafilaria* that may be found in deer.

Parafilaria bovicola (veterinary parasitology) A species of nematodes in the genus *Parafilaria* that causes haemorrhagic (*USA*: hemorrhagic) nodules in the skin of cattle and buffalo.

Parafilaria multipapillosa (veterinary parasitology) (synonym: *Filaria haemorrhagica*) A species of nematodes in the genus *Parafilaria* that causes nodules in the subcutaneous and intermuscular tissue, of horses. The nodules break open and discharge blood and then heal, but reappear each summer.

parafilariasis (parasitology) *See Parafilaria multi-papillosa.*

Parafilaroides (veterinary parasitology) A genus of nematodes in the metastrongyloid family Filaroididae that causes lesions in the lungs of aquatic mammals.

Parafilaroides decorus (veterinary parasitology) A species of nematodes in the genus *Parafilaroides* that may be found in sea lions.

Parafilaroides gymnurus (veterinary parasitology) A species of nematodes in the genus *Parafilaroides* that may be found in harbour (*USA*: harbor) seals.

Parafilaroides nanus (veterinary parasitology) A species of nematodes in the genus *Parafilaroides* that, similarly to *Parafilaroides prolificus*, may be found in Stellar's sea lion.

Parafilaroides prolificus (veterinary parasitology) A species of nematodes in the genus *Parafilaroides* that, similarly to *Parafilaroides nanus*, may be found in Stellar's sea lion.

paraformaldehyde (microscopy) An additive fixative used in the preparation of pathology slides of animal tissues. An alternative term for polymethanal.

paragonimiasis (parasitology) Infection with lung flukes of the genus *Paragonimus*. Characterized by lethargy, cough and dyspnea. On autopsy there is an eosinophilic peritonitis, pleurisy and myositis, and a chronic bronchitis and granulomatous pneumonia.

Paragonimus (parasitology) A genus of trematode parasites in the family Paragonimidae that causes paragonimiasis. Species includes *Paragonimus africanus*, *Paragonimus caliensis*, *Paragonimus iloktsuenensis*, *Paragonimus mexicanus*, *Paragonimus ohirai*, *Paragonimus peruvianus* and *Paragonimus uterobilateralis* which may be found in the lungs of a large number of animal species.

Paragonimus kellicotti (veterinary parasitology) A species of trematode parasites in the genus *Paragonimus* that may be found in the lungs of cats, dogs and pigs, with minks and muskrats being the probable primary hosts.

Paragonimus westermani (human/veterinary parasitology) A species of trematode parasites in the genus *Paragonimus* that may be found in the lungs and other organs of most animal species and humans.

Parahistomonas (parasitology) A genus of protozoa in the family Monocercomadidae.

Parahistomonas wenrichi (avian parasitology) A species of protozoa in the genus *Parahistomonas* that bears some similarity to *Histomonas* spp. that may be found in gallinaceous birds, especially pheasants, but is considered nonpathogenic

paraldehyde (chemistry) $(CH_3CHO)_3$ Cyclic trimer formed by the polymerization of acetaldehyde (ethanal), used as a sleep-inducing drug. Also termed: ethanal trimer.

parallel circuit (physics) Electrical circuit in which the voltage supply is connected to each side of all the components so that only a fraction of the total current flows through each of them.

See also series circuit.

paramagnetism (physics) Property of substances that possess a small permanent magnetic moment because of the presence of odd (unpaired) electrons; the substance becomes magnetized in a magnetic field as the magnetic moments align.

Paramecium (parasitology) A genus of ciliate protozoa.

Parametorchis (parasitology) A genus of digenetic nematodes in the family Opisthorchidae.

Parametorchis complexus (veterinary parasitology) A species of digenetic nematodes in the genus *Parametorchis* which may be found in the bile ducts of dogs and cats, and that may cause abdominal pain, ascites and jaundice.

parametric (statistics) Describing a statistical test that involves assumptions about the distribution and parameters of the population from which observations are sampled.

Paramoeba (fish parasitology) A genus of amoeba (*USA*: ameba) that are the cause of amoebic (*USA*: amebic) gill disease in salmonids.

Paramonostomum (avian parasitology) A genus of intestinal flukes (digenetic trematodes). Species includes *Paramonostomum alveatum* and *Paramonostomum parvum* which may be found in ducks but appear to cause little injury.

paramphistomes (parasitology) Members of the family Paramphistomatidae, belonging to genera such as *Paramphistomum*, *Calycophoron*, *Cotylophoron* and *Gigantocotyle*.

paramphistomiasis (parasitology) Infestation with members of the family Paramphistomatidae.

paramphistomosis (veterinary parasitology) The disease caused by infestation of ruminants with stomach flukes. Adult flukes cause weight loss, loss of production and anaemia (*USA*: anemia), while immature flukes additionally cause diarrhoea (*USA*: diarrhea). Also termed: stomach fluke disease.

See also Paramphistomum; *Calicophoron*; *Ceylonocotyle*.

Paramphistomum (veterinary parasitology) A genus of ruminal flukes (digenetic trematodes) in the family Paramphistomatidae. Species include *paramphistomum cervi*, *paramphistomum ichikawai* and *paramphistomum microbothrium* which are important parasites of cattle, sheep, goats and buffalo, and *paramphistomum explanatum*, *paramphistomum gotoi*, *paramphistomum hiberniae*, *paramphistomum liorchis*, *paramphistomum microbothrioides* and *paramphistomum scotiae* that are additional paramphistomes of cattle.

paramphistomum cotylophoron (parasitology) *See* paramphistomosis.

Paranaplasma (parasitology) A genus of organisms of uncertain validity in the family Anaplasmataceae, with a close resemblance to *Anaplasma* spp. They are often found in mixed infections with *Anaplasma marginale* in cattle. Species includes *Paranaplasma caudatum* and *Paranaplasma discoides*.

Paranoplocephala (veterinary parasitology) A genus of tapeworms in the family Anoplocephalidae that are parasitic in rodents.

See also Anoplocephaloides.

Paranoplocephala mamillana (parasitology) *See Anoplocephaloides*.

Parapygmephorus (insect parasitology) Mites in the Prostigmata that may be larval or egg parasites on the ants *Eciton*, *Neivamyrmex* and *Nomamyrmex*. They may also be phoretic on *Reticulitermes* and other various termites.

Parascaris (parasitology) A genus of roundworms in the family Ascarididae.

Parascaris equorum (veterinary parasitology) (synonym: *Ascaris equorum*) A species of roundworms in the genus *Parascaris* that may be found in the small intestine of horses and zebras. In foals up to 9 months of age heavy infestation with migrating larvae causes coughing. Heavy burdens of adult worms may cause diarrhoea (*USA*: diarrhea), debility, and potbelly in young animals.

parasitaemia (parasitology) The presence of parasites, including filariae and protozoa, in the blood. (*USA*: parasitemia).

parasite (parasitology) A plant or animal that lives upon or within another living organism at whose expense it obtains some advantage. Among the many parasites in nature, some feed upon animal hosts, causing diseases ranging from the mildly annoying to the severe and often fatal. Parasites include multicelled and single-celled animals, fungi and bacteria. Viruses are sometimes considered to be parasites. However, the most common use of the word refers to the multicellular helminth, arachnid, crustacean (copepod) and arthropod parasites.

See also symbiosis.

Parasitellus (insect parasitology) (formerly *Parasitus*) A species of mites in the Mesostigmata, that similarly to *Parasitus* are parasites of the bees Bombinae.

parasitic (parasitology) Pertaining to parasites.

parasitic bronchitis (parasitology) *See* lungworm.

parasitic crustaceans (fish parasites) Small crustaceans or copepods are common parasites on the gills of aquarium and pond fish where they appear as white spots. The gills become obstructed and the fish die of anoxia.

parasitic cyst (parasitology) A cyst forming around larval parasites, such as tapeworms, amoebae (*USA*: amebae), and trichinae, that enter the body.

parasitic otitis (parasitology) *See* otoacariasis.

parasitic tracheobronchitis (parasitology) *See Aelurostrongylus*; *Capillaria*; *Crenosoma*; *Filaroides*; *Paragonimus*; *Strongyloides*.

parasiticide (control measures) 1 Destructive to parasites. 2 A substance that is destructive to parasites.

parasitism (parasitology) Intimate relationship between two organisms in which one (the parasite) derives benefit from the other (the host), usually to obtain food or physical support. Parasitism can have minor or major effects on the survival of the host.

parasitize (parasitology) To live on or within a host as a parasite.

parasitogenic (parasitology) Due to parasites.

parasitoid (parasitology) An animal which is parasitic in one stage of the life history and subsequently free-living in the adult stage, as the parasitic Hymenoptera.

parasitological (parasitology) Pertaining to or emanating from parasitology.

parasitological examination (parasitology) To inspect and test for the presence of parasites. Procedures include examination of faeces (*USA*: feces) for protozoa, worm eggs or larvae and for tapeworm segments, skin scrapings for arthropod parasites, blood samples for protozoa, microfilariae, for plasma pepsinogen levels, and examination of gross specimens.

parasitologist (parasitology) A person skilled in parasitology.

parasitology (parasitology) The scientific study of parasites and parasitism.

parasitosis (parasitology) A disease caused by a parasitic infestation.

See also helminthiasis.

parasitotropic (parasitology) Having affinity for parasites.

Parasitus (insect parasitology) *See Parasitellus*.

Parasitus vesparum (insect parasitology) Mites in the Mesostigmata that may feed on nest associates of the wasps *Vespula vulgaris* and various social wasps.

Paraspidodera (parasitology) A genus of nematodes in the superfamily Heterakoidea.

Paraspidodera uncinata (veterinary parasitology) A species of nematodes in the genus *Paraspidodera* which may be found in the large intestine of guinea pigs and agouti and that appears to be nonpathogenic.

Parastrigea (parasitology) A genus of intestinal digenetic trematodes in the family Strigeidae.

Parastrigea robusta (avian parasitology) A species of intestinal digenetic trematodes in the genus *Parastrigea* which may be found in the intestine of domestic ducks and that may cause anaemia (*USA*: anemia) and haemorrhagic enteritis (*USA*: hemorrhagic enteritis).

parasympathetic nervous system (anatomy) Branch of the autonomic nervous system used in involuntary activities, for which acetylcholine is the transmitter substance. Effects of the parasympathetic nervous system generally counteract those of the sympathetic nervous system.

paratenic host (parasitology) An animal acting as a substitute intermediate host of a parasite, usually having acquired the parasite by ingestion of the original host; no development of the parasite takes place but the phenomenon aids in the transmission of infection. Also termed: transfer host; transport host.

parathyroid (anatomy) Four endocrine glands embedded in the thyroid in the neck which release a hormone that controls the levels of calcium in the blood.

Paratrichodorus (plant parasitology) A genus of nematodes in the order Dorylaimida and family Trichodoridae. The female body is stout, 0.5–1.4 mm long, and cigar-shaped especially when killed by heat. *Paratrichodorus* is similar to *Trichodorus* but the cuticle is somewhat looser and swells more in acid fixatives. Males are usually straight when killed by heat, a bursa is present, and spicular and copulatory muscles are much weaker than in *Trichodorus*. *See Paratrichodorus minor.*

Paratrichodorus minor (plant parasitology) A species of nematodes in the genus *Paratrichodorus* whose hosts cover a broad range including alfalfa, azalea, boysenberry, vegetables, corn, tomato, onion, wheat, sugarcane, rice, grasses, etc. Poor hosts include orchard grass, rye, spinach, radish, strawberry, pea and tobacco; non-hosts are asparagus, showy crotalaria, poinsettia and jimsonweed. They are widely distributed in temperate to sub-tropical climates, widespread in USA on many hosts but rare in Europe. They are migratory ectoparasites whose life cycle is completed in 16–17 days at 30°C (86°F), 21–22 days at 22°C (72°F). The temperature range for reproduction is 15 to 30°C (68–95)°F. They feed over the whole of the root surface, but usually close to the root tip, including the root cap, the meristematic region, and the region of elongation. The nematode pierces epidermal cell walls and root hairs with rapid thrusts of approximately 10 per second. Feeding causes root tips to stop growing and appear stubby. *Paratrichodorus minor* is often found in fields that have other species of plant parasitic nematodes and vectors tobacco rattle virus and causes corky ring spot of potatoes. Control measures includes the use of preplant nematicides, but this species is notorious for the speed at which it re-establishes itself. Also termed: stubby root nematode.

Paratylenchus (plant parasitology) A genus of nematodes in the order Tylenchida and family Tylenchulidae. The females are small, under 0.5 mm long, and often C-shaped when killed by heat, while males are more slender, esophagus degenerate, stylet reduced or absent, and juveniles similar to females but fourth stage may have a degenerate oesophagus (*USA*: esophagus) and no stylet. *See Paratylenchus hamatus.*

Paratylenchus hamatus (plant parasitology) A species of nematodes in the genus *Paratylenchus* whose hosts cover a wide range including celery, figs, grapes, and peaches, and distributed throughout Europe and North America. They are migratory ectoparasites which feed on mature parts of root on epidermal and outer cortical cells. The fourth stage juvenile shows some resistance to unfavorable environmental conditions. Frequently found in very large numbers on woody perennials without causing damage, they are however associated with crop decline in figs, may deform carrots and causes damage to celery in Northeastern USA and the Netherlands. This nematode can occur in very high numbers without apparently causing damage. Caution should be taken not to confuse pin nematode with pinworm. Also termed: pin nematode.

para-ureidobenzenearsonic acid (pharmacology) A substance that may be used for the treatment of avian histomoniasis.

parbendazole (pharmacology) A broad-spectrum benzimidazole anthelmintic used with good effects against the common parasites of cattle, pigs and horses. Like other benzimidazoles, it suffers from the problem of resistance developing in the resident worm population if it is used persistently.

parelaphostrongylosis (parasitology) *See Parelaphostrongylus tenuis.*

Parelaphostrongylus (parasitology) A genus in the worm family Protostrongylidae.

Parelaphostrongylus andersoni (veterinary parasitology) A species of parasite in the genus *Parelaphostrongylus* that may be found in the musculature, especially in the longissimus dorsi, in white-tailed deer.

Parelaphostrongylus odocoilei (veterinary parasitology) A species of parasite in the genus *Parelaphostrongylus* that may be found in connective tissue around blood vessels and in lymphatics of musculature below the vertebral column, abdomen and proximal parts of the limbs in mule and black-tailed deer and in moose.

Paraelaphostrongylus tenuis (veterinary parasitology) (Synonym: *Pneumostrongylus tenuis*, *Odocoileostrongylus tenuis*, *Elaphostrongylus tenuis*, *Neurofilaria cornelliensis*) A species of parasite in the genus *Parelaphostrongylus* that may be found in the cranial venous sinuses of white-tailed deer but is non-pathogenic in this species. Infection also occurs in moose, elks, caribous, red deer, black-tailed deer, llamas, sheep and goats. In these species the migrating larvae cause serious damage in the spinal cord and posterior paralysis, often in a number of animals at one time. Also termed moose sickness. Some infected goats also develop a local, linear dermatosis over the shoulders, thorax and flanks, believed to be caused by migrating *Parelaphostrongylus tenuis* larvae irritating nerve roots which leads to pruritus and self-trauma along dermatomes.

parenchyma (parasitological cytology) A loose collection of cells and fibres.

Parhadjelia neglecta (avian parasitology) A habronematid larva found in the proventricular submucosa of ducks in Brazil.

parietal cells (histology) Large cells in the lining of the stomach that secrete hydrochloric acid.

paroxysms (medical/parasitology) A term used in relation to malaria. Bouts of fever due to the liberation of merozoites during the erythrocytic schizogony.

parthenogenesis (biology) The development of an unfertilized egg into an adult organism. Virgin birth. This occurs naturally in bees and ants; in some animal species development of an ovum can be induced chemically or by pricking with a fine glass fibre (*USA*: glass fiber). The result is a clone of the mother cell identical in all respects. Only females can be produced by parthenogenesis, as no Y chromosome is present.

partial derivative (mathematics) Derivative of a function with respect to one of its variables, all other variables in the function being taken as constant. Also termed: partial differential.

partial fraction (mathematics) One of the component fractions into which another fraction can be separated (so that the sum of the partial fractions equals the original fraction).

partial pressure *See* Dalton's Law of partial pressures.

particle (physics) Minute portion of matter, often taken to mean an atom, molecule or elementary particle or subatomic particle.

particle physics (physics) Branch of science concerned with the properties of elementary particles.

partition coefficient (chemistry) Ratio of the concentrations of a single solute in two immiscible solvents, at equilibrium. It is independent of the actual concentrations.

parturition (physiology) Birth of a full-growth foetus (*USA*: fetus) at the completion of pregnancy (gestation).

parvaquone (pharmacology) An antiprotozoal agent.

pascal (Pa) (measurement) SI unit of pressure, equal to a force of 1 newton per square metre ($N\ m^{-2}$). It was named after the French physicist and mathematician Blaise Pascal (1623–62).

paspalitrems (plant parasitology) Tremorgens produced by *Claviceps paspali* which parasitize *Paspalum* spp. grasses.

Passalurus (parasitology) A genus of nematodes in the family Oxyuridae.

Passalurus ambiguus (veterinary parasitology) A species of nematodes in the genus *Passalurus* that may be found in the caecum (*USA*: cecum) and colon of rabbits, hares and other lagomorphs but that appear to be harmless even in very large numbers.

Passeromyia (avian parasitology) A genus of screw-worm flies of birds in Australia.

Passulurus ambiguus (veterinary parasitology) An oxyurid nematode of rabbits, cottontails and hares.

pasteurization (biology) Process of heating food or other substances under controlled conditions. It was developed by the French chemist Louis Pasteur (1822–95) to destroy pathogens. It is widely used in industry, *e.g.* milk production.

patella (anatomy) Bone in front of the knee joint. Also termed: kneecap.

path(o)- (terminology) A word element. [German] Prefix denoting *disease*.

path. (general terminology) An abbreviation of 'path'ology.

pathetic nerve (histology) An alternative term for trochlear nerve.

pathogen (biology) An organism, or other agent, capable of causing disease.

pathogenesis (medical) The origin and course of a disease.

pathogenic (medical) Producing disease.

pathognomonic (medical) Indicative of, or characteristic of, a particular illness.

pathognomy (1, 2 medical) 1 The study of how to recognise an illness from its symptoms. 2 Symptom or symptoms of an illness.

pathologic (medical) Due to or involving a morbid condition, as a pathologic state.

pathology (path.) (medical) The study of morphological changes characteristic of abnormal states, including both physiological (endogenous) effects, effects caused by pathogenic organisms and the adverse effects of erogenous physical and chemical agents. It includes changes at the level of gross anatomy.

-pathy (terminology) A word element. [German] Suffix denoting *morbid condition* or *disease*.

Pb (chemistry) The chemical symbol for Element 82, lead. The symbol is derived from Latin: plumbum.

PBI (biochemistry) An abbreviation of 'P'rotein-'B'ound 'I'odine.

p-block elements (chemistry) 30 nonmetallic elements that form Groups IIB, IVB, VB, VIB, VIIB and 0 of the periodic table (helium is usually excluded), so called because their 1 to 6 outer electrons occupy p-orbitals.

PCR (biochemistry) An abbreviation of 'P'olymerase 'C'hain 'R'eaction.

pd (physics) An abbreviation of 'p'otential 'd'ifference.

PDGF (physiology/biochemistry) An abbreviation of 'P'latelet-'D'erived 'G'rowth 'F'actor.

PE (statistics) An abbreviation for 'P'robable 'E'rror.

PEC (physics) An abbreviation of 'P'hoto 'E'lectric 'C'ell.

peccary flea (parasitology) *See Pulex porcinus.*

peck (measurement) Unit of dry capacity equal to a quarter of a bushel, or 2 gallons (= 9.092 litres).

pectin (chemistry) Complex polysaccharide derivative present in plant cell walls, to which it gives rigidity. It can be converted to a gel form in sugary acid solution.

pectoral (anatomy) Concerning the part of the front end of a vertebrate's body which supports the shoulders and forelimbs.

pectoral (medical) Relating to the chest.

pediat. (medical) An abbreviation of 'pediat'rics

pediatrics (medical) *See* paediatrics.

Pedicinus obtusus (veterinary parasitology) A species of lice in the genus *Pedicinus* that may be found on leaf and green monkeys and baboons.

Pedicinus (veterinary parasitology) A genus of lice in the family Pedliculidae that infests monkeys.

Pedicinus eurygaster (veterinary parasitology) A species of lice in the genus *Pedicinus* that may be found on macaques.

Pedicinus mjobergi (veterinary parasitology) A species of lice in the genus *Pedicinus* that may be found on howler monkeys.

Pedicinus patas (veterinary parasitology) A species of lice in the genus *Pedicinus* that may be found on colobus monkeys.

pediculation (parasitology) Infestation with lice.

pediculicide (control measures) 1 Destroying lice. 2 A substance that destroys lice.

Pediculoides ventricosus (parasitology) The grain itch mite that infests cereal crops and stored hay and grain, and that parasitizes animals causing a mild dermatitis but a severe pruritus.

pediculosis (parasitology) Infestation with lice. *See* louse.

pediculous (parasitology) Infested with lice.

pedigree (genetics) A diagram representing the genetic relationships between individuals. Usually drawn up with reference to one (or a few) specific traits, the pedigree uses conventional symbols to show the sex of the individuals (usually squares for males, circles for females); a diagonal line through their symbol shows that the person is deceased. Whether or not the individual is affected with the trait in question is shown by having their symbol solid or open, perhaps with heterozygotes (proven or surmised) shaded.

pedipalp (parasitological anatomy) A second pair of appendages which arise from the cephalothorax, found in Arachnida.

pedogenesis (parasitology) Metamorphotic phenomenon of production of a number of separate individuals in an intermediate host, *e.g.* a snail, by a single larval form.

peduncle (histology) Any bundle of nerve fibres (*USA*: nerve fibers) in the brain.

PEEP (physiology) An abbreviation of 'P'ositive 'E'nd-'E'xpiratory 'P'ressure breathing.

PEG (chemistry) An abbreviation of 'P'oly'E'thylene 'G'lycol.

PEL (toxicology) An abbreviation of 'P'ermissible 'E'xposure 'L'evel. *See* permissible dose.

Pelecitus roemeri (veterinary parasitology) A species of filaroid nematode in the genus *Pelecitus* that may be found in subcutaneous tissues, especially around the stifle joints, of wallabies and kangaroos. Also termed: *Dirofilaria roemeri.*

Pelodera (parasitology) A genus of nematodes in the family Rhabditida.

pelodera dermatitis (parasitology) Inflammation of the skin caused by larvae of the freeliving nematode *Pelecitus strongyloides* and characterized by alopecia, itching, thick, scurfy skin and 0.5 inch diameter pustules which contain the larvae.

Pelodera strongyloides (parasitology) A species of nematodes in the genus *Pelodera.* A freeliving worm which invades broken skin, usually from damp and infected bedding and that causes pelodera dermatitis.

pelvic (anatomy) Concerning the part of the rear end of a vertebrate's body which supports the hindlimbs.

pelvic mesocolon (anatomy) The peritoneum attaching the sigmoid colon to the dorsal abdominal wall. Also termed sigmoid megacolon.

pelvis (1, 2 anatomy) 1 Part of the skeleton (pelvic girdle) to which a vertebrate's hindlimbs are joined. 2 Cavity in the kidney which receives urine from the tubules and drains it into the ureter.

pemphigus (medical) A term used for skin diseases in which blisters develop.

penetrance (genetics) The frequency with which a dominant allele is expressed in the phenotype of the individual carrying it. A completely penetrant allele is expressed in every case, and this is normally the position with dominant alleles having major effects. An allele is said to have incomplete penetrance if it is not always expressed (this will depend on the genetic background and on the environment).

-penia (terminology) A word element. [German] Suffix denoting *deficiency*.

penis (anatomy) The male sex organ through which semen and urine are discharged by way of the urethra. The shaft of the penis contains spongy tissue which fills with blood during sexual excitement: the ensuing erection makes sexual intercourse possible.

pent(a)- (terminology) A word element. [German] Prefix denoting *five*.

pentahydrate (chemistry) Chemical containing five molecules of water of crystallization (*e.g.* $CUSO_4.5H_2O$).

pentamidine isethionate (pharmacology) A diamidine derivative effective against protozoa, that may be used in the treatment of *Babesia*, *Leishmania* and *Pneumocystis* spp.

pentane (chemistry) C_5H_{12} Liquid alkane, extracted from petroleum and used as a solvent and in organic synthesis. It has three isomers.

pentanoic acid (chemistry) $CH_3(CH_2)_3COOH$ Liquid carboxylic acid with a pungent odour (*USA*: odor), used in perfumes. Also termed: valeric acid.

pentastome (parasitology) Aberrant arthropod parasite belonging to the class Pentastomida, that includes *Linguatula*, *Porocephalus* and *Armillifer* spp.

pentastomiasis (veterinary parasitology) Infection by parasites of the phylum Pentastomida that may be found in the respiratory tract of reptiles.

Pentastomida (parasitology) *See* pentastome.

pentastomidiasis (parasitology) Infection with pentastomes.

Pentatrichomonas (parasitology) A genus of protozoan parasites with five flagella.

Pentatrichomonas hominis (human/veterinary parasitology) (synonym: *Trichomonas hominis*, *Trichomonas* intestinalis) A species of protozoan parasites in the genus *Pentatrichomonas* that may be found in dogs, primates and humans but that appears to be nonpathogenic.

pentavalent (chemistry) Having a valence of five.

pentose (chemistry) Monosaccharide carbohydrate (sugar) that contains five carbon atoms and has the general formula $C_5H_{10}O_5$; *e.g.* ribose and xylose. Also termed: pentaglucose.

pentyl group (chemistry) An alternative term for amyl group.

PEP (physiology) An abbreviation of 'P're'E'jection 'P'eriod.

pepsin (biochemistry) Enzyme produced in the stomach which, under acid conditions, brings about the partial hydrolysis of polypeptides which breaks down protein in the food.

peptic ulcer (medical) Ulcers in the stomach and duodenum are referred to as peptic ulcers. Duodenal ulcers are more common than those in the stomach, and more common in men than women, who are more liable to develop duodenal ulcers after menopause.

peptidase (biochemistry) Enzyme, often secreted in the body (*e.g.* by the intestine), which degrades peptides into free amino acids, thus completing the digestion of proteins.

peptide (biochemistry) Organic compound that contains two or more amino acid residues joined covalently through peptide bonds (-NH-CO-) by a condensation reaction between the carboxyl group of one amino acid and the amino group of another. Peptides polymerize to form proteins.

peptide bond (biochemistry) A bond joining amino acids to form peptides or proteins.

per- (terminology) A word element. [Latin] Prefix denoting *throughout*, *completely*, *extremly*.

percentage (statistics) Fraction expressed in hundredths (with the denominator omitted) *e.g.* one half = 1/2 = 50/100 = 50 per cent (often written 50%).

percentage composition (chemistry) Make-up of a chemical compound expressed in terms of the percentages (by mass) of each of its component elements *e.g.* ethane (C_2H_6), ethene (C_2H_4) and ethyne (C_2H_2) all consist of carbon and hydrogen. Their approximate percentage compositions are: ethane 80% carbon, 20% hydrogen; ethene 85% carbon, 15% hydrogen; ethyne 92% carbon 8% hydrogen.

percentile (statistics) Hundredth part of a range of statistics (data) of equal frequency *e.g.* the 80th percentile is the value below which 80 per cent of all the values fall. The 50th percentile is the median.

perdisulphuric(VI) acid (chemistry) H_2SO_5 White crystalline compound, used as a powerful oxidizing agent. Also termed: Caro's acid; persulphuric acid (*USA*: perdisulfuric(VI) acid; peroxomonosulphuric acid (*USA*: peroxomonosulfuric acid).

perfect gas (chemistry) An alternative term for ideal gas.

See also kinetic theory.

perfect number (mathematics) Number that equals the sum of all its factors (except the number itself *e.g.* 28 (= 1 + 2 + 4 + 7 + 14) and 496 (= 1 + 2 + 4 + 8 + 16 + 31 + 62 + 124 + 248) are perfect numbers.

peri- (terminology) A word element. [German] Prefix denoting *around*, *near*.

periarteritis nodosa (histology) Inflammation of the outer part of the wall of an artery and the tissues immediately roundabout.

periarthritis (histology) Inflammation of the tissues surrounding a joint.

pericarditis (medical) inflammation of the pericardium, the membrane surrounding the heart. It may occur in association with coronary thrombosis, virus infection, tuberculosis, rheumatic heart disease or bacterial infection, or as the result of injury. There is pain in the chest, and if there is an effusion of fluid between the two layers of the pericardium the action of the heart may be affected. It may then be necessary to aspirate the fluid.

pericardium (anatomy) The membranous sac which contains the heart. It has two layers, between which is a capillary space containing a film of fluid. This provides lubrication to assist the free movement of the heart in relation to neighbouring structures. The outermost part of the pericardium is fibrous.

pericentric (genetics) *See* inversion.

perichondrium (histology) The thin membrane which covers cartilage, except on freejoint surfaces.

pericranium (histology) The periosteal membrane covering the skull.

perikaryon (cytology) An alternative term for cell body.

perilymph (anatomy) Fluid that surrounds the cochlea of the ear.

perilymph (anatomy) The fluid found in the inner ear separating the membrane of the labyrinth from the surrounding bone. It does not communicate with the endolymph, the fluid inside the membranous labyrinth.

perinatal mortality rate (statistics) The number of stillbirths (babies born dead after 28 weeks) plus neonatal deaths (babies born alive but dying within a week, regardless of gestational age at delivery) expressed as a proportion of 1000 total births.

perinatal period (physiology) The developmental period preceding birth; the last third of gestation.

perinephric (medical) Surrounding the kidney.

perineum (anatomy) The region between the genital organs in front and the anus behind.

period (1 chemistry; 2 physics) 1 One of the seven horizontal rows of elements in the periodic table. 2 Time taken to complete a regular cycle.

periodic acid Schiffs orange G (histology) A histological staining technique used to make pituitary cell types visible under the microscope.

periodic acid, schiffs (histology) A histological staining technique used to make mucins and carbohydrate groups visible under the microscope.

periodic function (mathematics) Function that returns to the same value at regular intervals.

periodic law (chemistry) Properties of elements are a periodic function of their atomic numbers. Also termed: Mendeleev's law.

periodic ophthalmia (veterinary parasitology) A disease of horses with a possible causal relationship to infection wtih *Onchocerca cervicalis*, that usually terminates in blindness.

periodic parasite (parasitology) A parasite that parasitizes a host for short periods.

periodic table (chemistry) Arrangement of elements in order of increasing atomic number, with elements having similar properties, *i.e.* in the same family, in the same vertical column (group). Horizontal rows of elements are termed periods.

periodicity (chemistry) Regular increases and decreases of physical values for elements known to have similar chemical properties.

periosteum (histology) The thin membrane of connective tissue that surrounds the bones.

periostitis (medical) Inflammation of the periosteum.

peripheral (1 computing, 2 medical) 1 Any device attached to a computer but not forming part of the central processor or core store, *e.g.* a disk, printer or visual display unit. 2 A term applied to structures towards the outer parts or extremities, *e.g.* the peripheral as opposed to the central nervous system.

peripheral nervous system (anatomy) That portion of the nervous system consisting of the nerves and ganglia outside of the spinal cord and brain (central nervous system, CNS).

peripheral unit (computing) Equipment that can be linked to a computer, including input, output and storage devices.

peristalsis (physiology) The process by which the contents are propelled along tubular organs, as in the intestines; contractions of the muscle in the walls of the gut pass in a wave-like motion along its length, being controlled by the nervous plexus present there, *e.g.* in the oesophagus or intestine to push food along.

peristriate cortex (histology) An alternative term for parastriate cortex.

peritoneal (anatomy) Pertaining to the peritoneum.

peritoneal cysts (parasitological physiology) Cysts including those of cestode intermediate stages.

peritoneoscopy (laparoscopy) (medical) The examination of the peritoneal cavity and abdominal organs through an endoscope passed through the abdominal wall.

peritoneum (histology) The membrane covering the inner walls of the abdominal cavity and the organs it contains. There are two layers, the membrane covering the wall of the cavity is the parietal layer, and that covering the abdominal organs the visceral layer.

peritonitis (medical) Inflammation of the peritoneum.

peritrichous (parasitological anatomy) 1 Having flagella around the entire surface. 2 Having flagella around the cytostome only; said of *Ciliophora.*

peritrophic membrane (parasitological physiology) A membrane which is secreted from the anterior end of the midgut in some blood-feeding arthropods. This membrane encloses the blood meal.

periventricular (medical) Near the ventricles of the brain.

Perkinsus (veterinary parasitology) Protozoan parasite genus of molluscan tissues. Species includes *Perkinsus* marinus, a major pathogen of the American oyster, and *Perkinsus* olseni a pathogen of abalone in warm waters.

Perl's prussian blue ferric stain (histology) A histological staining technique used to make ferric iron visible under the microscope.

permanent gas (chemistry) Gas that is incapable of being liquefied by pressure alone; a gas above its critical temperature.

permanent magnet (physics) Ferromagnetic object that retains a permanent magnetic field and the magnetic moment associated with it after the magnetizing field has been removed.

permanent teeth (anatomy) Second set of teeth used by most mammals in adult life (after they have displaced the first set of milk or deciduous teeth).

permanganate bleach (histology) A histological staining technique used for removing melanin.

permeability (1, 2 physics)1 Rate at which a substance diffuses through a porous material. 2 Extent to which a substance can pass through a membrane. Membranes may be semi-permeable, *e.g.* plasma membrane, which allows small molecules such as those of water to pass through. 3 Magnetization developed in a material placed in a magnetic field, equal to the flux density produced divided by the magnetic field strength. Also termed: magnetic permeability.

permeable (physics) Porous; describing something (*e.g.* a membrane) that exhibits permeability.

permissible dose (toxicology) That dose of a chemical that may be received by an individual without expectation of an adverse effect.

permutation (mathematics) Number of ways a set of numbers can be arranged and ordered. For n numbers, there are $n!$ ways of arranging them n at a time, and $n!/(n - r)!$ ways of arranging them r at a time ($n!$ stands for factorial n).

pero- (terminology) A word element. [German] Prefix denoting *deformity, maimed.*

Peronia rostrata (parasitology) A tertiary blowfly parasitizing carcasses which are tending to dry out.

Perostrongylus (parasitology) A subgenus of nematodes in the family Filaroididae. *See Aelurostrongylus.*

peroxide (1, 2 chemistry) 1 Oxide of an element containing more oxygen than does the normal oxide of the element. 2 Oxide, containing the O_2^{2-} ion, that yields hydrogen peroxide on treatment with an acid. Peroxides are powerful oxidizing agents.

peroxomonosulphuric acid (chemistry) An alternative term for perdisulphuric (VI) acid (*USA*: perdisulfuric acid). (*USA*: peroxomonosulfuric acid).

Perperipes (insect parasitology) Mites in the Prostigmata that may be larval or egg parasites on the ants *Eciton, Neivamyrmex* and *Nomamyrmex.*

Perreyea lepida (parasitology) A South American sawfly whose larvae cause severe liver necrosis when ingested. The toxin involved is lophyrotomin.

Perspex (chemistry) Trade name for the plastic polymethyl methacrylate.

persulphate (chemistry) Salt of perdisulphuric(VI) acid (*USA*: persulfuric(VI) acid), used as an oxidizing agent. (*USA*: persulfate).

persulphuric acid (chemistry) An alternative term for perdisulphuric(VI) acid (*USA*: perdisulfuric(VI) acid). (*USA*: persulfuric acid).

Peru balsam (chemistry) *See* balsam.

pesticide (control measures) Compound used in agriculture to destroy organisms that can damage crops or stored food, especially insects and rodents. Pesticides include fungicides, herbicides and insecticides. The effects of some of them, *e.g.* organic chlorine compounds such as DDT, can be detrimental to the ecosystem.

pestilence (parasitology) A virulent contagious epidemic or infectious epidemic disease.

-petal (terminology) A word element. [Latin] Suffix denoting *directed, moving toward.*

petri dish (laboratory equipment) Sterilizable circular glass plate with a fitted lid used in microbiology for holding media on which micro-organisms may be cultured. It was named after the German bacteriologist Julius Petri (1852–1921).

petroleum wax (chemistry) An alternative term for paraffin wax.

-pexy (terminology) A word element. [German] Suffix denoting *surgical fixation.*

pH (chemistry) Hydrogen ion concentration (grams of hydrogen ions per litre) expressed as its negative logarithm; a measure of acidity and alkalinity. *E.g.* a hydrogen ion concentration 0 10^{-4} grams per litre corresponds to a pH of 4, and is acidic. A pH of 7 is neutral; a pH of more than 7 is alkaline.

pH scale (chemistry) Scale that indicates the acidity or alkalinity of a solution. *See* pH.

phac(o)- (terminology) A word element. [German] Prefix denoting *lens*.

phag(o)- (terminology) A word element. [German] Prefix denoting *eating, ingestion*.

phage *See* bacteriophage.

-phagia (terminology) A word element. [German] Suffix denoting *eating, swallowing*.

phagocyte (cytology) Cell that exhibits phagocytosis.

phagocytosis (cytology) Engulfment of external solid material by a cell, *e.g.* phagocyte. It is also the method by which some unicellular organisms (*e.g.* protozoa) feed.

-phagy (terminology) A word element. [German] Suffix denoting *eating, swallowing*.

phalanges (anatomy) Bones in the digits of the hand or foot.

pharmac(o)- (terminology) A word element. [German] Prefix denoting *drug, medicine*.

pharmacodynamics (pharmacology) The study of the effects of drugs upon the body.

pharmacokinetics (pharmacology) The study of how drugs are absorbed into, distributed and broken down in, and excreted from, the body.

pharmacological (pharmacology) Pertaining to pharmacology.

pharmacologist (pharmacology) A specialist in pharmacology.

pharmacology (pharmacology) The study of the action of drugs on living organisms; drug development, their use; their adverse effects and their beneficial effects.

pharmacotherapy (pharmacology) Treatment of disease with medicines.

pharmacy (general terminology) Preparation and dispensing of drugs, and the place where this is done.

pharyng(o)- (terminology) A word element. [German] Prefix denoting *pharynx*.

pharyngeal (anatomy) Pertaining to the pharynx.

Pharyngobolus (parasitology) A genus of parasitic flies in the family Oestridae, the larval stages of which resemble those of *Oestrus ovis*.

Pharyngobolus africanus (veterinary parasitology) A species of parasitic flies in the genus *Pharyngobolus* that may be found in the pharynx of African elephants. Also termed: elephant throat bot fly.

pharynx (anatomy) Area that links the buccal cavity (mouth) to the oesophagus (gullet) and the nares (back of the nostrils) to the trachea (windpipe). Food passes via the pharynx to the oesophagus when the epiglottis closes the entrance of the trachea.

phase (1 chemistry; 2 physics) 1 Any homogeneous and physically distinct part of a chemical system that is separated from other parts of the system by definite boundaries, *e.g.* ice mixed with water. 2 In physics, the part of a periodically varying waveform that has been completed at a particular moment (there being 360° or 2π radians in a full cycle).

phase contrast microscope (microscopy) Microscope that uses the principle that light passing through materials of different refractive indices undergoes a change in phase, transmitting these changes as different intensities of light given out by different materials.

phase microscope (microscopy) *See* phase-contrast microscope.

phase rule (physics) The number of degrees of freedom (F) of a heterogeneous system is related to the number of components (C) and of phases (P) present at equilibrium by the equation $P + F = C + 2$.

phasmid (parasitological anatomy) Sensory receptors situated near the posterior end of a nematode.

Phaulodinchyus (insect parasitology) *See Dinychus*.

Phe (biochemistry) An abbreviation of 'Phe'nylalanine.

Pheidoloxeton wheeleri (insect parasitology) Hymenoptera in the Eucharitidae that, similarly to *Pseudochalcura gibbosa*, *Pseudometagea schwarzii*, *Psilogaster antennatus* and *Psilogaster fasciventris*, may be found in larvae pupa and possibly adult workers of *Pheidole* spp., *Camponotus* spp., *Lasius* spp., and *Coelogyne*.

phenamidine isethionate (pharmacology) An aromatic diamidine used as an antiprotozoal, particularly in the treatment of infections by *Babesia*, *Leishmania* and *Pneumocystis* spp.

phenanthrene (chemistry) $C_{14}H_{10}$ Tricyclic aromatic organic compound consisting of three benzene rings fused together, the basic skeleton of steroids and many other biologically active compounds.

phenanthridium (pharmacology) A group of chemotherapeutic agents, including isometamidium, homidium, pyrithidium and quinapyramine, used in the treatment of trypanosomiasis. Also termed: phenanthridine.

phenol (chemistry) C_6H_5OH Colourless (*USA*: colorless) crystalline solid which turns pink on exposure to air and light. It has a characteristic, rather sweet odour (*USA*: odor). It is used as an antiseptic and disinfectant, and in the preparation of dyes, drugs, etc. Also termed: carbolic acid; hydroxybenzene. Other compounds with one or more hydroxy groups bound directly to a benzene ring are also known as phenols. They give reactions typical of alcohols (*e.g.* they form esters and ethers), but they are more acidic and form salts by the action of strong alkalis (*USA*: strong alkalies).

phenolphthalein (chemistry) Organic compound that is used as a laxative and as an indicator in volumetric analysis. It is red in alkalis (*USA*: alkalies) and colourless (*USA*: colorless) in acids.

phenothiazine (pharmacology) The first broad-spectrum veterinary anthelmintic which has been largely superseded by more efficient compounds. Its principal use is in mixtures with piperazine to inhibit egg-laying by resident worms in horses.

phenotype (genetics) Outward appearance and characteristics of an organism, or the way genes express themselves in an organism. Organisms of the same phenotype may possess different genotypes; *e.g.* in a heterozygous organism two alleles of a gene may be present with the expression of only one.

phenyl group (chemistry) C_6H_5 - Monovalent radical derived from benzene.

phenyl methyl ketone (chemistry) An alternative term for acetophenone.

phenylalanine (chemistry) $C_6H_5CH_2CH(NH_2)COOH$ Essential amino acid that possesses a benzene ring.

phenylamine (chemistry) An alternative term for aniline.

phenylethylene (chemistry) An alternative term for styrene.

phenylketonuria (biochemistry) An autosomal recessive inherited deficiency of the liver enzyme phenylalanine hydroxylase that converts the amino acid phenylalanine into tyrosine. Phenylalanine and its toxic derivatives accumulate in the body and can cause brain damage with mental retardation. In the UK babies are screened at birth by the Guthrie test and about 1 in 16000 is found to have the condition.

phenylmercuric (chemistry) Denoting a compound containing the radical C_6H_5Hg- , forming various antiseptic, antibacterial and fungicidal salts; compounds of the acetate salts are used as a herbicide.

phenylmercuric acetate (pharmacology) A compound used as a bacteriostatic preservative in pharmaceutical preparations, as a topical fungistatic agent, and as a herbicide.

3-phenylpropenoic acid (chemistry) An alternative term for cinnamic acid.

Pheromermis villosa (insect parasitology) A species of nematode in the Mermithidae that are parasites to the ants *Lasius* spp. causing morphological changes.

pheromone (biochemistry) Chemical substance produced by an organism which may influence the behaviour (*USA*: behavior) of another; *e.g.* in moths, pheromones act as sexual attractants; in social insects such as bees, pheromones have an important role in the development and behaviour of the colony.

phi (terminology) The twenty-first letter of the Greek alphabet, Φ.

Philaemon grandidieri (veterinary parasitology) A leech that invades the dorsal lymph sac of New Guinea frogs.

-philia (terminology) A word element. [German] Suffix denoting affinity for, morbid fondness of.

Philometra (fish parasitology) Genus of deep red-colored, large nematode of the Dracunculoid family Philometridae which may be found in the peritoneal cavity of fish.

Philophthalmus (parasitology) A genus of flukes (digenetic nematodes) in the family Philophthalmidae.

Philophthalmus gralli (avian parasitology) A species of the genus *Philophthalmus* which may be found in the conjunctival sacs of birds causing congestion and erosion of the conjunctiva. Also termed oriental avian eye fluke.

Phlebotomus argentipes (human parasitology) A species of biting fly of the genus *Phlebotomus* that is a vector of the human disease kala azar.

Phlebotomus caucasicus (human parasitology) A species of biting fly of the genus *Phlebotomus* that is a vector of the human disease cutaneous leishmaniasis.

Phlebotomus chinensis (human parasitology) A species of biting fly of the genus *Phlebotomus* that is a vector of the human diseases phlebotomus or sandfly fever, and kala azar.

Phlebotomus columbianum (human parasitology) A species of biting fly of the genus *Phlebotomus* that is a vector of the human illness Carrion's disease.

Phlebotomus major (human parasitology) A species of biting fly of the genus *Phlebotomus* that is a vector of the human disease cutaneous leishmaniasis.

Phlebotomus martini (human parasitology) A species of biting fly of the genus *Phlebotomus* that is a vector of the human disease kala azar.

Phlebotomus mongolensis (human parasitology) A species of biting fly of the genus *Phlebotomus* that is a vector of the human disease phlebotomus or sandfly fever.

Phlebotomus papatasi (human parasitology) A species of biting fly of the genus *Phlebotomus* that is a vector of the human diseases cutaneous leishmaniasis and phlebotomus or sandfly fever.

Phlebotomus perniciosus (human parasitology) A species of biting fly of the genus *Phlebotomus* that is a vector of the human disease kala azar.

Phlebotomus sergenti (human parasitology) A species of biting fly of the genus *Phlebotomus* that is a vector of the human disease cutaneous leishmaniasis.

Phlebotomus verrucarum (human parasitology) A species of biting fly of the genus *Phlebotomus* that is a vector of the human illness Carrion's disease.

Phlebotomus (human parasitology) A genus of biting flies, called sandflies, the females of which are blood-sucking. They are vectors of various human diseases.

Phlebotomus noguchi (human parasitology) A species of biting fly of the genus *Phlebotomus* that is a vector of the human illness Carrion's disease.

phlegm (medical) Viscid mucus excreted in abnormally large quantities from the respiratory tract.

phlegmasia (medical) Inflammation.

phlog(o)- (terminology) A word element. [German] Prefix denoting inflammation.

phlogogenic (medical) Producing inflammation.

Phocanema (fish/human parasitology) A genus of nematodes in the family Anisakidae which may parasitize codfish and be transmitted to humans who eat the fish.

Phocanema decipiens (veterinary parasitology) A species of the genus *Phocanema* which may be found in the South American sea lion and fur seals.

Phocitrema fusiforme (veterinary parasitology) A liver fluke found in fur seal, ringed seal and otter.

Pholcus phalangoides (parasitology) Daddy-long-legs; crane fly; harvestman.

phon (acoustics) Unit of loudness of sound (given as the number of decibels it appears to be above a reference tone of known frequency and intensity).

phon(o)- (terminology) A word element. [German] Prefix denoting sound, voice.

phorate (control measures) An organophosphorus compound used as an insecticide and capable of causing poisoning.

phorazetim (control measures) A very toxic organophosphorus compound used as a rodenticide.

-phore (terminology) A word element. [German] Suffix denoting a carrier.

-phoresis (terminology) A word element. [German] Suffix denoting transmission.

phoresy (parasitology) A method of dispersal, including for example transmission of a parasite by a parasite, *e.g.* *Histomonas meleagridis* by *Heterakis gallinarum*.

Phormia (parasitology) A genus of blowflies. Species includes *Phormia regina* and *Phormia terranovae* which are involved in cutaneous myiasis in North America.

phosgene (chemistry) $COCl_2$ Colourless (*USA*: colorless) gas with penetrating and suffocating smell. It was formerly used as a war gas, and is now used in organic synthesis. Also termed: carbonyl chloride.

phosmet (control measures) An organophosphorus insecticide used as a spray or pour-on to control ectoparasites.

phosphate (chemistry) Salt of phosphoric(V) acid, containing the ion PO_4^{3-}. Many phosphates occur naturally in minerals (*e.g.* apatite, calcium phosphate) and in biological systems, and some are of enormous commercial and practical importance, *e.g.* ammonium phosphate fertilizers and alkali phosphate buffers. Also termed: orthophosphate. Because phosphoric(V) acid is a tribasic acid, it also forms hydrogenphosphates, HPO_4^{2-} and dihydrogenphosphates, $H_2PO_4^-$ (*e.g.* superphosphate fertilizer is calcium dihydrogenphosphate).

phosphate buffer (biochemistry) Important phosphate-containing buffers.

phosphate buffered saline (pbs) (biochemistry) A special phosphate buffered saline used in tissue cultures.

phosphatidyl choline (biochemistry) An alternative term for lecithin, a phospholipid constituent of plasma membranes.

phospholipid (chemistry) Member of a class of complex lipids that are major components of cell membranes. They consist of molecules containing a phosphoric(V) acid ester of glycerol (*i.e.* phosphoglycerides), the remaining hydroxyl groups of the glycerol being esterified by fatty acids. Also termed: phosphoglyceride; phosphatide; glycerol phosphatide.

phosphonium ion (chemistry) Ion PH_4^+.

phosphor (chemistry) Substance capable of luminescence or phosphorescence, as used to coat the inside of a television screen or fluorescent lamp.

phosphorescence (chemistry) Emission of light (generally visible light) after absorption of light of another wavelength (usually ultraviolet or near ultraviolet) or electrons. It continues after the stimulating source is removed.

phosphoric acid (chemistry) A crystalline acid formed by oxidation of phosphorus; its salts are called phosphates.

See also phosphate.

phosphoric(V) acid (chemistry) H_3P0_4 Tribasic acid, a colourless (*USA*: colorless) crystalline solid, made by dissolving phosphorus(V) oxide in water. It is used to form a corrosion-resistant layer on steel. Its salts, the phosphates, are of great biological and commercial importance. Also termed: phosphoric acid; orthophosphoric acid.

phosphorolysis (chemistry) Cleavage of a chemical bond with simultaneous addition of the elements of phosphoric acid to the residues.

phosphorus (P) (chemistry) Nonmetallic element in Group VA of the periodic table. It exists as several allotropes, chief of which are red phosphorus and the poisonous and spontaneously inflammable white or yellow phosphorus. It occurs in many minerals (particularly phosphates) and all living organisms; it is an essential nutrient for plants. Phosphorus is made by heating calcium phosphate (with carbon and sand) in an electric furnace. It is used in matches and for making fertilizers. At. no. 15; r.a.m. 30.9738.

phosphorus(III) bromide (chemistry) PBr_3 Colourless (*USA*: colorless) liquid, used in organic synthesis to replace a hydroxyl group with a bromine atom. Also termed: phosphorus tribromide.

phosphorus(III) chloride (chemistry) PCl_3 Colourless (*USA*: colorless) fuming liquid, used to make organic compounds of phosphorus. Also termed: phosphorus trichloride.

phosphorus(III) oxide (chemistry) P_2O_3 White waxy solid which readily reacts with oxygen to form phosphorus(V) oxide. Also termed: phosphorus trioxide.

phosphorus(V) bromide (chemistry) PBr_5 Yellow crystalline solid, used as a brominating agent. Also termed: phosphorus pentabromide.

phosphorus(V) chloride (chemistry) PCl_5 Yellowish-white crystalline solid, used as a chlorinating agent. Also termed: phosphorus pentachloride.

phosphorus(V) oxide (chemistry) P_2O_5 Hygroscopic white powder which readily reacts with water to form phosphoric(V) acid. It is used as a desiccant. Also termed: phosphorus pentoxide.

phosphorylation (chemistry) Process by which a phosphate group is transferred to a molecule of an organic compound. In some substances, *e.g.* ATP, a high-energy bond may be formed by phosphorylation, which is essential for energy-transfer in living organisms. It is an important biochemical end-reaction which modifies the conformation of molecules such as enzymes, receptors, etc.

phosphosphingolipids (biochemistry) A subset of sphingolipids, that are membrane constituents of cells, constitutents of plasma lipoproteins, blood group substances and act as receptors for bacterial toxins.

photocell (physics) Device that converts light into an electric current. It can be used for the detection and measurement of light. Also termed: photoelectric cell; photoemissive cell.

photochemical reaction (chemistry) Chemical reaction that is initiated by the absorption of light. The most important phenomenon of this type is photosynthesis. It is also the basis of photography.

photochemistry (chemistry) Branch of chemistry concerned with the action of light in initiating chemical reactions.

photoconductivity (physics) Change in electrical conductivity of a substance when it is exposed to light; *e.g.* selenium, used in photoelectric light meters.

photoelectric cell (physics) An alternative term for photocell.

photoelectric effect (chemistry) Phenomenon that occurs with some semi-metallic materials. When photons strike them they are absorbed and the energized electrons produced flow in the material as an electric current. It is the basis of photocells and instruments that employ them, such as photographers' light meters.

photoelectron (chemistry) Electron produced by the photoelectric effect or by photoionisation.

photoemission (chemistry) Emission of photoelectrons by the photoelectric effect or by photoionisation.

photography (photography) Process of taking photographs by the chemical action of light or other radiation on a sensitive plate or film made of glass, celluloid or other transparent material coated with a light-sensitive emulsion. Light causes changes in particles of silver salts in the emulsion which, after development (in a reducing agent), form grains of dark metallic silver to produce a negative image. Unaffected silver salts are removed by fixing (in a solution of ammonium or sodium thiosulphate [*USA*: sodium thiosulfate]).

photoionisation (chemistry) Ionization of atoms or molecules by light or other electromagnetic radiation.

photoluminescence (chemistry) Light emission by a substance after it has itself been exposed to visible light or infra-red or ultraviolet radiation.

photolysis (chemistry) Photochemical reaction that results in the decomposition of a substance.

photometer (physics) Instrument for measuring the intensity of light.

photometry (measurement) Measurement of the intensity of light.

photomicrograph (microscopy/photography) Photograph obtained through a microscope.

photomultiplier (physics) Photocell of high sensitivity used for detecting very small quantities of light radiation. It consists of series of electrodes in an evacuated envelope, which are used to amplify the emission current by electron multiplication. Also termed: electron multiplier.

photon (physics) Quantum of energy in electromagnetic radiation, such as light or X-rays. The amount of energy per photon is $h\nu$, where h is Planck's constant and ν is the frequency of the radiation.

photoneutron (physics) Neutron resulting from the interaction of a photon with an atomic nucleus.

photophosphorylation (biology) Process during photosynthesis that results in the formation of ATP from energy derived from sunlight via chlorophyll. The reactions also produce hydrogen, which is used in combination with carbon dioxide to make sugars. Also termed: photosynthetic phosphorylation.

photoreceptor (physics) Receptor consisting of sensory cells that are stimulated by light, *e.g.* light-sensitive cells in the eye.

photosynthesis (biology) Type of autotrophic nutrition employed by green plants which involves the synthesis of organic compounds (mainly sugars) from carbon dioxide and water. Sunlight is used as a source of energy, which is trapped by chlorophyll present in chloroplasts. The process consists of a light stage, in which energy is converted into ATP and water is split into hydrogen and oxygen. Hydrogen is subsequently combined with carbon dioxide in the dark stage to form carbohydrates. Photosynthetic bacteria use different sources of hydrogen in the process.

photovoltaic cell (physics) An alternative term for a photocell.

phthalic acid (chemistry) $C_6H_4(COOH)_2$ White crystalline solid which on heating converts to its anhydride. It is used in organic synthesis and to make polyester resins. Also termed: benzene-1,2-dicarboxylic acid.

phylogeny (taxonomy) Relationship between groups of organisms (*e.g.* members of a phylum) based on the closeness of their evolutionary descent.

phylum (taxonomy) In biological classification, one of the groups into which the animal kingdom is divided. The members of the group, although often quite different in form and structure, share certain common features; *e.g.* the phylum Arthropoda (arthropods) includes all animals with jointed legs and an exoskeleton. Phyla are subdivided into classes. The equivalent of a phylum in the plant kingdom is a division.

physical change (chemistry) Reversible alteration in the properties of a substance that does not affect the composition of the substance itself (as opposed to a chemical change, which is difficult to reverse and in which composition is affected).

physical chemistry (chemistry) Branch of chemistry concerned with the physical properties of substances.

physical states of matter (chemistry) Three kinds of substances that make up matter: gases, liquids and solids.

physics (physics) Science concerned with the properties of matter, energy and radiation, particularly in processes involving no change of chemical composition.

physiology (physiology) Branch of biology that is concerned with the functioning of living organisms (as opposed to anatomy, which deals with their structure).

Pi π (mathematics) Symbol for the ratio between the circumference of a circle and its diameter. It is an irrational number with the value 3.1415926536.

Pi (biochemistry) An abbreviation of 'P'hosphate-'i'norganic.

picrate (chemistry) 1 Salt of picric acid 2 Charge-transfer complex of picric acid with aromatic hydrocarbons, amines and phenols; picrates are frequently used to identify these classses of compounds. *See* aromatic compound.

picric acid (chemistry) $C_6H_2(NO_3)_3OH$ Yellow crystalline solid obtained by nitrating phenol sulphonic acid (*USA*: phenol sulfonic acid). It has been used as an antiseptic, dye and explosive. Also termed: 2,4,6-trinitrophenol.

pictogram (statistics) Method of graphically representing statistical data that uses symbols, each representing a given number of items of information; *e.g.* if a barrel is used to symbolize a million gallons of oil, a row of 7 barrels represents 7 million gallons. It is thus a sort of pictorial bar chart.

pie chart (statistics) Method of graphically representing statistical data that consists of a circular diagram divided into sectors whose size (determined by the angle they subtend at the centre [*USA*: center]) represents a number of items of information, calculated as a percentage of the whole; *e.g.* an item allocated a quadrant (quarter of the pie chart) represents 25 per cent of the whole.

pigment (chemistry) Insoluble colouring (*USA*: coloring) material, used for imparting various colours to paints, paper, polymers, etc. (soluble colouring materials are dyes). Some naturally occurring coloured substances are also known as pigments; *e.g.* green chlorophyll in plants and red haemoglobin (*USA*: hemaglobin) in blood.

PIH (physiology/biochemistry) An abbreviation of 'P'rolactin-'I'nhibiting 'H'ormone.

pin nematode (plant parasitology) *See Paratylenchus hamatus.*

pine wood nematode (plant parasitology) *See Bursaphelenchus xylophilus.*

pineal gland (anatomy) Club-shaped, elongated outgrowth from the roof of the vertebrate forebrain. It may act as a third eye in some lower bony fishes; in other vertebrates it serves as a hormone-producing organ whose secretory function is regulated by light entering the body via the eyes. In human beings its role is not clear. Also termed: pineal body; epiphysis.

pinna (anatomy) The part of the ear that extends beyond the skull, consisting of a cartilaginous flap. It covers and protects the opening of the ear.

pinocytosis (cytology) Uptake of particles and macromolecules by living cells. Also termed: endocytosis.

pint (measurement) Unit of liquid measure equal to 20 fluid ounces or one-eighth of a gallon. 1 pint = 0.56826 litres.

pipette (laboratory equipment) Device for transferring a known volume of liquid. It consists of a glass tube, often with a swelling at its centre (*USA*: center), and may have a rubber bulb or glass 'cylinder' at one end. Pipettes are used in volumetric analysis.

pituitary (anatomy) Endocrine gland situated at the base of the brain in vertebrates, responsible for the production of many hormones. The anterior (front) lobe produces growth hormone, luteinizing hormone, follicle-stimulating hormone, thyrotrophic hormone, lactogenic hormone and acth. The posterior (rear) lobe secretes oxytocin and vasopressin produced in the hypothalamus. Also termed: pituitary gland; hypophysis.

pk value (mathematics) Negative logarithm of the equilibrium constant for the dissociation of an electrolyte in aqueous solution.

placenta (anatomy) Vascular organ that attaches the foetus (*USA*: fetus) to the wall of the uterus (womb).

Planck's constant (*h*) (physics) Fundamental constant that relates the energy of a quantum of radiation to the frequency of the oscillator that emits it. The relationship is $E = h\nu$, where E is the energy of the quantum and ν is its frequency. Its value is 6.62559×10^{-34} joule second. It was named after the German physicist Max Planck (1858–1947).

Planck's radiation law (physics) An object cannot emit or absorb energy, in the form of radiation, in a continuous manner; the energy can be taken up or given out only as integral multiples of a definite amount, known as a quantum. Also termed: Planck's law of radiation.

Planodiscidae (insect parasitology) Mites in the Meso-stigmata that are phoretic on legs of the ants *Eciton*.

plasma (1 physics; 2 haematology) 1 State of matter in which the atoms or molecules of a substance are broken into electrons and positive ions. All substances pass into this state of matter when heated to a very high temperature, *e.g.* in an electric arc or in the interior of a star. 2 Colourless (*USA*: colorless) fluid portion of blood or lymph from which all cells have been removed.

plasma cell (cytology) Large egg-shaped cell with granular, basophilic cytoplasm except for a clear area around the small eccentrically placed nucleus. Its function is believed to be antibody synthesis. Also termed: plasmacyte.

plasma membrane (histology) Thin layer of tissue consisting of fat and protein that forms a boundary surrounding the cytoplasm of eukaryotic cells and its organelles. It is a differentially permeable membrane that separates adjacent cells and cavities. Also termed: cytoplasmic membrane.

plasma pepsinogen (biochemistry) The presence of the zymogen, pepsinogen in the plasma, high levels of which are indicative of extensive mucosal damage in the abomasum, as in ostertagiasis in ruminants.

plasma protein (haematology/biochemistry) Protein in the plasma of blood (*e.g.* antibodies and various hormones).

plasmasol (cytology) An alternative term for endoplasm.

plaster of paris (chemistry) $CaSO_4.1/2H_2O$ Calcium sulphate hemihydrate (*USA*: calcium sulfate hemihydrate), obtained by heating gypsum. When water is added, it sets hard, re-forming gypsum. In doing so, it does not expand or contract much, and is therefore valuable as a moulding (*USA*: molding) material, particularly as a splint for broken bones and in the building industry.

platelet (haematology) Small non-nucleated oval or round fragment of cells from the red bone marrow found in mammalian blood. There are about 250,000 to 400,000 per mm^3 in human blood, which are required to initiate blood clotting by disintegrating and liberating thrombokinase. In some vertebrates platelets are represented by thrombocytes, which are small spindle-shaped nucleated cells.

platinum (Pt) (chemistry) Valuable silver-white metallic element in Group VIII of the periodic table (a transition element). It is used for making jewellery, electrical contacts and in scientific apparatus. At. no. 78; r.a.m. 195.09.

Platyhelminthes (fish parasitology) A phylum of flatworms. Also termed: flatworms. There are at least eight different schemes of classification of the phylum Platyhelminthes. Although the classification is controversial, a division of the phylum has been made into the Turbellaria, Digenea, Monogenea, Gyro-cotylidea, Amphilinidea, Caryophyllidea and Cestoda. *See* Turbellaria; Digenea; Monogenea; Gyrocotylidea; Amphilinidea; Caryophyllidea; Cestoda.

platymyarian (parasitological cytology) A term used in relation to nematodes. Somatic cells in which the muscle fibres lie only perpendicular to the hypodermis.

PLC (biochemistry) An abbreviation of 'P'hospho-'L'ipase 'C'.

pleomorphic (parasitological physiology) When a number of morphological types occur in one life cycle.

plerocercoid (parasitological physiology) The third stage larva of pseudophyllidea which has a solid body.

pleura (anatomy) Double membrane that covers the lungs and lines the chest cavity, with fluid between the membranes. Also termed: pleural membranes.

pleural (anatomy) To do with the lungs.

pleuron (parasitological anatomy) A term used in relation to arthropods. A lateral part of a segment.

plexus (anatomy) Network of interlacing nerves.

Plistophora (insect parasitology) Protozoal genus of microsporidia that are parasites to the termites *Odontotermes*, *Coptotermes* and *Reticulitermes*. *See Plistophora ganaptii*; *Plistophora weiseri*.

Plistophora ganaptii (insect parasitology) A protozoal species of microsporidia that, similarly to *Plistophora weiseri*, are parasites to the termites *Odontotermes*, *Coptotermes* and *Reticulitermes*. They may be found in the epithelial cells of the foregut or Malpighian tubules.

Plistophora weiseri (insect parasitology) A protozoal species of microsporidia that, similarly to *Plistophora ganaptii*, are parasites to the termites *Odontotermes*, *Coptotermes* and *Reticulitermes*. They may be found in the epithelial cells of the foregut or Malpighian tubules.

plumbic (chemistry) An alternative term for lead(IV).

plumbous (chemistry) An alternative term for lead(II).

plutonium (Pu) (chemistry) Radioactive element in Group IIIB of the periodic table (one of the actinides), produced from uranium-238 in a breeder reactor. It has several isotopes (with half-lives of up to 76 million years), some of which (*e.g.* Pu-239) undergo nuclear fission; all are very poisonous. At. no. 94; r.a.m. 244 (most stable isotope).

PMN (haematology) An abbreviation of 'P'oly-'M'orphonuclear 'N'eutrophilic leukocyte.

pneumatic (physics) operated by air pressure.

PNMT (biochemistry) An abbreviation of 'P'henyl-ethanolamine-'N'-'M'ethyl'T'ransferase.

poikilothermic (biology/physiology) Describing an animal that is cold-blooded and relies on the heat of the environment to warm its body; *e.g.* among vertebrates, fish, amphibians and reptiles.

See also homoiothermic.

poison (1 chemistry; 2 toxicology) 1 Substance that destroys the activity of a catalyst. 2 Substance that when introduced into a living organism in any way destroys life or causes injury to health; a toxin.

polar bond (chemistry) Covalent bond in which the bonding electrons are not shared equally between the two atoms.

polar molecule (chemistry) Molecule that is polarized even in the absence of an electric field. *See* polarization.

polarimeter (physics) Instrument for measuring the optical activity of a substance. Also termed: polariscope.

polarimetry (measurement) Measurement of optical activity, used in chemical analysis.

polarity (physics) Designation that something is either positively or negatively charged (*e.g.* the cathode in an electrolytic cell has negative polarity).

polarization (1 chemistry; 2,3 physics) 1 Separation of the positive and negative charges of a molecule. 2 Lining up of the electric and magnetic fields of an electromagnetic wave, *e.g.* as in polarized light. Only transverse waves can be polarized. 3 Formation of gas bubbles or a film of deposit on an electrode of an electrolytic cell, which tends to impede the flow of current.

polarized light (optics) Light waves (which normally oscillate in all possible planes) with fixed orientation of the electric and magnetic fields. It may be created by passing the light through a polarizer consisting of a plate of tourmaline crystal cut in a special way, through a Nicol prism or by using a Polaroid sheet. Substances that are optically active have the property of rotating the plane of polarized light.

polarography (chemistry) Method of chemical analysis for substances in dilute solution in which current is measured as a function of potential between mercury electrodes in an electrolytic cell containing the solution.

polaroid (photography) Trade name for a thin transparent film that produces plane polarized light when light is passed through it.

pole (physics) Electrode (particularly of a battery).

pollen (biology) Dust-like microspore of a seed plant produced by microsporangium cones in gymnosperms and by anthers in angiosperms. Each grain contains male gametes. If these gametes are carried by an external agent such as wind, insects or water to the ovules of gymnosperms or to the stigma of angiosperms, they produce fertilization. In susceptible (*i.e.* sensitive) people, pollen can be a powerful antigen that results in a vigorous allergic response. Also termed: farina. *See* allergy.

polonium (Po) (chemistry) Radioactive metallic element in Group VIA of the periodic table, used as a source of alpha-particles. It has 27 isotopes, more than any other element, with half-lives of up to 100 years. At. no. 84; r.a.m. 209 (most stable isotope).

poly- (general terminology) Prefix meaning many (*e.g.* polychaete, polygon).

polyamide (chemistry) Condensation polymer in which the units are linked by amide groups (- CONH -); *e.g.* proteins, hair, wool fibres (*USA*: wool fibers), nylon.

polybasic (chemistry) Describing an acid with two or more acidic (replaceable) hydrogen atoms in its molecule; *e.g.* phosphorus(V) (orthophosphoric) acid, H_3PO_4, with three replaceable hydrogens, is tribasic.

polycyclic (chemistry) Describing a substance that has more than one ring of atoms in its molecule.

polyhydric (chemistry) Containing a number of hydroxyl groups.

polyhydric alcohol (chemistry) Alcohol that contains three or more hydroxyl groups.

polymer (chemistry) Long-chain molecule built up of a number of smaller molecules called monomers, joined together by polymerization. Natural polymers include starch, cellulose and rubber. Synthetic polymers include all kinds of plastics.

polymerization (chemistry) Process of joining together of small molecules, called monomers, to form larger molecules, polymers (often in the presence of a catalyst). In condensation polymerization, two types of monomer molecules condense to form long chains, with the elimination of a small molecule (such as water). In addition to polymerization, long chains are formed by molecules of a single monomer joining together.

polymethanal (chemistry) Polymer formed from methanal (formaldehyde). Also termed: paraformaldehyde.

polymethyl methacrylate (chemistry/optics) Transparent colourless (*USA*: colorless) thermoplastic. Its optical properties of high transmission of light and high internal reflection, coupled with great strength, are responsible for its use in place of glass. Also termed: Perspex.

polymorphism (1 chemistry; 2 biology) 1 Occurrence of a substance in more than one crystalline form. 2 Occurrence of an organism in more than one structural form during its life cycle.

polymyarian (parasitological cytology) A nematode in which many muscle cells are seen between chords *e.g.* *Ascaris*, filarial worms and *Dracunculus*.

polynomial (mathematics) Algebraic expression with only one variable.

polynucleotide (chemistry) Polymer of many nucleotides.

Polyopisthocotylida (fish parasitology) An order of flatworms in class Monogenea and the phylum Platyhelminthes. Three suborders have been recognised: the Hexabothriidea on elasmobranchs, of which *Squalonchocotyle* is an example; the Mazocraeidea on teleosts, of which *Winkenthughesia*, *Microcotyle* and *Discocotyle* are examples; the Diclyobothriidea on Acipensiformes, of which *Diclybothrium* is an example.

polypeptide (chemistry) Chain of amino acids which is a basic constituent of proteins. It may be broken down by enzyme action (digestion) to form peptides. The functionally significant linking and folding of polypeptides makes up the three-dimensional structure of a protein.

polyploidy (genetics) Condition in which a cell or organism has three to four times the normal haploid or gametic number. It is often made use of in plant breeding because it results in the production of larger and more vigorous crops. Because it disturbs the sex-determining mechanism, polyploidy is rare in animals, and would result in sterility.

polypus Pendulous but usually benign tumour (*USA*: benign tumor) that grows from mucous membrane (*e.g.* in the nose or womb). Also termed: polyp.

polysaccharide (chemistry) High molecular weight carbohydrate, linked by glycoside bonds, that yields a large number of monosaccharide molecules (*e.g.* simple sugars) on hydrolysis or enzyme action. The most common polysaccharides have the general formula $(C_6H_{10}O_5)_n$; *e.g.* starch, cellulose, etc.

polyvalent (1, 2 chemistry) 1 Having a valence of more than one. 2 Having more than one valency.

POMC (biochemistry) An abbreviation of 'P'ro-'O'pio'M'elano'C'ortin.

population density (epidemiology) The number of animals per unit of area, which is important in relation to the rate of spread of disease.

population genetics (genetics) Study of the theoretical and experimental consequence of Mendelian inheritance on population levels, taking into account the genotypes, phenotypes, gene frequencies and the mating systems.

population inversion (chemistry) Condition in which a higher energy state in an atomic system is more populated with electrons than a lower energy state.

pork tapeworm (parasitology) *See Taenia solium.*

Porocephalus (veterinary parasitology) A genus of internal parasites in the class Pentastomida and the family Porocephalidae. Species includes *Porocephalus clavatus*, *Porocephalus crotali* and *Porocephalus subulifer*. The adults of these wormlike arthropods may be found in the respiratory passages of large snakes where heavy infestations may kill the snake.

porosity (physics) Property of substance that allows gases or liquids to pass through it.

porphyrin (chemistry) Member of an important class of naturally occurring organic pigments derived from four pyrrole rings. Many form complexes with metal ions, as in *e.g.* chlorophyll, haem (*USA*: heme), cytochrome, etc.

Porrocaecum (parasitology) A genus of roundworms in the family Ascarididae.

Porrocaecum angusticolle (avian parasitology) A species of roundworms in the genus *Porrocaecum* that, similarly to *Porrocaecum depressum*, may be found in birds of prey.

Porrocaecum aridae (avian parasitology) A species of roundworms in the genus *Porrocaecum* that may be found in herons.

Porrocaecum crassum (avian parasitology) A species of roundworms in the genus *Porrocaecum* which may be found in the intestine of ducks, and that may cause nasal discharge, diarrhoea (*USA*: diarrhea), anaemia (*USA*: anemia) and loss of weight.

Porrocaecum decipiens (veterinary parasitology) A species of roundworms in the genus *Porrocaecum* that may be found in the small intestine of walruses.

Porrocaecum depressum (avian parasitology) A species of roundworms in the genus *Porrocaecum* that, similarly to *Porrocaecum angusticolle*, may be found in birds of prey.

portal vein (anatomy) Any vein connecting two capillary networks, thus allowing for blood regulation from one network by the others *e.g.* the hepatic portal vein connects the intestine with the liver.

positive (1 mathematics; 2 physics) 1 Describing a number or quantity greater than zero. 2 Describing an electric charge or ion that is attracted by a negative one.

positive feedback (general terminology) Feedback in which the output adds to the input.

positive predictive value (statistics) The probability that a patient with a positive test result really does have the condition for which the test was conducted.

positron (physics) Elementary particle which has a mass equal to that of an electron, and an electrical charge equal in magnitude, but opposite in sign, to that of the electron.

posterior (biology) In bilaterally symmetrical animals, the end of the body directed backwards during locomotion; the rear or hind end. In bipedal animals (*e.g.* human beings), it corresponds to the dorsal side of quadrupeds.

posterior odds (statistics) Probability determined after consideration of the results of a study.

post-matum (medical) Over-developed; used of an infant born after its time.

potash (chemistry) Substance that contains potassium, particularly potassium carbonate.

potassium bicarbonate (chemistry) An alternative term for potassium hydrogenearbonate.

potassium bromide (chemistry) KBr White crystalline salt, used in medicine and photography.

potassium carbonate (chemistry) K_2CO_3 White granular solid, used in the manufacture of glass and soap. Also termed: potash.

potassium chloride (chemistry) KCl Colourless (*USA*: colorless) or white crystalline salt, used as a fertilizer and as a dietary salt (sodium chloride) substitute when sodium intake must be limited.

potassium cyanide (chemistry) KCN White poisonous solid, used in metallurgy and electroplating.

potassium ferricyanide (chemistry) $K_3Fe(CN)_6$ Red crystals, used as a chemical reagent and in the manufacture of pigments (*e.g.* Prussian blue). An alternative term: potassium hexacyanoferrate(III).

potassium ferrocyanide (chemistry) $K_4Fe(CN)_6.3H_2O$ Yellow crystals. Also termed: potassium hexacyanofer-rate(II).

potassium hydrogencarbonate (chemistry) $KHCO_3$ White granular solid, used in pharmaceuticals. Also termed: potassium bicarbonate.

potassium hydrogentartrate (chemistry) HOOC-$(CHOH)_2$ COOK White crystalline powder, used in baking powder. Also termed: cream of tartar.

potassium hydroxide (chemistry) KOH Strongly hygroscopic white solid. A strong alkali, it is used in the manufacture of soft soaps, and may be used for clearing skin scrapings in the diagnosis of ectoparasite infestation. Also termed: caustic potash.

potassium iodide (chemistry) KI Colourless (*USA*: colorless) crystalline salt, used in chemical analysis and organic synthesis. Its solution dissolves iodine.

potassium (K) (chemistry) Highly reactive silver-white metallic element in Group IA of the periodic table (the alkali metals). Its compounds occur widely (particularly the chloride) and have many uses; potassium is an essential nutrient for plants. The metal is used as a coolant in nuclear reactors. At. no. 19; r.a.m. 39.102.

potassium manganate(VII) (chemistry) An alternative term for potassium permanganate.

potassium nitrate (chemistry) KNO_3 Colourless (*USA*: colorless) crystalline salt, a powerful oxidizing agent. It is used in the manufacture of glass and explosives, and as a food preservative. Also termed: saltpetre; nitre.

potassium permanganate (chemistry) $KMnO_4$ Purple crystals, a powerful oxidizing agent. It is used in the manufacture of chemicals, as a disinfectant and fungicide, and in volumetric analysis. Also termed: permanganate of potash; potassium manganate(VII).

potassium thiocyanate (chemistry) KSCN Colourless (*USA*: colorless) hygroscopic solid, used in solution to test for iron(III) (ferric) compounds, which give a blood-red colour (*USA*: color).

potbellied (parasitology) Abnormal relative enlargement of the abdomen which may be caused due to parasitism.

potency (pharmacology) The strength of a drug based on its effectiveness to cause change.

potential (physics) An alternative term for potential difference or voltage.

potential difference (pd) (physics) Difference in electric potential between two points in a current-carrying circuit, usually expressed in volts (V). Also termed: voltage.

potentiometer (1, 2 physics) 1 Instrument for measuring potential difference or electromotive force. 2 Voltage divider.

potentiometric titration (chemistry) Method of quantitative analysis, using titration, that involves the measurement of changes in electrode potential of an electrode dipping into a solution.

Poteriostomum (parasitology) A genus of strongylid roundworms in the subfamily of Cyathostominae.

poultry flea (avian parasitology) *Echidnophaga gallinacea*.

poultry mite (avian parasitology) *Dermanyssus gallinae*.

pound (1 lb; 2 £) (1, 2 measurement) 1 Unit of weight equal to 16 ounces; 14 lb 1 stone, 112 lb = 1 hundredweight (cwt). 1 lb = 0.4536 kg. 2 Unit of currency equal to 100 pence.

pour-on (control measures) A technique for the application of insecticides, in which a small amount of liquid is poured onto an animal's back without any attempt to spread it over the surface. It is suited to systemic insecticides which are absorbed percutaneously.

power (1 physics; 2 optics; 3 mathematics) 1 Rate of doing work. The SI unit of power is the watt (equal to 10^7 erg s^{-1} or 1/745.7 horsepower). 2 The extent to which a curved mirror, lens or optical instrument can magnify an object. For a simple lens, power is expressed in dioptres. 3 An exponent or index, written as a small superior numeral; *e.g.* 3^2 is 3 to the power 2 (or 3 squared) = 9; 4^3 is 4 to the power 3 (or 4 cubed) = 256; x^5 is x to the power 5 (or x to the 5th).

power down (computing) Sequence of steps taken by a computer when power is shut off. The objective is to preserve the status of the computer and avoid damage to the peripheral appliances.

PPLO (biology) An abbreviation of 'P'leuro'P'neumonia-'L'ike 'O'rganisms.

Pr (chemistry) Chemical symbol, praseodymium.

PRA (physiology/biochemistry) An abbreviation of 'P'lasma 'R'enin 'A'ctivity.

prallethrin (control measures) A pyrethroid insecticide used to control fleas and ticks on dogs and cats.

praseodymium (Pr) (chemistry) Metallic element in Group IIIB of the periodic table (one of the lanthanides). At. no. 59; r.a.m. 140.9077.

Pratylenchus (plant parasitology) A genus of small (adults usually less than 1 mm long) nematodes in the order Tylenchida and family Pratylenchidae. *See Pratylenchus penetrans*.

Pratylenchus penetrans (plant parasitology) A species of nematodes in the genus *Paratylenchus* whose hosts cover a wide range with over 350 hosts and mainly found in temperate areas in the USA, Europe, Australia, Canada, Egypt, India, Japan, New Zealand, Peru, Philippines, Rhodesia, Russia, South Africa, Tunisia. Apple, cherry and other fruit trees; conifers; rose bushes; and tomato, potato, corn, sugarbeet, and many other plants are parasitized by this nematode. They are migratory endoparasites whose reproduction is sexual. Females lay eggs singly in roots or in soil. Second-stage juveniles hatch from eggs, feed, and undergo three moults (*USA*: molts) to the adult stage. The complete life cycle takes 30 to 86 days, depending on temperature, and is shortest at 30°C. If conditions in the root become unfavorable, any juveniles or adult nematodes may leave the root and invade other nearby roots. Invasion usually takes place in the region of elongation. *Pratylenchus penetrans* nematodes feed upon cells in the root cortex. Feeding results in cells being killed and, in many instances, small roots are killed also. The migratory parasitism of this nematode opens up roots to secondary invasion by other soil microorganisms such as fungi and bacteria. In some hosts such as walnut and stone fruits the nematodes may establish colonies in larger roots and large necrotic lesions are formed. Infested plants grow poorly, the foliage is frequently chlorotic and sparse and the terminal growth of branches may be suppressed, leading to reduction in crop yields. Seedlings planted in infested soil frequently fail to grow to normal size and may never produce satisfactory trees or vines. Control measures include the use of preplant nematicides, but because of its wide host range, crop rotation is usually not feasible. Also termed: lesion nematode; root-lesion nematode.

praziquantel (pharmacology) A widely used cestocide in dogs and cats that has a very high efficiency against *Echinococcus granulosus*, removing all of the worms with a single dose. Also termed Droncit.

PRC (biochemistry) An abbreviation of 'P'lasma 'R'enin 'C'oncentration.

pre- (terminology) A word element. [Latin] Prefix denoting *before*.

precipitate (chemistry) Solid that forms in and settles out from a solution.

precipitated sulphur (pharmacology) A scabicide and antiparasitic substance. (*USA*: precipitated sulfur). Also termed milk of sulphur (*USA*: sulfur).

precipitating cause (epidemiology) The trigger mechanism that initiates the commencement of the disease state.

precipitation (chemistry) Process of precipitate formation. *See* double decomposition.

preclinical (medical) Before a disease becomes clinically recognizable.

precursor (chemistry) Intermediate substance from which another is formed in a chemical reaction.

precyst (parasitological physiology) A term used in relation to amoebae (*USA*: amebae). The rounded form of trophozoite which precedes the cystic stage. It differs from the cyst in not having a cyst wall.

predilection host (parasitology) The host preferred by a parasite.

predisposing cause (parasitology) A mechanism that makes a patient more susceptible to the precipitating cause.

pregnancy (physiology) Time that elapses between fertilization or implantation of a fertilized ovum and an animal's birth; the time that an animal spends as an embryo or foetus (*USA*: fetus). Also termed: gestation. The time that an embryo reptile or bird spends in an egg between laying and hatching is usually termed incubation.

Premicrodispus (insect parasitology) Mites in the Prostigmata that may be phoretic on *Reticulitermes* and other various termites.

prepatent period (parasitology) A term used in relation to malaria. The minimum time between the entry of sporozoites and the first appearance of parasites in red cells.

presby- (terminology) A word element. [German] Prefix denoting *old age*.

presenting signs (medical) The signs or group of signs about which the patient complains or from which relief is sought.

pressure (p) (physics) Force applied to, or distributed over, a surface. Measured as force f per unit area a; $p = f/a$. At a depth d in a liquid, the pressure is given by $p = \varrho gd$, where ϱ is the liquid's density, d the depth and g is the acceleration of free fall. The SI unit of pressure is the pascal; other units include bars, millibars, atmospheres and millimetres of mercury.

pressure gauge (physics) Device for measuring fluid pressure (*e.g.* barometer, manometer).

prevalent (epidemiology) Widespread occurrence.

Pricetrema zalophi (veterinary parasitology) A liver fluke found in fur seals in large numbers without causing apparent clinical illness.

primaquine (pharmacology) An antiprotozoal agent used as the phosphate in the treatment of theileriosis, babesiosis, leishmaniasis and trypanosomiasis.

primary alcohol /amine (chemistry) alcohol or amine with only one alkyl group or aryl group.

primary cause (epidemiology) The principal factor in causing the disease.

primary cell (physics) Electrolytic cell (battery) in which the chemical reactions that cause the current flow are not readily reversible and the cell cannot easily be recharged, *e.g.* a dry cell.

See also secondary cell.

primary colour (1, 2 physics) 1 Red, green and violet, which give all other colours (*USA*: colors) when light producing them is combined in various proportions. All three mix to give white. 2 Pigment colours red, yellow and blue, which can also be combined to give pigments of all other colours. All three mix to give black. (*USA*: primary color).

primary conjunctivitis (medical) Inflammation of the conjunctiva caused by infectious agents, parasites or toxic agents in the first instance.

primary host (parasitology) Definitive host.

primary myocarditis (medical) A condition that is usually the result of a primary viral or protozoal infection of the myocardium.

prime number (mathematics) Integer (whole number) divisible only by itself and 1. All prime numbers except 2 are odd.

print queue (computing) A list of print jobs waiting in line at a printer.

printer (computing) Computer output device that produces hard copy as a printout. There are various kinds, including (in order of speed) daisy-wheel, dot-matrix, line, barrel and laser.

printer driver (computing) Software connecting the computer to the printer.

print-out (computing) Output (hard copy) from a computer printer.

Pristionchus lheritieri (insect parasitology) A species of nematode in the Rhabditidae that are parasites to the ants *Lasius* spp. They may be found in the head glands.

pro- (terminology) A word element. [Latin, German] Prefix denoting *before, in front of, favouring* (*USA*: favoring).

probability distribution of electrons (physics) Probability that an electron within an atom will be at a certain point in space at a given time. It predicts the shape of an atomic orbital.

probability, mathematical (mathematics) Expression of the extent to which an event is likely to occur, given a value between 0 (an impossibility) and 1 (a certainty); *e.g.* the probability of getting a 6 on one roll of a dice is 1/6 or 0.16666.

proboscis (biology) Tube-like organ of varying form and functions. In insects, a proboscis is a filamentous structure that projects outwards from the mouthparts, functioning as a piercing and sucking device for obtaining liquid food. In elephants, the proboscis is the trunk, and in some marine animals it is a tube-like pharynx that can be protruded.

Probstmayria (parasitology) A genus of roundworms in the family Kathlaniidae.

Probstmayria vivipara (veterinary parasitology) A species of minute worm in the genus *Probstmayria* that may be found in the colon of horses, often in enormous numbers but without apparent pathogenic effect.

procercoid (parasitological physiology) The first larval stage in the life cycle of a pseudophyllidean cestode,*e.g. Spirocerca erinacei*. The procercoid is a solid bodied stage with oncospheral hooks carried on the cercomer in the posterior region.

processing (methods) Exposure to a set of processes.

processor (computing) *See* computer.

proct(o)- (terminology) A word element. [German] Prefix denoting *rectum*.

Proctolaelaps (insect parasitology) A species of mites in the Mesostigmata, that are cleptoparasitic on the bees Bombinae and *Apis*.

pro-drug (pharmacology) A compound that, on administration, must undergo chemical conversion by metabolic processes before becoming an active pharmacological agent; a precursor of a drug.

product (1 mathematics; 2 chemistry) 1 The result of multiplying numbers together. 2 A substance formed as a result of a chemical change.

Prodynchius (insect parasitology) *See Dinychus.*

proenzyme (biochemistry) An alternative term for zymogen.

professional (education) 1 Pertaining to one's profession or occupation. 2 One whose income is derived from the practice of his/her profession.

professionalism (education) The upholding by individuals of the principles, laws, ethics and conventions of their profession.

Progamotoenia (veterinary parasitology) A genus of anoplocephalid cestodes that may be found in the bile ducts of kangaroos, wallabies and wombats.

progesterone (biochemistry) $C_{21}H_{30}O_2$ Steroid sex hormone secreted by the corpus luteum of the mammalian ovary, placenta, testes and adrenal cortex. In females it prepares the uterus for the implantation of a fertilized ovum (egg) and during pregnancy maintains nourishment for the embryo by developing the placenta, inhibiting ovulation and menstruation, and stimulating the growth of the mammary glands.

proglottid (parasitological anatomy) One of the segments making up the body of a tapeworm. Also termed: proglottis.

prognosis (medical) A forecast of the probable course and outcome of an attack of disease and the prospects of recovery as indicated by the nature of the disease and the clinical signs of the case.

program (computing) Sequence of instructions for a computer. Also termed: programme.

programme loop (computing) In a computer programme (*USA*: program) a sequence of commands which is repeated until a specified condition is reached.

progression (mathematics) Mathematical series of terms.

prohemistomiasis (human/veterinary parasitology) Disease caused by infection with *Prohemistomum vivax* which may occur in humans, dogs and cats in Israel, North Africa and Romania.

Prohemistomum vivax (human/veterinary parasitology) A species of digenetic trematode in the genus *Prohemistomum* that may be found in the intestines of humans, cats and dogs.

projectile (general terminology) Something thrown forward.

prokaryon (cytology) 1 Nuclear material scattered in the cytoplasm of the cell, rather than bounded by a nuclear membrane; found in some unicellular organisms. 2 Prokaryote.

prokaryote (biology) DNA-containing, single-celled organism with no proper nucleus or endoplasmic reticulum; *e.g.* bacteria, blue-green algae (cyanophytes).

See also eukaryote.

prokaryotic (biology) Describing or resembling a prokaryote.

prokaryotic cell (biology) *See* prokaryote.

prokinetics (pharmacology) Drugs which enhance the passage of intraluminal contents of the gastrointestinal tract.

prolactin (biochemistry) Protein hormone secreted by the anterior pituitary. In mammals it stimulates lactation and promotes functional activity of the corpus luteum. Also termed: luteotrophin; mammary stimulating hormone; mammogen hormone; mammotrophin.

proliferative kidney disease (fish parasitology) An important disease of salmonids, that is caused by a pre-spore stage of a myxosporean species presumably carried by other fish inhabiting the same waters.

proline (biochemistry) White crystalline amino acid that occurs in most proteins.

promastigote (parasitological physiology) One of the morphological stages in the development of certain protozoa, characterized by a free anterior flagellum and the kinetoplast and axoneme at the anterior end of the body. There is no undulating membrane. They are found in arthropods and plants. Also termed: 'leptomonad' stage.

promecarb (control measures) A carbamate pesticide.

promethium (Pm) (chemistry) Radioactive metallic element in Group IIIB of the periodic table (one of the lanthanides). It has several isotopes (none of which occurs naturally), with half-lives of up to 20 years. At. no. 61; r.a.m. 145 (most stable isotope).

promoter (chemistry) Substance used to enhance the efficiency of a catalyst. Also termed: activator.

prompt (computing) A message which indicates that the computer requires further information from the user before it can proceed.

pronotal comb (parasitological anatomy) Conspicuous spines seen on the pronotum of some fleas.

pronotum (parasitological anatomy) A term used in relation to arthropods. The dorsal surface of the first thoracic segment.

proof (measurement) Measure of the ethanol (ethyl alcohol) content of a solution (gunpowder moistened with a 100% proof spirit will just ignite). A 100% proof solution is 57. 1% ethanol by volume or 49.3% ethanol by weight.

propagation (biology) Reproduction.

propagative transmission (parasitology) Transfer, as of an infection from one patient to another, in which the agent multiplies in the transmission vehicle.

propanoic acid (chemistry) CH_3CH_2COOH Liquid carboxylic acid. Also termed: propionic acid.

propanol (chemistry) Alcohol that occurs as two isomers. 1 *n*-propanol C_3H_7OH is a colourless (*USA*: colorless) liquid, used as a solvent and in making toilet preparations. Also termed: n-propyl alcohol; propan-l-ol. 2 Isopropanol $(CH_3)_2CHOH$ is also a colourless liquid, used for preparing esters, acetone (propanone), and as a solvent. Also termed: isopropyl alcohol; propan-2-ol.

2-propanone (chemistry) An alternative term for acetone.

propellant (physics) Gas used in an aerosol to expel the contents through an atomizing jet.

2-propenal (chemistry) An alternative term for acrolein.

propene (chemistry) $CH_3CH = CH_2$ Colourless (*USA*: colorless) gaseous alkene (olefin), used in industry for the preparation of isopropanol, glycerol, polypropene, etc. Also termed: propylene.

propenoic acid (chemistry) An alternative term for acrylic acid.

propenonitrile (chemistry) An alternative term for acrylonitrile.

proper fraction (mathematics) Fraction whose numerator is less than its denominator (a fraction whose numerator is larger than its denominator is an improper fraction).

prophase (genetics) First stage of cell division in meiosis and mitosis. During prophase chromosomes can be seen to thicken and shorten and to be composed of chromatids. The spindle is assembled for division of chromosomes and the nuclear membrane disintegrates. In meiosis the first prophase is extended into several stages.

propionaldehyde (chemistry) An alternative term for propanal.

propionic acid (chemistry) An alternative term for propanoic acid.

propoxur (control measures) A carbamate insecticide used widely on companion animals to control ectoparasites.

pros(o)- (terminology) A word element. [German] Prefix denoting *forward, anterior.*

Prosarcoptes (veterinary parasitology) A genus of mange mites in the family Sarcoptidae. Species includes *Prosarcoptes faini, Prosarcoptes pitheci* which may be found on monkeys.

Prosobothrioidea (fish parasitology) An order of flatworms in subclass Eucestoda, class Cestoidea and the phylum Platyhelminthes. *Prosobothrium armigerum* is the only representative of the Prosobothrioidea. Adults are found only in the intestine of elasmobranchs. The four anteriorly-directed glandular suckers of the bothridia and a characteristic spiny neck distinguish them from the Tetraphyllidae and Proteocephalidae.

prosodemic (parasitology) Passing directly from one animal to another instead of reaching a large number at once, through such means as water supply; said of a disease progressing in that way.

prosop(o)- (terminology) A word element. [German] Prefix denoting *face.*

Prosostomata (fish parasitology) An order of flatworms in subclass Digenea, class Trematoda and the phylum Platyhelminthes. The Prosostomata are Digenea with a terminal or subterminal mouth.

prospective experiment (epidemiology) An experiment carried out to see what happens if certain influences are applied to an animal or a group of animals.

prospective study (epidemiology) A scholarly examination in which the data to be studied are yet to be generated, the events having not yet occurred.

prostaglandin (biochemistry) Member of a group of unsaturated fatty acids that contain 20 carbon atoms. They are found in all human tissue, and particularly high concentrations occur in semen. Their activities affect the nervous system, circulation, female reproductive organs and metabolism. Most prostaglandins are secreted locally and are rapidly metabolized by enzymes in the tissue.

prostate gland (anatomy) Gland located at the base of the urinary bladder that forms part of the male reproductive system. The size of the gland and the quantity of its secretion are controlled by androgens. Its function is secretion of a fluid containing enzymes and antiglutinating factor, which contributes to the production of semen.

Prosthenorchis (veterinary parasitology) A genus of parasites in the phylum Acanthocephala. Species includes *Prosthenorchis elegans* and *Prosthenorchis spirula* (synonym: *Prosthenorchis sigmoides*) that may be found in the terminal small intestine, colon and caecum (*USA*: cecum) of monkeys. Heavy infestations cause diarrhoea (*USA*: diarrhea), dehydration and death. There are yellow nodules on the serosal surface of the intestine which is swollen. There may be obstruction of the ileocaecal (*USA*: ileocecal) valve and perforation of the intestinal wall.

prosthetic group (biochemistry) Non-protein portion of a conjugated protein, *e.g.* haem (*USA*: heme) group in haemoglobin (*USA*: hemaglobin).

Prosthogonimus (avian/veterinary parasitology) A genus of flukes (digenetic trematodes) in the family Prosthogonimidae which parasitize birds but are found occasionally in mammals.

Prosthogonimus anatinus (avian parasitology) A species of flukes in the genus *Prosthogonimus* that may be found in domestic ducks.

Prosthogonimus cuneatus (avian parasitology) A species of flukes in the genus *Prosthogonimus* that may be found in swans.

Prosthogonimus macrorchis (avian parasitology) A species of flukes in the genus *Prosthogonimus* that, similarly to *Prosthogonimus ovatus*, may be found in the bursa of Fabricius and oviduct of domestic poultry and wild birds.***Prosthogonimus ovatus*** (avian parasitology) A species of flukes in the genus - *Prosthogonimus* that, similarly to *Prosthogonimus macrorchis*, may be found in the bursa of Fabricius and oviduct of domestic poultry and wild birds.

Prosthogonimus oviformis (avian parasitology) A species of flukes in the genus *Prosthogonimus* that may be found in ducks.

Prosthogonimus pellucidus (avian parasitology) A species of flukes in the genus *Prosthogonimus* that may be found in the bursa of Fabricius, oviduct and posterior intestine of fowl, ducks and wild birds.

Prosthorhyncus formosus (avian parasitology) An acanthocephalan parasite found in the small intestine of domestic fowl and some wild bird species, but that has doubtful pathogenicity.

Prot- (biochemistry) An abbreviation of 'Prot'ein anion.

protactinium (Pa) (chemistry) Radioactive element in Group IIIB of the periodic table (one of the actinides). It has several isotopes, with half-lives of up to 20,000 years. At. no. 91; r.a.m. 231.0319.

protean (biology) Changing form or assuming different shapes.

protease (biochemistry) Enzyme that breaks down protein into its constituent peptides and amino acids by breaking peptide linkages (*e.g.* pepsin, trypsin).

protected (biology) Not generally available; surrounded by some mechanism that prevents access.

protected field (computing) Information on the computer screen which cannot be overwritten.

protein (biochemistry) Member of a class of high molecular weight polymers composed of a variety of amino acids joined by peptide linkages. In conjugated proteins, the amino acids are joined to other groups. Proteins are extremely important in the physiological structure and functioning of all living organisms.

protein synthesis (biochemistry) Process by which proteins are made in cells. A molecule of messenger RNA decodes the sequence of copied DNA on ribosomes in the cytoplasm. A polypeptide chain is generated by the linking of amino acids in an order instructed by the base sequence of messenger RNA.

Proteocephalidea (fish parasitology) An order of flatworms in subclass Eucestoda, class Cestoidea and the phylum Platyhelminthes. Proteocephalideans are very common in freshwater teleost fish but may also be found in amphibians and reptiles. The scolex has four suckers and occasionally an apical sucker or armed rostellum.

Proteocephalus (parasitology) A genus of tapeworms in the order Proteocephalidea.

Proteocephalus ambloplitis (fish parasitology) A species of tapeworms in the genus *Proteocephalus* that may be found in cultured fish, especially bass. The parasite affects the gonads and causes sterility. Also termed: bass tapeworm.

Proteromonas (veterinary parasitology) Flagellate protozoa that may be found in the intestines of all lizards, but that is considered nonpathogenic.

prothorax (parasitological anatomy) A term used in relation to arthropods. The anterior segment of thorax.

protist (biology) Member of the Protista.

Protista (taxonomy) Kingdom that contains simple organisms such as algae, bacteria, fungi and Protozoa, although sometimes multicellular organisms are excluded.

proto- (terminology) A word element [German] Prefix denoting *first*.

Protolepsis tesselata (avian parasitology) A species of leech that commonly parasitizes the nasal cavity of aquatic birds.

proton (physics) Fundamental elementary particle with a positive charge equal in magnitude to the negative electron charge, and with a rest mass of 1.67252×10^{-27} kg (about 1,850 times that of an electron). Protons are constituents of the nucleus in every kind of atom.

proton number (chemistry) An alternative term for atomic number.

proton pump (physiology/biochemistry) The enzyme hydrogen-potassium ATP-ase which is involved in the final stage of acid production in the parietal cells.

protonic acid (chemistry) Compound that releases solvated hydrogen ions in a suitable polar solvent (*e.g.* water).

protoplasm (cytology) Material within a cell, *i.e.* the cytoplasm and nucleus.

protoscolex (parasitological anatomy) A term used in relation to cestodes. The scolex of a larval stage. Morphologically it resembles the adult scolex.

Protospirura (veterinary parasitology) A genus of spirurid nematodes in the family Spiruridae. Species include *Protospirura bestianum*, *Protospirura muricola* and *Protospirura numidia* that may be found in the stomach of various felids and rodents.

Protostrongylus (parasitology) A genus of hairlike lungworms found in the bronchioles, alveoli and parenchyma of the lungs of some mammals.

Protostrongylus boughtoni (veterinary parasitology) A species of lungworms in the genus *Protostrongylus* that may be found in cottontail rabbits and hares.

Protostrongylus brevispiculum (veterinary parasitology) A species of lungworms in the genus *Protostrongylus* that, similarly to *Protostrongylus hobmaieri*, *Protostrongylus stilesi*, *Protostrongylus rushi* and *Protostrongylus davtiani*, may be found in bighorn sheep.

Protostrongylus davtiani (veterinary parasitology) A species of lungworms in the genus *Protostrongylus* that, similarly to *Protostrongylus brevispiculum*, *Protostrongylus rushi*, *Protostrongylus stilesi* and *Protostrongylus hobmaieri*, may be found in bighorn sheep.

Protostrongylus hobmaieri (veterinary parasitology) A species of lungworms in the genus *Protostrongylus* that, similarly to *Protostrongylus brevispiculum*, *Protostrongylus rushi*, *Protostrongylus stilesi* and *Protostrongylus davtiani*, may be found in bighorn sheep.

Protostrongylus kochi (veterinary parasitology) A species of lungworms in the genus *Protostrongylus* that may be found in sheep and goats.

Protostrongylus oryctolagi (veterinary parasitology) A species of lungworms in the genus *Protostrongylus* that may be found in rabbits.

Protostrongylus pulmonalis (veterinary parasitology) A species of lungworms in the genus *Protostrongylus* that may be found in rabbits and hares.

Protostrongylus rufescens (veterinary parasitology) A species of lungworms in the genus *Protostrongylus* that may be found in sheep, goats and deer.

Protostrongylus rushi (veterinary parasitology) A species of lungworms in the genus *Protostrongylus* that, similarly to *Protostrongylus hobmaieri*, *Protostrongylus brevispiculum*, *Protostrongylus stilesi* and *Protostrongylus davtiani*, may be found in bighorn sheep.

Protostrongylus skrjabini (veterinary parasitology) A species of lungworms in the genus *Protostrongylus* that may be found in sheep and goats.

Protostrongylus stilesi (veterinary parasitology) A species of lungworms in the genus *Protostrongylus* that, similarly to *Protostrongylus hobmaieri*, *Protostrongylus brevispiculum*, *Protostrongylus rushi* and *Protostrongylus davtiani*, may be found in bighorn sheep.

Protostrongylus sylvilagi (veterinary parasitology) A species of lungworms in the genus *Protostrongylus* that may be found in cottontails and jackrabbits.

Protostrongylus tauricus (veterinary parasitology) A species of lungworms in the genus *Protostrongylus* that may be found in hares.

prototype (parasitology) The original type or form that is typical of later individuals or species.

Protozoa (taxonomy) Subkingdom or phylum of microscopic unicellular organisms which range from plant-like forms to types which feed and behave like animals. Most are freeliving, but some lead commensalistic, mutualistic, saprophytic or parasitic existences. They have no fundamental body shape but have specialized organelles. Basic mode of reproduction is by binary fission, although multiple fission and conjugation occur in some species. Some protozoans are colonial and many are parasitic, inhabiting freshwater, marine and damp terrestrial environments.

protozoacide (pharmacology) Destructive to protozoa; a substance destructive to protozoa.

protozoal (parasitology) Pertaining to or caused by protozoa.

protozoal encephalomyelitis (parasitology) Probably the same disease as equine protozoal myeloencephalitis caused by *Sarcocystis neurona*.

protozoal hepatitis (parasitology) A disease caused usually by *Toxoplasma*, *Neospora*, and *Leishmania*.

protozoan (parasitology) 1 Of or pertaining to protozoa. 2 An organism belonging to the Protozoa.

protozoiasis (medical) Any disease caused by protozoa.

protozoology (parasitology) The scientific study of protozoa.

protozoon (biology) Plural: *protozoa* [German] Any member of the Protozoa.

protozoonsis (medical) Any disease caused by protozoa.

protozoophage (cytology) A cell having phagocytic action on protozoa.

5-PRPP (biochemistry) An abbreviation of '5'-'P'hospho'R'ibosyl 'P'yro'P'hosphate.

pruritogenic (medical) Causing pruritus, or itching.

pruritus (medical) Itching, which is common in many types of skin disorders, especially allergic inflammation and parasitic infestations.

pruritus ani (medical) Intense chronic itching in the anal region that may be caused by intestinal parasitism.

prussic acid (chemistry) An alternative term for hydrocyanic acid.

Przevalskiana (veterinary parasitology) A parasitic fly that parasitizes goats, sheep and gazelles, with a life cycle resembling that of *Hypoderma* spp.

pseud(o)- (terminology) A word element. [German] Prefix denoting false.

Pseudacarapis (insect parasitology) A species of mites in the Prostigmata, that are phoretic on the bees *Apinae*.

Pseudamphistomum (parasitology) A digenetic trematode, parasitic in the liver, of the family Opisthorchiidae.

Pseudamphistomum truncatum (human/ veterinary parasitology) A species of digenetic trematode in the genus *Pseudamphistomum* that may be found in carnivores and humans and has a life cycle and pathogenicity similar to those of *Opisthorchis* spp.

Pseudaspidodera (avian parasitology) A genus of roundworms in the family Heterakidae. Species include *Pseudaspidodera jnanendre* and *Pseudaspidodera pavonis* which may be found in the caecum (*USA*: cecum) of peafowl but that do not appear to be pathogenic.

Pseudisobrachium terresi (insect parasitology) Hymenoptera in the Bethylidae that may be found in the larvae and pupa of the ants *Alphaenogaster* and *Solenopsis*. Its larvae is probably endoparasitic.

Pseudobilharziella (avian parasitology) A genus of blood flukes in the family Schistosomatidae that may be found in the blood vessels of some anatid birds.

pseudocelom (parasitological anatomy) A term used in relation to nematodes. Body cavity of a nematode which is filled with fluid and in which the internal organs are suspended. Also termed: pseudocoel.

Pseudochalcura gibbosa (insect parasitology) *See Pheidoloxeton wheeleri.*

pseudocysts (parasitology) A term used in relation to *Toxoplasma* and *Sarcocystis*. These consist of a large number of endozoites enclosed in a macrophage or some other host cell. The parasites are bound by host cell tissue.

Pseudodiplogasteroides sp. (insect parasitology) A species of nematode in the Diplogasteridae that are parasites to the termites *Mastotermes darwiniensis*, that may be found in the head.

Pseudodiscus (parasitology) A genus of intestinal flukes (digenetic trematodes) in the family Paramphistomatidae.

Pseudodiscus collinsi (veterinary parasitology) A species of intestinal flukes in the genus *Pseudodiscus* that may be found in the colon of horses.

Pseudohypocera kerteszi (insect parasitology) *See Pseudohypocera* spp.

***Pseudohypocera* spp.** (insect parasitology) Diptera in the Phoridae including *Pseudohypocera kerteszi* that are parasitic to larvae and possibly adults of the bees *Apis*, *Melipona* spp., *Cephalotrigona*, *Oxytrigona*, *Tetragona*, and *Trigona*; the parasite attacks larvae of entire colonies and can be very destructive.

Pseudolynchia (avian parasitology) A genus of flies in the family Hippoboscidae. Species includes *Pseudolynchia canariensis*, *Pseudolynchia capensis* and *Pseudolynchia lividicolor* which may be found on domestic pigeons and some wild birds. They are dark brown flies resembling a sheep ked and may transmit *Haemoproteus columbae* of pigeons and *Haemoproteus lophortyx* of quail.

Pseudometagea schwarzii (insect parasitology) *See Pheidoloxeton wheeleri.*

Pseudophyllidea (fish parasitology) An order of flatworms in subclass Eucestoda, class Cestoidea and the phylum Platyhelminthes. This is a large order of nine families of which the following are found in teleost fish: Ptychobothriida, Bothriocephalidae, Echinophallidae, Triaenophoridae, Amphicotylidae, and Parabothriocephalidae. The scolex has two bothria, with or without hooks.

pseudophyllidean (human/fish/avian/veterinary parasitology) Pertaining to tapeworms (cestodes) of the order Pseudophyllidae. Members within this order usually have a three-host aquatic life cycle. Adult tapeworms whose length ranges from 10 to 100 feet, are parasitic in the intestines of fish-eating mammals, birds, fish and humans.

pseudopodium (biology) Part of an amoeba (*USA*: ameba) or similar protozoan that bulges out of its single cell. Pseudopodia are used for locomotion and to engulf food particles (for digestion).

pseudopregnancy (physiology) In some female mammals, physiological state resembling pregnancy without the formation of embryos. Also termed: false pregnancy.

Pseudopygmephurs (insect parasitology) Mites in the Prostigmata that are parasites on the ants *Lasius flavus*.

Pseudostertagia bullosa (veterinary parasitology) A roundworm in the family Trichostrongylidae that may be found in the abomasum of sheep, bighorn and Barbary sheep and pronghorn antelopes.

pseudotracheae (parasitological anatomy) Small tubes on labella found in houseflies and Tabanidae through which fluid is sucked.

Psilogaster antennatus (insect parasitology) *See Pheidoloxeton wheeleri.*

Psilogaster fasciventris (insect parasitology) *See Pheidoloxeton wheeleri.*

psilostomatid (parasitology) A member of the family Psilostomatidae of small, globose flukes.

Psorergates (parasitology) A genus of parasitic mites in the family Cheyletidae. Also termed: *Psorobia* spp.

Psorergates bos (veterinary parasitology) A species of parasitic mites in the genus *Psorergates* that may be found on cattle but without obvious lesions.

Psorergates oettlei (veterinary parasitology) (synonym: *Psorergates simplex*) A species of parasitic mites in the genus *Psorergates* that may be found on laboratory mice and rats.

Psorergates ovis (veterinary parasitology) (synonym: *Psorobia ovis*) A species of parasitic mites in the genus *Psorergates* which may be found on the body of sheep and that causes a mild pruritus and scaly dermatitis. Also termed: itch mite.

Psorergates simplex (parasitology) *See Psorergates oettlei.*

psorergatic mange (parasitology) *See Psorergates ovis.*

Psorobia ovis (parasitology) *See Psorergates ovis.*

Psorophora (avian/veterinary parasitology) A genus of mosquitoes, which may be a cause of severe insect worry when in large numbers and that may cause deaths of poultry.

Psoroptes (parasitology) A genus of mange mites in the family Psoroptidae. *See* psoroptic mange.

Psoroptes cervinus (veterinary parasitology) A species of mange mites in the genus *Psoroptes* that may be found on bighorn sheep and American elk.

Psoroptes communis (parasitology) *See Psoroptes ovis.*

Psoroptes cuniculi (veterinary parasitology) A species of mange mites in the genus *Psoroptes* that may be found on the ears of rabbits, goats, horses, donkeys and mules.

Psoroptes equi (veterinary parasitology) A species of mange mites in the genus *Psoroptes* that may be found on horses, donkeys and mules.

Psoroptes hippotis (parasitology) *See Psoroptes cuniculi.*

Psoroptes natalensis (veterinary parasitology) A species of mange mites in the genus *Psoroptes* that may be found on the body of cattle (*Bos taurus* and *Bos indicus)* and Indian water buffalo.

Psoroptes ovis (veterinary parasitology) A species of mange mites in the genus *Psoroptes* that may be found on sheep where it causes sheep-scab, and on cattle and goats.

psoroptic mange (veterinary parasitology) A parasitic dermatitis of many species caused by *Psoroptes* spp. mites including *Psoroptes cervinus* on deer, *Psoroptes equi*, *Psoroptes natalensis* on cattle and water buffalo, and *Psoroptes ovis* on sheep, goats and cattle. The common ear mange mites are *Psoroptes* cuniculi. Also termed: sheep-scab; body mange; ear mange.

psych(o)- (terminology) A word element. [German] Prefix denoting *mind.*

psychodid (parasitology) A member of the family Psychodidae. The sandflies or owl midges that includes *Phlebotomus.*

psychr(o)- (terminology) A word element. [German] Prefix denoting *cold.*

PTA (haematology) An abbreviation of 'P'lasma 'T'hromboplastin 'A'ntecedent (clotting factor XI).

PTC (1, 2 physiology) 1 An abbreviation of 'P'lasma 'T'hromboplastin 'C'omponent (clotting factor IX); 2 An abbreviation of 'P'henyl'T'hio'C'arbamide.

Pterolichus (parasitology) A genus of mites in the family Dermoglyphidae.

Pterolichus bicaudatus (avian parasitology) A species of mites in the genus *Pterolichus* that may be found on the feathers of the South African ostrich.

Pterolichus obtusus (avian parasitology) A species of mites in the genus *Pterolichus* that may be found on the feathers of fowls.

Pteronyssus striatus (avian parasitology) A genus of feather mite found on sparrow, linnet and chaffinch.

Pterophagus (parasitology) A genus of mites in the family Proctophyllodidae.

Pterophagus strictus (avian parasitology) A species of mites in the genus *Pterophagus* that may be found on the feathers of pigeons.

PTH (physiology/biochemistry) An abbreviation of 'P'ara'T'hyroid 'H'ormone.

PTHRP (physiology/biochemistry) An abbreviation of 'P'ara'T'hyroid 'H'ormone-'R'elated 'P'rotein.

-ptosis (terminology) A word element. [German] Suffix denoting *downward displacement.*

ptyal(o)- (terminology) A word element. [German] Prefix denoting *saliva.*

Pulex (human/veterinary parasitology) A genus of fleas, several species of which transmit the microorganism causing bubonic plague in humans.

Pulex irritans (human/veterinary parasitology) A widely distributed species, known as the human flea, in the genus *Pulex*, which infests domestic animals as well as humans, and may act as an intermediate host of certain helminths.

Pulex porcinus (parasitology) A species of flea in the genus *Pulex*, the peccary flea.

pulicicide (control measures) A substance destructive to fleas.

pulmo- (terminology) A word element. [Latin] Prefix denoting *lung.*

pulmonary (anatomy) Concerning the lungs and breathing.

pulmonary artery (anatomy) In mammals, a paired artery that carries deoxygenated blood from the right ventricle of the heart to the lungs. It is the only artery that carries deoxygenated blood.

pulmonary cysts (parasitology) Cysts that may be caused by the parasites *Paragonimus* spp.

pulmonary vein (anatomy) A paired vein that carries oxygenated blood from the lungs to the right atrium of the heart. It is the only vein that carries oxygenated blood.

pulse (1 physics; 2 physiology) 1 Brief disturbance propagated in a similar way as a wave, but not having the continuous periodic nature of a wave. 2 Regular expansion of the wall of an artery caused by the blood pressure waves that accompany heartbeats.

pump (equipment) Mechanical device for transferring liquids or gases, or for compressing gases. A simple lift pump employs atmospheric pressure and cannot pump a liquid vertically more than about 10 m (32 feet); a force pump does not have this restriction.

punkies (veterinary parasitology) *See* biting midge.

pupa (parasitological physiology) Plural: pupae [Latin] The second stage in the development of an insect, between the larva and the imago, that is usually an inactive stage such as a coccoon.

puparium (parasitological physiology) The hard pupal case of the insect pupa.

pupate (parasitological physiology) To proceed to the stage of pupa in an insect life cycle.

purine (chemistry) $C_5H_4N_4$ Heterocyclic nitrogen-containing base from which the bases characteristic of nucleotides and DNA are derived; *e.g.* adenine, guanine. Other purine derivatives include caffeine and uric acid.

puromycin (pharmacology) An antibiotic that inhibits protein synthesis which may be used in the treatment of protozoal infections.

putrefaction (biology) Anaerobic decomposition of organic matter by microscopic organisms (*e.g.* bacteria, fungi, etc.) which results in the formation of incompletely oxidized products.

Pycnomonas (parasitology) A subgenus of trypanosomes that includes *Trypanosoma suis*.

pyel(o)- (terminology) A word element. [German] Prefix denoting *renal pelvis*.

Pyemotes (insect/veterinary parasitology) A genus of mites in the Prostigmata. Species includes *Pyemotes tritici* (straw-itch mite) and *Pyemotes ventricosus* (harvest mite) which can cause transitory dermatitis on animals eating infested feeds. They are parasites to the wasps *Polistes* spp. and similarly to *Pygmephorus* they are polyphagous ectoparasites on the bees *Apis* and *Bombus*.

Pyemotes ventricosus (insect parasitology) Mites in the Prostigmata that are ectoparasitic on immature forms of, and may be phoretic on, Reticulitermes and other various termites. They can also destroy colonies of the ants *Solenopsis* and other ant species.

pygidium (parasitological anatomy) A pincushion-like structure seen on the ninth segment of fleas. It is believed to have a sensory function.

Pygmephorus (insect parasitology) *See Pyemotes*.

pykn(o)- (terminology) A word element. [German] Prefix denoting *thick*, *compact*, *frequent*.

pyle- (terminology) A word element. [German] Prefix denoting *portal vein*.

pylor(o)- (terminology) A word element. [German] Prefix denoting *pylorus*.

pyo- (terminology) A word element. [German] Prefix denoting *pus*.

Pyramicocephalus phocarus (veterinary parasitology) A species of cestode parasite that may be found in seals.

pyranose (chemistry) Any of a group of monosaccharide sugars (hexoses) whose molecules have a six-membered heterocyclic ring of five carbon atoms and one oxygen atom.

See also furanose.

pyrantel (pharmacology) A broad-spectrum anthelmintic of low toxicity, which may be used as the pamoate and tartrate salts and that has extensive use in dogs and cats.

pyrazine (chemistry) $C_4H_4N_2$ Heterocyclic aromatic compound whose ring contains four carbon atoms and two nitrogen atoms. Also termed: 1,4-diazine.

pyrazole (chemistry) $C_3H_4N_2$ Heterocyclic aromatic compound whose ring contains three carbon atoms and two nitrogen atoms. Also termed: 1,2,-diazole.

pyrene (chemistry) $C_{16}H_{10}$ Aromatic compound consisting of four benzene rings fused together.

Pyrex (chemistry) Trade name for a heat-resistant borosilicate glass, used for domestic and laboratory glassware.

pyridine (chemistry) C_5H_5N Heterocyclic liquid organic base which occurs in the light oil fraction of coal-tar and in bone oil. It forms salts with acids and is important in organic synthesis.

pyridoxine (chemistry) Crystalline substance from which the active coenzyme forms of vitamin B_6 are derived. It is also utilized as a potent growth factor for bacteria.

pyrimethamine (pharmacology) A folic acid antagonist used in combination with sulphonamides (*USA*: sulfonamides) in the treatment of toxoplasmosis and avian coccidiosis.

pyrimidine (chemistry) $C_4H_4N_2$ Heterocyclic crystalline organic base from which bases found in nucleotides and DNA are derived; *e.g.* uracil, thymine and cytosine. Its derivatives also include barbituric acid and the barbiturate drugs.

pyro- (terminology) A word element. [German] Prefix denoting *fire*, *heat*.

pyrolysis (chemistry) Decomposition of a chemical compound by heat.

pyrometer (physics) Instrument for measuring high temperatures, above the range of liquid thermometers.

pyrometry (measurement) Measurement of high temperatures.

pyrrole (biochemistry) $(CH)_4NH$ Heterocyclic liquid organic compound whose ring contains four carbon atoms and one nitrogen atom. An aromatic compound, its derivatives are important biologically; *e.g.* haem (*USA*: heme), chlorophyll.

pyruvate (chemistry) Ester or salt of pyruvic acid.

pyruvic acid (biochemistry) $CH_3COCOOH$ Simplest keto-acid, important in making energy available from ingested food. It is the product of the first stage of respiration (glycolysis). If oxygen is available, the acid is broken down in the Krebs cycle (citric or tricarboxylic acid cycle) to yield energy. Also termed: 2-oxopropanoic acid.

PYY (biochemistry) An abbreviation of 'P'olypeptide 'YY'.

Q

Q (biochemistry) A symbol for glutamine.

q (genetics) Symbol used to designate the long arm of a chromosome. Thus if a gene's locus is given as 22q, this means that it is on the long arm of chromosome 22. The short arm is called p.

Q fever (medical) Query fever. An acute febrile illness caused by infection with the micro-organism *Coxiella burnetii,* a species of *Rickettsia.* The organism normally infects sheep, cattle and goats, passing to them from wild animals via ticks and lice, but humans acquire the infection by inhaling dust from dried excreta or drinking infected milk. Also termed queensland fever.

Q&A (general terminology) An abbreviation of 'Q'uestion 'and' 'A'nswer.

QC (quality standards) An abbreviation of 'Q'uality 'C'ontrol.

qty (general terminology) An abbreviation of 'q'uan't'it'y'.

quadr(i)- (terminology) A word element. [Latin] Prefix denoting *four.*

quadrate (anatomy) One of a pair of bones in the upper jaw of amphibians, birds, fish and reptiles that has evolved into the incus (an ear ossicle) in mammals.

quadratic equation (mathematics) Algebraic equation of the second order (degree) or square power, which has two possible solutions, known as the roots of the unknown. The roots may be real and different, real and the same (coincident) or imaginary. For the general quadratic equation $ax^2 + bx + c = 0$, the solutions are given by $x = [-b \pm (b^2 - 4ac)^{1/2}]/2a$.

quadriceps (anatomy) the large muscle covering the front of the thigh. It is a combination of four muscles, the vastus medialis, vastus intermediug, vastus lateralis and rectus femoris.

quadriplegia (medical) Paralysis of all four limbs.

quadrivalent (chemistry) Having a valence of four. Also termed: tetravalent.

qual. (1, 2, 3 general terminology) 1 An abbreviation of 'qual'ification. 2 An abbreviation of 'qual'itative. 3 An abbreviation of 'qual'ity.

qualitative (chemistry) Dealing with the qualities or appearance of something only.

qualitative analysis (chemistry) Identification of the constituents of a substance or mixture, irrespective of their amount.

quality assurance (quality standards) *See* quality control.

quality control (QC) (quality assurance/quality standards) A term used to describe mechanisms and procedures planned as a part of experimental protocols that are designed to reduce the possibility of error, particularly human error. Integral parts of quality control include the design of procedures in such a way as to minimise the possibility of human error, the collection of data not only on the results of the experiments, but also on the daily activities, personnel involved, etc; and the proper training of all laboratory and animal room personnel. Activity and results forms must be designed in such a way that omissions are immediately apparent.

quanta (measurement) Plural of quantum.

quantification (genetics) The application of measurement to experimental results.

quantitative (chemistry) Dealing with quantities of substances, *e.g.* mass, volume, etc., irrespective of their identity.

quantitative analysis (chemistry) Determination of the amounts of constituents present in a substance or mixture, often by weighing or manipulating volumes of solutions.

See also gravimetric analysis; volumetric analysis.

quantum (measurement) Unit quantity (an indivisible 'packet') of energy postulated in the quantum theory. The photon is the quantum of electromagnetic radiation (such as light) and in certain contexts the meson is the quantum of the nuclear field.

quantum electrodynamics (physics) Study of electromagnetic interactions, in accordance with the quantum theory.

quantum electronics (physics) Generation or amplification of microwave power, governed by quantum mechanics.

quantum evolution (genetics) Evolutionary change in which new species are formed very rapidly. The circumstance in which this is likely to happen is the colonisation of new habitats by small populations, in which both founder effect and genetic drift contribute to genetic change.

quantum mechanics (physics) Method of dealing with the behaviour (*USA*: behavior) of small particles such as electrons and nuclei. It uses the idea of the particle wave duality of matter. Thus an electron has a dual nature, particle and wave, but it behaves as one or the other according to the nature of the experiment.

quantum number (physics) Integer or half-integral number that specifies possible values of a quantitized physical quantity, *e.g.* energy level, nuclear spin, angular momentum, etc.

quantum state (physics) State of an atom, electron, particle, etc; defined by a unique set of quantum numbers.

quantum theory (physics) Theory of radiation. It states that radiant energy is given out by a radiating body in separate units of energy known as quanta; the same applies to the absorption of radiation. The total amount of radiant energy given out or absorbed is always a whole number of quanta.

quarantine (medical) The isolation of people or animals that are suspected of being infected with a particular infectious disease.

quart (qt) (measurement) Unit of liquid measure equal to 2 pints. 1qt (imperial) = 1.13652 litres and the American quart is 946 ml (0.946 litres).

quartan malaria (medical) Malaria caused by *Plasmodium malariae.*

quartile (statistics) Value below which one-quarter of a set of data lies; the 25th percentile. The second quartile equals the median.

quartz (chemistry) SiO_2 Natural crystalline silica (silicon dioxide), one of the hardest of common minerals. Its crystals (which can generate piezoelectricity) are frequently colourless (*USA*: colorless) and transparent. It is used as an abrasive and in mortar and cement.

quasi group (statistics) A collection of people, not yet a proper group, but with the capacity to become one.

quaternary ammonium compound (chemistry) Member of a group of white crystalline solids, soluble in water, and completely dissociated in solution. These compounds have the general formula $R_4N^+X^-$, where R is a long-chain alkyl group. They have detergent properties. Also termed: quaternary ammonium salt.

Queensland itch (parasitology) *See* sweet itch.

Queensland tick typhus (human/veterinary parasitology) A tickborne fever of humans, similar to Rocky Mountain spotted fever, caused by *Rickettsia australis* and transmitted by ixodid ticks. Dogs and cats may be unusual hosts.

quenching (chemistry) Rapid cooling.

query fever (parasitology) *See* Q fever.

Question and Answer (Q&A) (general terminology) Relating to speeches and lectures, the time when people in the audience can ask specific questions of the speaker or panel.

quicklime (chemistry) CaO Whitish powder prepared by roasting limestone, used in agriculture and in cements and mortar. Also termed: calcium oxide; lime.

quicksilver (chemistry) *See* mercury.

Quilonia (parasitology) A genus of roundworms in the subfamily Cyathostominae.

Quilonia africana (veterinary parasitology) A species of roundworms in the genus *Quilonia* that, similarly to *Quilonia ethiopica* and *Quilonia uganda*, may be found in the African elephant.

Quilonia ethiopica (veterinary parasitology) A species of roundworms in the genus *Quilonia* that, similarly to *Quilonia africana* and *Quilonia uganda*, may be found in the African elephant.

Quilonia renniei (veterinary parasitology) A species of roundworms in the genus *Quilonia* that, similarly to *Quilonia travancra*, may be found in the Indian elephant.

Quilonia travancra (veterinary parasitology) A species of roundworms in the genus *Quilonia* that, similarly to *Quilonia renniei*, may be found in the Indian elephant.

Quilonia uganda (veterinary parasitology) A species of roundworms in the genus *Quilonia* that, similarly to *Quilonia ethiopica* and *Quilonia africana*, may be found in the African elephant.

quinacrine (pharmacology) An antimalarial, anti-protozoal and anthelmintic used especially for suppressive therapy of malaria in humans and also in the treatment of giardiasis in dogs. Also termed mepacrine.

quinine (pharmacology) An alkaloid made from the bark of the South American cinchona tree. It is used against chloroquine-resistant strains of *Plasmodium falciparum*, the organism of malignant malaria.

quinoline (chemistry) C_9H_7N Colourless (*USA*: colorless) oily liquid heterocyclic base. It is an aromatic compound consisting of fused benzene and pyridine rings.

quinone (chemistry) Member of a group of cyclic unsaturated diketones in which the double bonds and keto groups are conjugated. Thus they are not aromatic compounds. Many quinones are used as dyes.

quint- (terminology) A word element [Latin] Prefix denoting *five*.

quinuronium (pharmacology) An antiprotozoal agent effective against *Babesia* spp. that has a low margin of safety because of its marked parasympath-omimetic effects.

quotient (mathematics) Result of division.

qwerty (computing) Description of the standard alphanumeric keyboard used on typewriters, typesetting machines, word processors and computers (named after the first six letters on the top rank of letters).

R

R plasmid (genetics) Any plasmid carying a gene for antibiotic resistance.

R technique (statistics) A factor analysis based on the correlations between tests that attempts to derive a limited number of factors underlying the correlations.

R unit (physiology) An abbreviation of unit of 'R'esistance in cardiovascular system.

R (radiology) An abbreviation of 'R'adiologist/ 'R'adiology.

R&D (education) An abbreviation of 'R'esearch 'and' 'D'evelopment.

r.a.m. (chemistry) An abbreviation of 'r'elative 'a'tomic 'm'ass (formerly called atomic weight).

Ra (chemistry) The chemical symbol for radium Element 88, radium.

rabbit tick (veterinary parasitology) *See Haemaphysalis leporispalustris.*

raccoon roundworm (veterinary parasitology) *Baylisascaris procyonis. See Baylisascaris.*

race (biology) In classification, an alternative term for subspecies.

racemic acid (chemistry) Racemic mixture of tartaric acid.

racemic mixture (chemistry) Optically inactive mixture that contains equal amounts of dextrorotatory and laevorotatory forms of an optically active compound.

racemization (chemistry) Transformation of optically active compounds into racemic mixtures. It can be effected by the action of heat or light, or by the use of chemical reagents.

rachi(o)- (terminology) A word element. [German] Prefix denoting spine.

rad (1 radiology; 2 measurement) 1 An abbreviation of 'r'adiation 'a'bsorbed 'd'ose. 2 Unit of absorbed dose of ionizing radiation, equivalent to 100 ergs per gram (0.01 J kg^{-1}) of absorbing material. The corresponding SI unit is the gray.

Rad. (radiology) An abbreviation of 'Rad'iologist; 'Rad'iology; 'Rad'iotherapist; 'Rad'iotherapy.

rada (radiology) An abbreviation of 'rad'io'a'ctive.

Radfordia (parasitology) A genus of mites in the family Myobiidae.

Radfordia affinis (veterinary parasitology) A species of mites of the genus *Radfordia* which may be found on mice.

Radfordia ensifera (veterinary parasitology) A species of mites of the genus *Radfordia* which may be found on wild and laboratory rats. The mites may cause sufficient pruritus to result in self-trauma.

radial immunodiffusion (methods) (Mancini technique) An immunodiffusion technique in which antigen diffuses into the agar which contains specific antibody and a ring of precipitate is formed, the diameter of which is directly proportional to the concentration of the antigen and can thereby be used to quantitate the amount of antigen. A reverse radial immunodiffusion test, in which antigen is incorporated in the agar, can be used to quantitate the amount of antibody in a sample.

radial symmetry (general terminology) Symmetry about any one of several lines or planes through the centre (*USA*: center) of an object or organism.

See also bilateral symmetry.

radian (rad) (measurement) SI unit of plane angle; the angle at the centre (*USA*: center) of a circle subtended by an arc whose length is equal to the radius of the circle. 1 radian = 57 degrees (approx.).

radiance (physics) (measurement) The amount of radiant energy (*e.g.* light) that passes through a given area per unit time. It is usually measured in watts square metre per steradian.

radiant (physics) Describing something that emits electromagnetic radiation (*e.g.* light, heat rays).

radiant energy (physics) Electromagnetic energy, which the part with wavelengths between about 380 and 750 nm is visible as light.

radiant flux (physics) The rate at which power is emitted or received by an object in the form of electromagnetic radiation.

radiant heat (physics) Heat that is transmitted in the form of infra-red radiation.

radiant intensity (physics) The intensity of energy emitted by a source per unit solid angle, usually measured in watt steradian.

radiation (physics) Energy that travels in the form of electromagnetic radiation, *e.g.* radio waves, infra-red radiation, light, ultraviolet radiation, X-rays and gamma rays. The term is also applied to the rays of alpha particles and beta particles emitted by radioactive substances. Particle rays and short-wavelength electromagnetic radiation may be harmful to tissues as they are ionizing radiation. Radiation is one of the causes of raised mutation rates and of increased incidence of cancers.

radiation biology (biology) Scientific study of the effects of ionizing radiation on living organisms.

radiation damage (physics) Alteration of properties of a material that results from exposure to ionizing radiation.

radiation hazard (physics) Radioactive material is on the premises and represents a potential threat to people and animals in that environment.

radiation pressure (physics) Minute force exerted on a surface by electromagnetic radiation that strikes it.

radiation unit (radiology) Activity of a radio-isotope expressed in units of disintegrations per second, called the becquerel in SI units. Formerly it was measured in curies.

radical (1 chemistry; 2 mathematics) 1 A group of atoms within a molecule that maintains its identity through chemical changes that affect the rest of the molecule, but is usually incapable of independent existence; *e.g.* alkyl radical. 2 Relating to the root of a number or quantity.

radiculitis (histology) Inflammation of a root, applied to inflammation of a nerve root.

radioactive (radiology) Possessing or exhibiting radioactivity.

radioactive decay (radiology) Way in which a radio-isotope spontaneously changes into another element or isotope by the emission of alpha or beta particles or gamma rays. The rate at which it does so is represented by its half-life.

radioactive equilibrium (radiology) Condition attained when a parent radioactive element produces a daughter radioactive element that decays at the same rate as it is being formed from the parent.

radioactive isotope (chemistry) A form of a chemical substance that emits radioactivity.

radioactive series (radiology) One of three series that describe the radioactive decay of 40 or more naturally occurring radioactive isotopes of high atomic number. They are known (after the element at the beginning of each sequence) as the thorium series, uranium series and actinium series.

radioactive standard (radiology) Radio-isotope of known rate of radioactive decay used for the calibration of radiation-measuring instruments.

radioactive tracers (radiology) Radioactive isotopes whose passage and behaviour (*USA*: behavior) through the body is monitored by recording their radiation, *e.g.* ^{14}C, ^{15}N.

radioactive waste (radiology) Hazardous radio-isotopes (fission products) that accumulate as waste products in a nuclear reactor. They have to be periodically removed and stored safely or reprocessed. The term is also applied to the waste ('tailings') produced by the processing of uranium ores.

radioactivity (radiology) Spontaneous disintegration of atomic nuclei, usually with the emission of alpha particles, beta particles or gamma rays.

radiobiology (radiology) Study of ionizing radiation in relation to living systems. It includes the effects of radiation on living organisms and the use of radio-isotopes in biological and medical work.

See also radiotherapy.

radiochemistry (radiology) Chemistry of radioactive elements and their compounds, and use of radio-isotopes (*e.g.* for 'labelling' in chemical analysis).

radiodiagnosis (radiology) Branch of medical radiology that is concerned with the use of X-rays or radio-isotopes in diagnosis.

radiograph (radiology) Photographic image that results from uneven absorption by an object being subjected to penetrating radiation. An X-ray photograph is a common example.

radiography (radiology) Photography using X-rays or gamma rays, particularly in medical applications.

radio-isotope (radiology) Isotope that emits radioactivity (ionizing radiation) during its spontaneous decay. Radio-isotopes are useful sources of radiation (*e.g.* in radiography) and are used as tracers for radioactive tracing.

radiology (radiology) Study of X-rays, gamma rays and radioactivity (including radio-isotopes), especially as used in medical diagnosis and treatment.

radioluminescence (radiology) Fluorescence caused by radioactivity.

radio-opaque (radiology) Resistant to the penetrating effects of radiation, especially X-rays, often used to describe substances injected into the body before a radiography examination.

radiotherapy (radiology) Treatment of disorders by the use of ionizing radiation such as X-rays or radiation from radio-isotopes.

radium (Ra) (chemistry) Silver-white radioactive metallic element in Group IIA of the periodic table (the alkaline earths). It has several isotopes, with half-lives of up to 1,620 years. It is obtained from pitchblende (its principal ore), and used in radiotherapy and luminous paints. At. no. 88; r.a.m. 226.0254.

radius (1 mathematics; 2 anatomy) 1 Distance from the centre (*USA*: center) of a circle to the circumference, equal to half the diameter or the circumference divided by 2π. 2 One of two bones in the forearm of a tetrapod vertebrate (the other is the ulna).

radius of curvature (mathematics) Of a point on a curve, the radius of a circle that touches the inside of the curve at that point.

radon (Rn) (chemistry) Radioactive gaseous element in Group 0 of the periodic table (the rare gases), a radioactive decay product of radium. It has several isotopes, with half-lives of up to 3.82 days. Radon coming out of the ground, particularly in hard-rock areas, is a source of background radiation that has been recognized as a health hazard. At. no. 86; r.a.m. 222 (most stable isotope).

Radopholus (plant parasitology) A genus of small, adults usually less than 1 mm long, nematodes in the order Tylenchida and family Pratylenchidae. Their body is straight or slightly curved ventrally when killed by heat, and there is marked sexual dimorphism in the anterior region. *See Radopholus similis.*

Radopholus citrophilus (plant parasitology) *See Radopholus similis.*

Radopholus similis (plant parasitology) A species of nematodes in the genus *Radopholus* whose hosts cover a wide range with over 200 hosts and widespread in tropical and subtropical regions of the world including West Africa, Central and South America, Hawaii, Florida, Puerto Rico, Cuba and Australia. They are migratory endoparasites spending their adult vermiform life in the root, but capable of emerging in adverse conditions. At 24–32°C the life cycle takes 20–25 days, 18–20 days at 24–27°C, and all larval stages and females are infective. Fertilization is usual but parthenogenesis does occur, with eggs hatching in three to seven days. Female produces an average of approximately two eggs/day. The male does not penetrate intact roots and may not feed, and the species survives less than six months in soil free of host roots. At least two biotypes have been demonstrated. They are similar morphologically but differ in host range. The 'banana race' attacks banana but not citrus. The 'citrus race' is pathogenic to both citrus and banana. In 1984, the citrus and banana race were described as sibling species with different chromosome numbers. There may also be a sugarcane race. *Radopholus similis* is the original banana race, attacking banana, but not citrus. *Radopholus citrophilus* is the citrus burrowing nematode attacking citrus and banana. The nematodes feed in cortex, seldomly in vascular tissue, resulting in lesions and cavities, root breakdown, and secondary decays; *e.g. Fusarium oxysporum* and *Rhizoctonia solani* in banana. They cause 'Blackhead' or toppling disease of bananas in which the root system is reduced and weakened so that the infested tree falls under weight of fruit or in the wind resulting in total crop loss. They may reduce the vigour (*USA*: vigor) of sucker growth for new trees and delay the rate of fruit development. Infestation with *Radopholus citrophilus* also causes 'spreading decline of citrus' an important disease of citrus crops and nematodes can be found up to 12 feet below the soil surface in citrus groves. Millions of black pepper trees have also been lost in Indonesia to 'Yellows disease' caused by *Radopholus similis*. The nematode is additionally a serious pest of sugarcane in Hawaii. Control measures includes the use of preplant nematicides, nematicides at planting and postplant application. Propping or guying of trees prevents total loss due to toppling, but the practice is expensive. Also termed: burrowing nematode.

raffinate (chemistry) Liquid that remains after a substance has been obtained by solvent extraction.

raffinose (chemistry) $C_{18}H_{32}O_{16}$ Colourless (*USA*: colorless) crystalline trisaccharide carbohydrate that occurs in sugar beet, which hydrolyses to the sugars galactose, glucose and fructose.

rafoxanide (pharmacology) A very efficient flukicide which is also effective against *Haemonchus contortus* and *Oestrus ovis*. Available for both oral and injectable administration. There is a good safety margin but poisoning can occur in sheep with heavy doses.

Railletina circumvalata (insect parasitology) A species of helminth in the Cestoda that, similarly to *Railletina echinobothrida*, are parasites to to the ants *Monomorium, Pheidole, Pheidologeton, Prenolepis* and *Tetramorium*. The final hosts of *Railletina echinobothrida* are poultry.

Railletina echinobothrida (insect parasitology) *See Railletina circumvalata.*

Railletina fedjushini (insect parasitology) A species of helminth in the Cestoda that, similarly to *Railletina friedbergeri, Railletina georgiensis, Railletina kashiwarensis, Railletina loeweni, Railletina tetragona* and *Railletina urogalli*, are parasites to to the ants *Brachyponera, Crematogaster* spp., *Dorymyrmex, Euponera, Formica, Iridomyrmex, Monomorium* spp., *Myrmecocystus, Myrmica* spp., *Pheidole, Prenolepis, Solenopsis* spp. and *Tetramorium*. The final hosts are turkeys, pheasants, and peacocks. For *Railletina loeweni* the final host is the rabbit.

Railletina friedbergeri (insect parasitology) *See Railletina fedjushini.*

Railletina georgiensis (insect parasitology) *See Railletina fedjushini.*

Railletina kashiwarensis (insect parasitology) *See Railletina fedjushini.*

Railletina loeweni (insect parasitology) *See Railletina fedjushini.*

Railletina tetragona (insect parasitology) *See Railletina fedjushini.*

Railletina urogalli (insect parasitology) *See Railletina fedjushini.*

Raillietia (parasitology) A genus of mites in the family Gamasidae.

Raillietia auris (veterinary parasitology) A species of mite of the genus *Raillietia* which may be found on the ears of cattle.

Raillietia australis (veterinary parasitology) A species of mite of the genus *Raillietia* which may be found in the ears of wombats.

Raillietia caprae (veterinary parasitology) A species of mite of the genus *Raillietia* which may be found in the ears of goats.

Raillietia hopkinsi (veterinary parasitology) A species of mite of the genus *Raillietia* which may be found on the ears of antelopes.

Raillietiella (avian/veterinary parasitology) A genus of pentastomid internal parasites in the class Pentastomida found in ophidians, lacertilians and birds.

Raillietina (parasitology) A genus of cyclophyllidean tapeworms in the family Davaineidae, containing a very large number of species, many of them uncommon.

Raillietina cesticillus (avian parasitology) A species of cyclophyllidean tapeworms in the genus *Raillietina* that are commonly found in domestic poultry.

Raillietina echinobothrida (avian parasitology) A species of cyclophyllidean tapeworms in the genus *Raillietina* that are commonly found in the small intestine of chickens and turkeys.

Raillietina georgiensis (avian parasitology) A species of cyclophyllidean tapeworms in the genus *Raillietina* that may be found in domestic and wild turkeys.

Raillietina magninumida (avian parasitology) A species of cyclophyllidean tapeworms in the genus *Raillietina* that may be found in guinea fowl.

Raillietina ransomi (avian parasitology) A species of cyclophyllidean tapeworms in the genus *Raillietina* that may be found in domestic and wild turkey.

Raillietina tetragona (avian parasitology) A species of cyclophyllidean tapeworms in the genus *Raillietina* that may be found in the posterior small intestine of many birds including domestic fowls.

Raillietina williamsi (avian parasitology) A species of cyclophyllidean tapeworms in the genus *Raillietina* that may be found in wild turkey.

rainbow trout fry syndrome (fish parasitology) Severe septicaemic (*USA*: septicemic) condition of rainbow trout fry caused by *Cytophaga psychrophila*.

rainworm (insect parasitology) *See Mermis subnigrescens*.

RAM (computing) An abbreviation of 'r'andom 'a'ccess 'm'emory of a computer.

Raman effect (physics) Scattering of monochromatic light, when it passes through a transparent homogeneous medium, into different characteristic wavelengths because of interaction with the molecules of the medium. It was named after the Indian physicist Chandrasekhara Raman (1888–1970).

random (general terminology) Unplanned, without direction or purpose.

random access memory (RAM) (computing) Part of a computer's memory that can be written to and read from.

See also rom.

random error (statistics) Error which occurs due to chance, such as sampling error.

random numbers (statistics) A list of numbers obtained by a standard randomization procedure.

random primer method (methods) A method of labelling (*USA*: labeling) DNA in vitro.

random sample (statistics) Sample taken from a large group in such a way that it is representative of the group as a whole. Random sampling is an important method of carrying out quality control for mass-produced articles.

Rankine scale (measurement) Temperature scale that expresses absolute temperatures in degrees Fahrenheit (absolute zero = 0°R). It was named after the British engineer and physicist William Rankine (1820–72).

rare earth (chemistry) Member of the series of elements, in Group IIIB of the periodic table, known also as the lanthanides. Also termed: rare earth element.

rare gas (chemistry) One of the uncommon, unreactive and highly stable gases in Group 0 of the periodic table. They are helium, neon, argon, krypton, xenon and radon. Also termed: inert gas; noble gas.

RAS (physiology) An abbreviation of 'R'eticular 'A'ctivating 'S'ystem.

rash (medical) A temporary eruption on the skin.

RAST (methods) An abbreviation of 'R'adio-'A'llergo'S'orbent 'T'est.

raster (computing) Display of information in the form of a grid, usually referring to the image produced by the parallel scanning action of a cathode-ray tube.

rat flea (parasitology) *Leptopsylla segnis*.

rate (measurement) The frequency with which an event or circumstance occurs per unit of time.

rate constant (chemistry) Constant of proportionality for the speed of a chemical reaction at a particular temperature. It can only be obtained experimentally. Also termed: velocity constant; specific rate constant.

rate equation (chemistry) An alternative term for rate law.

rate law (chemistry) Equation that relates the rate of a chemical reaction to the concentration of the individual reactants. It has the form rate = $k[X]^n$, where k is the rate constant, X is the reactant and n is the order of reaction. It can only be obtained experimentally. Also termed: rate equation.

rate of reaction (chemistry) Speed of a chemical reaction, usually expressed as the change in concentration of a reactant or product per unit time. It can be affected by temperature, pressure and the presence of a catalyst.

rate-determining step (chemistry) Slowest step of a chemical reaction which determines the overall rate, provided the other steps are relatively rapid. Thus the kinetics and order of reaction are basically those of the rate determining step. Also termed: limiting step.

rational number (mathematics) Number that may be expressed as a ratio of two integers, *i.e.* in the form of a fraction.

See also irrational number.

rat-tailed maggot (parasitology) Larva of *Eristalis* spp., a hover fly of no importance other than in confusing an identification. The maggots are found in areas with high concentrations of organic matter, *e.g.* stable drains. Also termed filth fly maggot.

raw data (statistics) Data as they are collected and before any calculation, ordering, etc. has been done.

ray (physics) Beam of any type of radiation, *e.g. light.*

rbc (haematology) An abbreviation of 'r'ed 'b'lood 'c'ell(s).

RDS (medical) An abbreviation of 'R'espiratory 'D'istress 'S'yndrome.

Re. (measurement) An abbreviation of 'Re'ynolds' number.

reactant (chemistry) Substance that reacts with another in a chemical reaction to form new substances).

reaction (chemistry) An alternative term for chemical reaction.

reactive dye (chemistry) Dye that forms a covalent bond with the fibre molecule (*USA*: fiber molecule) of the textile being dyed. This provides excellent fastness. Such dyes are used to dye cellulose fibres (*USA*: cellulose fibers) (*e.g.* rayon).

reactor (chemistry) Vessel in which a chemical reaction is carried out.

read-only memory (ROM) (computing) Part of a computer's memory that can only be read (and not written to).

See also random access memory.

reagent (1, 2 chemistry) 1 Substance that takes part in a chemical reaction; a reactant. 2 Common laboratory chemical used in chemical analysis and for experiments.

real gas (chemistry) Gas that never fully achieves 'ideal' behaviour (*USA*: behavior).

See also kinetic theory.

real image (optics) Image brought to a focus by a lens, mirror or optical system that can be displayed on a screen (as opposed to a virtual image, which can not).

real number (mathematics) Any number, positive or negative, from among all rational numbers and irrational numbers (as opposed to an imaginary number).

realgar (chemistry) As_2S_2 Natural red arsenic disulphide (*USA*: arsenic disulfide), used as a pigment and in pyrotechnics.

reboot (computing) Restarting the computer.

receptors (1 chemistry/physics; 2 pharmacology) 1 Sensory cells, which may be part of a group that form a sense organ capable of detecting stimuli. When a receptor is stimulated (*e.g.* by temperature or light), it produces electrical or biochemical changes that are relayed to the nervous system for processing. 2 Proteins through which many drugs and natural mediators act to exert their effects.

recessive (genetics) Describing a gene that is expressed in the phenotype when it is homozygous in a cell (*i.e.* there have to be two recessive genes for their effect to be apparent). The presence of a dominant allele masks the effect of a recessive gene (*i.e.* in a combination of a dominant gene and a recessive gene, the dominant gene manifests itself.

recidiva leishmaniasis (parasitology) Cutaneous leishmaniasis.

recipient (biology) Organism that receives material from another, *e.g.* as in the taking up of DNA by one bacterium from another.

reciprocal (mathematics) Quantity obtained by dividing a number into 1; *i.e.* the reciprocal of x is the number $1/x$. Zero has no reciprocal.

reciprocal proportions, law of (chemistry) An alternative term for the law of equivalent proportions.

reciprocal wavelength (physics) An alternative term for wave number.

recombinant DNA (molecular biology) Type of DNA that has genes from different sources, genetically engineered using recombination.

recombination (molecular biology) Process by which new combinations of characteristics not possessed by the parents are formed in the offspring. It results from crossing over during meiosis to form gametes that unite during fertilization to form a new individual. Genetic engineers have developed techniques for artificially recombining strands of DNA (to make recombinant DNA).

record (computing) A number of elements of data that together form one unit of stored information.

recrudescence (parasitology) A term used in relation to malaria. Renewed manifestation of infection due to the survival of erythrocytic forms.

recrystallization (1, 2 chemistry) 1 Change from one crystal structure to another; it occurs on heating or cooling through a critical temperature. 2 Purification of a substance by repeated crystallization from solution.

rectification (1 chemistry; 2 physics) 1 Purification of a liquid using distillation. 2 Conversion of alternating current (a.c.) into direct current (d.c.) using a rectifier.

rectified spirit (chemistry) Solution of ethanol (ethyl alcohol) that contains about 5–7% water. It is a constant-boiling mixture and the water cannot be removed by distillation.

rectifier (physics) Electrical device for converting an alternating current (a.c.) into a direct current (d.c.). It may take the form of a plate rectifier, a diode valve or a semiconductor diode.

rectilinear (mathematics) In or forming a straight line.

rectum (anatomy) Final part of the intestine, through which faeces (*USA*: feces) are passed after reabsorption of water.

recurrent (medical) Characterized by recurrence at intervals of weeks or months.

recurrent course (medical) Periods of normality of weeks to months.

recurring decimal (mathematics) Decimal that contains a number or block of numbers that repeat to infinity; *e.g.* the decimal for 2/3 = 0.666666 and the decimal for 2/11 = 0.181818.

See also terminal decimal.

red blood cell (haematology) An alternative term for an erythrocyte, also known as a red cell or red corpuscle.

red blood corpuscle (haematology) A erythrocyte.

red bug (parasitology) *See Eutrombicula alfreddugesi.*

red cell (haematology) *See* erythrocyte.

red lead (chemistry) Pb_3O_4 Bright red powdery oxide of lead. An oxidizing agent, it is used in anti-rust and priming paints. Also termed: minium; dilead(II); lead(IV) oxide; lead tetraoxide; triplumbic tetroxide.

red mange (veterinary parasitology) A synoym for sarcoptic mange in farm animals and demodectic mange in dogs.

red mite (parasitology) *See Dermanyssus gallinae.*

red ring disease (plant parasitology) *See Rhadinaphelenchus cocophilus.*

red ring nematode (plant parasitology) *See Rhadinaphelenchus cocophilus.*

red stomach worm (parasitology) *See Haemonchus.*

red urine (parasitology) *See* haematuria, haemoglobinuria.

redia (parasitological physiology) Plural: rediae [Latin]. A larval stage of certain trematode parasites, which develops in the body of a snail host and gives rise to daughter rediae, or to the cercariae.

red-legged tick (parasitology) *See Rhipicephalus evertsi.*

redox reaction (chemistry) Chemical reaction in which oxidation is necessarily accompanied by reduction, and vice versa; an oxidation-reduction reaction.

reduced pressure distillation (chemistry) An alternative term for vacuum distillation.

reducing agent (chemistry) Substance that causes chemical reduction, often by adding hydrogen or removing oxygen; *e.g.* carbon, carbon monoxide, hydrogen. Also termed: electron donor; reluctant.

reducing sugar (chemistry) Any sugar that can act as a reducing agent.

reductase (biochemistry) Enzyme that causes the reduction of an organic compound.

reduction (chemistry) Chemical reaction that involves the gain of electrons by a substance; the addition of hydrogen or removal of oxygen from a substance.

reduviid bug (parasitology) A member of the family Reduviidae. One of the Triatoma species that is involved in the transmission of Chagas' disease. Also termed cone-nose bug, assassin bug or kissing bug.

redwater fever (parasitology) *See* babesiosis.

redworm (parasitology) *See Strongylus.*

redworm infestation (parasitology) *See* strongylosis.

Reed-Frost model (epidemiology/statistics) A deterministic probability model of a theoretical epidemic.

REF (physiology/biochemistry) An abbreviation of 'R'enal 'E'rythropoietic 'F'actor.

reflectance (measurement) Ratio of the intensity of reflected radiation to the intensity of the incident radiation.

reflection (physics) Change in direction of an electromagnetic wave (*e.g.* light) or sound wave after it strikes a (smooth) surface (*e.g.* a mirror).

reflection of light laws (1, 2 physics) 1 The angle of reflection equals the angle of incidence. 2 The reflected ray is in the same plane as the incident ray and normal to the surface at the point of incidence.

reflection angle (optics) Angle between a reflected ray of light and the normal.

reflector (optics) Object or surface that reflects electromagnetic radiation (*e.g.* light, radio waves), particularly one around or inside a lamp to concentrate a light beam.

reflex (1 biology; 2 mathematics) 1 Sequence of nerve impulses that produce a fast involuntary response to an external stimulus. 2 Describing an angle that exceeds 180 degrees (but is less than 360 degrees).

reflux (chemistry) Boiling of a liquid for long periods of time. Loss by evaporation is prevented by using a reflux condenser.

reflux condenser (chemistry) Vertical condenser used in the process of refluxing. It is attached to a vessel that contains the liquid to be refluxed and condenses the vapour produced on boiling, which then runs back into the vessel.

reflux reduction division (genetics) An alternative term for meiosis.

reforming (chemistry) Production of branched-chain alkanes from straight-chain ones or the production of aromatic compounds (*e.g.* benzene) from alkenes, using cracking or a catalyst.

refraction (physics) Change in direction of an energy wave (*e.g.* light, sound) when it passes from one medium into another in which its speed is different.

refraction of light (optics) Change in direction of a light ray as it passes obliquely from one transparent medium to another of different refractive index.

refraction of light, laws of (1, 2 optics) 1 For two particular media, the ratio of the sine of the angle of incidence to the sine of the angle of refraction is constant (the refractive index). This is a statement of Snell's law. 2 The refracted ray is in the same plane as the incident ray and the normal at the point of incidence.

refraction, angle of (optics) Angle between a refracted ray of light and the normal.

refractive constant (physics) An alternative term for refractive index.

refractive index (*n*) (physics) Ratio of the speed of electromagnetic radiation (such as light) in air or vacuum to its speed in another medium. The speed depends on the wavelength of the radiation as well as on the density of the medium. For a refracted ray of light, it is equal to the ratio of the sine of the angle of incidence i to the sine of the angle of refraction r; *i.e.* $n = \sin i/\sin r$. Also termed: refractive constant.

refractometer (physics) Instrument for measuring the refractive index of a substance.

refrigerant (chemistry) Substance used as the working fluid in a refrigerator (*e.g.* ammonia, fluon, CFCS).

refrigeration (physics) Method of maintaining a cool temperature in a room or container by transferring heat from it to the exterior using a heat pump. Cooling is caused by the evaporation of a volatile liquid (refrigerant) into a vapour, which is compressed by a pump (compressor) to turn it back to a liquid.

regeneration (physiology) Regrowth of tissue to replace that which has been damaged or lost, *e.g.* wound healing.

Reighardia (avian parasitology) A genus of pentastomes in the family Porocephalidae which may be found in the air sacs of gulls and terns.

reindeer warble fly (veterinary parasitology) *See Oedemagena tarandi.*

reinfection (parasitology) A second infection by the same agent.

relapse (parasitology) The return of a disease weeks or months after its apparent cessation. In relation to malaria, renewed manifestation of infection due to the invasion of blood by merozoites from the late pre-erythrocytic stage (previously known as secondary exo-erythrocytic stage).

relationship (epidemiology) The state that exists when one variable is related to another variable in some way.

relative atomic mass (r.a.m.) (physics) Mass of an atom relative to the mass of the isotope carbon-12 (which is taken to be exactly 12). Former name: atomic weight.

relative density (physics) Ratio of the density of a given substance to the density of some reference substance. For liquids, relative densities are usually expressed with reference to the density of water at 4°C. Former name: specific gravity.

relative frequency (statistics) The number of observations of a particular, nominated value expressed usually as a proportion of the total frequency.

relative humidity (physics) Ratio of the pressure of water vapour present in air to the pressure the water vapour would have if the air were saturated at the same temperature (*i.e.* to the saturated water vapour pressure). It is expressed as a percentage.

relative molecular mass (chemistry) Sum of the relative atomic masses of all the atoms in a molecule of a substance. Also termed: molecular weight.

relativity (physics) Einstein's theory; scientific principle expressing in mathematical and physical terms the implications of the fact that observations depend as much on the viewpoint as on what is being observed.

relay, electrical (physics) Electromagnetic switching device that brings about changes in an independent circuit.

rem (radiology) An abbreviation of 'r'öntgen 'e'quivalent 'm'an, the quantity of ionizing radiation such that the energy imparted to a biological system per gram of living material has the same effect as one röntgen.

remote (general terminology) At a distance.

remote access (computing) Being able to access a computer by a telephone and a modem.

Renicola (parasitology) A genus of renal flukes (digenetic trematodes) in the family Renicolidae.

Renicola hayesanniae (avian parasitology) A species of renal flukes in the genus *Renicola* which may be found in turkeys and eider ducks and may cause sufficient damage to lead to renal failure.

renicolid (parasitology) A member of the family Renicolidae of flukes.

reniform nematode (plant parasitology) *See Rotylenchulus reniformis.*

renin (biochemistry) Enzyme produced by the kidney that constricts arteries and thus raises blood pressure.

rennin (biochemistry) Enzyme found in gastric juice that curdles milk. It is the active ingredient of rennet.

replicase (biochemistry) Enzyme that promotes the synthesis of DNA and RNA within living cells.

replication (molecular biology) The process by which a new complete molecule of DNA is made, by one strand being used as a template for the assembly of another strand.

replication (molecular biology) The production of exact copies of complex molecules during the growth of living organisms. *See* base pairing.

reproduction (biology) Procreation of an organism. Sexual reproduction involves the fusion of sex cells or gametes and the exchange of genetic material, thus bringing new vigour to a species. Asexual reproduction does not involve gametes, but usually the vegetative proliferation of an organism. *See* vegetative propagation.

resin (chemistry) Organic compound that is generally a viscous liquid or semi-liquid which gradually hardens when exposed to air, becoming an amorphous, brittle solid. Natural resins, found in plants, are yellowish in colour (*USA*: color) and insoluble in water, but are quite soluble in organic solvents. Synthetic resins (types of plastics) also possess many of these properties.

resistor (physics) Device that provides resistance in electrical circuits.

resmethrin (control measures) A synthetic pyrethroid insecticide used to control fleas and ticks on dogs and cats.

resonance (1 chemistry; 2 physics) 1 Movement of electrons from one atom of a molecule or ion to another atom of that molecule or ion to form a stable resonance hybrid structure (*e.g.* as in an aromatic compound). 2 Phenomenon in which a system is made to vibrate at its natural frequency as a result of vibrations received from another source of the same frequency.

resorantel (pharmacology) A safe, hydroxybenzanilide cestocide highly effective against *Moniezia* and *Thysaniezia* spp. and moderately effective against *Paramphistomum* spp.

resorb (biology) To take up or absorb again; to undergo resorption.

resorcinol (chemistry) $C_6H_4(OH)_2$ Crystalline dihydric phenol, used in the synthesis of drugs, dyes and plastics. Also termed: m-dihydroxybenzene; 1,3-benzenediol.

respiration (1, 2 physiology) 1 Release of energy by living organisms from the breakdown of organic compounds. In aerobic respiration, which occurs in most cells, oxygen is required and carbon dioxide and water are produced. Energy production is coupled to a series of oxidation reduction reactions, catalysed by enzymes. In anaerobic respiration (*e.g.* fermentation), food substances are only partly broken down, and thus less energy is released and oxygen is not required. 2 An alternative term for breathing.

respiratory movement (physiology) Movement by an organism to allow the exchange of respiratory gases, *i.e.* the taking up of oxygen and release of carbon dioxide. In mammals such as human beings this entails breathing, involving movements of the chest and diaphragm. In fishes, water is passed over the gills for gaseous exchange.

respiratory organ (anatomy) Organ in which respiration (breathing) takes place. In mammals (*e.g.* human beings), the process is carried out in the lungs; in fish, the gills. There gaseous exchange takes place (usually of oxygen and carbon dioxide).

respiratory pigment (biochemistry) Substance that can take up and carry oxygen in areas of high oxygen concentration, releasing it in parts of the organism with low oxygen concentrations where it is consumed, *i.e.* by respiration in cells. In vertebrates the respiratory pigment is haemoglobin (*USA*: hemaglobin); in some invertebrates it is haemocyanin (*USA*: hemocyanin).

respiratory quotient (RQ) (measurement) Ratio of carbon dioxide produced by an organism to the oxygen consumed in a given time. It gives information about the type of food being oxidized; *e.g.* carbohydrate has an RQ of approximately 1, but if the RQ becomes high (*i.e.* little oxygen is available), anaerobic respiration may occur.

response (physiology) Physical, chemical or behavioural (*USA*: behavioral) change in an organism initiated by a stimulus.

resting potential (physiology) Potential difference between the inner and outer surfaces of a resting nerve, which is about -60 to -80 mV. It occurs when the nerve is not conducting any impulse, and is in contrast to the action potential, which occurs during the application of a stimulus and brings about a rise in the potential difference to a positive value.

restriction enzyme (biochemistry) Enzyme (a nuclease) produced by some bacteria that is capable of breaking down foreign DNA. It cleaves double-stranded DNA at a specific sequence of bases, and the DNA of the bacteria is modified for protection against degradation. Restriction enzymes are used widely as tools in genetic engineering for cutting DNA. Also termed: restriction endonuclease.

retina (histology) Light-sensitive tissue at the back of the vertebrate eye, made up of a network of interconnected nerves. The first cells in the network are photo-receptors consisting of cones (which are sensitive to colour [*USA*: color]) or rods (which are sensitive to light). They contain visual pigments (*e.g.* rhodopsin) which ultimately cause impulses to be transmitted to the visual centre (*USA*: center) of the brain via the optic nerve.

retinol (biochemistry) Fat-soluble vitamin found in plants, in which it is formed from carotene. Also termed: vitamin A.

retort (chemistry) Heated vessel used for the distillation of substances, as in the separation of some metals.

Retortamonas (parasitology) A genus of nonpathogenic, pyriform, protozoan parasites in the family Retortamonadidae.

Retortamonas caviae (veterinary parasitology) A species of protozoan parasites in the genus *Retortamonas* which may be found in the caecum (*USA*: cecum) of guinea pigs.

Retortamonas cuniculi (veterinary parasitology) A species of protozoan parasites in the genus *Retortamonas* which may be found in the caecum (*USA*: cecum) of rabbits.

Retortamonas intestinalis (human/veterinary parasitology) A species of protozoan parasites in the genus *Retortamonas* which may be found in humans and nondomesticated primates.

Retortamonas ovis (veterinary parasitology) A species of protozoan parasites in the genus *Retortamonas* which may be found in sheep.

retro- (terminology) A word element [Latin] Prefix denoting *behind, backward.*

retrospective study (epidemiology) A scholarly examination based on examination of existing data, on events that have already occurred.

return (computing) The key on the keyboard of a computer terminal which carries out the commands previously typed in.

reverse zoonosis (parasitology) Infections or infestations transmitted from humans to animals.

reversible process (chemistry) Process that can theoretically be reversed by an appropriate small change in any of the thermodynamic variables (*e.g.* pressure, temperature). Real natural processes are irreversible.

reversible reaction (chemistry) Chemical reaction that can go either forwards or backwards depending on the conditions.

RFLP (physiology/biochemistry) An abbreviation of 'R'estriction 'F'ragment 'L'ength 'P'olymorphism.

Rh factor (haematology) An abbreviation of 'Rh'esus factor.

rhabd(o)- (terminology) A word element. [German]. Prefix denoting *rod, rod-shaped.*

Rhabdias (veterinary parasitology) A genus of nematodes which cause verminous pneumonia in reptiles and amphibians. The nematodes may also be found in their body cavities. Species includes *Rhabdias fuscovenosa.*

Rhabditis (parasitology) A genus of minute nematodes that are members of the family Rhabditidae, which are found mostly in damp earth but that may temporarily invade damaged skin. The Rhabditidae are parasites to the termites *Neotermes, Reticulitermes, Capritermes, Coptotermes, Macrotermes, Microtermes, Rhinotermes* and *Termes.*

Rhabditis axei (parasitology) A species of minute nematodes of the genus *Rhabditis* that are not recorded as pathogenic.

Rhabditis bovis (veterinary parasitology) A species of minute nematodes of the genus *Rhabditis* that causes bovine parasitic otitis and secondary myiasis.

Rhabditis clavopapillata (veterinary parasitology) A species of minute nematodes of the genus *Rhabditis* that may be found as a contaminant on the hair of dogs and monkeys.

Rhabditis gingivalis (veterinary parasitology) A species of minute nematodes of the genus *Rhabditis* that may be found in a granuloma of the gum of a horse.

Rhabditis janeti *(Caenorhabditis dolichura)* (insect parasitology) A species of nematode in the Rhabditidae that are parasites to the ants *Camponotus* spp., *Formica* spp., and *Lasius* spp where they may be found in the post-pharyngeal glands, and the termites *Reticulitermes flavipes* where they may be found in the head.

Rhabditis macrocerca (veterinary parasitology) A species of minute nematodes of the genus *Rhabditis* that may be found as a contaminant on the hair of dogs and monkeys.

Rhabditis strongyloides (parasitology) *See Pelodera strongyloides.*

Rhabpanus ossiculum (insect parasitology) A species of nematode in the Rhabditidae that are parasites to the termites *Reticulitermes flavipes.* They may be found in the termite's head.

Rhadinaphelenchus (plant parasitology) A genus of nematodes in the order Tylenchida and family Aphelenchoididae. The female body is approximately 1 mm long and very slender and arcuate to nearly straight when relaxed, while males are ventrally arcuate, and more strongly curved in the tail region. *See Rhadinaphelenchus cocophilus.*

Rhadinaphelenchus cocophilus (plant parasitology) A species of nematodes in the genus *Rhadinaphelenchus* whose hosts cover a narrow range, naturally infecting coconut and oil palms, but that can be artificially inoculated into cabbage palm and a few others. They are distributed over the Caribbean, Latin America, Grenada, Trinidad, Tobago, Guyana, Honduras, Mexico, Panama, Surinam, Venezuela, West Indies, and South and Central America. They are migratory endoparasites with a 10-day life cycle, and can migrate and survive in soil, especially in moist areas, but tree-to-tree spread seems minimal. Nematodes can be transmitted by putting infected tissue in soil near healthy trees, but nematode survives free in soil only three to four days. Their insect vector is the palm (coconut) weevil, *Rhynchophorus palmarum.* Nematodes are carried on body surface and also enter body through spiracles and mouth. Transmission to leaf axils occurs as the beetle feeds. Nematodes also aggregate around ovipositor and are injected into soft tissue as beetles deposit eggs. Beetle larvae hatch and tunnel into tissues, pupate, emerge, become infected and spread nematodes. There is some evidence to suggest that nematodes may persist in beetle larvae through moults (*USA*: molts), but this is unclear. Coconuts are especially susceptible to infestation for two years before and after the start of fruit bearing. Older and younger trees are more resistant. Older leaves become chlorotic and die, younger leaves turn yellow, and green nuts and unopen flowers are shed. Trees may die four months after the first symptoms appear. Symptoms include a band of discoloured (*USA*: discolored), reddish-brown tissue about 5 cm from the edge of leaf stems, and discolouration (*USA*: discoloration) extends into leaf petioles. Red ring (3–4 cm wide) may appear up to 8 feet (2.4 m) above soil line. Nematodes are numerous in and around the discoloured (*USA*: discolored) tissues, and adult nematodes are usually located at the inner edge of the red ring. Up to 5000 nematodes can be found per gram of tissue, the greatest numbers occur 6–12 inches (15–30 cm) below upper limit of ring. Roots are similarly discoloured (*USA*: discolored), with soft, spongy cortex. There is apparent phytotoxin in the red ring. Damage causes reduction in water uptake by the tree. The nematode has been associated with both red ring disease and little leaf disease of coconut and oil palms. Control measures include palm weevil control by various techniques. Also termed: red ring nematode.

-rhaphy (terminology) A word element. [German] Suffix denoting *seam*, *suture*; *surgical repair of*.

rhenium (Re) (chemistry) Rare metallic element of Group VIIB of the periodic table (a transition element), used in making thermocouples. At no 75; r.a.m. 186.20.

rheo- (terminology) A word element. [German] Prefix denoting *electric current*, *flow* (as of fluids).

rheostat (physics) Variable electrical resistor, generally used to control current flow.

rhin(o)- (terminology) A word element. [German] Prefix denoting *nose*, *noselike structure*.

rhinitis (medical) Inflammation of the mucous membrane of the nose. It may be mild and chronic, or acute. There are signs of wheezing, sneezing and respiratory stertor at all levels. There is a strong nasal discharge which may be serous to purulent.

Rhinoestrus (parasitology) A genus of flies, resembling *Oestrus* spp., and obligate parasites of *Eguus* spp.

Rhinoestrus purpureus (veterinary parasitology) A species of flies of the genus *Rhinoestrus* whose larvae parasitize larynx and nasal sinuses of Equidae. Also termed Russian gad fly.

rhinogenous (medical) Arising in the nose.

rhinograph (radiography) A contrast radiograph of the nasal cavity.

rhinologist (medical) A specialist in rhinology.

rhinology (medical) The sum of knowledge about the nose and its diseases.

Rhinonyssus (avian parasitology) A genus of mites which parasitize birds without being important pests.

Rhinonyssus rhinolethrum (avian parasitology) A species of mites of the genus *Rhinonyssus* which may be found in ducks and geese.

rhinopathy (medical) Any disease of the nose.

Rhipicentor (parasitology) A genus of ticks in the family Ixodidae that resemble *Dermacentor* spp.

Rhipicentor bieorinis (veterinary parasitology) A species of ticks of the genus *Rhipicentor* which may be found on a wide range of wild and domesticated animals.

Rhipicentor nuttalli (veterinary parasitology) A species of ticks of the genus *Rhipicentor* which may be found on a wide range of wild and domesticated animals.

Rhipicephalus (parasitology) A genus of ticks in the family Ixodidae.

Rhipicephalus appendiculatus (veterinary parasitology) A three-host tick of the genus *Rhipicephalus* which may be found on most animal species. The tick transmits *Theileria parva*, *Babesia* spp. and other protozoan and viral diseases including Nairobi sheep disease and louping ill. It is the principal vector of East Coast fever. Also termed brown ear tick.

Rhipicephalus ayrei (parasitology) A tick of the genus *Rhipicephalus* which transmits *Theileria parva*.

Rhipicephalus bursa (veterinary parasitology) A tick of the genus *Rhipicephalus* which transmits *Babesia*, *Theileria*, *Anaplasma*, *Rickettsia*, and *Coxiella* spp.

Rhipicephalus capensis (veterinary parasitology) A three-host tick of the genus *Rhipicephalus* which is parasitic on cattle, and that transmits East Coast fever.

Rhipicephalus evertsi (parasitology) A two-host tick of the genus *Rhipicephalus* which transmits *Babesia*, *Theileria*, *Borrelia*, and *Rickettsia* spp. Also termed red-legged tick.

Rhipicephalus jeanelli (parasitology) A tick of the genus *Rhipicephalus* which transmits *Theileria parva*.

Rhipicephalus neavei (parasitology) A tick of the genus *Rhipicephalus* which transmits *Theileria parva*.

Rhipicephalus pulchellus (parasitology) A tick of the genus *Rhipicephalus* which transmits *Theileria parva* and Nairobi sheep disease.

Rhipicephalus sanguineus (avian/veterinary parasitology) A three-host tick of the genus *Rhipicephalus* which is mainly a parasite of dogs but occurs on all species of mammals and birds. It transmits *Babesia*, *Borrelia*, *Coxiella*, *Rickettsia*, *Anaplasma* and *Pasteurella* spp. Also causes tick paralysis.

Rhipicephalus simus (parasitology) A three-host tick of the genus *Rhipicephalus* which transmits *Theileria parva*.

Rhipipallus affinis (insect parasitology) Hymenoptera in the Eucharitidae that, similarly to *Tricoryna*, may be found in pupa of *Odontomachus*, *Myrmecia*, *Chalcoponera* and *Ectatomma*, and ocassionally superparasitism found.

***Rhipiphorus* sp.** (insect parasitology) Arthropods in the Coleoptera that are a parasitic to the bees *Apis florea*, and whose larva are endoparasitic in bee larva.

rhizo- (terminology) A word element. [German] Prefix denoting *root*.

Rhizoglyphus (insect parasitology) A species of mites in the Astigmata, that similarly to *Sancassania*, are cleptoparasitic to the bees *Apis*, and *Halictinae*.

Rhizopoda (parasitology) A class of protozoa of the subphylum Sarcodina, having pseudopodia, and including the amoebae (*USA*: amebae).

rhod(o)- (terminology) A word element. [German] Prefix denoting *red*.

rhodium (Rh) (chemistry) Silver-white metallic element in Group VIII of the periodic table (a transition element), used as a catalyst and in making thermocouples. At. no. 45; r.a.m. 102.9055.

Rhodnius prolixus (insect parasitology) A triatomine bug, vector of trypanosomiasis. Member of the insect family Reduviidae of assassin or kissing bugs.

rhodopsin (biochemistry) Protein (derived from vitamin A) in the rods of the retina of the eye which acts as a light-sensitive pigment; the action of light brings about a chemical change that results in the production of a nerve impulse. Also termed: visual purple.

rhoptries (parasitological anatomy) A term used in relation to *Toxoplasma* and *Sarcocystis*. These are electron dense club-like structures located at the anterior end of endozoites, cystozoites and merozoites. It is postulated that these structures produce enzymes required for the penetration of the parasite into the host. Also termed: paired organelles.

RIA (immunology) An abbreviation of 'R'adio-'I'mmuno'A'ssay. *See* immunoassay.

Ribeiroia (avian parasitology) A genus of alimentary canal flukes (digenetic trematodes) in the family Cathaemasiidae; found in fowl, domestic ducks and geese in which they cause proventricular inflammation and ulceration.

Ribeiroia ondatrae (avian/fish parasitology) A species of alimentary canal flukes of the genus *Ribeiroia* which may be found in fish-eating birds, chicken and geese.

riboflavin (biochemistry) Orange water-soluble crystalline solid, member of the vitamin B complex. It plays an important role in growth. Also termed: riboflavine; lactoflavin; vitamin B_2.

ribonuclease (biochemistry) An enzyme that catalyzes the breakdown of ribonucleic acid.

ribonucleic acid (molecular biology) *See* RNA.

ribose (chemistry) $C_5H_{10}O_5$ Optically active pentose sugar, a component of the nucleotides of RNA (ribonucleic acid).

ribosome (cytology) Particle present in the cytoplasm of cells, often attached to the endoplasmic reticulum, that is essential in the biosynthesis of proteins. Ribosomes are composed of protein and RNA, and are the site of attachment for messenger RNA during protein synthesis. They may be associated in chains called polyribosomes.

Rickettsiae (human/insect parasitology) Group of micro-organisms, often classified as being part way between bacteria and viruses, that are parasitic on the cells of arthropods (lice, mites and ticks) and vertebrates. Some can cause serious disorders (*e.g.* typhus in human beings).

RID (immunology) Abbreviation of 'R'adial 'I'mmuno'D'iffusion. *See* immunoassay.

ring nematode (plant parasitology) *See Criconemella xenoplax*.

ringer's fluid (chemistry) Physiological saline solution used for keeping tissues and organs alive outside the body (in vitro). It is similar in composition to the fluid that naturally bathes cells and tissues, maintaining a constant internal environment. It contains chlorides of sodium, potassium and calcium. It was named after the British physiologist Sydney Ringer (1835–1910).

risk (statistics) The chance of an unfavourable (*USA*: unfavorable) event occurring.

risk factor (statistics) An attribute or exposure which increases the probability of occurrence of a disease or other outcome.

RMV (physiology) An abbreviation of 'R'espiratory 'M'inute 'V'olume.

RNA (molecular biology) An abbreviation of 'R'ibo'N'ucleic 'A'cid, one of the nucleic acids present in cells, the other being DNA. It is composed of nucleotides that contain ribose as the sugar. RNA contains the bases adenine, guanine, cytosine and uracil. Messenger RNA takes part in transcription or copying of the genetic code from a DNA template. Transfer RNA and ribosomal RNA take part in translation or protein synthesis, all of which occur in prokaryotes and eukaryotes.

robenidine (pharmacology) A guanidine derivative used at one time as an anticoccidial in poultry little used now because of the rapid development of resistance to it by the protozoa.

ROC curve (statistics) An abbreviation of 'R'eceiver 'O'perating 'C'haracteristic curve. A graphical method of assessing the characteristic of a diagnostic test.

Rochalimaea (human/insect parasitology) A genus of the family Rickettsiaceae resembling the genus *Rickettsia*, but usually found extracellularly in the arthropod host, including *Rochalimaea quintana*, the aetiological (*USA*: etiological) agent of trench fever of humans and voles, transmitted by the body louse *Pediculus humanus*.

Rocky Mountain wood tick (parasitology) *See Dermacentor andersoni.*

rod (cytology) Type of sensory cell present in the retina of the vertebrate eye. It is stimulated by light and is concerned with vision in low illumination. The absorption of light energy (photons) by the visual pigment rhodopsin present in the rod causes a nervous impulse, which travels along the optic nerve to the brain.

roentgen (chemistry) An alternative term for röntgen.

rolling disease (veterinary parasitology) A nervous system disease of mice caused by *Mycoplasma neurolyticum*.

ROM (computing) An abbreviation of 'r'ead-'o'nly 'm'emory of a computer.

Roman numerals (mathematics) Number system, originally used by the Romans, based on letters: I = 1; V = 5; X = 10; L = 50; C = 100; D = 500; M = 1,000. Other numbers are written using combinations of these; there is no zero.

Romanowsky stains (parasitology) A group of eosin-methylene blue stains generally used for blood smears, protozoa and bacteria, and includes Giemsa, Wright and Leishman stains.

Rondanioestrus apivorus (insect parasitology) Diptera in the Tacinidae that are parasitic to the bees *Apis mellifera*, attacking workers in flight and depositing larva on the host.

röntgen (R) (measurement) Unit of radiation; the amount of X-rays or gamma rays that produce a charge of 2.58×10^{-4} coulomb of electricity in 1 cm^3 of dry air. It was named after the German physicist Wilhelm Röntgen (1845–1923). Also termed: roentgen.

röntgen rays (radiology) An alternative term for X-rays.

roost mite (parasitology) *See Dermanyssus gallinae.*

root (1, 2 mathematics; 3, 4 anatomy) 1 Number or quantity that when multiplied by itself some specified times gives the number again; *e.g.* the square root of a number is that when multiplied by itself gives the number. 2 Solution of an algebraic equation. 3 The descending and subterranean part of a plant. 4 That portion of an organ, such as a tooth, hair or nail, that is buried in the tissues, or by which it arises from another structure, or the part of a nerve that is adjacent to the centre (*USA*: center) to which it is connected.

root directory (computing) The top of a hierarchical directory structure for a computer disk.

root lesion nematode (plant parasitology) *See Pratylenchus penetrans.*

root-knot nematode (plant parasitology) *See Meloidogyne incognita.*

root-mean-square (rms) (mathematics) Average equal to the square root of the sum of the squares of a number of values divided by the total number of values.

rostellum (parasitological anatomy) Apical projection of the scolex of a tapeworm, which may or may not bear hooks.

rotary microtome (equipment) A microtome in which wheel action is translated into a back-and-forth movement of the specimen being sectioned.

rotatory (chemistry/optics) Optically active; capable of rotating the plane of polarized light.

rotenone (pharmacology/control measures) A compound derived from the root and rhizomes of derris (*Derris elliptica*) and lonchocarpus (*Lonchocarpus utilis*) which may be used as an insecticide and as an acaricide, mainly on dogs, cats, cattle, birds and fish.

rotoscope (physics) Alternative term for stroboscope.

rotund tick (parasitology) *See Ixodes kingi.*

Rotylenchulus (plant parasitology) A genus of nematodes in the order Tylenchida and family Hoplolamidae. They are 0.3–0.5 mm long, sexually dimorphic, and immature females may be found free in the soil. They are C-shaped when killed by heat. Mature females are endoparasitic in roots, greatly swollen, irregular to kidney shaped, enlarged gonads occupying much of the body. *See Rotylenchulus reniformis.*

Rotylenchulus reniformis (plant parasitology) A species of nematodes in the genus *Rotylenchulus* whose hosts cover a very broad range including fruit trees, cotton, cowpea, tea, soybean and pineapple. They are distributed over tropical and subtropical countries including West and Central Africa, Central and South America, Southeast Asia, and USA. They are sedentary endoparasite, bisexual, and amphimictic. Eggs are laid in a gelatinous matrix, 40–60 eggs per mass, and hatch in 8–10 days. Juveniles then undergo 3 moults (*USA*: molts) in the soil without feeding. Young females are the infective stage and only females are parasitic. Females enter root with posterior outside, feed and swell to reniform shape. The life cycle is completed in 25 days at 25°C (80°F). Nurse cells, 100–200 per female, form near pericycle. Feeding causes hypertrophy of pericycle and endodermis cells, increased cytoplasm density, but cells remain uninucleate with large nucleolus, and walls may rupture to form a syncytium. Infestation causes a reduction in cotton yield, with a concomitant increase in *Fusarium* wilt. In the presence of this nematode, *Fusarium* wilt-resistant varieties of cotton also become susceptible. Control measures include the use of preplant nematicides. Also termed: reniform nematode.

rough endoplasmic reticulum (anatomy) Parts of the endoplasmic reticulum to which ribosomes are attached on the cytoplasmic side; involved in the biosynthesis of proteins for export to the outside of the cell and enzymes to be incorporated into cellular organelles such as lysosomes.

round window (histology) Lower of two membranous areas on the cochlea of the inner ear (the other is the oval window). Also termed: fenestra rotunda.

roundworm (parasitology) Any of the parasitic, unsegmented, cylindrical in cross-section, elongated in shape, nematode worms which invade principally the gastrointestinal tract although almost any organ can be involved. Comprises the class Nematoda and its large number of genera.

roup (veterinary parasitology) Any disease of poultry manifested by signs of coryza and involvement of the nasal chambers.

See also avian trichomoniasis.

roxarsone (pharmacology/control measures) An organic arsenical compound used as a growth promoter and coccidiostat in swine and poultry. Also termed 4-hydroxy-3-nitrophenylarsonic acid.

RPF (physiology) An abbreviation of 'R'enal 'P'lasma 'F'low.

RQ (physiology) An abbreviation of 'R'espiratory 'Q'uotient.

-rrhagia (terminology) A word element. [German] Suffix denoting excessive flow.

-rrhea (terminology) A word element. [German] Suffix denoting profuse flow.

-rrhexis (terminology) A word element. [German] Suffix denoting rupture of a vessel or organ.

rRNA (molecular biology) An abbreviation of 'r'ibosomal RNA. *See* RNA.

Ru (chemistry) The chemical symbol denoting ruthenium.

rubidium (Rb) (chemistry) Reactive silver-white metal in Group IA of the periodic table (the alkali metals), with a naturally occurring radioactive isotope (Rb-87). At. no. 37; r.a.m. 85.4678.

Rugopharynx (parasitology) A gastrointestinal nematode found in free-living macropods.

Rugopharynx australis (veterinary parasitology) A species of nematode of the genus *Rugopharynx* which may be found in free-living macropods.

Rugopharynx rosemariae (veterinary parasitology) Gastrointestinal nematodes which may be found embedded in nodules in the saccular stomachs of kangaroos.

rumen (veterinary anatomy) Plural: rumens, rumina. The largest of the compartments of the forestomach of ruminant animals that serves as a fermentating vessel.

rumenal (anatomy) Pertaining to the rumen.

ruminal (anatomy) Pertaining to the rumen.

ruminal flukes (veterinary parasitology) *See* paramphistomiasis.

Russian gad fly (parasitology) *Rhinoestrus purpureus*.

rust (fish parasitology) A disease of tropical fish in aquariums caused by the protozoa *Oodinium limneticum* and characterized by loss of luster of the skin surface. It causes heavy mortality.

ruthenium (Ru) (chemistry) Silver-white metallic element in Group VIII of the periodic table (a transition element), used to add hardness to platinum alloys. At. no. 44; r.a.m. 101.07.

rutherfordium (Rf) (chemistry) Element no. 104 (a post-actinide). It is a radioactive metal with at least three very short-lived isotopes (half-lives up to 70 seconds), made by bombarding an actinide with carbon, oxygen or neon atoms. Also termed: kurchatovium.

S

S (1 chemistry; 2 biochemistry) 1 The chemical symbol for sulphur (*USA*: sulfur). 2 An abbreviation of 'S'erine.

SA (physiology) An abbreviation of 'S'pecific 'A'ctivity.

SA node (histology) An abbreviation of 'S'ino'A'trial node.

SAB (governing body) An abbreviation of 'S'cience/ 'S'cientific 'A'dvisory 'B'oard.

saccharide (chemistry) Simplest type of carbohydrate, with the general formula ($C_6H_{12}O_6$), common to many sugars. Also termed: saccharose.

saccharimetry (measurement) Measurement of the concentration of sugar in a solution from its optical activity, by using a polarimeter.

saccharin (chemistry) $C_6H_4SO_2CONH$ White crystalline organic compound that is about 550 times sweeter than sugar; an artificial sweetener. It is almost insoluble in water and hence it is used in the form of its soluble sodium salt. Also termed: 2-sulphobenzimide (*USA*: 2-sulfobenzimide); saccharine.

saccharose (chemistry) An alternative term for saccharide.

sacr(o)- (terminology) A word element. [Latin] Prefix denoting *sacrum*.

SADR (biochemistry) An abbreviation of 'S'uspected 'A'dverse 'D'rug 'R'eaction.

safranine O/fast green stain (histology) A histological staining technique used to make cartilage visible under the microscope.

sagittal fissure (histology) The fissure separating the cerebral hemispheres.

SAIMR (institution) An abbreviation of 'S'outh 'A'frican 'I'nstitute of 'M'edical 'R'esearch.

sal ammoniac (chemistry) Old name for ammonium chloride.

sal volatile (chemistry) Old name for ammonium carbonate.

salicylanilides (pharmacology) A group of ant-helmintics which exert their action by uncoupling mitochondrial reactions which are critical to electron transport and associated phosphorylation in the metabolic system of the parasite. They are effective against cestodes and trematodes but not nematodes. Some are active against *Haemonchus contortus*, *e.g.* rafoxanide and closantel.

salicylate (chemistry) Ester or salt of salicyllic acid.

salicyllic acid (chemistry) $C_6H_4(OH)COOH$ White crystalline organic compound, a carboxylic acid. It is used as an antiseptic, in medicine, and in the preparation of azo dyes. Its acetyl ester is aspirin. Also termed: 2-hydroxybenzoic acid.

saline (chemistry) Salty; describing a solution of sodium chloride (common salt).

salinomycin (pharmacology) An ionophore coccidiostat with a broad spectrum of efficiency and closely related to monensin.

saliva (biochemistry) Neutral or slightly alkaline fluid secreted by the salivary glands in the mouth. It lubricates food during chewing and aids digestion. It consists of a mixture of mucus and the enzyme amylase (ptyalin), which breaks down starch to maltose.

Salivaria (parasitology) The anterior station group or Group B trypanosomes which transmit the protozoa with their mouthparts.

salivarian (parasitology) Said of trypanosomes which belong to the subgenera of *Duttonella*, *Trypanozoon*, *Pycnomonas* and *Nannomonas*. These trypanosomes are passed to the recipient in the saliva of the tsetse fly (*Glossina* spp.) during the act of biting.

See also Salivaria.

salivary (biology) Pertaining to the saliva.

Salmincola (fish parasitology) Parasitic crustaceans which attach to the gills of salmonid fish and cause obstruction of respiration and blood loss resulting in delayed sexual maturity, reduced growth rate and some mortality.

salmon louse (fish parasitology) *See Lepeophtherius salmonis.*

salmon poisoning (fish/veterinary parasitology) A disease of dogs and other canids which eat salmon from streams in the Pacific Northwest of the USA, and caused by *Neorickettsia helminthoeca*. The infection is transmitted by the fluke, *Nanophyetus salmincola*, parasitic in the salmon. The disease in dogs is characterized by fever, ocular discharge and oedema (*USA*: edema) of the eyelids, followed by vomiting, then diarrhoea (*USA*: diarrhea) and later severe dysentery and death in untreated cases. Also termed: salmon disease.

See also Elokomin fluke fever.

salping(o)- (terminology) A word element. [German] Prefix denoting *tube*.

salt (1, 2 chemistry) 1 Product obtained when a hydrogen atom in an acid is replaced by a metal or its equivalent (*e.g.* the ammonium ion NH_4^+). It results from the reaction between an acid and a base.

saltcake (chemistry) An alternative term for crude sodium sulphate (*USA*: sodium sulfate).

salting out (chemistry) Precipitation of a colloid (*e.g.* gelatine) by the addition of large amounts of a salt.

saltpetre (chemistry) An alternative term for potassium nitrate.

salvage statistics (statistics) Statistical technique used in an attempt to derive some useful information from a poorly designed or poorly executed experiment.

salvation (chemistry) Attachment between solvent and solute molecules. The greater the polarity of the solvent, the greater is the attraction between solute and solvent molecules.

samarium (Sm) (chemistry) Metallic element in Group IIIB of the periodic table (one of the lanthanides). It is slightly radioactive and arises from fission fragments in a nuclear reactor, where it acts as a 'poison'. At. no. 62; r.a.m. 150.35.

samore (parasitology) Trypanosomiasis. Also termed: nagana.

sample (1 parasitology; 2 statistics) 1 A specimen of fluid, blood or tissue collected for analysis on the assumption that it represents the composition of the whole. 2 Part of a population, usually randomly selected to be representative of the whole population.

sample bias (statistics) Any way in which a sample is not representative of the population from which it was drawn.

sampling (parasitology) The process of selecting a sample.

sampling error (statistics) The difference between any value of a statistic obtained from a sample of a population.

sampling population (statistics) The population from which a sample is drawn.

sampling theory (statistics) The principles that govern the selection of representative samples.

sampling validity (statistics) The extent to which a test appears to sample those traits under test, and only those traits.

sampling with replacement (statistics) Sampling from a finite population while replacing each item sampled.

sampling without replacement (statistics) Sampling from a finite population without replacing the items sampled.

sanatorium (medical) A hospital that specialised in the treatment of tuberculosis. Since the introduction of chemotherapy for the disease, such hospitals are no longer necessary.

Sancassania (insect parasitology) *See Rhizoglyphus.*

Sancassania berlessei (veterinary parasitology) A mite of stored products that occasionally infests sheep.

sand flea (veterinary parasitology) *See Tunga penetrans.*

sandfly (parasitology) *Phlebotomus* spp. In some countries *Culicoides*, *Simulium* and *Austrosimulium* spp. are also called sandflies. Also termed: owl midges.

sandfly zieria (parasitology) *Zieria smithii.*

sandwich compound (chemistry) Orgaometallic compound whose molecules consist of two parallel planar rings with a metal atom centred (*USA*: centered) between them (*e.g.* ferrocene).

sangui- (terminology) A word element. [Latin] Prefix denoting *blood.*

Sanguinicola (fish parasitology) A genus of the family Sanguinicolidae, digenetic flukes of the vascular system of freshwater and marine fish. Species includes *Sanguinicola inermis* and *Sanguinicola klamanthensis* that may be found in cyprinid and salmonid fish in which it is a serious pathogen, especially in cultured carp and trout. The gills and kidneys are most affected and mortality may be high.

sanguinopurulent (medical) Containing both blood and pus.

sanguivorous (parasitology) Blood-eating; said of female mosquitoes that prefer blood to other nutrients.

SAP (biochemistry) An abbreviation of 'S'erum 'A'lkaline 'P'hosphatase.

saponification (chemistry) Hydrolysis of an ester, using an alkali, to produce a free alcohol and a salt of the organic acid. It is the process by which soap is made.

saponification value (chemistry) Number of milligrams of potassium hydroxide required for the complete saponification of 1 g of the substance being tested. Also termed: saponification number.

sapr(o)- (terminology) A word element. [German] Prefix denoting *rotten*, *putrid*, *decay*, *decayed material.*

saprocyclozoonoses (medical) A disease with features of both the saprozoonoses and cyclozoonoses, *e.g.* tick paralysis.

saprophyte (biology) Organism that feeds on dead or decaying organic matter. Saprophytic activity is the first step in the decomposition of dead animals and plants, and consequently is important in the recycling of elements. Also termed: saprotroph.

saprotroph (biology) An alternative term for saprophyte.

saprozoic (biology) Living on decayed organic matter; said of animals, especially protozoa.

sarc(o)- (terminology) A word element. [German] Prefix denoting *flesh*.

Sarcocystis (avian/veterinary parasitology) A genus of parasitic protozoa in the family Sarcocystidae. The definitive hosts are domestic and wild carnivores in which they appear as a form of coccidosis. Birds and reptiles are also infected. It appears to be part of any prey-predator system, *e.g.* snakes, rats, owls, mice, etc. The effects of their intermediate stages are manifested as the disease sarcocystosis.

Sarcocystis bertrami (veterinary parasitology) (synonym: *Sarcocystis equicanis*) A species of parasitic protozoa in the genus *Sarcocystis* that has a dog-horse cycle and produces cysts in muscles but without clinical disease.

Sarcocystis bovifelis (parasitology) *See Sarcocystis hirsuta.*

Sarcocystis capracanis (veterinary parasitology) A species of parasitic protozoa in the genus *Sarcocystis* which has a dog-goat cycle but that is not considered pathogenic.

Sarcocystis cervi (veterinary parasitology) A species of parasitic protozoa in the genus *Sarcocystis* whose cysts may be found in deer.

Sarcocystis cruzi (veterinary parasitology) A species of parasitic protozoa in the genus *Sarcocystis* that has a dog (and other feral canids)-cattle cycle. They cause fever, anorexia, anaemia (*USA*: anemia) and weight loss in cattle but the microscopic cysts found in dogs are not considered to be pathogenic.

See also sarcocystosis.

Sarcocystis cuniculi (veterinary parasitology) A species of parasitic protozoa in the genus *Sarcocystis* whose cysts may be found in wild and domesticated rabbits but without apparent pathogenicity in rabbits. It has a cat-rabbit cycle and the macroscopic cysts in cats are pathogenic.

Sarcocystis equicanis (parasitology) *See Sarcocystis bertrami.*

Sarcocystis fayeri (veterinary parasitology) A species of parasitic protozoa in the genus *Sarcocystis* that has a dog-horse cycle but appears not to be pathogenic to the horse.

Sarcocystis fusiformis (veterinary parasitology) A species of parasitic protozoa in the genus *Sarcocystis* that has a cat-buffalo cycle but appears not to be pathogenic in the water buffalo.

Sarcocystis gigantea (veterinary parasitology) (synonym: *Sarcocystis ovifelis*) A species of parasitic protozoa in the genus *Sarcoonym:cystis* that has a cat-sheep cycle. The cysts in sheep are very large and visible with the naked eye but are not considered to be pathogenic.

Sarcocystis hemionilatrantis (veterinary parasitology) A species of parasitic protozoa in the genus *Sarcocystis* that has dog or coyote-mule deer cycle and the disease in young deer may be fatal.

Sarcocystis hirsuta (veterinary parasitology) (synonym: *Sarcocystis bovifelis*) A species of parasitic protozoa in the genus *Sarcocystis* that has a cat-cattle cycle and is not considered to be pathogenic in cattle.

Sarcocystis hominis (human/veterinary parasitology) A species of parasitic protozoa in the genus *Sarcocystis* that has a human-cattle cycle, the enteric infection in humans causing diarrhoea (*USA*: diarrhea), and the cysts in the muscles of cattle having no observable effect.

Sarcocystis kortei (veterinary parasitology) A species of parasitic protozoa in the genus *Sarcocystis* whose definitive host is unknown. The intermediate host is the rhesus monkey but the cysts in its muscles appear to cause no apparent disability.

Sarcocystis levinei (veterinary parasitology) A species of parasitic protozoa in the genus *Sarcocystis* that has a dog-buffalo cycle but causes no apparent illness.

Sarcocystis lindemanni (human/veterinary parasitology) A species of parasitic protozoa in the genus *Sarcocystis* whose intermediate host is humans but whose final host is unknown.

Sarcocystis miescheriana (veterinary parasitology) A species of parasitic protozoa in the genus *Sarcocystis* that has a dog-pig cycle but no apparent pathogenicity.

Sarcocystis muris (veterinary parasitology) A species of parasitic protozoa in the genus *Sarcocystis* that has a cat-mouse, rat, vole, etc., cycle but the cysts of which are not considered to be pathogenic.

Sarcocystis nesbitti (veterinary parasitology) A species of parasitic protozoa in the genus *Sarcocystis* whose intermediate hosts are rhesus monkeys but the muscle cysts are clinically silent. The final host has not been identified.

Sarcocystis neurona (veterinary parasitology) A species of parasitic protozoa in the genus *Sarcocystis* that causes equine protozoal encephalomyelitis.

Sarcocystis orientalis (veterinary parasitology) A species of parasitic protozoa in the genus *Sarcocystis* whose intermediate hosts are goats which remain apparently unaffected. Dogs are probably the final host.

Sarcocystis ovicanis (parasitology) *See Sarcocystis tenella.*

Sarcocystis ovifelis (parasitology) *See Sarcocystis gigantea.*

Sarcocystis porcifelis (veterinary parasitology) A species of parasitic protozoa in the genus *Sarcocystis* that has a cat-pig cycle and is pathogenic for pigs, causing diarrhoea (*USA*: diarrhea), myositis and lameness.

Sarcocystis porcihominis (parasitology) *See Sarcocystis suihominis.*

Sarcocystis rileyi (avian parasitology) A species of parasitic protozoa in the genus *Sarcocystis* whose cysts occur in the muscles of many species of domestic and wild birds without causing any apparent ill-effects.

Sarcocystis suihominis (human/veterinary parasitology) A species of parasitic protozoa in the genus *Sarcocystis* that has a human-pig cycle without any apparent pathogenic effects.

Sarcocystis tenella (veterinary parasitology) (synonym: *Sarcocystis ovicanis*) A species of parasitic protozoa in the genus *Sarcocystis* which has a dog-sheep cycle and that causes mortality in lambs if the infection is heavy.

sarcocystosis (veterinary parasitology) A rare clinical disease in all food animal species caused by the intermediate stage of the protozoan parasite *Sarcocystis* spp. The terminal stages are passed in the dog or cat. Clinically the disease is manifested by emaciation, lameness, hypersalivation, loss of tail switch, anaemia (*USA*: anemia) and abortion. The subclinical infection with the intermediate stage of cysts in muscles is very common in all species. The common sites are the oesophageal (*USA*: esophageal), cardiae and lingual muscles. Abnormally there are localizations in brain, uterus and lungs. Also termed: rat-tail syndrome; Dalmeny disease.

Sarcodina (parasitology) A subphylum of protozoa, including all the amoebae (*USA*: amebae), both freeliving and parasitic, characterized by the ability to produce pseudopodia during most of the life cycle. Flagella, when present, develop only during the early stages.

Sarcophaga (parasitology) A genus of flesh flies in the family Sarcophagidae that deposit their larvae in wounds or sores. Species from varying geographical distribution includes *Sarcophaga carnaria*, *Sarcophaga dux*, *Sarcophaga fuscicauda* and *Sarcophaga haemorrhoidalis.*

Sarcophaga sarracenioides (insect parasitology) Diptera in the Sarcophagidae that simlarly to *Sarcophaga surrubea* and *Senotainia tricuspis* are parasitic to adults of the bees *Bombus* spp., and *Apis mellifera*. *Senotainia tricuspis* larva are deposited externally and burrow into the host.

Sarcophaga surrubea (insect parasitology) *See Sarcophaga sarracenioides.*

Sarcopterinus (parasitology) A genus of mites in the family Myobiidae.

Sarcopterinus nidulans (avian parasitology) A species of mites in the genus *Sarcopterinus* that may be found in the feather follicles of pigeons and other birds.

Sarcopterinus pilirostris (avian parasitology) A species of mites in the genus *Sarcopterinus* that may be found in the skin on the head of the sparrow.

Sarcoptes (parasitology) A widely distributed genus of mites in the family Sarcoptidae which causes sarcoptic mange and that includes *Sarcoptes scabiti*, the itch mite which causes scabies. Their nomenclature is confused but the most widely used system of nomenclature is *Sarcoptes scabiei* var. *canis*, var. *suis*, var. *equi*, var. *ovis*, var. *bovis*, etc.

Sarcoptes tapiri (veterinary parasitology) A species of mites in the genus *Sarcoptes* that are potentially pathogenic for tapirs.

sarcoptic mange (parasitology) An intensely pruritic dermatitis caused by the acarid mite *Sarcoptes scabiei*. Although there is some species specificity with subspecies of the mite this is not complete. Lesions commence as erythema and small red papules. Extensive self trauma leads to loss of hair and secondary infection. In long-standing cases, debilitation is also common. The lesions are usually widespread but are most easily seen on the abdominal skin and inside the thighs. In dogs, the elbows, hocks and pinnae are most commonly affected. Besides the common infections in domestic species, the disease occurs frequently in captive and freeliving primates, monkeys, rodents, canids and ungulates. Also termed: red mange; fox mange.

See also scabies.

sarcosporidian cysts (parasitology) Cylindrical cysts (schizonts) containing bradyzoites, found in the muscles of those infected with *Sarcocystis* spp.

SAT (methods) An abbreviation of 'S'erum 'A'gglutination 'T'est.

sat. (chemistry) An abbreviation of 'sat'urated.

satellite chromosome (genetics) A part of a chromosome that is joined to the end of the main chromosome by a very fine thread of non-condensed chromatin.

satellite DNA (molecular biology) Fraction of DNA with significantly different density and thus base composition from most of the DNA in an organism.

saturated compound (chemistry) Organic compound that contains only single bonds; all the atoms in the compound exert their maximum combining power (valency) with other atoms, so that a chemical change can be effected only by a substitution reaction and not in an addition reaction.

saturated solution (chemistry) Solution that cannot take up any more solute at a given temperature.

See also supersaturated solution.

saturated vapour (physics) Vapour that can exist in equilibrium with its parent solid or liquid at a given temperature.

saturated vapour pressure (physics) Pressure exerted by a saturated vapour. It is temperature dependent.

saturation (chemistry) Point at which no more of a material can be dissolved, absorbed or retained by another.

sawfly (parasitology) Member of the insect family Pergidae. Species include *Arge pullata*, found in Denmark; *Lophyrotoma interrupta*, in Australia; and *Perreyea lepida*, in South America. The larvae of these leaf-eating insects collect in piles under the trees they parasitize and cattle may eat them avidly, developing acute hepatitis due to the ingestion of lophyrotomin in the larvae.

SB (education) An abbreviation of '*S'cientiae 'B'accalaureus*. Latin, meaning 'Bachelor of Science'.

s-block elements (chemistry) Metallic elements that form Groups IA and IIA of the periodic table, which include the alkali metals, alkaline earths and the lanthanides and actinides (together also known as the rare earths); hydrogen is usually included as well. They are so called because their 1 or 2 outer electrons occupy s-orbitals.

Sc (chemistry) Chemical symbol for scandium.

sc (medical) An abbreviation of 's'ub'c'utaneously.

scab (medical) 1 A crust composed of coagulated serum, blood, pus and skin debris covering a skin lesion. 2 Used colloquially to mean psoroptic mange.

scab mites (parasitology) *See Chorioptes*; *Psoroptes*; *Sarcoptes*.

scabicide (control measures) 1 Fatal to *Sarcoptes scabiei*. 2 A substance fatal to *Sarcoptes scabiei*.

scabies (parasitology) Infestation by mites of the genus *Sorcoptes*.

See also sarcoptic mange.

scabies incognito (veterinary parasitology) A variant of sarcoptic mange in dogs in which mites are difficult or impossible to recover in skin scrapings, presumably because of the extensive grooming and generally high level of skin hygiene that lacks only the use of a scabicide. Also there are usually only a few mites present once an immune response develops. Further infection may cause a hypersensitivity to develop but the mites present are usually still in small numbers.

scabietic (parasitology) Pertaining to scabies.

scaly face (parasitology) *See Cnemidocoptes pilae*.

scaly leg (parasitology) *See Cnemidocoptes pilae*.

scaly leg mite (parasitology) *See Cnemidocoptes*.

scaly skin (medical) Condition characterized by scales; scalelike.

scandium (Sc) (chemistry) Silvery-white metallic element in Group IIIB of the periodic table (a rare earth element); its oxide, Sc_2O_3, is used as a catalyst and to make ceramics. At. no. 21; r.a.m. 44.9559.

scanner (computing) An alternative term for optical character reader.

scanning electron microscope (microscopy) Electron microscope that scans the sample to be examined with a beam of electrons.

scapula (anatomy) An alternative term for the shoulder blade.

scatoscopy (parasitology) Examination of the faeces (*USA*: feces).

scattering of light (physics) Irregular reflection or diffraction of light rays that occurs when a beam of light passes through a material medium.

SCE (genetics) An abbreviation of 'S'ister 'C'hromatid 'E'xchange.

Schiff's base (chemistry) Organic compound formed when an aldehyde or ketone condenses with a primary aromatic amine with the elimination of water. Also termed: aldimine; azomethine. It was named after the German chemist Hugo Schiff (1834–1915).

Schiff's reagent (chemistry) Reagent for testing for the presence of aliphatic aldehydes, which quickly restore its magenta colour (*USA*: color). It is prepared by dissolving rosaniline hydrochloride in water and passing sulphur dioxide (*USA*: sulfur dioxide) through it until the magenta colour is discharged.

schist(o)- (terminology) A word element. [German] Prefix denoting *cleft*, *split*.

Schistocephalus (parasitology) A genus of tapeworms in the family Diphyllobothriidae.

Schistocephalus solidus (avian/fish parasitology) A species of tapeworms in the genus *Schistocephalus* the adults of which may be found in the intestine of birds and can cause mortality in ducks. The intermediate stages in cyprinid fishes cause marked swelling of the abdomen.

Schistosoma (parasitology) A genus of elongated dioecious trematodes which inhabit blood vessels of the host. The eggs are found in the wall of the bladder, uterus and urethra.

Schistosoma bovis (veterinary parasitology) A species of trematodes in the genus *Schistosoma* that may be found in ruminants.

Schistosoma curassoni (veterinary parasitology) A species of trematodes in the genus *Schistosoma* that may be found in ruminants.

Schistosoma haematobium (human parasitology) A species of trematodes in the genus *Schistosoma* that may be found in humans.

Schistosoma incognitum (veterinary parasitology) A species of trematodes in the genus *Schistosoma* that may be found in pigs and dogs.

Schistosoma indicum (veterinary parasitology) A species of trematodes in the genus *Schistosoma* that may be found in ruminants and horses.

Schistosoma intercalatum (human/veterinary parasitology) A species of trematodes in the genus *Schistosoma* that may be found in humans, ruminants and horses.

Schistosoma japonicum (human/veterinary parasitology) A species of trematodes in the genus *Schistosoma* that may be found in humans and many other species.

Schistosoma lieperi (veterinary parasitology) A species of trematodes in the genus *Schistosoma* that may be found in wild artiodactyls.

Schistosoma magrebowiei (veterinary parasitology) A species of trematodes in the genus *Schistosoma* that may be found in ruminants and zebra.

Schistosoma makongi (human/veterinary parasitology) A species of trematodes in the genus *Schistosoma* that may be found in humans and dogs.

Schistosoma mansoni (human/veterinary parasitology) A species of trematodes in the genus *Schistosoma* that may be found in humans and wild animals.

Schistosoma mattheei (parasitology) A species of trematodes in the genus *Schistosoma* that may be found in most species.

Schistosoma nasalis (veterinary parasitology) A species of trematodes in the genus *Schistosoma* that may be found in ruminants and horses.

Schistosoma rodhaini (veterinary parasitology) A species of trematodes in the genus *Schistosoma* that may be found in dogs and rodents.

Schistosoma spindale (veterinary parasitology) A species of trematodes in the genus *Schistosoma* that may be found in ruminants and dogs.

Schistosoma suis (parasitology) *See Schistosoma incognitum*.

Schistosomatium (veterinary parasitology) A genus of flukes in the family Schistosomatidae that may be found in rodents.

Schistosomatium douthitti (veterinary parasitology) A species of flukes in the genus *Schistosomatium* that may be found in mesenteric veins and may cause dermatitis in rodents.

schistosome (parasitology) A member of the family Schistosomatidae, which includes the genera *Austrobilharzia*, *Bilharziella*, *Bivitellobilharzia*, *Dendritobilharzia*, *Gigantobilharzia*, *Heterobilharzia*, *Ornithobilharzia*, *Pseudobilharziella*, *Schistosoma*, *Schistosomatium* and *Trichobilharzia*.

schistosome dermatitis (human parasitology) A disease of humans caused by invasion of the skin by the cercariae of nonhuman schistosomes, especially avian ones. Also termed clam-digger's itch; swimmer's itch; rice-paddy-itch; swamp itch.

schistosomiasis (human/veterinary parasitology) The disease caused by the trematode *Schistosoma* spp. The commonest syndrome is one of haemorrhagic (*USA*: hemorrhagic) enteritis, anaemia (*USA*: anemia) and emaciation with many affected animals dying after an illness of several months. Necropsy lesions include distended mesenteric vessels filled with flukes, haemorrhagic (*USA*: hemorrhagic) enteritis with granuloma formation in some cases. Granulomatous lesions occur in the liver in a hepatic form of the disease. Nasal schistosomiasis, caused by *Schistosoma nasalis*, is characterized by nasal discharge, snoring and dyspnea. *Shistosoma haematobium* causes haematuria (*USA*: hematuria). Human infestation with cercariae causes schistosome dermatitis.

schistosomicide (pharmacology/control measures) A substance that destroys schistosomes.

schiz(o)- (terminology) A word element. [German] Prefix denoting *divided*, *division*.

Schizaspidia convergens (insect parasitology) Hymenoptera in the Eucharitidae that, similarly to *Schizaspidia doddi*, *Schizaspidia polyrhachicida*, *Schizaspidia tenuicornis*, *Stilbula cynipiformis*, *Stilbula* (*Schizaspidia*) *polyrhachicida* and *Stilbula tenuicornis*, may be found in pupa of *Calomyrmex*, *Camponotus* spp., *Odontomachus* and *Polyrhachis*.

Schizaspidia doddi (insect parasitology) *See Schizaspidia convergens*.

Schizaspidia polyrhachicida (insect parasitology) *See Schizaspidia convergens*.

Schizaspidia tenuicornis (insect parasitology) *See Schizaspidia convergens*.

schizodemes (parasitology) Parasites with mitochondrial DNA of similar characteristics.

schizogony (biology) Form of asexual reproduction employed by an apicomplexan parasite by multiple fission, in which a parent cell divides into more than two independent cells, within the body of the host, giving rise to merozoites.

See also binary fission.

schizont (parasitological physiology) The asexual reproductive stage in the development of the *Eimeria* spp. and in many other coccidians, *e.g.* *Toxoplasma* spp. in the cat, *Cystoisospora*, *Hammondia*, *Frenkelia*, *Isospora* spp. and *Plasmodium* spp, following the trophozoite whose nucleus divides into many smaller nuclei. This process of division is known as schizogony.

See also malaria.

Schizotrypanum (parasitology) *See Trypanosoma*.

Schmorl stain (histology) A histological staining technique used to make lipofuscin visible under the microscope.

Schoengastia (parasitology) *See* harvest mites.

Schuffner's dots (parasitology) A term used in relation to malaria. Pinkish small and round stippling seen on *Plasmodium vivax* and *Plasmodium ovale* infected red cells, in Romanowsky stained films. They appear earlier and in greater numbers in *Plasmodium ovale* than in *Plasmodium vivax.*

Schwann cell (cytology) Cell that produces the myelin sheath that surrounds a nerve cell (neurone). Schwann cells are in close contact with the axon of the neurone and are separated by gaps called nodes of Ranvier. It was named after the German physiologist Theodor Schwann (1810–82).

scintillation counter (physics) Device that counts the incidence of photons upon a material by the visible or near-visible light which is emitted.

scirrho- (terminology) A word element. [German] Prefix denoting *hard.*

scler(o)- (terminology) A word element. [German] Prefix denoting *hard, sclera.*

scleroprotein (biochemistry) Member of a group of fibrous proteins that provide organisms with structural materials (*e.g.* collagen, keratin).

sclerotic (histology) Outermost of the three layers that form the eyeball (outside the choroid and retina).

SCN (histology) An abbreviation of 'S'upra'C'hiasmatic 'N'ucleus.

scolex(parasitological anatomy) Plural: scoleces. [German] The attachment organ of a tapeworm, generally considered the anterior, or cephalic, end. Also termed: holdfast.

scoli(o)- (terminology) A word element. [German] Prefix denoting *crooked, twisted.*

-scopy (terminology) A word element. [German] Suffix denoting *examination of.*

scoto- (terminology) A word element. [German] Prefix denoting *darkness.*

scour worm (parasitology) *See Trichostrongylus; Ostertagia; Cooperia; Nematodirus* spp.

scraping (methods) A scraping of the superficial elements of the skin for laboratory examination for parasitic elements.

scratch test (immunology) A test for hypersensitivity in which a minute amount of the substance in question is inserted in small scratches made in the skin. A positive reaction is swelling and reddening at the site within 30 minutes. Used in allergy testing. In animals, intradermal testing is used more commonly.

See also skin test.

screening (methods) 1 Examination of a large sample of animals in a population in order to detect the presence of disease or to ascertain the prevalence of certain diseases. 2 In diagnostic tests, the use of a test which has a high sensitivity but often only a moderate specificity.

screening test (methods) Any test used to screen a population.

screw-worm (parasitology) The larvae of *Callitroga, Chrysomya bezziana* and *Passeromyia.*

screw-worm flies (parasitology) The flies that produce screw-worms.

screw-worm myiasis (human/veterinary parasitology) Invasion of normal skin wounds by maggots of *Callitroga, Chrysomya bezziana* and *Passeromyia* flies. Large masses of tissue may be destroyed and the case fatality rate is very high. All animal species, including humans, are susceptible.

scrotal (anatomy) Pertaining to scrotum.

scrotal mange (veterinary parasitology) *See* chorioptic mange.

scrotal myiasis (veterinary parasitology) Blowfly strike of a recently castrated ram lamb.

scrotum (anatomy) Sac present in males of some mammals that contains the testes. Positioned outside the body cavity so that their temperature is cool enough for sperm production.

scrub tick (parasitology) *See Ixodes holocyclus.*

scrub typhus (human parasitology) A disease of humans transmitted by *Trombicula akamushi* and resident in rodents which serve as reservoirs. Also termed: Japanese river fever; tsutsugamushi disease.

scrub-itch mite (parasitology) *See Acomatacarus; Trombicula minor.*

Scutacaridae (insect parasitology) Mites in the Prostigmata that are phoretic on legs of adults of the ants *Eciton* and *Atta sexdens.*

Scutacarus (insect parasitology) A species of mites in the Prostigmata, that similarly to *Siteroptes* are parasites on the bees *Apinae, Halictinae* and Bombinae.

Scutacarus acarorum (insect parasitology) Mites in the Prostigmata that may be parasites to social wasps.

Scutacarus attae (insect parasitology) Mites in the Prostigmata that are phoretic on legs of adults of the ants *Eciton* and *Atta sexdens.*

Scutacarus minutus (insect parasitology) Mites in the Prostigmata that are phoretic on legs of adults of the ants *Eciton* and *Atta sexdens.*

scutellum (parasitological anatomy) A term used in relation to arthropods. A small posterior section of the tergum.

scutum (parasitological anatomy) A protective covering or shield, *e.g.* a chitin plate in the exoskeleton of hard-bodied ticks

SD (statistics) An abbreviation of 'S'tandard 'D'eviation.

SDA (physiology) An abbreviation of 'S'pecific 'D'ynamic 'A'ction.

SDS (computing) An abbreviation of 'S'cientific 'D'ata 'S'ystems.

SDS-PAGE (chemistry) An abbreviation of 'sodium 'd'odecyl 's'ulphate – 'p'oly 'a'crylamide 'g'el 'e'lectrophoresis.

sebaceous gland (anatomy) Small gland found in large numbers in the skin of mammals, usually alongside a hair follicle, that secretes the protective skin oil sebum.

sebum (biochemistry) Waxy material secreted by sebaceous glands, which helps to keep skin waterproof.

sec (measurement) An abbreviation of 'sec'ond. A unit of time equal to 1/60 of a minute.

Secernentea (fish parasitology) A subclass of roundworms in the phylum Nematoda. Three orders of this subclass, the Oxyurida, Ascaridida and Spirurida, contain a number of important genera found in fish. In the Secernentea, there are no stichosome or trophosome, caudal papillae are almost always numerous (the basic number being 21), phasmids are present and eggs are without polar plugs. *See* Nematoda.

second (measurement) 1 SI unit of time, defined as the duration of 9,192,631,770 periods of the radiation between the two hyperfine levels of the ground state of the caesium-133 (*USA*: cesium-133) atom. Abbreviated to sec or s. 2 Angle equal to 1/60 of a minute or 1/360 of a degree.

secondary cell (physics) Electrolytic cell that must be supplied with electric charge before use by passing a direct current through it, but it can be recharged over and over again. Also termed: accumulator; storage cell.

See also primary cell.

secondary colour (optics) Colour (*USA*: color) obtained by mixing primary colours. (*USA*: secondary color).

secondary exo-erythrocytic stage (parasitology) A term used in relation to malaria. It was used to denote schizonts developing in the liver as a result of invasion by merozoites from pre-erythrocytic schizont. It is now believed that re-infection of liver cells does not occur by pre-erythrocytic merozoites.

secondary host (parasitology) Intermediate host.

secondary sexual characteristics (physiology) Features that develop in some animals after the onset of puberty, distinguishing males from females but not required for sexual function. They result from the actions of sex hormones, principally testosterone and oestrogen (*USA*: estrogen).

secretion (biochemistry) Release of a substance by a cell or gland with a specialized function, *e.g.* secretion of digestive enzymes by cells of the small intestine, or secretion of hormones by the pituitary.

section (histology) A slice of tissue cut for examination under a microscope.

sedimentation (physics) Removal of solid particles from a suspension by gravitational force or in a centrifuge.

seed and leaf gall nematode (plant parasitology) *See Anguina tritici.*

seed tick (parasitological physiology) Larval form; the stage prior to the nymph.

segregation (genetics) Separation of a pair of alleles in a diploid organism during meiosis in the formation of gametes. A gamete receives one of the two alleles in a diploid organism because it receives only one of a pair of homologous chromosomes.

selection pressure (genetics) The intensity with which natural selection is acting upon a population to change the gene frequencies from one generation to the next.

selective advantage (genetics) The increase in fitness of one genotype compared to others in the same population.

selectively permeable membrane (physics) An alternative term for semipermeable membrane.

selenium (Se) (chemistry) Nonmetallic element in Group VIA of the periodic table, obtained from flue dust in refineries that use sulphide (*USA*: sulfide) ores. One of its allotropes conducts electricity in the presence of light, and is used in photocells and rectifiers. At. no. 34; r.a.m. 78.96.

self-absorption (radiology) Decrease in radiation from a large radioactive source due to absorption by the material itself of some of the radiation produced. Also termed: self-shielding.

self-cure (parasitology) A phenomenon in sheep in which a hypersensitivity to an adult worm load develops and the worms are discharged. The hypersensitivity is induced by a second larval infestation. The apparent cure of the sheep is misleading in the assessment of a control programme (*USA*: program).

self-induced electromotive force (physics) Production of an electromotive force (e.m.f.) in an electric circuit when the current is varied.

self-induction (physics) Resistance to a change in electric current in a circuit by the creation of a back electromotive force.

selfish DNA (molecular biology) Those parts of a DNA sequence in a species that serve no apparent useful function, and which are thought to have survived evolution only because they do not actually harm the organism in any way.

self-limited (medical) Limited by its own peculiarities, and not by outside influence; said of a disease that runs a definite limited course and resolves itself without intervention.

self-limited disease (medical) *See* self-limited.

SEM (statistics) An abbreviation of 'S'tandard 'E'rror of the 'M'ean.

semen (biochemistry) Fluid produced in male reproductive organs of many animals. It contains sperm, and, in mammals, secretions from the accessory sex glands.

semi- (terminology) A word element. [Latin] Prefix denoting *half*.

semicarbazone (chemistry) Crystalline organic compound formed when an aldehyde or a ketone reacts with semicarbazide ($NH_2NHCONH_2$) with the elimination of water. Semicarbazones are used to identify the original aldehyde or ketone.

semicircular canal (anatomy) Part of the ear that is involved in maintaining balance.

semiferous tubule (anatomy) One of many tubes within the testes in which sperm are made.

seminal receptacle (parasitological anatomy) A term used in relation to trematodes. A dilated organ in the female genital tract which stores sperms.

seminal vesicle (anatomy) 1 Organ in the testes that is used for storing sperm. 2 A term used in relation to trematodes. The dilated lower part of the vas deferens which opens into cirrus.

semipermeable membrane (physics) Porous membrane that permits the passage of some substances but not others; *e.g.* plasma membrane, which permits entry of small molecules such as water but not large molecules, allowing osmosis to occur. Such membranes are extremely important in biological systems and are used in dialysis. Also termed: selectively permeable membrane.

Senotainia tricuspis (insect parasitology) *See Sarcophaga sarracenioides.*

sense organ (histology) Group of receptor cells specialized to react to (detect) a certain stimulus (*e.g.* the eye to light, the ear to sound, and chemoreceptors in the tongue and nose to tastes and smells).

senses (physiology) The five primary senses, common to most vertebrates but sometimes lacking in less highly evolved animals, are sight, hearing, taste, smell and touch, to which may be added the sense of balance. They are effected by various sense organs.

sensilla (parasitological anatomy) A term used in relation to arthropods. Hairs which have sensory function.

septic fever (medical) Fever associated with infection either as a local abscess or cellulitis or as a septicaemia (*USA*: septicemia) or bacteremia. The infective agent may be a protozoa.

septicaemia (medical) Disorder that results from the presence of microorganisms, or their toxins, in the bloodstream. Also termed: blood poisoning. (*USA*: septicemia).

septivalent (chemistry) Having a valency of seven. Also termed: heptavalent.

septum (anatomy) Dividing wall found in biological systems, *e.g.* between the nostrils or between the two halves of the heart.

sequela (medical) Any abnormality caused by an illness that has ended.

SER (histology) An abbreviation of 'S'mooth 'E'ndoplasmic 'R'eticulum.

Ser. (biochemistry) An abbreviation of 'Ser'ine.

serial passage (methods) Repeated passage through a series of experimental animals or media, often with the objective of altering the virulence of the agent or adapting it to grow better.

series (1 chemistry; 2 physics) 1 Systematically arranged succession of chemical compounds (*e.g.* homologous series) or of numbers or algebraic terms (*e.g.* arithmetic series, exponential series, geometric series). 2 Describing the arrangement of components in a series circuit.

series circuit (physics) Electrical circuit in which the components are arranged one after the other so that the same current flows through each of them. For a series of resistors, the total resistance R is equal to the sum of the individual resistors; *i.e.* for three individual resistors $R = R_1 + R_2 + R_3$. For a series of capacitors, the reciprocal of the total capacitance C is equal to the sum of the reciprocals of the individual capacitances; *i.e.* $1/C = 1/C_1 + 1/C_2 + 1/C_3$.

See also parallel circuit.

serine (chemistry) $CH_2OHCHNH_2COOH$ White crystalline amino acid, present in many proteins. Also termed: 2-amino-3-hydroxypropanoic acid.

seroconversion (immunology) The point at which an individual exposed to a virus or other agent becomes serologically positive.

serodiagnosis (parasitology) Diagnosis of disease based on serum reactions.

seroepidemiology (epidemiology) A system of epidemiological surveillance and examination based on mass and serial testing of sera of samples of the animal populations.

serological (immunology) Pertaining to or emanating from serology.

serological test (methods) A test involving examination of blood serum usually for antibody.

serologist (immunology) A specialist in serology.

serology (immunology) Branch of immunology concerned with reactions between antibodies of one organism with antigens of the serum of another.

seronegative (immunology) Showing a negative serum reaction; i.e. an animal with no detectable serum antibodies to a specified microorganism.

seropositive (immunology) Showing positive results on serological examination, i.e. an animal with detectable serum antibodies to a particular microorganism.

seroprognosis (medical) Prognosis of disease based on serum reactions.

seropurulent (medical) Both serous and purulent.

seropus (medical) Serum mingled with pus.

seroreaction (immunology). Any reaction taking place in serum, or as a result of the action of a serum.

seroresistant (immunology) Showing a seropositive reaction to a pathogen after treatment.

serosa (anatomy) Any serous membrane.

serosanguineous (medical) Composed of serum and blood.

serosurvey (immunology/epidemilogy) A screening test of the serum of animals at risk to provide data about specific diseases.

serous (biology) 1 Pertaining to serum; thin and watery, like serum. 2 Producing or containing serum.

Serratospiculum amaculatum (avian parasitology) An air sac nematode that may be found infecting falcons.

serum (haematology) The clear portion of any animal or plant fluid that remains after the solid elements have been separated out. The term usually refers to the constituent of plasma of blood, which contains all the substances in plasma except for fibrinogen.

server (computing) A computer which can be accessed by other computers in a network and is dedicated to that task. Used widely by providers of network services. It is the critical feature, supported by imaginative software, which makes the Internet possible.

set (mathematics) A group of things (elements) that have at least one property in common.

seta (parasitological anatomy) A term used in relation to arthropods. Hair-like structures which are hollow internally.

Setaria (parasitology) A genus of filarioid worms in the family Onchocercidae found usually in the peritoneal cavity of ungulates. They cause no apparent clinical illness unless they invade abnormal tissues such as the central nervous system.

Setaria altaica (veterinary parasitology) A species of filarioid worms in the genus *Setaria* that may be found in the deer Cervus canadensis asiaticus.

Setaria cervi (veterinary parasitology) A species of filarioid worms in the genus *Setaria* that may be found in deer, occassionally in the spinal cord causing spinal nematodiasis, and possibly buffaloes.

Setaria congolensis (veterinary parasitology) A species of filarioid worms in the genus *Setaria* that may be found in pigs.

Setaria cornuta (veterinary parasitology) A species of filarioid worms in the genus *Setaria* that may be found in antelopes.

Setaria digitata (veterinary parasitology) A species of filarioid worms in the genus *Setaria* that may be found in cattle and buffalo. They may also occur in the urinary bladder in these species and in the central nervous system of sheep, goats and horses, causing cerebrospinal nematodiasis.

Setaria equina (veterinary parasitology) A species of filarioid worms in the genus *Setaria* that may be found in horses in the eyes, scrotum, pleural cavity and lungs as well as the peritoneal cavity. They may also be found in the eyes of cattle.

Setaria labiato-papillosa (veterinary parasitology) A species of filarioid worms in the genus *Setaria* that may be found in cattle, deer, giraffes and antelopes.

Setaria marshalli (veterinary parasitology) A species of filarioid worms in the genus *Setaria* that may be found in sheep.

Setaria tundrae (veterinary parasitology) A species of filarioid worms in the genus *Setaria* that may be found in reindeer.

Setaria yehi (veterinary parasitology) A species of filarioid worms in the genus *Setaria* that may be found in deer, moose, caribous and bison.

sex cell (cytology) An alternative term for gamete.

sex chromosome (genetics) Chromosome that carries the genes determining sex. In mammals, the female possesses two identical sex chromosomes or X-chromosomes; whereas in the male, the two sex chromosomes differ, one being an X- and the other a Y-chromosome.

sex determination (genetics) Inheritance of particular combination of sex chromosomes, which is the deciding factor in whether an organism is male or female. Inheritance of a homologous pair of sex chromosomes predisposes the organism to one sex (*e.g.* in mammals, the female). Inheritance of a pair of dissimilar sex chromosomes determines the other sex (in mammals, the male).

sex hormone (biochemistry) Hormone that determines secondary sexual characteristics and regulates the reproductive behaviour (*USA*: behavior) of an organism. *See* oestrous cycle.

sex linkage (genetics) Distribution of genes according to the sex of an organism because they are carried on the sex chromosomes. In human males, a recessive gene carried on the X-chromosome will be expressed because no corresponding allele is present on the Y-chromosome to mask it. In the female, the corresponding allele will be present on the other X-chromosome, and for this reason human males have a predisposition to recessive sex-linked disorders, *e.g.* haemophilia (*USA*: hemophilia), colour blindness (*USA*: color blindness).

sex ratio (measurement) Ratio of the number of males to the number of females in a population. It may be expressed as the number of males to every 100 females.

sexagesimal (mathematics) Describing a number system with the base 60.

sexual dimorphism (biology) The existence of physical differences between the two sexes, other than the differences in the reproductive organs.

sexual reproduction (physiology) Reproduction of an organism that involves the fusion of specialized sex cells or gametes (which are haploid) to form diploid progeny. It is important in bringing new vigour to a species by the mixing of genetic material from the parents to give a genetically different organism.

sexually transmitted disease (medical) Previously called venereal disease. The currently generally preferred term is genitourinary medicine.

SFO (anatomy) An abbreviation of 'Sub'F'ornical 'O'rgan.

sg (physics) An abbreviation of 's'pecific 'g'ravity.

SGLT 1 (physiology/biochemistry) An abbreviation of 'Sodium-dependent 'GL'ucose 'T'ransporter.

SGOT (biochemistry) An abbreviation of 'S'erum 'G'lutamic-'O'xaloacetic 'T'ransaminase.

SH (biochemistry) An abbreviation of 'S'ulp'H'ydryl (*USA*: sulfhydryl).

shaft (anatomy) A long slender part, such as the portion of a long bone between the wider ends or extremities, the shaft of a hair and the central shaft of a feather.

shaft louse (parasitology) *See Menopon gallinae.*

shampoo (control measures) A cleaning agent, usually liquid, for hair; usually consists of a detergent and perfume. Some, usually referred to as medicated shampoos, contain therapeutic substances such as parasiticides.

sharps (equipment) Needles, scalpel blades, broken ampoules, etc. Anything in a hospital or clinic which has been used on patients, and which may be contaminated with infectious material; to be discarded into special containers for disposal without any risk to disposal personnel.

sheath nematode (plant parasitology) *See Hemicycliophora arenaria.*

Sheather's flotation method (methods) A method for examining faeces (*USA*: feces) for the presence of worm eggs or larvae by mixing with a saturated solution of sodium chloride or sugar and collecting a sample from the top of a column for microscopic examination.

shedding (parasitology) Excretion of an infectious agent from the body of an infected host.

shedding agent (parasitology) An agent, *e.g.* a wild animal, shedding an infectious agent.

sheep itch mite (veterinary parasitology) *Psorergates ovis.*

sheep ked (veterinary parasitology) *See Melophagus ovinus.*

sheep nasal bot (veterinary parasitology) *See Oestrus ovis.*

sheep tick (veterinary parasitology) Sheep ked. *See Melophagus ovinus.*

sheep-head fly (veterinary parasitology) *Hydrotoea irritans. See Hydrotoea.*

sheep-scab (veterinary parasitology) *See* psoroptic mange.

sheet feeder (computing) A device fitted to a printer which feeds single sheets or envelopes into the printer one at a time.

shingle tick (parasitology) *See Dermacentor albipictus.*

shoe-leather epidemiology (epidemiology) Epidemiology conducted as a field study. Also termed: gum-boots epidemiology.

short-nosed cattle lice (veterinary parasitology) *Haematopinus eurysternus.*

shortnosed louse (parasitology) *Haematopinus eurysternus.*

shotty eruption (veterinary parasitology) Round, cystic papules around the tailhead and buttocks of pigs. The lesions are in sweat glands and are thought to be caused by *Eimeria fusca*, a protozoan parasite.

shoulder blade (anatomy) An alternative term for scapula.

shoulder tick (parasitology) *See Ixodes scapularis.*

shunt (1 physics; 2 medical) 1 Device that directs an electric current in a known way. 2 A surgically implanted tube or vessel that diverts the flow of fluid (*e.g.* to by-pass an obstruction).

shuttle vectors (parasitology) Vectors which contain both prokaryotic and eukaryotic replication signals, thus allowing replication of the vector in both kinds of cells.

SI units (measurement) An abbreviation for 'S'ystème 'I'nternational d' Unités, an international system of scientific units. It has seven basic units: metre (m), kilogram (kg), second (s), kelvin (K), ampere (A), mole (mol) and candela (cd), and two supplementary units radian (rad) and steradian (sr). There are also 18 derived units.

sial(o)- (terminology) A word element. [German] Prefix denoting *saliva, salivary glands*.

Siberian tick typhus (human/veterinary parasitology) A disease of humans and many species of feral mammals, especially rodents, caused by *Rickettsia siberica* and transmitted by the ticks *Dermacentor* and *Haemaphysalis*.

sibling (medical) A brother or sister.

sibship (medical) All the brothers and sisters in a family.

side reaction (chemistry) Chemical reaction that takes place at the same time as the main reaction.

sidero- (terminology) A word element. [German] Prefix denoting *iron*.

SIDS (medical) An abbreviation of 'S'udden 'I'nfant 'D'eath 'S'yndrome.

siemens (S) (measurement) SI unit of electric conductance, formerly expressed in reciprocal ohms (mhos). It was named after the German physicist Ernst Werner von Siemens (1816–92).

sievert (Sv) (measurement) SI unit of radiation dose equivalent.

SIF cells (histology) An abbreviation of 'S'mall, 'I'ntensely 'F'luorescent cells in sympathetic ganglia.

sigmoid mesocolon (anatomy) *See* pelvic megacolon.

significant figure (mathematics) Digit that gives information about a number containing it, and not a zero used simply to indicate a vacant place at the beginning or end of the number; a digit that makes an actual contribution to the number. Also termed: significant digit.

silica gel (chemistry) Porous amorphous variety of silica (SiO_2) which is capable of absorbing large quantities of water and other solvents. It is used as a desiccant and adsorbent, and has been used topically on dogs and cats for flea control.

silicon (Si) (chemistry) Nonmetallic element in Group IVA of the periodic table, which exists as amorphous and crystalline allotropes. It is the second most abundant element, occurring as silicates in clays and rocks. Sand and quartz consists of silica (silicon dioxide, SiO_2). It is used in making refractory materials and temperature-resistant glass. At. no. 14; r.a.m. 28.086.

Silpha (avian parasitology) Genus of the family of Silphidae of carrion beetles which infest pigeon droppings and whose larval maggots may invade the skin of live squabs.

silver (Ag) (chemistry) Silver-white metallic element in Group IB of the periodic table (a transition element). It occurs as the free element (native) and in various sulphide (*USA*: sulfide) ores. It is used in jewellery, electrical contacts, batteries and mirrors. Silver halides are used in photographic emulsions. At. no. 47; r.a.m. 107.868.

silver bromide (chemistry) AgBr Pale yellow insoluble crystalline salt, used for making light-sensitive photographic emulsions.

silver chloride (chemistry) AgCl White insoluble crystalline salt, used in the manufacture of pure silver and in photographic emulsions.

silver iodide (chemistry) AgI Pale yellow insoluble crystalline salt, used in photographic emulsions.

silver nitrate (chemistry) $AgNO_3$ Colourless (*USA*: colorless) crystalline salt, used in volumetric analysis and as a caustic in medicine (*e.g.* for removing warts).

silver oxide (chemistry) Ag_2O Brown amorphous solid, only slightly soluble in water but soluble in ammonia solution. Also termed: silver(I) oxide.

Simondsia paradoxa (veterinary parasitology) A member of the nematode family Thelaziidae. A thick white stomach worm of pigs that causes chronic gastritis.

simple microscope (microscopy) A microscope that consists of a single lens.

simulation study (epidemiology) A scholarly examination in which the real circumstances are simulated, either in fact, or by means of a set of mathematical formulae each of which expresses the probability of each outcome in a series of consequential events that mirror the possible pathways in a real-life situation.

Simulium (insect parasitology) A genus of insects in the family Simuliidae causing insect worry with livestock. Cutaneous lesions may include vesicles and wart-like papules, and oedema (*USA*: edema) and petechiation of thin-skinned, ventral areas. Poultry may be affected by anaemia (*USA*: anemia). The species also transmits a number of animal diseases including the protozoa, *Leucocytozoon* and *Haemoproteus* spp., and are intermediate parasites for some *Onchocerca* spp. There are a large number of species distinguished largely by their geographic distribution and includes: *Simulium arcticum, Simulium callidum, Simulium columbaczense, Simulium damnosum, Simulium erythrocephalum, Simulium indicum, Simulium metallicum, Simulium neavei, Simulium ochraceum, Simulium ornatum, Simulium pecuarum, Simulium rugglesi* and *Simulium venustum*. Also termed: black fly; buffalo gnat.

simultaneous equations (mathematics) Set of algebraic equations that are all true for the same particular values of their variables.

sine wave (physics) Waveform that represents the periodic oscillations of constant amplitude as given by the sine of a linear function. Also termed: sinusoidal wave; sine curve.

Singhfilaria hayesi (avian parasitology) A tissue-dwelling nematode found in tissues around the crop and trachea of quail and turkeys in USA.

single bond (chemistry) Covalent bond formed by the sharing of one pair of electrons between two atoms.

sinistr(o)- (terminology) A word element. [Latin] Prefix denoting *left, left side.*

Sinostrongylus (parasitology) *See Caballonema.*

sinus (anatomy) Irregular cavity or depression that forms part of an animal's anatomy; *e.g.* sinuses in the bones of the face in mammals.

siphon (physics) Device consisting of an inverted U-shaped tube that moves a liquid from one place to another place at a lower level. The tube has to be initially filled with liquid in order to function.

Siphonaptera (parasitology) The order of fleas.

SIR (medical) An abbreviation of 'S'ystemic 'I'nflammatory 'R'esponse.

-sis (terminology) A word element. [German] Suffix denoting *state, condition.*

Siteroptes (insect parasitology) *See Scutacarus.*

skeleton (anatomy) Structure that supports the tissues and organs of an animal and is attached to muscles to allow locomotion. An endoskeleton is internal, made of bone or cartilage, and possessed by vertebrates. Exoskeletons lie outside the muscles, *e.g.* in arthropods. Some invertebrates possess a hydrostatic skeleton which consists of fluid under pressure.

skewed distribution (statistics) A distribution in which the curve illustrating it is not symmetrical but has a long tail on one or the other side of the graph.

skia- (terminology) A word element. [German] Prefix denoting *shadow.*

skin (anatomy) Organ that protects the body from invasion by pathogens. In warm-blooded animals the skin also takes part in temperature regulation, *e.g.* through sweating and the constriction and dilation of its blood vessels. It consists of epithelial tissue and connective tissue arranged in two major layers, the thin outer epidermis and the thicker underlying dermis.

skin test (methods) Application or intradermal injection of a substance to the skin to test the body's reaction to it. There are several types of skin tests, including the patch test, the scratch test, and the intradermal test.

skin-maggot fly (parasitology) *See Cordylobia anthropophaga.*

Skrjabinagia (parasitology) A genus of worms in the subfamily Ostertaginae and the family Trichostrongylidae.

Skrjabinagia boevi (veterinary parasitology) A species of worms in the genus *Skrjabinagia* that may be found in the large and small intestines of water buffaloes.

Skrjabinagia dagestanica (veterinary parasitology) A species of worms in the genus *Skrjabinagia* that may be found in the small intestine of sheep.

Skrjabinagia kolchida (veterinary parasitology) A species of worms in the genus *Skrjabinagia* that may be found in the abomasum of deer.

Skrjabinagia lyrata (parasitology) *See Ostertagia lyrata.*

Skrjabinagia popovi (veterinary parasitology) A species of worms in the genus *Skrjabinagia* that may be found in the small intestine of sheep.

Skrjabinema (veterinary parasitology) A genus of worms in the family Oxyuridae that inhabit the caeca (*USA*: ceca) of ruminants.

Skrjabinema africana (veterinary parasitology) A species of worms in the genus *Skrjabinema* that may be found in steinbock in Africa.

Skrjabinema alata (veterinary parasitology) A species of worms in the genus *Skrjabinema* that may be found in sheep.

Skrjabinema caprae (veterinary parasitology) A species of worms in the genus *Skrjabinema* that may be found in goats.

Skrjabinema ovis (veterinary parasitology) A species of worms in the genus *Skrjabinema* that may be found in sheep, goats and antelopes.

Skrjabinema tarandi (veterinary parasitology) A species of worms in the genus *Skrjabinema* that may be found in reindeer and caribou.

Skrjabingylus (parasitology) A genus of worms in the metastrongyloid family Skrjabingylidae found in the nasal sinuses of their hosts.

Skrjabingylus chitwoodorum (veterinary parasitology) A species of worms in the genus *Skrjabingylus* that may be found in skunks.

Skrjabingylus magnus (veterinary parasitology) A species of worms in the genus *Skrjabingylus* that may be found in skunks.

Skrjabingylus nasicola (veterinary parasitology) A species of worms in the genus *Skrjabingylus* that may be found in nasal sinuses of minks, polecats and foxes.

Skrjabingylus petrowi (veterinary parasitology) A species of worms in the genus *Skrjabingylus* that may be found in pine marten.

Skrjabinoptera phrynosome (veterinary parasitology) A physalopterid roundworm that may be found in the stomach of lizards.

Skrjabinotrema ovis (veterinary parasitology) A digenetic trematode in the family Hasstilesiidae, that may be found in the small intestine of sheep.

skull (anatomy) Bones that form the head and face, including the cranium and jaws.

sleeping sickness (medical) Any disease characterized by drowsiness or somnolence.

See also Cryptobia; Trypanosomiasis.

slender goose louse (avian parasitology) *Anaticola anseris. See Anaticola.*

slender guinea fowl louse (avian parasitology) *Lipeurus numidae.*

slide (microscopy) A piece of glass or other transparent substance on which material is placed for examination under the microscope.

sliding microtome (equipment) A microtome in which the specimen being sectioned is made to slide on a track.

slime balls (parasitological physiology) Form in which the cercariae of *Dicrocoelium dendriticum* are discharged from the intermediate host snail's lung. The cercariae become infective only if the next intermediate stage, the ant, ingests the slime ball.

slimy skin disease (fish parasitology) A disease of aquarium fish caused by the flagellated protozoan *Costia necatrix* and characterized by a copious exudate on the skin of mucous and epithelial debris.

slow decline of citrus (plant parasitology) *See Tylenchulus semipenetrans.*

small intestine (anatomy) Part of the digestive tract in mammals which is composed of the duodenum and ileum, and is the main site of digestion and absorption in the gut. Bile, pancreatic juice and intestinal juice are liberated in it, to supply many digestive enzymes.

See also intestine.

small liver fluke (parasitology) *See Dicrocoelium dendriticum.*

small stomach worm (parasitology) *See Trichostrongylus axei.*

small strongyles (parasitology) *See Caballonema; Cyathostomum; Cylicocylus; Cylicodontophorus; Cylicostephanus; Cylindropharynx; Gyalocephalus; Poteriostomum* spp.

smear (parasitology) Specimen for microscopic study, the material being spread thinly and unevenly across the slide with a swab or loop, or with the edge of another slide, *e.g.* blood smear for trypanosome sighting, faecal smear (*USA*: fecal) for the presence of worm eggs or coccidial oocysts.

smell (physiology) One of the primary senses that enables animals to detect odours (*USA*: odors), using chemoreceptors usually located (in mammals) in olfactory bulbs in the nasal cavity.

smg (biochemistry) An abbreviation of 'sm'all 'g'TP-binding protein.

smooth muscle (histology) Type of muscle in internal organs and tissues, not under voluntary control. Also termed: involuntary muscle.

snail (parasitology) Gastropod mollusc with a spiral, coiled shell, some species of which act as intermediate hosts for flukes.

SND (statistics) An abbreviation of 'S'tandardized 'N'ormal 'D'eviation.

snoring disease (veterinary parasitology) Is an enzootic rhinitis of cattle caused by *Helminthosporium* spp.

soap (chemistry) Sodium or potassium salt of a fatty acid of high molecular weight (*e.g.* palmitic acid, stearic acid). Soaps are made by the hydrolysis or saponification of fats with hot sodium hydroxide or potassium hydroxide, giving glycerol as a by-product. They emulsify grease and act as wetting agents.

See also detergent.

Soboliphyme baturini (veterinary parasitology) A nematode in the family Soboliphymatidae that may be found in the intestines of foxes, sables and cats.

soda (chemistry) Imprecise term for a compound of sodium, usually referring to sodium carbonate.

See also caustic soda; soda ash.

soda ash (chemistry) Common name for anhydrous sodium carbonate.

sodium acetate (chemistry) CH_3COONa White crystalline solid, used in photography and in the manufacture of ethyl ethanoate (acetate) and various pigments. Also termed: sodium ethanoate.

sodium arsenate (pharmacology) Like the arsenite, a toxic compound that may be used as an acaricide.

sodium arsenite (pharmacology) A substance used as a topical acaricide.

sodium azide (chemistry) NaN_3 White poisonous crystalline solid, used in the manufacture of detonators.

sodium bicarbonate (chemistry) An alternative term for sodium hydrogencarbonate.

sodium bisulphate (chemistry) An alternative term for sodium hydrogensulphate (*USA*: sodium hydrogensulfate). (*USA*: sodium bisulfate).

sodium bisulphite (chemistry) An alternative term for sodium hydrogensulphite (*USA*: sodium hydrogensulfite). (*USA*: sodium bisulfite).

sodium borate (chemistry) An alternative term for borax.

sodium bromide (chemistry) NaBr White crystalline solid, used in medicine.

sodium carbonate (chemistry) $Na_2CO_3.10H_2O$ White crystalline solid which exhibits efflorescence and forms an alkaline solution in water. It is used in glass making, as a water softener, and for the preparation of sodium chemicals. Also termed: washing soda; soda; soda ash.

sodium chlorate (chemistry) $NaClO_3$ White soluble crystalline solid. It is a powerful oxidizing agent, used as a weed-killer and in the textile industry. Also termed: sodium chlorate (V).

sodium chloride (chemistry) NaCl White soluble crystalline salt, extracted from seawater or underground deposits. It is used for seasoning and preserving food. Industrially, it is used in the manufacture of a wide variety of chemicals, including chlorine, sodium carbonate, sodium hydroxide and hydrochloric acid. Also termed: common salt; salt; sea salt; table salt.

sodium dihydrogenphosphate(V) (chemistry) NaH_2PO_4 White solid, used in detergents and certain baking powders. Also termed: sodium dihydrogen orthophosphate.

sodium ethanoate (chemistry) An alternative term for sodium acetate.

sodium hydrogencarbonate (chemistry) $NaHCO_3$ White soluble powder, used in making baking powder, powder-based fire extinguishers and antacids. Also termed: sodium bicarbonate; baking soda.

sodium hydrogensulphate (chemistry) $NaHSO_4.H_2O$ White solid, used in the dyeing industry and in the manufacture of sulphuric acid (*USA*: sulfuric acid). Also termed: sodium bisulphate (*USA*: sodium bisulfate). (*USA*: sodium hydrogensulfate).

sodium hydrogensulphite (chemistry) $NaHSO_3$ White powder, used in medicine as an antiseptic, and as a preservative. Also termed: sodium bisulphite (*USA*: sodium bisulfite). (*USA*: sodium hydrogensulfite).

sodium hydroxide (chemistry) NaOH White deliquescent solid; a strong base. It is made by the electrolysis of brine (sodium chloride solution). It is used in the manufacture of soaps, rayon and paper and many other sodium compounds. Also termed: caustic soda; soda.

sodium (Na) (chemistry) Soft, silvery-white metallic element in Group IA of the periodic table (the alkali metals). It occurs widely, principally as its chloride (common salt, NACI) in seawater and as underground deposits, from which it is extracted by electrolysis. The metal is used as a coolant in some nuclear reactors; its many compounds are important in the chemical industry, particularly in addition to the chloride, sodium hydroxide (caustic soda, NAOH) and sodium carbonate (soda, Na_2C0_3). At. no. 11; r.a.m. 22.9898.

sodium nitrate (chemistry) $NaNO_3$ White crystalline salt, used as a food preservative and in the manufacture of explosives and fireworks. Also termed: Chilean saltpetre; soda nitre.

sodium peroxide (chemistry) Na_2O_2 Pale yellow powdery solid that reacts readily with water to give sodium hydroxide and oxygen. It is an oxidizing agent, used as a bleach.

sodium pump (physiology) Process by which potassium and sodium ions are transported across membranes that surround animal cells.

sodium silicate (chemistry) $Na_2SiO_3.5H_2O$ Colourless (*USA*: colorless) crystalline solid, used in various types of detergents and cleaning compounds, and as a bonding agent in many ceramic cements and in various refractory applications. Also termed: sodium metasilicate.

sodium silicate solution (chemistry) Concentrated solution of sodium silicate in water, used to prepare silica gel and precipitated silica. Also termed: water glass.

sodium sulphate (chemistry) $NaSO_4.10H_2O$ White crystalline salt, used in the manufacture of paper, glass, dyes and detergents. Also termed: Glauber's salt; saltcake. (*USA*: sodium sulfate).

sodium sulphide (chemistry) NaS_2 Reddish-yellow deliquescent amorphous solid, used in the manufacture of dyes. (*USA*: sodium sulfide).

sodium sulphite (chemistry) Na_2SO_3 White soluble crystalline solid, used in bleaching and photography. (*USA*: sodium sulfide).

sodium thiosulphate (chemistry) $Na_2S_2O_3$ White soluble crystalline solid. It is a strong reducing agent, used as a photographic fixing agent (when it reacts with unexposed silver halides) and in dyeing and volumetric analysis. Also termed: hypo. (*USA*: sodium thiosulfate).

soft copy (computing) The text displayed on the computer monitor. The opposite of hard copy.

soft palate (anatomy) Rear part of the roof of the mouth, consisting of muscle tissue covered by mucous membrane. The uvula hangs from the back of the soft palate.

soft radiation (radiology) Radiation of relatively long wavelength whose penetrating power is very limited.

soft ticks (parasitology) Ticks lacking a dorsal shield or scutum. Members of the family Argasidae, *e.g.* *Argas persicus*. Also termed: nonscutate.

soft tissue (histology) Tissue other than bone and cartilage.

soft water (chemistry) Water that lathers immediately with soap. Water from which most of the calcium and magnesium compounds have been removed.

See also hardness of water.

softening of water (chemistry) *See* ion exchange.

software (computing) Program that can be used on a computer; *e.g.* executive programs, operating systems and utility programs.

See also hardware.

sol (chemistry) Type of colloid consisting of a solid dispersed in a liquid. It is usually liquid (unlike a gel, which is a jelly-like solid).

solar cell (physics) Photocell that converts solar energy directly into electricity.

solenoid (physics) Cylindrical coil of wire, carrying an electric current, used to produce a magnetic field. It may have an iron core that moves, often to work a switch.

Solenopotes capillatus (veterinary parasitology) A species of sucking louse of cattle in the family Linognathidae.

solid (1 chemistry; 2 mathematics) 1 A state of matter in which the constituent molecules or ions possess no translational motion, but can only vibrate about fixed mean positions. A solid has a definite shape and resists deforming forces. 2 A figure having three dimensions.

solidifying point (physics) An alternative term for freezing point.

solubility (chemistry) Amount of a substance (solute) that will dissolve in a liquid (solvent) at a given temperature, usually expressed as a weight per unit volume (*e.g.* gm per litre) or a percentage.

See also concentration.

solubility product (chemistry) When a solution is saturated with an electrolyte, the product of the concentrations of its constituent ions.

soluble (chemistry) Describing a substance (solute) that will dissolve in a liquid (solvent).

solute (chemistry) Substance that dissolves in a solvent to form a solution.

solution (1 chemistry; 2 general terminology) 1 Homogeneous mixture of solute and solvent. 2 Result of solving a problem (*e.g.* finding the unknown quantity in a mathematical equation).

solvent (chemistry) Substance in which a solute dissolves; the component of a solution which is in excess.

solvent extraction (chemistry) Removal of a substance from a (usually aqueous) solution by dissolving it in a (usually organic) solvent. The resulting liquid containing the substance is called a raffinate. Also termed: liquid-liquid extraction.

solvolysis (chemistry) Chemical reaction between solvent and solute molecules.

See also hydrolysis.

Somaphantus lusius (avian parasitology) A species of louse of guinea fowl.

somat(o)- (terminology) A word element. [German] Prefix denoting *body*.

somatic (histology) Of the body. Somatic cells include all cells of an organism except for the gametes or sex-cells.

somatic cell (histology) Any cell in the body other than its gametes, the sperm in its testes or the eggs in its ovary.

somatotrophin (biochemistry) An alternative term for growth hormone.

sorbitol (chemistry) Alcohol formed by the reduction of glucose, used as a sweetening agent.

sore head (parasitology) *See* elaeophoriasis.

Soret effect *See* thermal diffusion.

sort (computing) To rearrange data, *e.g.* into alphabetical or numerical order.

SOT (society) An abbreviation of 'S'ociety 'O'f 'T'oxicology.

Southern blotting (molecular biology) DNA blotting. Named after Edwin Mellor Southern (1938–). *See* Western blotting; Northern blotting.

Southgates mucicarmine stain (histology) A histological staining technique for mucin.

sp (biology) An abbreviation of 'sp'ecies.

Spaniopsis (parasitology) A genus of biting flies in the family Rhagionidae. They are blood-suckers but do not appear to transmit disease.

Spanish blister fly (parasitology) *See Epicauta vittata.*

sparganosis (human/veterinary parasitology) Infection with *Spirometra mansoni* (synonym: *Spirometra erinacei*), which invades the subcutaneous tissues of pigs causing inflammation and fibrosis. If the lymphatics are involved, there is oedematous (*USA*: edematous) enlargement of the part. It is transmissible to humans who eat infected meat.

sparganum (human/veterinary parasitology) Plural: spargana. [German] A migrating plerocercoid of a tapeworm. Usually refers specifically to larvae of *Spirometra.* They cause sparganosis in humans and pigs.

Spathebothriidea (fish parasitology) An order of flatworms in subclass Eucestoda, class Cestoidea and the phylum Platyhelminthes. In this order, the scolex may be a funnel-shaped adhesive organ as in the Cyathocephalidae, two adhesive cups, or a proboscis. Some spathebothriidean tapeworms can mature and produce eggs in crustaceans and others in teleost fish. They may represent a fundamental ancestral type of the segmented tapeworms. On the one hand they may have affinities with Caryophyllidea and, on the other, with both Pseudophyllidea and Tetraphyllidea.

species (sp) (taxonomy) Smallest group commonly used in biological classification and into which a genus is divided. Species are sometimes further divided into subspecies (races). Generally, no more than one type of organism is present in one species. Members of a species may breed with one another, but cannot generally breed with members of another species. Rarely, very closely related species interbreed to produce a hybrid.

See also binomial nomenclature; variety.

specific activity (physics) Number of disintegrations of a radio-isotope per unit time per unit mass.

specific charge (physics) Ratio of electric charge to unit mass of an elementary particle.

specific gravity (physics) Former name for relative density.

specific latent heat (physics) Amount of latent heat per unit mass of a substance.

specific rate constant (physics) *See* rate constant.

specimen (parasitology) A small sample or part taken to show the nature of the whole, such as a small quantity of urine for urinalysis, or a small fragment of tissue for microscopic study.

specimen artifacts (parasitology) Changes in tissues or other samples for laboratoy examination, caused by the collection, transport, fixing, section cutting, staining or other procedural manipulations.

SPECT (physics) An abbreviation of 'S'ingle 'P'hoton 'E'mission 'C'omputed 'T'omography.

spectrochemistry (chemistry) Branch of chemistry concerned with the study of spectra of substances.

spectrometer (physics) Spectroscope that has some form of photographic or electrical detection device.

spectrometry (measurement) Measurement of the intensity of spectral lines or spectral series as a function of wavelength.

spectrophotometer (physics) Instrument that measures the intensity of electromagnetic radiation absorbed or transmitted by a substance as a function of wavelength, usually in the visible, infra-red and ultraviolet regions of the electromagnetic spectrum.

spectroscope (physics) Instrument for splitting various wavelengths of electromagnetic radiation into a spectrum, using a prism or diffraction grating.

spectroscopy (physics) Study of the properties of light, using a spectroscope; the production and analysis of spectra.

spectrum (1, 2 physics) 1 Band, continuous range, or lines of electromagnetic radiation emitted or absorbed by a substance under certain circumstances. 2 Coloured (*USA*: colored) band of light or bands of colours produced by splitting various wavelengths of electromagnetic radiation, using a prism or diffraction grating.

spectrum colours (physics) Visible colours that are observed in the spectrum of white light (*e.g.* in a rainbow). (*USA*: spectrum colors).

sperical aberration (optics) Type of aberration in a lens or mirror.

sperm (cytology) An abbreviation of 'sperm'atozoan, the gamete (sex-cell) produced by the male in many species of animals (in mammals, in the testes). It consists of a head containing genetic material in the nucleus (which is haploid) and usually possesses cilia or flagella for movement.

sperm(o)- (terminology) A word element. [German] Prefix denoting *seed.*

spermatheca (parasitological anatomy) Accessory organ opening into the common oviduct in female insects; stores spermatozoa delivered by the male during copulation; the sperm may remain viable for the life of the female.

spermato- (terminology) A word element. [German] Prefix denoting *seed.*

spermatocyte (cytology) Cell from which sperm (spermatozoa) are derived through spermatogenesis. Primary spermatocytes are diploid; after meiosis, secondary spermatocytes which are haploid are formed. These further divide to produce spermatids, which differentiate to form sperm.

spermatogenesis (physiology) Formation of sperm in the testis. It commences with the repeated mitosis of primordial germ cells to form spermatogonia, which grow to form a primary spermatocyte. This ultimately forms haploid spermatozoa after meiosis.

spermatozoon *See* sperm.

SPF (parasitology) An abbreviation of 'S'pecific 'P'athogen 'F'ree.

Sphaeridiotrema (parasitology) A genus of small, globular digenetic trematodes in the family Psilostomatidae.

Sphaeridiotrema globulosus (avian parasitology) A species of digenetic trematodes in the genus *Sphaeridiotrema* which may be found in the intestine of wild and domestic ducks and in swans and that causes severe enteritis. It has caused severe mortalities in swans.

Sphaeridiotrema spinoacetabulum (parasitology) A species of digenetic trematodes in the genus *Sphaeridiotrema* which inhibits the caeca (*USA*: ceca) and causes severe typhlitis.

Sphaerolaelaps holothyroides (insect parasitology) Mites in the Mesostigmata that are cleptoparasitic on the ants Solenopis and Lasius.

Sphaerularia bombi (insect parasitology) A species of nematode in the Tylenchidae that are parasites to the bees *Bombus* spp. *and Psithyrus* spp., that may be found in overwintering queens resulting in behavioural (*USA*:

behavioral) changes, the queen seeking overwintering sites instead of nesting sites. They are also parasites of the wasps *Vespula* spp.

spheno- (terminology) A word element. [German] Prefix denoting *wedge-shaped, sphenoid bone.*

sphincter (histology) Circular muscle that controls the flow of a liquid or semi-solid through an orifice (*e.g.* the anal sphincter, round the anus).

sphygmo- (terminology) A word element. [German] Prefix denoting *the pulse.*

sphygmomanometer (physics) Instrument for measuring blood pressure.

spicule (parasitological anatomy) Part of the male genital apparatus in nematodes which is elongated and protrusible and that engages the female genital orifice during copulation. Its length, shape and number vary in different species.

Spiculocaulus (veterinary parasitology) A genus of nematodes in the family Protostrongylidae that may be found in the lungs of small ruminants.

Spiculocaulus austriacus (veterinary parasitology) A species of nematodes in the genus *Spiculocaulus* that may be found in goats and ibex.

Spiculocaulus kwongi (veterinary parasitology) A species of nematodes in the genus *Spiculocaulus* that may be found in sheep and goats.

Spiculocaulus leuckarti (veterinary parasitology) A species of nematodes in the genus *Spiculocaulus* that may be found in sheep and ibex.

Spiculocaulus orloffi (veterinary parasitology) A species of nematodes in the genus *Spiculocaulus* that may be found in sheep, goats and ibex.

Spiculopteragia (veterinary parasitology) A genus of worms in the family Trichostrongylidae, that may be found in the abomasum.

Spiculopteragia boehmi (veterinary parasitology) A species of worms in the genus *Spiculopteragia* that may be found in moufflon and deer.

Spiculopteragia peruviana (veterinary parasitology) A species of worms in the genus *Spiculopteragia* that may be found in llamas, alpacas and vicunas.

Spiculopteragia spiculoptera (veterinary parasitology) A species of worms in the genus *Spiculopteragia* that may be found in sheep and deer.

spike (quality control) Prepare samples with a known amount of substance for testing laboratory proficiency.

Spilopsyllus cuniculi (veterinary parasitology) A flea in the order Siphonaptera which is a parasite of rabbits and hares and occasionally their predators, and transmits myxomatosis. A cause of severe irritation on the pinnae of domestic cats. Also termed: European rabbit flea.

spinal column (anatomy) *See* spine.

spinal cord (anatomy) Part of the central nervous system (CNS) in vertebrates that is enclosed within the spine. It consists of a hollow nerve tube containing many interconnecting neurones and connected to the spinal nerves.

spinal cord myelitis (parasitology) A condition that may be caused by protozoa and helminth parasites. The signs range from weakness to complete flaccid paralysis, including paralysis of the anus and the tail.

spinal fluid (anatomy) The fluid within the spinal canal.

spinal meningitis (medical) Inflammation of the meninges of the spinal cord.

spinal nerve (histology) Any of several peripheral nerves arising from the spinal cord which are connected to receptors and effectors in other parts of the body.

spine (anatomy) Backbone; dorsally situated bony column composed of vertebrae, which enclose the spinal cord. Also termed: spinal column.

spinose ear tick (parasitology) *See Otobius megnini.*

spinous ear tick (parasitology) *See Otobius megnini.*

spiracle (parasitological physiology) Small, circular openings in the exoskeleton of insects that are the portal of entry for air into the insect body.

spiral worm (parasitology) *Dispharynx nasuta. See also Synhimantus spiralis.*

spirit (1, 2 chemistry) 1 Volatile liquid obtained by distillation; a volatile distillate (*e.g.* aviation spirit). 2 Solution that consists of a volatile substance dissolved in ethanol (ethyl alcohol).

See also methylated spirits.

Spirocerca (parasitology) A genus of spiruroid nematodes in the family Spirocercidae.

Spirocerca lupi (veterinary parasitology) A species of spiruroid nematodes in the genus *Spirocerca* that may be found in domestic and wild Canidae and wild Felidae. The worms are located in the walls of the oesophagus (*USA*: esophagus), the aorta and the stomach, and sometimes in other organs, persisting in nodules. The nodules may be large enough to cause obstruction of the oesophagus (*USA*: esophagus) and the aorta. The oesophageal (*USA*: esophageal) lesion converts to a fibrosarcoma or an osteosarcoma in a number of cases.

Spirocerca sanguinolenta (parasitology) *See Spirocerca lupi.*

spirocercosis (parasitology) *See Spirocerca lupi.*

spirochetosis (avian parasitology) An infectious disease of many species of fowl caused by *Borrelia anserina* and characterized by fever, cyanosis of the head and diarrhoea (*USA*: diarrhea). It is transmitted by the fowl ticks *Argas persicus*, *Argas miniatus* and *Argas reflexus*. Also termed: rabbit syphilis; treponematosis; vent disease.

Spirometra (human/veterinary parasitology) A pseudophyllidean tapeworm similar to *Diphyllobothrium* spp. that may be found in the intestines of wild carnivores and domestic cats and dogs. The larvae (plerocercoids) infest amphibians but humans can also be infected, the resulting disease being known as sparganosis.

Spirometra erinacei (veterinary parasitology) A species of pseudophyllidean tapeworm in the genus *Spirometra* that may be found in cats and dogs.

Spirometra felis (veterinary parasitology) A species of pseudophyllidean tapeworm in the genus *Spirometra* that may be found in big zoo cats.

Spirometra mansoni (veterinary parasitology) A species of pseudophyllidean tapeworm in the genus *Spirometra* that may be found in cats and dogs.

Spirometra mansonoides (veterinary parasitology) A species of pseudophyllidean tapeworm in the genus *Spirometra* that may be found in cats, dogs, and raccoons.

Spirometra reptans (veterinary parasitology) A species of pseudophyllidean tapeworm in the genus *Spirometra* that may be found in New World primates.

Spironucleus (fish parasitology) A genus of protozoa that is one of the causes of hole-in-the-head disease of ornamental fish.

Spironucleus muris (veterinary parasitology) A protozoon in the genus *Spironucleus* which may be found in mice and that causes weight loss, diarrhoea (*USA*: diarrhea) and occassionally death.

Spiroptera incerta (parasitology) *See Habronema incertum.*

Spirura (veterinary parasitology) A genus of nematodes in the family Spiruridae that may be found in the alimentary canal of carnivores, rodents and insectivores.

Spirura rytipleurites (veterinary parasitology) A species of nematodes in the genus *Spirura*. There are two varieties, one found in the stomach of cats and foxes and one in hedgehogs.

Spirura talpae (veterinary parasitology) A species of nematodes in the genus *Spirura* that may be found in the European common mole, *Talpa europea*, and the black rat, *Rattus rattus*.

Spirurida (fish parasitology) An order of roundworms in subclass Secernentea, and the phylum Nematoda. Six superfamilies and about 20 genera of the Spirurida are important: the superfamily Habronematoidae contains one large family, the Cystidicolidae, with 12 genera; the superfamily Camallanoidea contains *Camallanus*, a member of the Camallanidae; the superfamily Physalopteroidea has four genera of the family Physalopteridae, *Proleptus*, *Paraleptus*, *Heliconema* and *Bulbocephalus*, which are nematodes of fish; the superfamily Dracunculoidea has three families, the Anguillicolidae, Guyanemidae and Philometridae, all of which are found in fish; members of the superfamily Thelazoidea, family Rhabdochonidae, are parasites of the intestine or of various other organs of fish; and the superfamily Gnathostomatoidea of which *Ancyracanthus* and *Echinocephalus* are members of the Gnathostomatidae. *Ancyracanthus* larvae occur in teleosts whilst *Echinocephalus* adults occur in elasmobranchs.

spiruroid (parasitology) Roundworms of the order Spirurida that includes the genera of *Cyrnea*, *Draschia*, *Habronema*, *Hartertia*, *Mastophorus*, *Protospirura*, *Spirura* and *Streptopharagus*.

splanchn(o)- (terminology) A word element. [German] Prefix denoting *viscus* (viscera), *splanchnic nerve*.

splanchnology (biology) Branch of biology and medicine concerned with the organs within the central body cavity of vertebrates.

spleen (anatomy) Organ present in the abdomen of some vertebrates that aids in defence against invading organisms. It produces lymphocytes and also stores and removes red blood cells (erythrocytes) from the blood system.

spodo- (terminology) A word element. [German] Prefix denoting *waste material.*

spondyl(o)- (terminology) A word element. [German] Prefix denoting *vertebra*, *vertebral column*.

spongi(o)- (terminology) A word element. [Latin, German] Prefix denoting *sponge*, *spongelike*.

spontaneous generation (philosophy) Theory (now disproved) that living matter can arise from non-living matter. Also termed: abiogenesis.

See also biogenesis.

sporadic (medical) Occurring irregularly, usually infrequently.

sporadic disease (medical) Occurring singly and haphazardly; widely scattered; not epidemic or endemic.

spore (biology). The reproductive element, produced sexually or asexually, of organisms, such as protozoa.

sporicide (control measures) A substance that kills spores.

sporoblast (parasitological physiology) An asexual reproductive phase in the development of coccidial parasites.

See also sporogony.

sporocyst (parasitological physiology) 1 Any cyst or sac containing spores or reproductive cells; contained in the oocyst of coccidia in which sporozoites develop. 2 The larval stages of flukes in snails.

sporogenic (biology) Producing spores.

sporogony (parasitological physiology) An asexual stage in the life cycle of an apicomplexan parasite,

with development of sporocysts and sporozoites within the oocysts. The oocysts are extraintestinal, in for example *Eimeria* and *Isospora*, or intraintestinal, as in *Sarcocystis* and *Cryptosporidium*.

sporont (biology) A mature protozoon in its sexual cycle.

sporoplasm (parasitological anatomy) The protoplasm of a spore.

Sporozoa (parasitology) A subphylum of endoparasitic protozoa in the phylum Apicomplexa which are marked by the lack of locomotor organs in adult stages and a complex life cycle including schizogony, gametogony and sporogony.

sporozoa (biology) Plural of sporozoon.

sporozoan (biology) 1 Pertaining to the Sporozoa. 2 An individual of the Sporozoa.

sporozoite (parasitological physiology) Spore formed after fertilization; a sickleshaped nucleated germ formed by division of the protoplasm of a spore of an apicomplexan organism during sporogony.

sporozoon (biology) Plural: sporozoa. An individual organism of the Sporozoa.

spot film device (equipment) A device attached to the X-ray machine which moves an X-ray cassette into position for exposure during a fluoroscopic examination.

spot-on (control measures) A means of delivering medication topically, usually in a small area of the skin, where the active ingredient is absorbed percutaneously. A method used for flea control agents.

spotted (medical) Characterized by spots.

spotted fever (human parasitology) A febrile disease characterized by a skin eruption, such as Rocky Mountain spotted fever, boutonneuse fever, and other human infections due to tickborne rickettsiae.

spotted fever tick (parasitology) *See Dermacentor andersoni.*

Sprague-Dawley rat (parasitology) Albino laboratory rat.

spray (control measures) Material applied in liquid form by pressure through a fine orifice creating a mist of fine droplets, *e.g.* insecticidal spray in a spray dip, pressure pack spray of insecticide or wound treatment.

spray race (control measures) A lane with high fences, used to restrain animals while they are being sprayed, usually with insecticide.

spraydip (control measures) Device for applying insecticide or other solution to the coat of animals, replacing the plunge dip. The animal walks through a compartment which carries spray nozzles on the sides, below and above. Excess spray drains into a sump and is reused.

spreading decline of citrus (plant parasitology) *See Radopholus citrophilus.*

spreadsheet (computing) A computer software consisting of a programme (*USA*: program) used for quick analysis of figures based on a grid of cells created by the intersection of vertical columns and horizontal rows, each cell containing data or formulae.

spring rise (veterinary parasitology) A phenomenon in some nematode infestations, *e.g.* Haemonchus spp., in ruminants in which there is an increase in the number of eggs excreted in the faeces (*USA*: feces) during the spring months.

sq cm (measurement) An abbreviation of 'sq'uare 'c'enti'm'eter(s) (*USA*: centimeter[s]).

squamous (histology) Scaly or plate-like; a type of cell.

squamous cell carcinoma (histology) A malignant neoplasm derived from squamous epithelium.

square root (mathematics) Of a number, another number that when multiplied by itself gives the original number. It is indicated by the symbol √ or the index (power) ½; *e.g.* $\sqrt{16}$ (or $16^{1/2}$) = 4.

SR (haematology) An abbreviation of 'S'edimentation 'R'ate.

Sr (chemistry) Chemical symbol, strontium.

SRC (governing body) An abbreviation of 'S'cience 'R'esearch 'C'ouncil (UK).

SRIF (physiology/biochemistry) An abbreviation of 'S'omatotropin 'R'elease-'I'nhibiting 'F'actor.

SRNA (biochemistry) An abbreviation of 'S'oluble (or transfer) 'R'ibo'N'ucleic 'A'cid.

SRY (genetics) An abbreviation of product of 'S'ex-determining 'R'egion of 'Y' chromosome.

SS (society) An abbreviation of Royal 'S'tatistical 'S'ociety.

ssp. (biology) An abbreviation of 's'ub 'sp'ecies.

stabilate (parasitology) A mixture of blood and tick tissues, expected to contain infective elements of rickettsia or protozoa, in a stable state suitable for storage.

stabilization (parasitology) The process of making firm and steady.

stable (1 chemistry; 2, 3 physics) 1 In chemistry, relatively inert (*e.g.* describing a rare gas) or hard to decompose (*e.g.* describing an oxide such as silica). 2 In atomic physics, describing an isotope or nucleus that shows no tendency to decompose (*e.g.* by emitting radioactivity). 3 Describing a type of equilibrium that does not have a tendency to shift (as opposed to unstable equilibrium).

stable fly (parasitology) *See Stomoxys calcitrans.*

stack (computing) The work area in a computer's memory that expands and contracts dynamically as required to store temporary data.

stain (microscopy) A substance used to impart colour (*USA*: color) to tissues or cells, to facilitate microscopic study and identification.

stainable iron (microscopy) A method of staining a bone marrow smear to determine the amount of body storage of iron.

staining (parasitology) Artificial colouration (*USA*: coloration) of a substance to facilitate examination of tissues, parasites or other cells under the microscope.

stand-alone computer (computing) *See* computer.

standard (quality control) Something established as a measure or model to which other similar things should conform.

standard deviation (statistics) Measure of the spread of a set of numbers about their arithmetic mean. It is the square root of the mean (average) of the squares of all the deviations from the mean value of the set (root mean square value). It is a measure of the compactness of a set of numbers. Also termed: standard error.

standard electrode potential (physics) Electrode potential specified by comparison with a standard electrode.

standard error (statistics) An alternative term for standard deviation.

standard error of mean (statistics) The sampling variability of the mean.

standard population (epidemiology) A population not yet divided into classes; the population against which each of its constituent classes can be compared.

standard solution (chemistry) Solution of definite concentration, *i.e.* having a known weight of solute in a definite volume of solution.

standard state (chemistry) Element in its most stable physical form at a specified temperature and a pressure of 101,325 pascals (760 mm Hg).

standard temperature and pressure (s.t.p. or STP) (physics) Set of standard conditions of temperature and pressure. By convention, the standard temperature is 273.15 K (0°C) and the standard pressure is 101,325 pascals (760 mm Hg).

standardized (quality control) Pertaining to data that have been submitted to standardization procedures.

stannic chloride (chemistry) An alternative term for tin(IV) chloride.

stannous (chemistry) Also termed: tin(IV).

stannous chloride (chemistry) An alternative term for tin(II) chloride.

staphyl(o)- (terminology) A word element. [German] Prefix denoting *uvula, resembling a bunch of grapes, staphylococci.*

starch (chemistry) $(C_6H_{10}O_5)_n$ Complex polysaccharide carbohydrate, a polymer of glucose, that occurs in all green plants, where it serves as a reserve energy material. It forms glucose on complete hydrolysis. Also termed: amylum.

-stasis (terminology) A word element. [German] Suffix denoting maintenance of (or maintaining) a constant level; preventing increase or multiplication.

stat (general terminology) (Do) immediately.

state (medical) Condition or situation.

state of matter (physics) An alternative term for physical state of matter.

statistic (statistics) A numerical value calculated from a number of observations in order to summarize them.

statistical (statistics) Pertaining to or emanating from statistics.

statistical analysis (statistics) Evaluation of data by the use of statistics.

statistical significance (statistics) *See* significance.

statistics (statistics) Branch of science concerned with the collection and classification of numerical data and facts, and their interpretation in mathematical terms, especially the determination of probabilities.

STD (medical) An abbreviation of 'S'exually 'T'ransmitted 'D'isease.

steam point (physics) Normal boiling point of water; it is taken to be a temperature of 100°C (at normal pressure).

stear(o)- (terminology) A word element. [German] Prefix denoting *fat.*

stearate (chemistry) Ester or salt of stearic acid.

See also soap.

stearic acid (chemistry) $CH_3(CH_2)_{16}COOH$ Long-chain fatty acid (a carboxylic acid) that occurs in most fats and oils. Its sodium and potassium salts are constituents of soaps. Also termed: octadecanoic acid.

steat(o)- (terminology) A word element. [German] Prefix denoting *fat, oil.*

Stegomyia fasciata (parasitology) A mosquito vector of the virus of fowl pox.

stem (anatomy) Stalk; a stalklike supporting structure.

stem and bulb nematode (plant parasitology) *See Ditylenchus dipsaci.*

stem cell (cytology) A cell that gives rise to a particular type of cell, as occurs in haematopoiesis (*USA*: hematopoiesis).

Stempellia odontotermi (insect parasitology) A protozoal species of microsporidia that are parasites to the termites *Odontotermes* species. They may be found in the epithelial cells of the foregut of workers.

steno- (terminology) A word element. [German] Prefix denoting *narrow*, *contracted*, *constriction*.

Stephanofilaria (veterinary parasitology) A genus of nematodes in the family Filariidae that causes chronic dermatitis of cattle termed cascado.

Stephanofilaria assamensis (veterinary parasitology) A species of nematodes in the genus *Stephanofilaria* that causes chronic dermatitis in buffaloes, goats and cattle. In cattle the disease is termed humpsore.

Stephanofilaria dedoesi (veterinary parasitology) A species of nematodes in the genus *Stephanofilaria* that may be found in the skin of cattle.

Stephanofilaria dinniki (veterinary parasitology) A species of nematodes in the genus *Stephanofilaria* that causes filarioid dermatitis in black rhinoceroses.

Stephanofilaria kaeli (veterinary parasitology) A species of nematodes in the genus *Stephanofilaria* that may be found on the legs of cattle.

Stephanofilaria okinawaensis (veterinary parasitology) A species of nematodes in the genus *Stephanofilaria* that causes dermatitis of the muzzle and teats of cattle.

Stephanofilaria stilesi (veterinary parasitology) A species of nematodes in the genus *Stephanofilaria* that causes dermatitis on the ventral abdominal wall of cattle.

Stephanofilaria zaheeri (veterinary parasitology) A species of nematodes in the genus *Stephanofilaria* that may be found on the inner side of the ear pinna of buffaloes.

stephanofilariasis (parasitology) *See* stephanofilarosis.

stephanofilarosis (parasitology) Infestation with filariid worms.

Stephanostomum (parasitology) A genus of intestinal flukes in the family Acanthocolpidae.

Stephanostomum baccatum (fish parasitology) A species of intestinal flukes in the genus *Stephanostomum* which may be found in the intestines of marine fish but that is not considered to be pathogenic.

stephanuriasis (veterinary parasitology) The disease of pigs caused by the kidney worm *Stephanurus dentatus* that is characterized by poor feed utilization and growth, and later emaciation, paralysis and ascites.

Stephanurus (parasitology) A genus of nematodes in the family Stephanuridae.

Stephanurus dentatus (veterinary parasitology) A species of nematodes in the genus *Stephanurus* that may be found in the kidneys of pigs and rarely as aberrant parasites in cattle and horses. They may also be found in the liver, pancreas and other organs of pigs.

stepwise (general terminology) Incremental; additional information is added at each step.

stercoraceous (parasitology) Of faecal (*USA*: fecal) origin; said of trypanosomes in the subgenus *Schizotrypanum*. These trypanosomes are passed to the recipient in the faeces (*USA*: feces) of the tsetse fly *Glossina* spp.

stercoracic (parasitology) *See* stercoraceous.

stercoral (parasitology) *See* stercoraceous.

Stercoraria (parasitology) The posterior station group of trypanosomes transmitted by contamination through the faeces (*USA*: feces) of the insect vector. Includes *Trypanosoma cruzi*, *Trypanosoma lewisi*, *Trypanosoma melophagium*, *Trypanosoma nabiasi*, *Trypanosoma rangeli*, *Trypanosoma theileri* and *Trypanosoma theodori*.

stercorian trypanosome (parasitology) A term used in relation to flagellates. Infective forms which develop in the faeces (*USA*: feces) of the insect vector and enter the vertebrate host by the contamination of the bite area. This is also known as the posterior station development.

stereo- (terminology) A word element. [German] Prefix denoting *solid*, *firm*, *three-dimensional*.

stereochemistry (chemistry) Branch of chemistry concerned with the study of the spatial arrangement of the atoms within a molecule and the way that these affect the properties of the molecule.

stereoisomer (chemistry) One of two or more isomers with the same molecular formula, but different configurations (arrangements of atoms).

stereoisomerism (chemistry) Isomerism of compounds of the same molecular formula that results when the spatial arrangement of the atoms within the molecules are different.

See also geometric isomerism; optical isomerism.

stereoregular (chemistry) Describing a compound that has a regular spatial arrangement of atoms within its molecule.

stereoscope (physics) Optical instrument that gives a three-dimensional illusion of depth, normally from a pair of flat photographs.

stereoscopic (physics) Describing an instrument or system of vision that allows objects to be viewed in three dimensions.

stereoscopic microscope (microscopy) A binocular microscope modified to give a three-dimensional view of the specimen.

stereospecific Describing a chemical reaction in which a different product is formed from each geometric isomer of the reactant.

steric effect (chemistry) Phenomenon in which the shape of a molecule affects its chemical reactions.

sterilant (control measures) A sterilizing agent, i.e. a substance that destroys microorganisms.

sterile (biology) 1 Aseptic; not producing microorganisms; free from living microorganisms. 2 Not fertile; barren; not producing young.

sterile insect release method (control measures) A method of insect control of particular use in insects in which mating occurs only once, *e.g.* as with screwworm fly. Artificially bred flies are sterilized by irradiation and released; if sufficient sterile flies are released most of the wild flies mate with a sterile fly and the population is sufficiently diminished to lead to a control situation. Also termed: SIRM.

sterility (biology) The state of being sterile.

sterilization (control measures) Treatment of an apparatus or substance so that it contains no microorganisms that could cause disease or spoilage, usually by means of high temperatures, gamma radiation, etc.

stern(o)- (terminology) A word element. [Latin, German] Prefix denoting *sternum.*

Sternostoma (avian parasitology) A genus of mites of the family Rhinonyssidae that may be found in the trachea of companion birds, causing respiratory difficulty.

Sternostomum tracheacolum (avian parasitology) A genus of tracheal mite that may be found in free-living finches.

sternum (anatomy) Bone in tetrapod vertebrates on the ventral side of the thorax, parallel to the spine, to which most of the ribs are attached. Also termed: breastbone.

steroid (biochemistry) Any of a group of naturally occurring tetracyclic organic pounds, widely found in animal tissues. Most have very important physiological activities (*e.g.* adrenal hormones, bile acids, sex hormones, sterols). Some can be made synthetically (*e.g.* for use as contraceptive pills).

sterol (biochemistry) Subgroup of steroids or steroid alcohols. They include cholesterol, abundant in animal tissues, which is the precursor of many other steroids.

steth(o)- (terminology) A word element. [German] Prefix denoting *chest.*

stethoscope (medical) An instrument through which the physician listens to sounds made by internal organs (*e.g.* heart, lungs).

stibogluconate (pharmacology) An antimony compound used in the treatment of leishmaniasis.

stibophen (pharmacology) A toxic compound used at one time as a treatment for heartworm in dogs.

stichosome (parasitological cytology) A term used in relation to nematodes. Glandular cells (stichocytes) arranged in a row along the oesophagus. Seen in Trichinelloidea.

stickfast flea (parasitology) *Echidnophaga gallinacea.*

Stilbula (Schizaspidia) polyrhachicida (insect parasitology) *See Schizaspidia convergens.*

Stilbula cynipiformis (insect parasitology) *See Schizaspidia convergens.*

Stilbula tenuicornis (insect parasitology) *See Schizaspidia convergens.*

Stilesia (parasitology) A genus of tapeworms in the family Thyanosomatidae.

Stilesia globipunctata (veterinary parasitology) A species of tapeworms in the genus *Stilesia* which may be found in the small intestine of ruminants and that causes sufficient injury to kill some animals.

Stilesia hepatica (veterinary parasitology) A species of tapeworms in the genus *Stilesia* that may be found in the bile ducts in all ruminant species and causes thickening of the duct walls but with no clinically apparent disease.

still (chemistry) Apparatus for the distillation of liquids.

stimulated emission (physics) Process by which a photon causes an electron in an atom to drop to a lower energy level and emit another photon. It is the principle of the laser.

stimulus (physics) Environmental factor that is detected by a receptor and induces a response from an effector.

sting nematode (plant parasitology) *See Belonolaimus longicaudatus.*

stoichiometric compound (chemistry) Chemical whose molecules have the component elements present in exact proportions as demanded by a simple molecular formula.

stoichiometric mixture (chemistry) Mixture of reactants that in a chemical reaction yield a stoichiometric compound with no excess reactant.

stoichiometry (chemistry) Branch of chemistry that deals with the relative quantities of atoms or molecules taking place in a reaction.

Stoll's method (methods) A technique for counting nematode and trematode eggs in a faecal (*USA*: fecal) sample diluted in a 0.1 M caustic soda solution.

stomach (anatomy) Muscular sac present in vertebrates in which food is partly digested and stored after passage through oesophagus (gullet). Hydrochloric acid is secreted in the stomach, as is the enzyme pepsin, which begins the digestion of proteins.

stomach bot (parasitology) *See Gasterophilus.*

stomach fluke (parasitology) *See Paramphistomum.*

stomach worm (parasitology) *See Gnathostoma spinigerum*; *Haemonchus*; *Ostertagia.*

stomat(o)- (terminology) A word element. [German] Prefix denoting *mouth.*

Stomoxys calcitrans (veterinary parasitology) The ubiquitous blood-sucking stable fly, which is a pest for horses and that may transmit a number of try-panosomes including *Trypanosoma evansi* (surra), *Trypanosoma equinum* (mal de caderas), *Trypanosoma brucei* and *Trypanosoma vivax* (nagana of cattle) and is an intermediate host for *Hebronema majus.* It is probably also involved in the transmission of *Dermatophilus congolensis* (mycotic dermatitis).

-stomy (terminology) A word element. [German] Suffix denoting *creation of an opening into* or *a communication between.*

stool (biology) The faecal (*USA*: fecal) discharge from the bowels.

See also faeces.

storage (computing) Computer memory capacity.

storage cell (physics) An alternative term for secondary cell.

storage device (computing) An alternative term for a computer memory or store.

See also random access memory (RAM); read-only memory (ROM).

storage mites (parasitology) Many species, *e.g.* *Acarus siro*, that may be found in stored cereals.

STP or s.t.p. (physics) An abbreviation of 's'tandard 't'emperature and 'p'ressure.

STPD (physics) An abbreviation of 'S'tandard 'T'emperature and 'P'ressure (0°C, 760mm Hg), 'D'ry.

stratum (histology) Layer of cells or tissue.

straw-itch mite (parasitology) *Pyemotes tritici. See Pyemotes.*

strepto- (terminology) A word element. [German] Prefix denoting *twisted.*

Streptopharagus (veterinary parasitology) A genus of roundworms in the family Spirocercidae. Species includes *Streptopharagus armatus* and *Streptopharagus pigmentatus* that may be found in the stomach of apes and monkeys.

stress (biology) Any environmental factor, or combination of factors, that has adverse effects on the structure or behaviour (*USA*: behavior) of an organism.

Stretocara (avian parasitology) A genus of spiruroid worms which may be found in the gizzard of many bird species and whose intermediate host is a crustacean.

striated muscle (histology) Muscle that contains well-aligned threads of protein, which enable it to contract strongly in a particular direction. Also termed: voluntary muscle.

strigeids (avian parasitology) Intestinal trematodes in the family Strigeidae that may be found in birds.

string (computing) A sequence of alphanumeric characters.

string variable (computing) A sequence of characters, numerical or alphabetical, which are treated by a computer programme (*USA*: program) as a single entity.

strobila (parasitological anatomy) Plural: strobilae [Latin] The chain of proglottides constituting the bulk of the body of an adult tapeworm.

strobilisation (parasitological physiology) A term used in relation to cestodes. The process of producing or growing new segments (proglottids). This happens near the neck region.

strobilocercus (parasitological physiology) The larval stage of a tenid cestode. A fluid-filled cyst containing a scolex with a pseudosegmented elongated region connecting to the bladder, *e.g. Taenia taeniae* forms.

stroboscope (physics) Instrument consisting of a rapidly flashing lamp, employed for measuring speeds of rotation. It can also be used, by controlling the rate of flashing, to view objects that are moving rapidly with periodic motion and to see them as if they were at rest. Also termed: rotoscope.

stroke (medical) Sudden damage to the vascular system in the brain, caused by the rupture or blockage of a blood-vessel and usually resulting in impairment or loss of some functions.

stroma (histology) The tissue that forms the supporting framework of an organ, as distinct from the parenchyma, the functioning tissue.

strong acid (chemistry) Acid that is completely dissociated into its component ions (*e.g.* hydrochloric acid).

strong base (chemistry) Base that is completely dissociated into its component ions (*e.g.* sodium hydroxide).

strongyle (parasitology) Any roundworm in the superfamily Strongyloidea.

Strongylogaster (*Tamiclea*) *globula* (insect parasitology) Diptera in the Tacinidae that are endoparasitic to young queens of the ants *Lasius*, later being cared for by workers.

Strongyloides (parasitology) A genus of nematode parasites in the family Strongyloididae, the larvae of which are able to penetrate the intact skin of the host then migrate to the intestine via the bloodstream, lung, trachea and pharynx.

See also strongyloidosis.

Strongyloides avium (avian parasitology) A species of nematode parasites in the genus *Strongyloides* that may be found in the caeca (*USA*: ceca) and small intestine of fowl, turkeys and wild birds.

Strongyloides cati (veterinary parasitology) A species of nematode parasites in the genus *Strongyloides* that may be found in the small intestine of cats.

Strongyloides felis (veterinary parasitology) A species of nematode parasites in the genus *Strongyloides* that may be found in cats.

Strongyloides fuelleborni (veterinary parasitology) A species of nematode parasites in the genus *Strongyloides* that may be found in the small intestine of primates.

Strongyloides papillosus (veterinary parasitology) A species of nematode parasites in the genus *Strongyloides* that may be found in the small intestine of ruminants and rabbits.

Strongyloides planiceps (veterinary parasitology) A species of nematode parasites in the genus *Strongyloides* that may be found in cats.

Strongyloides procyonis (veterinary parasitology) A species of nematode parasites in the genus *Strongyloides* that may be found in raccoons.

Strongyloides ransomi (veterinary parasitology) A species of nematode parasites in the genus *Strongyloides* that may be found in the small intestine of pigs.

Strongyloides ratti (veterinary parasitology) A species of nematode parasites in the genus *Strongyloides* that may be found in rats.

Strongyloides simiae (parasitology) *See Strongyloides fuelleborni.*

Strongyloides stercoralis (human/veterinary parasitology) A species of nematode parasites in the genus *Strongyloides* that may be found in the intestine of humans and other mammals, primarily in the tropics and subtropics, usually causing diarrhoea (*USA*: diarrhea) and intestinal ulceration.

Strongyloides tumefaciens (veterinary parasitology) A species of nematode parasites in the genus *Strongyloides* that is associated with tumours (*USA*: tumors) of the large intestine of cats.

Strongyloides venezuelensis (veterinary parasitology) A species of nematode parasites in the genus *Strongyloides* that may be found in rats.

Strongyloides westeri (veterinary parasitology) A species of nematode parasites in the genus *Strongyloides* that may be found in the small intestine of horses, pigs and zebras.

strongyloidiasis (parasitology) Infestation with *Strongyloides* spp. *See* strongyloidosis.

strongyloidosis (veterinary parasitology) Infestation with the nematode *Strongyloides*, a parasite of the small intestine that can cause dermatitis and balanoposthitis due to percutaneous entry, or diarrhoea (*USA*: diarrhea) when the intestinal infection is very heavy. In kangaroos it is a stomach parasite causing gastritis. *Strongyloides papillosus* may be associated with the introduction of organisms into the skin of the feet, causing footrot. Also termed: strongyloidiasis.

strongylosis (veterinary parasitology) Infestation of horses with large strongyles (*Strongylus edenatus*, *Strongylus equinus*, *Strongylus vulgaris*), *Triodontophorus* spp., *Oesophagodontus* spp. and some members of the genera *Cyclostephanus*, *Cyathostomum*, *Cylicocyclus*, and the genera *Cylicodontophorus*, *Cylindropharynx*, *Gyalocephalus*, and *Poteriostomum*. *Strongylus vulgaris* sucks blood and also causes verminous arteritis and thromboembolic colic. The other *Strongylus* spp. and the *Triodontophorus* spp. cause anaemia (*USA*: anemia) and debility. The small strongyles can cause severe enteritis and diarrhoea (*USA*: diarrhea). Also termed: redworm infestation.

Strongylus (parasitology) A genus of roundworm, in the family Strongylidae. Also termed: redworm.

See also strongylosis.

Strongylus asini (veterinary parasitology) A species of roundworm in the genus *Strongylus* that may be found in the large intestine of asses and wild equids.

Strongylus edentatus (veterinary parasitology) A species of roundworm in the genus *Strongylus* that may be found in the large intestine of horses.

Strongylus equinus (veterinary parasitology) A species of roundworm in the genus *Strongylus* that may be found in the caecum (*USA*: cecum) and colon of equids.

Strongylus tremletti (veterinary parasitology) A species of roundworm in the genus *Strongylus* that may be found in rhinoceroses.

Strongylus vulgaris (veterinary parasitology) A species of roundworm in the genus *Strongylus* that may be found in the large intestine of equids; larval stages are found in the anterior mesenteric and other arteries.

strontium (Sr) (chemistry) Silvery-white metallic element in Group IIA of the periodic table (one of the rare earth elements). Strontium compounds impart a bright red colour (*USA*: color) to a flame, and are used in flares and fireworks. At. no. 38; r.a.m. 87.62.

structural formula (chemistry) Shorthand description of a chemical compound that indicates the arrangement of the atoms in its molecules as well as its composition; *e.g.* C_2H_5COOH and $CH_2 = CH_2$ are the structural formulae of ethanol and ethene.

See also empirical formula; molecular formula.

structural isomerism (chemistry) Isomerism of chemical compounds that have the same molecular formula but different structural formulae. Structural isomers have different physical and chemical properties.

strychnine (chemistry) White crystalline insoluble alkaloid with a bitter taste, one of the most powerful poisons known. Also termed: vauqueline.

stubby root nematode (plant parasitology) *See Paratrichodorus minor.*

study (epidemiology) A scholarly examination. Specific types of study are also detailed under blind, case-control, cohort, cross-sectional.

sturdy (veterinary parasitology) A neurological disease in sheep caused by the pressure of a *Taenia multiceps* metacestode. Also termed: gid.

styl(o)- (terminology) A word element. [Latin, German] Prefix denoting *stake, pole, styloid process of the temporal bone.*

subatomic particle (physics) Particle that is smaller than an atom or forms part of an atom (*e.g.* electron, neutron, nucleus, proton). Sometimes also called an elementary particle.

subclinical infections (medical) Infections with minimal or no apparent symptoms; said of the early stages or a very mild form of a disease, *e.g.* sub-clinical disease, infection, parasitism, or when a disease is detectable by clinicopathological tests but not by a clinical examination.

subcosta (parasitological anatomy) A term used in relation to arthropods. A vein lying posterior to the costa.

subcutaneous (medical) Beneath the surface of the skin, but not necessarily beneath all the layers of skin, usually applied to injections.

subcutaneous mite (parasitology) *See Laminosioptes cysticola.*

subcutaneous tissue (histology) Layer of tissue below the dermis of the skin. It often contains deposits of fat.

subgenus (taxonomy) A taxonomic category sometimes established where the number of specimens is large, but not generally used; subordinate to a genus and superior to a species.

sublimate (chemistry) Solid formed by the process of sublimation.

sublimation (chemistry) Direct conversion of a solid substance to its vapour state on heating without melting taking place, *e.g.* solid carbon dioxide (dry ice). The vapour condenses to give a sublimate. The process is used to purify various substances.

sublimed sulphur (control measures) A parasiticide and scabicide. (*USA*: sublimed sulfur). Also termed sulphur (*USA*: sulfur) sublimatum; flowers of sulphur (*USA*: sulfur). (*USA*: sublimed sulphur).

subroutine (computing) In computer usage a self-contained subsidiary programme (*USA*: program) that is called upon by one programme (*USA*: program) to carry out a specific task.

sub-shell (physics) Subdivision of an electron shell.

subspecies (taxonomy) Group of organisms within a species that have certain characteristics not possessed by other members of the species. Breeding may occur between members of different sub-species. Also termed: race.

subspecies (taxonomy) A subdivision of a species; a variety or race.

substage (microscopy) The part of a microscope underneath the stage.

substance (histology) The material constituting an organ or body.

substituent (chemistry) Atom or group that replaces another atom or group in a molecule of a compound.

substitution product (chemistry) Product formed from substitution.

substitution reaction (chemistry) Chemical reaction that involves the direct replacement of an atom or group in a molecule of a compound (usually an organic compound) by some other atom or group; *e.g.* the reaction in which an atom of chlorine replaces an atom of hydrogen in a molecule of benzene (C_6H_6) to form a molecule of chlorobenzene (C_6H_5Cl). Also termed: displacement reaction.

substrate (1 biochemistry; 2 biology) 1 The molecule or compound upon which an enzyme acts. Each substrate is specific to its own enzyme, and vice versa. 2 Substance upon which an organism grows or is attached to.

subtribe (taxonomy) A taxonomic category sometimes established, subordinate to a tribe and superior to a genus.

Subtriquetra (fish/veterinary parasitology) A genus of pentastomes in the family Porocephalidae whose nymphs may be found in fish, and adult forms in crocodiles.

Subulura (parasitology) A genus of roundworms in the family Subuluridae.

Subulura brumpti (avian parasitology) A species of roundworms in the genus *Subulura* which may be found in the caeca (*USA*: ceca) of fowl, turkeys, guinea fowl and related wild birds but that has low pathogenicity.

Subulura differens (avian parasitology) A species of roundworms in the genus *Subulura* which may be found in fowls and guinea fowl.

Subulura minetti (avian parasitology) A species of roundworms in the genus *Subulura* which may be found in chickens.

Subulura strongylina (avian parasitology) A species of roundworms in the genus *Subulura* which may be found in chickens, guinea fowl and wild birds.

Subulura suctoria (avian parasitology) A species of roundworms in the genus *Subulura* which may be found in the caeca (*USA*: ceca) of chickens and wild birds.

succinic acid (chemistry) $(CH_2COOH)_2$ White crystalline dicarboxylic acid, used in the manufacture of dyes and in organic synthesis. It is an intermediate in the Krebs cycle (citric acid or tricarboxylic acid cycle). Also termed: butanedioic acid; ethylenedicarboxylic acid.

suckers (parasitological anatomy) A term used in relation to trematodes. Adhesive organs. There are generally two suckers, oral and ventral.

sucking (parasitology) The application of suction to an object by the mouth.

sucking lice (parasitology) Lice in the order Anoplura which includes the families Echinophthiriidae, Haematopinidae, Hoplopleuridae, Lignognathidae and Pediculidae.

sucrase (biochemistry) Enzyme that breaks down sucrose into simpler sugars. Also termed: invertase.

sucrose (chemistry) $C_{12}H_{22}O_{11}$ White optically active soluble crystalline disaccharide which is obtained from sugar cane and sugar-beet; ordinary sugar, used to sweeten food. It is hydrolysed to fructose and glucose. Also termed: cane-sugar; beet-sugar; sugar.

sudden infant death syndrome (SIDS) (medical) A term applied to babies who die unexpectedly and of undetermined cause, usually during sleep. Also called crib death.

sugar (1, 2 chemistry) 1 Crystalline soluble carbohydrate with a sweet taste; usually a monosaccharide or disaccharide. 2 Common name for sucrose.

sugar faecal centrifugation (methods) Using Sheather's solution, the standard procedure for examination of faeces (*USA*: feces) for parasite eggs. (*USA*: sugar fecal centrifugation).

sugarbeet cyst nematode (plant parasitology) *See Heterodera schachtii.*

Suidasia nesbitti (parasitology) A grain mite in the family Acaridae which causes wheat pollard itch.

Suifilaria (parasitology) A genus of filarioid worms in the family Filariidae.

Suifilaria suis (veterinary parasitology) A species of filarioid worms in the genus *Suifilaria* which may be found in the intermuscular and subcutaneous connective tissue of pigs but that cause no apparent problems other than the physical presence of the worms in the meat.

sulfur (chemistry) *See* sulphur.

sulphamethazine (pharmacology) A sulfonamide which is rapidly absorbed from the alimentary canal, slowly excreted with minimal precipitation in renal tubules that may be used as an coccidiostat. (*USA*: sulfamethazine). Also termed: sulfadimidine, sulfamezathine.

sulphanitran (pharmacology) A coccidiostat used in fowls. (*USA*: sulfanitran).

sulphaquinoxaline (pharmacology) A cocciodistat which was used extensively in chickens but has been superseded. (*USA*: sulfaquinoxaline).

sulphate (chemistry) Ester or salt of sulphuric acid (*USA*: sulfuric acid). (*USA*: sulfate).

sulphation (chemistry) Conversion of a substance into a sulphate (*USA*: sulfate). (*USA*: sulfation).

sulphide (chemistry) Binary compound containing sulphur (*USA*: sulfur); a salt of hydrogen sulphide (*USA*: hydrogen sulfide). Organic sulphides (*USA*: sulfides) are called thioethers. (*USA*: sulfide).

sulphite (chemistry) Ester or salt of sulphurous acid (*USA*: sulfurous acid). (*USA*: sulfite).

2-sulphobenzimide (chemistry) An alternative term for saccharin. (*USA*: sulfobenzimide).

sulphonate (chemistry) Ester or salt of a sulphonic acid (*USA*: sulfonic acid). (*USA*: sulfonate).

sulphonation (chemistry) Substitution reaction that involves the replacement of a hydrogen atom by the sulphonic acid group (*USA*: sulfonic acid group). (*USA*: sulfonation).

sulphonic acid (chemistry) Acid that contains the group SO_3H. Organic sulphonic acids (*USA*: sulfonic acids) are used in the manufacture of dyes, detergents and drugs. (*USA*: sulfonic acid).

sulphonium compound (chemistry) Compound of the empirical formula R_3SX, where R is an organic radical and X is an electronegative element or radical. (*USA*: sulfonium compound).

sulphoxide (chemistry) Compound of empirical formula RSOR', where R and R' are organic radicals. (*USA*: sulfoxide).

sulphur dichloride oxide (chemistry) An alternative term for thionyl chloride. (*USA*: sulfur dichloride oxide).

sulphur dioxide (chemistry) SO_2 Colourless (*USA*: colorless) poisonous gas with a strong pungent odour (*USA*: odor), made by burning sulphur (*USA*: sulfur), roasting sulphide (*USA*: sulfide) ores or by the action of acids on sulphites (*USA*: sulfites). It is also produced when sulphur containing compounds (such as fossils fuels) are burned, and from this source is a major atmospheric pollutant. It is used to make sulphuric acid (*USA*: sulfuric acid) and, in aqueous solution, as a bleach (*e.g.* for straw and paper). Also termed: sulphur(IV) oxide (*USA*: sulfur(IV) oxide). (*USA*: sulfur dioxide).

sulphur dye (chemistry) Dye made by heating certain organic compounds with sulphur (*USA*: sulfur) or alkali polysulphides (*USA*: polysulfides), used for dyeing industrial fabrics. (*USA*: sulfur dye).

sulphur (S) (chemistry) Yellow nonmetallic solid element in Group VIB of the periodic table, which forms several allotropes including alpha- (rhombic) sulphur (*USA*: sulfur) and beta- (monoclinic) sulphur. It occurs as the free element in volcanic regions and as underground deposits and as sulphates and sulphides, which include important minerals (*e.g.* galena, PbS, and pyrites, FeS_2). Chemically it behaves like oxygen, and can replace it in organic compounds (*e.g.* thioethers and thiols). It is also fed as an oldfashioned worm prophylaxis and coccidiostat, to make sulphuric acid, matches, gunpowder, drugs, fungicides and dyes, and in the vulcanization of rubber. At. no. 16; r.a.m. 32.06. (*USA*: sulfur).

sulphur trioxide (chemistry) SO_3 Volatile white solid made by the catalytic oxidation of sulphur dioxide (*USA*: sulfur dioxide), usually stored in sealed tubes. It reacts with water to form sulphuric acid (*USA*: sulfuric acid). Also termed: sulphur (VI) oxide (*USA*: sulfur(VI) oxide). (*USA*: sulfur trioxide).

sulphur(IV) oxide (chemistry) An alternative term for sulphur dioxide (*USA*: sulfur dioxide). (*USA*: sulfur(IV) oxide).

sulphur(V1) oxide (chemistry) An alternative term for sulphur trioxide (*USA*: sulfur trioxide). (*USA*: sulfur(VI) oxide).

sulphuric acid (chemistry) H_2SO_4 Corrosive, colourless (*USA*: colorless), oily liquid acid, made mainly from sulphur dioxide (*USA*: sulfur dioxide) by the contact process. It is a desiccant, and when hot a powerful oxidizing agent. It is produced in large quantities and used in the manufacture of other acids, fertilizers, explosives, accumulators, petrochemicals, etc. Its salts are sulphates (*USA*: sulfates). Also termed: vitriol; oil of vitriol; hydrogen sulphate (*USA*: hydrogen sulfate). (*USA*: sulfuric acid).

sulphurous acid (chemistry) H_2SO_3 Colourless (*USA*: colorless) aqueous solution of sulphur dioxide (*USA*: sulfur dioxide), used as a bleach and a reducing agent. Its salts are sulphites (*USA*: sulfites). (*USA*: sulfurous acid).

sumicidin (pharmacology) A synthetic, toxic, pyrethroid anthelmintic.

summer dermatitis (parasitology) *See* sweet itch.

summer eczema (parasitology) *See* sweet itch.

summer mange (parasitology) *See* onchocercosis.

summer sores (parasitology) *See* sweet itch; swamp cancer.

super- (terminology) A word element. [Latin] Prefix denoting *above*, *excessive*.

superconductivity (physics) Large increase in electrical conductivity exhibited by certain metals and alloys at a temperature a few degrees above absolute zero. Also termed: supraconductivity.

superconductor (physics) Metal that exhibits superconductivity.

supercooling (chemistry) Metastable state of a liquid in which its temperature has been brought below the normal freezing point without any solidification or crystallization occurring.

superficial pustular dermatitis (veterinary parasitology) A condition occurring in immature dogs who may develop pustules on the inguinal or axillary skin, often in association with parasitism.

superheated steam (chemistry) Steam (under pressure) at a temperature above the boiling point of water (100°C). *See* superheating.

superheating (chemistry) Metastable state of a liquid or gas that has been heated above its boiling point in the liquid state, by increasing the pressure above that of the atmosphere. Also termed: overheating.

supernatant liquid (chemistry) Clear liquid that lies above a sediment or precipitate.

superoxide (chemistry) 1 Compound that yields the free radical $O_2^{\cdot-}$, which is highly toxic to living cells. 2 Oxide that yields both hydrogen peroxide and oxygen on treatment with an acid.

See also peroxide.

supersaturated solution (chemistry) Unstable solution that contains more solute than a saturated solution would contain at the same temperature. It easily changes to a saturated solution when the excess solute is made to crystallize.

suppurate (medical) Produce pus.

suppuration (medical) Formation or discharge of pus.

suppurative (medical) Pertaining to or emanating from suppuration.

supra- (terminology) A word element. [Latin] Prefix denoting *above*.

supravital stain (parasitology) A stain introduced in living tissue or cells that have been removed from the body.

surface (anatomy) The outer part or external aspect of a body.

surface active agent (chemistry) An alternative term for surfactant.

surface biopsy (parasitology) Sample of cells scraped from the surface of a lesion or obtained by impression smears.

surface tension (physics) Force per unit length acting along the surface of a liquid at right-angles to any line drawn in the surface. It has the effect of making a liquid behave as if it has a surface skin (which can support *e.g.* small aquatic insects), and is responsible for capillarity and other phenomena. It is measured in newtons per metre ($N\ m^{-1}$).

surfactant (chemistry) Substance that reduces the surface tension of a liquid, used in detergents, wetting agents and foaming agents. Also termed: surface active agent.

surveillance (epidemiology) Keeping a watch over.

susceptible (immunology) Readily affected or acted upon; lacking immunity or resistance.

suspension (chemistry) Mixture of insoluble small solid particles and a fluid in which the insoluble substance stays evenly distributed throughout the fluid (because of molecular collisions and the fluid's viscosity, which prevent precipitation). Also termed: suspensoid.

suspensory ligament (anatomy) One of the structures that hold the lens of the eye in position.

swab (equipment) 1 A small piece of cotton or gauze wrapped around the end of a slender wooden stick or wire for applying medications or obtaining specimens of secretions, etc., from body surfaces or orifices. 2 Use of a swab to collect a specimen.

swamp cancer (veterinary parasitology) A common lesion of the skin and mucosae of horses in tropical and subtropical regions that may be caused by *Habronema megastoma* larvae. The lesions are dense granulation tissue in the form of an ulcer which may rapidly extend to 8 inches diameter. The tissue may contain cores of necrotic yellow or black, sometimes calcified, material. Also termed: cutaneous habronemiasis, summer sore.

swamp itch (parasitology) *See* schistosome dermatitis.

sweat (physiology) Watery fluid containing salts secreted from glands in the skin. Evaporation of sweat aids in cooling the body. Also termed: perspiration.

sweat fly (parasitology) *See Morrelia.*

sweating sickness (veterinary parasitology) A tick toxicosis of young cattle caused by the bites of *Hyalomma truncatum* (synonym: *Hyalomma transiens*) that is characterized by hyperaemia (*USA*: hyperemia) of the mucosae and an extensive moist eczema which may lead to sloughing. Also termed: dyshidrosis tropicale.

sweet itch (veterinary parasitology) An intensely itchy dermatitis of horses caused by hypersensitivity to the bites of *Culicoides* spp. The lesions are worst in the summer months and most evident along the middle of the back. The skin is thickened and the hair is missing. Also termed queensland itch. *See* equine allergic dermatitis.

swimmer's itch (parasitology) *See* schistosomiasis.

symbiosis (biology) Association between two organisms of different species in which both partners benefit.

See also commensalism; parasitism; saprophyte.

symbiotic mange (parasitology) *See* chorioptic mange.

symbol, chemical (chemistry) Letter or letters that represent the name of a chemical element, or one atom of it in a chemical formula. The symbol may be based on the element's Latin, not English, name; *e.g.* Au for gold (Latin *aurum*), Fe for iron (ferrum).

symmetry (mathematics) The property of being symmetrical, *i.e.* having the same shapes on each side of or around a point, axis or plane.

sympathetic nervous system (anatomy) Branch of the autonomic nervous system that is structurally different to the parasympathetic nervous system. Noradrenaline (*USA*: norepinephrine) is produced at the end of sympathetic nerve fibres (*USA*: sympathetic nerve fibers), unlike the parasympathetic system. Effects produced by each system are generally antagonistic.

Symphoromyia (parasitology) A genus of flies which is ordinarily predatory on other flies but which are blood-suckers and do inflict a painful bite. They are members of the family Rhagionidae.

syn- (terminology) A word element. [German] Prefix denoting *union*, *association*.

synapse (histology) Point of connection between neurones (nerve cells). It consists of a gap between the membranes of two cells, across which impulses are transmitted by a transmitter substance (*e.g.* acetylcholine). Specialized synapses occur at nerve-muscle junctions.

See also neurotransmitter.

syndrome (medical) Set of symptoms occurring together all thought to be produced by the same illness. (*e.g.* nephrotic syndrome).

syngamiasis (avian parasitology) Infestation with *Syngamus trachea* in fowl, turkeys, pheasants, guinea fowl, geese and wild birds, which causes pneumonia while migrating through the lungs. In the trachea they cause tracheitis and anaemia (*USA*: anemia) because of heavy blood-sucking. The predominant clinical signs are dyspnea, head shaking and spasmodic gaping.

Syngamus (parasitology) A genus of nematodes in the family Syngamidae.

Syngamus ierei (veterinary parasitology) A species of nematodes in the genus *Syngamus* that is a cause of chronic nasopharyngitis in cats in Puerto Rico.

Syngamus laryngeus (veterinary parasitology) A species of nematodes in the genus *Syngamus* that may be found in the larynx of cattle.

Syngamus nasicola (veterinary parasitology) A species of nematodes in the genus *Syngamus* that may be found in the nasal cavities of ruminants.

Syngamus skrjabinomorpha (avian parasitology) A species of nematodes in the genus *Syngamus* that may be found in domestic geese and in chickens.

Syngamus trachea (avian parasitology) A species of nematodes in the genus *Syngamus* that may be found in fowl, turkeys, pheasants, guinea fowl, geese and wild birds. They cause pneumonia while migrating through the lungs and in the trachea cause trachcitis and anaemia (*USA*: anemia) because of heavy blood-sucking. The predominant clinical signs are dyspnea, head shaking and spasmodic gaping.

syngamy (parasitological physiology) A method of reproduction in which two individuals (gametes) unite permanently and their nuclei fuse; sexual reproduction. A common form of reproduction in protozoa.

Synhimantus (avian parasitology) A genus of nematodes in the family Acuariidae that infect birds. Species includes *Synhimantus spiralis* and *Synhimantus nasula* that may be found in the walls of the proventriculus, oesophagus (*USA*: esophagus), and sometimes intestine of most domestic and wild bird species, mainly in water birds, *e.g.* pelicans, but also in hawks and owls. It causes severe lesions in the proventriculus and affected birds waste away and die. Losses may be heavy.

synnecrosis (biology) Symbiosis in which the relationship between populations (or individuals) is mutually detrimental.

synonym (taxonomy) An alternative name for the same disease, sign, bacteria, etc. A key word or sign may have a number of synonyms.

synovial fluid (anatomy) Liquid secreted by a synovial membrane.

synovial membrane (anatomy) Lining of the capsule that encloses a joint between bones. It secretes synovial fluid, which acts as a lubricant to prevent friction in the joint during movement.

synthetic (chemistry) Formed by artificial means; describing a chemical compound that has been produced by synthesis.

Syphacia obvelata (veterinary parasitology) The pinworm found in mice and hamsters, a common parasite in laboratory colonies.

syring(o)- (terminology) A word element. [German] Prefix denoting *tube*, *fistula*.

Syringophilus (avian parasitology) A genus of trombidiform mites that may be found inside the quills of feathers of birds.

Syringophilus bipectinatus (avian parasitology) A species of trombidiform mites in the genus *Syringophilus* that may be found in fowl feather quills.

Syringophilus columbae (avian parasitology) A species of trombidiform mites in the genus *Syringophilus* that may be found in pigeon feather quills.

Syringophilus uncinatus (avian parasitology) (synonym: *Cheyletoides uncinata*) A species of trombidiform mites in the genus *Syringophilus* that may be found in peacock feather quills.

systemic (medical) Pertaining to or affecting the body as a whole.

systemic disease (medical) Sufficiently widespread in the body to cause clinical signs referable to any organ or system, and in which localization of infection may occur in an organ.

systole (physiology) The contraction phase of the cardiac cycle, as opposed to the diastole.

syzygy (parasitology) Temporary adherence of male and female protozoa of the subclass Gregarinomorpha before numerical increase occurs by sporogony.

synergy (biology) Collective action of two or more things (*e.g.* drugs, muscles) that is more effective than it would be if they acted on their own. Also termed: synergism.

synthesis (chemistry) Formation of a chemical compound by combining elements or simpler compounds; the building of compounds through a planned series of steps.

T

T (biochemistry) A symbol for threonine.

t (general terminology) An abbreviation of 't'ime.

T helper cells (immunology) A subset of T cells that carry the T4 marker and are essential for turning on antibody production, activating cytotoxic T cells and initiating many other immune responses.

T lymphocytes (immunology) A group of white blood cells involved in the immune system. They have on their surfaces receptors that recognise histocompatibilty antigens on the surfaces of cells, so that they bind on to these antigens on any foreign cell that enters the body. They are therefore the main part of the body's response to transplants. T lymphocytes (so called because they originate in the thymus) recognise the antigens by sequence (unlike B lymphocytes). T lymphocytes also release factors that induce proliferation of T lymphocytes and B lymphocytes.

T4 (1 genetics; 2 biochemistry) 1 One of the most intensively studied of the bacteriophages. It only attacks *Escherichia coli*. The genome is double-stranded DNA, linear, but with matching ends that can join up to make a circular DNA molecule after entry into the host cell. 2 Thyroxine. *See* thyroid.

tabanid (parasitology) A fly of the family Tabanidae, including the genera *Chrysops*, *Haematopota*, *Pangonia* and *Tabanus*.

Tabanus (parasitology) A genus of blood-sucking biting flies (horse flies, deer flies or march flies) in the family Tabanidae which transmit trypanosomes to various animals and have a painful bite.

tachy- (terminology) A word element. [German] Prefix denoting *rapid*, *swift*.

tachycardia (medical) Rapid heartbeat, above 100 beats per minute.

Tachygonetria (veterinary parasitology) A genus of oxyurid worms, in the superfamily Oxyuroidea, that may be found in the large intestine of tortoises.

tactical treatment (pharmacology) Treatment at times when the activities of a disease are at their worst, *e.g.* when diarrhoea (*USA*: diarrhea) caused by *Ostertagia* spp. in cattle is most severe.

Taenia (human/veterinary parasitology) A genus of cyclophyllidean tapeworms of the family Taeniidae. The adult tapeworm inhabits the intestine of carnivores, the larval stage (metacestode) invades the tissues of a variety of animals, in some cases humans. Tapeworms and their hosts where known are listed individually, but species whose intermediate hosts are unknown include: *Taenia bubesi* in lions, *Taenia crocutae* in spotted hyenas, *Taenia erythraea* in black-backed jackals, *Taenia gongamai* and *Taenia hlosei* in lions and cheetahs, *Taenia lycaontis* in hunting dogs and *Taenia regis* in lions. *See Taenia brauni*; *Taenia crassiceps*; *Taenia hydatigena*; *Taenia hyenae*; *Taenia laticollis*; *Taenia krabbei*; *Taenia macrocystis*; *Taenia martis*; *Taenia multiceps*; *Taenia mustelae*; *Taenia omissa*; *Taenia ovis*; *Taenia parva*; *Taenia pisiformis*; *Taenia polyacantha*; *Taenia rileyi*; *Taenia saginata*; *Taenia serialis*; *Taenia serrata*; *Taenia solium*; *Taenia taeniaeformis*; *Taenia twitchelli*.

Taenia brauni (veterinary parasitology) A species of tapeworms in the genus *Taenia* that is probably a subspecies of *Taenia serialis*. Adult tapeworms of this species may be found in dogs and jackals and the larval stage, a coenurus, in rats, mice and porcupines.

Taenia crassiceps (veterinary parasitology) A species of tapeworms in the genus *Taenia*. Adult tapeworms of this species may be found in foxes and coyotes and the larval stage, a cysticercus, in rodents.

Taenia hydatigena (veterinary parasitology) A species of tapeworms in the genus *Taenia* that may be found in the small intestine of dogs, wolves and wild Carnivora, and the larval stage, *Cysticercus tenuicollis*, in sheep and other ruminants, pigs and occasionally primates.

Taenia hyenae (veterinary parasitology) A species of tapeworms in the genus *Taenia* that may be found in hyenas and the cysticerci in antelopes.

Taenia krabbei (veterinary parasitology) A species of tapeworms in the genus *Taenia*. Adult tapeworms of this species may be found in dogs and in wild carnivores and the larval cestode *Cysticercus tarandi*, in the muscles of wild ruminants, especially deer.

Taenia laticollis (veterinary parasitology) A species of tapeworms in the genus *Taenia* that may be found in carnivores and larval forms in rodents.

Taenia macrocystis (veterinary parasitology) A species of tapeworms in the genus *Taenia*. Adult tapeworms of this species may be found in lynx and coyotes, and the intermediate stage in snowshoe lagomorphs.

Taenia martis (veterinary parasitology) A species of tapeworms in the genus *Taenia*. Adult tapeworms of this species may be found in martens and the cysticercus in voles.

Taenia multiceps (veterinary parasitology) (synonym: *Multiceps multiceps*) A species of tapeworms in the genus *Taenia*. Adult tapeworms of this species may be found in dogs and wild canids, the larvae, *Coenurus cerebralis*, in the brain and spinal cord of sheep and goats.

Taenia mustelae (veterinary parasitology) A species of tapeworms in the genus *Taenia*. Adult tapeworms of this species may be found in martens, weasels, otters, skunks, and badgers, and larval stages in voles and other rodents.

Taenia omissa (veterinary parasitology) A species of tapeworms in the genus *Taenia*. Adult tapeworms of this species may be found in cougars and larvae in deer.

Taenia ovis (veterinary parasitology) A species of tapeworms in the genus *Taenia*. Adult tapeworms of this species may be found in dogs and wild carnivores and the larval stage, *Cysticercus ovis*, in the skeletal and cardiac muscles of sheep and goats.

Taenia parva (veterinary parasitology) A species of tapeworms in the genus *Taenia*. Adult tapeworms of this species may be found in genets, and the larval stage in rodents.

Taenia pisiformis (veterinary parasitology) A species of tapeworms in the genus *Taenia*. Adult tapeworms of this species may be found in the small intestine of dogs, foxes, some wild carnivores, and very rarely in cats. The metacestode stage, *Cysticercus pisiformis*, may be found in the liver and peritoneal cavity of lagomorphs.

Taenia polyacantha (veterinary parasitology) A species of tapeworms in the genus *Taenia*. Adult tapeworms of this species may be found in the intestine of foxes and the metacestodes in microtine rodents.

Taenia rileyi (veterinary parasitology) A species of tapeworms in the genus *Taenia*. Adult tapeworms of this species may be found in lynx, and the larvae in rodents.

Taenia saginata (human/veterinary parasitology) A species of tapeworms in the genus *Taenia*. Adult tapeworms of this species may be found in humans, and the metacestode, *Cysticercus bovis*, in cattle and some wild ruminants.

Taenia serialis (veterinary parasitology) A species of tapeworms in the genus *Taenia*. Adult tapeworms of this species may be found in dogs and foxes and the metacestode, *Coenurus serialis*, in the subcutaneous and intramuscular tissues of lagomorphs.

Taenia serrata (veterinary parasitology) *See Taenia pisiformis*.

Taenia solium (human/veterinary parasitology) A species of tapeworms in the genus *Taenia*. Adult tapeworms of this species may be found in the small intestine of humans and some apes, the metacestode, *Cysticercus cellulosae*, in the skeletal and cardiac muscle of pigs and in the brain of humans.

Taenia taeniaeformis (veterinary parasitology) A species of tapeworms in the genus *Taenia*. Adult tapeworms of this species may be found in the small intestine of cats and other related carnivores and the metacestode, *Cysticercus fasciolaris*, in the livers of rodents.

Taenia twitchelli (veterinary parasitology) A species of tapeworms in the genus *Taenia*. Adult tapeworms of this species may be found in wolverines, and the larvae in lungs and pleural cavity of porcupines.

taeniacide (pharmacology) 1 Fatal to tapeworms. 2 A substance fatal to tapeworms. Also termed: teniacide.

taeniafuge (pharmacology) A medicine for expelling tapeworms. Also termed: teniafuge.

taeniasis (parasitology) Infection with tapeworms of the genus *Taenia*. The adult tapeworms do not cause a clinically recognizable disease, but the larval stages may cause clinical signs described under coenurosis or cysticercosis. (*USA*: teniasis).

taeniid (parasitology) Pertaining to or emanating from cestodes in the family Taeniidae. (*USA*: teniid).

Taeniorhyncus (parasitology) The modern genus for *Mansonia* mosquitoes.

tail (statistics) The extreme part of a distribution, which when represented graphically looks like a tail.

tail louse (parasitology) *See Haematopinus quadripertusus*.

tail mange (parasitology) *See* chorioptic mange.

tail switch louse (parasitology) *See Haematopinus quadripertusus*.

talc (chemistry) $3MgO.4SiO_2.H_2O$ Finely powdered hydrous magnesium silicate. Soft white or grey-green mineral. Its purified form is a white powder, used in talcum powder, in medicine and in ceramic materials. Also termed: French chalk; magnesium silicate monohydrate.

talus (anatomy) The ankle bone.

tampan (parasitology) A tick. *See Ornithodorus*.

Tanaisia (parasitology) A genus of flukes in the family Eucotylidae.

Tanaisia bragai (avian parasitology) A species of flukes in the genus *Tanaisia* that may be found in the kidneys and ureters of chickens, turkeys and pigeons but with little apparent pathogenicity.

Tanaisia zarudnyi (avian parasitology) A species of flukes in the genus *Tanaisia* that may be found in ruffed grouse.

tandum duplication (genetics) A form of duplication in which part of a chromosome duplicates itself, the second copy being inserted in line with the first.

tanin (chemistry) Yellow substance that is a member of a class of organic compounds, of vegetable origin (*e.g.* in tree bark, oak galls and tea), that are derivatives of polyhydric benzoic acids. They are used in tanning hides to make leather.

See also tannic acid.

tannic acid (chemistry) White amorphous solid organic acid, a member of the class of compounds called tannins. It is used in tanning and for making inks and dyes. Also termed: tannin.

tantalum (Ta) (chemistry) Hard blue-grey metallic element in Group VA of the periodic table (a transition element), used in electronic and chemical equipment, and for making surgical instruments. At. no. 73; r.a.m. 180.9479.

tapeworm (parasitology) *See* taenia.

target dose (biochemistry) The amount or concentration of a chemical that gets to the site of action, causing a measurable effect.

targeting vector (molecular biology) A vector carrying a DNA sequence that is able to take part in a specified chromosomal crossover in the host.

tarone test (statistics) A statistical test used in carcinogenicity studies to evaluate the significance of the time of observation of tumor rather than death as an endpoint. This test is only appropriate under certain defined conditions such as when the number of animals dying early with a lethal tumor is large and control survival has been large.

tars(o)- (terminology) A word element. [German] Prefix denoting *edge of eyelid, tarsus of the foot.*

tarsal (anatomy) Bone that occurs in the feet of tetrapods. Human beings have seven tarsals in each foot, one of them modified to form the heel bone (calcaneum).

Tarsenomidae (insect parasitology) Mites in the Prostigmata that feed on excretions of ants' anus in the ants *Atta* and *Solenopsis.*

Tarsocheylus atomarius (insect parasitology) Mites in the Prostigmata that feed on excretions of ants' anus in the ants *Atta* and *Solenopsis.*

Tarsonemus (insect parasitology) A species of mites in the Prostigmata, that similarly to *Tetranychus* are phoretic on the bees *Apinae* and *Apis.*

tartar emetic (control measures/chemistry) Potassium antimonyl tartrate, $K.SbO.C_4H_4O_6.1/2H_2O$, a poisonous compound used as an insecticide.

tartaric acid (chemistry) HOOC.CH(OH)-CH-(OH).COOH White crystalline hydroxycarboxylic acid that occurs in grapes and other fruits. It is used in dyeing and printing. Its salts, the tartrates, are used as buffers and in medicine. Also termed: 2,3- dihydroxybutanedioic acid; dihydroxysuccinic acid.

tartrate (chemistry) Ester or salt of tartaric acid.

taste (physiology) Sense that enables animals to detect flavours, which in mammals involves taste buds.

taste bud (physiology) Small sense organ containing chemoreceptors for the sense of taste, located in the mouth (particularly on the upper surface of the tongue).

Tata box (genetics) A sequence found in eukaryote genes. Part of the promoter region, it occurs about 35 base-pairs upstream from the point at which RNA transcription is to start. Prokaryotes have a 'Pribnow box'.

tautomerism (chemistry) Equilibrium between two organic isomers. It usually involves a shift in the point of attachment of a mobile hydrogen atom and a shift in the position of a double bond in a molecule. Also termed: dynamic isomerism.

taxis (biology) Orientation of an organism, involving movement, with respect to a stimulus from a specific direction.

See also chemotaxis.

taxon (genetics) Any group of living things, from the viewpoint of classification. The key taxon is the species; the higher taxa are for animals: genus, family, order, class, phylum, kingdom. (pl. taxa).

taxonomy (genetics) Study of the classification of living organisms. The arrangement of taxa into a system of classification. A modem technique is numerical taxonomy, in which all attributes of the phenotype metric, biochemical or whatever, are given equal weighting, and groupings are worked out by computer.

Tb (chemistry) The chemical symbol for terbium.

TBG (physiology/biochemistry) An abbreviation of 'T'hyroxine-'B'inding 'G'lobulin.

TBPA (physiology/biochemistry) An abbreviation of 'T'hyroxine-'B'inding 'P're'A'lbumin.

T-butylaminoethanol (pharmacology) A coccidiostat which may induce a choline deficiency leading to depressed growth in chickens.

TBW (physiology) An abbreviation of 'T'otal 'B'ody 'W'ater.

Tc (chemistry) The chemical symbol for Element 43, technetium.

TCA (chemistry) An abbreviation of 'T'ri'C'hloroacetic 'A'cid.

T-cell (immunology) An abbreviation of 'T'hymus-derived 'cell'.

TDN (biochemistry) An abbreviation of 'T'otal 'D'igestible 'N'utrients.

Te (chemistry) The chemical symbol for Element 52, tellurium.

TEA (biochemistry) An abbreviation of 'Tetra-'E'thyl'A'mmonium.

tech. (general terminology) Technical.

technetium (Tc) (chemistry) Artificial radioactive metallic element in Group VIIA of the periodic table (a transition element), which occurs among the fission products of uranium. It has several isotopes, with half-lives of up to 2.12×10^5 years. At. no. 43; r.a.m. 99 (most stable isotope).

teflon (chemistry) Trade name for polytetrafluoroethene (PTFE).

Teladorsagia (veterinary parasitology) A genus of nematodes in the family Trichostrongylidae, which are sometimes considered as a synonym of *Ostertagia*. Species includes *Teladorsagia* circumcincta (*Teladorsagia davtiani*, *Teladorsagia trifurcata*), that may be found in sheep and goats.

telangiectasis (histology) A red spot on the skin formed by dilated capillary or terminal arterial blood vessels.

tele- (terminology) A word element. [German] Prefix denoting *far away*, *operating at a distance*, *an end.*

telluride (chemistry) Binary compound that contains tellurium.

tellurium (Te) (chemistry) Silvery-white semi-metallic element in Group VIB of the periodic table, obtained as a by-product of the extraction of gold, silver and copper. It is used as a catalyst and to add hardness to alloys of lead or steel. At. no. 52; r.a.m. 127.60.

telo- (terminology) A word element. [German] Prefix denoting *end.*

telodendria (histology) The very fine terminal branches of an axon.

telophase (cytology) Final phase of cell division that occurs in mitosis and meiosis. During telophase chromatids, on reaching the poles of the cell, become densely packed and the cell divides. In animal cells, the plasma membrane constricts. In plants, a wall divides the cell in two.

temp. (measurement) An abbreviation of 'temp'erature.

temperature (measurement) The degree of hotness or coldness, a measure of the average kinetic energy of its atoms or molecules. The average body temperature in health is 37°C (98.6°F). It is a little lower in the morning and higher in the evening.

temperature coefficient (physics) Change in a physical quantity with change in temperature, usually given as per degree rise of temperature.

See also temperature scale.

temperature gradient (physics) Degree or measured rate of the temperature change between two points of reference in a substance or in an area.

temperature scale (physics) Method of expressing temperature. There are various scales, based on different fixed points. The Fahrenheit scale has largely been replaced by the Celsius scale (formerly centigrade), all of which express temperatures in degrees (°F and °C). In science, temperatures are frequently expressed in kelvin (K), the thermodynamic unit of temperature in SI units.

temporary parasite (parasitology) A parasite that lives free of its host during part of its life cycle.

tendon (histology) Strong connective tissue that attaches muscles to bones. It consists of collagen fibres (*USA*: collagen fibers).

See also ligament.

Tenebrio molitor (avian parasitology) A pest of stored grain which may attack the claws of setting hens and tissues of newly hatched birds. Other mealworm species are known to have the same effect. Also termed: yellow mealworm.

tenesmus (medical) Straining painfully but ineffectively to pass a motion or pass urine.

teniacide (pharmacology) *See* taeniacide.

teniafuge (parasitology) *See* taeniafuge.

teno- (terminology) A word element. [German] Prefix denoting *tendon.*

tenonto- (terminology) A word element. [German] Prefix denoting *tendon.*

tenosynovitis (histology) Inflammation of the sheath of a tendon.

ter- (terminology) A word element. [Latin] Prefix denoting *three*, three-fold.

tera- (terminology) A word element. [German] Prefix denoting *monster.*

terato- (terminology) A word element. [German] Prefix denoting *monster*, *monstrosity.*

teratogen (toxicology) Any substance, natural or artificial, that causes abnormal development of an embryo.

terbium (Tb) (chemistry) Silvery-grey metallic element in Group IIIA of the periodic table (one of the lanthanides), used in making semiconductors and phosphors. At. no. 65; r.a.m. 158.9254.

tergum (parasitological anatomy) A term used in relation to arthropods. A dorsal plate on the thoracic segment.

term (medical) A definite period of time, in medicine usually applied to the duration of a pregnancy; full term is 282 days from the first day of the last menstrual period.

terminal (computing) An input device or output device that can handle data.

terminal decimal (mathematics) Decimal quantity that has a finite number of digits after the decimal point.

See also recurring decimal.

terminal fronsferase (genetics) An enzyme that synthesises a 'tail' on to a DNA molecule. It is used in genetic engineering to add a poly-A tail (one that consists only of repeated As) to the end of one DNA molecule and a poly-T tail (one that consists only of repeated Ts) to the end of another so that they join up.

termination (medical) Commonly used to mean termination of pregnancy, or abortion.

Termirhabditis fastidiosus (insect parasitology) A species of nematode in the Rhabditidae that are parasites to the termites *Reticulitermes flavipes*. They may be found in the termites head.

Termitacarius cuneiformis (insect parasitology) *See Histiostoma formosana.*

Ternidens (parasitology) A genus of strongylid worms in the family Chabertiidae.

Ternidens deminatus (human/veterinary parasitology) A species of strongylid worms in the genus *Ternidens* that may be found in the large intestine of primates and occasionally humans. They cause anaemia (*USA*: anemia) and the development of nodules in the intestinal wall similar to those of *Oesophagostomum* spp.

Terranova (fish parasitology) A genus of nematodes in the family Anisakidae that parasitize elasmobranch fish.

tesla (measurement) SI unit of magnetic flux density, named after the Croatian-born American physicist Nikola Tesla (1857–1943).

testes (anatomy) Plural of testis.

testis (anatomy) The male gonad whose primary structures are the seminiferous tubules in which spermatogenesis occurs and the interstitial cells (Leydig cells) which secrete androgens.

testosterone (biochemistry) Sex hormone. The most active androgen hormone. It is produced by the testes in the male, in small quantities by the ovaries in the female, and by the adrenal cortex in both sexes. It stimulates the development of the male reproductive organs and of male secondary sex characteristics like body hair and strong muscles. It also raises the sexual drive in males, and increases aggression in some species, probably including man.

tetmosol (pharmacology) An acaricide that may be toxic to birds if used to excess. Also termed: monosulfiram.

tetra- (terminology) A word element. [German] Prefix denoting *four*.

tetrachlorodifluorethane (pharmacology) A fasciolicide, that because of toxicity is not recommended for general use. Also termed: Freon-112; difluorotetrachloroethane.

tetrachloroethene (chemistry) $CCl_2 = CCl_2$ A liquid halogenoalkene, used as a solvent (especially as a de-greasing agent). Also termed: ethylene tetrachloride; tetrachloroethylene.

tetrachloroethylene (chemistry) An alternative term for tetrachloroethene.

tetrachloromethane (chemistry) CCl_4 Colourless (*USA*: colorless) liquid halogenoalkane, used as a solvent, cleaning agent and in fire extinguishers. Also termed: carbon tetrachloride.

tetrachlorvinphos (control measures) An organophosphorus insecticide used for ectoparasites.

tetrad (1, 2 genetics) 1 Four cells formed after the second division in meiosis is complete. 2 Four spores formed by meiosis in a spore mother cell, often seen in fungi. *See* tetrad analysis.

tetradifon (control measures) An acaricide used in horticulture.

Tetradonema solenopsis (insect parasitology) A species of nematode in the Tetradonematidae that are parasites to the ants *Solenopsis invicta*.

tetrahedral compound (chemistry) Compound that has a central atom joined to other atoms or groups at the four corners of a tetrahedron. It is the configuration of carbon in many saturated compounds.

tetrahydrate (chemistry) Chemical containing four molecules of water of crystallization.

tetrahydropyrimidines (pharmacology) A group of broad-spectrum anthelmintics including pyrantel and morantel.

tetramastigote (parasitological anatomy) 1 Having four flagella. 2 An organism having four flagella.

Tetrameres (avian parasitology) A genus of roundworms in the family Tetrameridae, that may be found in the proventriculus of birds and may kill young birds. They show marked sexual dimorphism, the female being globular and the male slender and filiform.

tetrameres (parasitology) Spiruroid nematodes in the family Tetrameridae.

Tetrameres americana (avian parasitology) A species of roundworms in the genus *Tetrameres* that may be found in fowl and turkeys.

Tetrameres confusa (avian parasitology) A species of roundworms in the genus *Tetrameres* that may be found in fowl and pigeons.

Tetrameres crami (avian parasitology) A species of roundworms in the genus *Tetrameres* that may be found in domestic and wild ducks.

Tetrameres fissispina (avian parasitology) A species of roundworms in the genus *Tetrameres* that may be found in fowl, ducks, turkeys, pigeons and wild aquatic birds.

Tetrameres mohtedai (avian parasitology) A species of roundworms in the genus *Tetrameres* that may be found in fowl.

Tetrameres pattersoni (avian parasitology) A species of roundworms in the genus *Tetrameres* that may be found in quail.

tetrameriasis (parasitology) Infestation with tetrameres.

tetrameric (biology) Having four parts.

tetramethylthiuram disulphide (pharmacology/control measures) A relatively nontoxic acaricide. Also termed: thiram. (*USA*: tetramethylthiuram disulfide).

Tetramitus rostratus (human/veterinary parasitology) A coprophilic, flagellate protozoan found in stagnant water and human and rat faeces (*USA*: feces) but which is considered to be nonpathogenic. A member of the family Tetramitidae.

Tetranychus (insect parasitology) *See Tarsonemus.*

tetrapeptide (biochemistry) A peptide which, on hydrolysis, yields four amino acids.

Tetrapetalonema (veterinary parasitology) A filarioid nematode found in the peritoneal cavity of primates and insectivores but even in very large numbers do not apparently cause any damaging effect.

Tetraphyllidea (fish parasitology) An order of flatworms in subclass Eucestoda, class Cestoidea and the phylum Platyhelminthes. Adult tetraphyllideans are found only in the intestine of elasmobranchs. The scolex is very variable: the four bothridia may be sessile, pedunculate, simple in outline, folded or loculate: suckers may be present and the scolex may be armed with hooks and/or spines.

tetraploid (genetics) Describing an organism that has four sets of homologous chromosomes.

See also diploid; haploid.

tetrathyridium (parasitological physiology) One of the forms of metacestodes in the life cycles of the cestodes of domestic animals, *e.g. Mesocestoides* spp.

Tetratrichomonas (parasitology) A genus of protozoa with four anterior flagella in the family Trichomonadidae.

Tetratrichomonas anatis (avian parasitology) A species of protozoa of the genus *Tetratrichomonas* that may be found in ducks but that does not appear to be pathogenic.

Tetratrichomonas anseri (avian parasitology) A species of protozoa of the genus *Tetratrichomonas* that may be found in the caecum (*USA*: cecum) of geese but that does not appear to be pathogenic.

Tetratrichomonas bovis (parasitology) *See Tetratrichomonas pavlovi.*

Tetratrichomonas buttreyi (veterinary parasitology) A species of protozoa of the genus *Tetratrichomonas* that may be found in the caecum (*USA*: cecum) and colon of pigs but that does not appear to be pathogenic.

Tetratrichomonas gallinarum (avian parasitology) A species of protozoa of the genus *Tetratrichomonas* that may be found in fowl, ducks, guinea fowl, quail, pheasants, partridges, and probably Canada geese, which causes hepatic lesions similar to those caused by *Histomonas meleagridis.*

Tetratrichomonas microti (veterinary parasitology) A species of protozoa of the genus *Tetratrichomonas* that may be found in the caecum (*USA*: cecum) of rodents.

Tetratrichomonas ovis (veterinary parasitology) A species of protozoa of the genus *Tetratrichomonas* that may be found in the caecum (*USA*: cecum) of sheep but that does not appear to be pathogenic.

Tetratrichomonas pavlovi (synonym: T. Bovis) A species of protozoa of the genus *Tetratrichomonas* that may be found in the caecum (*USA*: cecum) of calves with diarrhoea (*USA*: diarrhea).

tetravalent (chemistry) Having a valence of four. Also termed: quadrivalent.

TG (biochemistry) 1 An abbreviation of 'T'hyro-'G'lobulins. 2 An abbreviation of 'T'riacyl'G'lycerol.

TGE (medical) An abbreviation of 'T'ransmissible 'G'astro'E'nteritis.

TGFA (physiology) An abbreviation of 'T'ransforming 'G'rowth 'F'actor 'A'lpha.

Th (chemistry) The chemical symbol for thorium.

thalamus (anatomy) Part of the forebrain of vertebrates that is concerned with the routing of nervous impulses to and from the spinal cord.

thallium (Tl) (control measures/chemistry) Silver-grey metallic element in Group IIIB of the periodic table, used in electronic equipment and to make pesticides. At. no. 81; r.a.m. 204.39.

thallous (biochemistry) Pertaining to thallium.

thallous acetate (control measures) A rodenticide.

thallous sulphate (control measures) A rodenticide. (*USA*: thallous sulfate).

thanato- (terminology) A word element. [German] Prefix denoting death.

thanatology (medical) The medicolegal study of death and conditions affecting human bodies.

***Thaumatopelvis* sp.** (insect parasitology) Mites in the Prostigmata that are parasites on various ants.

Theileria (parasitology) A genus of protozoan parasites in the family Theileriidae. They are transmitted by ticks, multiply in leukocytes and then invade erythrocytes.

Theileria annulata (veterinary parasitology) (synonym: *Theileria dispar*) A species of protozoan parasites of the genus *Theileria* which may be found in cattle and water buffalo, is transmitted by *Hyalomma* spp. ticks and causes a clinical disease similar to East Coast fever.

Theileria buffali (veterinary parasitology) A species of protozoan parasites of the genus *Theileria* which may be found in cattle and buffalo in Australia and that is transmitted by *Haemaphysalis longicornis* and *Haemaphysalis bancrofti*. Similar to *Theileria mutans*. Only sporadic cases of clinical disease have been reported and are seen mostly in splenectomized calves.

Theileria camalensis (veterinary parasitology) A species of protozoan parasites of the genus *Theileria* which may be found in camels, and transmission is thought to be by *Hyalomma* spp. ticks.

Theileria cervi (veterinary parasitology) A species of protozoan parasites of the genus *Theileria* which may be found in splenectomized deer, but that does not appear to be pathogenic.

Theileria dispar (parasitology) *See Theileria annulata.*

Theileria hirci (veterinary parasitology) A species of protozoan parasites of the genus *Theileria* which may be found in sheep and goats, which may cause a disease similar to East Coast fever in cattle. The vector is probably the tick *Rhipicephalus bursa*.

Theileria lawrenci (veterinary parasitology) A species of protozoan parasites of the genus *Theileria* which may be found in cattle, buffaloes and water buffaloes, is transmitted by the tick, *Rhipicephalus appendiculatus*, and that may cause fatal corridor disease.

Theileria mutans (veterinary parasitology) A species of protozoan parasites of the genus *Theileria* which may be found in ailing cattle, is transmitted by ticks including *Rhipicephalus* and *Haemaphysalis* spp. and that may cause a benign bovine theileriasis.

Theileria orientalis (veterinary parasitology) A species of usually benign protozoan parasites of the genus *Theileria* which may be found in cattle where they may cause severe anaemia (*USA*: anemia).

Theileria ornithorhynci (veterinary parasitology) A species of protozoan parasites of the genus *Theileria* which may be found in platypus.

Theileria ovis (veterinary parasitology) A species of protozoan parasites of the genus *Theileria* which may be found in sheep and goats, is transmitted by ticks of the genera *Rhipicephalus*, *Dermacentor*, *Haemaphysalis* and *Ornithodoros* spp., and which may cause a mild form of theileriasis.

Theileria parva (veterinary parasitology) A species of protozoan parasites of the genus *Theileria* which may be found in cattle, African buffaloes and Indian water buffaloes, is transmitted by *Rhipicephalus appendiculatus* and possibly other ticks, and that may cause the widespread and serious disease East Coast fever.

Theileria sergenti (veterinary parasitology) A species of protozoan parasites of the genus *Theileria* of doubtful identity which may be found in cattle where it is mostly benign.

Theileria tarandi (veterinary parasitology) A species of protozoan parasites of the genus *Theileria* which may be found in ailing reindeer, is transmitted by *Ixodes persulcatus* and may cause an acute disease.

Theileria taurotragi (veterinary parasitology) A species of protozoan parasites of the genus *Theileria* which may be found in cattle in Africa and Asia and may be mildly pathogenic.

Theileria velifera (veterinary parasitology) A species of protozoan parasites of the genus *Theileria* which may be found in cattle, and may be mildly pathogenic.

theileriasis (parasitology) Infestation with the protozoan parasites of the genus *Theileria*. In general the diseases may be acute and severe or mild. The severe disease is characterized by high fever, nasal discharge, jaundice, petechiation of mucusae, enlargement of the spleen and lymph nodes, enlargement of the kidneys and transient haemoglobinuria (*USA*: hemoglobinuria). Diagnosis is based on finding protozoa in the blood or in the lymph nodes. A feature, considered to be diagnostic, is the presence of Koch's blue bodies or spots, lymphocytes containing macroschizonts, and detectable in smears stained with common blood-stain.

See also corridor disease.

theileriosis (parasitology) *See* theileriasis.

Thelazia (avian/veterinary parasitology) A genus of spiruroid worms in the family Thelaziidae that cause thelaziasis. They are all parasites of the lacrimal duct or conjunctival sac of mammals and birds. The larvae are deposited in the conjunctival sac by the intermediate host, *Musca* spp. flies.

Thelazia alfortensis (veterinary parasitology) A species of spiruroid worms in the genus *Thelazia* which may be found in cattle.

Thelazia bubalis (veterinary parasitology) A species of spiruroid worms in the genus *Thelazia* which may be found in water buffaloes.

Thelazia californiensis (human/veterinary parasitology) A species of spiruroid worms in the genus *Thelazia* which may occur in cats, dogs, humans, sheep and deer.

Thelazia callipaeda (human/veterinary parasitology) A species of spiruroid worms in the genus *Thelazia* which may occur in dogs, rabbits and humans.

Thelazia erschowi (veterinary parasitology) A species of spiruroid worms in the genus *Thelazia* which may be found in pig.

Thelazia gulosa (veterinary parasitology) A species of spiruroid worms in the genus *Thelazia* which may be found in cattle.

Thelazia lacrymalis (veterinary parasitology) A species of spiruroid worms in the genus *Thelazia* which may be found in horses.

Thelazia leesi (veterinary parasitology) A species of spiruroid worms in the genus *Thelazia* which may be found in dromedary.

Thelazia rhodesii (veterinary parasitology) A species of spiruroid worms in the genus *Thelazia* which may be found in cattle, sheep, goats, and buffaloes.

Thelazia skrjabini (veterinary parasitology) A species of spiruroid worms in the genus *Thelazia* which may be found in cattle.

thelaziasis (parasitology) Infestation of the conjunctival sac with *Thelazia* spp. which causes conjunctivitis, lacrimation, blepharospasm and keratitis.

Thelohanellus (fish parasitology) A genus of myxosporid protozoa in the class Myxosporea that are parasites of freshwater fish.

Thelohanellus piriformis (fish parasitology) A species of myxosporid protozoa in the genus *Thelohanellus* that causes cutaneous abscesses in cyprinid and coregonid fish.

Thelohania (parasitology) A genus of parasitic protozoa in the phylum Microspora.

Thelohania apodemi (veterinary parasitology) A species of parasitic protozoa in the genus *Thelohania* which may be found in the brain of field mice.

Thelohania baueri (fish parasitology) A species of parasitic protozoa in the genus *Thelohania* which may be found in the ovary of the stickleback fish.

Thelohania solenopsae (insect parasitology) A protozoal species of microsporidia that are parasites to the ants *Solenopsis geminata*, *Solenopsis invicta*, *Solenopsis quinquecuspis*, *Solenopsis richteri*, and *Solenopsis saevissima*-complex.

thenium closylate (veterinary parasitology) A very effective anthelmintic against hookworms in dogs, commonly in combination with piperazine to include control of ascarids.

Theobaldia (parasitology) One of the genera of mosquitoes which are the definitive hosts for *Plasmodium* spp., the cause of avian malaria.

theorem (philosophy) General conclusion in science or mathematics which makes certain assumptions in order to explain observations.

See also hypothesis.

theoretical epidemiology (epidemiology) The use of mathematical models to explain and examine aspects of epidemiology, *e.g.* computer simulation models of outbreaks.

therap. (general terminology) An abbreviation of 'therap'eutic.

therapeutic immunosuppression (medical) Treatment which suppresses immune function where it is contributing to the disease process.

therapeutic index (TI) (pharmacology) A numerical estimate of the relationship between the toxic dose of a drug and its therapeutic dose.

therapeutic window (pharmacology) The range of plasma levels of a drug within which optimal therapeutic effects are obtained; below the range the drug has too little beneficial effect, while too high concentrations may either produce serious side effects or reduce the therapeutic effect.

therapeutics (pharmacology) The study of the science of treating disease.

therapy (pharmacology) The treatment of disease.

therm(o)- (terminology) A word element. [German] Prefix denoting *heat*.

thermal conduction (physics) Transmission of heat by materials.

thermal conductivity (physics) Thermal conducting power of a material.

thermal diffusion (physics) Process of forming a concentration gradient in a fluid mixture by the application of a temperature gradient. Also termed: Soret effect.

thermic (physics) Pertaining to heat.

thermistor (physics) Temperature-sensitive semiconductor device whose resistance decreases with an increase in temperature, used in electronic thermometers and switches.

thermobarograph (physics) Instrument for measuring and recording the temperature and pressure of the atmosphere. It consists of a thermograph and a barograph.

thermochemistry (chemistry) Branch of chemistry that deals with the study of heat changes in relation to chemical reactions. The measurement of heat of reaction, specific heat capacities and bond energies falls within its scope.

thermocouple (physics) Device for measuring temperature which relies on the Seebeck effect: a heated junction between two dissimilar metals produces a measurable electromotive force (e.m.f.) which depends on the temperature of the junction.

thermodynamic temperature (physics) An alternative term for absolute temperature.

thermodynamics (physics) Branch of physics concerned with the study of the effects of energy changes in physical systems and the relationship between various forms of energy, principally heat and mechanical energy.

thermoelectricity (physics) Electricity produced from heat energy, as in a thermocouple.

thermograph (physics) Thermometer that records variations in temperature over a period in time on a graph; a self-registering thermometer.

thermography (physics) Variations of temperature over the surface of a body can be recorded by using photographic film sensitive to infra-red radiation. The technique is called thermography, and the record a thermogram. Because the surface temperature of the body is determined by the state of the local circulation, and variations in the blood supply to a part can be produced by underlying disease processes.

thermolabile (biology) Adversely affected by heat (as opposed to thermostable, not affected by heat).

thermometer (measurement) An instrument for measuring temperature. The common liquid in-glass thermometer relies on the expansion of the liquid (*e.g.* mercury or dyed alcohol) in a calibrated sealed glass capillary tube.

thermometer, clinical thermometer (physics) Used in medicine for the accurate measurement of body temperature. It measures only a small range of temperatures.

thermometer, maximum and minimum (physics) Thermometer that records the maximum and minimum temperatures attained during a given period of time.

thermopile (physics) Temperature-measuring device consisting of several thermocouples connected in a series, with one set of junctions blackened so as to absorb thermal radiation.

thermoreceptors (measurement) Sensory receptors sensitive to temperature.

thermoregulation (physiology) The control of the body's temperature. The main regulating centre (*USA*: center) is in the hypothalamus, which contains thermoreceptors and can control vasodilation, sweating, etc.

thermostabile (physics) Not affected by heat.

thermostasis (biology) Maintenance of temperature, as in warm-blooded animals.

Theromyzon tessulatum (avian parasitology) A small leech that normally parasitizes the nasal sinuses of geese, but which may migrate to the conjunctival sacs and cause keratoconjunctivitis.

THI (physics) An abbreviation of 'T'emperature 'H'umidity 'I'ndex.

thi(o)- (terminology) A word element. [German] Prefix denoting *sulphur* (*USA*: sulfur).

thiacetarsamide sodium (pharmacology) A parenteral anthelmintic used against adult heartworms in dogs with significant hepatic and renal toxicity. Also termed: Caparasolate.

thiamine (chemistry) White water-soluble crystalline B vitamin, found in cereals and yeast. Also termed: thiamin; aneurin; vitamin B_1.

thiazine (chemistry) Member of a group of heterocyclic compounds that contain sulphur (*USA*: sulfur) and nitrogen (in addition to carbon) in the ring.

thick stomach worms (parasitology) *See Ascarops.*

thin-layer chromatography (TLC) (biochemistry) A chromatographic separation technique using a liquid mobile phase that moves by capillary action through a thin layer of sorbent coated on an inert, rigid backing material or plate. Once separated, they can be identified.

thio- (chemistry) Prefix denoting the presence of sulphur (*USA*: sulfur) in a compound.

thioarsenites (pharmacology) Organic arsenicals that may be used as anthelmintics.

thiocarbamide (chemistry) An alternative term for thiourea.

thioether (chemistry) Member of a class of organic compounds, analogous to ethers, in which sulphur (*USA*: sulfur) takes the place of oxygen. Also termed: alkyl or aryl sulphide (*USA*: aryl sulfide).

thionine (microscopy) A dark green powder, purple in solution, used as a metachromatic stain in microscopy.

thiourea (chemistry) NH_2CSNH_2 Colourless (*USA*: colorless) crystalline organic compound, the sulphur (*USA*: sulfur) analogue of urea. Its conversion to its isomer ammonium thiocyanate on heating was the first demonstration of an organic compound being changed directly into an inorganic one. It is used in medicine and as a photographic sensitizer. Also termed: thiocarbamide.

thorac(o)- (terminology) A word element. [German] Prefix denoting *chest.*

thoracoscopy (medical) The inspection of the interior of the thoracic cavity through an endoscope.

thoracotomy (medical) The operation of opening the wall of the chest.

thorax (anatomy) Region of the body that contains the heart and lungs; the chest. It is separated from the abdomen by the diaphragm.

thoria (chemistry) An alternative term for thorium dioxide.

thorium dioxide (chemistry) ThO_2 White insoluble powder, used as a refractory and in non-silica optical glass. Also termed: thoria.

thorium (Th) (chemistry) Silvery-white radioactive element in Group IIIA of the periodic table (one of the actinides). It has several isotopes, with half lives of up to 1.39×10^{10} years. Thorium-232 captures slow, or thermal, neutrons and is used to 'breed' the fissile uranium-233. Its refractory oxide (thoria, ThO_2) is used in gas mantles. At.no. 90; r.a.m. 232.0381.

thorn-headed worm (parasitology) *See* thorny-headed worm.

thorny-headed worm (parasitology) Thorn-headed worm member of the phylum Acanthocephala, a group of parasitic worms related to the nematodes. *See Corynosoma*; *Filicollis*; *Macracanthorhynchus*; *Oncicola*.

Thr (biochemistry) An abbreviation of 'Thr'eonine.

threadworm (pinworm) (parasitology) *Oxyuris vermicularis*.

threonine (biochemistry) Amino acid that is essential in the diet of animals. Also termed: 2-amino-3-hydroxybutanoic acid.

threshold (physics) The value of a physical stimulus at which it becomes detectable, 'absolute threshold', or the minimum difference in the values of two stimuli lying on the same dimension at which they can be discriminated, 'difference threshold'. Within limits the value assigned to a threshold is arbitrary; it is usually taken to be that at which the stimulus (or the difference between two stimuli) is detected on 50 per cent of presentations.

-thrix (terminology) A word element. [German] Suffix denoting *hair*.

thromb(o)- (terminology) A word element. [German] Prefix denoting *clot*, *thrombus*.

thrombin (biochemistry) An enzyme concerned with the conversion of fibrinogen into fibrin in the clotting of blood following injury. *See* coagulation of the blood.

thrombocytes (platelets) (haematology) The smallest of the formed elements in the blood. They are fragments of megakaryocytes, giant precursor cells. Platelets adhere to collagen fibres (*USA*: collagen fibers) exposed at points of blood vessel injury. Additional platelets aggregate at this point to form a platelet plug, which is the first step in haemostasis (*USA*: hemostasis). They release factors to promote coagulation. A typical platelet count is 250,000 per cubic millimetre (*USA*: cubic millimeter). Aspirin interferes with platelet aggregation and may be useful in preventing clot formation.

thromboembolic colic (parasitology) A syndrome caused by stimulation by migrating strongyle larvae or infarction of a section of gut wall and may appear as intermittent spasmodic colic or subacute colic for a number of days followed by development of peritonitis.

See also strongylosis.

thromboembolic parasitism (parasitology) *See* thromboembolic colic.

thrombophlebitis (haematology) Inflammation of a vein with consequent thrombosis, the formation of a clot. It is quite common in varicose veins, which become red and tender and sometimes painful enough to make walking difficult.

thrombosis (haematology) The formation of a clot or thrombus in a blood vessel. It may occur in arteries, particularly when the wall has been roughened by atheroscierosis, or in the veins, especially when the blood flow is stagnant or sluggish.

thrombus (medical) A solid mass formed from the constituents of blood within the blood vessels or the heart. Thrombi that form within the rapidly moving arterial circulation are composed largely of fibrin and platelets with only a few trapped red and white cells. (pl. thrombi).

thulium (Tm) (chemistry) Silvery-white metallic element in Group IIA of the periodic table (one of the lanthanides). Its radioactive isotopes emit gamma rays and X-rays and are used in portable radiography equipment. At. no. 69; r.a.m. 168.9342.

thym(o)- (terminology) A word element. [German] Prefix denoting *thymus*; *mind*, *soul*, *emotions*.

-thymia (terminology) A word element. [German] Suffix denoting *condition of mind*.

thymine (chemistry) $C_5H_6N_2O_2$ Colourless (*USA*: colorless) crystalline heterocyclic compound, one of the pyrimidines. It is found in the nucleotides of DNA (along with cytosine, adenine and guanine). In RNA the corresponding base is uracil. Also termed: 5-methyluracil; 5-methyl-2,4-dioxopyrimidine.

thymol (chemistry) $C_{10}H_{14}O$ Colourless (*USA*: colorless) crystalline organic compound found in the oils of thyme and mint. It is used in antiseptic mouth washes. Altemative names: 2-hydroxy-p-cymene, 2-hydroxy-1-isopropyl-4-methylbenzene.

thymus (anatomy) Twin-lobed endocrine gland, situated in the chest near the heart, that plays an important role in the immune response. After birth it produces many lymphocytes and induces them to develop into antibody-producing cells. It declines after puberty and atrophies in older adults.

thymus-derived cell (T-cell) (haematology) A lymph cell of the body that fights against infection and foreign particles.

Thynnascaris (fish parasitology) A genus of nematodes in the family Anasakidae whose adult forms parasitize fish.

Thynnascaris aduncum (fish parasitology) A species of nematodes in the genus *Thynnascaris* that may be found in fish.

thyro- (terminology) A word element. [German] Prefix denoting *thyroid*.

thyroid (biochemistry) An endocrine gland, located in the neck region, that produces three hormones: thyroxine (tetraiodothyronine, T4); triiodothyronine (T3); thyrocalcitonin (calcitonin). The former two are the classic 'thyroid hormones'. Thyroxine is the cleavage product of thyroglobulin found in the thyroid gland colloid. Thyroxine represents the predominant circulating form. The L-form is the thyroid hormone, whereas the D-form has an anticholesteremic action. Thyroxine is metabolized within target cells to T3, which is five times more active than T4. T3 binds to nuclear receptors where it activates genes. Its primary actions are to increase metabolism, to increase protein synthesis, stimulate basal metabolic rate and increase heat production. Thyroid hormone synthesis and secretion are under the control of thyroid-stimulating hormone from the adenohypophysis. The third hormone from the thyroid gland, calcitonin, decreases blood calcium levels and acts antagonistically to parathyroid hormone in calcium homeostasis. Calcitonin secretion is controlled by blood calcium levels and not by thyroid-stimulating hormone.

thyroid-stimulating hormone (TSH) (biochemistry) Hormone produced by the pituitary gland which stimulates the thyroid gland into activity.

thyrotrophin (biochemistry) (thyroid-stimulating hormone, TSH) A tropic hormone from the adenohypophysis that causes the thyroid gland to proliferate, absorb more iodine and synthesise and secrete the thyroid hormones thyroxine and triiodothyronine, but not thyrocalcitonin. TSH secretion is under the control of thyrotropin-releasing factor (TRF) from the hypothalamus. Thyrotropin is a glycoprotein, consisting of two large peptide units (one of which is similar to LH and FSH). (*USA*: thyrotropin).

thyroxin (biochemistry) White crystalline organic compound, an iodine-containing amino acid derived from tyrosine. It is a hormone secreted by the thyroid, which promotes growth in immature organisms and increases metabolic rates in periods of increased activity.

thyroxine (T4) (biochemistry) *See* thyroid.

Thysaniezia (parasitology) A genus of cyclophyllidean cestodes in the family Thysanosomatidae.

Thysaniezia giardi (veterinary parasitology) A species of cestodes in the genus *Thysaniezia* that may be found in the small intestine of sheep, goats and cattle, that appears to be of little pathogenic significance. Also termed: *Helictometra giardi.*

Thysanosoma actinioides (veterinary parasitology) A cyclophyllidean cestode in the family Thysanosomatidae which may be found in the small intestine and biliary and pancreatic ducts of sheep, cattle and deer, and that may cause unthriftiness. Also termed: fringed tapeworm.

Ti (chemistry) The chemical symbol for titanium.

TI(chemistry) The chemical symbol for Element 81, thallium.

tiamulin (pharmacology) A carboxypenicillin antibiotic effective against *Mycoplasma.*

tibia (anatomy) One of the two bones below the knee in a tetrapod vertebrate (the other is the fibula); the shinbone.

tick (parasitology) A blood-sucking insect. They may be separated into hard, *Ixodidae*, and soft, *Argasidae*, ticks. Ticks carry several diseases, among them encephalitis, relapsing fever, Lyme disease, typhus, spotted lever, Q fever, and babesiosis which is a protozoan parasitic disease common in wild and domestic animals in tropical and subtropical countries, passed on to man by hard ticks. *See Dermacentor variabilis*; *Amblyomma hebraeum*; *Ixodes canisuga*; *Rhipicephalus sanguineus*; *Rhipicephalus appendiculatus*; *Dermacentor nigrolineatus*; *Amblyomma cajennense*; *Amblyomma maculatum*; *Amblyomma americanum*; *Ornithodorus coriaceus*; *Rhipicephalus everti*; *Dermacentor andersoni*; *Dermacentor albipictus*; *Otobius megnini*; *Amblyomma variegatum*; *Haemaphysalis leachi leachi.*

tick collar (control measures) A neck collar made of a PVC resin which releases particles of insecticide over a period of several months and aids in the control of tick infestations in companion animals.

tick fever (parasitology) *See* babesiosis; anaplasmosis.

tick paralysis (human/veterinary parasitology) The female of several species of ticks but most commonly *Ixodes* or *Dermacentor* spp. elaborates a neurotoxin that typically causes an ascending flaccid paralysis in many animal species and humans but particularly in companion animals and young food animals. Affected dogs first develop weakness and paralysis of the hindlimbs, then forelimb, and ultimately respiratory paralysis unless the tick is removed and, in some cases, treatment with hyperimmune serum is given.

tick pyemia (veterinary parasitology) An infection of lambs caused by the bacteria Staphylococcus aureus and transmitted by the bites of ticks. Also termed: staphylococcal pyemia.

tick toxicosis (parasitology) *See* sweating sickness.

tick vectors (parasitology) Ticks act as vectors of protozoa, bacteria, viruses, and rickettsia.

tick worry (parasitology) An all-embracing term to describe the debilitating effects of heavy tick infestations. It includes anaemia (*USA*: anemia), irritation by the ticks, local infection as a result of bites, secondary blowfly and screw-worm infestation.

tick-stained (veterinary parasitology) Said of wool or fleece that is heavily discolored by the faeces (*USA*: feces) of sheep ked (*Melophagus ovinus*).

tidal volume (physiology) The volume of gas inspired or expired during each respiratory cycle. It is an indicator of the depth of breathing.

time series (statistics) Data tabulated by the time which successive observations are made.

time-lapse photography (photography) Technique for producing a speeded-up film, achieved by introducing a delay between each frame exposed during filming. It is used to film very slow-moving or slow-developing processes.

tin (Sn) (chemistry) Metallic tin and inorganic tin compounds are not toxic to humans. Some organo-tin compounds are toxic.

tin dioxide (chemistry) An alternative term for tin(IV) oxide.

tin disulphide (chemistry) An alternative term for tin(IV) sulphide (*USA*: tin(IV) sulfide). (*USA*: tin disulfide).

tin hydroxide oxide (chemistry) An alternative term for stannic acid.

tin salt (chemistry) An alternative term for tin(II) chloride.

tin (Sn) (chemistry) Soft silvery-white metallic element in Group IVB of the periodic table, which forms three allotropes. It occurs mainly as tin(IV) oxide, SnO_2, in ores such as cassiterite (tinstone). It is used mainly as a protective coating for steel (tin plate) and in making alloys with lead. Its compounds are used as catalysts, fungicides and mordants. At. no. 50; r.a.m. 118.69.

tin(II) (chemistry) An alternative term for stannous.

tin(II) chloride (chemistry) $SnCl_2$ White soluble solid. A reducing agent, it is used as a catalyst in organic reactions and as an anti-sludge agent for oils. Also termed: stannous chloride; tin salt.

tin(IV) (chemistry) An alternative term for stannic.

tin(IV) hydride (chemistry) SnH_4 Unstable gas, used as a reducing agent in organic chemistry. Also termed: stannane.

tin(IV) oxide (chemistry) SnO_2 White crystalline solid, used as a pigment and as a refractory material. Also termed: tin dioxide.

tin(IV) sulphide (chemistry) SnS_2 Yellow insoluble solid, used as a pigment. Also termed: tin disulphide (*USA*: tin disulfide). (*USA*: tin(IV) sulfide).

tincal (chemistry) Naturally occurring crude borax.

tinnitis (medical) An abnormality of hearing in which an apparent persistent ringing sound is heard when in fact no such sound is present.

tis. (general terminology) An abbreviation of 'tis'sue.

tissue (histology) The tissues are the substance of the body; particular tissues such as connective or fatty tissue are collections of cells of the same type specialised in the same way to carry out a particular function. In higher organisms tissues may combine to form a highly specialized organ.

tissue culture (biology) Process by which cells or tissues are maintained outside the body (in vitro) in a suitable medium. The material is kept at a suitable temperature, pH and osmotic pressure. The composition of the medium depends on the type of tissue cultured. Depending on the type of cell, a culture may be able to go on indefinitely, with cells dividing from time to time. Cancer cells (or normal cells that have undergone transformation in culture) will go on dividing infinitely often, but most differentiated cells seem to lose the ability to divide after 30–40 divisions; this is known as the Hayflick phenomenon.

tissue fluid (biology) An alternative term for lymph.

tissue typing (histology) The identification of histocompatibility antigens in donor and recipient prior to transplant surgery.

titania (chemistry) An alternative term for titanium(IV) oxide.

titanic chloride (chemistry) An alternative term for titanium(IV) chloride.

titanium tetrachloride (chemistry) An alternative term for titanium(IV) chloride.

titanium (Ti) (chemistry) Silvery-white metallic element in Group IVA of the periodic table (a transition element). Its corrosion-resistant lightweight alloys are employed in the aerospace industry. Naturally occurring crystalline forms of titanium(IV) oxide (titania, TiO_2) are the semi-precious gemstone rutile. The powdered oxide is used as a white pigment and a dielectric in capacitors. At. no. 22; r.a.m. 47.90.

titrant (chemistry) Chemical solution of known concentration, *i.e.* a standard solution, which is added during the course of a titration.

titration (chemistry) Technique in volumetric analysis in which one chemical solution of known concentration is added (using a burette) to a known volume of another chemical solution of unknown concentration (measured by a pipette), and the chemical reaction followed by observing changes in colour (*USA*: color), pH, etc. An indicator may be added to indicate the end-point of the reaction, which allows the unknown concentration to be determined.

titre (measurement) Reciprocal value of the highest possible dilution that illicits a response or reaction. (*USA, titer*)

TLC (biochemistry) An abbreviation of 'T'hin-'L'ayer 'C'hromatography.

TLV (toxicology) An abbreviation of 'T'hreshold 'L'imit 'V'alue.

Tm (1 chemistry; 2 physiology) 1 The chemical symbol for Element 69, thulium. 2 An abbreviation of renal 'T'ubular 'm'aximum.

Tn (chemistry) The chemical symbol for thoron.

toco- (terminology) A word element. [German] Prefix denoting *parturition, labour* (*USA*: labor).

tocopherol (biochemistry) Vitamin isolated from plants that increases fertility in rats. Deficiency of it causes wasting of muscles in animals. It has been found to have antioxidant activity, and it is important in maintaining membranes. Also termed: vitamin E.

toko- (terminology) A word element. [German] Prefix denoting *parturition*, *labour* (*USA*: labor).

tolerance (1 pharmacology; 2 immunology) 1 An adaptational state when, after repeated exposure, a given dose of an agent produces a decreased effect or, conversely, when increasingly larger doses are necessary to obtain the effects observed with the original dose. Two mechanisms of acquired pharmacological tolerance are generally recognized: (a) dispositional; (b) pharmacodynamic. Dispositional tolerance results from alterations in the pharmacokinetic properties of the agent. Pharmacodynamic tolerance results from adaptive changes within affected systems, such that the response is reduced in the presence of the same concentration of the agent. Tolerance may not develop uniformly to all the actions of an agent. The toxicological manifestation of tolerance development is typically expressed as a progressive increase in the LD50 for a given agent, although it should be recognised that tolerance development is not absolute. 2 A state of nonresponsiveness to a particular antigen or group of antigens.

Tollens' reagent (chemistry) Ammoniacal solution of silver oxide used as a test for aldehydes, which reduce it to deposit a mirror of silver. It was named after the German chemist Bernhard Tollens (1841–1918).

toltrazuril (pharmacology) A triazinon drug with anticoccidial and antiprotozoal activity.

toluene (chemistry) $C_6H_5CH_3$ Colourless (*USA*: colorless) aromatic organic liquid that occurs in coaltar, used as an anthelmintic against roundworms and hookworms in dogs, as an industrial solvent and starting point for making explosives. Also termed: methylbenzene.

toluidine (chemistry) $CH_3C_6H_4NH_2$ One of three isomeric aromatic amines, used in the manufacture of dyes and drugs. The ortho- and meta- forms are colourless (*USA*: colorless) liquids; the para- isomer is a colourless crystalline solid. Also termed: aminotoluene; methylaniline.

toluidine blue in sorensons pH 6.8 buffer (histology) A histological staining technique used to make helicobacter pylori visible under the microscope.

toluidine blue stain (histology) 1 A histological staining technique used to make amyloid visible under the microscope. **2** Histological staining technique for mast cells.

-tome (terminology) A word element. [German] Suffix denoting *an instrument for cutting*, *a segment*.

tomo- (terminology) A word element. [German] Prefix denoting *a cutting*, *a segment*.

tomogram (radiology) An X-ray taken to show structures lying in a selected plane in the body.

tomography (medical) Any non-invasive technique that yields information about the different spatial parts of the brain, particularly about successive slices through it.

-tomy (terminology) A word element. [German] Suffix denoting *incision*, *cutting*.

ton (measurement) Unit of mass equal to 20 cwt or 2,240 lb (an imperial, or long, ton); a short ton equals 2,000 lb.

See also tonne.

tongue (anatomy) Muscular organ located in the buccal cavity (mouth) of some animals, used for manipulating food (and, in human beings, involved in speech).

tongue worm (parasitology) *See Linguatula serrata.*

tonne (measurement) Unit of mass equal to 1,000 kilograms. 1 tonne = 23204.62 lb, slightly less than the imperial ton (2,240 lb). Also termed: metric ton.

tono- (terminology) A word element. [German] Prefix denoting *tone*, *tension*.

tonsil (anatomy) One of a pair of lymphoid tissue regions at the back of the mouth which help to prevent infection by producing lymphocytes.

top(o)- (terminology) A word element. [German] Prefix denoting *particular place or area*.

toppling disease of bananas (plant parasitology) A disease caused by the nematode *Radopholus similis*. Initial entry of the nematode into the root produces a reddish, elongate fleck parallel to the root axis. The fleck or discoloured (*USA*: discolored) area enlarges as the nematode and progeny feed. The older parts of the lesion turn black and shrink, with the advancing margin remaining red. Neither the nematode nor the eggs are found beyond the red margin, and they are also rare in the older portions of the lesion. Continued feeding causes extensive, deep lesions on roots and rhizomes weakening trees which fall under the weight of fruit or in the wind resulting in total crop loss.

torr (measurement) Unit of pressure equivalent to that produced by a 1 mm column of mercury. It is equal to 133.3 newtons m^{-2}.

torsalo grub (parasitology) Larvae of *Dermatobia hominis*.

total worm count (parasitology) *See* worm count.

tox(o)- (terminology) Word element. [German] Prefix denoting *toxin*, *poison*.

toxascariasis (parasitology) Disease caused by infestation with *Toxascaris* spp.

Toxascaris (parasitology) A genus of roundworms in the family Ascarididae.

Toxascaris leonina (veterinary parasitology) A species of roundworms in the genus *Toxascaris* which may be found in the small intestine of dogs, cats, foxes and wild carnivora, and that causes toxascariasis.

toxic (toxicology) Poisonous; pertaining to poisoning.

toxic(o)- (terminology) A word element. [German] Prefix denoting *poison, poisonous*.

toxicant (toxicology) Any chemical, of natural or synthetic origin, capable of causing a deleterious effect on a living organism. The term toxin should never be used as an alternative term for toxicant, being properly reserved for only those toxicants synthesised metabolically by a living organism.

toxicara (parasitology) An infestation with *Toxicara canis,* a worm that lives in the intestines of dogs and foxes. The eggs are passed in the excreta, and if humans eat contaminated food the larvae travel through the lungs and liver but normally do no harm and die in about a year. Occasionally however they reach the eye, where they can set up a granulomatous inflammation and affect the retina, causing a squint and loss of vision.

toxicol. (toxicology) An abbreviation of 'toxicol'ogical; 'toxicol'ogist; 'toxicol'ogy.

toxicology (toxicology) The science that deals with poisons (toxicants) and their effects. A poison is defined as any substance that causes a harmful effect, either by accident or design, when administered to a living organism.

toxicosis (toxicology) The state of having been poisoned.

toxin (1, 2 toxicology) 1 A poison; usually applied to those produced by a living organism. 2 Any harmful substance.

toxinology (toxicology) The science dealing with the toxins produced by certain higher plants, animals and humans and by pathogenic bacteria.

Toxocara (parasitology) A genus of nematode parasites in the family Ascarididae.

Toxocara canis (human/veterinary parasitology) A species of nematodes in the genus *Toxocara*, the adult forms of which may be found in the small intestine of dogs and foxes. Infection is via oral, transmammary and transplacental routes, the larvae migrating through tissues, including to the foetus (*USA*: fetus) where they establish a prenatal infection, eventually passing through the lungs and then to the alimentary canal. In hosts other than the dog and fox, migration in abnormal tissues occurs and the life cycle is not completed. It is the cause of ocular and visceral larva migrans in humans.

Toxocara cati (veterinary parasitology) A species of nematodes in the genus *Toxocara*, the adult forms of which may be found in the small intestine of domestic and wild cats. Larvae pursue a migratory course through tissues but intrauterine infection of the foetus (*USA*: fetus) does not occur.

Toxocara pteropodis (veterinary parasitology) A species of nematodes in the genus *Toxocara* that may be found in flying fox bats.

Toxocara vitulorum (veterinary parasitology) A species of nematodes in the genus *Toxocara* that may be found in the small intestine of cattle, buffaloes and in sheep and goats. Calves are infected via the milk of the dam.

toxocariasis (veterinary parasitology) Infection by worms of the genus *Toxocara*. Heavy infestations in young puppies and kittens may be responsible for abdominal distension, signs of colic, diarrhoea (*USA*: diarrhea) and poor growth. Somatic tissue migration of larvae in neonatal puppies may cause respiratory and nervous signs.

toxoid (toxicology) A bacterial toxin so modified that it has lost its poisonous properties, but can still act as an antigen to provoke the formation of antibodies *e.g.* tetanus toxoid is used by injection to induce immunity to tetanus, but it does not produce symptoms of the disease.

Toxoplasma (parasitology) A genus of sporozoan parasites in the family Sarcocystidae.

toxoplasma encephalitis (parasitology) *See toxoplasmosis*.

Toxoplasma gondii (human/avian/veterinary parasitology) A species of coccidian parasite in the genus *Toxoplasma* that may be found in the intestine of all felids, including especially the domestic cat, jaguarundi, ocelot, mountain lion, leopard cat, and bobcat, which are definitive hosts. Most vertebrates, including humans and birds, can be infected with the intermediate stages and experience one or other forms of the disease toxoplasmosis.

Toxoplasma hammondi (parasitology) *See Hammondia hammondi*.

toxoplasmin (parasitology) An antigen prepared from mouse peritoneal fluids rich with *Toxoplasma gondii* which may be injected intracutaneously as a test for toxoplasmosis.

toxoplasmosis (human/veterinary parasitology) Infection with the protozoan parasite *Toxoplama gondii*. Human beings may become infected from cats or by eating meat containing tissue cysts which has been badly cooked; the disease is common in sheep. In the vast majority of cases the infection does no harm and passes unrecognised, but in some there may be enlargement of the lymph nodes and rarely a rash, with general malaise which may last weeks; the disease resembles infectious mononucleosis. It usually resolves without trouble, but it may infect those suffering from immunosuppressive disorders such as AIDS or those who are taking immunosuppressive drugs, when the disease may affect the brain and produce encephalitis. Most importantly, however, mothers who become infected during pregnancy may pass the infection on to the child. This may result in miscarriage, or the child may be born with abnormalities of the central nervous system which are apparent at birth or become apparent later, perhaps after several years. The principal manifestation in animals is as abortion in ewes, and it is also a cause of sporadic cases of pneumonia, central nervous system disease, and less often retinochoroiditis, and hepatitis in dogs and cats.

TPA (physiology/biochemistry) An abbreviation of 'T'issue 'P'lasminogen 'A'ctivator.

TPN (biochemistry) An abbreviation of 'T'ri'P'hosphopyridine 'N'ucleotide.

TPNH (biochemistry) An abbreviation of reduced 'T'ri'P'hosphopyridine 'N'ucleotide.

Tr (chemistry) Chemical symbol for Terbium.

trace element (biochemistry) Element essential to metabolism, but necessary only in very small quantities (*e.g.* copper and cobalt in animals, molybdenum in ants). Such elements are usually poisonous if large quantities are ingested.

trachea (anatomy) Tube through which air is drawn into the lungs; windpipe.

trachel(o)- (terminology) A word element. [German] Prefix denoting *neck, necklike structure*, especially the cervix uteri.

tracheo- (terminology) A word element. [German] Prefix denoting *trachea.*

tracheobronchitis (medical) Inflammation of the trachea and bronchi.

Tracheophilus (avian parasitology) A genus of trematodes in the family Cyclocoelidae, that are parasites of aquatic birds.

Tracheophilus cucumerinum (avian parasitology) A species of trematodes in the genus *Tracheophilus* which may be found in the trachea, air sacs and oesophagus (*USA*: esophagus) of domestic and wild ducks, and that causes obstruction leading to dyspnea and asphyxia.

Tracheophilus cymbius (avian parasitology) (synonym: *Tracheophilus sisowi*) A species of trematodes in the genus *Tracheophilus* which may be found in the trachea and bronchi of domestic and wild ducks, and that may cause tracheal obstruction and asphyxia.

tracheotomy (medical) The operation of cutting an opening into the trachea.

Trachytidae (insect parasitology) *See Trachyuropoda.*

Trachyuropoda (insect parasitology) Mites in the Mesostigmata that, similarly to Trachytidae and Trhypachthonius, are parasitic to the ants *Atta, Formica,* and *Solenopsis* spp., and are phoretic on the heads of workers of the termites *Odontotermes, Glyptotermes,* and other termite genera.

tranquillizer (pharmacology) Drug that acts on the central nervous system (CNS), used for calming people and animals without affecting consciousness.

trans- (terminology) A word element. [Latin] Prefix denoting *through, across, beyond.*

trans configuration (genetics) One of the two possible arrangements for the alleles in an individual that is heterozygous for mutations at two linked loci. *Trans* (from the Latin meaning 'across') means that one of the mutant alleles is on one chromosome and the other is on the homologous chromosome, *i.e.* the two chromosomes are a + and + b. The opposite is *cis.* Also termed repulsion.

transamination (chemistry) Removal and transference of an amino group from one compound (usually an amino acid) to another.

transcendental (1, 2 mathematics) 1 Irrational number that is not the root of a polynomial equation. 2 Function that is not a finite polynomial equation (*e.g.* logarithmic or exponential function).

transcription (molecular biology) The process by which one strand of DNA is copied into a single strand of RNA; the first step in protein synthesis. The RNA molecule is synthesized by RNA polymerase, copying the DNA message by following the base-pairing rules. The polymerase first recognizes the promoter region, and actual transcription begins at the start codon a few bases further downstream (going from the 5' towards the 3' end of the DNA strand). *See* genetic code.

transducer (physiology) A device that receives a signal in one physical form and outputs it in another; *e.g.* sensory receptors signal the stimulus received by a change in their membrane potential.

transduction (genetics) The transfer of genetic information from one bacterium to another when it is carried by a bacteriophage.

transfer factor (tf) (immunology) A factor released from sensitized lymphocytes that has the capacity to transfer delayed hypersensitivity to a normal (nonsensitized) animal.

transfer host (parasitology) A host that is used until the appropriate definitive host is reached, but that is not necessary to complete the life cycle of the parasite. Also termed: transport host.

transfer RNA (molecular biology) Small molecule of RNA that acts as a carrier of specific amino acids in the synthesis of proteins. Amino acids are placed in a specific order by the transfer RNA molecules according to instructions in the messenger RNA, to form a polypeptide chain.

transference number (physics) An alternative term for transport number.

transformation (genetics) 1 A permanent change in the genetic characteristics of one bacterium by exposure to DNA of a different origin. (It was the phenomenon of transformation that led Avery to the discovery that DNA was the molecule responsible for carrying genetic information). 2 A change in an animal cell in tissue culture so that it grows and divides in the same way as a cancer cell, possibly due to activation of a viral gene.

transformation constant (radiology) An alternative term for disintegration constant.

transition (molecular biology) A mutation in which a purine base (adenine, guanine) is replaced with another purine, or a pyrimidine (cytosine, thymine) with another pyrimidine.

See also transversion.

transition element (chemistry) Member of a large group of elements that have partly filled inner electron shells, which gives them their distinctive physical and chemical properties (particularly variable valency and the tendency to form coloured [*USA*: colored] compounds). They occupy Groups IIIA, IVA, VA, VIA, VIIA, VIII, IB and IIB of the periodic table. Many of these elements and their compounds are used as catalysts.

transition point (1, 2 chemistry) 1 Temperature at which the transformation of one form of a substance into another form can occur (usually one crystalline modification into another). 2 Temperature at which two solid phases exist at equilibrium. 3 Temperature at which a change happens in a substance.

transition temperature (chemistry) Temperature above and below which different allotropes are stable.

translation (biochemistry) Process by which protein is synthesized in cells. It occurs by the action of messenger RNA, which attaches to a ribosome in the cytoplasm. Transfer RNA molecules which are attached to a specific amino acid then line up according to the sequence of amino acids encoded in the messenger RNA to form a polypeptide chain. Also termed: protein synthesis.

translocation (genetics) A mutation consisting of the transfer of part of a chromosome to another part of that chromosome or of a different chromosome.

translucent (chemistry) Describing a substance that transmits and diffuses light, but does not allow a well-defined image to be seen through it.

See also transparent.

transmissible infection (parasitology) An infection capable of being transmitted from one animal to another. Also termed contagious.

transovarial (parasitology) Via the ovary. The infectious agent is passed to the foetus (*USA*:fetus) in the ovum, having been infected from the parents circulation. A common occurrence in ticks and other arthropods, *e.g.* with babesiosis in ticks.

transparent (chemistry) Describing a substance that allows light (or other radiation) to pass through it with little or no diffusion.

See also translucent.

transport host (parasitology) *See* paratenic host; transfer host.

transport number (physics) In electrolysis, fraction of the total current carried by a particular ion in the electrolyte. Also termed: transference number.

transposition (genetics) The movement of a transposer or other movable sequence of DNA from one place in the genome to another.

transposon (genetics) A movable genetic element similar to a jumping gene. They are found in the genome of bacteria, and are involved in the resistance to antibiotics. Unlike jumping genes, transposers leave a copy of themselves in the original position when they move to a new site. They make it possible for antibiotic resistance to be spread very rapidly, not only within a single strain of bacteria, but from one genus to another.

trans-stadial (medical) Across or between stages of a process or disease.

trans-stadial vector transmission (parasitology) Occurs when an infection is picked up by one stage in the vector's life cycle and transmitted to succeeding stages in its metamorphosis.

transthoracic (parasitology) Through the thoracic cavity or across the chest wall.

transtracheal aspiration (medical) Passage of needle and plastic catheter through the trachea for obtaining lower respiratory tract secretions.

transudate (cytology) Similar to exudate, but with low protein content.

transversal (general terminology) Line that intersects another set of lines.

transversion (molecular biology) A mutation in which a purine base (adenine, guanine) is substituted for a pyrimidine one (cytosine, thymine), or vice versa.

See also transition.

TRAP (physiology/biochemistry) An abbreviation of 'T'hyroid 'R'eceptor 'A'uxiliary 'P'rotein.

trauma (medical) An injury or wound.

traumat(o)- (terminology) A word element. [German] Prefix denoting *trauma*.

traveller's diarrhoea (medical) A term applied to the short attacks of diarrhoea which afflict many people when they first arrive in a warm country. Most cases are due to a type *of Escherichia coli* which produces a toxin which affects the bowel. (*USA*: traveller's diarrhea)

treatment (medical) Management and care of a patient or the combating of disease or disorder.

trematocides (pharmacology) Drugs effective in the treatment of immature and adult flukes, *e.g.* rafoxanide, closantel, triclabendazole, diamphenethide. Bromsalans, oxyclozanide, nitroxynil and carbon tetrachloride are effective only against mature flukes.

Trematoda (fish parasitology) A class of flatworms in the phylum Platyhelminthes. Two subclasses of trematodes have been recognised, the Aspidogastrea and the Dignea. *See* Aspidogastrea; Dignea.

trematode (fish/veterinary parasitology) Parasitic worm in the class Trematoda. There are three subclasses: Monogenea, containing parasites of fish, amphibians and mammals; Digenea, also termed digenetic trematodes, containing the flukes of domestic animals which cause parasitic disease of most systems, including the blood, eye, liver, reproductive tract, respiratory system, skin and urinary system; Aspidogastrea, containing parasites of molluscs, fish and reptiles.

trematodiasis (parasitology) Infestation by trematodes.

Trematurella (insect parasitology) *See Dinychus.*

trend (statistics) A generally consistent movement in the same direction over a long period in a time series.

TRH (physiology/biochemistry) An abbreviation of 'T'hyrotropin-'R'eleasing 'H'ormone.

Trhypachthonius (insect parasitology) *See Trachyuropoda.*

tri- (terminology) A word element. [German] Prefix denoting three.

Triaenophorus (parasitology) A genus of tapeworms in the family Triaenophoridae.

Triaenophorus nodulosus (fish parasitology) A species of tapeworms in the genus *Triaenophorus*, the adult forms of which may be found in pike and other predatory fish. Intermediate hosts are first copepods, and second fish, especially trout. The plerocercoids in the trout tissues cause loss of value and liver damage may cause deaths.

triamcinolone acetonide (pharmacology) A synthetic corticosteroid with anti-inflammatory and anti-allergic properties.

Triatoma (human/veterinary parasitology) A genus of bugs in the order Hemiptera, the cone-nosed bugs, important in human medicine as vectors of *Trypanosoma cruzi* from its natural vectors, dogs, cats, foxes, monkeys and others. Species include *Triatoma dimidiata*, *Triatoma infestans*, *Triatoma protracta*, and *Triatoma sanguisuga* (the vector for equine encephalomyelitis).

triatome (parasitology) A member of the genus *Triatoma* of true bugs.

triatomic molecule (chemistry) Molecule of an element that consists of three atoms, *e.g.* ozone, O_3.

triatomine (parasitology) Pertaining to the genus *Triatoma.*

triazine (chemistry) $C_3H_3N_3$ One of a group of isomeric heterocyclic organic compounds with three nitrogen atoms and three carbon atoms in the ring. Triazine derivatives are used as plastics, dyes and herbicides.

tricarboxylic acid cycle (biochemistry) An alternative term for Krebs cycle.

Tricephalobus gingivalis (parasitology) *See Rhabditis gingivalis.*

trich(o)- (terminology) A word element. [German] Prefix denoting *hair.*

trichina (parasitology) Plural: trichinae [German] A single larval stage worm of those in the genus *Trichinella.*

Trichinella (human/veterinary parasitology) A genus of nematode parasites in the family Trichinellidae. Adult worms of the following species are amongst those found in the intestines and the encapsulated, first stage larvae in the striated muscle of various animals: *Trichinella spiralis spiralis*, *Trichinella pseudospiralis*, *Trichinella spiralis domestica* (*Trichinella spiralis*), *Trichinella britovi*, *Trichinella nativa* and *Trichinella nelsoni*. *Trichinella spiralis* is a common cause of infection in humans as a result of ingestion of poorly cooked pork.

trichinellosis (parasitology) *See* trichinosis.

trichinosis (human/veterinary parasitology) Infection with the parasitic roundworm *Trichinella spiralis*, which enters the human body in infected meat eaten raw or insufficiently cooked. Found in most parts of the world with the exception of Australia and the Pacific Islands. The larvae, or early forms, of *Trichinella spiralis* live embedded in tiny capsule-like cysts of muscle tissue of infected pork. When the meat is properly cooked, the larvae are killed by the high temperature. If, however, the pork is undercooked, they survive, and when the meat is eaten, digestive juices dissolve the cyst capsules and free the larvae in the intestines, where they grow to maturity.

trichinous (parasitology) Affected with or containing trichinae.

trichinous myositis (parasitology) Inflammation of a voluntary muscle caused by the presence of *Trichinella spiralis.*

trichlorfon (pharmacology/control measures) An organophosphorus insecticide and anthelmintic, used in horses, often in combination with other anthelmintics, for treatment of endoparasites and cutaneous habronemiasis. Also used in dogs against whipworms and as a pour-on in cattle for control of warble flies. It is used in fishponds and commercial fish farms to control anchorworms, gill flukes and lice. Also termed: metrifonate.

trichloroacetaldehyde (chemistry) An alternative term for trichloroethanal.

trichloroacetic acid (TCA) (chemistry) A colourless (*USA*: colorless), hygroscopic solid, used as a herbicide (sodium trichloroacetate) and as an intermediate in pesticide manufacture. TCA is also used in *in vitro* laboratory studies to stop enzyme reactions by precipitation of proteins. It is corrosive to the skin and eyes, but is not otherwise hazardous.

trichloroethanal (chemistry) CCl_3CHO Pungent colourless (*USA*: colorless) oily liquid aldehyde, which forms a solid hydrate (trichloroethanediol). Also termed: chloral; trichloroacetaldehyde.

trichloroethanediol (chemistry) $Cl_3CCH(OH)_2$ White crystalline organic compound, used as a sedative. Also termed: chloral hydrate.

trichloromethane (chemistry) $CHCl_3$ Colourless (*USA*: colorless) volatile liquid haloform, used as an anaesthetic (*USA*: anesthetic) and as a solvent. Also termed: chloroform.

Trichobilharzia (avian parasitology) A genus of blood flukes (digenetic trematodes) in the family Schistosomatidae. Species include *Trichobilharzia ocellata*, *Trichobilharzia physellae* and *Trichobilharzia stagnicolae* which may be found in the portal veins of birds, especially waterfowl.

Trichocephalus (parasitology) *See Trichuris.*

Trichodectes (parasitology) A genus of biting lice in the subfamily of Ischnocera.

Trichodectes canis (veterinary parasitology) A species of biting lice in the genus *Trichodectes* that may be found on dogs.

Trichodectes equi (veterinary parasitology) A species of biting lice in the genus *Trichodectes* that may be found on zebra and wild equids.

Trichodectes pinquis euarctidos (veterinary parasitology) A species of biting lice in the genus *Trichodectes* that may be found on wild black bear.

Trichodina (fish parasitology) A genus of protozoa in the subclass Peritricha which parasitizes the skin and gills of fish and causes local irritation. They have a characteristic disc-like (*USA*: disk-like) appearance and are in constant revolving motion, but generally only pathogenic when in large numbers and or in stressed fish.

Trichodinella (fish parasitology) A protozoan parasite of fish similar to and having similar effects to *Trichodina.*

Trichoecius romboutsi (veterinary parasitology) A mite that causes hair loss and dermatitis in mice. Also termed: Myocoptes romboutsi.

Trichomitus (parasitology) A genus of trichomonads in the family Trichomonadidae.

Trichomitus fecalis (human parasitology) A species of trichomonads in the genus *Trichomitus* that may be found in human faeces (*USA*: feces).

Trichomitus rotunda (veterinary parasitology) A species of trichomonads in the genus *Trichomitus* that may be found in the caecum (*USA*: cecum) and colon of pigs, but is not considered to be pathogenic.

Trichomitus wenyoni (veterinary parasitology) A species of trichomonads in the genus *Trichomitus* that may be found in the caecum (*USA*: cecum) and colon of rodents and rhesus monkeys.

trichomonacide (pharmacology) A substance destructive to trichomonads.

trichomonad (parasitology) A parasite of the genera *Trichomonas*, *Tritrichomonas*, and *Tetratrichomonas*.

trichomonal (parasitology) Pertaining to trichomonads.

trichomonal enteritis (parasitology) Trichomonads may be a cause of enteric disease, but it is likely that the organisms may be opportunistic pathogens only.

Trichomonas (human/avian/veterinary parasitology) A genus of flagellate protozoa in the family Trichomonadidae which is characterized usually by the presence of a single flagellum and parasitic in animals, birds and humans.

Trichomonas caballi (veterinary parasitology) A species of flagellate protozoa in the genus *Trichomonas* that may be found in the colon of horses.

Trichomonas canistomae (veterinary parasitology) A species of flagellate protozoa in the genus *Trichomonas* that may be found in the mouth of dogs.

Trichomonas equi (parasitology) (synonym: *Trichomonas faecalis*) *See Tritrichomonas equi.*

Trichomonas equibuccalis (veterinary parasitology) A species of flagellate protozoa in the genus *Trichomonas* that may be found in the mouth of horses.

Trichomonas felistomae (veterinary parasitology) A species of flagellate protozoa in the genus *Trichomonas* that may be found in the mouth of cats.

Trichomonas foetus (parasitology) *See Tritrichomonas foetus.*

Trichomonas gallinae (avian parasitology) A species of flagellate protozoa in the genus *Trichomonas* that may be found in the upper digestive tract of many birds but mostly in pigeon squabs where it causes avian trichomoniasis.

Trichomonas gallinarum (parasitology) *See Tetratrichomonas gallinarum.*

Trichomonas hominis (parasitology) *See Pentatrichomonas hominis.*

Trichomonas intestinalis (parasitology) *See Pentatrichomonas hominis.*

Trichomonas macacovaginae (veterinary parasitology) A species of flagellate protozoa in the genus *Trichomonas* that may be found in the vagina of the rhesus monkey.

Trichomonas phasioni (avian parasitology) A species of flagellate protozoa in the genus *Trichomonas* that may cause diarrhoea (*USA*: diarrhea) and dehydration of pheasant poults.

Trichomonas tenax (human/veterinary parasitology) A species of flagellate protozoa in the genus *Trichomonas* that may be found in the mouth of monkeys and humans but has no apparent pathogenic effect.

Trichomonas vaginalis (human parasitology) A species of flagellate protozoa in the genus *Trichomonas* that causes inflammation of the vagina in women and urethritis in men, and is commonly transmitted via in sexual intercourse.

trichomoniasis (parasitology) Disease caused by infection with the protozoan parasite *Trichomonas* spp.

Trichonema (parasitology) A now obsolete genus name of strongylid worms that has been replaced by the genera *Cyathostomum*, *Cylicocyclus*, *Cylicodontophorus*, and *Cylicostephanus*.

Trichophora (fish parasitology) A protozoan parasite of fish.

Trichosomoides crassicauda (veterinary parasitology) A nematode in the family Trichuridae which may be found in the urinary bladder of Norway and black rats and that may cause granulomatous lesions.

trichostrongyliasis (veterinary parasitology) The disease caused by the infestation of the intestine and abomasum of ruminants by *Trichostrongylus* spp. Manifested by poor growth, wasting and persistent diarrhoea (*USA*: diarrhea). Also termed: trichostrongylosis.

trichostrongylid (parasitology) A worm of the family Trichostrongylidae.

Trichostrongylus (human/veterinary parasitology) A genus of nematode parasites belonging to the family Trichostrongylidae, which infects animals and humans.

Trichostrongylus affinis (veterinary parasitology) A species of nematodes in the genus *Trichostrongylus* that may be found in the small intestine of rabbits and occasionally in sheep.

Trichostrongylus axei (human/veterinary parasitology) A species of nematodes in the genus *Trichostrongylus* that may be found in the abomasum of cattle, sheep, goats, deer, antelope and in the stomach of pigs, horses, donkeys and rarely humans.

Trichostrongylus capricola (veterinary parasitology) A species of nematodes in the genus *Trichostrongylus* that may be found in the small intestine of sheep and goats.

Trichostrongylus colubriformis (human/veterinary parasitology) A species of nematodes in the genus *Trichostrongylus* that may be found in the small intestine and sometimes abomasum, in cattle, sheep, goats, antelopes, camels, and occasionally in pigs, humans, dogs and rabbits.

Trichostrongylus drepanoformis (veterinary parasitology) A species of nematodes in the genus *Trichostrongylus* that may be found in the small intestine of sheep.

Trichostrongylus falculatus (veterinary parasitology) A species of nematodes in the genus *Trichostrongylus* that may be found in the small intestine of sheep, goats and antelopes.

Trichostrongylus hamatus (veterinary parasitology) A species of nematodes in the genus *Trichostrongylus* that may be found in the intestine of sheep and steinboks.

Trichostrongylus longispicularis (veterinary parasitology) A species of nematodes in the genus *Trichostrongylus* that may be found in sheep and cattle.

Trichostrongylus orientalis (human/veterinary parasitology) A species of nematodes in the genus *Trichostrongylus* that may be found in the small intestine of humans and rarely sheep.

Trichostrongylus probolurus (human/veterinary parasitology) A species of nematodes in the genus *Trichostrongylus* that may be found in the small intestine of sheep, goats, camels and rarely humans.

Trichostrongylus retortaeformis (veterinary parasitology) A species of nematodes in the genus *Trichostrongylus* that may be found in the small intestines of rabbits and hares.

Trichostrongylus rugatus (veterinary parasitology) A species of nematodes in the genus *Trichostrongylus* that may be found in the small intestines of sheep and goats.

Trichostrongylus skrjabini (veterinary parasitology) A species of nematodes in the genus *Trichostrongylus* that may be found in sheep, moufflons and roe deer.

Trichostrongylus tenuis (avian parasitology) A species of nematodes in the genus *Trichostrongylus* that may be found in the small intestine and caeca (*USA*: ceca) of domestic and wild birds.

Trichostrongylus vitrinus (human/veterinary parasitology) A species of nematodes in the genus *Trichostrongylus* that may be found in the small intestine of sheep, goats, deer, and occasionally in pigs, rabbits, camels and humans.

trichuriasis (parasitology) The disease caused by the infestation of the caecum (*USA*: cecum) by *Trichuris* species. The most obvious clinical feature is diarrhoea, (*USA*: diarrhea) sometimes with mucus and blood.

Trichuris (parasitology) A genus of nematodes in the family Trichuridae, found in the large intestine of most species. Also termed: whipworms.

Trichuris cameli (veterinary parasitology) A species of nematodes in the genus *Trichuris* that may be found in camels.

Trichuris campanula (veterinary parasitology) A species of nematodes in the genus *Trichuris* that may be found in cats.

Trichuris discolor (veterinary parasitology) A species of nematodes in the genus *Trichuris* that may be found in buffaloes, cattle, goats, sheep and zebras.

Trichuris globulosa (veterinary parasitology) A species of nematodes in the genus *Trichuris* that may be found in ruminants.

Trichuris leporis (veterinary parasitology) A species of nematodes in the genus *Trichuris* that may be found in rabbits, hares and coypu.

Trichuris ovis (veterinary parasitology) A species of nematodes in the genus *Trichuris* that may be found in ruminants.

Trichuris raoi (veterinary parasitology) A species of nematodes in the genus *Trichuris* that may be found in dromedaries.

Trichuris serrata (veterinary parasitology) A species of nematodes in the genus *Trichuris* that may be found in cats.

Trichuris skrjabini (veterinary parasitology) A species of nematodes in the genus *Trichuris* that may be found in sheep, goats, and camels.

Trichuris suis (veterinary parasitology) A species of nematodes in the genus *Trichuris* that may be found in pigs.

Trichuris syvilagi (veterinary parasitology) A species of nematodes in the genus *Trichuris* that may be found in rabbits, hares and coypu.

Trichuris tenuis (veterinary parasitology) A species of nematodes in the genus *Trichuris* that may be found in dromedaries.

Trichuris trichiura (human/veterinary parasitology) A species of nematodes in the genus *Trichuris* that may be found in humans and simian primates.

Trichuris vulpis (veterinary parasitology) A species of nematodes in the genus *Trichuris* that may be found in dogs and foxes.

trichurosis (parasitology) Infection with the worm *Trichuris*.

triclabendazole (pharmacology) A highly inffective fasciolicide against liver flukes.

Tricoryna (insect parasitology) *See Rhipipallus affinis.*

triglycerides (biochemistry) Neutral fat comprises three fatty acids esterified to glycerol. The fatty acids may be saturated or unsaturated, and the most common ones are stearic, oleic and palmitic acids. This is the primary storage form of lipid as an energy reserve. Triglycerides are synthesised in adipose tissue from glycerol phosphate, derived from glucose, and from fatty acids, either derived from acetyl CoA formed from glucose or absorbed from the plasma after they are released from lipoproteins. Triglyceride synthesis in adipose cells is dependent upon insulin-stimulated uptake of plasma glucose. Whereas the amount of glycogen that can be stored as an energy reserve is limited, the amount of triglyceride stored appears to be essentially unlimited.

triiodomethane (chemistry) CHI_3 Yellow crystalline solid haloform, used as an antiseptic. Also termed: iodoform.

Trimenopon hispidum (veterinary parasitology) An amblycerid louse that may be found on guinea pigs.

trimer (chemistry) Chemical formed by the combination of three similar (monomer) molecules.

trinitrophenol (chemistry) An alternative term for picric acid.

trinomial (chemistry) Polynomial with only three terms.

Trinoton anserinum (avian parasitology) An amblycerid mite that may be found on ducks and swans.

Triodontophorus (veterinary parasitology) One of the genera of large strongyles of horses in the family Strongylidae. They are all parasites of the large intestine of equids.

Triodontophorus brevicauda (veterinary parasitology) A species of large strongyles in the genus *Triodontophorus* that may be found in horses and asses.

Triodontophorus minor (veterinary parasitology) A species of large strongyles in the genus *Triodontophorus* that may be found in donkeys.

Triodontophorus serratus (veterinary parasitology) A species of large strongyles in the genus *Triodontophorus* that may be found in horses, asses, mules and zebras.

Triodontophorus tenuicollis (veterinary parasitology) A species of large strongyles in the genus *Triodontophorus* that may be found in the right dorsal colon of horses where it causes deep, haemorrhagic (*USA*: hemorrhagic) ulcers.

triple bond (chemistry) Covalent bond formed by the sharing of three pairs of electrons between two atoms.

triploidy (genetics) The condition of having three copies of every chromosome, a form of polyploidy. Triploidy does not occur in nature as a permanent feature of a species, as triploid organisms have very low fertility. However, triploid organisms are just as capable of normal growth as any others (mitosis is not impeded at all, since the process consists of each chromosome dividing itself in half, irrespective of any partner). Artificially produced triploids can be commercially useful. One example of this is the banana, the trees of which are bred as triploids from stocks of a tetraploid and a diploid strain. The triploid hybrids grow well, but their seeds never develop properly, so that the inside of the banana is not full of inedible seeds as it is in a normal banana.

trisaccharide (chemistry) Carbohydrate consisting of three joined monosaccharides.

trisomy (genetics) The state of having three representatives of a given chromosome instead of the usual pair, as in trisomy 21 (Down's syndrome).

tritium (chemistry) A radioactive isotope of hydrogen, with mass number 3 and atomic mass 3.016. Its half-life is 12.5 years. Tritium is used extensively as a radiolabel tracer in toxicity studies.

triton (chemistry) Atomic nucleus of tritium, consisting of two neutrons and one proton.

Tritrichomonas (parasitology) A genus of protozoan parasites with three anterior flagella in the family Trichomonadidae.

Tritrichomonas caviae (veterinary parasitology) A species of protozoan parasites in the genus *Tritrichomonas* that may be found in the caecum (*USA*: cecum) and colon of guinea pigs but is apparently nonpathogenic.

Tritrichomonas eberthi (avian parasitology) A species of protozoan parasites in the genus *Tritrichomonas* that may be found in the caeca (*USA*: ceca) of the chicken, turkey and duck.

Tritrichomonas enteris (veterinary parasitology) A species of protozoan parasites in the genus *Tritrichomonas* that may be found in the caecum (*USA*: cecum) and colon of *Bos indicus* and *Bos taurus*, but that has no apparent pathogenic effects.

Tritrichomonas equi (veterinary parasitology) A species of protozoan parasites in the genus *Tritrichomonas* that may be found in horses but is apparently nonpathogenic.

Tritrichomonas foetus (veterinary parasitology)A species of protozoan parasites in the genus *Tritrichomonas* that may be found in cattle, pig, horse and deer but that causes trichomoniasis only in cattle.

Tritrichomonas minuta (veterinary parasitology) A species of protozoan parasites in the genus *Tritrichomonas* that may be found in the caecum (*USA*: cecum) and colon of rat, mouse and hamster.

Tritrichomonas muris (veterinary parasitology) A species of protozoan parasites in the genus *Tritrichomonas* that may be found in the caecum (*USA*: cecum) and colon of rat, hamster and wild rodent.

Tritrichomonas suis (veterinary parasitology) A species of protozoan parasites in the genus *Tritrichomonas* that may be found in the stomach, small intestine, caecum (*USA*: cecum) and nasal passages of pigs.

4-tritylmorpholine (control measures) A molluscicide used in the control of bilharziasis. Also termed: trifenmorph; Frescon.

trivalent (chemistry) Having a valence of three. Also termed: tervalent.

Trixacarus (parasitology) A genus of mange mites in the family Sarcoptidae.

Trixacarus caviae (veterinary parasitology) (synonym: *Caviacoptes caviae*) A species of mite in the genus *Trixacarus* that causes mange in guinea pigs characterized by alopecia, pruritus and keratinization.

Trixacarus diversus (veterinary parasitology) A species of mite in the genus *Trixacarus* that causes mange in rats, mice and hamsters.

tRNA (molecular biology) An abbreviation of 't'ransfer (or soluble) 'R'ibo'N'ucleic 'A'cid.

Trochometridium (insect parasitology) A species of mites in the Prostigmata, that are parasitic to the larvae of the bees *Halictinae*.

Troglodytella (veterinary parasitology) A genus of ciliated protozoa isolated from cases of diarrhoea (*USA*: diarrhea) in recently captured great apes, chimpanzees and gorillas. Species includes *Troglodytella abrassarti* and *Troglodytella gorillae*.

Troglostrongylus (veterinary parasitology) A genus of nematodes in the family Crenosomatidae that may be found in the lungs of members of the family Felidae.

Troglostrongylus brevior (veterinary parasitology) A species of nematodes in the genus *Troglostrongylus* that may be found in the respiratory tract of cats.

Troglostrongylus subcrenatus (veterinary parasitology) A species of nematodes in the genus *Troglostrongylus* that may be found in the lungs of domestic and wild cats.

Troglotrema (parasitology) A genus of trematodes in the family Troglotrematidae.

Troglotrema acutum (veterinary parasitology) A species of trematodes in the genus *Troglotrema* which may be found in the frontal and ethmoidal sinuses of fox, mink and polecat and that may destroy the walls of the sinuses.

Troglotrema salmincola (parasitology) *See Nanophyetus salmincola.*

Trombicula (human/veterinary parasitology) A genus of mites in the family Trombiculidae, whose larvae are parasitic on all animal species and cause dermatitis. Some also transmit diseases from rodents, their natural hosts, to humans, *e.g.* scrub typhus. The larvae are also termed: chiggers.

Trombicula akamushi (human parasitology) A species of mites in the genus *Trombicula* that transmits scrub typhus to humans.

Trombicula alfreddugesi (parasitology) *See Eutrombicula alfreddugesi.*

Trombicula autumnalis (human/avian/veterinary parasitology) A species of mites in the genus *Trombicula* that are distinctively red and may be found on all domestic animal species including poultry and humans. They may cause dermatitis. Also termed: harvest mite, aoutat, lepte automnale.

Trombicula batatas (parasitology) A species of mites in the genus *Trombicula* that causes dermatitis.

Trombicula delhiensis (human/veterinary parasitology) A species of mites in the genus *Trombicula* that transmits scrub typhus of humans from rodents.

Trombicula minor (parasitology) A species of mites in the genus *Trombicula* that is also termed the scrub-itch mite.

Trombicula sarcina (parasitology) *Eutrombicula sarcina.*

Trombicula spendens (parasitology) A species of mites in the genus *Trombicula* that causes dermatitis.

trombiculiasis (parasitology) *See* trombiculidiasis.

trombiculid (parasitology) A member of the Trombiculidae family of mites whose parasitic larvae (chiggers) infest vertebrates causing trombiculidiasis.

trombiculidiasis (avian/veterinary parasitology) Dermatitis in all pastoral animal species and birds caused by mites of the family Trombiculidae. The bites produce wheals and intense pruritus followed by the development of moderate to severe dermatitis. This is mostly on the lower part of the face and the distal extremities. The disease is most likely to occur in autumn when the parasites are active and is often confined to particular fields that provide the best ecological niche for the mite.

trombiculosis (parasitology) Infestation with *Trombicula.*

trombidiform mite (parasitology) *See Trombicula.*

trop. med. (medical) An abbreviation of 'trop'ical 'med'icine.

troph(o)- (terminology) A word element. [German] Prefix denoting *food, nourishment.*

-trophic (terminology) A word element. [German] Suffix denoting *nourishing, stimulating.*

-trophin (terminology) A word element. [German] Suffix denoting *nourishing, stimulating.*

trophont (fish parasitology) Sperm of the protozoon *Oodinium* spp. that may be found attached to the skin or gills of fish.

trophozoite (parasitological physiology) The active, motile feeding stage of an apicomplexan parasite, and the motile stage of flagellate protozoa, *e.g. Giardia* spp.

-tropic (terminology) A word element. [German] Suffix denoting *turning toward, changing, tending to turn or change.*

tropical bont tick (parasitology) *See Amblyomma variegatum.*

tropical cattle tick (veterinary parasitology) *See Boophilus microplus.*

tropical fowl mite (avian parasitology) *See Ornithonyssus bursa.*

tropical horse tick (veterinary parasitology) *See Dermacentor nitens.*

tropical rat mite (veterinary parasitology) *See Ornithonyssus bacoti.*

tropical sore (medical) a slow-healing ulcer in the skin caused by *Leishmania major* or *Leishmania tropica. See* Leishmaniasis.

tropical theileriosis (parasitology) *See* Mediterranean coast fever.

Tropilaelaps (insect parasitology) A species of mites in the Mesostigmata, that are parasites of the bees Apinae.

Tryp. (biochemistry) An abbreviation of 'Tryp'tophan.

trypan blue (biochemistry) A vital dye that is used to determine the viability of isolated cells. Living cells exclude the dye, whereas nonviable cells do not. Also, a largely superseded trypanocide.

trypanocidal (pharmacology) Destructive to trypanosomes. The common drugs used are diminazene aceturate, homidium bromide, quinapyramine and suramin.

trypanocide (pharmacology) A trypanocidal drug.

trypanolysis (pharmacology) The destruction of trypanosomes.

Trypanoplasma cyprini (parasitology) A species of fusiform protozoan parasite with a flagellum at each pole, that is a member of the Cryptobiidae family. They may invade the vascular system and cause depression and emaciation.

See also Cryptobia.

trypanoplasmiasis (fish parasitology) Disease of finfish caused by *Trypanoplasma* spp.

Trypanorrhyncha (Tetrarhynchidea) (fish parasitology) An order of flatworms in subclass Eucestoda, class Cestoidea and the phylum Platyhelminthes. This order is found in elasmobranchs only. Except for *Aporhynchus* *norvegicus*, the scolex has four eversible armed tentacles and two or four bothridia.

Trypanosoma (human/insect/veterinary parasitology) A multispecies genus of protozoa in the family Trypanosomatidae, that are parasitic in the blood, lymph and tissues of invertebrates and vertebrates, including humans. Most species live part of their life cycle in the intestines of insects and other invertebrates, the flagellate stage being found only in the vertebrate host.

Trypanosoma avium (avian parasitology) A species of protozoa that are of minor pathogenicity in the genus *Trypanosoma*, which may be found in birds.

Trypanosoma binneyi (veterinary parasitology) A species of protozoa that are of minor pathogenicity in the genus *Trypanosoma*, which may be found in platypus.

Trypanosoma brucei (veterinary parasitology) (synonym: *Trypanosoma pecaudi*) A species of protozoa in the genus *Trypanosoma* that causes a severe disease in all species, including horses, cattle, sheep, dogs and cats.

Trypanosoma calmetti (avian parasitology) A species of protozoa that are of minor pathogenicity in the genus *Trypanosoma*, which may be found in ducklings.

Trypanosoma congolense (human/veterinary parasitology) (synonym: *Trypanosoma pecorum*, *Trypanosoma nanum*, *Trypanosoma montgomeryi*) A species of protozoa in the genus *Trypanosoma* that causes diseases in all domestic animals but most serious in humans, where it causes sleeping sickness, and cattle. Reservoir hosts are wild ruminants.

Trypanosoma cruzi (human/veterinary parasitology) (synonym: *Trypanosoma escomeli*) A species of protozoa in the genus *Trypanosoma* that causes Chagas' disease, American trypanosomiasis in humans, and has a reservoir in pigs, dogs, cats and many wild animals. It also causes disease in these hosts and may be fatal to dogs.

Trypanosoma diazi (veterinary parasitology) A species of protozoa that are of minor pathogenicity in the genus *Trypanosoma*, which may be found in capuchin monkeys.

Trypanosoma dimorphon (veterinary parasitology) A species of protozoa that are of minor pathogenicity in the genus *Trypanosoma*, which may be found in domestic animals generally.

Trypanosoma equinum (veterinary parasitology) A species of protozoa in the genus *Trypanosoma* that may be found in all species but is most serious in equids, in which it is characterized by posterior paralysis, called mal de Caderas.

Trypanosoma equiperdum (veterinary parasitology) A species of protozoa in the genus *Trypanosoma* that may be found in all species but a common infection and a serious disease only in equids in which it is transmitted venereally and is known as dourine.

Trypanosoma evansi (veterinary parasitology) A species of protozoa in the genus *Trypanosoma* that causes infection in many species especially in camels, horses and dogs. The disease in horses is surra. In cattle and buffalo the disease is subclinical but these species act as reservoirs.

Trypanosoma gallinarum (avian parasitology) A species of protozoa that are of minor pathogenicity in the genus *Trypanosoma*, which may be found in fowls.

Trypanosoma gambiense (human/veterinary parasitology) (synonym: *Trypanosoma hominis*, *Trypanosoma nigeriense*, *Trypanosoma ugandense*) A species of protozoa in the genus *Trypanosoma* that is a chronic disease of humans which can occur also in cattle, goats, sheep, horses, dogs and cats.

Trypanosoma lewisi (veterinary parasitology) A species of protozoa in the genus *Trypanosoma* that may be found in rats and may cause death in ratlings.

Trypanosoma melophagium (veterinary parasitology) A species of protozoa that are of minor pathogenicity in the genus *Trypanosoma*, which may be found in sheep.

Trypanosoma minasense (veterinary parasitology) A species of protozoa that are of minor pathogenicity in the genus *Trypanosoma*, which may be found in monkeys, such as the marmosets.

Trypanosoma nabiasi (veterinary parasitology) A species of protozoa that are of minor pathogenicity in the genus *Trypanosoma*, which may be found in rabbits.

Trypanosoma primatum (veterinary parasitology) A species of protozoa that are of minor pathogenicity in the genus *Trypanosoma*, which may be found in chimpanzees and gorillas.

Trypanosoma rangeli (human/veterinary parasitology) A species of protozoa that are of minor pathogenicity in the genus *Trypanosoma*, which may be found in humans, cats and dogs. Also termed: *Trypanosoma ariarii.*

Trypanosoma rhodesiense (human/veterinary parasitology) A species of protozoa in the genus *Trypanosoma* that causes a serious disease in humans but only a mild one in ruminants and other domestic animals and monkeys.

Trypanosoma saimiriae (veterinary parasitology) A species of protozoa that are of minor pathogenicity in the genus *Trypanosoma*, which may be found in squirrel monkeys.

Trypanosoma sanmartini (veterinary parasitology) A species of protozoa that are of minor pathogenicity in the genus *Trypanosoma*, which may be found in squirrel monkeys.

Trypanosoma suis (veterinary parasitology) A species of protozoa in the genus *Trypanosoma* that may be found in pigs in which it causes a fatal disease.

Trypanosoma theileri (veterinary parasitology) A species of protozoa in the genus *Trypanosoma* that may be considered to be nonpathogenic in cattle, in which it occurs almost universally but that may cause illness in stressed animals.

Trypanosoma theodori (veterinary parasitology) A species of protozoa that are of minor pathogenicity in the genus *Trypanosoma*, which may be found in pigs.

Trypanosoma uniforme (veterinary parasitology) A species of protozoa in the genus *Trypanosoma* that may be found in most ruminants and which is similar to *Trypanosoma vivax* in pathogenicity.

Trypanosoma vivax (veterinary parasitology) (synonym: *Trypanosoma caprae*; *Trypanosoma angolense*) A species of protozoa in the genus *Trypanosoma* that may be found in ruminants and horses but not pigs, dogs or cats, and that causes a serious and fatal disease in cattle and goats, especially in animals under stress.

trypanosome (parasitology) A protozoan of the genus *Trypanosoma.*

trypanosomiasis (human/veterinary parasitology) Infestation with trypanosomes. Clinically a nondescript disease which may be peracute, acute or chronic. In humans *Trypanosoma brucei gambiewe* causes sleeping sickness in West Africa, and *Trypanosoma brucei rhoesiense* causes the disease in East Africa. The diagnosis in humans and animals is based on a positive blood smear and the presence of an insect vector, often a tsetse fly, or a history of mating in the case of equids in which it is transmitted venereally and is known as dourine.

See also Chagas' disease.

trypanosomicide (pharmacology) 1 Lethal to typanosomes. 2 A substance lethal to trypanosomes.

trypanosomid (parasitology) A skin eruption occurring in trypanosomiasis.

trypanotolerance (parasitology) Resistance to infection with trypanosomes, inherent in some breeds of cattle, *e.g.* the N'dama, Nigerian shorthorn, Lagune and others.

trypomastigote (parasitological physiology) Developmental stage of trypanosomes, usually in the vertebrate host but may be in the invertebrate. A stage in the life cycle that is elongated with the kinetoplast posterior and distal to the nucleus. A leaf-like form with an undulating membrane and often a free flagellum.

tryptophan (biochemistry) Essential amino acid which is a precursor of serotonin, that contains an aromatic group, needed in animals for proper growth and development.

tsetse (parasitology) An African fly of the genus *Glossina*, which transmits trypanosomiasis.

tsetse fly (parasitology) *See* trypanosomiasis.

TSF (haematology) An abbreviation of 'T'hrombopoietic 'S'timulating 'F'actor, thrombopoietin.

TSH (physiology/biochemistry) An abbreviation of 'T'hyroid-'S'timulating 'H'ormone.

TSI (physiology/immunology) An abbreviation of 'T'hyroid-'S'timulating 'I'mmunoglobulins.

t-test (statistics) A parametric test for assessing hypotheses about population means. It is most commonly used when the null hypothesis is that two populations have the same mean value on some variable of interest. The form of the test used in this case depends on whether independent samples are drawn from each population or the samples are matched in some way (*e.g.* thereby having the same subjects each perform under two conditions). Though the test is based on assumptions of normality and homogeneity of population variances, it is relatively robust against departures from these assumptions.

Tu (chemistry) The chemical symbol for thulium.

Tubellaria (fish parasitology) A class of flatworms in the phylum Platyhelminthes. Tubellaria are mostly free-living flatworms in terrestrial, freshwater and marine environments. *See* Platyhelminthes.

tubo- (terminology) A word element. [Latin] Prefix denoting *tube.*

tumbu fly (parasitology) *See Cordylobia anthropophaga.*

tumour (histology) An overt neoplasm, either benign (*e.g.* a polypus) or malignant (*i.e.* cancerous). (*USA*: tumor).

Tunga (parasitology) A genus of fleas native to tropical and subtropical USA and Africa.

Tunga penetrans (human/avian/veterinary parasitology) A species of flea in the genus *Tunga.* The chigoe or jigger flea, which attacks humans, dogs, pigs and other animals, as well as poultry, and causes intense skin irritation.

See also chigoe.

tungiasis (parasitology) Infestation with *Tunga* fleas.

tunicamycin (parasitology) Bacterial toxin produced by *Clavibacter rathayi* which infects galls on grass caused by the nematode *Anguina funesta.* Also termed: corynetoxin.

turbid (general terminology) Cloudy.

turgid (biology) Swollen and congested.

turkey louse (avian parasitology) Lice which includes *Chelopistes meleagridis* (large turkey louse), *Colpocephalum tausi*, *Oxylipeurus polytrapezius* (slender turkey louse), and *Oxylipeurus corpelentus*.

turnaround time (computing) The time elapsed between submission of data to a computer bureau and the receipt of the result.

Turnbull stain (histology) Histological staining technique for ferrous iron.

turning sickness (parasitology) An aberrant form of theileriasis in which parasitized lymphocytes cause emboli and haemorrhagic (*USA*: hemorrhagic) infarcts in central nervous tissue. *Theileria parva* is credited with causing the disease which is characterized by convulsive attacks of spinning followed by collapse and unconsciousness, or by a more chronic sydrome of circling, head pressing, incoordination and blindness.

Turraea robusta (pharmacology) An African tree in the plant family Meliaceae, the juice of the leaves of which may be used as a treatment for diarrhoea (*USA*: diarrhea) and as an anthelmintic, but the effective agent has not been identified.

twist disease (fish parasitology) A disease of salmonid fish, especially rainbow trout, caused by infection with the protozoan parasite *Myxosoma cerebralis* which erodes the skeleton, especially the cranium. Affected fish lose their balance and chase their tails when startled. Survivors are usually badly deformed.

two-tailed test (statistics) A statistical test of an hypothesis whose regions of rejection are placed at both ends (or tails) of the distribution of the test statistic, *e.g.* when the alternative to the null hypothesis is that the mean of one population differs from that of the other, regardless of whether it is higher or lower. *Contrast* one-tailed test.

Tylenchulus (plant parasitology) A genus of nematodes in the order Tylenchida and family Tylenchulidae. They are sexually dimorphic. Immature females are migratory, vermiform, and small, under 0.5 mm long. Mature females are sedentary endoparasites, the anterior third of the body embedded in root tissue, the posterior body swollen, and irregular in shape. The male body is short and slender, with the stylet and oesophageal (*USA*: esophageal) region much weaker than in the female. Males and male juveniles do not feed, but female juveniles are ectoparasitic and endoparasitic on roots. *See Tylenchulus semipenetrans*.

Tylenchulus semipenetrans (plant parasitology) A species of nematodes in the genus *Tylenchulus* whose hosts cover a narrow range including citrus, olive, grape, lilac, and persimmon. Currently, there are four known biotypes on various hosts; these can be distinguished by their ability to parasitize citrus rootstocks. They are distributed worldwide on citrus; Australia, California and Chile on grapes; and USA on olives. They are sedentary endoparasites. Reproduction occurs by parthenogenesis; males are present but are not required. Second-stage juveniles hatch from eggs and these and the third and fourth stage juveniles feed upon root cells of the hypodermis. The young adult females which are vermiform penetrate more deeply into the cortical parenchyma, leaving the posterior half of the body outside the root. The exposed part of the body becomes saccate and a gelatinous matrix is produced. Egg masses contain about 100 eggs deposited in the matrix; these hatch and the second-stage juveniles attack roots. The feeding site consists of 8–10 nurse cells which have thick walls, a large nucleus and nucleolus. The area may become invaded by other micro-organisms. Males do not penetrate roots and they are about the same size as the second-stage juveniles. The male passes through three moults (*USA*: molts) without feeding, and the stylet becomes progressively less distinct; males reach maturity in one week. The second-stage female juveniles are the persistent stage, and these have been recovered from stored soil after two and a half years and from field soil four years after pulling lemon trees. Trees heavily infected with *Tylenchulus semipenetrans* exhibit symptoms of 'slow decline', dieback of small branches, yellow foliage, reduced fruit size and numbers of feeder roots. Feeding causes the formation of necrotic areas and cells are killed. Heavy infections cause disintegration of the cortical parenchyma cells and the feeder roots eventually are killed. Root destruction causes plant decline over three to five years. Nematodes occur in very high numbers and the female body's outside root swells, and eggs are produced in a gelatinous matrix. The soil adheres to this matrix causing a 'dirty root' symptom. Control measures include the use of preplant nematicides and postplant nematicides on citrus and grapes. Also termed: citrus nematode.

Tylenchus (plant parasitology) A genus of nematodes in the order Tylenchida and family Tylenchidae. They prefer aquatic or semi-aquatic habitats and were first recovered from moss on boulders in freshwater streams. They have been found on more than 40 plant hosts, but no apparent damage has been observed and it has never been considered to be a true parasite. Cosmopolitan, they occur in every continent except Antartica, but seem to prefer the Northern Hemisphere.

type 1 error (statistics) Rejection of the null hypothesis when it is in fact true.

type 11 error (statistics) Acceptance of the null hypothesis when it is in fact false.

type species (taxonomy) The original species from which the description of the genus is formulated.

typhl(o)- (terminology) Word element [German] Prefix denoting *caecum* (*USA*: cecum), *blindness*.

typhus (human parasitology) Acute infectious diseases caused by rickettsia which are usually transmitted from infected rats and other rodents to humans by lice, fleas, ticks and mites. Also termed: Gail fever.

Tyr (biochemistry) An abbreviation of 'Tyr'osine.

Tyroglyphus (parasitology) A genus of mites in the family Acaridae. Parasites of grain and other vegetable matter that only parasitizes animals accidentally. They cause itching and occasionally a mild dermatitis, but the infestation is usually self-limiting.

Tyroglyphus farinae (parasitology) (synonym: *Acarus farinae*) A species of mites in the genus *Tyroglyphus* that may be found in cheeses and grain.

Tyroglyphus siro (parasitology) (synonym: *Acarus siro*) The cheese mite. A species of mites in the genus *Tyroglyphus* that may also be found in grain and that may cause diarrhoea (*USA*: diarrhea).

Tyrophagus (parasitology) A genus of mites in the family Acaridae.

Tyrophagus farinae (human/veterinary parasitology) Housedust mites. A species of mites in the genus *Tyrophagus* that are common in stored cereal and thought to be associated with allergic dermatitis.

Tyrophagus longior (human/veterinary parasitology) A species of mites in the genus *Tyrophagus* that may be found in grains and copra, and that is the cause of copra itch in humans.

Tyrophagus palmarum (veterinary parasitology) A species of pasture mite in the genus *Tyrophagus* that may be found in the nostrils of cattle, especially those with nasal granuloma.

tyrosine (chemistry) White crystalline organic compound, a naturally occurring essential amino acid that is a precursor for adrenergic transmitters, *e.g.* Adrenaline (*USA*: epinephrine) and noradrenaline (*USA*: norepinephrine).

Tyzzeria (parasitology) A genus of protozoa in the family Eimeriidae that are intracellular parasites in the intestinal epithelium.

Tyzzeria alleni (avian parasitology) A species of protozoa in the genus *Tyzzeria* that may be found in teal.

Tyzzeria anseris (avian parasitology) A species of protozoa in the genus *Tyzzeria* that may be found in domestic and wild geese.

Tyzzeria pellerdyi (avian parasitology) A species of protozoa in the genus *Tyzzeria* that may be found in wild birds.

Tyzzeria perniciosa (avian parasitology) A species of protozoa in the genus *Tyzzeria* that may be found in domestic duck.

tzaneen disease (veterinary parasitology) A mild disease of cattle without apparent clinical signs, caused by *Theileria mutans*.

tzetze (parasitology) *See* tsetse.

U

U (1 chemistry; 2 measurement) 1 The chemical symbol for uranium, Element 92. 2 An abbreviation of 'U'nit(s).

UCHD (medical) An abbreviation of 'U'sual 'C'hild'H'ood 'D'iseases.

UCL (physics) An abbreviation of 'U'pper 'C'ontrol 'L'imit.

UDPG (biochemistry) An abbreviation of 'U'ridine 'D'i'P'hospho'G'lucose.

UDPGA (biochemistry) An abbreviation of 'U'ridine 'D'i'P'hospho'G'lucuronic 'A'cid.

UFA (physiology/biochemistry) An abbreviation of 'U'nesterified free 'F'atty 'A'cid.

UHF (physics) An abbreviation of 'U'ltra'H'igh 'F'requency.

uitpeuloog (veterinary parasitology) A contagious disease of the eyes in ruminants and horses caused by the larvae of the fly *Gedoelstia*.

UL (histology) An abbreviation of 'U'nstirred 'L'ayer.

ulf (1, 2 physics) 1 An abbreviation of 'u'ltra 'l'ow 'f'requency. 2 An abbreviation of 'u'pper 'l'imiting 'f'requency.

ulna (anatomy) Rearmost (and usually larger) of the two bones in the lower forelimb of a tetrapod vertebrate (the other bone is the radius).

ultra- (terminology) A word element. [Latin] Prefix denoting beyond, excess.

ultracentrifugation (methods) Subjection of material to an exceedingly high centrifugal force, which will separate and sediment the molecules of a substance or subcellular components.

ultracentrifuge (equipment) The centrifuge used in ultracentrifugation.

ultrafilter (equipment) The filter used in ultra-filtration.

ultrafiltrate (methods) Substances which pass through an ultrafilter, *i.e.* a semipermeable membrane through which the filtrate passes under pressure.

ultrafiltration (methods) Filtration through a filter capable of removing colloidal particles from a dispersion medium, as in the filtration of plasma at the capillary membrane.

ultraheat (physics) Heating to a very high temperature for a very brief period.

ultrahigh frequency (UHF) (physics) A radio frequency between 300 and 3,000 megahertz.

ultramicroscope (microscopy) Instrument for viewing sub-microscopic objects. *e.g.* particles of smoke and fog.

ultramicroscopic (microscopy) Too small to be seen with the ordinary light microscope.

ultrasonic (physics) Describing a band of sound frequencies of about 2×10^9 hertz, which are just above the upper limit of normal human hearing. Ultrasonic energy is used in sonar, for degreasing (in conjunction with a suitable solvent) and for scanning soft tissues in medical diagnosis. Also termed: supersonic.

ultrasonic (physics) Beyond the audible range; relating to sound waves having a frequency of more than 20,000 cycles per second.

ultrasonogram (methods) The record obtained by ultrasonography.

ultrasonography (methods) An imaging technique in which deep structures of the body are visualized by recording the reflections (echoes) of ultrasonic waves directed into the tissues. Frequencies in the range of 1 million to 10 million hertz are used in diagnostic ultrasonography.

ultrasound (physics) Mechanical radiant energy of a frequency greater than 20,000 cycles per second; used in the technique of urtrasonography.

ultrasound-guided biopsy (methods) Use of ultrasonography to guide the passage of a needle or biopsy instrument into an internal organ or lesion.

ultrastructure (microscopy) The structure visible under the electron microscope.

ultraviolet irradiation (methods) The projection of ultraviolet light from a generator which may be used for the treatment of skin disease and for sterilization of materials.

ultraviolet microscope (microscopy) A microscope that uses an ultraviolet light source; only photographic images are available.

ultraviolet radiation (UV) (physics) Electromagnetic radiation with wavelengths in the range 4×10^{-7} to 4×10^{-9} m, the region between visible light and X-rays. Also termed: ultraviolet light.

ultraviolet rays (physics) *See* ultraviolet radiation.

ultronics (physics) Sound waves which are of a frequency above the range of audible sound, which lies between 20 and 20,000 cycles per second

umbilical cord (anatomy) In embryology, vascular structure that contains the umbilical arteries and veins, connecting the foetus (*USA*: fetus) to the placenta.

umbilicus (anatomy) The navel.

unarmed tapeworm (parasitology) *See Taenia saginata.*

uncertainty principle (physics) It is impossible to determine simultaneously with accuracy both the position and the momentum of a moving particle. The limit of accuracy may be given by the relation $p_x x >$ h/2_, where p_x is the uncertainty in the momentum, x the uncertainty in position and h is Planck's constant. Also termed: Heisenberg uncertainty principle.

Uncinaria (veterinary parasitology) A genus of canine hookworms in the order Strongylida.

Uncinaria criniformis (veterinary parasitology) A species in the genus *Uncinaria* that occurs in the badger and fox.

Uncinaria lucasi (veterinary parasitology) A species in the genus *Uncinaria* that occurs in the fur seal.

Uncinaria stenocephala (veterinary parasitology) A species in the genus *Uncinaria* that commonly occurs in the dog, cat and fox.

Uncinaria yukonensis (veterinary parasitology)) A species in the genus *Uncinaria* that occurs in all ursids.

uncinariasis (veterinary parasitology) The disease caused by *Uncinaria* in cats and dogs. It is similar to, but less severe than ancylostomiasis, with only mild blood loss and enteritis.

undulating membrane (parasitological anatomy) A fold of the protozoa's cell membrane formed when the flagellum of the protozoa beats and pulls up the membrane along the full length of the parasite's body.

undulation (biology) A wavelike motion in any medium; a vibration.

uni- (terminology) A word element. [Latin] Prefix denoting one.

unicellular (biology) Describing an organism that consists of only one cell (*e.g.* protozoans, bacteria).

uniform distribution (statistics) A frequency distribution in which all classes or values have the same frequency or probability.

unilateral (medical) Pertaining to one side of the body or one hemisphere of the brain.

unimodal (statistics) Of a distribution, having only one mode, which, if the data are represented graphically, corresponds to there being only one peak.

unimolecular reaction (chemistry) Chemical reaction that involves only one molecule as the reactant.

unit (1, 2 measurement) 1 A standard quantity in which something is measured, *e.g.* decibel or wavelength. 2 A single unit. 3 An alternative term for kilowatt-hour, the unit that measures consumption of electricity.

univalent (1 cytology; 2 chemistry) 1 Single chromosome that separates during the meiotic division. *See* meiosis. 2 Monovalent.

univariate (statistics) Having only one variable.

universal indicator (chemistry) Mixture of chemical indicators that give a definite for various values of pH.

unkn (general terminology) An abbreviation of 'unk'now'n'.

unpub (literary terminology) An abbreviation of 'unpub'lished.

unsaturated (1, 2 chemistry) 1 Describing a solution that can dissolve more solute before reaching saturation. 2 Describing an organic compound with doubly or triply bonded carbon atoms.

See also saturated compound.

up. (general terminology) An abbreviation of 'up'per.

update (computing) To modify a master file by replacing old information with new.

upper respiratory tract (URT) (anatomy) Comprises the nasal cavities, pharynx and larynx. Some anatomists also include the upper segments of the bronchial tree.

upstream (molecular biology) Further back on a DNA molecule, in respect of the direction in which the sequence is being read. *See* replication.

upstream activating (molecular biology) Site DNA sequences that are upstream from the promoter and have a regulatory role in transcription.

Ur (chemistry) An abbreviation of 'Ur'anium.

Ura (chemistry) An abbreviation of 'Ura'cil.

uracil (Ura) (molecular biology) $C_4H_4N_2O_2$ Pyrimidine base; one of the four nucleotide bases in RNA; in DNA the corresponding base is thymine. Also termed: 2,6-dioxypyrimidine.

uraemia (biochemistry) Excess of urea and other nitrogen-containing compounds, principally creatinine, in the blood. The condition is the result of kidney failure, except in cases where there is failure of the circulation from any cause, when the state is described as pre-renal uraemia; such cases may recover when the circulatory collapse is remedied. Also termed 'azotaemia'.

uran(o)- (terminology) A word element. [German] Prefix denoting *palate*.

uranium (U) (chemistry) Radioactive grey metallic element in Group IIIA of the periodic table (one of the actinides), obtained mainly from its ore uraninite (which contains uranium(IV) oxide, UO_2). It has three natural and several artificial isotopes with half-lives of up to 4.5 $\times 10^9$ years. Uranium-235 undergoes nuclear fission and is used in nuclear weapons and reactors; uranium-238 can be converted into the fissile plutonium-239 in a breeder reactor. At. no. 92; r.a.m. 238.029.

urates (biochemistry) Salts of uric acid, the result of the breakdown of purines, which are compounds which have an important role in metabolism. Urates are normally present in the blood, but when the blood level is too high crystals may form; in the joints they give rise to gout, and in the kidneys to stones.

urea (biochemistry) H_2NCONH_2 White crystalline organic compound, found naturally in the urine of mammals as the natural end-product of the metabolism of proteins. It is also manufactured commercially from carbon dioxide and ammonia under high pressure. It is used in plastics, adhesives, fertilizers and animal-feed additives. Also termed:carbamide.

Ureaplasma (human/veterinary parasitology) A genus in the family Mycoplasmataceae. There are two species, *Ureaplasma urealyticum*, which may be found in humans, and *Ureaplasma* diversum, which is associated with genial disease in cattle. Also termed: t-strain mycoplasma. *See* ureaplasmosis.

ureaplasmosis (veterinary parasitology) Infection with *Ureaplasma* spp.; occurs in the vagina and vulva of cows and ewes and there is speculation that they are causally associated with granular vaginitis and possibly with transitory endometritis. They may also play a role in seminal vesiculitis in bulls, and in pneumonia in calves.

urease (chemistry) Enzyme that occurs in plants (*e.g.* soya beans) and acts as a catalyst for the hydrolysis of urea to ammonia and carbon dioxide.

uredofos (pharmacology) A broad-spectrum anthelmintic introduced for the treatment of canine and feline hookworms, ascarids, tapeworms and whipworms but withdrawn because of a large number of adverse reactions with many deaths.

ureter (anatomy) One of a pair of ducts that carry urine from the kidneys to the bladder.

ureter(o)- (terminology) A word element. [German] Prefix denoting *ureter*.

urethane (chemistry) $CO(NH_2)OC_2H_5$. Highly toxic, inflammable organic used in veterinary medicine, biochemical research and as a chemical intermediate. Also termed: ethyl carbamate; ethyl urethane.

urethr(o)- (terminology) A word element [German]. Prefix denoting *urethra*.

urethra (anatomy) Tube through which urine is discharged to the exterior from the urinary bladder of most mammals.

urethritis (medical) Inflammation of the urethra, the canal through which urine is discharged.

URF (physiology) An abbreviation of 'U'terine-'R'elaxing 'F'actor; relaxin.

URI (medical) An abbreviation of 'U'pper 'R'espiratory 'I'nfection.

-uria (terminology) A word element. [German] Suffix denoting condition of the urine.

uric acid (biochemistry) $C_5H_4N_4O_3$ White crystalline organic acid of the purine group, the end-product and the principal excretory product of purine metabolism. Defects in uric acid metabolism and excretion appear to be associated with a number of disease states, and it frequently occurs as a component of renal calculi. Uric acid deposition in the joints is the principal cause of gout. Also termed: 2,6,8-trihydroxypurine.

urin(o)- (terminology) A word element. [German; Latin] Pertaining to urine.

urinary output (biology) The amount of urine secreted by the kidneys.

urine (biochemistry) Liquid, produced in the kidneys and stored in the urinary bladder, that contains urea and other excretory products. It is discharged to the outside via the urethra.

urine sediment (parasitology) A centrifuged deposit suitable for microscopic examination for the presence of parasites, eggs, cells, casts, bacteria, crystals, etc.

uro- (terminology) A word element. [German] Prefix pertaining to urine.

urobilinogen (biochemistry) A derivative of bilirubin formed by the gut microflora; this compound is colourless (*USA*: colorless).

Urobovella (insect parasitology) *See Dinychus*; Macrochelidae; *Urodiscella*.

Urodiscella (insect parasitology) A species of mites in the Mesostigmata, that similarly to *Urobovella*, *Uroplitana and Urozercon*, are parasites of the bees Meliponinae and Bombinae. *See Dinychus*; Macrochelidae.

urogenital system (veterinary parasitology) Genitourinary system.

urogenous (biochemistry) 1 Producing urine. 2 Produced from or in the urine.

urogram (radiography) A radiograph obtained by urography.

urography (radiography) Radiography of any part of the urinary tract.

urol. (medical) An abbreviation of 'urol'ogy.

urologist (medical) A specialist in urology.

urology (medical) The branch of medicine that deals with diseases of the urinary tract in both sexes, and those of the genital organs in the male.

Uronema (fish parasitology) Ciliated protozoa within the order Scuticociliatida. These or closely related genera cause severe losses in marine aquacultured finfish and in ornamental species.

uropathogen (medical) Pathogenic organisms in the urinary tract.

Uroplitana (insect parasitology) *See Urodiscella.*

Uropodidae (1 parasitology; 2 insect parasitology) 1 A family of scavenger mites that frequent poultry litter but do not infest birds. 2 *See* Macrochelidae.

uroradiology (radiography) Radiology of the urinary tract.

uroscopy (medical) Diagnostic examination of the urine.

Uroseius (insect parasitology) *See Dinychus.*

Urozercon (insect parasitology) *See* Macrochelidae; *Urodiscella.*

Ursicoptes americanus (veterinary parasitology) American sarcoptiform mange mite which causes pruritus and alopecia in bears.

ursicoptic (parasitology) Emanating from or pertaining to *Ursicoptes americanus.*

ursicoptic mange (veterinary parasitology) Pruritus and alopecia caused by *Ursicoptes americanus,* a sarcoptiform mite, which may be found in the follicular sinuses of bears.

URT (anatomy) An abbreviation of 'U'pper 'R'espiratory 'T'ract.

URTI (anatomy) An abbreviation of 'U'pper 'R'espiratory 'T'ract 'I'nfection.

urticaria (immunology) Nettle-rash. Swelling and redness of the skin, usually with itching, caused by an allergic reaction which results in the liberation of histamine in the skin.

urticaria papulosa (medical) An allergic reaction to the bite of various insects, with appearance of lesions that evolve into inflammatory, increasingly hard, red or brownish, persistent papules.

US (physiology) An abbreviation of 'U'nconditioned 'S'timulus.

user (computing) A person who uses either a computer or the information generated by one.

user friendly (computing) A computer programme (*USA*: program) or system which is easy to understand, to learn and to use.

USG (biochemistry) An abbreviation of 'U'rine 'S'pecific 'G'ravity.

u-shaped curve (statistics) Any graph of a distribution that is shaped like a U.

USPHS (governing body) An abbreviation of 'U'nited 'S'tates 'P'ublic 'H'ealth 'S'ervice.

usu. (general terminology) An abbreviation of 'usu'ally.

uter(o)- (terminology) A word element. [Latin] Prefix denoting uterus.

uterography (radiography) Radiographic examination of the uterus; hysterography.

uteroscope (equipment) An instrument for viewing the interior of the uterus; hysteroscope.

uterus (anatomy) Muscular organ located in the lower abdomen of female mammals, in which a fertilized ovum develops into a foetus (*USA*: fetus) prior to birth. Also termed: womb.

UTI (medical) An abbreviation of 'U'rinary 'T'ract 'I'nfection.

UTP (biochemistry) An abbreviation of 'U'ridine 'T'ri'P'hosphate.

UV (physics) An abbreviation of 'U'ltra'V'iolet.

uve(o)- (terminology) A word element. [Latin] Prefix denoting uvea.

UVL (physics) An abbreviation of 'U'ltra'V'iolet 'L'ight.

V

V (1 chemistry; 2 biochemistry) 1 The chemical symbol for Element 23, vanadium. 2 A symbol for valine.

V region (immunology) Variable region. Part of the heavy and light chains of an immunoglobulin molecule. It is this region that interacts with the antigen.

v.v. (literary terminology) An abbreviation of '*v'ice 'v'ersa.* Latin, meaning 'reversal of order'.

v/v (measurement) An abbreviation of 'v'olume (of solute) per 'v'olume (of solvent).

vac. pmp (physics) An abbreviation of 'vac'uum 'p'u'mp'.

vaccinate (immunology) To inoculate with vaccine to produce immunity.

vaccination (immunology) The artificial production of active immunity by the injection of a vaccine. The word 'vaccination' is derived from the Latin for 'cow', and was coined by Edward Jenner at the end of the 18th century to describe his idea of injecting patients with cowpox virus to protect them against smallpox (a very similar virus). A vaccine now means any suspension of dead or non-virulent viruses or bacteria used in immunisation.

vaccination schedules (immunology) Specified ages and intervals for administration of vaccines to ensure the best immunological response.

vaccine (immunology) A preparation of dead organisms, attenuated live organisms, live virulent organisms, or parts of micro-organisms which are either injected or ingested into the body, where they stimulate the production of antibodies and so confer immunity against infection. Less commonly, vaccines are used in treating a disease.

vaccinotherapy (medical) Therapeutic use of vaccines.

Vacor (control measures) A single dose rodenticide that acts as an antagonist of B vitamins, particularly nicotinamide, which is useful against warfarin-resistant rodents. Also termed pyriminil.

vacuity (anatomy) Any gap between the bones of a skull.

vacuolar (biology) Containing, or of the nature of, vacuoles.

vacuolated (biology) Containing vacuoles.

vacuolation (biology) The process of forming vacuoles; the condition of being vacuolated.

vacuole (cytology) A space or cavity in the cytoplasm of a cell. *See* contractile vacuole.

vacuum (physics) Space containing no matter. A good laboratory vacuum still contains about 10^{14} molecules of air per cubic metre; intergalactic space may have an almost perfect vacuum (although it does contain some subatomic particles).

vacuum collection (parasitology) Use of a handheld vacuum to recover ectoparasites from the coat of animals.

vacuum pump (equipment) An alternative term for diffusion pump.

vagina (anatomy) The canal which leads from the external female genitalia to the cervix.

vaginal (parasitology) Pertaining to the vagina.

vaginal aspiration (parasitology) Use of a suction apparatus to collect a sample of vaginal fluid for culture, cytological, immunological or parasitological examination.

vaginal biopsy (parasitology) Collection of a sample of mucosa by a pinch biopsy instrument for histopathogical examination.

vaginitis (medical) Inflammation of the vagina. One of the commonest causes is infection with *Trichomonas vaginalis.*

vagus nerve (anatomy) In vertebrates, 10th cranial nerve, which forms the major nerve of the parasympathetic nervous system, supplying motor nerve fibres (*USA*: nerve fibers) to the stomach, kidneys, heart, liver, lungs and other organs.

Vairmorpha invictae (insect parasitology) A protozoal species of microsporidia that are parasites to the ants *Solenopsis geminata*, *Solenopsis invicta*, *Solenopsis quinquecuspis*, *Solenopsis richteri*, and *Solenopsis saevissima*-complex.

Val. (biochemistry) An abbreviation of 'Val'ine.

valence (chemistry) Positive number that characterizes the combining power of an atom of a given element to the number of hydrogen atoms or their equivalent (in a chemical reaction). For an ion, the valence equals the charge on the ion. Also termed: valency.

valence band (1, 2 chemistry) 1 Highest energy level in an insulator or semiconductor that can be filled with electrons. 2 Region of electronic energy level that binds atoms of a crystal together.

valence bond (chemistry) Chemical bond formed by the interaction of valence electrons between two or more atoms.

valence electron (chemistry) Electron in an outer shell of an atom which participates in bonding to other atoms to form molecules.

valency (chemistry) *See* valence.

validation (biology) The process of establishing that a theory or test is valid.

validity (biology) The extent to which a test or experiment genuinely measures what it purports to measure.

valine (biochemistry) $C_5H_{11}NO_2$ One of the essential amino acids required for normal growth in animals. Also termed: 2-aminoisovaleric acid; 2-amino-3-methylbutyric acid.

Valone (control measures) An insecticide and rodenticide. A indandione compound with actions similar to that of warfarin.

valor (control measures) A rodenticide no longer marketed because of toxicity in horses. Also termed N-3-pyridyl methyl N^1-*p*-nitrophenyl urea.

valve (anatomy) Flap of tissue that controls movement of fluid through a tube, duct or aperture in one direction, *e.g.* as between the chambers of the heart or in the veins.

van der Waals forces (physics) Very weak forces acting between the nucleus of one atom and the electrons of another atom (*i.e.* between dipoles and induced dipoles). The attractive forces arise from slight distortions induced in the electron clouds surrounding each nucleus as two atoms are brought close together.

van Gieson stain (histology) A histological staining technique used to make connective tissue visible under the microscope.

van't Hoff's law (physics) Osmotic pressure of a solution is equal to the pressure that would be exerted by the solute if it were in the gaseous phase and occupying the same volume as the solution at the same temperature. It was named after the Dutch chemist Jacobus van't Hoff (1852–1911).

vanadium (V) (chemistry) Silvery-grey metallic element in Group VA of the periodic table (a transition element), used to make special steels. Vanadium(V) oxide, V_2O_5, is used as an industrial catalyst and in ceramics. At. no. 23; r.a.m. 50.9414.

vaporize (chemistry) To convert into vapour (*USA*: vapor) or to be transformed into vapour (*USA*: vapor).

vapour (chemistry) A gas when its temperature is below the critical temperature; a vapour can thus be condensed to a liquid by pressure alone. (*USA*: vapor).

vapour density (physics) Density of a gas relative to a reference gas, such as hydrogen, equal to the mass of a volume of gas divided by the mass of an equal volume of hydrogen at the same temperature and pressure. It is also equal to half the relative molecular mass. (*USA*: vapor density).

vapour pressure (physics) Pressure under which a liquid and its vapour (*USA*: vapor) coexist at equilibrium. Also termed: saturation vapour pressure. (*USA*: vapor pressure).

var. (1 statistics; 2 general terminology) 1 An abbreviation of 'var'iable. 2 An abbreviation of 'var'ious.

variability (statistics) The spread of scores in a sample or the extent to which they differ from the mean.

variable (1 mathematics; 2 computing) 1 Something that is not constant. 2 Block of data that is stored at different locations during the operation of a program.

variable length record (computing) A computer record whose length is only as long as it needs to be. Many data storage programmes (*USA*: programs) have fixed length records which can waste much disk space especially if most of the records are likely to be short.

variable number tandem repeats (VNTRs) (molecular biology) The repeated sequences of DNA that vary from one individual to another and are the basis for genetic fingerprinting.

variable region (immunology) *See* v region.

variance (statistics) For a set of numbers, the mean of the squares of the deviations of each number from the arithmetic mean of the set. Its square root is the standard deviation.

variance ratio distribution (statistics) *See* F distribution.

variant (biology) An organism or tissue that is different from the majority of the population but is still sufficiently similar to the common mode to be considered to be one of them.

variation (biology) Differences between members of the same species, which may be either continuous (having a normal distribution about a species mean, *e.g.* height and weight) or discontinuous (having different specific characteristics with no intermediate forms, *e.g.* blood types).

variegated tick (parasitology) *See Amblyomma variegatum.*

variety (taxonomy) Any sub-division of a species, *e.g.* breed, race, strain, etc.

Varroa (insect parasitology) A species of mites in the Mesostigmata, that are parasites of the bees Apinae and *Bombus.*

vas deferens (anatomy) One of a pair of ducts that carry sperm from the testes. In mammals it joins the urethra and passes along the penis. Plural: vasa deferentia. Also termed: sperm duct.

vas efferens (parasitological anatomy) A canal extending from a testis to the vas deferens.

vas(o)- (terminology) A word element [Latin]. Prefix denoting vessel, duct.

vasc. (histology) An abbreviation of 'vasc'ular.

vascular system (anatomy) A system of interlinked fluid-filled vessels, *e.g.* the blood vascular system.

vasculitis (medical) Inflammation of a vessel.

vasoconstriction (physiology) Reduction in diameter of a blood vessel due to contraction of the smooth muscles in its walls. It may be induced by the secretion of adrenaline (*USA*: epinephrine) in response to pain, fear, decreased blood pressure, low external temperature, etc. or result from stimulation by vasoconstrictor nerve fibres (*USA*: nerve fibers).

vasodilation (physiology) Increase in diameter of small blood vessels due to relaxation of the smooth muscles in their walls. It is induced in response to exercise, high blood pressure, high external temperature, etc. or results from stimulation by vasodilator nerve fibres (*USA*: nerve fibers). Also termed: vasodilatation.

vasodilator (medical) An agent that causes dilatation of the blood vessels.

vasomotor (physiology) Pertaining to the motor control of the diameter of blood-vessels.

vasomotor nerve (physiology) Nerve of the autonomic nervous system that controls the variation in the diameter of blood vessels, *e.g.* causing them to become constricted or dilated.

vasopressin (biochemistry) Peptide hormone, secreted by the pituitary gland and hypothalamus, that stimulates water resorption in the kidney tubules and contraction of the smooth muscles in the walls of blood vessels. It is secreted in response to low blood pressure. A lack of vasopressin results in diabetes insipidus. Also termed: antidiuretic hormone (ADH).

vauqueline (chemistry) An alternative term for strychnine.

vcm (physics) An abbreviation of 'v'a'c'uu'm'.

Vd (chemistry) Chemical symbol for vanadium.

vd (physics) An abbreviation of 'v'apour 'd'ensity.

VDH (medical) An abbreviation of 'v'alvular 'd'isease of 'h'eart.

VDT (computing) An abbreviation of 'V'isual 'D'isplay 'T'erminal.

VDU (computing) An abbreviation of 'V'isual 'D'isplay 'U'nit.

vector (1 molecular biology; 2 parasitology; 3 mathematics) 1 A plasmid or other self-replicating DNA molecule that transfers DNA between cells in nature or in recombinant DNA technology. In the latter case it may be called a cloning vector or cloning vehicle. 2 A carrier, especially the animal, frequently an arthropod, which transfers an infective agent from one host to another, *e.g.* the tsetse fly, which carries trypanosomes from animals to humans. 3 A quantity having both magnitude and direction, which can be used to characterise forces.

vector transmission (parasitology) *See* vector.

vegetable (biology) 1 Pertaining to or derived from plants. 2 Any plant or species of plant, especially one cultivated as a source of food.

vegetal (biology) 1 Pertaining to plants or a plant. 2 Vegetative.

vegetation (biology) Plant growth.

vegetative propagation (plant biology) A form of asexual reproduction in plants in which vegetative organs are able to produce new individuals.

vein (anatomy) Blood vessel that, with the exception of the pulmonary vein, carries deoxygenated blood away from cells and tissues.

velogenic (parasitology) 1 A host-parasite relationship in which the parasite dominates and the host frequently dies. 2 Highly virulent. Compare lentigenic.

velvet disease (fish parasitology) *See Oodinium, Amyloodinium.*

velvet mite (parasitology) *See Trombicula.*

ven- (terminology) A word element. [Latin] Prefix denoting vein.

vena (anatomy) Plural: venae [Latin] Vein.

vena cava (anatomy) Collective term for the precaval (anterior vena cava) and postcaval (posterior vena cava) vein. The precaval vein is paired, and carries oxygenated blood away from the head and forelimbs; the postcaval vein is single and carries deoxygenated blood away from most of the body and hind limbs (or legs) to the heart.

vene- (terminology) A word element. [Latin] Prefix denoting vein.

venectomy (medical) Phlebectomy.

venemous (toxicology) Venomous.

venereal disease (medical) *See* sexually transmitted disease.

venesection (medical) The cutting of a vein to draw off blood or, more usually, to insert a cannula for intravenous therapy.

veni- (terminology) A word element. [Latin] Prefix denoting vein.

veno- (terminology) A word element. [Latin] Prefix denoting vein.

venom (toxicology) Poison, especially a toxic substance normally secreted by an insect, snake or other animal.

venom kinin (toxicology) A peptide found in the venom of insects.

venomous (biology) Secreting poison; poisonous.

ventr(i)- (terminology) A word element. [Latin] Prefix denoting belly, front (anterior) aspect of the body, ventral aspect.

ventr(o)- (terminology) A word element. [Latin] Prefix denoting belly, front (anterior) aspect of the body, ventral aspect.

ventral (general terminology) Describing something that is on or near the surface of an organism and, in a tetrapod, directed downwards (on a human being it is directed forwards).

ventral midline dermatitis (medical) A type of dermatitis in which small ulcers with haemorrhagic (*USA*: hemorrhagic) crusts and hair loss, is located on the abdomen, particularly around the umbilicus, of horses; caused by biting flies and gnats.

ventricle (1, 2 anatomy) 1 In mammals, thick-walled muscular lower chamber of the heart. Contraction of the right ventricle pumps deoxygenated blood into the pulmonary artery, and contraction of the left ventricle forces oxygenated blood into the aorta. 2 In vertebrates, one of the fluid-filled interconnected cavities within the brain.

ventricular (anatomy) Pertaining to a ventricle.

Venturi tube (physics) Cylindrical pipe with a constriction at its centre (*USA*: center). When a fluid flows through the tube, its rate of flow increases and fluid pressure drops in the constriction. The rate can be calculated from the difference in pressure between the ends of the tube and at the constriction. It was named after the Italian physicist G. Venturi (1746–1822).

venule (anatomy) Small vein located close to capillary blood vessels, where it collects and conveys deoxygenated blood from the capillary network to a vein.

verdigris green (chemistry) Basic copper(II) carbonate, $CUC0_3.CU(OH)_2$, formed by corrosion of metallic copper or its alloys. The term is also used for the similar basic copper(II) acetate used as a pigment, fungicide and mordant in dyeing.

vermicide (control measures) A substance fatal to worms or intestinal animal parasites.

vermicular (biology) Wormlike in shape or appearance.

vermiculation (biology) Peristaltic motion; peristalsis.

vermiculous (parasitology) 1 Wormlike. 2 Infested with worms.

vermiform (biology) Worm-shaped.

vermiform appendix (anatomy) An alternative term for the appendix.

vermifugal (parasitology) Expelling worms or intestinal animal parasites.

vermifuge (pharmacology) Any agent that expels the worms or intestinal animal parasites; an anthelmintic.

vermin (biology) Any vertebrate or invertebrate animals of an objectional kind.

vermination (parasitology) Infestation with worms or vermin.

verminous (parasitology) Pertaining to, due to, or abounding in worms or in vermin.

verminous bronchitis (parasitology) *See* lungworm.

verminous encephalitis (parasitology) A condition caused by migration of *Strongylus vulgaris* or larvae, resulting in paralysis due to destruction of nervous tissue.

See also brain trauma, neurofilariasis.

verminous mesenteric arteritis (parasitology) A condition caused by migrating *Strongylus vulgaris* larvae, resulting in a defective blood supply to the intestine which causes intermittent colic, and sometimes terminal infarction.

See also thromboembolic colic.

verminous pneumonia (parasitology) *See* lungworm.

Vermipsylla (veterinary parasitology) A genus of fleas in the order Siphonaptera. Species include *Vermipsylla alacurt*, *Vermipsylla dorcadia*, *Vermipsylla ioffi*, and *Vermipsylla perplexa* which may be found on sheep, goats, other ruminants and horses.

Vermipsylla alacurt (veterinary parasitology) A species of flea whose principal hosts are ruminants and horses.

Vermipsylla dorcadia (veterinary parasitology) A species of flea whose principal hosts are ruminants and horses.

Vermipsylla ioffi (veterinary parasitology) A species of flea whose principal hosts are ruminants and horses.

Vermipsylla perplexa (veterinary parasitology) A species of flea whose principal hosts are ruminants and horses.

vermis (biology) [Latin] A worm, or wormlike structure.

vertebr(o)- (terminology) A word element. [Latin] Prefix denoting vertebra, spine.

vertebra (anatomy) One of the hollow bones or pieces of cartilage that form the vertebral column.

vertebral column (anatomy) Flexible column of closely arranged vertebrae that form an axial skeleton running from the skull to the tail. It provides a protective channel for the spinal cord. The vertebral column becomes larger and stronger towards the posterior, which is the major weightbearing region. Also termed: spinal column; backbone.

Vertebrata (taxonomy) Major subphylum of Chordata that contains all animals with a vertebral column, *i.e.* mammals, birds, fishes, reptiles and amphibians. Vertebrates are characterized by a well-developed brain, complex nervous systems and a flexible endoskeleton of bone and cartilage. Also termed: Craniata.

vertebrate (taxonomy) Backboned animal; a member of the subphylum Vertebrata.

vertical scrolling (computing) The ability to move up or down through a page of script so that it can be accessed for editing.

vertical transmission (genetics) The usual type of genetic inheritance, from one generation to the next. It is contrasted with horizontal transmission.

very high frequency (VHF) (physics) A radio frequency between 30 and 300 kilohertz.

vesic(o)- (terminology) A word element. [Latin] Prefix denoting blister, bladder.

vesical (medical) Pertaining to the bladder.

vesicle (1 biology; 2 medical) 1 A small fluid-filled sac of variable origin, *e.g.* Golgi apparatus, pinocytotic vesicle. Also termed: vacuole; air sac; bladder. 2 Small blister on the skin.

vessel (anatomy) Tubular structure that transports fluid (*e.g.* blood, lymph).

vet (veterinary science) An abbreviation of 'vet'erinary surgeon or 'vet'erinarian.

veterinary (veterinary science) 1 Pertaining to domestic animals and their diseases. 2 Vernacular for veterinary surgeon or veterinarian.

veterinary dermatology (veterinary science) The study of the diseases of the skin of animals.

veterinary investigation centres (veterinary science) The system in the UK of regional veterinary laboratories dedicated to the study and diagnosis of the diseases of animals in the region.

veterinary medicine (veterinary science) The study of the diseases of animals including their diagnosis, prevention and treatment.

rotational (physics) Characterized by rotation.

veterinary pharmacology (veterinary science) The study of medicines used in the treatment of animals.

veterinary science (veterinary science) The study of the diseases and health maintenance of animals.

veterinary surgeon (veterinary science) A person trained and authorized to practice veterinary medicine and surgery; a doctor of veterinary medicine. Also termed veterinarian.

VGA (computing) An abbreviation of 'V'ideo 'G'raphics 'A'rray.

VHF (physics) An abbreviation of 'V'ery 'H'igh 'F'requency.

viability (biology) The state or quality of being viable.

viable (biology) Capable of living.

vial (equipment) A small bottle.

vibration receptor (physiology) The skin receptor that responds to vibration, thought to be the Pacinian corpuscle.

video graphics array (VGA) (computing) A type of computer graphics circuitry that drives a computer monitor with very high resolution, or the monitor itself.

villus (histology) Plural: villi [Latin]. A small protrusion from the surface of a membrane. One of many finger-like structures that line the inside of the small intestine. Villi increase the surface area for absorption. Each villus contains a central lacteal and a network of blood capillaries, which absorb the soluble products of digestion into the body.

VIP (physiology/biochemistry) An abbreviation of 'V'asoactive 'I'ntestinal 'P'olypeptide.

viroid (biology) A small disease-causing agent, a tight loop of RNA lacking any form of capsid (outer coat).

virological (virology) Pertaining to viruses.

virologist (virology) A microbiologist specializing in virology.

virology (virology) The study of viruses and viral diseases.

virulence (biology) The ability of a pathogen (a virus, bacterium or other micro-organism) to produce a disease. It depends partly on its capacity to invade the host cell and multiply, and partly on its ability to produce toxins. Virulence tends to be favoured by natural selection, as virulent strains are often the most successful at reproducing themselves; but beyond a certain point virulence is a disadvantage to a strain of pathogens, as they may kill their hosts too soon for their own good.

viscera (histology) The large internal organs of the body, *e.g.* lungs, liver and intestines.

visceral (anatomy) Relating to the viscera.

visceral larva migrans (human/veterinary parasitology) A condition due to prolonged migration of larvae of animal nematodes in human tissues other than skin, commonly caused by larvae of the roundworms *Toxocara canis* and *Toxocara cati*.

viscer(o)- (terminology) A word element. [Latin] Prefix denoting viscera.

viscid (parasitology) Glutinous or sticky.

viscidity (physics) The property of being viscid.

viscometer (physics) Instrument for measuring viscosity.

viscosity (physics) Property of a fluid (liquid or gases) that makes it resist flow, resulting in ditterent velocities of flow at different points in the fluid. Also termed: internal friction.

viscous (physics) Sticky or gummy; having a high degree of viscosity.

viscus (anatomy) Plural: viscera [Latin] Any large interior organ in any of the great body cavities, especially those in the abdomen.

visible light (physics) Light that can be seen by the human eye (the visible spectrum), as opposed to infra-red and ultraviolet radiation.

visible spectrum (physics) Range of wavelengths of visible electromagnetic radiation (light), between about 780 and 380 nm.

vision (biology) The faculty of seeing; sight.

visual (biology) Pertaining to vision.

visual display (computing) The television screen attached to a computer which allows data to be viewed and editing to be performed on screen. Includes visual display unit, VDU, and visual display terminal, VDT.

visual purple (biochemistry) An alternative term for rhodopsin.

vital (biology) Pertaining to life; necessary to life.

vital dye (microscopy) *See* vital stain.

vital index (statistics) The ratio of births to deaths within a given time in a population.

vital signs (medical) The signs of life, i.e. pulse, respiration and temperature.

vital stain (methods) A stain introduced into the living organism, and taken up selectively by various tissue or cellular elements. Often used to determine the live/dead cell ratio in a cell population.

vital statistic rate (statistics) Vital statistics presented as a proportion of a population, *e.g.* foetal (*USA*: fetal) deaths as a percentage of total births.

vitamins (biochemistry) A group of substances that are necessary for metabolism but only in minute quantities. There are two major groups, water-soluble (*e.g.* vitamins C, B) and fat-soluble (*e.g.* A, D, E, K), which are present in foodstuffs and must be taken as part of a balanced diet. Vitamins work as co-enzymes within the enzyme system, enabling the various essential chemical reactions within cells to take place, *e.g.* the production of energy, the synthesis of proteins (for tissue-building, hormones, etc). Excessive doses of vitamins often have a deleterious effect, such toxic overdoses being known by the general term, hypervitaminosis.

vitamin A (biochemistry) Retinol.

vitamin B_1 (biochemistry) Thiamine.

vitamin B_2 (biochemistry) Riboflavin.

vitamin C (biochemistry) Ascorbic acid.

vitamin D (biochemistry) Calciferol.

vitamin E (biochemistry) Tocopherol.

vitamin H (biochemistry) Biotin.

vitamin K_1 (biochemistry) Phytomenadione.

vitelline glands (parasitological anatomy) A term that may be used in relation to trematodes. The glands which provide substances for the development of the egg and the formation of the shell.

vitreous humour (anatomy) In the vertebrate eye, firm transparent gel-like substance that fills the space behind the lens, thus maintaining the shape of the eyeball.

See also aqueous humour.

vitriol (chemistry) An alternative term for sulphuric acid (*USA*: sulfuric acid).

Vitula edmandsii (insect parasitology) Arthropods in the Lepidoptera that is a parasite of the bees *Bombus* spp., and which may be considered the North American equivalent to *Aphomia*.

vivi- (terminology) A word element. [Latin] Prefix denoting alive, life.

viviparous (parasitological physiology) A term that may be used in relation to nematodes. Larviparous. A species which discharges larvae instead of eggs.

VLDL (biochemistry) An abbreviation of 'V'ery 'L'ow 'D'ensity 'L'ipoprotein.

VLDLP (chemistry) An abbreviation of 'V'ery-'L'ow-'D'ensity 'L'ipo'P'rotein. *See* lipoproteins.

VLN (physics) An abbreviation of 'V'ery 'L'ow 'N'itrogen.

VMA (biochemistry) An abbreviation of 'V'anillyl'M'andelic 'A'cid (3-methoxy-4-hydroxymandelic acid).

vocal cord (anatomy) One of a pair of membranous flaps in the larynx that are vibrated by air from the lungs to produce sounds.

Vogeloides (veterinary parasitology) A genus of nematodes in the family Pneumospiruridae that are parasites of carnivores and primates. Species include *Vogeloides massinoi* and *Vogeloides ramanujacharii* which may be found in the lungs of cats.

void (biology) To cast out as waste matter, especially the urine.

voiding (biology) A euphemism for urination, defecation.

vol. (terminology) An abbreviation of 'vol'ume.

volatile (1 chemistry; 2 computing) 1 Describing any substance that is readily changed to a vapour and hence lost through evaporation. Volatile liquids have low boiling points. 2 Describing stored information that is lost through a power cut.

volt (V) (measurement) SI unit of potential difference (p.d) or electromotive force (e.m.f.), which equals the p.d. between two points when one coulomb of electricity produces one joule of work in going from one point to the other. It was named after the Italian physicist Alessandro Volta (1745–1827).

voltage (physics) Value of a potential difference, or the potential difference itself.

voltage divider (physics) Resistor that can be tapped at a point along its length to give a particular fraction of the voltage across it. Also termed: potentiometer.

voltaic cell (physics) Any device that produces an electromotive force (e.m.f.) by the conversion of chemical energy to electrical energy, *e.g.* a battery or accumulator. Also termed: galvanic cell.

voltmeter (physics) Instrument for measuring voltage or potential difference.

Volucella inanis (insect parasitology) Diptera in the Syrphidae that simlarly to *Volucella pellucens* and *Volucella zonaria* are parasitic to the larva and pupa of the wasps *Paravespula* spp., and whose larva actively seeks out the host.

Volucella pellucens (insect parasitology) *See Volucella inanis.*

Volucella zonaria (insect parasitology) *See Volucella inanis.*

volume (physics) Amount of space occupied by a solid object, or the capacity of a hollow vessel.

volumetric analysis (chemistry) Method of chemical analysis that relies on the accurate measurement of the reacting volumes of substances in a solution (*e.g.* by carrying out a titration).

volumetric solution (chemistry) One that contains a specific quantity of solvent per stated unit of volume.

voluntary muscle (histology) Type of muscle, connected to the bones, that is under conscious control. It is responsible for most body movements. Also termed: striated muscle.

volunteer sample (statistics) Sample donated by interested parties; a biased sample because it does not represent all sections of the population. Also termed self-selection.

vomit (medical) 1 Matter expelled from the stomach via the mouth. 2 To eject stomach contents through the mouth.

vomition (medical) The act of vomiting.

vomitus (medical) 1 Vomiting. 2 Vomited material.

VOR (physiology) An abbreviation of 'V'estibulo-'O'cular 'R'eflex.

Vs (medical) An abbreviation of 'V'ene's'ection.

vulgar fraction (mathematics) Fraction in which both numerator and denominator are integers (whole numbers).

vulva (anatomy) In female mammals, the external opening of the vagina.

W (1 chemistry; 2 physics; 3 biochemistry) 1 The chemical symbol for the element tungsten (from its German name: 'Wolfram'). 2 Symbol for the unit watt. 3 A symbol for tryptophan.

w/o (literature terminology) An abbreviation of 'w'ith'o'ut.

w/v (measurement) Abbreviation of 'w'eight (of solute) per 'v'olume (of solvent).

Wagner-nelson method (chemistry) Technique to characterise the rate and extent of absorption of a substance.

wahi (parasitology) Dermatitis caused by *Onchocerca gutturosa.*

Walchia americana (veterinary parasitology) One of the harvest mites that cause trombiculidiasis in domestic animals.

warble fly (parasitology) *See Hypoderma*; warbles.

warble myiasis (parasitology) *See Hypoderma*; warbles.

warbles (veterinary parasitology) The disease caused by *Hypoderma,* which may result in damage to the hides where the larvae emerge, some cases of choke caused by perioesophagitis (*USA*: periesophagitis), posterior paresis or paralpsis in a small percentage of infested cattle due to a reaction to dead *Hypoderma bovis* in the spinal canal, and some deaths due to invasion of the brain or to anaphylaxis.

warfarin (pharmacology/control measures) Organic compound used (as its sodium derivative) as an anticoagulant drug and as a pesticide for killing rats and mice.

Waring blender syndrome (parasitology) The shearing of erythrocytes by obstructions of the vascular bed, such as heartworms, resulting in the formation of schistocytes.

warm haemagglutinin (immunology) A haemagglutinin that acts only at temperatures near 98.6°F(37°C). (*USA*: warm hemagglutinin).

warm-blooded (biology) An alternative term for homoiothermic.

washed red cells (haematology) Blood component in which platelet and leucocyte (*USA*: leukocyte) antigens have been removed.

washing soda (chemistry) An alternative term for hydrated sodium carbonate.

washout (methods) To disperse or empty by flooding with water or other solvent.

washout period (pharmacology) In drug trials, the period allowed for all of the administered drug to be eliminated from the body.

wasp (biology) Stinging insect of the order Hymenoptera. There is local irritation at the site of the sting. Wasp stings contain histamine, serotonin and 'wasp kinin' plus hyaluronidase and phospholipase. Unlike bees, wasps may sting several times.

wasting (medical) Used in a general sense to indicate serious loss of body weight, or locally to indicate atrophy.

wasting disease (medical) Any disease marked especially by progressive emaciation and weakness.

WAT (physics) An abbreviation of 'W'eight, 'A'ltitude and 'T'emperature.

water (chemistry) H_2O Colourless (*USA*: colorless) liquid, one of the oxides of hydrogen (the other is hydrogen peroxide, H_2O_2) and the most common substance on Earth. It can be made by burning hydrogen or fuels containing it in air or oxygen, or by the action of an acid on an alkali or alcohol. It is a good (polar) solvent, particularly for ionic compounds, with which it may form solid hydrates. It can be decomposed by the action of certain reactive metals (*e.g.* the alkali metals) or by electrolysis. Water is essential for life and forms the major part of most body fluids (*e.g.* blood, lymph). It freezes at 0°C and boils at 100°C (at normal atmospheric pressure), and has its maximum density at 3.98°C.

See also hardness of water.

water balance (biochemistry) Fluid balance.

water flea (parasitology) *See Cyclops*; *Diaptomus gracilis.*

water of crystallization (chemistry) Definite amount of water retained by a compound (usually a salt) when crystallized from solution. The chemical formula of the resulting hydrate shows the number of molecules of water of crystallization associated with each molecule of hydrate; *e.g.* $Na_2SO_4.7H_2O$. The water can usually be removed by heating, and the resulting compound is termed anhydrous. Also termed: water of hydration.

water of hydration (chemistry) The water present in hydrated compounds. These compounds when crystallized from solution in water retain a definite amount of water, *e.g.* copper (II) sulphate (*USA*: copper (II) sulfate), $CuSO_4.5H_2O$.

waterborne infection (parasitology) Infection by microorganisms transmitted in water.

waterbrash (medical) Regurgitation of a watery secretion containing acid fluids from the stomach into the oesophagus (*USA*: esophagus) and mouth.

water-in-oil emulsion adjuvant (immunology) Adjuvant in which the antigen, dissolved or suspended in water, is enclosed in tiny droplets within a continuous phase of mineral oil. The antigen solution constitutes the dispersed phase, stabilised by an emulsifying agent such as mannitol mono-oleate.

waters (anatomy) Popular name for amniotic fluid. Also termed waterbag.

watery faeces (parasitology) Diarrhoeic faeces (*USA*: diarrheic feces) with a high water content.

watt (W) (measurement) SI unit of power, equal to 1 joule per second (j s^{-1}). 745.70 watts = 1 horsepower. Wattage is the power of an electrical circuit determined by multiplying the voltage by the amperage. It was named after the British engineer james Watt (1736–1819).

watt-hour (measurement) Measure of electric power consumption. Also termed: unit.

wattmeter (physics) Instrument for measuring power consumption in an electric circuit. Power consumption is usually expressed in watt hours or units.

wave (physics) Regular (periodic) disturbance in a substance or in space; *e.g.* in an airborne sound wave, alternate regions of high and low pressure travel through the air, although the air itself does not move along. In an electromagnetic wave, such as light, electric and magnetic waves at right angles to each other and the direction of movement travel through a medium or through space.

wave function (mathematics) Equation that expresses time and space variation in amplitude for a wave system.

wave number (physics) Reciprocal of the wavelength of an electromagnetic wave. Also termed: reciprocal wavelength.

wavelength (λ) (physics) Distance between two successive points at which a wave has the same phase, *e.g.* visible light has a wavelength of between 400nm (violet) to 750nm (red).

wax (chemistry) Solid or semi-solid organic substance that is an ester of a fatty acid: produced by a plant or an animal (*e.g.* beeswax, tallow) or : a high molecular weight hydrocarbon (*e.g.* paraffin wax, from petroleum), also called mineral wax.

WBC (1, 2 haematology) 1 An abbreviation of 'W'hite 'B'lood 'C'ell. 2 An abbreviation of 'W'hite 'B'lood 'C'ount.

weak acid (chemistry) Acid that shows little ionization or dissociation in solution, *e.g.* carbonic acid, (ethanoic) acetic acid.

weak electrolyte (chemistry) An electrolyte which is only slightly ionised in moderately concentrated solutions.

weaning index (toxicology) An expression of survival of offspring to weaning in reproductive toxicity tests. The number of offspring surviving to weaning (*i.e.* 21 days in the rat) as a percentage of those alive at four days.

weber (Wb) (measurement) SI unit of magnetic flux, named after the German physicist Wilhelm Weber (1804–91).

weed (plant biology) A plant growing out of place.

Wehrdikmansia (parasitology) *See Onchocerca.*

weight (physics) The force of gravity (9.8 m s^{-2}) acting on an object at the Earth's surface; *i.e.* weight = mass X acceleration of free fall (acceleration due to gravity). It is measured in newtons, pounds-force or dynes.

weights and measures *See* SI units.

Weil-Felix reaction (immunology) An agglutination test used in the diagnosis of rickettsial infections (typhus etc.) which depends upon a carbohydrate cross-reacting antigen shared by *Rickettsiae* and certain strains of *Proteus.* The agglutination pattern of patients with rickettsial disease against O-agglutinable strains of *Proteus* OX19, OX2 and OXK is diagnostic of the various rickettsial diseases. Named after Viennese physician E. Weil (1880–1916) and London bacteriologist A. Felix (1887–1956).

Weil's myelin stain (histology) A histological staining technique used to make myelin visible under the microscope.

wen (histology) A sebaceous cyst.

Wenyonella (parasitology) A genus of coccidia of the family Eimeriidae.

Wenyonella anatis (avian parasitology) A species of coccidia of the genus *Wenyonella* which may be found in domestic ducks.

Wenyonella gagari (avian parasitology) A species of coccidia of the genus *Wenyonella* which may be found in domestic ducks.

Wenyonella philiplevinei (avian parasitology) A species of coccidia of the genus *Wenyonella* which may be found in domestic ducks and causes severe inflammation of the mucosa of the ileum and rectum.

Western blotting (biochemistry) A technique for the analysis and identification of proteins in complex mixtures. The proteins are separated by polyacrylamide gel electrophoresis and the separated bands then transferred by 'blotting' onto a polymer membrane by electrophoresis at 90° to the gel surface. The proteins bound on the polymer membrane can be then be identified by reaction with specific reagents such as radioactive or fluorescent labelled antibodies. The method is very specific and can be used to confirm the results of the ELISA test. It is named by analogy with Southern blotting.

See also Northern blotting; Southern blotting.

western chicken flea (avian parasitology) *See Ceratophyllus niger.*

western hen flea (avian parasitology) *See Ceratophyllus niger.*

western transfer (methods) Western blotting.

wet preparations (parasitology) A method of preparing specimens for examination in which they are kept in their liquid state or suspended in a liquid, rather than being dried and then examined. Used for example in the diagnosis of trichomoniasis.

whale lice (parasitology) *See Isocyamus delpini.*

wheal (medical) A localized area of oedema (*USA*: edema) on the body surface, often attended with severe itching and usually evanescent. It is the typical lesion of urticaria.

wheal-flare (medical) *See* wheal-flare reaction.

wheal-flare reaction (medical) A cutaneous sensitivity reaction to skin injury or administration of antigen, due to histamine production and marked by oedematous (*USA*: edematous) elevation and erythematous flare.

wheat pollard itch (parasitology) Dermatitis caused by the acarid mite *Suidasia nesbitti.*

wheat weevil disease (medical) An immediate immune complex-mediated hypersensitivity pneumonitis of humans caused by inhalation of flour infested with *Sitophilus granarius.*

Wheatstone bridge (physics) Electric circuit for measuring the resistance of a resistor (by comparing it with three other resistors of known values). It was named after the British physicist Charles Wheatstone (1802–75).

wheeze (medical) A whistling respiratory sound.

wheezing (medical) Breathing with a rasp or whistling sound. It results from constriction or obstruction of the throat, pharynx, trachea or bronchi.

whewellite (biochemistry) Hydrated calcium oxalate $CaC_2O_4.H_2O$. It occurs uncommonly in the mineral world but is abundant in human calculi.

whipworm (parasitology) A popular name for *Trichuris trichiura.*

whirling disease (fish parasitology) Important disease of juvenile rainbow trout. Caused by the myxosporean *Myxobolus cerebralis* which parasitizes the cartilage of the head.

white ant (parasitology) *See* Isoptera.

white arsenic (chemistry) An alternative term for arsenic(III) oxide (arsenious oxide). *See* arsenic.

white blood cell *See* leucocyte.

white blood corpuscle (haematology) *See* leucocyte.

white grub (parasitology) *Posthodiplostomum minimum.*

white lead (chemistry) Basic lead carbonate.

white light (physics) Light that is composed of a mixture of wavelengths in the visable spectrum.

white matter (histology) Those parts of the central nervous system containing mainly myelinated fibres (*USA*: myelinated fibers), which appear white; *e.g.* the myelinated fibres of the cortex which run below the grey matter.

white petrolatum (chemistry) Vaseline.

white phosphorus (control measures) Pure phosphorus, used at one time as a rodenticide.

white spot disease (fish parasitology) A disease of freshwater fish caused by the protozoan *Ichthyophthirius multifiliis* and manifested by white pustules on the skin and gills and in severe cases sloughing of extensive affected areas. Common in aquariums but occurs also in cultured fish in large groups in hatcheries and on fish farms.

white spot liver (parasitology) *See Ascaris lumbricoides.*

white vitriol (chemistry) An alternative term for zinc sulphate (*USA*: zinc sulfate).

WHO (governing body) An abbreviation of 'W'orld 'H'ealth 'O'rganisation.

whole blood (haematology) Blood which has not been broken down into its various components. It is used in transfusion therapy in neonatal total blood exchange and in cases of acute massive blood loss. Whole blood is usually collected in citrate-phosphate-dextrose (CPD) anticoagulant preservative solution. Whole blood is suitable for transfusion up to a 21 day storage period.

wide-spectrum (pharmacology/control measures) Of antibiotics, etc., effective against a wide range of micro-organisms. The term 'broad-spectrum' is also used.

Wilcoxon rank-sum test (statistics) A distribution-free, non-parametric test for two independent groups. The data from both groups are combined and ranked with the lowest value assigned a rank of 1. The ranks assigned to each group are then summed, with the test statistic T' being the sum of the ranks for the smaller group. For small groups ($n < 10$), the critical value for T' can be obtained from special tables. If $n > 10$ for both groups, the distribution approximates normal, and the obtained value for T' can be assessed using the normal curve table.

See also non-parametric.

Wilcoxon signed ranks test (statistics) *See* Wilcoxon rank-sum test.

wild card (computing) Characters (usually * or ?) used to replace one or a number of characters in a search string.

wild strain (biology) A strain of a particular type of micro-organism that has not become laboratory adapted.

wild type (genetics) The allele that is deemed to be the one that occurs in the species in its original state, as opposed to subsequent mutations, either spontaneous or induced. The wild-type allele may be either dominant or recessive in relation to a mutant allele. Its symbol is +.

Willis technique (parasitology methods) A flotation method, utilizing saturated common salt solution to float helminth eggs, and hence separate eggs from faecal (*USA*: fecal) debris. The eggs can then be counted in a faecal (*USA*: fecal) sample. Suitable for most nematodes but not for cestodes or trematodes.

winchester (equipment) Bottle, commonly used for liquid chemicals, with a capacity of about 2.25 litres. It was named after the city of Winchester, Hampshire, United Kingdom.

window (computing) Portion of a computer screen which is an entity in its own right. The size of the window can often be changed. Each application running on the computer has one or more windows in which to communicate to the user.

windpipe (anatomy) The trachea.

wing (avian anatomy) A modified limb suitable for generating aerodynamic lift. Wing membranes or patagia are stretched between bony elements. In birds the wing surface is increased by large flight feathers (remiges) borne on the hand (primaries) or ulna (secondaries). In bats the patagia are more extensive than in birds through enlargement of the bones of the hand.

wing louse (avian parasitology) *See Lipeurus caponis.*

winter tick (parasitology) *See Dermacentor albipictus.*

wire worm (parasitology) *See Haemonchus placei.*

Wistar rat (parasitology) A white laboratory rat.

within-group variance (statistics) The amount of variance caused by differences within a group.

Wohlfahrtia (parasitology) A genus of flesh flies in the family Sarcophagidae. Species that deposit their larvae anywhere on the bodies of animals and cause tissue loss and some disfigurement includes *Wohlfahrtia magnifica*, *Wohlfahrtia meigini*, *Wohlfahrtia nuba*, and *Wohlfahrtia vigil.*

wolfram (chemistry) Old name for tungsten.

womb (anatomy) Uterus.

wood alcohol (chemistry) *See* methyl alcohol.

wood spirit (chemistry) An alternative term for methanol.

wood sugar (chemistry) An alternative term for xylose.

wood tick (parasitology) *See Dermacentor andersoni.*

Wood's filter (equipment) *See* Wood's lamp.

Wood's lamp (equipment) Ultraviolet radiation from a mercury vapor source, transmitted through a nickel oxide filter (Wood's filter), which holds back all but a few violet rays and passes ultraviolet wavelengths of about 367 nm.

wool fat (chemistry) An alternative term for lanolin.

wool maggot (parasitology) *See* cutaneous myiasis.

wool rubbing (parasitology) An occurrence when a sheep rubs its fleece against a hard object, which is frequently an indication of itching caused by external parasites.

wool wax (chemistry) *See* lanolin.

word processor (computing) Microcomputer that is programmed to help in the preparation of text, for data transmission or printing.

work (W) (physics) The measurement of a force multiplied by the distance moved by the point of application of the force in the direction of the force. It is measured in joules.

working (mathematics) Pertaining to something assumed or approximated, *e.g.* 'working hypothesis'.

work-up (parasitology) The procedures done to arrive at a diagnosis, including history taking, laboratory tests, and so on.

World Health Organisation (WHO) (governing body) The health agency of the United Nations, WHO is based in Geneva, Switzerland. Set up in 1948 it co-ordinates international health activities, and aims to improve health, particularly of developing countries, through education and information and practical assistance with mass vaccination programmes, public health schemes and medical facilities.

world scourge (parasitology) *See* schistosomiasis.

world wide web (computing) A group of Internet servers using protocols such as HTTP which provides ready access to documents which contain text, sound, video.

worm count (parasitology) A numerical computation or indication of the total number of worms. A total worm count requires a freshly slaughtered cadaver, collection of intestinal or other fluid in an aliquot sample, counting actual worms and, by multiplication, measuring the total worm burden. In the case of lungs it is necessary to digest the tissue.

worm egg count (parasitology) *See* egg count.

worm nodule disease (parasitology) *See* oesophagostomiasis.

worm resistance (pharmacology) A significant increase in the ability of worms to tolerate doses of individual drugs which have previously been fatal to the worms.

worms (parasitology) An imprecise term applied to elongated invertebrates with no appendages. Those that infest human beings can be divided into three groups: roundworms or nematodes, tapeworms or cestodes and flukes or trematodes.

wound (medical) A bodily injury caused by physical means, with disruption of the normal continuity of structures.

wound drain (medical) Any device by which a channel or open area may be established for the exit of material from a wound or cavity.

wound healing (medical) The restoration of integrity to injured tissues by replacement of dead tissue with viable tissue.

wounds (medical) Disruption of the tissues by external agents of a mechanical nature which may be classified as incised, punctured, contused, lacerated, perforated and penetrated.

WP (computing) An abbreviation of 'W'ord 'P'rocessing.

wraparound (computing) A feature of most word-processing programmes (*USA*: programs) which automatically places a word that will not fit on a line or page onto the next line or page.

wrinkles (parasitology) Small folds of skin in sheep, especially merino, that are susceptible to staining and wetness and therefore to blowfly strike.

write-protect (computing) A physical method of preventing anyone from over-writing information on computer disk, or other computer media, which has valuable data or systems on it.

wt (measurement) An abbreviation of 'w'eigh't'.

Wuchereria (parasitology) *See Brugia.*

Wyominia tetoni (veterinary parasitology) A cestode of the family Thysanosomatidae that may be found in bighorn sheep.

X

X (statistics) The arithmetic mean.

x coordinate (mathematics) A value on the horizontal axis of a graph.

xanth(o)- (terminology) A word element. [German] Prefix denoting *yellow*.

xanthene (chemistry) $CH_2(C_6H_4)_2O$ Yellow crystalline organic compound, used as a fungicide and in making dyes. Also termed: tricyclicdibenzopyran.

xanthic (chemistry) 1 Yellow. 2 Pertaining to xanthine.

xanthine (chemistry) $C_5H_4N_2O_2$ Organic compound that occurs in potatoes, coffee beans, blood and urine, used industrially as a chemical intermediate. Also termed: 3 7-dihydro-1H-purine-2,6-dione; 2,6-dihydroxypurine.

xanthone (control measures/chemistry) $CO(C_6H_4)_2O$ Plant pigment that occurs in gentian and other flowers, used commercially as an insecticide and dye intermediate. Also termed: 9H-xanthen-9-one.

xanthophyll (chemistry) $C_{40}H_{56}O_2$ Yellow to orange pigment present in the normal chlorophyll mixture of green plants. Also termed: lutein.

x-axis (mathematics) The horizontal axis on a graph.

X chromosome (genetics) One of the sex chromosomes. In humans the female has two X chromosomes and the male an X and a Y. The human X chromosome is the seventh largest in the human karyotype, nearly three times the size of the Y chromosome. It carries a large number of genes, which are said to be X-linked. *See* X linkage.

Xe (chemistry) The chemical symbol for the element xenon.

xenobiotic (toxicology) A general term used to describe any chemical interacting with an organism that does not occur in the normal metabolic pathways of that organism.

xenodiagnosis (parasitology) 1 Procedure involving the feeding of laboratory-reared triatomid bugs on patients suspected of having Chagas' disease; after several weeks the faeces (*USA*: feces) of the bugs are checked for intermediate stages of *Trypanosoma cruzi*. 2 Diagnosis of trichinosis by means of feeding laboratory-bred rats or mice on meat suspected of being infected with *Trichinella*, and then examining the animals for the parasite.

xenogenous (biology) Caused by a foreign body, or originating outside the organism.

xenoimmunization (immunology) Development of antibodies in response to antigens derived from an individual of a different species.

xenoma (fish parasitology) A massive hypertrophic lesion caused in fish by the microsporidian protozoan parasites *Nosema* and *Pleistophora* spp.

xenon (Xe) (chemistry) Unreactive gaseous element in Group 0 of the periodic table (the rare gases) which occurs as traces in the atmosphere, from which it is extracted. It is used in electronic flash tubes and high-intensity arc lamps. The isotope xenon-135 is a uranium fission product and a troublesome 'poison' in nuclear reactors (because it captures slow neutrons). At. no. 54; r.a.m. 131.30.

xenoparasite (parasitology) An organism not usually parasitic on a particular species, but becomes so because of a weakened condition of the host.

Xenopsylla (parasitology) A genus of fleas, including more than 30 species, many of which transmit disease-producing microorganisms.

Xenopsylla cheopis (veterinary parasitology) The rat flea, which transmits *Pasteurella pestis*, the causative organism of plague and *Richettsia typhi*, the causative organism of murine typhus.

xenoreactivity (immunology) The reaction of lymphocytes or antibodies with xenoantigens.

xer(o)- (terminology) A word element. [German] Prefix denoting *dry, dryness*.

xiph(o)- (terminology) A word element. [German] Prefix denoting *xiphoid process*.

xiphidiocercaria (parasitology) Cercaria with a stylet in the anterior rim of its oral sucker.

Xiphinema (plant parasitology) A genus of nematodes in the order Dorylaimida and family Longidoridae. They are somewhat similar to *Longidorus* but the amphid ampertures are wide slits and the amphid pouch is a short funnel shape. *See Xiphinema index*.

Xiphinema index (plant parasitology) A species of nematodes in the genus *Xiphinema* whose hosts cover a narrow range including grape, fig, apple, rose, mulberry and woody perennials. Their distribution is closely related to that of its most important host, grapevine, and covers Argentina, Australia, Chile, France, Germany, Greece, Hungary, Iran, Iraq, Italy, North Africa, Portugal, South Africa, Spain, Turkey and the USA. They are migratory root ectoparasites, all stages feeding at root tips. Reproduction is by meiotic parthenogenesis and males are very rare. Attacked roots show necrosis, lack of lateral roots, terminal swelling or galling, cellular hypertrophy and multinucleate condition of cortical cells near the feeding sites. The nematode transmits grapevine fanleaf virus. The virus is intimately associated with the oesophageal (*USA*: esophageal) lining. It is acquired in five to fifteen minutes of feeding, and persists for up to nine months when the nematode is not feeding. The virus is lost at moult (*USA*: molt). Grapevine fanleaf virus causes reduced vigour (*USA*: vigor), lack of fruit set, and reduced yield. Distinctive leaf symptoms consisting of vein banding and misshapen leaves are typically present in the fall. Control measures include removal of virus-infected vines and initiation of a five-to-ten year rotation for roots to die followed by preplant fumigation. Also termed: dagger nematode.

X linkage (genetics) The situation in which a gene is located on the X chromosome. In males, with sex chromosomes XY, any allele on the X chromosome behaves as a dominant, because there is no corresponding allele to mask it on the other chromosome (the Y does not behave as homologous to the X). A male passes on to all his daughters any gene that he carries on the X chromosome, but his sons will not inherit it. A female who carries the same allele on one of her X chromosomes will not be affected, if the gene is recessive as most X-linked genes are. Such a female is known as a carrier, and passes on the gene to half of her offspring, sons and daughters alike: males who receive the gene will inherit the condition, females will be carriers. It is possible for a female to inherit an X-linked gene simultaneously from her father (he will have the trait in question) and from her carrier mother, but this is rare (how rare depends on the frequency of the X-linked allele in the population). Over 100 genes are known to be X-linked in humans, including colour blindness (*USA*: color blindness), Duchenne muscular dystrophy and haemophilia (*USA*: hemophilia). Occasionally an X-linked gene is dominant; its pattern of transmission is similar, in that a father cannot pass it on to his sons.

X-linked (genetics) Traits transmitted by genes on the X chromosome; sex-linked; the categories are X-linked dominant, X-linked recessive.

XO (genetics) Symbol for the karyotype in which there is only one sex chromosome, an X chromosome.

XPS (physics) An abbreviation of 'X'-ray 'P'hoto-electron 'S'pectroscopy.

X-ray (physics) An electromagnetic radioactive wave.

X-ray crystallography (biochemistry) The technique used for investigating the three-dimensional structure of large molecules such as proteins and nucleic acids. The wavelength of the X-rays used is similar to the inter-atomic distances in these molecules, and the position of atoms can be deduced from the patterns of diffraction seen when X-rays are passed through them.

X-ray diffraction (physics) The pattern of variable intensifies produced by diffraction of X-rays when passed through a diffraction grating consisting of spacings of about 10^{-8}cm, in particular that formed by the lattice of a crystal.

X-ray emulsion (radiography) Radiation-sensitive coating of an X-ray film consisting of a suspension of finely divided grains of silver halide in gelatin.

X-ray fluorescence (physics) Less penetrating, secondary X-rays emitted by a substance when subjected to primary X-rays or high-energy electrons. The secondary X-rays are characteristic of the bombarded substance.

X-rays (radiology) Electromagnetic radiation produced in a partial vacuum by the sudden arrest of high-energy bombarding electrons as they collide with the heavy atom nuclei of a target metal. The X-rays produced are thus characteristic of the target's atoms. X-rays have very short wavelengths (10^{-3} to 1 nm) and can penetrate solids to varying degrees; this characteristic has made them useful in medicine, dentistry and X-ray crystallography. Also termed: rontgen (roentgen) rays; X-radiation.

X-ray spectrum (physics) Line spectrum of the intensity of X-rays emitted when a solid target is bombarded with electrons. It consists of sharp superimposed lines, which are characteristic of the target atoms, on a continuous background.

X-ray tube (radiology) Vacuum tube designed to produce X-rays by using an electrostatic field which accelerates and directs electrons on to a target.

xylene (chemistry) $C_6H_4(CH_3)_2$ Aromatic liquid organic compound that exists in three isomeric forms (ortho-, meta- and para-xylene), obtained from coaltar and petroleum. They are used as solvents in polyester synthesis, in microscopy for preparation of specimens and as cleaning agents. Also termed: dimethylbenzene.

xylose (chemistry) $C_5H_{10}O_5$ Naturally occurring pentose sugar, found in the form of xylan or as glycosides in many plants (*e.g.* cherry and maple wood, straw, pecan shell, corn cobs and cotton-seed hulls). Also termed: wood sugar.

xylulose (chemistry) A pentose sugar occurring as D-xyulose and as L-xylulose, one of the few L-sugars found in nature; it is sometimes excreted in the urine.

xysma (parasitology) Material resembling bits of membrane in stools of diarrhoea (*USA*: diarrhea).

Y

Y (1 chemistry; 2 biochemistry) 1 The chemical symbol for yttrium. 2 The symbol for tyrosine.

Y chromosome (genetics) One of the sex chromosomes and the third smallest in the human karyotype. In humans the male has an X and a Y, while the female has two X chromosomes. It is the presence of the Y chromosome that causes maleness to develop. The sex-determining gene is on the short arm of the Y chromosome, and individuals who are XY but have lost the short arm of the Y develop as female, while the longer arm of Y chromosomes can apparently be lost with no effect on the phenotype.

y coordinate (mathematics) A value on the vertical axis of a graph.

Y1 adrenal cells (biology) A commonly used continuous cell line used in tissue culture systems.

Yangtze River fever (parasitology) *See* schistosomiasis.

yard (measurement) Unit of length equal to 3 feet; there are 1,760 yards in 1 mile. 1 yard = 0.9144 m.

Yates correction (statistics) A correction for small samples in the calculation of chi-square in a 2-by-2 table, in which 0.5 is deducted from each figure exceeding expectation, and 0.5 is added to each figure that is less than the expected value. Assuming fixed marginal totals, which is rarely appropriate, the effect is to bring the distribution of the calculated chi-square nearer to the continuous distribution from which the usual chi-square tables are derived.

y-autosome (methods) Chromosomal aberration created artificially by irradiation in the process of producing sterile flies by the sterile insect release method.

y-axis (mathematics) The axis perpendicular to and in the horizontal plane through the *x-axis* in any type of graph.

Yb (chemistry) The chemical symbol for ytterbium.

yeast artificial chromosome (YAC) (genetics) A system used for producing segments of DNA of more than 200,000 bases. Genetic engineering techniques have been used to construct DNA molecules that contain a yeast centromere and several yeast genes. Segments of DNA can be inserted into such molecules and, when introduced into yeast cells, they behave like chromosomes and divide in step with the cell. YACs have become important host vectors for cloning large regions of genomes.

yellow body louse (parasitology) *See Menacanthus stramineus.*

yellow dog tick (veterinary parasitology) *See Haemaphysalis leachi leachi.*

yellow fever (medical) An acute and severe disease endemic in tropical America and Africa. It is caused by a flavivirus of the togavirus family transmitted between humans by the bite of the mosquito *Aedes aegypti.*

yellow grub (fish parasitology) Metacercariae of digenetic flukes in skin and or musculature of finfish. *See Clinostomum marginatum.*

yellow grub disease (fish parasitology) Metacercariae of *Clinostomum marginatum* cause cyst development of the skin and viscera of freshwater fish.

yellow mealworm (parasitology) *See Tenebrio molitor.*

yolk (biology) Part of an ovum that stores the nutritive materials, or the yellow central portion of the egg of birds and reptiles.

yolk sac (biology) A tiny bag attached to the embryo that provides early nourishment before the placenta is formed. *See* yolk sac isolation.

yolk sac isolation (methods) A technique used for harvesting and increasing yields of chlamydiae. All known Chlamydiae grow in the yolk sac of the embryonated hen egg. However, these techniques are relatively slow and will not provide an aetiologic (*USA*: etiologic) diagnosis quickly enough to be clinically relevant.

yomesan (pharmacology) The trade name for niclosamide. A medication that is used to treat parasitic tapeworm infections. It works by irreversibly damaging the head of the worm so it can no longer attach itself to the gut wall. The worm is therefore separated from the gut wall and expelled in the faeces (*USA*: feces). Niclosamide does not kill the eggs or the larvae of the tapeworm. The damaged segments from tapeworm infections have the potential to release eggs, which are not affected by the medication. A laxative is often given in conjunction with niclosamide in these infections to flush out any eggs or larvae which may cause problems at a later date.

ytterbium (Yb) (chemistry) Silvery-white metallic element in Group IIIA of the periodic table (one of the lanthanides), with no commercial uses. At. no. 70; r.a.m. 173.04.

yttrium (Y) (chemistry) Grey metallic element in Group IIIA of the periodic table). It is used in alloys for superconductors and magnets, and yttrium(VI) oxide, Y_2O_6, is employed in lasers and phosphors. At. no. 39; r.a.m. 88.9059.

Z

Zalaphotrema (veterinary parasitology) A genus of digenetic trematodes which includes *Zalaphotrema hepaticum* found in the liver of the sea lion.

z-axis (mathematics) The vertical axis in any three-dimensional co-ordinate system.

Z DNA (genetics) DNA in which the double helix is wound left-handed rather than right-handed as normal. The sugars do not fit together so well in this configuration, giving the molecule a jagged appearance. Z DNA has been studied in *Drosophila* chromosomes, and it has been suggested that it has a role in gene regulation.

Zener diode (physics) Semiconductor diode which at a certain negative voltage produces a sharp breakdown of current and hence may be used as a voltage control device. Also termed: avalanche diode; breakdown diode.

zeolite (chemistry) Hydrated aluminosilicate mineral, from which the water is easily removed, used for making molecular sieves and for ion exchange columns.

zero (measurement) The point on a thermometer scale from which the graduations begin. The zero of the Celsius (centigrade) scale is the ice point; on the Fahrenheit scale it is 32° below the ice point.

zero method (statistics) An alternative term for null method.

zero-order correlation (statistics) A correlation performed on the raw data without first removing the effects of any related variables. *Compare* partial correlation.

zero-order reactions (biochemistry) Generally pertaining to enzyme reactions where the rate of product formation is independent of substrate concentration.

Ziehl/Neelsen stain (methods) Microbiological staining technique for alcohol fast organisms, especially *Mycobacterium tuberculosis*, in smears and tissues. A modified method using mild acid to decolorize is used for staining *Nocardia asteroides* and *Chlamydia psittaci*.

See also acid-fast.

Ziemann's dots (parasitology) Pinkish small dots sometimes seen in *Plasmodium malariae* infected red cells in Romanowsky stained films.

zinc (Zn) (chemistry) Bluish-white metallic element in Group IIB of the periodic table (a transition element), used to give a corrosion-resistant coating to steel (galvanizing), to make dry batteries and in various alloys (*e.g.* brass, bronze). It is an essential trace element which is needed for enzyme systems in the body. Although a sufficient amount is normally present in the diet, deficiency results in stunted growth, skin disease and inadequacy of the immune system. At. no. 30; r.a.m. 65.38.

zinc cadmium sulphide (equipment) Used in the preparation of fluoroscopic screens; is fluorescent and emits yellow-green light when excited by X-rays. (*USA*: zinc cadmium sulfide).

zinc chloride (chemistry) $ZnCl_2$ White hygroscopic salt produced commercially by heating metallic zinc in dry chlorine gas. It is used to fireproof timber, in battery making, vulcanizing, galvanizing, oil refining, and as a fungicide and catalyst.

zinc oxide (chemistry) ZnO White crystalline solid (yellow when hot) which can be produced directly by heating zinc in air. It is used as a white pigment (Chinese white), in ceramics, cosmetics, pharmaceuticals and floor coverings, and in the manufacture of tyres. It dissolves in alkalis (*USA*: alkalies) to form zincates.

zinc phosphide (control measures) Used at one time as a rodenticide. When ingested, the poisonous gas phosphine is liberated and kills the animal without diagnostic signs or lesions.

zinc sulphate (chemistry) $ZnSO_4.7H_2O$ Colourless (*USA*: colorless) crystalline salt prepared by dissolving metallic zinc in dilute sulphuric acid (*USA*: sulfuric acid). It is used in the manufacture of rayon, glue, fertilizers, fungicides, wood preservatives, rubber, paint and varnishes. Also termed: white copperas; white vitriol; zinc vitriol. (*USA*: zinc sulfate).

zinc sulphate flotation test (methods) A method used to demonstrate nematode eggs and protozoan cysts, and larvae in faeces (*USA*: feces) and bronchial secretions. (*USA*: zinc sulfate flotation test).

zinc sulphide (chemistry) ZnS Occurs naturally as blende (an important zinc ore) and can be prepared as a white precipitate by adding ammonium sulphide (*USA*: ammonium sulfide) or hydrogen sulphide (*USA*: hydrogen sulfide) to a solution of a zinc salt. It is used as the pigmentary base for white zinc sulphide (*USA*: zinc sulfide) (lithopone), which contains up to 60% zinc sulphide and a balance of barium sulphate (*USA*: barium sulfate). It is also used in fungicides and phosphors. (*USA*: zinc sulfide).

zincate (chemistry) Compound formed by the reaction of metallic zinc or zinc oxide with an alkali; *e.g.* Na_2ZnO_2.

zipper worm (parasitology) *See Spirometr erinacei.*

zirconium (Zr) (chemistry) Silvery-grey metallic element in Group IVA of the periodic table (a transition element). It is used to clad uranium fuel rods in nuclear reactors. Naturally occurring crystalline zirconium(IV) oxide, ZrO_2, is the semi-precious gemstone zircon; the oxide is also used as an electrolyte in fuel cells. At. no. 40; r.a.m. 91.22.

Zn (chemistry) The chemical symbol for zinc.

ZN (microscopy) An abbreviation of 'Z'iehl/'N'eelsen stain.

zo(o)- (terminology) A word element. [German] Prefix denoting *animal.*

zoetic (biology) Pertaining to life.

zonifugal (epidemiology) Passing outward from a zone or region.

zonipetal (epidemiology) Passing toward a zone or region.

zoo- (terminology) A word element. [German] Prefix denoting *animal.*

zoogenous (veterinary parasitology) 1 Acquired from animals. 2 Viviparous.

zoogeny (biology) The development and evolution of animals.

zoogeography (epidemiology) Defining the location and numbers of animal populations, and their variability with time.

zoogony (parasitology) The production of living young from within the body.

zooid (parasitology) 1 Animal-like. 2 An animal-like object or form. 3 An individual in a united colony of animals.

zool. (biology) An abbreviation of 'zool'ogy.

zoology (biology) Systematic study of the animal kingdom.

Zoomastigophorea (parasitology) A class of flagellated protozoa, many of which cause diseases such as Leishmaniasis, trichomoniasis and trypanosomiasis.

zoonosis (parasitology) Any disease of animals which may be transmitted to human beings. Examples are plague and rabies; anthrax, Q fever and salmonella infections; all kinds of worm infestations; and various virus, rickettsial, leptospiral and fungus infections; and brucellosis, ornithosis and glanders.

zoonotic disease (parasitology) Disease capable of spread from animals to humans.

See also zoonosis.

zooparasite (parasitology) Any parasitic animal, organism or species.

zoopathology (terminology) The science of the diseases of animals.

zootoxicosis (parasitology) A disease caused by a zootoxin.

zootoxin (parasitology) A toxic substance of animal origin, *e.g.* venom of snakes, spiders and scorpions.

Zr (chemistry) The chemical symbol for Element 40, zirconium.

z-score (statistics) A score expressed as the number of units of standard deviation above or below the mean, according to the formula, $z = (X\text{-}M)/SD$, where X is the score, M the mean and SD the standard deviation. They have the following properties: (i) their sum is zero; (ii) their standard deviation is 1.0.

zwitterion (chemistry) An ion carrying both a positive and a negative charge, *e.g.* present in solid and liquid amino acids such as glycine, $N^+H_3CH_2COO^-$. Also termed dipolar ion.

zyg(o)- (terminology) A word element. [German] Prefix denoting *yoked, joined, a junction.*

Zygocotyle lunata (avian parasitology) A common parasite of ducks and other water fowl. The parasite is found in the small intestine and eggs are passed in the faeces (*USA*: feces). Adults of this species can measure up to 9 mm in length. The first intermediate host is a snail, and the cercariae that are liberated from the snail encyst on the surfaces of various objects in the water (*e.g.* plants, branches, etc.). The definitive host is infected when it ingests the metacercariae.

zygosis (genetics) The union of two gametes.

zygote (genetics) The individual as it exists at the moment of fertilization, *i.e.* the diploid cell that results from the fusion of two gametes. *See* hybrid.

zygotene (genetics) One of the stages in meiosis.

zym(o)- (terminology) A word element. [German] Prefix denoting *enzyme, fermentation.*

zymase (chemistry) Enzyme that catalyses the fermentation of carbohydrates to ethanol (ethyl alcohol).

zymic (terminology) Pertaining to enzymes or fermentation.

zymodemes (parasitology) Populations of parasites with identical isoenzymes.

zymogen (biochemistry) Inactive precursor of an enzyme. It is activated by the action of a kinase. Also termed: proenzyme.

zymogen cells (biochemistry) The secretory cells that secrete the zymogens.

zymogenesis (biochemistry) The conversion of an enzyme precursor to the active state.

zymotic (medical) Describing an agent that causes an infectious disease.

List of Appendices:

Appendix 1 Metric Conversions

Distance

1 inch = 2.54 centimetres
1 foot = 0.3048 metre
1 yard = 0.9144 metre
1 rod = 5.0292 metres
1 chain = 20.117 metres
1 furlong = 201.17 metres
1 mile = 1.6093 kilometre
1 nautical mile = 1.8532 kilometre

1 millimetre = 0.03937 inch
1 centimetre = 0.3937 inch
1 decimetre = 0.3281 foot
1 metre = 3.281 feet
1 metre = 1.094 yard
1 decametre = 10.94 yards
1 kilometre = 0.6214 mile
1 kilometre = 0.539 nautical mile

Surface or Area

1 square inch = 6.4516 square centimetres
1 square foot = 929.03 square centimetres
1 square yard = 0.8361 square metre
1 acre = 4046.9 square metres
1 square mile = 259.0 hectares
1 square centimetre = 0.1550 square inch
1 square metre = 1550 square inches
1 acre = 119.6 square yards
1 hectare = 2.4711 acres
1 square kilometre = 0.3861 square mile

Capacity

1 cubic inch = 16.387 cubic centimetres
1 cubic foot = 0.0283 cubic metre

1 cubic yard = 0.7646 cubic metre
1 cubic centimetre = 0.061 cubic inch
1 cubic decimetre = 0.035 cubic foot
1 cubic metre = 1.308 cubic yard

Dry Measure (Imperial)

1 pint = 0.5506 litre
1 quart = 1.136 litre
1 gallon = 4.546 litres
1 peck = 9.092 litres
1 bushel = 36.369 litres

Liquid Measure (USA)

1 pint = 0.473 litre
1 quart = 0.9463 litre
1 gallon = 3.785 litres
1 peck = 8.809 litres
1bushel = 35.24 litres

Avoirdupois Weight

1 ounce = 28.35 grams
1 pound = 453.59 grams
1 hundredweight = 50.802 kilograms
1 ton = 907.18 kilograms
1 gram = 0.035 ounce
1 hectogram = 3.527 ounces
1 kilogram = 2.205 pounds
1 ton = 1.102 ton (short)

Temperature conversion

Celsius° = 5/9 (Fahrenheit° - 32°)
Fahrenheit° = 9/5 Celsius° + 32°

Appendix 2 Human Vector-Borne Infections

Infection (disease)	**Causative Agent**	**Vector** (common name)
Viral Infections		
	(Toga viridae)	
Yellow fever	Yellow fever virus	Mosquitoes
Dengue haemorrhagic fever	Dengue virus	Mosquitoes
Japanese encephalitis	JE virus Mosquitoes	
Murray valley encephalitis	MVE virus	Mosquitoes
Western equine encephalomyelitis	WEE virus	Mosquitoes
Eastern equine encephalomyelitis	EEE virus	Mosquitoes
Chikungunya	Chikungunya virus	Mosquitoes
O'Nyong-Nyong	O'Nyong-Nyong virus	Mosquitoes
Tick-borne encephalitis (Central European)	CET virus	Ticks
Tick-borne encephalitis (Russian Far East)	FER virus	Ticks
Kyasanur forest disease	KFD virus	Ticks
Omsk haemorrhagic fever	OHF virus	Ticks
	(Reo viridae)	
Colorado tick fever	CTF virus	Ticks
	(Unclassified Arboviruses)	
Rift valley fever	Rift valley virus	Mosquitoes
Epidemic haemorrhagic fever	EHF virus	Mite (?)
Sandfly fever	Sandfly fever virus	Sandflies
Bacterial and Rickettsial Infections		
Tick typhus	*Rickettsia rickettsi*	Ticks
Queensland tick typhus	*Rickettsia australis*	Ticks
Q fever	*Coxiella burnetti*	Ticks
Relapsing fever	*Borrelia duttoni*	Ticks
Tularaemia	*Pasteurella tularensis*	Ticks
Scrub typhus	*Rickettsia tsutsugamushi*	Mites
Epidemic typhus	*Rickettsia prowazeki*	Lice
Relapsing fever	*Borrelia recurrentis*	Lice
Epidemic (murine) typhus	*Rickettsia mooseri*	Fleas
Plague	*Yersinia pestis*	Fleas
Orayafever	*Bartonella bacilliformis*	Sandflies

Infection (disease)	Causative Agent	Vector (common name)
Protozoal Infections		
Malaria	*Plasmodium* spp	Mosquitoes
Kala-azar	*Leishmania donovani*	Sandflies
Oriental sore	*Leishmania tropica*	Sandflies
Chaga's disease	*Trypanosoma cruzi*	Triatomid bugs
East African trypanosomiasis	*Trypanosoma rhodesiense*	Tsetse flies
West African trypanosomiasis	*Trypanosoma gambiense*	Tsetse flies
Babesiosis	*Babesia* spp	Ticks
Helminthic Infections		
Filariasis	*Wuchereria bancrofti*	Mosquitoes
Filariasis	*Brugia malayi*	Mosquitoes
Filariasis	*Dirofilaria spp*	Mosquitoes
Filariasis	*Dipetalonema perstans*	Biting midges
Filariasis	*Dipetalonema streptocerca*	Biting midges
Filariasis	*Mansonella ozzardi*	Biting midges
Onchocerciasis	*Onchocerca volvulus*	Black flies
Loiasis	*Loa loa*	Deer flies

Appendix 3 Worldwide Societies of Parasitology

Parasitology incorporates information from a large number of disciplines, and worldwide there are many hundreds of associations, societies and other organisations dedicated to various aspects. The following information contains the e-mail contact addresses of key organisations worldwide relevant to parasitology from which specialised information and assistance may be sought. These are listed under geographical headings. It is hoped that the list provides a useful guide, as a starting point, in identifying sources of information being sought.

Australia

- Australian Society for Parasitology
  http://parasite.org.au/

Austria

- Austrian Society of Tropical Medicine and Parasitology
  http://www.vu-wien.ac.at/i116/OeGTPhome.html

Belgium

- Societé Belge de Parasitologie
  http://www.icp.ucl.ac.be/bsp/belgian_society_proto.htm

Canada

- Parasitology Section, Canadian Society of Zoologists
  http://www.biology.ualberta.ca/parasites/home.htm

Czech Republic

- Czech Society for Parasitology
  http://www.natur.cuni.cz/hydrobiology/parpages/Society-us.htm

Denmark

- Danish Society for Parasitology
  http://www.dsp.kvl.dk/

Germany

- Deutsche Gesellschaft fur Parasitologie
  http://www.dgparasitologie.de

Hungary

- Hungarian Society of Parasitologists
  http://bio.univet.hu/mpt/angol.htm

India

- Indian Society for Parasitology
  http://www.parasitologyindia.org

Japan

- Japanese Society of Parasitology
  http://jsp.tm.nagasaki-u.ac.jp/~parasite

Korea

- Korean Society for Parasitology
  http://www.parasitol.or.kr/eng

Netherlands

- Netherlands Society for Parasitology
  http://www.parasitologie.nl

New Zealand

- New Zealand Society for Parasitology
  http://nzsp.rsnz.govt.nz/

Southern Africa

- Parasitological Society of Southern Africa
  http://www.parsa.ac.za

Switzerland

- Swiss Society for Tropical Medicine and Parasitology
  http://www.sstmp.unibe.ch

United Kingdom

- British Society for Parasitology
  http://www.abdn.ac.uk/bsp/

- Society of Protozoologists – British Section
  http://www.bssp.org/

- Royal Society of Tropical Medicine and Hygiene
  http://www.rstmh.org

United States of America

- American Association of Veterinary Parasitologists
  http://www.vetmed.ufl.edu/aavp/

- American Society of Parasitologists
  http://www-museum.unl.edu/asp/

- American Society of Tropical Medicine and Hygiene
  http://www.astmh.org

- Society for Inverterbrate Pathology
  http://www.sipweb.org
- Society of Nematologists
  http://www.ianr.unl.edu/son/
- Society of Protozoologists
  http://www.uga.edu/~protozoa/

Appendix 4 Select Bibliography and Further Reading List

Academic Press Dictionary of Science and Technology, Academic Press Inc, 1992.
Acha PN, Szyfres B, Zoonoses and communicable diseases common to man and animals, Pan American Health Organization, 1987.
Anderson KN, Anderson L, Glanze WD, Mosby's Medical, Nursing, and Allied Health Dictionary, Mosby, 1994.
Ash LR, Orihel TC, Parasites: A Guide to Laboratory Procedures and Identification, ASCP Press, 1987.
Beaver PC, Jung RC, Cupp EW, Clinical Parasitology, Lea and Febiger, 1984.
Blackwell's Dictionary of Nursing, Blackwell Science (UK), 1997.
Borowski EJ, Borwein JM, Dictionary of Mathematics, HarperCollins, 1999.
Cheesbrough M, Medical Laboratory Manual for Tropical Countries, Butterworths, 1987.
Cook GC, Manson's Tropical Diseases, WB Saunders, 1996.
Daintith J, Oxfords' Dictionary of Chemistry, Oxford University Press, 1996.
Despommier DD, Gwadz RW, Hotez PJ, Parasitic diseases, Springer-Verlag, 2000.
Dictionary of Science, Brockhampton Press, 1997.
Dictionary of the Sciences, Harcourt Publishers Ltd, 2000.
Dupayrat J, Dictionary of Biomedical Acronyms and Abbreviations, John Wiley & Sons, 1990.
Everitt B, Cambridge Dictionary of Statistics, Cambridge University Press,1998.
Everitt BS, The Cambridge Dictionary of Statistics in the Medical Sciences, Cambridge University Press, 1995.
Farr AD, Dictionary of Medical Laboratory Sciences, Blackwell Science (UK), 1988.
Fraser, CM, Airello, SE, eds, Merck Veterinary Manual: A Handbook of the Diagnosis, Therapy, and Disease Prevention and Control for the Veterinarian, 8th ed. Merck and Co., Rahway, NJ, 1998.
Fukui S, Schmid R, Dictionary of Biotechnology, Springer-Verlag Berlin and Heidelberg GmbH & Co. KG, 1986.
Ganong WF, Review of Medical Physiology, Appleton and Lange, 1995.
Garcis LS, Bruckner DA, Diagnostic Medical Parasitology, ASM Press, 1997.

Gard P, Human Pharmacology, Taylor & Francis, 2001.
Gosling PJ, Dictionary of Biomedical Sciences, Taylor & Francis, 2002.
Gray H, Gray's Anatomy, Parragon, 1998.
Hale WG, Margham JP, Saunders VA, Dictionary of Biology, HarperCollins, 1999.
Harrap's Dictionary of Science and Technology, Harrap, 1991.
Harrison P, Waites G, Cassell Dictionary of Chemistry, Ward Lock, 1999.
Hausmann K, Hulsman N, Protozoology, Thieme Medical Publications, 1997.
Heister R, Dictionary of Abbreviations in Medical Sciences, Springer-Verlag Berlin and Heidelberg GmbH & Co. KG, 1989.
Herbert WJ, Wilkinson PC, Dictionary of Immunology, Blackwell Science (UK), 1985.
Higgins SJ, Turner AJ, Wood EJ, Biochemistry for the Medical Sciences, Longman Group Ltd, 1994.
Hodson A, Essential Genetics, Bloomsbury, 1992.
Hutchinson Dictionary of Science, Helicon, 1998.
Lafferty P, Rowe J, The Dictionary of Science, Prentice Hall, 1994.
Lee JJ, Leedale GF, Bradbury P, Lawrence KS, Illustrated Guide to the Protozoa, Society of Protozoologists, 2000.
Leventhal R, Cheadle RF, Medical Parasitology, F. A. Davis Company, 1985.
Lewis RJ Sr, Hawley's Condensed Chemical Dictionary, John Wiley & Sons, 1997.
MacPherson G, Black's Medical Dictionary, A & C Black, 1999.
Marquardt WC, Demaree RS, Grieve RB, Parasitology and Vector Biology, Academic Press, 2000.
McGraw-Hill, Dictionary of Bioscience, McGraw-Hill Publishing Company, 1996.
Melvin DM, Brooke MM, Laboratory Procedures for the Diagnosis of Intestinal Parasites, US Department of Health and Human Services, 1982.
Merrell S, ed, Medicines, Bloomsbury Publishing Plc, 1995.
Molecular Biology, Oxford University Press, 1997.
Muir H, Walker P, Larousse Dictionary of Science and Technology, Kingfisher Chambers Harrap, 1995.
Noble ER, Noble GA, Parasitology: The Biology of Animal Parasites, Lea and Febiger, 1989.
Olsen OW, Animal Parasites: Their Life Cycles and Ecology, University Park Press, 1974.

Oxfords' Science Dictionary, Oxford University Press, 1999.
Oxfords' Concise Medical Dictionary, Oxford University Press, 1998.
Pearce EC, Pearce's Medical and Nursing Dictionary and Encyclopaedia, for the Social Sciences, Sage Publications Ltd, 1993.
Reuter P, Birkhauser Pocket Dictionary of Biochemistry, Birkhauser Verlag AG, 2000.
Roper N, Churchill Livingstone Pocket Medical Dictionary, Harcourt Publishers Ltd, 1988.
Roper N, New American Pocket Medical Dictionary, Prentice Hall, 1988.
Ross JS, Wilson KJW, Foundations of Anatomy and Physiology, Churchill Livingstone, 1984.
Smith AD, Datta SP, Smith G, et al, Oxford Dictionary of Biochemistry and Molecular Biology, Oxford University Press, 1997.
Smyth JD, Introduction to Parasitology, Cambridge University Press, 1994.
Stenesh J, Dictionary of Biochemistry and Molecular Biology, John Wiley & Sons, 1989.
Strickland GT, Hunter's Tropical Medicine and Emerging Infectious Diseases, WB Saunders, 2000.
Uvarov EB, Isaacs A, Penguin Dictionary of Science, Penguin Books, 1993.
Vogt WP, Dictionary of Statistics and Methodology, A Non-Technical Guide, Mosby, 1983.
Walker P, Chambers Science and Technology Dictionary, Kingfisher Chambers Harrap, 1991.
Webster's New American Dictionary, Merriam-Webster Inc, 1995.
WHO, Control of Lymphatic Filariasis: A Manual for Health Personnel, World Health Organization, 1987.
WHO, Manual of Basic Techniques for a Health Laboratory, World Health Organization, 1980.
Wilson J, Hunt T, Molecular Biology of the Cell, Garland Publishing Inc, 1994.
Wordsworth Dictionary of Science and Technology, Wordsworth Editions Ltd, 1995.
Wyler DJ, Modern Parasite Biology: Cellular, Immunological, and Molecular Aspects, WH Freeman, 1990.
Youngson RM, Dictionary of Medicine, HarperCollins, 1999.